*Electronics
for
Modern Communications*

Prentice-Hall Series in Electronic Technology

Dr. Irving L. Kosow, *editor*
Charles M. Thomson, Joseph L. Gershon, and Joseph A. Labok,
consulting editors

Electronics for Modern Communications

George J. Angerbauer

Department of Electronics
College of San Mateo

PRENTICE-HALL, INC., *Englewood Cliffs, New Jersey*

Library of Congress Cataloging in Publication Data

ANGERBAUER, GEORGE J.
  Electronics for modern communications.

  (Prentice-Hall series in electronic technology)
  1. Electronics. 2. Radio. I. Title.
  TK7815.A53   621.38   72-13521
  ISBN 0-13-252338-8

© 1974 by Prentice-Hall, Inc., Englewood Cliffs, N.J.

All rights reserved. No part of this book may be reproduced in any form, by mimeograph or any other means, without permission in writing from the publisher.

10 9 8 7 6 5 4 3 2 1

Printed in the United States of America.

PRENTICE-HALL INTERNATIONAL, INC., *London*
PRENTICE-HALL OF AUSTRALIA, PTY. LTD., *Sydney*
PRENTICE-HALL OF CANADA, LTD., *Toronto*
PRENTICE-HALL OF INDIA PRIVATE LIMITED, *New Delhi*
PRENTICE-HALL OF JAPAN, INC., *Tokyo*

Preface

This text provides, in one volume, the basic electronic and radio theory necessary for a person to pass successfully the several Federal Communications Commission (FCC) license examinations. This includes both commercial and radio amateur licenses. Important laws, rules, and regulations governing the operation of radio stations have also been included.

This book has been written to meet the needs of junior colleges, technical institutes, and correspondence schools whose students desire to learn the essentials of electronic communication. High school students studying electronics should also be able to profit from it. Anyone should be able to study the text on his own and still be able to pass the desired FCC license examinations.

All the essential material has been arranged in appropriate chapters, from the simple to the more complex, thus making the volume suitable as a text for nearly all beginning electronics students as well as those wishing to acquire any of the FCC licences. The author has chosen this approach, rather than the question-and-answer method, because of the continuity provided from subject to subject.

Numerous FCC-type questions have been included at the end of each chapter, compiled from the latest publications provided by the FCC. These are intended to indicate the *nature* and *scope* of the questions asked in the various FCC examinations. The actual FCC examination questions are of the *multiple-choice* type from which the applicant selects the best answer. These sample questions have served as a basis for writing this comprehensive text on electronic communication for applicants for the several FCC commercial and radio amateur licenses.

Solid-state devices and circuits have been included, where possible, to reflect the changing state of the art. Appropriate questions are included in the respective chapters to acquaint the reader with likely changes in future FCC questions.

The author has also included a number of problems at the end of most chapters to be used at the discretion of the instructor or for self-study. These are intended to supplement the FCC-type questions. A solution manual for these problems is available to instructors upon request.

Chapters 25 and 26 cover the essential information needed to answer the FCC questions for Elements 1 and 2, which are required for all classes of FCC licenses.

The author is indebted to the following individuals who have made various contributions and suggestions in the preparation of this volume. Mr. Leon Savage, instructor in commercial and radio amateur license classes; Mr. John Hecomovich, telecommunications and electronics instructor at College of San Mateo; Mr. Sam Cooper, General Electric Company; and Mr. Jack Gittings, telecommunications.

For the final credit I wish to thank my wife, Gerry, for her excellent work in typing the manuscript.

GEORGE J. ANGERBAUER

Contents

PREFACE   *v*

1 CIRCUITS REVIEW   *1*

    1-1   The Atom   *1*
    1-2   Potential Difference   *2*
    1-3   Current   *2*
    1-4   Resistance   *3*
    1-5   Resistor Color Code   *3*
    1-6   Ohm's Law   *3*
    1-7   Power   *5*
    1-8   Voltage-Dropping Resistors   *6*
    1-9   Parallel Circuits   *7*
    1-10   Conductance   *8*
    1-11   Series–Parallel Circuits   *8*
    1-12   Maximum Power Transfer Theorem   *10*
    1-13   Prefixes Used in Electronics   *10*
    1-14   Miscellany   *11*
    Practice Problems   *12*
    Commercial License Questions   *14*

2 CONDUCTORS AND INSULATORS   *17*

    2-1   Conductors   *17*
    2-2   Specific Resistance or Resistivity   *18*
    2-3   Wire Sizes and Fuses   *19*
    2-4   Conduction in Gases and Liquids   *20*
    2-5   Insulators   *20*
    2-6   Skin Effect   *21*
    2-7   Soldering   *21*
    Practice Problems   *22*
    Commercial License Questions   *22*

3 MAGNETISM REVIEW 24

3-1 Permanent Magnetism 24
3-2 Magnetic Flux 25
3-3 Permeability 26
3-4 Ampere Turns 27
3-5 *BH* Magnetization Curve 29
3-6 Magnetic Circuit 29
3-7 Hysteresis Loss 29
3-8 Induced Current 31
3-9 Earphones 33
Practice Problems 33
Commercial License Questions 34

4 ALTERNATING CURRENT 36

4-1 Generation of Alternating Current 36
4-2 Waveshapes 37
4-3 Sine Wave 37
4-4 Peak and Instantaneous Values 38
4-5 Average Value 39
4-6 Effective or Root-Mean-Square 40
4-7 Frequency 41
4-8 Phase Relationships 41
4-9 Radians 43
Practice Problems 44
Commercial License Questions 45

5 INDUCTANCE AND TRANSFORMERS 46

5-1 Inductance 46
5-2 Unit of Inductance 48
5-3 Mutual Inductance 48
5-4 Series- and Parallel-Connected Inductors 49
5-5 Energy in a Magnetic Field 50
5-6 Time Constant of Inductance–Resistance Circuit 50
5-7 Inductive Reactance 52
5-8 Phase Relationships 53
5-9 Basic Transformer 53
5-10 Turns Ratio 55
5-11 Transformer Losses 57
5-12 Autotransformers 59
5-13 Impedance Matching 60
5-14 Output Transformers 62
5-15 Radio-Frequency Transformers 62
5-16 Three-Phase Transformers 63
5-17 Shorted Turns 64
Practice Problems 65
Commercial License Questions 66

6 CAPACITANCE 68

6-1 Basic Capacitor 68
6-2 Unit of Capacitance 69

- 6-3 Charging a Capacitor *70*
- 6-4 Factors Affecting Capacity *71*
- 6-5 Types of Capacitors *72*
- 6-6 Capacitors in Parallel and Series *75*
- 6-7 Capacitor Color Codes *76*
- 6-8 Resistance–Capacitance Time Constant *78*
- 6-9 Capacitive Reactance *79*
- 6-10 Losses in a Capacitor *81*
- 6-11 Voltage Ratings of Capacitors *81*
- Practice Problems *82*
- Commercial License Questions *82*

7 ALTERNATING-CURRENT CIRCUITS 84

- 7-1 Series Resistance–Inductance Circuit *84*
- 7-2 Series Resistance–Capacitance Circuit *86*
- 7-3 Series Inductance–Capacitance–Resistance Circuit *88*
- 7-4 The j Operator *90*
- 7-5 Power in Alternating-Current Circuits *90*
- 7-6 Power Factor *95*
- 7-7 Parallel Alternating-Current Circuits *96*
- 7-8 Complex Circuits *99*
- Practice Problems *102*
- Commercial License Questions *102*

8 RESONANT CIRCUITS 104

- 8-1 Resonance *104*
- 8-2 Series Resonant Circuits *106*
- 8-3 Parallel Resonant Circuits *108*
- 8-4 Circuit Q *111*
- 8-5 Bandwidth *112*
- 8-6 Low Pass Filters *114*
- 8-7 High Pass Filters *115*
- 8-8 Band Rejection and Band-Pass Filters *117*
- Practice Problems *119*
- Commercial License Questions *120*

9 RECTIFIERS 121

- 9-1 Semiconductor Materials *121*
- 9-2 *pn* Junctions *125*
- 9-3 Biasing the *pn* Junction *126*
- 9-4 Diode As a Rectifier *129*
- 9-5 Vacuum-Tube Diode *129*
- 9-6 Vacuum-Tube-Diode Characteristics *131*
- 9-7 Volt–Ampere Characteristic *132*
- 9-8 Plate Resistance *133*
- 9-9 Zener Diodes *135*
- Practice Problems *136*
- Commercial License Questions *136*

10 TRANSISTOR FUNDAMENTALS 138

10-1 Basic Transistor 138
10-2 *npn* Transistor 139
10-3 *pnp* Transistor 141
10-4 Collector Family of Curves 141
10-5 Load Lines 143
10-6 Biasing Techniques 145
10-7 Transistor Circuit Configurations 149
10-8 *h* Parameters 151
10-9 Transistor Equivalent Circuit 152
10-10 Leakage Currents 152
10-11 Other Transistor Types 153
Practice Problems 156
Commercial License Questions 157

11 REVIEW OF VACUUM-TUBE FUNDAMENTALS 159

11-1 Triode 159
11-2 Tube Parameters 160
11-3 Relation Between μ, g_m, and r_p 162
11-4 Tetrodes 162
11-5 Pentodes 163
11-6 Miscellany 164
Practice Problems 165
Commercial License Questions 166

12 POWER SUPPLIES 167

12-1 Half-Wave Rectifiers 167
12-2 Basic Full-Wave Rectifier Circuit 169
12-3 Bridge Rectifier 170
12-4 Waveform Analysis 172
12-5 Capacitance Input Filter 173
12-6 Inductance Input Filter 174
12-7 Pi-Section Filter 175
12-8 Miscellaneous Filter Circuits 176
12-9 Filter Chokes 177
12-10 Vacuum Rectifiers 178
12-11 Mercury-Vapor Rectifiers 180
12-12 Peak Inverse Voltage 183
12-13 Series-Connected Filter Capacitors 184
12-14 Bleeder Resistors and Voltage Dividers 185
12-15 Voltage Regulation 187
12-16 Regulated Power Supplies 189
12-17 Voltage-Doubler Circuits 192
12-18 Vibrator Power Supplies 192
12-19 Three-Phase Power Supplies 194
Practice Problems 196
Commercial License Questions 196

13 AUDIO AMPLIFIERS *199*

- 13-1 Classification of Amplifiers *199*
- 13-2 Class A Amplifiers *199*
- 13-3 Class B Amplifiers *201*
- 13-4 Class AB Operation *202*
- 13-5 Class C Amplifiers *203*
- 13-6 Biasing *204*
- 13-7 Coupling Circuits *207*
- 13-8 Distortion *211*
- 13-9 Amplifier Gain *211*
- 13-10 Comparison of Transistor Configurations *213*
- 13-11 Paraphase Amplifiers *213*
- 13-12 Push–Pull Power Amplifiers *215*
- 13-13 Transistor Amplifier *217*
- 13-14 Complementary amplifiers *218*
- 13-15 Inverse Feedback *219*
- 13-16 Cathode and Emitter Followers *221*
- Practice Problems *222*
- Commercial License Questions *223*

14 RADIO-FREQUENCY AMPLIFIERS *226*

- 14-1 Radio-Frequency Transformer Coupling *226*
- 14-2 Class C Amplifiers *229*
- 14-3 Grid Current Loading *232*
- 14-4 Series and Shunt Feed *233*
- 14-5 Interstage Coupling Techniques *234*
- 14-6 Neutralization—General *236*
- 14-7 Plate Neutralization (Hazeltine) *236*
- 14-8 Grid Neutralization (Rice) *238*
- 14-9 Coil Neutralization *239*
- 14-10 Neutralization of Push–Pull Stages *239*
- 14-11 Neutralization Procedures *240*
- 14-12 Parallel Operation of Radio-Frequency Amplifiers *240*
- 14-13 Push–Pull Operation of Radio-Frequency Amplifiers *242*
- 14-14 Frequency Multipliers *242*
- 14-15 Grounded-Grid Amplifiers *245*
- 14-16 Very High and Ultrahigh-Frequency Amplifiers *246*
- 14-17 Coupling the Radio-Frequency Amplifier to the Load *247*
- 14-18 Troubleshooting Procedures *249*
- Commercial License Questions *250*

15 MEASURING INSTRUMENTS *253*

- 15-1 Basic Direct-Current Meter Movement *253*
- 15-2 Direct-Current Ammeter Connections and Shunts *256*
- 15-3 Direct-Current Voltmeters *259*
- 15-4 Ohmeters *262*
- 15-5 Volt–Ohm–Milliammeters *263*
- 15-6 Direct-Current Vacuum-Tube Voltmeters *263*

15-7 Alternating-Current Vacuum-Tube Voltmeters *265*
15-8 Alternating-Current Rectifier Meters *265*
15-9 Peak-Reading Voltmeters *268*
15-10 Electrodynamometers *268*
15-11 Wattmeters *269*
15-12 Watthour Meters *270*
15-13 Radio-Frequency Ammeters *270*
15-14 Volume-Unit Meters *272*
15-15 Oscilloscopes *273*
15-16 Absorption Frequency Meters *275*
15-17 Grid-Dip Meters *278*
15-18 Primary Frequency Standards *279*
15-19 Secondary Frequency Standards *281*
15-20 Heterodyne Frequency Meters *283*
15-21 Carrier-Shift Detector *289*
15-22 Frequency Measurements with Lecher Wires *290*
Practice Problems *290*
Commercial License Questions *291*

16 OSCILLATORS *294*

16-1 Requirements for Oscillation *294*
16-2 Two-Stage Amplifier Using Regenerative Feedback *295*
16-3 Armstrong Oscillator *296*
16-4 Oscillator Biasing *298*
16-5 Tune-Plate Tuned-Grid Oscillator *299*
16-6 Hartley Oscillator *301*
16-7 Colpitts Oscillator *302*
16-8 Electron-Coupled Oscillators *304*
16-9 Crystal-Controlled Oscillators *305*
16-10 Crystals and Their Characteristics *307*
16-11 Crystal Ovens *310*
16-12 Other Crystal-Controlled Oscillator Circuits *313*
16-13 Audio Oscillators *314*
16-14 Phase-Shift Oscillator *315*
16-15 High-Frequency Oscillators *316*
16-16 Dynatron Oscillators *318*
16-17 Resistance–Capacitance Oscillators *320*
16-18 Unijunction Transistor Oscillator *321*
16-19 Multivibrators *322*
16-20 Parasitic Oscillations *324*
16-21 Methods Used to Detect Oscillation *325*
16-22 Oscillator Stabilization *326*
Commercial License Questions *327*

17 BASIC TRANSMITTERS *330*

17-1 Fundamental Concepts *330*
17-2 Single-Stage Transmitter *331*
17-3 Master Oscillator Power Amplifier *332*
17-4 Break-In Operation *334*
17-5 Buffer Amplifiers *334*

CONTENTS xiii

    17-6   Power Amplifiers *335*
    17-7   Dummy Antennas *338*
    17-8   Bias Methods *339*
    17-9   Keying Requirements *342*
    17-10 Key-Click Filters *343*
    17-11 Oscillator Keying *344*
    17-12 Blocked-Grid Keying *344*
    17-13 Vacuum-Tube Keying *346*
    17-14 Cathode Keying *347*
    17-15 Frequency-Shift Keying (F1) *349*
    17-16 Modulated Continuous Wave Signals (A2) *350*
    17-17 Variable-Frequency Oscillator Operation *351*
    17-18 Reducing Harmonic Radiation *351*
    17-19 Shielding *354*
    17-20 Typical Continuous Wave Transmitter *356*
    17-21 Tuning a Continuous Wave Transmitter *356*
    17-22 Locating Troubles *358*
    Commercial License Questions *359*

18 AMPLITUDE MODULATION *362*

    18-1   Basic Modulation Concepts *362*
    18-2   Bandwidth of Amplitude-Modulated Wave *365*
    18-3   Basic Modulation System *365*
    18-4   Plate Modulation *367*
    18-5   Percentage of Modulation *369*
    18-6   Sideband Power *371*
    18-7   Modulator Power Requirements *371*
    18-8   Modulation Level *374*
    18-9   Plate and Screen-Grid Modulation *374*
    18-10 Control-Grid Modulation *375*
    18-11 Suppressor-Grid Modulation *377*
    18-12 Heising Modulation *378*
    18-13 Collector Modulation *380*
    18-14 Base Modulation *381*
    18-15 Checking Percentage of Modulation *381*
    18-16 Linear Radio-Frequency Amplifiers *384*
    18-17 Carrier Shift *386*
    18-18 Antenna Ammeter Indications *387*
    18-19 Characteristics of Radiotelephone Transmitters *388*
    18-20 Principles of Single-Sideband Transmission *391*
    18-21 Balanced Modulator Circuits *393*
    18-22 Sideband Filters *394*
    18-23 Phase-Shift Method of Generating Single Sidebands *395*
    Commercial License Questions *396*

19 FREQUENCY MODULATION *400*

    19-1   Reasons for Frequency Modulation *400*
    19-2   Fundamental Principles *401*
    19-3   Modulation Index *403*
    19-4   Percentage of Modulation *404*

19-5 Sidebands and Bandwidth *405*
19-6 Preemphasis and Deemphasis *407*
19-7 Reactance-Tube Modulators *408*
19-8 Solid-State Frequency-Modulated Modulator *409*
19-9 Phase Modulation *410*
19-10 Other Methods of Producing Phase Modulation *411*
19-11 Block Diagram of a Frequency-Modulated Transmitter *413*
19-12 Public Safety Radio Service *414*
19-13 Stereo Multiplex *417*
Practice Problems *421*
Commercial License Questions *421*

20 ANTENNAS 425

20-1 Characteristics of Radio Waves *425*
20-2 Ground Waves *427*
20-3 Ionosphere *428*
20-4 Sky Waves *429*
20-5 Atmospheric Disturbances *431*
20-6 Transmission Lines *432*
20-7 Half-Wave Antenna *434*
20-8 Antenna Impedance *437*
20-9 Wave Polarization *437*
20-10 Hertz Antennas *438*
20-11 Marconi Antennas *439*
20-12 Counterpoise *440*
20-13 Long-Wire Antennas *441*
20-14 Folded Dipoles *442*
20-15 Parasitic Arrays *443*
20-16 Collinear Arrays *446*
20-17 Loop Antennas *447*
20-18 Methods of Antenna Feed *448*
20-19 Tuning or Loading the Antenna *452*
20-20 Harmonic Suppression *456*
20-21 Antenna Radiation Patterns *458*
20-22 Calculating the Power in an Antenna *460*
20-23 Antenna Gain *463*
20-24 Phase Monitors *464*
20-25 Miscellany *464*
Practice Problems *465*
Commercial License Questions *466*

21 RECEIVERS 471

21-1 Basic Functions of a Receiver *471*
21-2 Crystal Detectors *472*
21-3 Tuned Radio-Frequency Receiver *473*
21-4 Diode Detectors *473*
21-5 Grid-Leak Detectors *474*

CONTENTS   XV

    21-6 Plate Detectors *475*
    21-7 Regenerative Detectors *476*
    21-8 Superregenerative Detectors *477*
    21-9 Superheterodyne Receivers *478*
    21-10 Radio-Frequency Amplifier or Preselector *479*
    21-11 Mixers and Converters *481*
    21-12 Image Frequency *484*
    21-13 Intermediate-Frequency Amplifier *485*
    21-14 Crystal Filters *486*
    21-15 Second Detector and Automatic Volume Control Circuit *488*
    21-16 Delayed Automatic Volume Control *490*
    21-17 Beat-Frequency Oscillators *490*
    21-18 Signal-Strength Meters *491*
    21-19 Noise Limiters *493*
    21-20 Squelch Circuits *494*
    21-21 Double-Conversion Superheterodyne Receivers *496*
    21-22 Diversity Receiving System *497*
    21-23 Wave Traps *498*
    21-24 Frequency-Modulated Receivers *499*
    21-25 Limiters *500*
    21-26 Foster–Seeley Discriminator *501*
    21-27 Ratio Detector *503*
    21-28 Maintenance Procedures *505*
    Commercial License Questions *506*

22 BROADCAST STATIONS   *510*

    22-1 Standard Broadcast Station *510*
    22-2 Frequency-Modulated Broadcast Stations *511*
    22-3 Components of a Broadcast System *511*
    22-4 Remote Broadcast Facilities *514*
    22-5 Broadcast Automation *515*
    22-6 Volume-Unit Meters and Audio Levels *516*
    22-7 Attenuator Networks *517*
    22-8 Automatic Gain Control Amplifiers *520*
    22-9 Frequency Monitors *521*
    22-10 Modulation Monitors *522*
    22-11 Broadcast Transmitters *523*
    22-12 Operating Powers *524*
    22-13 Remote-Control Systems *525*
    22-14 Proof-of-Performance Tests *526*
    22-15 Log Requirements *526*
    22-16 Emergency Broadcast System *528*
    22-17 Operator License Requirements *528*
    22-18 Broadcast Microphones *529*
    22-19 Transmission Lines *529*
    22-20 Antenna Systems *530*
    22-21 Miscellaneous Requirements *532*
    Commercial License Questions *532*

23 TELEVISION 536

23-1 Basic Principles 536
23-2 Magnetic Deflection and Focusing 537
23-3 Interlaced Scanning 538
23-4 Iconoscope 538
23-5 Image Orthicon 539
23-6 Vidicon 541
23-7 Television Station 542
23-8 Synchronizing-Pulse Generator 544
23-9 Video Modulation 545
23-10 Sound Transmitter 546
23-11 Operating Power of the Transmitter 546
23-12 Camera Chains 547
23-13 Vestigial Sideband Transmission 547
23-14 Log Requirements 549
23-15 Channel Allocations 549
23-16 Receiving Functions—The Antenna-Tuner Circuits 551
23-17 Intermediate-Frequency Amplifier 552
23-18 Video Detector 554
23-19 Automatic Gain Control 554
23-20 Video Amplifier 555
23-21 Synchronizing-Pulse Separation 555
23-22 Vertical Deflection 555
23-23 Horizontal Deflection 556
23-24 Color 559
23-25 Transmitting Color Signals 561
23-26 Color Synchronization 564
23-27 Color Television Receiver 565
23-28 Three-Gun Color Picture Tube 567
Commercial License Questions 569

24 GENERATORS, MOTORS, AND BATTERIES 571

24-1 Basic Direct-Current Generator 571
24-2 Practical Generators 573
24-3 Generator Types 574
24-4 Direct-Current Motors 575
24-5 Motor Generators 578
24-6 Alternators 579
24-7 Induction and Synchronous Motors 582
24-8 Single-Phase Motors 583
24-9 Maintenance of Motors and Generators 585
24-10 Primary Cells 586
24-11 Cell Connections 588
24-12 Secondary Cells 590
24-13 Battery Charging 591
24-14 Lead–Acid Battery Maintenance 594
24-15 Edison Storage Battery 595
Commercial License Questions 596

25 FEDERAL COMMUNICATIONS COMMISSION LICENSES AND LAWS, ELEMENT 1   601

25-1   Role of the Federal Communications Commission   *601*
25-2   Types of Federal Communications Commission Licenses and Permits   *602*
25-3   License and Permit Requirements   *604*
25-4   Types of Communication   *605*
25-5   Suspension of Operator Licenses   *606*
25-6   Notice of Violations   *607*
25-7   Restrictions on Operating, Servicing, and Adjusting Transmitters   *609*
25-8   Logs   *609*
Commercial License Questions   *610*

26 BASIC OPERATING PRACTICE, ELEMENT 2   613

26-1   Message Priority for Mobile Service   *613*
26-2   Distress, Urgent, and Safety Messages   *614*
26-3   Microphone Techniques   *615*
26-4   Phonetic Alphabet   *615*
26-5   Commonly Used Radiotelephone Abbreviations   *616*
26-6   Calling Another Station   *616*
26-7   Calling and Working Frequencies   *617*
26-8   Standard Operating Practices   *618*
26-9   Miscellaneous   *619*
Commercial License Questions   *620*

27 LOGARITHMS AND DECIBELS— MICROPHONES AND LOUDSPEAKERS   622

27-1   Logarithms   *622*
27-2   Decibels   *623*
27-3   Microphones   *626*
27-4   Loudspeakers   *629*
27-5   Earphones   *630*
Commercial License Questions   *631*

Appendix A   Copper-Wire Table   **633**

Appendix B   Natural Sines, Cosines, and Tangents   **634**

Appendix C   Greek Alphabet   **640**

Appendix D   Classification of Emissions   **641**

Appendix E   Federal Communications Commission Field Offices   **643**

Appendix F   Commonly Used Q Signals   **644**

Appendix G   Miscellaneous Abbreviations and Signals   **646**

Appendix H   Common Logarithm Table   **648**

Appendix I   Standard Fixed Resistor Values   **650**

*Electronics
for
Modern Communications*

1

Circuits Review

1-1 THE ATOM

Matter may be defined as any substance that has weight (mass) and occupies space. It is found in any one of three states: solid, liquid, or gaseous. All matter consists of one or more basic materials called elements, of which there are over 100.

All elements are made up of atoms, which in turn are made up of electrons, protons, and other minute particles. The center, or *nucleus*, of an atom contains positively charged particles called *protons*. Orbiting the nucleus are negatively charged particles called *electrons*. Normally, there are as many electrons as protons in the atom.

In some elements the outermost electrons are not tightly bound to their nucleus and can move from one atom to another. These are called *free electrons* and constitute the flow of electricity in a wire when an electrical pressure, or voltage, is applied.

When an atom loses one or more of its electrons, it becomes a *positive ion*. If it acquires an extra electron, it becomes a *negative ion*. Under certain conditions electricity can flow between two points as a result of these charges.

An *electrostatic* (or *dielectric*) field surrounds any charged body. Its magnitude is a function of the strength of the electric charge. This charge may by represented by electric lines of force, as shown in Fig. 1-1.

The fundamental law of electrostatic charges states: *like charges repel and unlike charges attract.*

When one of the charged bodies has a polarity opposite to the other, there is said to be a *difference of potential* between them. If a wire were connected between these two charged bodies, the excess of electrons on the negatively charged body would flow through the wire to the positively charged body. This flow would continue until the charges were equalized.

Lines of Force

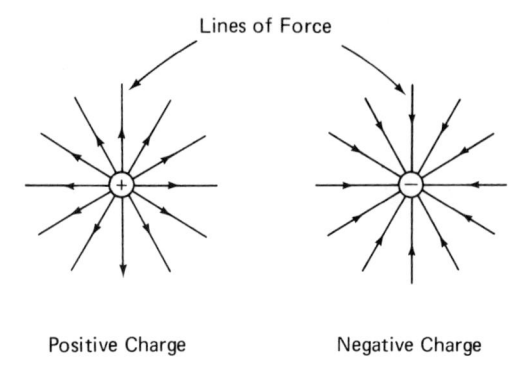

Positive Charge   Negative Charge

FIG. 1-1 Representation of electrostatic field surrounding positively and negatively charged bodies.

1-2 POTENTIAL DIFFERENCE

Potential difference, sometimes called *electromotive force*, is measured in *volts*. Its symbol is E or V. Electromotive force, abbreviated emf, is the potential difference that forces electrons through a wire or circuit. Potential difference is usually called voltage. A potential difference may be obtained by several means, such as

1. Static (friction).
2. Chemical (batteries).
3. Electromagnetism (generator).
4. Photoelectric (light-sensitive devices).
5. Thermal (heating the junction of dissimilar metals).
6. Piezoelectric (vibration of certain crystals).

1-3 CURRENT

When a potential difference is applied to an electrical conductor, electric charges flow, as shown in Fig. 1-2. The switch must be closed to complete

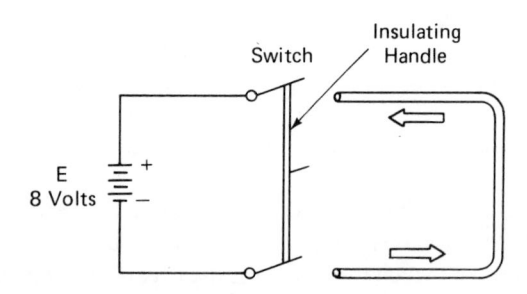

FIG. 1-2 Current flows through circuit when switch is closed.

Ohm's Law

the circuit before current flows. Current will continue to flow as long as the battery maintains a potential difference across its terminals. Current flow is measured in *amperes* (A) and is represented by the letter I. When 6.28×10^{18} electrons [1 coulomb (C)] flow past a given point in 1 second (s), the current is 1 A.

1-4 RESISTANCE

The amount of current flowing in a conductor with a given applied voltage depends upon its *resistance*. Resistance opposes current flow. The unit of resistance is the *ohm*, which is defined as the amount of resistance that allows 1 A of current to flow when 1 V is applied. The symbol for resistance is R. The unit abbreviation is the Greek letter omega (Ω). The schematic symbol for resistance is a zigzag line, as shown in Fig. 1-3.

FIG. 1-3 Resistor schematic symbol.

1-5 RESISTOR COLOR CODE

Customarily, resistors have color bands to indicate their ohmic value and tolerance. Figure 1-4 shows a typical axial carbon composition resistor and color chart to determine its value. Reading from left to right, the color bands indicate the first and second significant figures, the multiplier, and the tolerance. Hence, the value of the resistor shown in Fig. 1-5 is 6300 Ω, ± 10 per cent.

1-6 OHM'S LAW

A circuit may be defined as a path for current flow. A simple *series circuit* is shown in schematic form in Fig. 1-6. Ohm's law states that the current in an electrical circuit is proportional to the voltage and inversely proportional to the resistance. This may be expressed as

$$I = \frac{E}{R}$$

where I = intensity of current flow in amperes
 E = potential difference in volts
 R = resistance in ohms

By simple algebra we may solve for E by obtaining the product of IR:

$$E = IR$$

Resistance is determined by a rearrangement of the basic formula:

$$R = \frac{E}{I}$$

CIRCUITS REVIEW

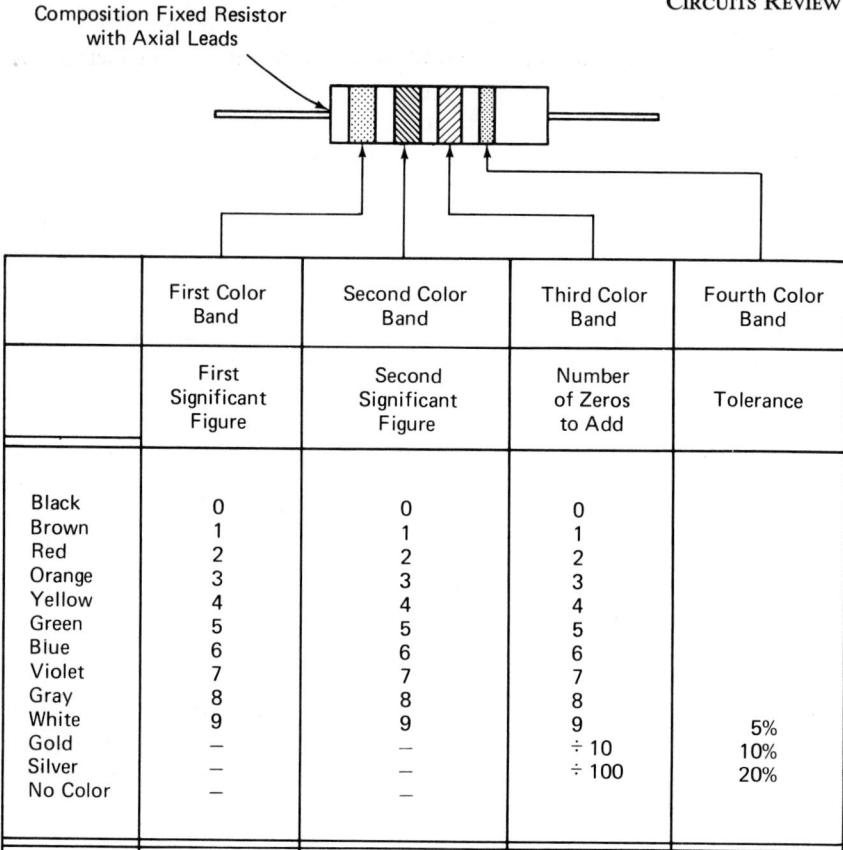

Composition Fixed Resistor with Axial Leads

	First Color Band	Second Color Band	Third Color Band	Fourth Color Band
	First Significant Figure	Second Significant Figure	Number of Zeros to Add	Tolerance
Black	0	0	0	
Brown	1	1	1	
Red	2	2	2	
Orange	3	3	3	
Yellow	4	4	4	
Green	5	5	5	
Blue	6	6	6	
Violet	7	7	7	
Gray	8	8	8	
White	9	9	9	5%
Gold	—	—	÷ 10	10%
Silver	—	—	÷ 100	20%
No Color	—	—		

FIG. 1-4 Resistor color code.

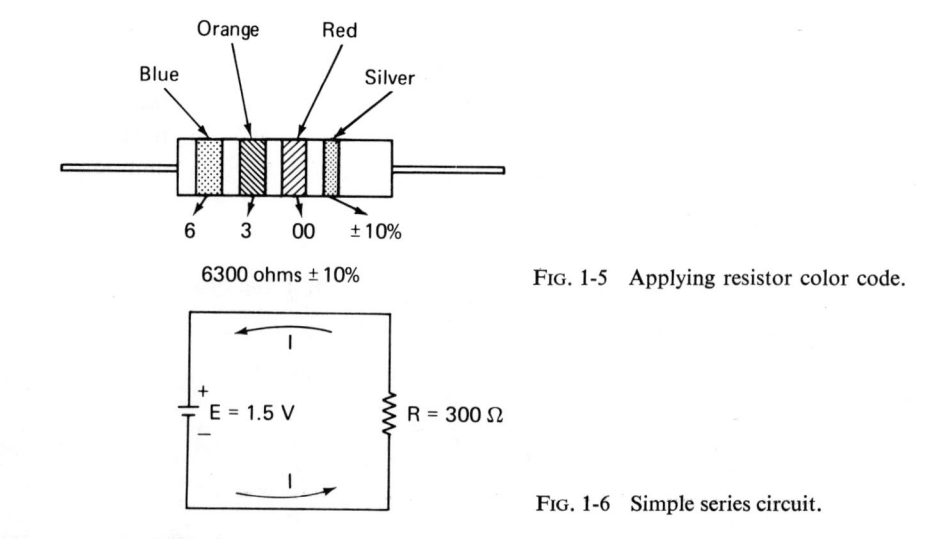

Blue Orange Red Silver
6   3   00   ±10%
6300 ohms ± 10%

FIG. 1-5 Applying resistor color code.

$E = 1.5$ V $R = 300\ \Omega$

FIG. 1-6 Simple series circuit.

Power 5

To illustrate how the current in a simple series circuit can be determined, refer to Fig. 1-7:

$$I = \frac{E}{R} = \frac{12}{3} = 4 \text{ A}$$

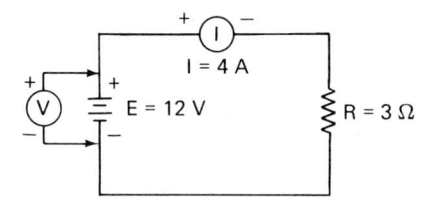

FIG. 1-7 Schematic diagram of simple series circuit.

Notice that the ammeter is in *series* with the battery and resistor. Its positive terminal connects to the positive side of the battery and its negative terminal is connected, via the load resistor, to the negative terminal of the battery. *Ammeters are always connected in series.*

When several resistors are connected in series, the total resistance can be calculated as

$$R_T = R_1 + R_2 + R_3 + R_4$$

1-7 POWER

Electrical power is the *time rate* at which a charge is forced to move by a voltage and is measured in *watts* (W). One watt of electrical power equals the work done in 1 s by 1 V of potential difference in moving 1 C of charge. Therefore, watts equal the product of volts and amperes:

$$P = EI$$

It is possible to calculate electrical power by two other formulas:

$$P = \frac{E^2}{R} \quad \text{or} \quad P = I^2 R$$

The rate at which electrical power is used is measured in *watthours* (Wh). If 10 V were impressed across a certain resistor and 0.1 A flowed, the power would be 10 V × 0.1 A = 1 W. If this amount of power flowed for 1 hour (h), the total power consumption would be 1 Wh. When larger amounts of power are used, it is customary to express the total consumption in *kilowatt hours* (kWh).

Example 1-1

What is the maximum rated current-carrying capacity of a resistor marked 5000 Ω, 200 W?

Solution

$P = I^2 R$. Solving for I,

$$I = \sqrt{\frac{P}{R}} = \sqrt{\frac{200}{5000}} = \sqrt{0.04} = 0.2 \text{ A}$$

Example 1-2

What should be the minimum power dissipation rating of a resistor of 20,000 Ω to be connected across of potential of 500 V?

Solution

$$P = \frac{E^2}{R} = \frac{500^2}{20,000} = 12.5 \text{ W}$$

The total power dissipated in a series circuit is the *sum* of the power being dissipated across each circuit or resistance element, as shown by the formula

$$P_T = P_1 + P_2 + P_3 + \text{etc.}$$

1-8 VOLTAGE-DROPPING RESISTORS

In many communications and industrial applications it is necessary to operate a device from some potential higher than that required for proper operation. Under these conditions it is essential that a *voltage-dropping resistor* be added in series with the device to drop the excessive voltage.

Example 1-3

A relay with a coil resistance of 500 Ω is designed to operate when 0.2 A flows through the coil. What value of resistance must be connected in series with the coil if operation is to be made from a 110-V dc line? The circuit is shown is Fig. 1-8.

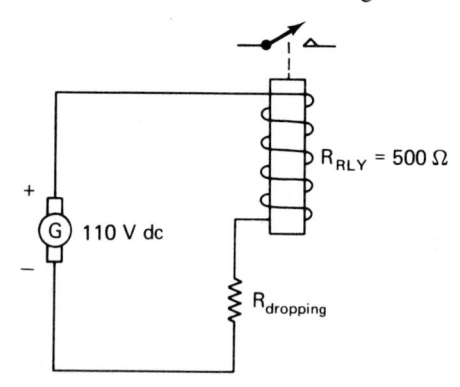

FIG. 1-8 Circuit showing use of voltage-dropping resistor.

Solution

First determine the *IR* drop across the coil.

$$E_{coil} = IR = 0.2 \times 500 = 100 \text{ V}$$

Inasmuch as the supply is 110-V direct current, it is apparent that the difference emf, or 10 V, must be dropped across the series resistor.

Parallel Circuits

Therefore,
$$R = \frac{E}{I} = \frac{10}{0.2} = 50\ \Omega$$

Example 1-4

What value of resistance should be connected in series with a 6-V battery that is to be charged at a 3-A rate from a 115-V dc line?

Solution

The series dropping resistor must be of such an ohmic value as to permit a 109-V drop across it (115 V − 6 V = 109 V). Hence,
$$R = \frac{109}{3} = 36.33\ \Omega$$

1-9 PARALLEL CIRCUITS

A parallel circuit differs from a series circuit in that *two or more* circuit elements are *connected directly to the same source of voltage.* An arrangement of this type, therefore, provides more than one path for the current to flow through. The more paths, or *branches*, added in parallel, the *less* opposition there is to the flow of current from the supply source.

From Ohm's law we know that the amount of current flowing through each branch depends upon the impressed voltage and the resistance of the branch. Therefore, the branch with the lowest resistance will have the most current (see Fig. 1-9).

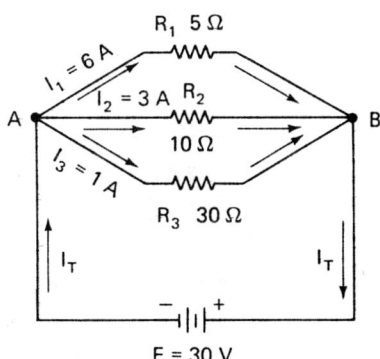

FIG. 1-9 Resistors in parallel.

The total current I_T of the parallel circuit is equal to the sum of the currents through the individual branches.
$$I_T = I_1 + I_2 + I_3$$
The total resistance of two resistors in parallel may be calculated as
$$R_T = \frac{R_1 \times R_2}{R_1 + R_2}$$

When there are three or more resistors parallel connected, the following formula is used:

$$\frac{1}{R_T} = \frac{1}{R_1} + \frac{1}{R_2} + \frac{1}{R_3}$$

1-10 CONDUCTANCE

Another method used to find R_T of a parallel resistive circuit involves calculating the *conductance* of the circuit. The conductance, G, is the *reciprocal* of its resistance and is expressed as $G = 1/R$. The unit of conductance has been the *mho*, which is ohm spelled backward. More recently this has been changed by international agreement to the *siemen*, which is abbreviated as S.

The conductance of a circuit is a measure of how *easily* current can flow through it. High conductance means considerable current will flow when only a relatively small emf is applied. Hence, the circuit having *high* conductance has *low* resistance, and vice versa. The total conductance of a parallel circuit can be found by calculating the conductance of each branch and then adding them up. Expressed mathematically, $G_T = G_1 + G_2 + G_3$. Once the total conductance is determined, the circuit's resistance can be found by obtaining the reciprocal of G_T.

1-11 SERIES–PARALLEL CIRCUITS

Many electronics communications circuits contain combinations of series and parallel resistors, such as shown in Fig. 1-10. Suppose that it is desired to find the total current and the voltage drop across each resistor,

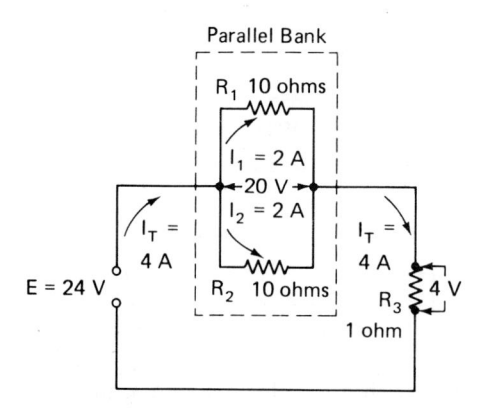

FIG. 1-10 Simple series-parallel circuit.

using the supply voltage and resistors shown. Our first step is to find the *equivalent* resistance of the parallel branch. Two 10 Ω resistors in parallel have an equivalent resistance of 5 Ω. This, added to R_3, gives $R_T = 6$ Ω. Then

$$I_T = \frac{24}{6} = 4\text{A}$$

Series–Parallel Circuits

The IR drop across $R_3 = 4\text{ A} \times 1\text{ }\Omega = 4\text{ V}$. The voltage drop across the parallel combination of R_1 and R_2 is $24\text{ V} - 4\text{ V} = 20\text{ V}$. This voltage causes 2 A to flow through each parallel resistor.

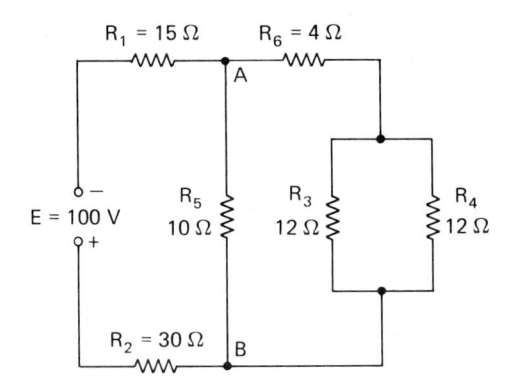

FIG. 1-11 Series-parallel circuit.

As a further study of series–parallel circuits, refer to Fig. 1-11. We begin by reducing or simplifying the branch *farthest* from the supply source and working toward the applied voltage. R_3 and R_4 reduce to an equivalent 6 Ω, which is in series with R_6. This adds up to 10 Ω, which is in parallel with R_5. The equivalent circuit thus far is shown in Fig. 1-12. The equivalent resistance of this combination is 5 Ω, which is in series with R_1 and R_2, which are 15 and 30 Ω, respectively. Hence, the R_T of the entire circuit is 50 Ω.

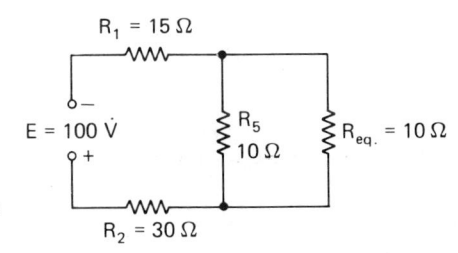

FIG. 1-12 Simplification of schematic diagram shown in Fig. 1-11.

With the voltage indicated, the total current will be 2 A. This 2 A of current will cause a 30-V drop across R_1, a 60-V drop across R_2, and a 10-V drop across R_5.

All sources of emf have some *internal resistance* that acts in series with the load. This resistance, designated R_s, is generally indicated in circuit diagrams as a separate resistor connected in series with the source. Both the voltage and power made available to the load may be increased if the internal resistance of the source is *reduced*.

1-12 MAXIMUM-POWER-TRANSFER THEOREM

The maximum-power-transform theorem states that maximum power is transferred from the source to the load when the *resistance of the load is equal to the internal resistance of the source.*

The efficiency of power transfer (ratio of output to input power) from the source to the load increases as the load resistance is increased. The efficiency approaches 100 per cent as the load resistance approaches a relatively large value compared with that of the source, since less power is lost in the source.

The problem of *high efficiency* and *maximum power transfer* is resolved as a *compromise* somewhere between the low efficiency of maximum power transfer and the high efficiency of the high-resistance load. When the amounts of power involved are large and the efficiency is important, the load resistance is made large relative to the source resistance so that the losses are kept small. In this case the efficiency will be high. When the problem of matching a source to a load is of paramount importance, as in communications circuits, a strong signal may be more important than the losses within the source, in which case the efficiency of transmission will be of the order of 50 per cent and at the same time the power in the load will be the maximum that the source is capable of supplying.

1-13 PREFIXES USED IN ELECTRONICS

The usual units of amperes, volts, and ohms are either too large or too small for most electronic applications. It is generally more convenient to use new units of measurement. These new units are formed by placing a special word or *prefix* in front of the unit. Each prefix has a definite meaning. Four

TABLE 1-1 PREFIXES

Multiples and Submultiples	Prefixes	Symbols
10^{12}	tera	T
10^9	giga	G
10^6	mega	M
10^3	kilo	k
10^2	hecto	h
10	deka	da
10^{-1}	deci	d
10^{-2}	centi	c
10^{-3}	milli	m
10^{-6}	micro	μ
10^{-9}	nano	n
10^{-12}	pico	p
10^{-15}	femto	f
10^{-18}	atto	a

Miscellany

of the more commonly used prefixes are

1. Milli-unit means $\frac{1}{1000}$ of the unit.
2. Kilo-unit means 1000 of these units.
3. Micro-unit means $\frac{1}{1,000,000}$ of the unit.
4. Meg-unit means 1,000,000 of these units.

A complete list of prefixes has been adopted by the International Committee on Weights and Measures and is shown in Table 1-1.

To illustrate the use of prefixes and how conversions can be made from one electrical unit to another, the following examples are given:

$$2.3 \text{ V} = 2300 \text{ mV (millivolts)}$$

$$0.045 \text{ A} = 45 \text{ mA (milliamperes)}$$

$$0.025 \text{ mA} = 25 \text{ } \mu\text{A (microamperes)}$$

$$1{,}150{,}000 \text{ hertz (Hz) (formerly cycles per second)}$$
$$= 1150 \text{ kHz (kilohertz)}$$

$$10{,}700 \text{ kHz} = 10.7 \text{ MHz (megahertz)}$$

$$0.25 \text{ megohm (M}\Omega\text{)}$$
$$= 250 \text{ kilohms (k}\Omega\text{)} = 2.5 \times 10^5 \text{ }\Omega$$

$$27 \text{ nanoseconds (ns)}$$
$$= 27 \times 10^{-9} \text{ s} = 27 \text{ billionths of a second}$$

1-14 MISCELLANY

It is sometimes necessary to make up a series–parallel combination of resistors to provide a desired value. For example, show how three resistors of equal value may be connected so that the total resistance is one half times the resistance of one unit. The solution is shown in Fig. 1-13.

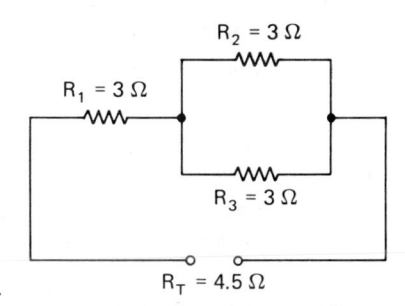

FIG. 1-13 Series-parallel circuit. $R_T = 4.5 \text{ }\Omega$

Consider another example. Draw a simple schematic diagram showing how three resistors of equal value may be connected so that the total resistance is two thirds of one unit. This is accomplished as shown in Fig. 1-14.

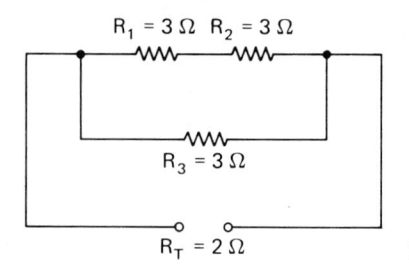

FIG. 1-14 Series-parallel circuit.

Practice Problems

1. One coulomb past a given point in 0.5 s is equal to how many amperes?
2. What is the ohmic value of a resistor with the following color code: yellow, violet, gold?
3. Within what resistance values should an axial resistor be that is marked blue, green, and red?
4. How much current will flow through a resistance of 20 Ω if connected across a potential of 120 V?
5. A small soldering iron has 0.4 A of current flowing through its heating element when connected to a 120-V outlet. What is the heater resistance?
6. A certain radio tube requires 0.3 A. Its resistance is 21 Ω. What voltage is required?
7. The transmitting tube type 801 requires a filament current of 1.25 A and a filament voltage of 7.5 V. Calculate the hot resistance of the filament.
8. An electric toaster draws 10.5 A from the power line when the voltage across the heater element is 117 V. What is the hot resistance of the toaster?
9. A certain resistor in a television receiver is 2200 Ω. Determine the current through the resistor if the impressed potential is 45 V.
10. A transistor has an 8.5-V drop across its collector resistance of 680 Ω. Calculate the current flowing through the resistor.
11. A plate load resistor in a vacuum-tube circuit is 56,000 Ω. If 0.004 A of current flows through it, what is the *IR* drop across the resistor?
12. What is the wattage rating of a soldering iron if it is designed for 120-V operation and has 303-Ω resistance?
13. A certain voltage-dropping resistor of 1200 Ω has 0.065 A of current flowing through it. What is its power dissipation?
14. What is the hot resistance of the heater of a 12 AU7 radio tube that has 0.15 A flowing through it when connected to a 12.6-V potential?
15. What is the voltage drop across a 470-Ω cathode biasing resistor of a certain radio tube if 0.008 A flows through it?
16. Calculate the current through a 10-kΩ resistor, connected in the collector circuit of a transistor, when the voltage across it is 9.4 V.

Practice Problems

17. A certain resistor in a voltage-divider circuit has 90 V across it. What wattage will it dissipate if its resistance is 6800 Ω?

18. If the voltage impressed across a series circuit is increased three times and the resistance is doubled, how many times greater (or less) will the resultant current be compared to the original value?

19. A 270-Ω emitter resistor in a transistor circuit has 0.01 A of current flowing through it. What is the power dissipation in the resistor?

20. If a certain transistor has a base current of 0.00012 A, what is the base current in microamperes?

21. The carrier frequency of an FM radio station is 108,200 kHz. What is its frequency in megahertz?

22. What is the total resistance of resistors 500, 1000, and 1500 Ω when parallel connected?

23. What is the total circuit current of a 180- and 240-Ω resistor parallel connected across a 100-V source?

24. A 55- and a 23-Ω resistor are connected in parallel. What is the source voltage if the source current is 2.5 A?

25. A 400-Ω, 10-W resistor and a 1500-Ω, 50-W resistor are connected in parallel. What is the maximum voltage that can be connected across the combination without exceeding the wattage rating of either resistor?

26. In the circuit shown in Fig. 1-15, what is the value of E_T when $E_3 = 17.5$ V?

Fig. 1-15

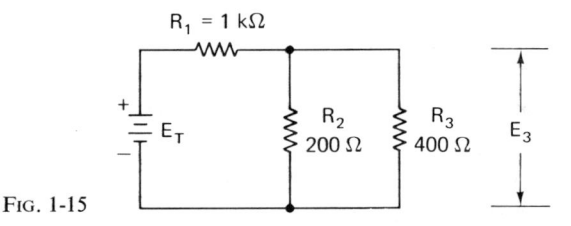

27. What is the power dissipated in resistor R_2 (Fig. 1-15)?

Problems 28 through 31 refer to Fig. 1-16.

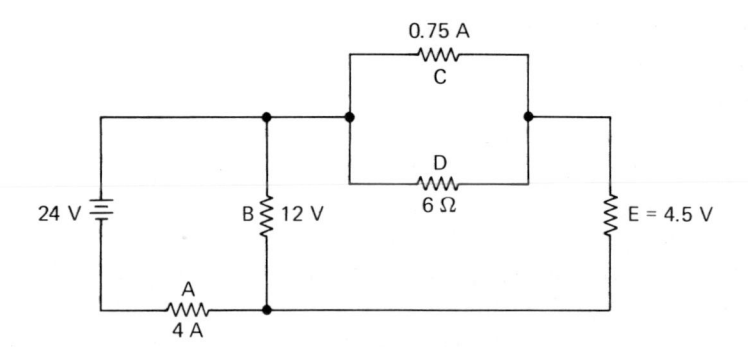

Fig. 1-16

28. What is the voltage drop across resistor A?
29. What is the resistance of resistor E?
30. What is the current through resistor D?
31. Determine the voltage drop across resistor C.

Problems 32 through 36 refer to Fig. 1-17.

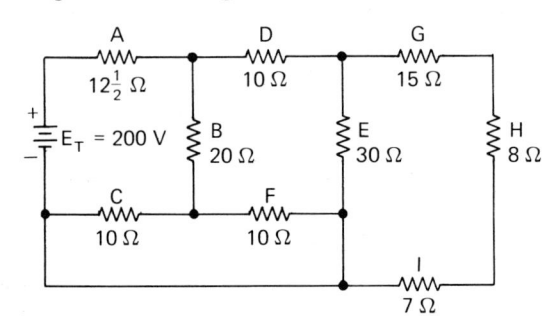

FIG. 1-17

32. What is the total resistance of the circuit?
33. Determine the current through resistor A.
34. What is the voltage drop across resistor B?
35. Calculate the voltage across resistor H.
36. Find the total power dissipated in the circuit.

Commercial License Questions

Sections in which answers to questions are given appear in parentheses. A bracketed number following a question implies that it applies only to that element.

1. What is an electron? An ion? (1–1)
2. By what other expression may a difference of potential be described? (1–2)
3. Name four methods by which an electrical potential may be generated. (1–2)
4. Define the term *coulomb*. (1–3)
5. By what other expression may an electric current flow be described? (1–3)
6. What is the unit of resistance? (1–4)
7. What is the ohmic value and tolerance of a resistor if its left-to-right color code is green, brown, yellow, and silver? (1–5)
8. What is the ohmic value and tolerance of a resistor if its left-to-right color code is red, black, orange, and gold? (1–5)

9. State the three ordinary mathematical forms of Ohm's law. (1-6)
10. What instrument is used to measure current flow? (1-6)
11. If the voltage applied to a circuit is doubled and the resistance of the circuit is increased to three times its former value, what will be the final current value? (1-6)
12. Show by a diagram how a voltmeter and ammeter should be connected to measure power in a dc circuit. (1-6, 1-7)
13. Draw a simple schematic diagram showing the method of connecting three resistors of equal value so that the total resistance will be three times the resistance of one unit. (1-6)
14. What is the sum of all voltage drops around a simple dc series circuit, including the source? (1-6)
15. What is the unit of electrical power? (1-7)
16. What is the formula for determining the power in a dc circuit when the voltage and resistance are known? (1-7)
17. What is the formula for determining the power in a dc circuit when the current and resistance are known? (1-7)
18. What is the formula for determining the power in a dc circuit when the current and voltage are known? (1-7)
19. What is the difference between electrical power and electrical energy? (1-7)
20. If the value of a resistance to which a constant emf is applied is halved, what will be the resultant proportional power dissipation? (1-7)
21. If the value of a resistance across which a constant emf is applied is doubled, what will be the resultant proportional power dissipation? (1-7)
22. What will be the heat dissipation, in watts, of a resistor of 20 Ω having a current of 0.25 A passing through it? (1-7)
23. What is the maximum rated current-carrying capacity of a resistor marked 5000 Ω, 200 W? (1-7)
24. What should be the minimum power dissipation rating of a resistor of 20,000 Ω to be connected across a potential of 500 V? (1-7)
25. How much energy is consumed in 20 h by a radio receiver rated at 60 W? (1-7)
26. A 6-V storage battery has an internal resistance of 0.01 Ω. What current will flow when a 3-W, 6-V lamp is connected? (1-7)
27. If a vacuum tube having a filament rated at 0.25 A and 5 V is to be operated from a 6-V battery, what is the value of the necessary series resistor? (1-8)
28. A relay with a coil resistance of 500 Ω is designed to operate when 0.2 A flows through the coil. What value of resistance must be connected in series with the coil if operation is to be made from a 110-V dc line? (1-8)

29. What value of resistance should be connected in series with a 6-V battery that is to be charged at a 3-A rate from a 115-V dc line? (1–8)
30. What is the total resistance of a parallel circuit consisting of one branch of 10-Ω resistance and one branch of 25-Ω resistance? (1–9)
31. If resistors of 5, 3, and 15 Ω are connected in parallel, what is the total resistance? (1–9)
32. Indicate by a diagram how the total current in three branches of a parallel circuit can be measured by one ammeter. (1–9)
33. Draw a simple schematic diagram showing the method of connecting three resistors of equal value so that the total resistance will be one third of one unit. (1–9)
34. What is the unit of conductance? (1–10)
35. Define the term *conductance*. (1–10)
36. What is the conductance of a circuit if 6 A flows when 12 V_{dc} is applied to the circuit? (1–10)
37. Explain the effects of the internal resistance of a power source on the amount of power it can deliver to a load. (1–12)
38. Discuss the maximum power transfer theorem. (1–12)
39. Discuss the relationship between maximum efficiency and maximum power transfer in a circuit. (1–12)
40. What is the meaning of the prefix kilo? (1–13)
41. What is the meaning of the prefix micro? (1–13)
42. Explain the meaning of the prefix micromicro. (1–13)
43. How many micromicrofarads are there in 1 microfarad? What is the difference between a milliwatt and a kilowatt? (1–13)
44. What is the meaning of the prefix meg? (1–13)
45. Draw a simple schematic diagram showing the method of connecting three resistors of equal value so that the total resistance will be one and one half times the resistance of one unit. (1–14)
46. Draw a simple schematic diagram showing the method of connecting three resistors of equal value so that the total resistance will be two thirds the resistance of one unit. (1–14)

2

Conductors and Insulators

2-1 CONDUCTORS

The various components of an electronic circuit are interconnected by means of *conductors*. These may be small insulated wires or the conductive pattern on printed wiring boards. In some applications one of the conductors may be the metal chassis of a receiver or transmitter. In other cases it may be the metallic structure of a missile or the frame of an aircraft or automobile.

The amount of current the conductor is to carry dictates the *wire size*. Also, the operating voltage of the system will regulate the type and thickness of the insulated material around the wire. In high-power applications, where the current may reach several hundred amperes, *bus bars* are used.

The conductors listed in Table 2-1 are arranged in order of their *conductivity*. All are presumed to have the same cross-sectional area and length.

TABLE 2-1

Conductors
Silver
Copper
Aluminum
Zinc
Brass
Iron

To compare the resistance of one conductor with another, a standard size must be established. The standard unit for diameter is the *mil* [0.001 inch (in.)] and for length is the foot. Hence, the *mil-foot* has become the standard unit of size. The resistance in ohms of a unit size of conductor is called its *specific resistance* and is abbreviated ρ (Greek letter rho).

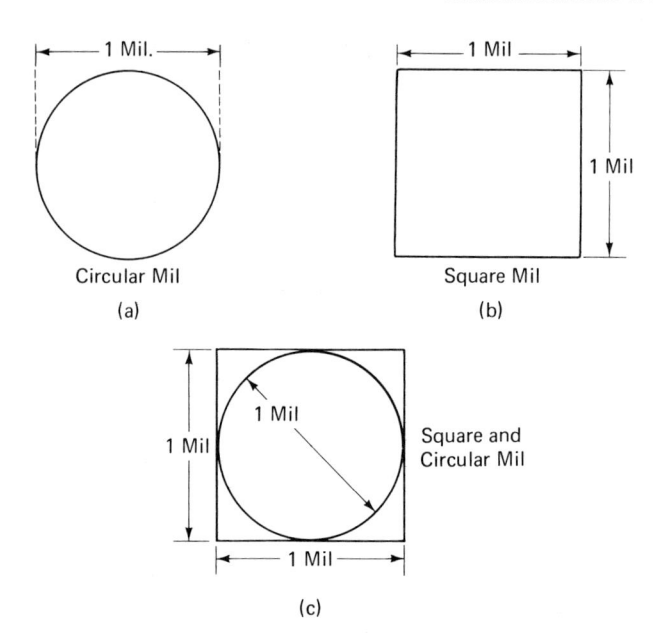

FIG. 2-1 (a) Circular mil; (b) square mil; (c) comparison of circular mil to square mil.

A *circular mil* is the area of a circle having a diameter of 1 mil, as shown in Fig. 2-1a. The area in circular mils of a round conductor is obtained by squaring the diameter measured in mils. Thus, a wire having a diameter of 25 mils has an area of 25^2 or 625 circular mils.

The *square mil* is a convenient unit of cross-sectional area for square or rectangular conductors. A square mil is the area of a square the sides of which are 1 mil, as shown in Fig. 2-1b. To obtain the cross-sectional area in square mils of a rectangular conductor, multiply the length of one side by that of the other, each length being expressed in mils.

In comparing square and round conductors the circular mil is the *smaller* unit of area. The comparison is shown in Fig. 2-1c. The area of a circular mil is equal to 0.7854 of a square mil. Therefore, to determine the circular mil area when the square-mil area is given, divide the area in square mils by 0.7854. Conversely, to determine the square-mil area when the circular-mil area is given, multiply the area in circular mils by 0.7854.

2-2 SPECIFIC RESISTANCE OR RESISTIVITY

The specific resistance of a substance is the resistance of a unit volume of that substance. Tables of specific resistance are based on the resistance in ohms of a volume of the material 1 foot (ft) long and 1 circular mil in cross-sectional area, usually at 20°C. The specific resistances of some common materials are given in Table 2-2.

Wire Sizes and Fuses

TABLE 2-2 SPECIFIC RESISTIVITIES
(Resistivities of various materials in ohms per circular mil-foot at 20°C)

Material	Resistivity
Silver	9.56
Copper (annealed)	10.4
Aluminum	17.
Tungsten	34.
Brass	42.
Nickel	60.
Iron	61.
Nichrome	675.
Carbon	22,000

The resistance of a conductor may be calculated if the length, cross-sectional area, and specific resistance are known, as follows:

$$R = \rho \frac{L}{A}$$

where ρ is the specific resistance in ohms per circular mil-foot, L the length in feet, and A the cross-sectional area in circular mils.

The resistance of pure metals, such as silver, copper, and aluminum, *increases* as the temperature increases. However, the resistance of some alloys, such as constantan and manginin, changes very little as the temperature changes. Some measuring instruments use these alloys because the resistance of the circuits must remain constant if accurate measurements are to be achieved.

2-3 WIRE SIZES AND FUSES

Wires are manufactured in sizes known as the American Wire Gage (AWG). The wire diameters become *smaller* as the gage numbers become larger. A table of wire sizes is shown in Appendix A.

It is customary to protect circuits against *overloads* in order to prevent possible damage to electronics components. *Fuses* are employed for this purpose. Excessive current melts the fusible element and opens the circuit.

The resistance of a fuse is normally very low. When a fuse blows, it becames an open circuit. The full applied voltage of the circuit then appears across the open terminals of the fuse. For this reason fuses have *voltage ratings* to indicate the maximum voltage that may appear across the fuse without arcing.

In many cases circuits are protected against excessive current by means of *circuit breakers* or thermal cutout devices because they are more convenient. They contain a thermal element, generally in the form of a spring, that expands with heat and trips open the circuit. After cooling, the thermal element may be reset to restore normal operation, assuming that the short circuit has been eliminated.

2-4 CONDUCTION IN GASES AND LIQUIDS

Gases may become conductors of electricity under certain conditions. If the voltage across a small volume of gas is sufficiently large, one or more electrons may be pulled from the various atoms. These will be attracted toward the positive potential. The atoms that have lost electrons become *positive ions* and will be attracted toward the negative potential. If the voltage is increased, the electron flow and ionic currents will increase. These currents are influenced by the pressure and temperature of the gas.

The potential difference through which an electron must pass to acquire sufficient speed to break up a gas molecule is called the *ionizing voltage*. This varies for different gases.

Under some conditions it is possible for a molecule of gas to acquire an extra electron, causing it to have an overall negative charge. Under these conditions we say that the gas molecule is a *negative ion*. After ionization takes place the resistance drops to a low value, and a large amount of current can flow.

Liquids that are good conductors because of ionization are called *electrolytes*. Acid and salt solutions in water may be classified as electrolytes. The process of producing a chemical change in a material by allowing ionization current to flow in an electrolyte is called *electrolysis*.

2-5 INSULATORS

Any substance that has a very high resistance, ranging in the order of megohms, may be classified as an *insulator*. Basically, insulators are used for two functions: (1) to isolate conductors to prevent conduction between them, and (2) to store an electric charge. An insulator maintains this charge because electrons cannot flow to neutralize the charge. Insulators are commonly referred to as *dielectric* materials.

Table 2-3 makes a comparison of the *dielectric strength* of several types of insulators, The dielectric strength, also known as the *voltage breakdown rating*, is the amount of voltage that may be placed across a material before breakdown occurs. The thickness of the dielectric material is assumed to be 1 mil.

Corona discharge is a result of excessively high potentials on an object. Generally, if the voltage is raised to a higher level, a spark discharge may occur. To reduce corona effect, the conductors that are carrying high potentials should be smooth, rounded, and thick. This tends to equalize the potential difference from all parts on the conductor to the surrounding air. Any

TABLE 2-3 VOLTAGE BREAKDOWN OF INSULATORS

Material	Dielectric Strength (V/mil)	Material	Dielectric Strength (V/mil)
Air or vacuum	20	Paraffin wax	200–300
Bakelite	300–550	Phenol, molded	300–700
Fiber	150–180	Polystyrene	500–760
Glass	335–2000	Porcelain	40–150
Mica	600–1500	Rubber, hard	450
Paper	1250	Shellac	900
Paraffin oil	380		

sharp point may have a more intense field developed, making it more susceptible to corona and eventual spark discharge.

Materials that are good insulators at one frequency may not be at another. At low frequencies, such as power or audio, materials like rubber fiber, porcelain, cloth, and so forth, are satisfactory. At radio frequencies (RF) these insulators have excessive dielectric losses, and may allow currents to flow through them.

For RF and other high-frequency applications it is necessary that such materials as Pyrex, Micalex, isolantite, steatite, polyethylene, and Teflon be used. These materials have excellent insulation qualities because they keep dielectric losses low.

2-6 SKIN EFFECT

Skin effect is the tendency of RF currents to travel at or near the surface of a conductor rather than uniformly throughout the entire cross-sectional area. Although skin effect is present at all frequencies its magnitude *increases* with frequency. At ultrahigh frequencies (UHF) the depth of current penetration is very small, resulting in virtually all the current existing at the surface of the conductor. Because of this phenomenon, it is common practice to use tubular conductors of relatively large diameters so that the surface area will be large.

2-7 SOLDERING

Ordinary soft solder is a *fusible alloy* consisting essentially of *tin* and *lead*. It is used for the purpose of joining together two or more metals (electrical components and their associated conductor wires) at temperatures below their melting points.

All common metals are covered with a nonmetallic film known as an *oxide*, which forms an effective insulating barrier that prevents metals from touching each other. As long as this nonmetallic barrier is present on the surface of metals, the metals themselves cannot make actual metal-to-metal contact; as a result, intermetallic action (soldering) cannot take place. It is the function of the soldering flux to *remove* the nonmetallic oxide film from the surface of the metals and keep it removed during the soldering operation in order that the clean, free metals can make mutual metallic contact.

Rosin fluxes are the only type used in electronic work. Never should an acid-core solder be used when soldering electrical connections. Ultimately, the acid will cause corrosion and the joint becomes one having a high resistance.

Practice Problems

1. Find the circular-mil area of a conductor 0.032 in. in diameter.
2. Calculate the diameter in mils of a wire whose cross-sectional area is 168,000 circular mils.
3. Find the diameter in inches of a wire whose cross-sectional area is 6530 circular mils.
4. What is the square-mil area of a conductor whose circular-mil area is 1624?
5. A rectangular conductor is 0.5 by 2.0 in. What is its circular-mil area?
6. What is the resistance of a copper wire whose diameter is 25.3 mils and whose length is 125 ft?
7. What is the approximate resistance of an aluminum wire whose dimensions are equal to that of the copper wire in problem 6?
8. A certain piece of nickel wire has a specific resistance of 34 Ω/mil-foot. If the length is 1.5 ft and the diameter is 20.1 circular mils, what is its resistance?
9. What is the length of a copper wire whose resistance is 0.45 Ω and whose area is 160 circular mils?
10. Calculate the length of a piece of nichrome wire of 8-mil diameter whose specific resistance is 675 Ω/mil-foot?

Commercial License Questions

Sections in which answers to questions are given appear in parentheses. A bracketed number following a question implies that it applies only to that element.

1. Name four conducting materials in the order of their conductivity. (2–1)
2. What effect does the cross-sectional area of a conductor have upon its resistance per unit length? (2–2)
3. Explain the factors that influence the resistance of a conductor. (2–2)
4. If the diameter of a conductor of given length is doubled, how will the resistance be affected? (2–2)

Commercial License Questions

5. Does the resistance of a copper conductor vary with variations in temperature and, if so, in what manner? (2–2)
6. With respect to electrons, what is the difference between conductors and non-conductors? (2–5)
7. Name four materials that are good insulators at radio frequencies. (2–5)
8. Name four materials that are not good insulators at radio frequencies, but are satisfactory for use at commercial power frequencies. (2–5)
9. What is the meaning of skin effect in conductors of RF energy? (2–6)
10. Why is rosin used as soldering flux in radio construction work? (2–7)

3

Magnetism Review

3-1 PERMANENT MAGNETISM

Magnets may be divided into two groups: (1) permanent magnets, and (2) electromagnets.

Hardened steel and certain alloys when subjected to strong magnetic fields become magnetized and retain their magnetism over long periods of time. The amount of magnetism remaining in the material after the magnetizing force has been removed in called *residual magnetism*.

A magnet has two poles: one *north* and the other *south*. If a small permanent bar magnet is suspended so that it is free to swing in a horizontal plane, it will come to rest with the same end always pointing to the earth's magnetic north pole. This north-seeking pole is frequently stamped with an N; the south seeking pole is marked S.

If unlike poles of magnets are brought close together, they are attracted to each other. Like poles would repel each other. This indicates the law of magnetic forces: *Like poles repel, unlike poles attract*.

The attractive or repulsive forces between two magnets vary with their *strength* and *inversely* as the *square of the distance* between them. The following equation expresses this relationship:

$$F = \frac{m_1 m_2}{d^2}$$

where F = force between two poles
m_1 and m_2 = magnitudes of the poles
d = distance between poles

Magnetic Flux

3-2 MAGNETIC FLUX

The field surrounding a magnet is conventionally represented by lines, as shown in Fig. 3-1, called *lines of force*. The arrowheads on these lines imply that they leave the magnet at or near the north pole and return to the south pole.

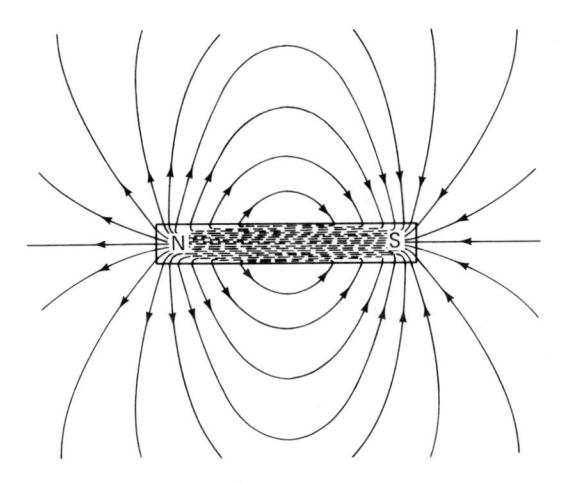

FIG. 3-1 Representation of magnetic field.

The entire group of lines of force is called *magnet flux* and is represented by the symbol ϕ (Greek letter phi). The unit of flux is the *maxwell* (Mx), which is equal to one magnetic line of force. In Fig. 3-2 the flux illustrated is equal to 6 Mx, because there are six lines of force shown.

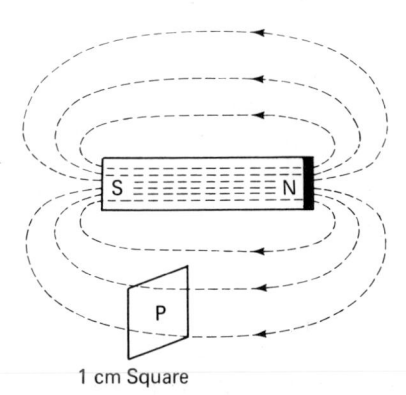

FIG. 3-2 Flux density at *P* equals two gauss (2 lines per cm²). 1 cm Square

The number of magnetic lines passing through an area 1 centimeter (cm) square is referred to as *flux density* and is represented by the letter *B*. If one line passed through an area of 1 cm² is it called a *gauss* (G). From the illustration in Fig. 3-2 we see that the flux density *B* is 2 G at point *P* in the field.

3-3 PERMEABILITY

The ability of a material to concentrate magnetic lines of force is called *permeability*. Any material that is easily magnetized, such as soft iron, tends to concentrate the magnetic flux. The numerical values of permeability for different materials are assigned in comparison with *air* or *vacuum*, which have a reference value of 1. The permeability of magnetic materials therefore equals the number of times greater the flux density is in the magnetic material as compared to air.

The symbol for permeability is the Greek letter μ. Typical values of μ for iron vary from as low as 100 to as high as 5000, depending upon the grade used. Assume that the flux density, in air, of a given magnet is 1 G. If an iron core is placed in the same position in the magnetic field, it increases the flux density to 400 G. Therefore, the permeability of the iron core is 400.

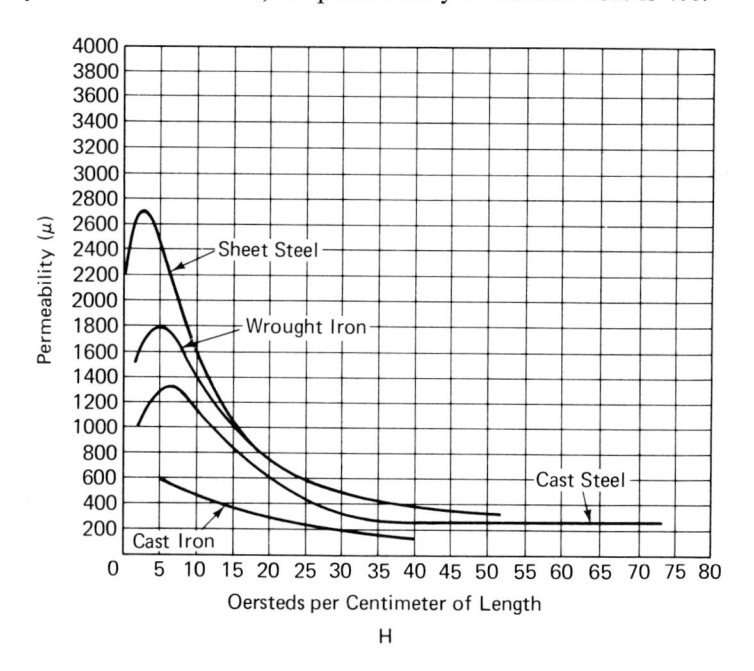

FIG. 3-3 Permeability curves.

The permeability of magnetic materials also varies according to the degree of magnetization. Consider the curve for sheet steel shown in Fig. 3-3. This shows how the permeability of a magnetic material varies with magnetizing force, which is measured in *oersteds* (Oe) and whose symbol is H. If too large a magnetizing force is used for a given size and type of magnetic material, the permeability reduces and the efficiency is impaired.

The relationship between the magnetizing force H, flux density B, and permeability μ is indicated by

$$B = \mu \times H$$

Because of the high permeability of some materials, the lines of force surrounding a magnet may be distorted if this material is placed in a magnetic field. An example of this appears in Fig. 3-4. It is easier for the lines of force to take a more devious route because the *reluctance* of the path is reduced. This same principle is utilized to protect sensitive instruments, such as meters and watches, from external magnetic fields.

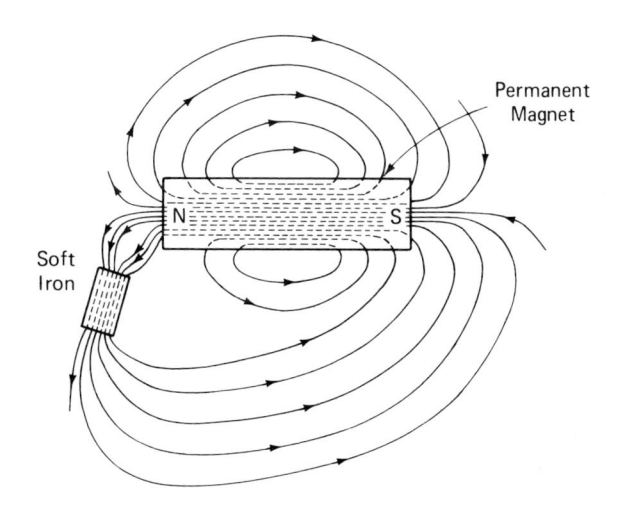

FIG. 3-4 Distortion of magnetic field by addition of soft iron bar.

A number of ceramic-like materials, called *ferrites*, have been developed having ferrromagnetic properties. They are characterized by very high permeability and yet have high values of electrical resistance.

3-4 AMPERE TURNS

A conductor carrying an electric current has a magnetic field built up around it. The greater the current, the larger the field. The field is at right angles to the current producing it and extends the complete length of the conductor, as shown in Fig. 3-5. If the conductor is bent into a loop, as in Fig. 3-6, the lines of force are concentrated inside the loop. If the conductor is wound into a number of turns forming a coil whose length is considerably longer than its diameter, it is called a *solenoid*. This results in a much stronger field inside the solenoid.

The direction of the magnetic field can be determined when the direc-

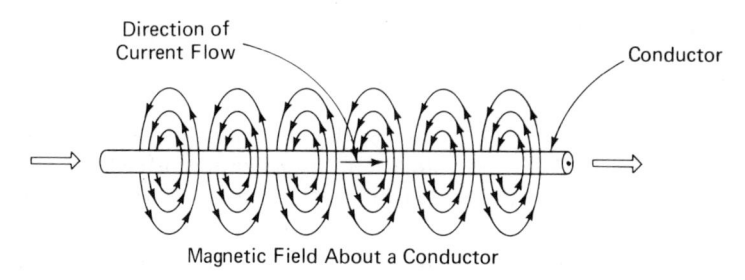

FIG. 3-5 Magnetic fielding surrounding a conductor.

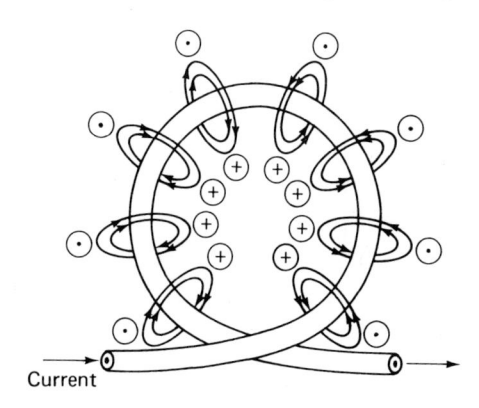

FIG. 3-6 Concentration of magnetic field around a single loop.

tion of the current is known. It is known as the left-hand rule and may be stated as follows:

> Grasp the conductor in the left hand with the thumb extended, in the direction of current flow. The fingers will then point in the direction of the magnetic field. Using this rule, if either the direction of the current or the direction of the field is known, the other may be obtained.

The strength of the magnetic field surrounding a coil depends upon the *number of turns* and the *current* flowing through them. The product of the current, in amperes, and the number of turns is therefore a measure of the field strength of an electromagnet. This may be expressed by the simple formula

$$\text{ampere turns} = NI$$

If a piece of soft iron is placed inside the solenoid, an *electromagnet* is formed. When direct current flows through the coil, the core will become magnetized with the same polarity that the coil would have without the core. If the current is reversed, the polarity of both the coil and the soft-iron core is reversed.

A soft-iron bar is attracted by a current-carrying coil, if the coil and bar

Hysteresis Loss

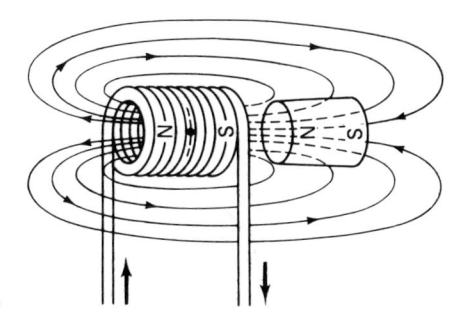

FIG. 3-7 Solenoid with iron core.

are orientated as in Fig. 3-7. The lines of force extend through the soft iron and magnetize it by induction. Because unlike poles attract, the iron bar is pulled toward the coil. If the bar is free to move, it will be drawn into the coil to a position near the center where the field is the strongest. This is the principle of operation of the solenoid, which finds widespread application in electromechanical systems.

3-5 BH MAGNETIZATION CURVE

The relationship between the number of gauss (B) flowing in a magnetic circuit and the oersteds (H) producing the flux can be shown graphically. Several curves showing this relationship for different magnetic materials are shown in Fig. 3-8. These are called BH curves. They indicate how much magnetizing force is needed to produce a given flux density in the core of the device (i.e., transformer, solenoid, etc.).

3-6 MAGNETIC CIRCUIT

The path that the flux takes is called the magnetic circuit. To produce a magnetic flux, a *magnetomotive force* (mmf) is required. The flux also depends on the amount of opposition, which is called *reluctance*. This is expressed by

$$\phi = \frac{F}{\mathcal{R}}$$

where ϕ = flux
 F = magnetomotive force
 $\mathcal{R}$ = reluctance

This formula has often been referred to as Ohm's law for the magnetic circuit.

The unit for mmf is the *gilbert* (Gb):

$$\text{mmf} = 1.26\,NI \quad \text{Gb}$$

3-7 HYSTERESIS LOSS

Hysteresis is a lagging of the magnetizing effect behind the magnetizing cause. In the core of a power transformer, for example, the flux lags behind

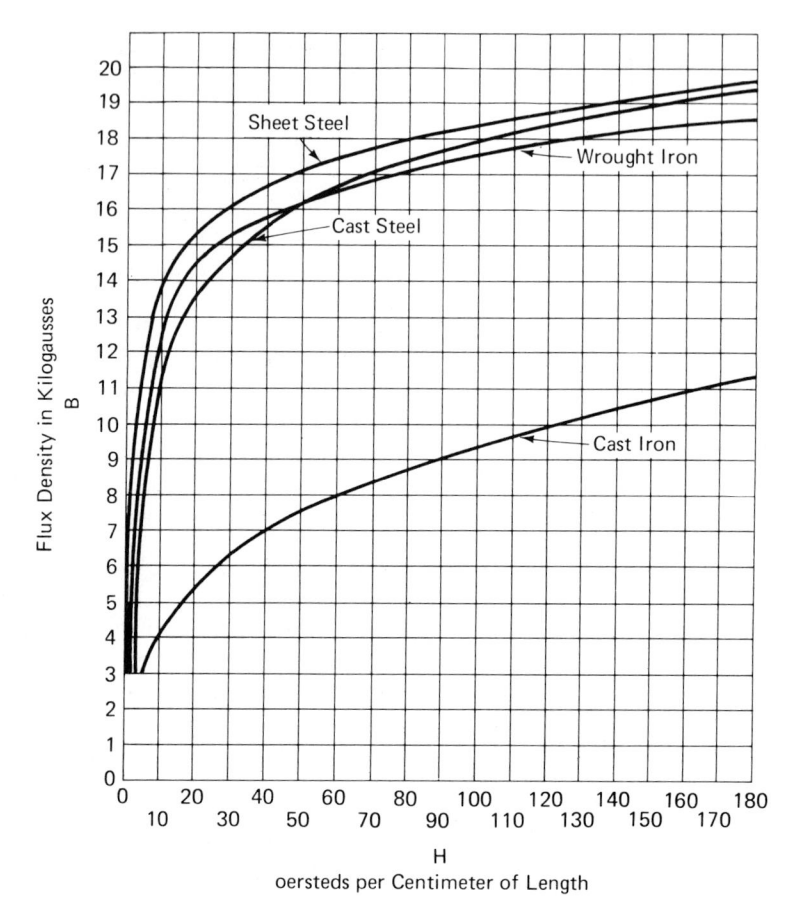

FIG. 3-8 *BH* magnetization curve.

the increase or decrease of the magnetizing force. This results from the fact that the magnetic characteristics of the metal are not capable of immediately aligning themselves with the changing external magnetizing force. Also, they do not return exactly to their original positions when the magnetic forces are removed.

If the magnetizing force varies at a slow rate, the hysteresis loss is negligible. When the magnetizing force reverses thousands or even millions of times per second, the hysteresis may cause a considerable amount of energy to be dissipated in the core. Heat is produced in the process and is called *hysteresis loss*.

By using a graph having *BH* coordinates, it is possible to plot the hysteresis characteristics of a given metal, as shown in Fig. 3-9, which is called a *hysteresis loop*. The area contained within the loop indicates the amount of loss.

Induced Current 31

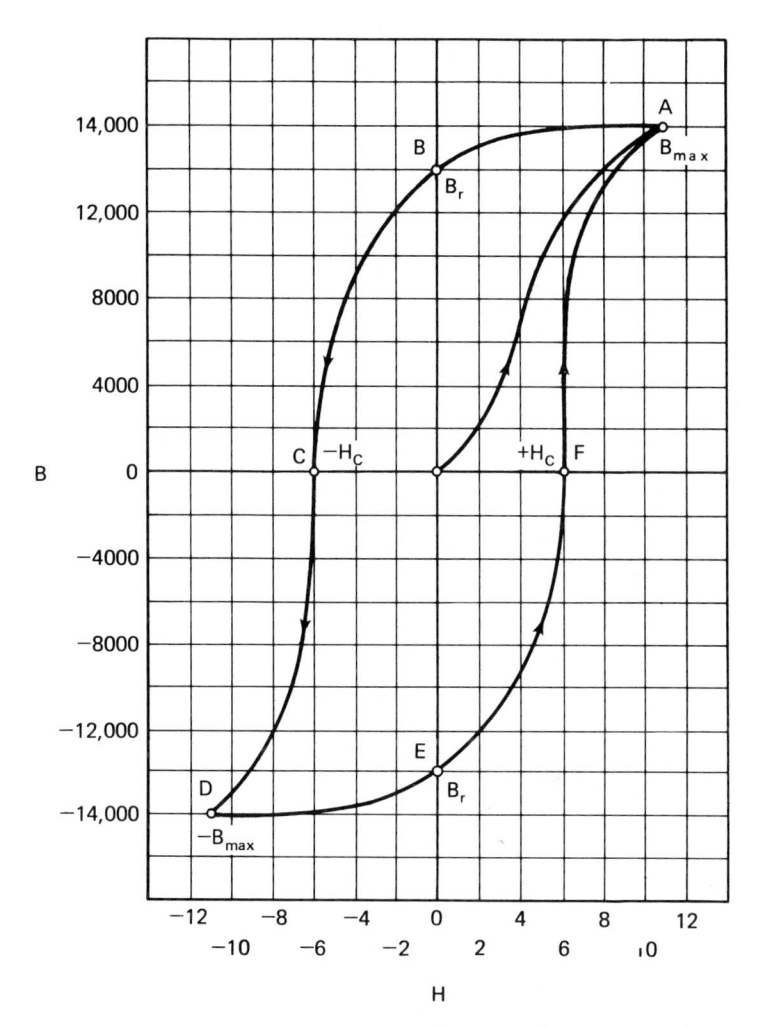

FIG. 3-9 Hysteresis loop.

3-8 INDUCED CURRENT

When a conductor is moved through a magnetic field, a voltage is developed in the conductor (see Fig. 3-10). The *galvanometer* (a sensitive current-indicating meter) will deflect momentarily in one direction as the conductor is moved downward and in the opposite direction when pulled out. No emf is produced when the conductor moves parallel to the lines of force.

The factors that determine the magnitude of the induced emf are

1. The number of *turns* of the conductor passing through the magnetic field.

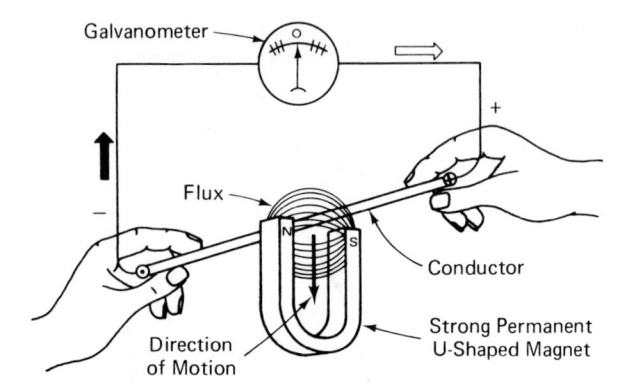

FIG. 3-10 Inducing a current in a conductor.

2. The *strength* of the magnetic field.
3. Relative *speed* between the conductor and the field.
4. The *angle* at which the conductor passes through the field. The voltage is greatest when cutting at right angles and zero when moving parallel to the field.

The induced voltage may be calculated from the formula

$$E_{ind} = \frac{N \times \phi}{T \times 10^8} \text{ V}$$

where ϕ = maxwells or number of lines
T = time in seconds
N = number of turns

It is possible for the magnetic field around one conductor carrying a current to *induce* a voltage and cause current to flow in another conductor (circuit) that is not electrically connected to the first. This is known as *mutual induction*, as shown in Fig. 3-11. When the switch is closed, current flows

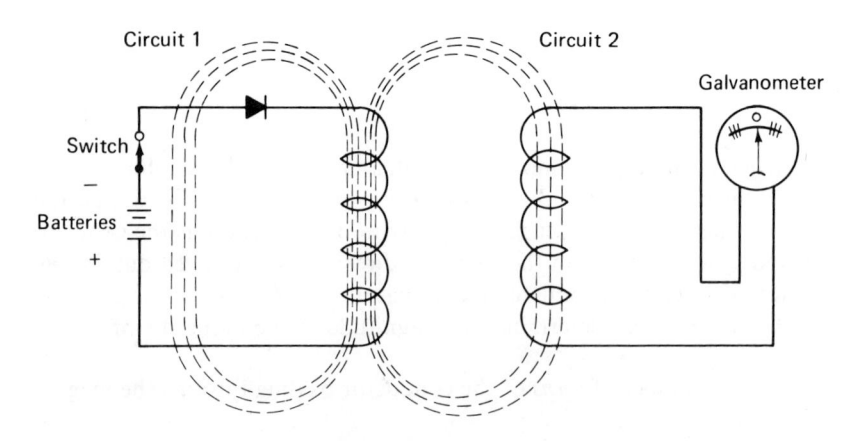

FIG. 3-11 Mutual induction between two coils closely spaced.

through the first coil (circuit 1) and induces a current in circuit 2, causing the galvanometer to momentarily deflect and then return to zero.

When the switch is opened, the current flowing through coil 1 *collapses* very quickly, inducing a voltage of *opposite polarity* in circuit 2. This causes the galvanometer to deflect in the opposite direction for just an instant and then return to zero.

3-9 EARPHONES

Earphones utilize permanent magnets and electromagnets. The basic components of earphones are shown in Fig. 3-12. When no signal currents

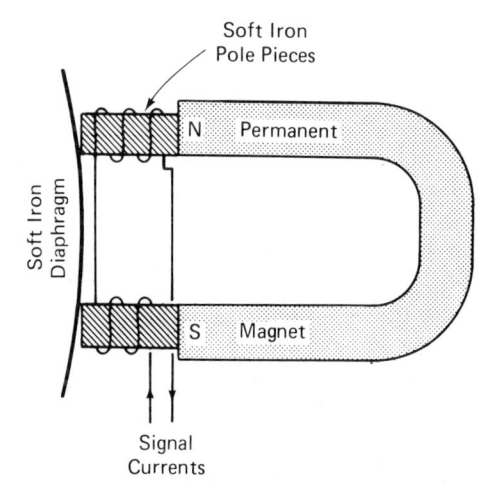

FIG. 3-12 Basic components of earphones.

are present, the permanent magnet exerts a steady pull on the soft-iron diaphragm. Signal currents flowing through the coils wound on the soft-iron pole pieces develop a magnetomotive force that either *adds* to or *subtracts* from the field of the permanent magnet. The diaphragm thus moves in or out according to the resultant field. Sound waves that have amplitudes and frequencies similar to the amplitudes and frequencies of the signal currents (within the capabilities of the reproducer) will then be produced.

A summary of the relationships of the various magnetic units is shown in Table 3-1, on p. 34.

Practice Problems

1. A certain magnet produces 7500 Mx. If the cross-sectional area of the magnet is 4 cm², what is the flux density?
2. What is the strength of a magnet, in maxwells, if it has a cross-sectional area of 9 cm² and its flux density is 850 G?

TABLE 3-1 MAGNETIC UNITS AND DEFINITIONS

Term	Description	Symbol	Metric Unit	Notes
Flux	Total lines	ϕ	1 Mx = 1 line	1 kiloline = 1000 lines
Permeability	Ratio of ϕ in material compared with air	μ	No units	μ of air or vacuum is 1
Flux density	Lines per unit area, $B = \mu H$	B	1 G = 1 line/cm^2	Also in kilolines per in.^{2}
Magnetomotive force	Total force producing flux	mmf	1 Gb = 1.26 amp-turns	Corresponds to voltage independent of length
Magnetizing force, or field intensity	Force per cm of flux path	H	1 Oe $= \dfrac{1.26 \text{ amp-turn}}{\text{cm}}$	Corresponds to voltage per cm
Reluctance	Opposition to flux	$\mathcal{R}$	$\dfrac{\text{Gb}}{\text{mx}}$	Corresponds to resistance

3. What is the flux density in the core of a transformer of 7-cm^2 cross-sectional area when a total of 22,000 Mx is flowing through the entire core?

4. A magnetic circuit having an air gap 7 cm long has a flux density of 27 G. When a certain piece of ferromagnetic material is placed in the gap, the flux density increases to 41,850. What is the permeability of the material?

5. The ϕ inside an air-core coil is 15-G. What is the μ of a ferromagnetic material that increases ϕ to 13,500 when inserted in the air core?

6. A solenoid 10 cm long is wound with 600 turns of No. 22 wire. What is the ampere turns if 24 mA of current passes through it?

7. What is the magnetizing force of a solenoid that is 3.5 cm long, has 850 turns of No. 28 wire, and has 24 mA of current flowing through it?

8. A certain toroid (iron ring) requires a magnetizing force of 1.35 Oe for proper magnetic operation. What must the current be through the winding if it has 500 turns and the mean magnetic path is 12 cm long?

9. What is the magnetomotive force in a circuit where 20,000 Mx flows through a reluctance of 0.0065?

10. Determine the reluctance of a magnetic circuit if the flux is 5360 Mx and the magnetomotive force is 320 Gb.

Commercial License Questions

Sections in which answers to questions are given appear in parentheses. A bracketed number following a question implies that it applies only to that element.

1. Define the term *residual magnetism*. (3–1)

Commercial License Questions 35

2. Name several pieces of electronic equipment that use magnets. (3–1)
3. What factors influence the force existing between two magnets? (3–1)
4. What is meant by the term *flux density*? (3–2)
5. Define the term *gauss*. (3–2)
6. Define the term *permeability*. (3–3)
7. Define the term *reluctance*. (3–3, 3–6)
8. Of what importance is magnetic shelding? (3–3)
9. What are ferrites? (3–3)
10. What is meant by *ampere turns*? (3–4)
11. What factors determine the strength of a magnetic field surrounding an electromagnet? (3–4)
12. How can the direction of flow of dc electricity in a conductor be determined? (3–4)
13. What factors affect the strength of the magnetic field surrounding an energized solenoid? (3–4)
14. Neglecting temperature coefficient of resistance and using the same gauge of wire and the same applied voltage in each case, what would be the effect upon the field strength of a single-layer solenoid of a small increase in the number of turns? (3–4)
15. What factors influence the direction of magnetic lines of force generated by an electromagnet? (3–4)
16. What is a solenoid? (3–4)
17. What does a *BH* magnetization curve show? (3–5)
18. What is Ohm's law for the magnetic circuit? (3–6)
19. What type of core materials are used in transformers to minimize hysteresis losses? (3–7)
20. Explain the meaning of the term *hysteresis loss*. (3–7)
21. Which factors determine the amplitude of the emf induced in a conductor which is cutting magnetic lines of force? (3–8)
22. What is mutual induction? (3–8)
23. Describe the principle of operation of an earphone. (3–9)

4

Alternating Current

4-1 GENERATION OF ALTERNATING CURRENT

Our studies thus far have dealt only with dc circuits. While direct current has many applications, alternating currents (ac) have even more. One advantage is that alternating can be easily transformed to higher or lower voltages to fit the needs of a particular situation. A large percentage of communications circuits use alternating current, and therefore a thorough knowledge of their characteristics is important.

If a coil is *rotated* between the poles of a magnet as shown in Fig. 4-1, an emf will be generated in the coil. The output voltage appearing at the rotating contacts or *slip rings* will rise and fall according to the rate at which the lines of force are cut. The *waveshape* appearing in Fig. 4-1a is representative of the voltage and current produced by such an ac generator. Each complete revolution of the coil produces one *cycle* of emf. By international

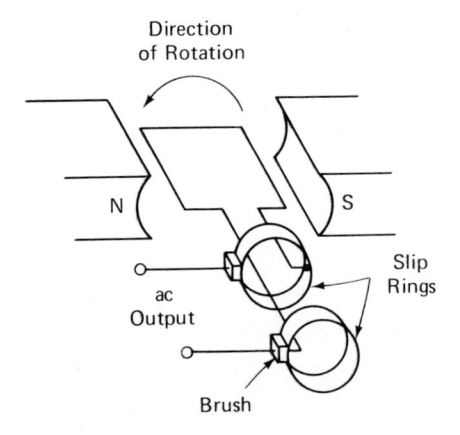

Fig. 4-1 Simple ac generator.

Sine Wave 37

agreement the unit cycles per second is being replaced by *hertz*, abbreviated Hz. The time required to complete 1 Hz is naturally dependent upon the *speed* of the coil. The number of hertz produced (each second) is known as the *frequency* of the generator. For large generators, such as used by utility companies, 50 to 60 Hz is commonly used. Such machines are called *alternators*. For radio communications alternating currents up to millions of hertz are used.

4-2 WAVESHAPES

Many types of waveshapes are encountered in the electronics and communications fields. These vary from the simple *sine wave* to complex waveforms. Several typical waveforms are shown in Fig. 4-2. The sine wave (Fig. 4-2a) is the most common and will be the one studied in this chapter.

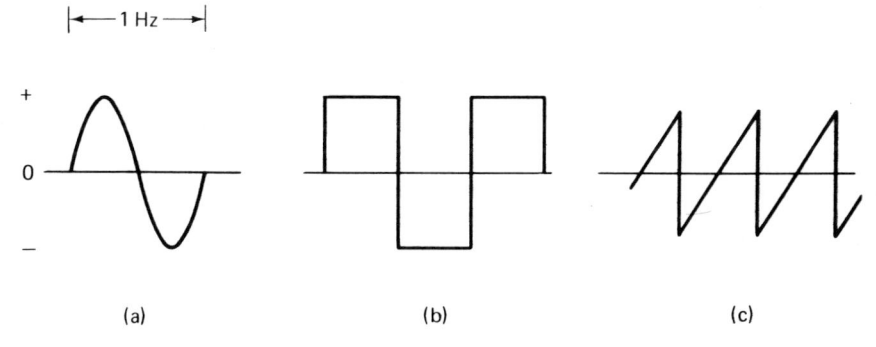

FIG. 4-2 Typical waveshapes: (a) sine, (b) square, (c) triangular.

The waveform shown in Fig. 4-2b rises abruptly to a maximum value and remains there for a short period of time, then suddenly reverses. This is called a *square wave*. Waveshapes of this type are generated electronically and are used in pulse and digital circuits. Figure 4-2c represents a *triangular* or *sawtooth* wave. Waveshapes of this type are generated electronically and find widespread application in timing circuits and test instruments, such as *oscilloscopes*, where they control the horizontal sweep.

4-3 SINE WAVE

The wave of Fig. 4-2a is characterized by a periodic rising and falling above and below the reference line, called a *time base*. The rate at which the wave is constantly changing conforms to the sine function of an angle and therefore is called a sine wave. For communications purposes high-frequency sine waves are needed and are generated by special circuits called *oscillators*.

The graph of a sine wave appears in Fig. 4-3. This represents one complete cycle or 360°. Observe that when the sine wave has gone from 0° to 30° it has reached 0.5 of its maximum amplitude. At 60° the wave has an in-

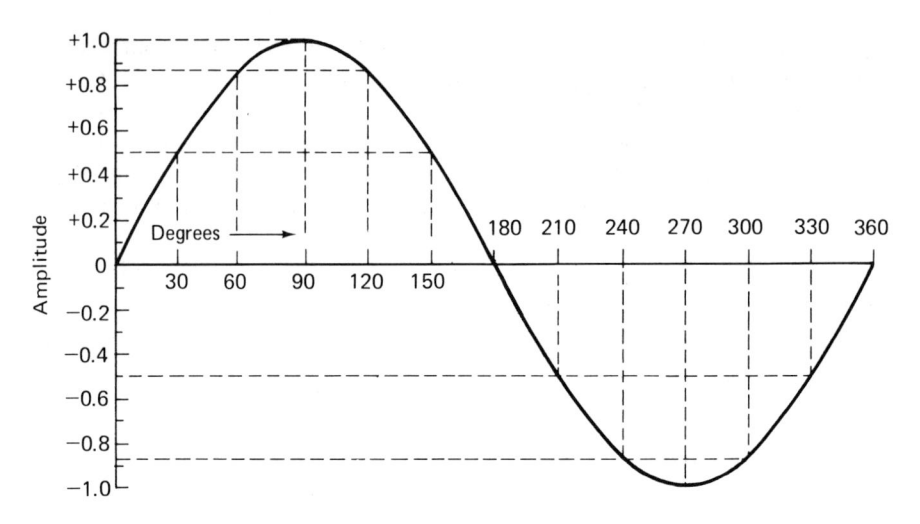

FIG. 4-3 Graph of sine function.

stantaneous value of 0.86 of maximum. Of course, at 90° and 270° the wave has reached its maximum instantaneous values in the positive and negative directions, respectively. Note that at 120° the amplitude is the same as at 60°, and at 150° the same as at 30°.

It should be evident that the negative half of the cycle is inverted and displaced 180° in time with respect to the first half-cycle. Observe that the amplitude of the wave at 210° and 330° is the same as at 30° and 150°, but the polarity is reversed. Likewise, at 240° and 300° the amplitude is the same as at 60° and 120°, but in the opposite direction. From this graph it is possible to estimate the sine of any angle.

4-4 PEAK AND INSTANTANEOUS VALUES

Because of the *symmetrical* nature of the sine wave, the overall amplitude, called *peak to peak*, is twice the peak amplitude, as shown in Fig. 4-4.

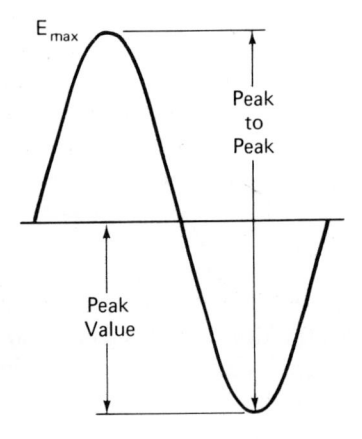

FIG. 4-4 Peak and peak-to-peak values.

Average Value

Peak-to-peak voltage is generally measured with an oscilloscope, although many voltmeters have a special scale so calibrated. The peak value of the voltage is designated as E_{max}.

Frequently, it is desirable to know the *instantaneous* value of a voltage (or current) at some particular time in the cycle. This can be done graphically as outlined in Section 4-3. Usually, more accuracy is required than possible by graphical analysis. When this is the case, we must refer to a table of *trigonometric functions*, such as found in Appendix B. To calculate the instantaneous value of voltage, we use the formula

$$e = E_{max} \sin \theta$$

where e = instantaneous voltage
E_{max} = maximum or peak voltage
$\sin \theta$ = sine of the desired angle

Example 4-1

If the peak value of a sine wave is 169 V, what is the instantaneous value at 45°?

Solution

By reference to Appendix B we find, opposite the angle 45°, 0.707 in the vertical column marked sine. Hence,

$$e = E_{max} \sin \theta$$
$$= 169 \text{ V} \times 0.707 = 120 \text{ V}$$

Similarly, the equation for the instantaneous value of a sinusoidal current is

$$i = I_{max} \sin \theta$$

4-5 AVERAGE VALUE

The *average* value of a complete cycle of a sine wave is zero, since the positive alternation is identical to the negative alternation. In certain types of circuits, however, it is necessary to compute the average value of one alternation. This could be accomplished by adding together a series of instantaneous values of the wave between 0 and 180°, and then dividing the sum by the number of instantaneous values used. Such a computation would show one alternation of a sine wave to have an average value equal to 0.637 of the peak value. In terms of an equation,

$$E_{avg} = E_{max} \times 0.637$$

where E_{avg} = average voltage of one alternation
E_{max} = maximum or peak voltage

Similarly,

$$I_{avg} = I_{max} \times 0.637$$

4-6 EFFECTIVE OR ROOT-MEAN-SQUARE VALUE

Neither the instantaneous nor average value of an ac voltage or current is used in calculating the amount of power delivered to a resistive load. It is necessary to have some way of *comparing* the amount of power that an alternating current will produce compared to a direct current. Since the power developed in the load will vary with the changing current ($P = I^2R$), it becomes necessary to find the *effective* value of alternating current that is equivalent to direct current. It can be proved mathematically that the effective value can be found by taking the square root of the average of the sum of the squares of a number of instantaneous values throughout 180° of the cycle. This is known as the *root mean square* or simply *rms* value. This turns out to be 0.707 of the maximum value. When referring to ac voltages or currents, it is assumed that they are rms values unless specifically indicated otherwise.

Example 4-2

What is the effective value of an ac voltage whose peak value is 8.9 V?

Solution
$$E_{\text{eff}} = 0.707 \times 8.9 = 6.3 \text{ V}$$

If the rms value of a current is known, the peak value can be determined by multiplying by the reciprocal of 0.707 (1/0.707), or 1.414. It should be apparent that an rms value of voltage or current can be converted to a peak-to-peak value by multiplying by 2 times 1.414, or 2.828.

Figure 4-5 graphically shows the relationships between peak-to-peak (p-p), peak, rms or effective, and average values.

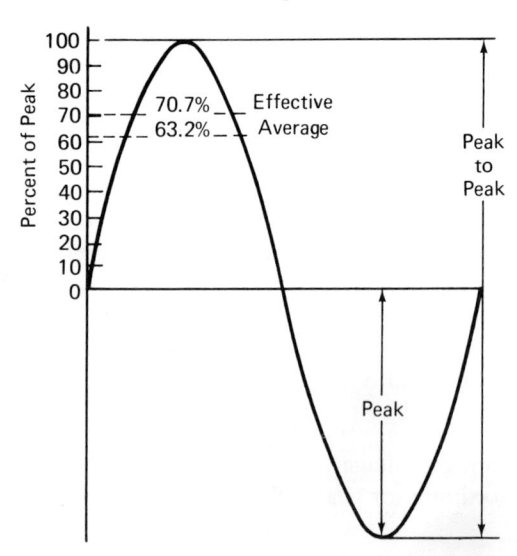

FIG. 4-5 Relationship between various ac values.

4-7 FREQUENCY

Alternating voltages and currents can and do vary over an extremely broad frequency spectrum. At the low end are the *audio* frequencies (AF), so called because they can be heard by the human ear. These frequencies range from approximately 20 Hz to 20 kHz. High audio frequencies, about 300 Hz and above, can be considered to provide treble tones. Low audio frequencies, about 300 Hz and below, provide bass tones. The higher the frequency, the higher the pitch or tone of the sound.

Loudness is determined by amplitude. The greater the amplitude of the AF variation, the louder is its corresponding sound.

Alternating current and voltage above the audio range provide radiofrequency (RF) variations, since electrical variations of high frequency can be transmitted by electromagnetic radio waves. Some of the more common frequency bands are listed in Table 4-1. The velocity of transmission through space for electromagnetic waves equals the speed of light, which is 186,000 miles/s.

TABLE 4-1 FREQUENCY SPECTRUM

Designations and Abbreviations	Frequency Range
Power frequencies	50 to 400 Hz
Audio frequencies (AF)	20 to 20,000 Hz
Very low frequencies (VLF)	15 to 30 kHz
Low radio frequencies (LF)	30 to 300 kHz
Medium frequencies (MF)	300 to 3000 kHz
High frequencies (HF)	3 to 30 MHz
Very high frequencies (VHF)	30 to 300 MHz
Ultrahigh frequencies (UHF)	300 to 3000 MHz
Superhigh frequencies (SHF)	3 to 30 GHz
Extremely high frequencies (EHF)	30 to 300 GHz

4-8 PHASE RELATIONSHIPS

Figure 4-6 shows a sine wave of voltage and the resulting current superimposed on the same time axis. The voltage and current pass through the same relative parts of their respective cycles at the same time. When two waves, such as these, are precisely in step, they are said to be *in phase*.

Figure 4-7 shows a voltage wave E_1 considered to start at 0° (t_1). As voltage wave E_1 reaches its positive peak, a second voltage wave E_2 starts its rise (t_2). Since these waves do not go through their maximum and minimum points at the same instant of time, a *phase difference* exists between them. The two waves are said to be *out of phase*, in this case 90°.

To further describe the phase relationship between two waves, the terms *lead* and *lag* are used. The amount by which one wave leads or lags another is measured in degrees. In Fig. 4-7 wave E_2 is seen to start 90° later in time than wave E_1; thus, wave E_2 lags E_1 by 90°. This relationship could also be de-

FIG. 4-6 Voltage and current waves in phase.

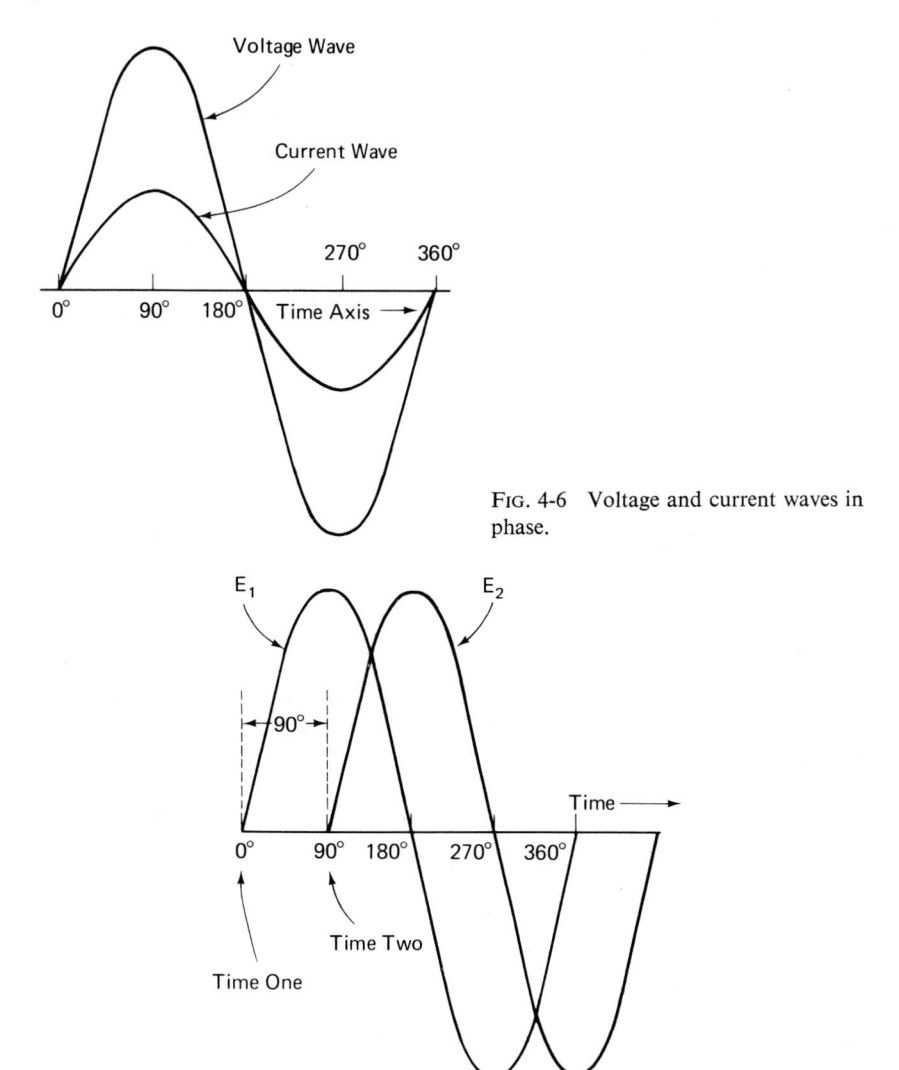

FIG. 4-7 Voltage waves 90° out of phase.

scribed by stating that wave E_1 leads E_2 by 90°. It is possible for one wave to lead or lag another by any number of degrees, except 0 or 360°, in which condition the two waves are in phase.

Another phase relationship is shown in Fig. 4-8. The waves illustrated have a phase difference of 180°. Notice that although the waves pass through their maximum and minimum values at the same time, their instantaneous voltages are always of opposite polarity. If two such waves existed across the same component, they would have a canceling effect. If they are equal in amplitude, the resultant would be zero. However, if they have different ampli-

Radians

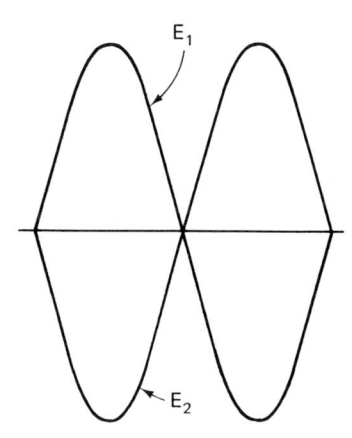

FIG. 4-8 Two waves 180° out of phase.

tudes, the resultant wave would have the polarity of the larger and be the difference of the two.

4-9 RADIANS

In electronic calculations the number of hertz is expressed in *radian* measure. A radian (rad) is a unit of angular measurement equal to that angle which, when its vertex is upon the center of a circle, intercepts an arc that is equal in length to the radius of the circle. Thus, in Fig. 4-9 central angle AOB is equal to 1 rad because arc AB is equal to radius OA.

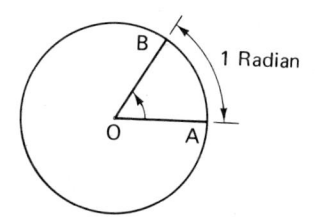

FIG. 4-9 The radian or circular system of measurement.

If the arc subtended by the angle AOB is divided into the circumference of the circle, it goes in 6.28 times. Consequently, the circumference of a circle—or one cycle—equals 2π, or 6.28 rad. The natural system of angular measurement, called the radian system, is used extensively in electronic formulas because it is more convenient than the arbitrary 360° system.

It is frequently important to know how many degrees constitute an angle of 1 rad. Because

$$2\pi \text{ rad} = 360°$$

$$1 \text{ rad} = \frac{360°}{2\pi} = \frac{180°}{\pi} = 57.3°$$

4-10 VECTORS

A line used to represent a quantity that has *direction* and *magnitude* is a *vector*. The length of the line indicates the magnitude and the arrowhead the direction. The quantity represented by a vector is called a *vector quantity*. For example, the instantaneous value of a voltage or current can be represented by a vector. The radius of the circle in Fig. 4-10 may be called a vector and can represent the value of a current at 0° in the cycle. If the vector is rotated counterclockwise to position A, the side *i* of the right triangle represents the instantaneous value of the assumed current at that moment.

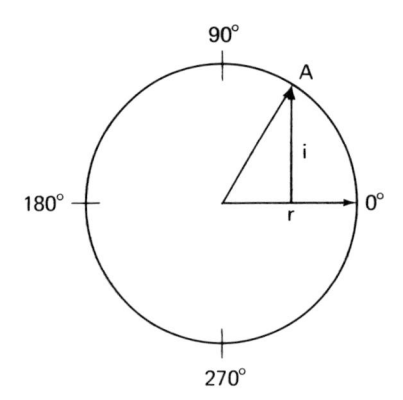

FIG. 4-10 The use of vectors to illustrate the maximum and instantaneous values of current.

Practice Problems

1. What is the instantaneous value of a sine wave at 75° if the peak value is 170 V?
2. The peak value of a sinusoidal current is 23 mA. What is its instantaneous value at 17°?
3. A sine wave has a peak value of 200 V. What is its value at 270°?
4. What is the average value of a sine wave of voltage if its peak value is 67 V?
5. What is the effective value of an ac sine wave whose peak value is 155 V?
6. What is the peak value of a sine wave if its effective value is 117 V?
7. The effective value of a sinusoidal current is 60 mA. What is its maximum value?
8. An ammeter connected in an ac circuit reads an effective current of 10 A. Calculate the maximum value of the current.
9. If the rms value of a voltage is 6.3 V, what is its peak-to-peak value?
10. How many degrees are there in 1.7 rad?

Commercial License Questions

Sections in which answers to questions are given appear in parentheses. A bracketed number following a question implies that is applies only to that element.

1. What is the meaning of *frequency*? (4-1)
2. What is a sine wave? (4-3)
3. How many degrees are there in one cycle? (4-3)
4. What is the ratio of peak to rms values of a sine wave? (4-4)
5. What is the relationship between the following ac values: rms, peak, peak-to-peak? (4-4, 4-5, 4-6)
6. What is the effective value of a sine wave in relation to its peak value? (4-6)
7. What is the relationship between the effective value of a radio-frequency current and the heating value of the current? (4-6)
8. What are audio frequencies and what approximate band of frequencies is referred to as the audio-frequency range? (4-7)
9. What is the meaning of *phase difference*? (4-8)
10. What is the meaning of *in phase* with regard to alternating currents? (4-8)

5

Inductance and Transformers

5-1 INDUCTANCE

Inductors are one of the main building blocks in electronics circuits. An *inductor* is simply a coil of wire with or without a magnetic core. This chapter deals with the characteristics of such a device when alternating and direct current are applied.

All coils have *inductance*, which is the property of *opposing any change of current* flowing through it. If a coil offers a large opposition to the current flowing through it at a certain frequency, it is said to have a large inductance. A small inductance would provide less opposition at the same frequency.

The opposition to changing current flow can be explained as follows. Assume that a dc potential is applied to an inductor. The inrush of current causes a magnetic field to be built up around each turn, which links with the adjacent turns, causing an emf to be *induced* in them. The polarity of this voltage is opposite to the applied voltage and is called *counter emf* (cemf) or *back emf*. The amount of cemf is proportional to the *rate of change* of current. Consequently, the applied voltage, which is trying to force current through the coil, is being bucked by the cemf. The difference between these two voltages is what is actually forcing the current to increase exponentially through the inductor, as shown in Fig. 5-1. The cemf can never equal the applied emf; otherwise no current would flow.

The net result of these actions is that the direct current cannot rise instantaneously to its maximum value, as it would do in a purely resistive circuit.

There is a law, known as *Lenz's law*, that explains this action as follows: An induced emf always has such a direction as to *oppose* the action that produces it. Thus, when a current flowing through a circuit is varying

Inductance

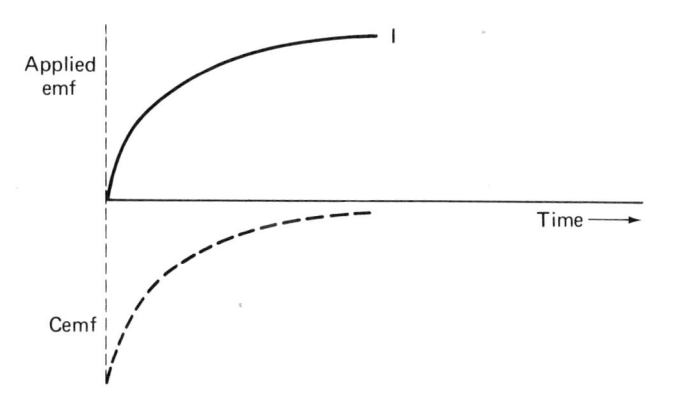

FIG. 5-1 Relationship between applied emf, cemf, and current plotted against time.

in magnitude, it produces a varying magnetic field which sets up an induced emf that opposes the current change producing it.

The schematic symbols for different types of inductors are shown in Fig. 5-2. An air-core inductor is represented in Fig. 5-2a; in b an iron core is shown; and in c a *permeability tuned* coil is shown. This usually consists of a powdered iron slug attached to the end of an adjustment screw that can be moved in or out, thereby controlling the permeability of the flux path within the coil. Sometimes the symbol is represented with dotted lines or with a single arrow passing diagonally through the coil.

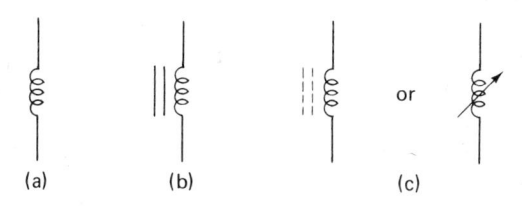

FIG. 5-2 Schematic symbols for different types of inductors: (a) air core, (b) iron core, (c) slug or permeability tuned.

The effect of cemf may be observed experimentally in that an alternating current through an inductor is opposed by a force much greater than its dc resistance. For example, assume that a certain coil has 5-Ω dc resistance and is connected to a 110-V line, as shown in Fig. 5-3. From Ohm's law we would assume that the current would be

$$I = \frac{E}{R} = \frac{110}{5} = 22 \text{ A}$$

However, we observe that the ac ammeter reads approximately 1 A. It becomes apparent that some opposition other than the 5-Ω resistance is present in an ac circuit. This opposition is the cemf.

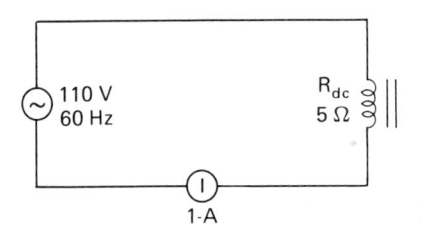

FIG. 5-3 Effect of cemf on current flow.

5-2 UNIT OF INDUCTANCE

The unit of inductance is the *henry* (H) and is represented by the letter L. A coil is said to possess an inductance of 1 H if an emf of 1 V is induced in the coil when the current is changing at the rate of 1 A/s.

The amount of inductance a coil possesses is determined by several factors. These are the square of the number of turns, spacing of turns, core material, core area, and coil geometry.

5-3 MUTUAL INDUCTANCE

When two coils are positioned so that flux lines from one cut the turns of the other, they are said to exhibit *mutual inductance*. The action of mutual inductance in producing an emf is basically the same as that of self-inductance, the main difference being that self-inductance is the property of a single coil, whereas mutual inductance is the property of two (or more) coils acting together. Mutual inductance is measured in henrys and is designated by M.

Mutual inductance is dependent on the physical dimensions of the two coils, number of turns on each coil, permeability of the cores, and the *coefficient of coupling*. The last factor, coefficient of coupling, is a new term. It is dependent upon the distance between two coils and on the position of the coil axes with respect to each other. Coefficient of coupling is a measure of how much of the flux from one coil cuts the turns of the other coil. When all the flux from one coil cuts all the turns on the other, the coefficient of coupling is 1.

If two coils in close proximity are arranged so that the axis of one is perpendicular to the other, the flux developed by one coil will not induce an emf in the other. This condition establishes a *null point* between them. The coefficient of coupling is zero, and therefore the mutual inductance is zero. The relationship between self-inductance, mutal inductance, and coefficient of coupling (designated by the letter K) is shown mathematically as

$$M = K\sqrt{L_1 L_2}$$

where L = self-inductance of each coil in henrys
K = coefficient of coupling expressed as a decimal fraction
M = mutual inductance in henrys

5-4 SERIES- AND PARALLEL-CONNECTED INDUCTORS

When series-connected inductors are well shielded, or located far enough apart to make the effects of mutual inductance negligible, the equivalent of total inductance of the circuit is the algebraic sum of the individual inductances. Expressed mathematically,

$$L_t = L_1 + L_2 + L_3$$

where L_t = total inductance in henrys
L_1 = individual inductances in henrys, etc.

Example 5-1

Assuming perfect shielding, find the total inductance of the following four coils, series connected:

$L_1 = 1.5$ H
$L_2 = 500$ millihenrys (mH)
$L_3 = 0.2$ H
$L_4 = 25,000$ microhenrys (μH)

Solution

First convert all inductance to the same unit.

$L_1 = 1.5$ H $\quad = 1500$ mH
$L_2 = 500$ mH $\quad = 500$ mH
$L_3 = 0.2$ H $\quad = 200$ mH
$L_4 = 25,000\ \mu$H $= 25$ mH

Substitute values in the equation:

$L_t = 1500 + 500 + 200 + 25$
$\quad = 2225$ mH or 2.225 H

When two inductors in series are so positioned that there is magnetic coupling between them, the total inductance is expressed mathematically as

$$L_t = L_1 + L_2 \pm 2M$$

where L_t = total inductance of the two coils
L = self-inductance of each coil
M = mutual inductance

The signs $\pm$ in this equation must be considered due to the fact that the coils can be arranged so that their fields are either *series aiding* or *series opposing*. When the fields are series aiding, the plus sign is used, and when opposing, the minus sign is used.

The total inductance of inductors in parallel is calculated in the same manner that the total resistance of resistors in parallel is calculated, provided the coefficient of coupling is zero. Mathematically, this is

$$\frac{1}{L_t} = \frac{1}{L_1} + \frac{1}{L_2} + \frac{1}{L_3}$$

When there are only two inductors in parallel, the product-over-the-sum method may be employed. Mathematically, this is

$$L_t = \frac{L_1 L_2}{L_1 + L_2}$$

5-5 ENERGY IN A MAGNETIC FIELD

The field built up around an inductor represents *stored energy* that came from the source supplying the voltage and current. If the circuit should be suddenly opened, the magnetic energy would be converted to electrical energy, appearing as a spark across the open connection. The voltage produced in this case may be many times greater than the original supply potential. The reason for this is that the field collapses faster than it originally built up. The stored energy can be calculated as

$$E = \frac{LI^2}{2}$$

where L = inductance in henrys
I = current in amperes
E = energy in joules (watt/second)

5-6 TIME CONSTANT OF INDUCTANCE–RESISTANCE CIRCUIT

When a resistance is added in series with an inductance, the time required for the current to reach its maximum value is changed. The relationship between inductance and resistance so far as rise time is concerned is expressed as

$$T = \frac{L}{R}$$

where L = inductance in henrys
R = resistance in ohms
T = time in seconds

This indicates the length of time required for the current to reach approximately 63 per cent of maximum value. This is called *one time constant*. In one more time constant the current will rise another 63 per cent of the remaining amount of maximum. This goes on for a total of five time constants at which time the current is essentially at maximum value (within 1 per cent).

Time Constant of Inductance–Resistance Circuit

It is apparent that the current in a series *LR* circuit does not rise in a *linear* manner. The instantaneous current magnitude with respect to time follows what is called an *exponential curve*, such as shown in Fig. 5-4.

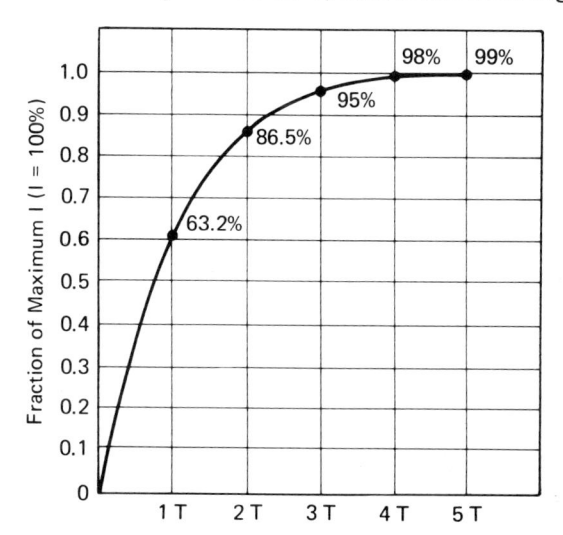

FIG. 5-4 Universal time constant chart.

Example 5-2

Find the time constant of a circuit in which the inductance is 2 H and the resistance is 10 Ω.

Solution

$$T = \frac{L}{R} = \frac{2}{10} = 0.2 \text{ s}$$

When an inductive circuit is opened, the resistance suddenly becomes very large—in the order of megohms. The characteristics of the circuit are now greatly changed. The previous time constant is now shortened to a small fraction of a second.

The energy contained in the collapsing magnetic field must be dissipated somewhere within the circuit. The voltage developed in the inductor is sufficient to create an arc across the switch contacts. The energy in the magnetic field is dissipated in the heat of this arc. This can seriously burn an individual, damage the switch contacts, or break down the insulation of the coil.

The development of a large voltage pulse from a low voltage source (inductive kick) is not always a disadvantage, and is commonly used in the ignition systems of most gasoline engines.

5-7 INDUCTIVE REACTANCE

When an alternating current passes through an inductance, the field will be constantly changing due to the changing current. The cemf therefore is continually opposing the varying current. This opposition is called *inductive reactance* and is measured in ohms. The symbol for inductive reactance is X_L.

The inductive reactance of a coil can be calculated as

$$X_L = 2\pi f L$$

where X_L = inductive reactance in ohms
2π = number radians in one cycle (6.28)
f = frequency of current in hertz
L = inductance in henrys

This formula indicates that X_L varies directly with frequency and inductance. It can be shown graphically as a slanting, straight line, as in Fig. 5-5. The steepness of this line is a function of the amount of inductance involved. Larger inductances would cause a steeper line, and small ones a shallow line.

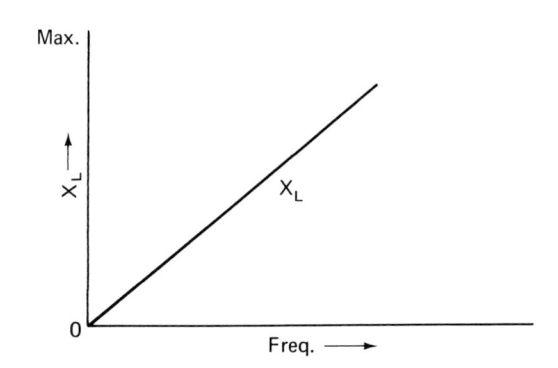

FIG. 5-5 Relationship between X_L and f in an inductive ac circuit.

Example 5-3

A pure inductance of 0.2 H is connected across a 100-V, 60-Hz source. Find the amount of current that will flow.

Solution

First calculate X_L.

$$X_L = 6.28 \times 60 \times 2 \times 10^{-1}$$
$$= 75.4 \, \Omega$$

Use Ohm's law by substituting X_L for R:

$$I = \frac{E}{X_L} = \frac{100}{75.4} = 1.33 \text{ A}$$

Basic Transformer 53

5-8 PHASE RELATIONSHIPS

The induced voltage, or cemf, is caused by the current. The rate of change of current is greatest as it is *crossing the zero axis*. Therefore, we can expect the greatest cemf to be produced at this moment. By referring to Fig. 5-6, we see that this is the case. Remembering that the cemf is 180° out of phase with the applied voltage, we observe that the current is *lagging* the applied emf by 90°. This is true of all pure inductances (no resistance present).

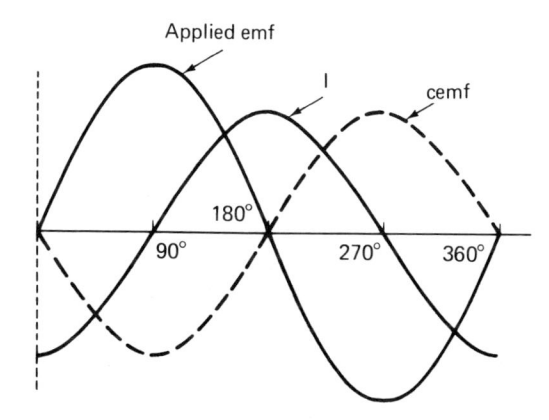

FIG. 5-6 Phase shift in an inductive ac circuit.

5-9 BASIC TRANSFORMER

Transformers are widely used in electronics and communications circuits. Essentially, it is a device that *transforms energy* from one circuit to another. The basic transformer consists of two separate windings, a primary and secondary, on a common iron core. The transformer does not change the amount of power available between secondary and primary, but rather *converts* electrical power from one voltage–current level to another.

Figure 5-7 shows a basic transformer connected between an ac generator and a resistive load. The coil connected to the source of power is called the *primary winding*, and the coil connected to the load is called the *secondary winding*. The power delivered by the generator passes through the transformer

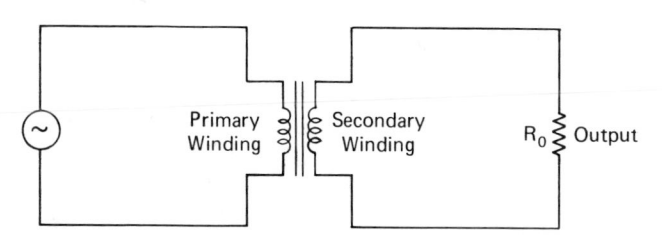

FIG. 5-7 Basic transformer.

and is delivered to the load, although no direct connection exists between the primary and the secondary winding. The connection that does exist is the *flux linkage* between the coils, and power is effectively transferred by induction. Thus, the power consumed by the primary is equal to the power delivered by the secondary, or $P_p = P_s$.

For maximum transfer of power from primary to secondary, the flux linkage must be complete; that is, all the lines of force set up by the primary must link the secondary. For this reason, the secondary is often wound directly on the primary with only protective insulation separating the two. The introduction of a soft-iron core of high permeability increases the flux linkage between the coils and makes possible a high percentage of power transfer. Even then a few of the flux lines fail to link the secondary winding and are effectively lost, constituting a flux leakage. However, a well-designed iron-core transformer may effect a 98 per cent flux linkage, which means that K, the coefficient of coupling between the coils, is 0.98.

Figure 5-8 is a cross-sectional view of a typical transformer core showing the relationship of primary and secondary. Notice that the windings are located on the central leg of the core. This results in increased efficiency, because fewer flux linkages are lost. Each layer of wire is separated from the

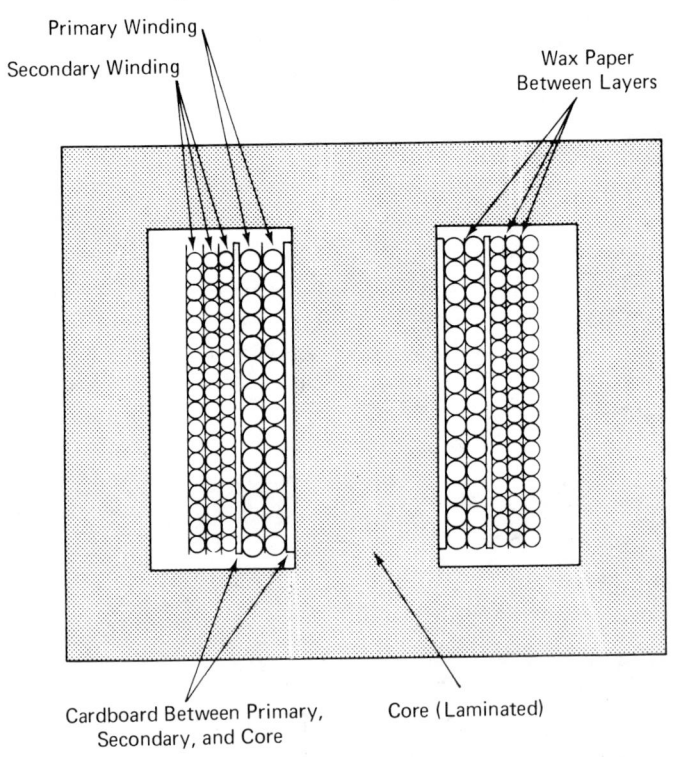

FIG. 5-8 Cross section of iron core transformer.

Turns Ratio

other by sheets of waxed paper, and the primary winding is separated from the secondary winding by thicker varnished paper or cardboard.

Because of the tight coupling between primary and secondary, the primary power is controlled by the load on the secondary. If no load is connected to the secondary, the primary power is zero, except for small losses, to be discussed. Under these conditions the primary acts like an iron-core inductor. When a resistive load is connected to the secondary, as in Fig. 5-7, the following action takes place:

1. The secondary voltage, E_s, causes a secondary current, I_s, to flow through the load.
2. I_s produces a magnetic field which generates a voltage in the primary (mutual induction) that cancels part of the primary cemf.
3. With the primary cemf reduced more I_p flows.

Therefore, the secondary load affects the primary current. As the load increases (R_o becomes *smaller*), more I_p flows, creating a stronger magnetic field, which induces more electrical energy into the secondary.

5-10 TURNS RATIO

The amount of voltage induced in the secondary winding is determined by the ratio of the number of turns in the secondary to the number in the primary. Expressed mathematically,

$$\frac{E_p}{N_p} = \frac{E_s}{N_s}$$

where E_p = voltage applied to the primary
N_p = number of turns in the primary
E_s = voltage induced in the secondary
N_s = number of turns in the secondary

This equation may be rewritten

$$E_p N_s = E_s N_p \quad \text{or} \quad E_s = \frac{E_p N_s}{N_p}$$

The expression N_s/N_p or N_p/N_s is called the *turns ratio*. Figure 5-9 shows the schematic of a transformer with 1000 turns in the secondary and 250 turns in the primary. The turns ratio is 4:1 or 4. If 110-V_{ac} is applied to the primary, the induced secondary voltage is

$$E_s = \frac{E_p N_s}{N_p} = \frac{110 \times 1000}{250} = 440 \text{ V}$$

This is a *step-up* transformer. If the ratio N_s/N_p is less than 1, the transformer is step-down. Whether a transformer is step-up or step-down always refers to *voltage level*.

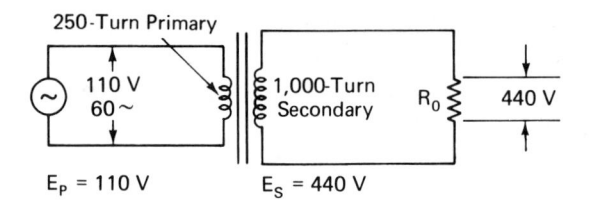

56   INDUCTANCE AND TRANSFORMERS

FIG. 5-9 Step-up transformer.

Since the voltage induced in the secondary is directly proportional to the number of turns in the secondary compared with the number of turns in the primary, the voltage ratio of secondary to primary may be found by determining the number of volts per turn for a given transformer. Thus, 110 V applied to the primary of 250 turns results in a volts-per-turn ratio of 0.44, or

$$\text{volts/turn} = \frac{110}{250} = 0.44$$

Since the number of volts per turn is a constant for any given transformer, the voltage in the secondary may be determined by multiplying this constant by the number of turns in the secondary,

$$E_s = 1000 \times 0.44 = 440 \text{ V}$$

The volts-per-turn ratio is also convenient for determining a number of secondary voltages when a transformer carries more than one secondary winding, as shown in Fig. 5-10. The primary consists of 200 turns, secondary

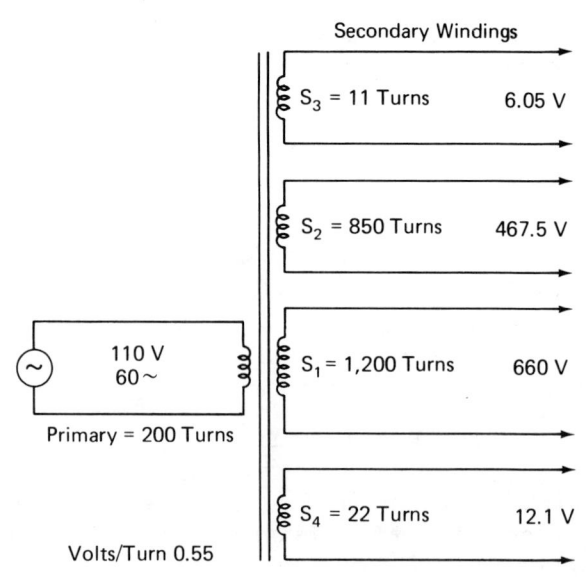

FIG. 5-10 Multi-winding transformer.

Transformer Losses

winding S_1 is 1200 turns, S_2 is 850 turns, S_3 is 11 turns, and S_4 is 22 turns. The volts per turn of this transformer is E_p/N_p, or 110/200, and is equal to 0.55. Winding S_1, then, has an induced voltage of

$$E_{s1} = 1200 \times 0.55 = 660 \text{ V}$$
$$E_{s2} = 850 \times 0.55 = 467.5 \text{ V}$$
$$E_{s3} = 11 \times 0.55 = 6.05 \text{ V}$$
$$E_{s4} = 22 \times 0.55 = 12.1 \text{ V}$$

The ratio of current in the primary to current in the secondary can be determined by the formula

$$\frac{I_p}{I_s} = \frac{N_s}{N_p} \quad \text{or} \quad I_p = I_s\left(\frac{N_s}{N_p}\right)$$

Thus, the current ratio varies inversely as the number of turns. For example, Fig. 5-11 shows a power transformer of 300 turns on the primary

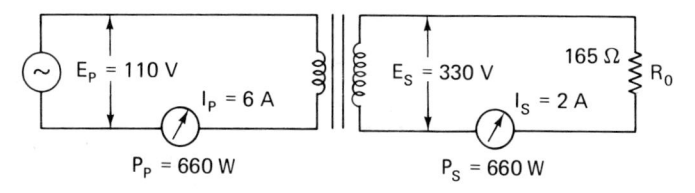

FIG. 5-11 Current ratio in a step-up transformer.

and 900 on the secondary connected to a 110-V ac line and a 165-Ω load. The turns ratio is 3:1, and E_s is equal to 330 V. Then, since the load determines the energy used, the current in the secondary is

$$I_s = \frac{E_s}{R_o} = \frac{330}{165} = 2 \text{ A}$$

The current in the primary is

$$I_p = \frac{I_s N_s}{N_p} = 2 \times 3 = 6 \text{ A}$$

A turns ratio, then, of 3:1 increases the applied voltage from 110 to 330 V, at the same time decreasing current from 6 to 2 A. From these observations it will be noted that the *product of current and voltage* in one side of a transformer is equal to the product of current and voltage in the other side. Thus, the power in the primary circuit is $110 \times 6 = 660$ W; the power consumed by the secondary circuit is $330 \times 2 = 660$ W.

5-11 TRANSFORMER LOSSES

The principal losses in a transformer are *flux leakage, hysteresis, eddy currents,* and *copper losses.* These normally amount to about 2 to 5 per cent of the total power handled by the transformer. The first two have been discussed in previous sections.

The rapidly changing magnetic field set up around the primary not only induces an emf in the secondary winding but also in the *iron core* of the transformer. The core behaves as a single loop of conducting material having a large cross-sectional area. The voltage induced in the core is small, because it is essentially a winding having but one turn. However, because the resistance of the core is low, the resultant currents are high. These currents circulate throughout the volume of the core and, because of their circulatory nature, are called *eddy currents*. An example of a simple transformer core is shown in Fig. 5-12a.

Figure 5-12b shows a cross-sectional view of the core and the resulting eddy currents. The magnitude of these currents can be large, and the resulting power loss (heating of the core material) can be considerable. To reduce eddy currents to a minimum, transformer cores are *laminated*, that is, made up of thin strips of soft annealed steel pressed together. Each strip is sprayed with

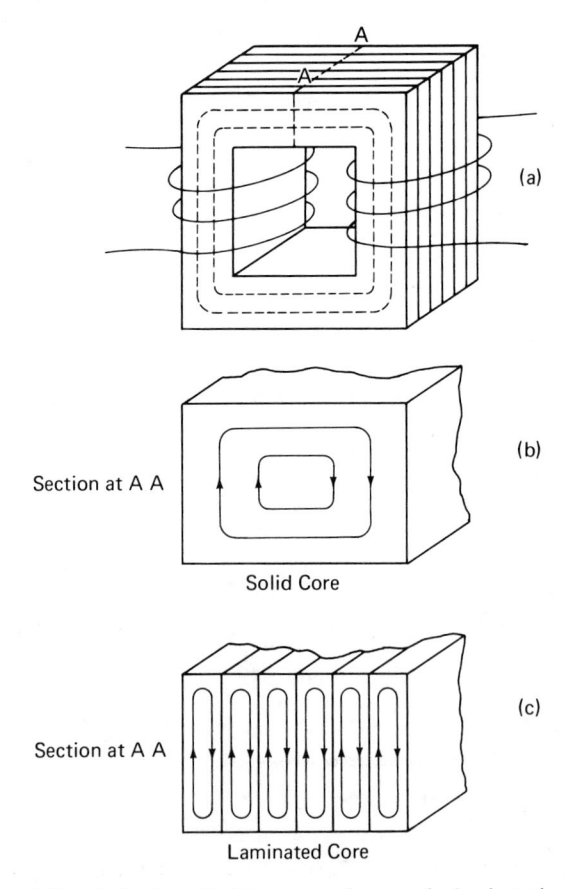

FIG. 5-12 Reduction of eddy currents by use of a laminated core.

Impedance Matching 59

an insulating coating so that the dc resistance between laminations is very high. Thus, eddy currents find a series of high-resistance paths, whereas the magnetic permittivity of the core is unaffected, as shown in Fig. 5-12c.

The current in the primary and secondary windings of a transformer must flow through the dc resistance of the wire. A certain amount of power is lost through heat, since this is a true power (I^2R) loss. Transformers carrying considerable amounts of power use wire of large cross-sectional area to cut down this heat loss. Since a high percentage of flux linkage at power-line frequencies requires large inductance, some compromise between the size of the core and the number of windings must be made. Thus, a large core and small winding for a given inductance would show little copper loss, but the transformer would be heavy and awkward. On the other hand, a small core reduces core loss and increases copper losses because of the increased number of turns. For most applications, the dc resistance of the wire may be ignored if the ratio of inductive reactance to resistance is 10 : 1 or greater.

It is desirable to operate power transformers at or near their rated capacities. For instance, any power transformer operated at one fifth its proper load shows considerable primary and secondary inductive reactance, and the induced voltages across it are very high, causing arcing and breakdown of insulation with resultant burning out of the transformer. Under full load the inductive reactance of the primary circuit is almost completely canceled by the opposing magnetizing force set up by the secondary current.

Some power transformers employ a thin copper sheet between primary and secondary windings. This serves as an *electrostatic shield* and prevents RF currents being coupled between the windings.

5-12 AUTOTRANSFORMERS

The *autotransformer*, or self-transformer, is a special type of power transformer designed for good voltage regulation under varying loads. A single tapped winding characterizes the autotransformer, as shown in Fig. 5-13. When used as a step-up transformer, as in Fig. 5-13a, all the primary winding is part of the secondary winding, and when used as a step-down transformer, all the secondary winding is part of the primary winding, as in Fig. 5-13b. Voltages across the individual windings follow the turns ratios. A disadvantage of this arrangement is that the *secondary circuit is not isolated electrically from the primary circuit*, as it is in a conventional transformer.

The chief advantage of the autotransformer is that the load may be varied without arcing and with little change in output voltage.

5-13 IMPEDANCE MATCHING

Besides stepping voltage up or down, transformers are extensively used as *impedance-matching* devices. *Impedance* is an electrical term used to indi-

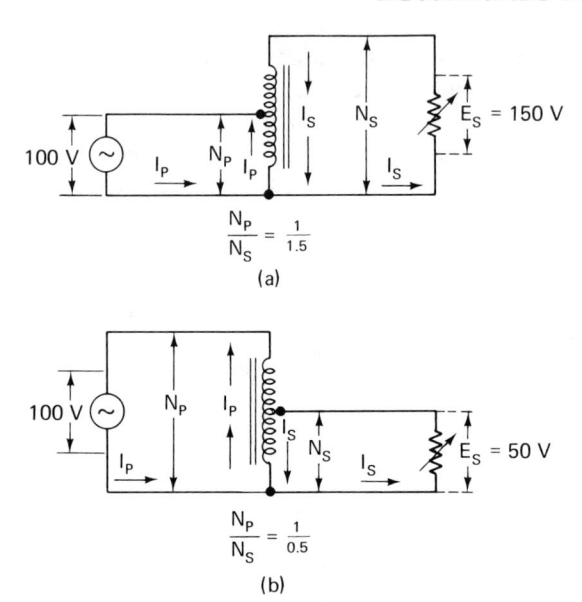

FIG. 5-13 Autotransformer: (a) providing step-up voltage, (b) stepping-down voltage.

cate the *total opposition* a circuit offers to the flow of alternating current through it. It is sometimes referred to as "ac resistance," but this is an improper term. Briefly stated, impedance, represented by the letter Z, is the vector sum of the dc resistance and reactance in a circuit and is measured in ohms. Impedance will be treated in more detail in Chapter 7.

The *impedance ratio* of a transformer is equal to the square of the turns ratio. Expressed mathematically,

$$\frac{Z_p}{Z_s} = \left(\frac{N_p}{N_s}\right)^2$$

Example 5-4

Find the primary impedance of a step-down transformer whose secondary is connected to a 1.5-Ω impedance and whose turns ratio is 15:1.

Solution

Solving for Z_p,

$$Z_p = Z_s \left(\frac{N_p}{N_s}\right)^2$$
$$= 1.5\left(\frac{15}{1}\right)^2 = 337.5 \ \Omega$$

In the transfer of power from any electrical source to its load, the impedance of the load must equal, or match, the internal impedance of the source

Impedance Matching

for maximum transfer of power. The importance of matching the load to the source for maximum transfer of power makes the transformer useful as an impedance-matching device. For example, it often becomes necessary in electronic circuits to match a low-impedance load to a high-impedance source. Figure 5-14a shows such a circuit. The impedance of the source R_G is 10 kΩ and of the load 400 Ω. The primary impedance is to match the source, and the secondary impedance is to match the load. The turns ratio of the transformer is

$$\frac{N_p}{N_s} = \sqrt{\frac{Z_p}{Z_s}} = \sqrt{\frac{10{,}000}{400}} = \sqrt{\frac{25}{1}}$$

Then

$$\frac{N_p}{N_s} = \frac{5}{1}$$

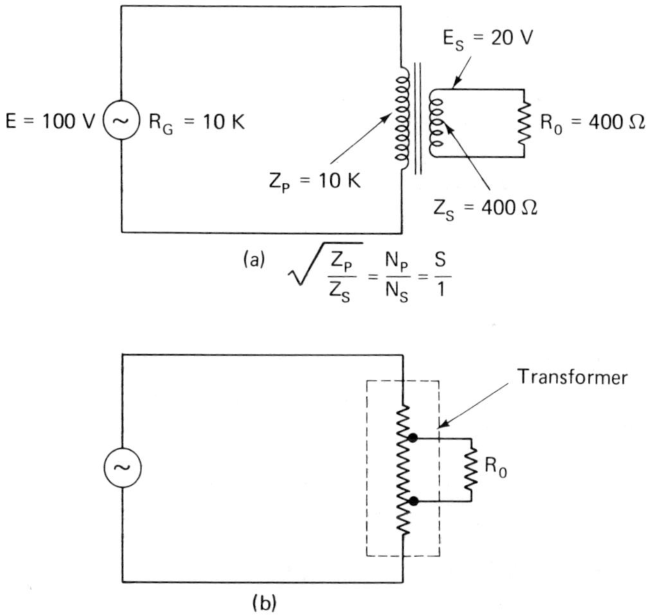

FIG. 5-14 (a) Impedance matching transformer. (b) Equivalent circuit.

If 100 V is applied to the primary, the secondary voltage is 20 V. The current in the secondary is $\frac{20}{400}$, or 0.05 A, the current in the primary is $\frac{100}{10{,}000}$ or 0.01 A. Figure 5-14b is the equivalent circuit. Since the power in the primary (1 W) is equal to the power in the secondary, the transformer has matched a 400-Ω load to a 10,000-Ω source with maximum transfer of energy. It may be said then that the source sees the primary impedance as a matching impedance, and that the secondary, which by transformer action receives the power of the primary, sees the load impedance as a matching impedance.

5-14 OUTPUT TRANSFORMERS

The problem of matching the output of an amplifier to a loudspeaker is once again a problem in the transfer of power, not of voltage alone. Since the cone of the speaker has physical mass and moves against air, power is required to move it. Therefore, the power output of the amplifier must be delivered to the cone of the speaker with as little loss as possible. Figure 5-15 shows the schematic of a circuit. The output impedance of the transistor is 1.5 kΩ, and the impedance of the coil attached to the cone is 8 Ω. The turns ratio of the transformer should be

$$\frac{N_p}{N_s} = \sqrt{\frac{Z_p}{Z_s}} = \sqrt{\frac{1500}{8}} = 13.7:1$$

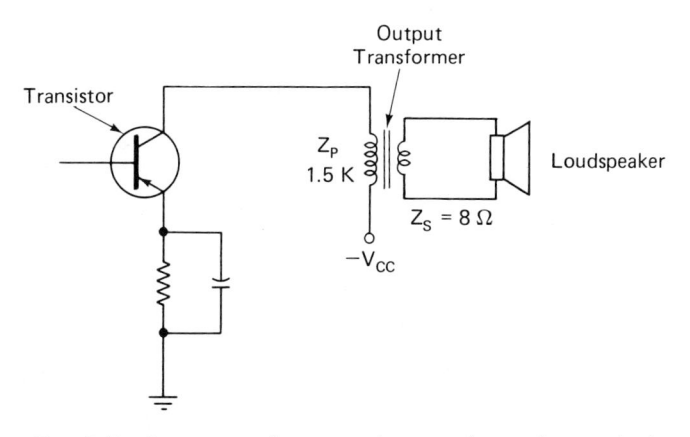

FIG. 5-15 Output transformer used to match transistor output impedance to loudspeaker.

At this turns ratio, the 8-Ω impedance of the speaker appears as a 1500-Ω impedance to the output of the transistor.

5-15 RADIO-FREQUENCY TRANSFORMERS

Power and audio-frequency transformers are primarily power-transfer or impedance-matching devices in which very high mutual coupling is obtained by means of the large number of turns and the soft steel core. At radio frequencies, eddy current and hysteresis losses in a magnetic core become prohibitive; therefore, RF transformers are generally of *air-core* construction. Eddy current loss is the more important factor, since it increases as the square of the frequency; hysteresis loss, being constant for each cycle, increases directly with frequency. With air-core construction, the tightest possible coupling results in a coefficient of coupling of about 0.65 compared to the 0.98 of well-designed iron-core transformers.

A large part of the primary and secondary windings of these partially coupled transformers constitutes a leakage reactance; therefore, voltage,

Three-Phase Transformers 63

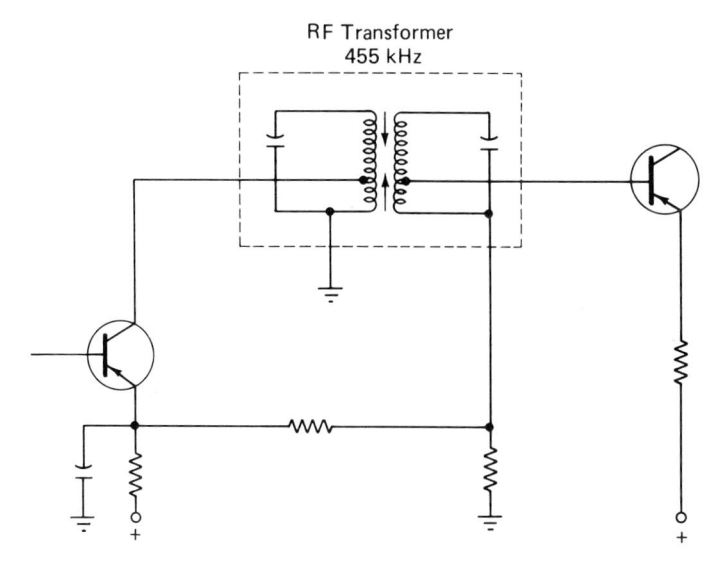

FIG. 5-16   RF transformer.

current, and impedance ratios cannot be calculated by the turns ratio equations, which are based on an assumption of almost perfect coupling. Hence, the turns ratios of RF transformers have no exact significance. In fact, the windings of these transformers are generally very loosely coupled (K may be as small as 0.005), and they function primarily as devices coupling two circuits, rather tnan as voltage- or current-level transformers.

A typical application of an RF transformer coupling the output of one transistor stage to the input of another is shown in Fig. 5-16. The dotted rectangle around the transformer implies that it is located inside a small can, usually aluminum.

5-16   THREE-PHASE TRANSFORMERS

Three-phase transformers are usually used in high-power applications. They have definite advantages when used in power supplies of large transmitters. All connections of three-phase transformers are usually made internally, and only the primary and secondary leads are brought out.

Three-phase transformers may be connected in various ways, depending on the systems with which they are used. The primaries and secondaries may all be connected in *delta* or in *wye*, as shown in the delta–delta and wye–wye connections in Fig. 5-17. In both these cases the secondary line voltage is equal to the primary voltage times the step-up or step-down ratio of the transformer.

If the primary windings of a three-phase transformer are connected in delta and the secondary windings in wye, the connection is a *delta–wye* type. In this case the secondary line voltage is equal to the primary line voltage times the turns ratio times 1.732. This is due to the fact that the wye connection of the secondary causes the secondary line voltage to be equal

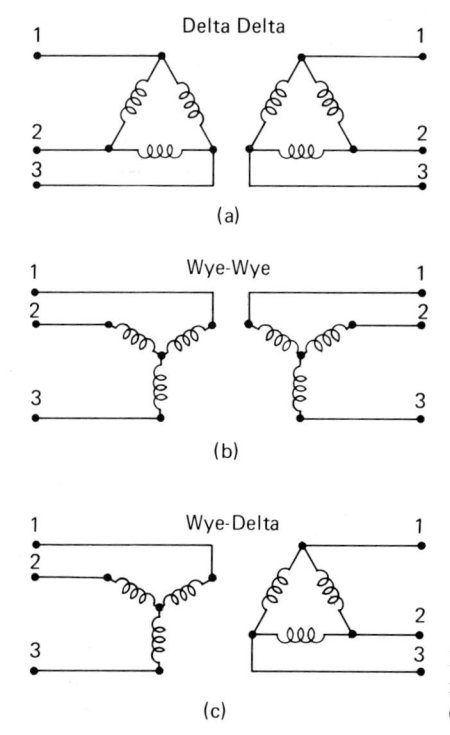

FIG. 5-17 Three phase transformer connections: (a) delta-delta, (b) wye-wye, (c) wye-delta.

to 1.732 times the voltage of each phase of the secondary. One advantage of this type of transformer connection is that the insulation requirements for the secondaries are reduced. This is particularly important in high-voltage systems.

If the primary windings are connected in wye and the secondary windings in delta, the connection is a *wye–delta* type. In this case the secondary line voltage is equal to the primary line voltage times the turns ratio times the factor 1/1.73. This is due to the fact that the output is equal to the voltage across two coils that are 120° out of phase. An advantage of this type of transformer connection is in stepping down high voltages and stepping up current outputs proportionately. Therefore, the secondary windings need not be as well insulated; however, they are usually of larger diameter.

5-17 SHORTED TURNS

When the current flowing through an inductor or transformer is suddenly turned off, a very large voltage, called a *transient*, is developed. This may be large enough to puncture the insulation of the wire on adjacent turns, causing a short. With alternating current flowing through the coil, voltage is induced into the shorted turn by the surrounding turns, causing excessive current to flow within the shorted turn. This current is in opposition to that flowing in the balance of the winding. The field set up by the current in the shorted turn *counteracts* the main field, materially reducing the total inductance. The reduced L is much greater than if just one turn had been removed from the coil. Generally, the inductor overheats under these conditions and must be replaced. There are some applications in which a shorted turn in the form of a loop of wire or a brass disc is brought near a coil, effectively reducing the inductance of the coil. The inductance is reduced by an amount determined by the mutual inductance.

Practice Problems

1. What is the mutual inductance of two coils if one has an inductance of 2 H and the other has 8 H when the coefficient of coupling is 1?
2. One coil of 81.6 mH is wound close to another having 448.75 mH. If the coefficient of coupling is 1, what is the mutual inductance?
3. What is the coefficient of coupling of two coils whose mutual inducance is 1.0 H and whose self-inductances are 2 and 1.2 H?
4. Two coils, each having 3-H inductance, are connected series aiding and have a coupling coefficient of 0.5. What is their total inductance?
5. What is the time constant of a series LR circuit where the inductance is 53 mH and the resistance is 2.7 Ω?
6. A series circuit containing 8-H inductance and 52.5-Ω resistance is connected to a 12.6-V dc source. How long will it take for the current to reach 63 per cent of full value? What current is flowing in the circuit at one time constant?
7. Calculate the inductive reactance of a coil having an inductance of 40 mH at 450 kHz.
8. What is the inductive reactance of a 17-μH coil at 10 MHz?
9. What is the inductance of a coil whose reactance is 2500 Ω at 456 kHz?
10. What is the voltage drop across a pure inductance of 0.68 H if a current of 37 mA at 800 Hz flows through it?
11. A transformer has 15 turns on its primary and 110-V_{ac} applied to it. How many turns are required on the secondary to produce 1650 V to a load?
12. If 10 V_{ac} at 5 A is fed to the primary winding of a transformer, what will the output current be if the secondary voltage across the load is 100-V_{ac}? (assume 100 per cent efficiency)?
13. In a transformer (100 per cent efficiency) the primary voltage and current are 60 V and 3 A. What is the secondary voltage if the secondary current is 15 A?
14. A transformer is 90 per cent efficient. What will the primary current be if the primary voltage is 117 V and the transformer delivers 54 W to its load?
15. What is the I_p of a step-up transformer with a 1:5 turns ratio when $e_p = 120$ V, $Z_L = 600$ Ω, and efficiency is 100 per cent?

16. A certain power transformer delivers 200 mA at 375 V either side of the center-tapped secondary. What is the primary current if the unit is 90 per cent efficient and 117-V_{ac} is applied to the primary?
17. What is the turns ratio of an output transformer designed to match a load resistance of 4.5 kΩ to an 8-Ω voice coil of a loud speaker?
18. What is the I_p of an autotransformer with a 2.5 : 1 step-up ratio when 120-V_{ac} is connected to the primary and the secondary load is 470 Ω?
19. The power transformer of a certain mobile transmitter is connected via a vibrator to a 12-V dc source. The secondary winding develops 720-V_{ac} and has a 0.005-microfarad (μF) buffer capacitor across it. Calculate the reflected capacitance across the primary.
20. A three-phase, delta–wye power transformer has a turns ratio of 16.5 : 1, step-up. Calculate the ac voltage appearing across one of the three phases of the wye if 440 V_{ac} is applied to each phase of the delta-connected primary.

Commercial License Questions

Sections in which answers to questions are given appear in parentheses. A bracketed number following a question implies that it applies only to that element.

1. Define the term *inductance*. (5–1)
2. What is the unit of inductance? (5–2)
3. What is the relationship between the number of turns and the inductance of a coil? (5–2)
4. What is the effect of adding an iron core to an air-core inductance? (5–2)
5. If the mutual inductance between two coils is 0.1 H and the coils have inductances of 0.2 and 0.8 H, respectively, what is the coefficient of coupling? [4] (5–3)
6. When two coils of equal inductance are connected in series, with unity coefficient of coupling and their fields in phase, what is the total inductance of the two coils? [4] (5–3, 5–4)
7. What is the total reactance of two inductances connected in series with zero mutual inductance? [4] (5–4, 5–7)
8. What is the reactance of a 5-mH choke coil at a frequency of 1000 kHz? (5–7)
9. What is the current and voltage relationship when inductive reactance predominates in an ac circuit? (5–8)
10. What would be the effect if direct current were applied to the primary of an ac transformer? (5–9)
11. What would be the effects of connecting 110 V at 25 cycles to the primary of a transformer rated at 110 V and 60 cycles? (5–9, 5–11)
12. If a power transformer having a voltage step-up ratio of 1 : 5 is placed under load, what will be the approximate ratio of the primary to secondary current? (5–10)

Commercial License Questions 67

13. If a power transformer has a primary voltage of 4400 V and a secondary voltage of 220 V, and the transformer has an efficiency of 98 per cent when delivering 23 A of secondary current, what is the value of primary current? [4]
(5–10)
14. What factors determine the core losses in a transformer? [4] (5–11)
15. Why are laminated iron cores used in audio and power transformers?
(5–11)
16. What circuit constants determine the copper losses of a transformer? [4]
(5–11)
17. Why are electrostatic shields used between windings in coupling transformers?
(5–11)
18. What factor(s) determine the ratio of impedances that a given transformer can match? [4] (5–13)
19. If a transformer, having a turns ratio of 10:1, is working into a load impedance of 2000 Ω and out of a circuit having an impedance of 15 Ω, what value of resistance may be connected across the load to effect an impedance match? [4]
(5–13, 5–14)
20. Three single-phase transformers, each with a ratio of 220 to 2200 V, are connected across a 220-V three-phase line, primaries in delta. If the secondaries are connected in wye, what is the secondary line voltage? [4] (5–16)
21. Draw a schematic wiring diagram of a three-phase transformer with delta-connected primary and wye-connected secondary. [4] (5–16)
22. What will be the effect of a shorted turn in an inductance? (5–17)

6

Capacitance

6-1 BASIC CAPACITOR

Capacitors, sometimes called condensers, are one of the principal building blocks in electronics circuits. Their basic function is to *store* electrical energy for a period of time and then release it. This time interval may be as short as a few picoseconds or as long as minutes. The more *electrostatic energy* a capacitor can store, the larger the capacitance.

The essential elements of a capacitor are shown in Fig. 6-1. It consists of two metallic plates separated by an insulating material called the *dielectric*.

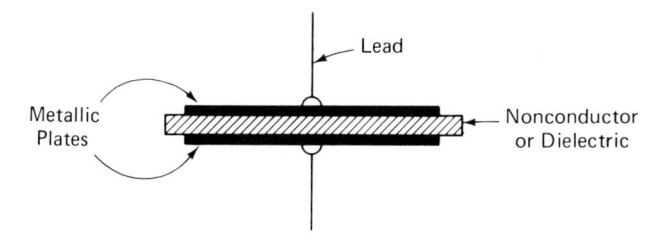

FIG. 6-1 Simple capacitor.

Leads are attached to the plates so that they can be connected to the circuit. The simplest type of capacitor consists of two metal plates separated by air, which is the dielectric in this case. Schematic symbols for capacitors are shown in Fig. 6-2. Capacitance is represented by the letter *C*.

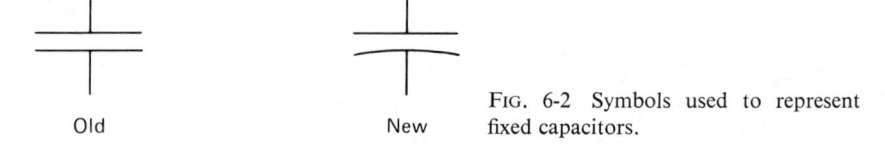

FIG. 6-2 Symbols used to represent fixed capacitors.

Unit of Capacitance

If an insulating material such as mica is placed between the plates of a capacitor that have no potential applied to them, the planetary electrons revolve about their nuclei in undistorted paths, as shown in Fig. 6-3a. If a positive potential is applied to one plate and a negative to the other, the electron orbits are distorted. The positive plate provides an attractive force; the negative one provides a repelling force. The result is as shown in Fig. 6-3b. If the charge remains on the plates, the orbits stay as shown. A small amount of energy is required to move the electrons, and this energy is stored in the dielectric. Under proper conditions this stored energy can be returned to the circuit.

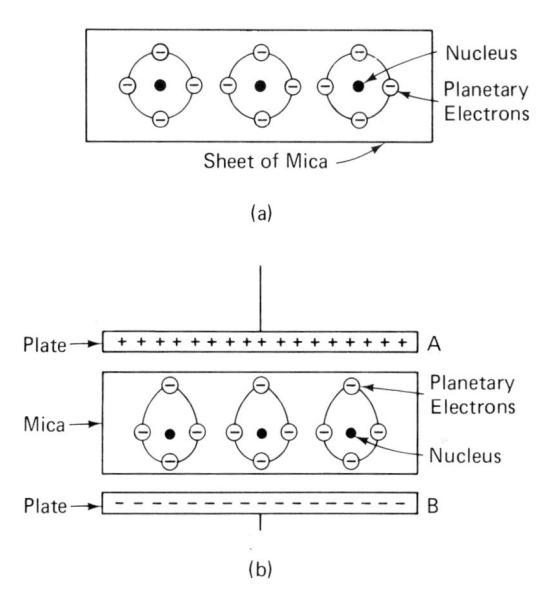

FIG. 6-3 Electron orbits in a dielectric: (a) without an electrical potential applied to the plates, (b) with a potential.

6-2 UNIT OF CAPACITANCE

Capacitance is measured in *farads*, abbreviated F. A capacitance is equal to 1 F when a voltage changing at the rate of 1 V/s causes a charging current of 1 A to flow.

The farad can also be defined in terms of charge and voltage. A capacitor has a capacitance of 1 F if it will store 1 *coulomb* of charge when connected across a potential of 1 V. This relationship can be expressed mathematically as

$$C = \frac{Q}{E}$$

where C = capacitance in farads
        Q = charge in coulombs
        E = applied potential in volts

Example 6-1

What is the capacitance of two metal plates separated by 1 cm of air if 0.001 C of charge is stored when a potential of 200 V is applied to the capacitor?

Solution

$$C = \frac{Q}{E} = \frac{10 \times 10^{-4}}{2 \times 10^2} = 5 \times 10^{-6} \text{ F}$$

Although this capacitance might appear rather small, many electronic circuits require capacitors of much smaller value. Consequently, the farad is a cumbersome unit, far too large for most applications. The *microfarad* is a more convenient unit. The symbols used to designate microfarad are μF or mfd. In high-frequency circuits even the microfarad becomes too large, and the *picofarad* is used. The symbols for picofarad are pF or $\mu\mu$fd.

6-3 CHARGING A CAPACITOR

To understand the charging action taking place in a capacitor, assume that an uncharged unit is suddenly placed across a battery and that the circuit has no resistance, as shown in Fig. 6-4. The positive terminal of the battery extracts electrons from the bottom plate and forces electrons into the top plate. Thus, in every part of the circuit a clockwise displacement of electrons occurs.

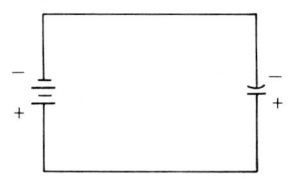

FIG. 6-4 Charging action in a capacitor.

This builds up a potential difference across the capacitor that *opposes* the source voltage. The emf developed across the capacitor, however, has a tendency to oppose the current flowing into it. As the capacitor continues to charge, the voltage across it rises until it is equal to the source voltage. Once the capacitor voltage equals the source voltage, the two balance one another and current ceases to flow.

In the charging process *no current* flows through the capacitor because of the insulating qualities of the dielectric. However, if a milliammeter had been inserted in the circuit, it would have momentarily registered a current flow and then returned to zero. The current that appears to flow in a capacitive circuit is called *displacement current*.

When a capacitor is fully charged, the electrostatic field between the plates will be maximum, which indicates that maximum energy is stored. If an excessive amount of voltage is applied to a capacitor, the dielectric

Factors Affecting Capacity

will break down and short the capacitor. The dielectric material in a practical capacitor is not perfect, and a small leakage current will flow through it. This will eventually dissipate the charge.

If a charged capacitor is connected across a discharged capacitor, part of the charge on the one will be delivered to the other. The resultant voltage across the capacitors will be less than the original charge.

Example 6-2

Given two identical capacitors of 0.1 μF each, one is charged to a potential of 125 V and disconnected from the charging source. It is then connected across the uncharged capacitor. What voltage will appear across the two?

Solution

The original charge divides equally so that 62.5 V appears across each. This can be proved by the formula

$$C = \frac{Q}{E}$$

and solving for E

$$E = \frac{Q}{C}$$

If C is doubled, then

$$E = \frac{Q}{2C}$$

which shows that the voltage is halved.

6-4 FACTORS AFFECTING CAPACITY

Several factors affect the ability of a capacitor to store a charge. These are size of plates, type of dielectric material used, and the spacing between them. The relationship between these can be expressed by the formula

$$C = 0.225 \frac{KA}{d}$$

where C = charge in picofarads
 A = area of one plate in square inches
 d = distance between plates in inches
 K = dielectric constant of insulating material

This formula shows that capacitance is a direct function of the dielectric constant and the area of the capacitor plates, and an inverse function of the distance between the plates. Several materials and their dielectric constants are given in Table 6-1.

TABLE 6-1 DIELECTRIC CONSTANTS

Vacuum	1.0
Air	1.0006
Aluminum oxide	7–10
Glass	4–7
Paper	2–3.5
Rubber	2–3
Mica	7–7.3
Water	81
Porcelain	6–7
Ceramic	20 to over 1000

6-5 TYPES OF CAPACITORS

Capacitors are classified into two general types: *variable* and *fixed*. Variable capacitors are constructed so that their capacitance can be varied. There are two types of variable capacitors: the *rotor–stator* type and the *trimmer* type.

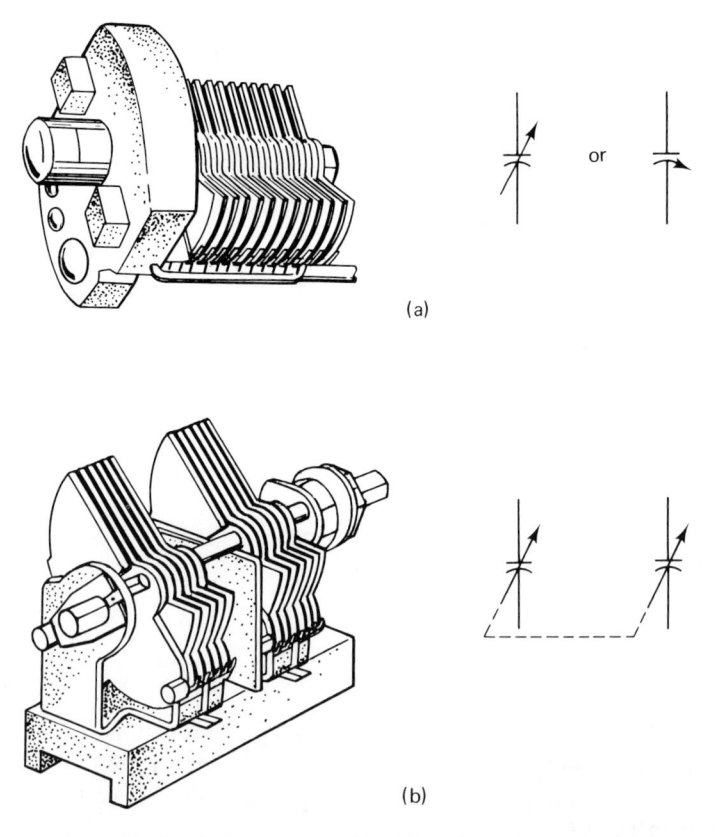

FIG. 6-5 Variable capacitors (a) midget single (b) split-stator.

Types of Capacitors 73

The rotor–stator type uses air as the dielectric. The amount of capacitance is varied by changing the position of the rotor plates (movable plates). This changes the effective plate area of the capacitor. When the rotor plates are fully meshed between the stator plates, the capacitance is maximum. Two rotor–stator types are illustrated in Fig. 6-5 with their symbols adjacent.

The trimmer type of variable capacitor consists of two plates separated by a dielectric other than air. The capacitance is varied by changing the distance between the plates. This is ordinarily accomplished by means of a screw, which forces the plates closer together, as shown in Fig. 6-6.

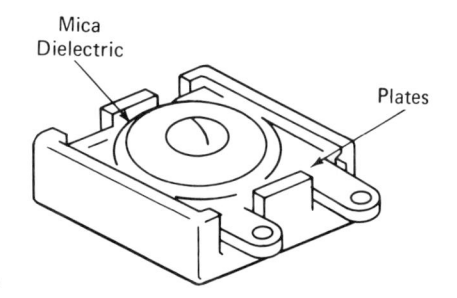

FIG. 6-6 Trimmer capacitor.

Fixed capacitors are categorized by the type of dielectric used. Some of the more common types are paper, mica, Mylar, and ceramic.

Figure 6-7 shows the details of the internal construction of a paper capacitor. The rolled strips of metal foil project slightly on either side so that the connecting bond may be soldered to each turn and not to the end of the strip only. Such an arrangement prevents the metal-foil conductors from becoming inductances (by shunting the turns), since physically the capacitor resembles two coils wound one within the other. A capacitor so constructed is called a *noninductive* capacitor.

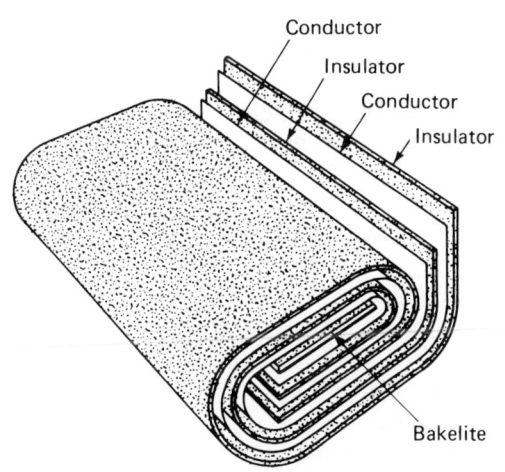

FIG. 6-7 Internal construction of a paper capacitor.

A word of caution about fixed capacitors. They have a nominal voltage rating in addition to a *surge* voltage rating. The surge voltage rating should never be exceeded, not even during tests. To test a capacitor near or at its surge rating is practically asking for a failure. Always test capacitors at their rated or working voltage, never above it.

Electrolytic capacitors are used where a large amount of capacitance is required. As the name implies, electrolytic capacitors contain an *electrolyte*. This can be in the form of either a liquid (wet electrolytic capacitor) or a paste (dry electrolytic capacitor). Wet electrolytic capacitors are no longer in popular use due to the care needed to prevent spilling of the electrolyte.

Dry electrolytic capacitors consist essentially of two metal plates between which is placed the electrolyte. In most cases the capacitor is housed in a cylindrical aluminum container, which acts as the negative terminal of the capacitor, as shown in Fig. 6-8. The positive terminal (or terminals if the capacitor is of the multisection type) is in the form of a lug on the bottom end of the container. The size and voltage rating of the capacitor are generally printed on the side of the aluminum case.

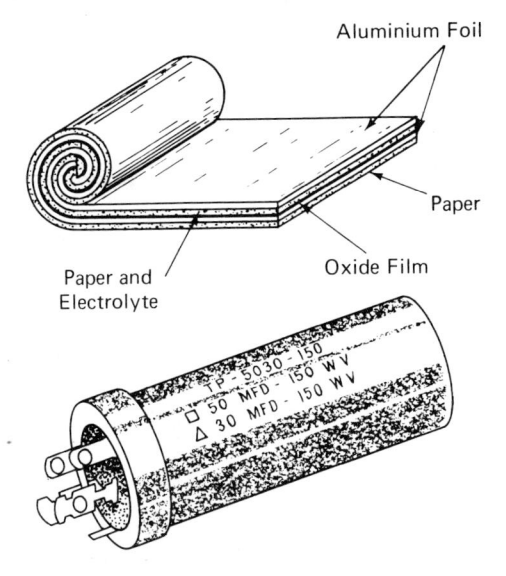

FIG. 6-8 Construction of an electrolytic capacitor.

Internally, the electrolytic capacitor is constructed similarly to the paper capacitor. The positive plate consists of aluminum foil covered with an extremely thin film of *oxide*. This acts as the dielectric. Next to, and in contact with, the oxide is placed a strip of paper or gauze that has been impregnated with a pastelike electrolyte. The electrolyte acts as the negative plate of the capacitor. A second strip of aluminum foil is then placed against the elec-

Capacitors in Parallel and Series

trolyte to provide electrical contact to the negative electrode (electrolyte). When the three layers are in place, they are rolled up into a cylinder.

Electrolytic capacitors are polarized and have a low leakage resistance. Should the positive plate be accidentally connected to the negative terminal of the source, the thin oxide film dielectric will dissolve, and the capacitor will become a conductor (i.e., it will short). Since electrolytic capacitors are polarity sensitive, their use is ordinarily restricted to dc circuits or circuits in which a small ac voltage is superimposed on a dc voltage.

6-6 CAPACITORS IN PARALLEL AND SERIES

When capacitors are connected in parallel, the total capacity increases. The reason is that the effective capacity is made up of the total plate area of all capacitors. The formula for parallel capacitors is

$$C_T = C_1 + C_2 + C_3 + \cdots$$

Example 6-3

What is the total capacity of the following capacitors parallel connected: 0.03, 2.0, and 0.25 μF?

Solution

$$C_T = C_1 + C_2 + C_3$$
$$= 0.03 + 2.0 + 0.25 = 2.28 \ \mu F$$

The maximum voltage that can be impressed across the combination is dictated by the capacitor having the *lowest* voltage rating. If this should be exceeded, all units would be short circuited by the shorted capacitor. If electrolytic capacitors are parallel, it is very important that polarities be observed. An example of parallel-connected capacitors appears in Fig. 6-9.

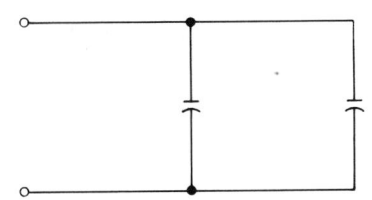

FIG. 6-9 Parallel connected capacitors.

When two capacitors are connected in series, the total capacity is reduced. For example, two 8-μF units will give 4 μF when connected in series. An example of this arrangement is shown in Fig. 6-10. A study of this simple drawing will reveal that there is twice the thickness of dielectric material between points X and Y as with one capacitor. From Section 6-4 we learned that capacity varies inversely with dielectric thickness. Hence, if the effective plates of the capacitor are twice as far apart, the capacity is halved.

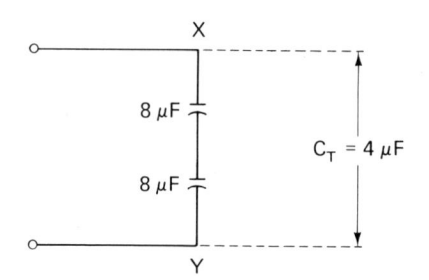

FIG. 6-10 Series capacitors.

To find the equivalent capacitance of series-connected capacitors, the following formula can be used:

$$\frac{1}{C_T} = \frac{1}{C_1} + \frac{1}{C_2} + \frac{1}{C_3} + \cdots$$

Note that this is similar to the formula for parallel-connected resistors. If the circuit contains only two capacitors, the product-over-the-sum formula can be used:

$$C_T = \frac{C_1 C_2}{C_1 + C_2}$$

The total capacity of series-connected units will always be *smaller* than the smallest of the individual capacitors.

If the voltage is impressed across two series-connected capacitors, it will divide in a manner *inverse* to their capacities. Consider 0.5 and 1.0 μF in series across a 600-V supply. Four hundred volts will be dropped across the 0.5 μF and 200 V across the 1.0 μF, whether alternating or direct current is supplied. This is true so long as no leakage is present in the dc voltage case.

In an emergency it is possible to series connect several low-voltage capacitors to obtain the required voltage rating. Suppose that a number of 2-μF, 400-V capacitors are available and it is necessary to connect them in such a manner as to form 1.5 μF at 1600 V. If four of them are connected in series, they will each be able to handle 400 V, but the capacity would be only one fourth of 2 μF or 0.5 μF. However, if two additional series combinations are connected in parallel with the first, a total of 1.5 μF of capacity will result. A total of 12 capacitors would be needed to make up this series–parallel arrangement. To ensure equal voltage drops across each series-connected capacitor (their leakage resistances are not always the same), *equalizing resistors* should be used. These resistors would likely be at least 40-kΩ each.

6-7 CAPACITOR COLOR CODES

Capacitors are coded in a manner somewhat similar to the method used with resistors (Section 1-6). There are two color-coding systems for

Capacitor Color Codes 77

capacitors. One is the JAN (Joint Army Navy) system by which all capacitors produced for military use are marked. The other is the RMA (Radio Manufacturers Association) system, renamed the EIA (Electronic Industries Association) system.

The simplest method is the three-dot code used with mica capacitors having 20 per cent tolerance shown in Fig. 6-11. If the dots were orange, white, and brown, reading in the direction of the arrow, the capacitance would be 390 pF.

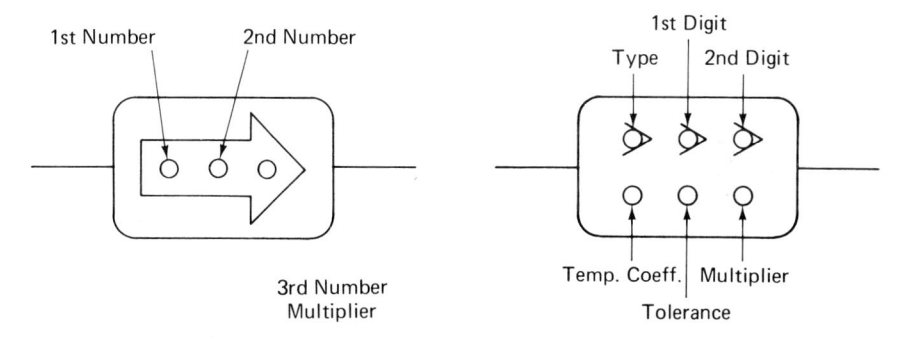

Fig. 6-11 Typical 3-dot EIA mica capacitor.

Fig. 6-12 Six-dot EIA and JAN color coding system for mica and molded paper capacitors.

In the six-dot EIA and JAN system shown in Fig. 6-12 a 5600-pF, 10 per cent mica capacitor would be coded (reading clockwise from upper left dot) white (EIA) or black (JAN), green, blue, red, and silver. The last dot represents the *temperature coefficient*, which is the degree of capacitance change with temperature. If there is no change, it has a *zero coefficient*. If it increases with temperature, it has a *positive coefficient*. A -750 coefficient means the capacitance will decrease by 750 parts per million (ppm) per each degree rise in temperature.

A typical color-coded tubular ceramic capacitor is shown in Fig. 6-13. A 270-pF, 10 per cent tolerance, zero-temperature-coefficient capacitor would be coded as follows: black, red, violet, brown, and silver.

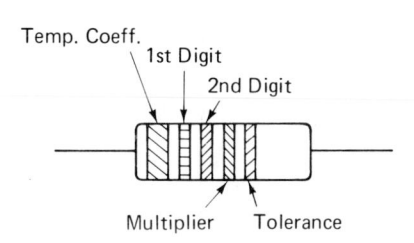

Fig. 6-13 Tubular ceramic capacitor.

6-8 RESISTANCE–CAPACITANCE TIME CONSTANT

A definite amount of time is required for a capacitor to charge through a series resistor. The time required to charge a capacitor to 63 per cent of its maximum voltage is called the *time constant* of the circuit. The time constant is equal to the product of the resistance and capacitance. Thus,

$$T = RC$$

where T = one time constant in seconds
 R = circuit resistance in megohms
 C = circuit capacitance in microfarads

Example 6-4

Determine the time constant of a circuit containing a 0.1-μF capacitor and a 100-kΩ resistor.

Solution

$$T = RC$$
$$= 1 \times 10^{-1} \times 10^{-1} = 1 \times 10^{-2} \text{ s}$$

This means that the capacitor will reach 63 per cent of the supply voltage in 10 milliseconds (ms). If the supply potential is assumed to be 100 V, the capacitor will have 63 V across it in one time constant.

During the second time constant the capacitor will charge to 63 per

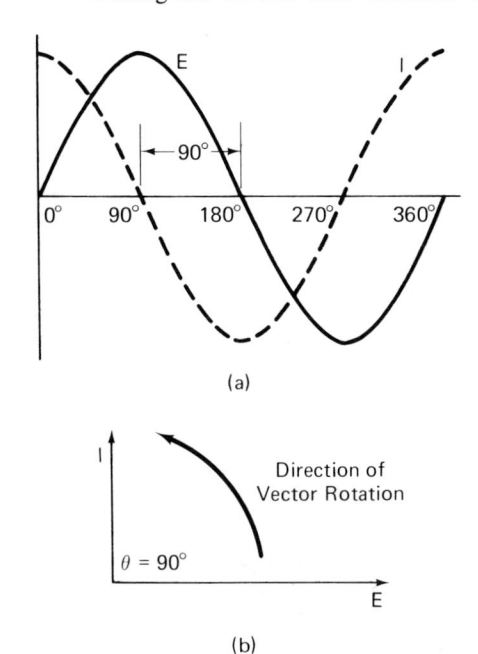

FIG. 6-14 (a) Phase relationship of E and I in a purely capacitive circuit. (b) Vector representation of same.

Capacitive Reactance

cent of the remaining value of voltage between 100 V and its present charge. Therefore, in the second time constant the capacitor will gain an additional charge of

$$e = 0.63 \times (100 - 63)$$
$$= 23.3 \text{ V}$$

The charge on the capacitor will now equal 86.3 V. This process continues for three more time constants until the charge is approximately 100 V. The shape of the charging curve is the same as the time-constant curve shown in Fig. 6-14.

6-9 CAPACITIVE REACTANCE

If a voltage is applied to an inductance, opposition is immediate, and there is a delay in current rise through it. If a voltage is applied to a circuit containing capacitance, current flows at a *maximum* almost instantaneously and then *gradually* falls to zero as opposition to it builds up. Thus, when the applied voltage is changed, the capacitor charges or discharges until the voltage on the capacitor equals the new value of applied emf.

At the time when the capacitor voltage is equal to the source voltage, no more current flows. Since a capacitor reacts to a voltage change by producing a cemf, a capacitor is said to be *reactive*.

The opposition offered to a changing voltage is called *capacitive reactance*, and is measured in ohms. The symbol for this reactance is X_c.

Although no current actually flows through a capacitor, circuit current will exist whenever a capacitor charges or discharges. If a capacitor is connected across an alternating voltage source, an alternating current will flow as the capacitor charges and discharges. If a sine wave of voltage is applied to a capacitor, a sine wave of current will result. Since current in a capacitive circuit is maximum when the *rate of change* of voltage is maximum, the current waveform will be offset 90° from the voltage waveform. This is illustrated in Fig. 6-15a. Notice that when the voltage is passing through zero (maximum rate of change) the current is maximum. When the voltage is at its peak value (minimum rate of change), the current is zero; thus, in a capacitive circuit the *current leads the voltage by 90°*. This phase relationship is shown vectorally in Fig. 6-15b.

Capacitive reactance is an inverse function of frequency and capacitance. Expressed mathematically,

$$X_c = \frac{1}{2\pi f C}$$

where X_c = capacitive reactance in ohms
2π = a constant
f = frequency in hertz
C = capacitance in farads

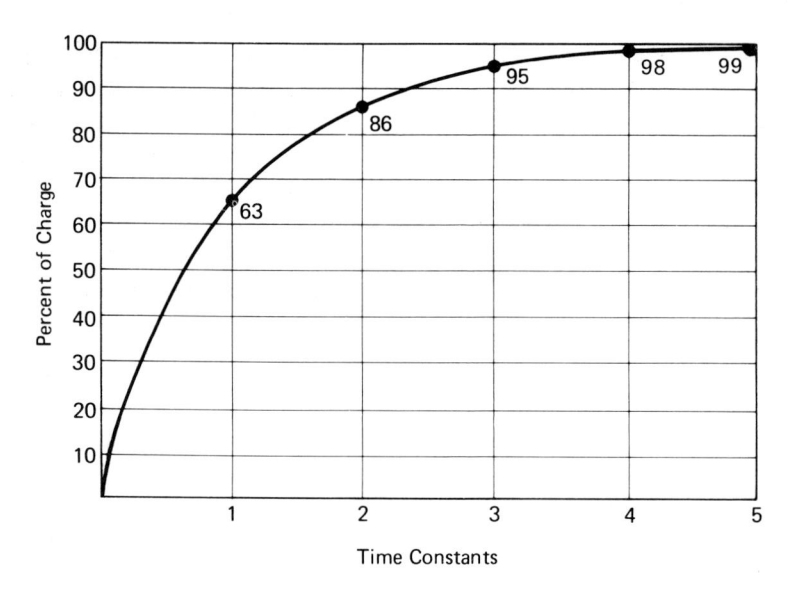

FIG. 6-15 X_c, I, and f relationships in a capacitive ac circuit.

Example 6-5

What is the capacitive reactance of a 2-μF capacitor at a frequency of 4 kHz?

Solution

$$X_c = \frac{1}{2\pi f C}$$

$$= \frac{1}{6.28(4 \times 10^3)(2 \times 10^{-6})} = 19.8 \ \Omega$$

Capacitive reactances in series or parallel are calculated in the same way as resistances in series or parallel.

Figure 6-16 illustrates the relationships between capacitive reactance, current, and frequency in an ac capacitive circuit.

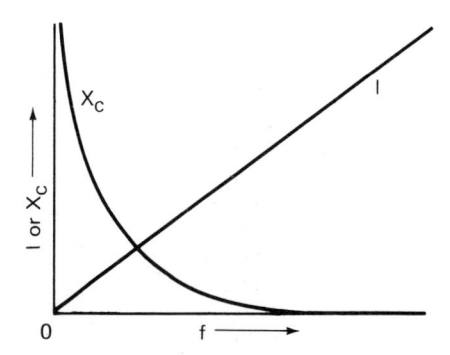

FIG. 6-16

Voltage Ratings of Capacitors

From analyzing Fig. 6-16 it can be seen that, as the frequency increases, X_c decreases and the current increases. At high frequencies a capacitor has a low opposition to current. Thus, as frequency is increased a capacitor's characteristics approach those of a *short circuit*. At the point marked zero frequency the capacitor exhibits the characteristics of an *open circuit*.

To solve a capacitive network, the same general laws and equations for direct current are used, except that the notation in the formulas is changed to comply with the quantities under consideration. For example, Ohm's law for a capacitive circuit is expressed mathematically as

$$I_c = \frac{E_c}{X_c}$$

6-10 LOSSES IN A CAPACITOR

There are two losses worth considering in a capacitor. One is the leakage of charges from one plate to another either through the dielectric or over the surface of the capacitor. These will be small due to the very high resistance of the dielectric and leakage path over the surface. Heat is produced as a result ($P = I^2R$).

The second loss is due to hysteresis in the dielectric, which also produces heat. As the charge on the capacitor increases and decreases, particularly at high frequencies, the orbital electrons are strained, first in one direction and then another, as shown in Fig. 6-3. The energy used to produce this action results in the generation of heat. For this reason some capacitors may operate well at low frequencies, but fail at very high frequencies. Paper capacitors would likely have much more leakage at high frequencies than ceramic, mica, or vacuum.

6-11 VOLTAGE RATINGS OF CAPACITORS

Two factors affect the voltage rating of a capacitor. One is the thickness of the dielectric, and the other is the type of material used for the dielectric. For any given dielectric the voltage rating (amount of voltage that can

TABLE 6-2 DIELECTRIC STRENGTHS

Material	Volts per Mil
Air	3000
Bakelite	10,000–28,000
Glass (commercial)	20,000–60,000
Glass (electrical)	80,000–330,000
Isolantite	12,600
Mica	50,000–225,000
Paper (kraft)	30,000–40,000
Porcelain	5700
Vinyl plastic	15,800

be safely connected across the capacitor) increases as the thickness increases. Of course, increasing the thickness of the dielectric increases the separation of the plates and, consequently, reduces the capacity.

The type of dielectric material significantly affects the voltage rating. For a given thickness some materials can withstand considerably more voltage before breakdown than others. When high-voltage operation is required, as in certain transmitter circuits, it is particularly important to select the right voltage-rating capacitor. The dielectric strengths of several materials in volts per millimeter are given in Table 6-2.

Practice Problems

1. What is the total capacitance of 40 and a 20 μF connected in series?
2. Calculate the total capacitance of two 200-μF capacitors connected in series, which are in turn connected across a 50-μF capacitor.
3. What is the time constant of a 40-μF capacitor in series with a 2-kΩ resistor?
4. If a 200-V dc potential is suddenly connected across the circuit in problem 3, what will the voltage across the capacitor be in one time constant? In two time constants?
5. Calculate the X_c of a 415-pF capacitor at 1410 kHz.
6. What capacitance is required to have a reactance of 100 Ω at 1 MHz?
7. A capacitor having 2-kΩ reactance at 1 kHz is connected across a 20-V source. How much current flows in the circuit?
8. What reactive voltage appears across a capacitor of 0.047 μF when 36 mA at 2.5 kHz is flowing in the circuit?

Commercial License Questions

Sections in which answers to questions are given appear in parentheses.
A bracketed number following a question implies that it applies only to that element.

1. The charge in a condenser is stored in what portion of the condenser?
(6–1)
2. What is the unit of capacitance? (6–2)
3. State the formula for determining the quantity or charge of a condenser and the energy stored in a condenser. (6–2)
4. Given two identical mica condensers of 0.1-μF capacitance each. One of these is charged to a potential of 125 V and disconnected from the charging circuit. The charged condenser is then connected in parallel with the uncharged condenser. What voltage will appear across the two condensers connected in parallel? (6–3)

Commercial License Questions

5. What are the properties of a series condenser acting alone in an ac circuit? (6-3, 6-9)
6. What factors determine the charge stored in a condenser? (6-4)
7. Explain the effect of increasing the number of plates upon the capacitance of a condenser. (6-4)
8. What effect does a change in the dielectric constant of a condenser dielectric material have upon the capacitance of a condenser? (6-4)
9. If the specific inductive capacity of a condenser dielectric material between the condenser plates were changed from 1 to 2, what would be the resultant change in capacitance? (6-4)
10. What precaution should be observed when connecting electrolytic condensers in a circuit? (6-5)
11. If condensers of 1, 3, and 5 μF are connected in parallel, what is the total capacitance? (6-6)
12. What is the formula used to determine the total capacitance of three or more capacitors connected in series? (6-6)
13. If condensers of 5, 3, and 7 μF are connected in series, what is the total capacitance? (6-6)
14. Having available a number of condensers rated at 400 V and 2 μF each, how many of these condensers would be necessary to obtain a combination rated at 1600 V and 1.5 μF? (6-6)
15. The voltage drop across an individual condenser of a group of condensers connected in series across a cource of potential is proportional to what factors? (6-6)
16. What is meant by the *time constant* of a resistance–capacitance circuit? (6-8)
17. What is the reactance of a condenser at the frequency of 1200 kHz if its reactance is 300 Ω at 680 kHz? (6-9)
18. What is the reactance value of a condenser of 0.005 μF at a frequency of 1000 kHz? (6-9)

7

Alternating-Current Circuits

7-1 SERIES RESISTANCE–INDUCTANCE CIRCUIT

If an alternating voltage is applied to a series circuit containing inductive reactance and resistance, the current that flows is *not* determined by the mere addition of X_L and R divided into the voltage. The reason is that the total opposition that a circuit offers to the flow of an alternating current is the *vector sum* of the reactance and resistance. This is called *impedance* and is measured in ohms. The impedance of a circuit may be calculated from the formula $Z = \sqrt{R^2 + X^2}$. Example 7-1 will serve to illustrate how the impedance of a simple LR circuit may be calculated.

Example 7-1

What is the impedance of a circuit containing an inductive reactance of 8 Ω in series with 6 Ω of resistance? See Fig. 7-1a.

Solution

$$Z = \sqrt{R^2 + X_L^2}$$
$$= \sqrt{6^2 + 8^2} = \sqrt{100} = 10 \text{ Ω}$$

The vector representation of this circuit appears in Fig. 7-1b. Observe the 90° relationship between X_L and R. The current flowing in the circuit is determined by Ohm's law for the ac circuit:

$$I = \frac{E}{Z} = \frac{100}{10} = 10 \text{ A}$$

Series Resistance–Inductance Circuit

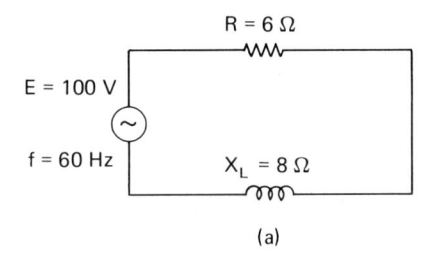

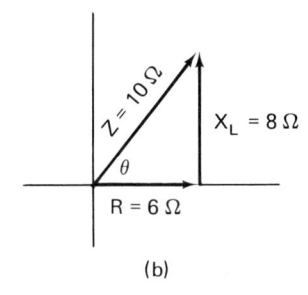

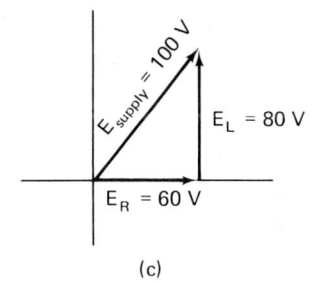

FIG. 7-1 Series *LR* circuit.

The voltage drop across each component is

$$E_R = IR = 10 \times 6 = 60 \text{ V}$$
$$E_L = IX_L = 10 \times 8 = 80 \text{ V}$$

For a vector representation of this, see Fig. 7-1c. Notice the similarity between the right triangles in Figs. 7-1b and c. The relationship between the ac voltages is expressed by

$$E_{\text{supply}} = \sqrt{E_R^2 + E_L^2}$$

Because the circuit contains inductance, the current will lag the voltage. The angle is usually designated by the Greek letter θ (theta) and is called the *phase angle*. This angle can be determined by the ratio of R to Z and is known as the *cosine*. The cosine of θ in Fig. 7-1b is R/Z or $\frac{6}{10} = 0.60$. By referring to a table of trigonometric functions (Appendix B), we find under the column marked cosine the number 0.602, which is closest to 0.60, and that this number corresponds to 53°.

It is possible to find the phase angle by the ratio of X_L/Z, called the *sine*, and also X_L/R, called the *tangent*. For example, the sine of $\theta = \frac{8}{10} = 0.80$. The closest number to this under the sine column (Appendix B) is 0.799, which corresponds to 53°. Using the tangent method, $\theta = X_L/R = \frac{8}{6} = 1.33$. The closest number to this in the tangent is 1.327, which again represents 53°.

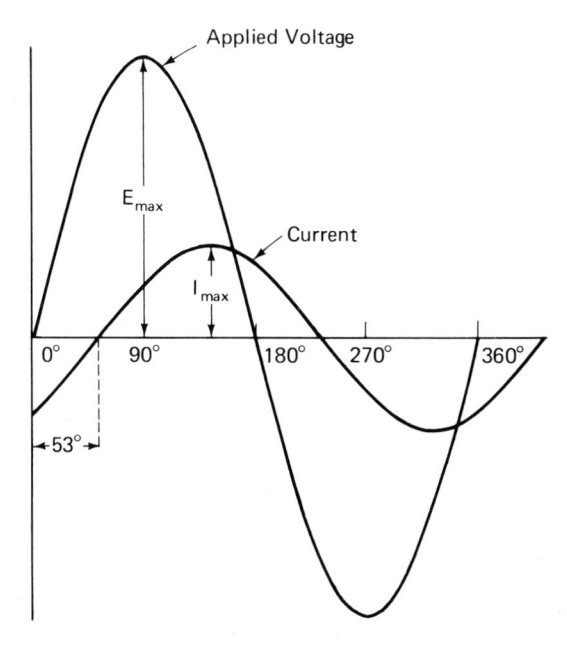

FIG. 7-2 *RL* circuit.

The sinusoidal relationship between the applied emf and the circuit current is shown graphically in Fig. 7-2. The current wave passes through the zero axis approximately 53° after the voltage. If the circuit's resistance were increased but the reactance left unchanged, the circuit would look *less inductive;* that is, the phase angle would diminish. The current and voltage would come closer to being in phase. Obviously, the reverse would be true if the circuit's resistance were reduced and the reactance remained unchanged.

7-2 SERIES RESISTANCE–CAPACITANCE CIRCUIT

The series circuit shown in Fig. 7-3a can be analyzed in much the same way as the *LR* circuit. To find current flowing in the circuit, it is necessary to first determine the impedance. By substituting X_c for X_L, we can use the same impedance formula as in Section 7-1, where

Series Resistance–Capacitance Circuit

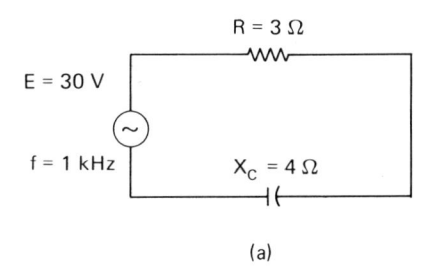

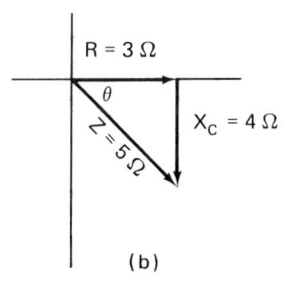

FIG. 7-3 Series *RC* circuit.

$$Z = \sqrt{R^2 + X_c^2}$$

Substituting the values shown in Fig. 7-3a, we have

$$Z = \sqrt{3^2 + 4^2} = \sqrt{25} = 5\ \Omega$$

The vector diagram in Fig. 7-3b graphically shows the angular relationships between *R*, X_c, and *Z*. The phase angle θ is calculated in precisely the same manner as outlined in Section 7-1. However, the current is *leading* the

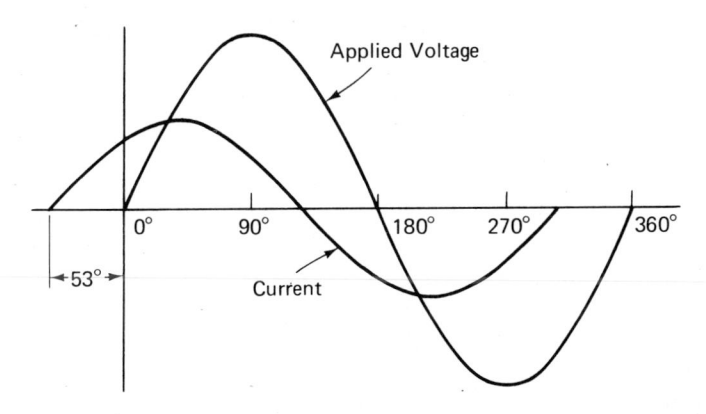

FIG. 7-4 *RC* circuit.

applied voltage by the angle indicated, and therefore the vectors are drawn in the fourth quadrant. The phase angle θ by which the current leads the voltage is

$$\cos \theta = \frac{R}{Z} = \frac{3}{5} = 0.600$$

$$\theta = 53°$$

The relationship between emf and current is shown in Fig. 7-4. Observe that the current waveform leads the voltage.

7-3 SERIES INDUCTANCE–CAPACITANCE–RESISTANCE CIRCUIT

When inductance, capacitance, and resistance are brought together in a series circuit, the voltage drops, current, and phase angle may be determined by combining the methods previously described. Figure 7-5 shows a series *LCR* circuit having $R = 6\,\Omega$, $X_L = 8\,\Omega$, and $X_c = 16\,\Omega$ connected to a 60-Hz, 30-V source.

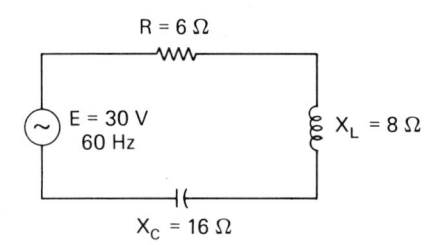

FIG. 7-5 Series *RLC* circuit.

Figures 7-6a and b are the graph and vector diagrams, respectively, for this circuit. Since current is the same in all parts of a series circuit, it is the *reference* vector. Then E_R is in phase with I, E_L leads I by 90°, and E_c lags I by 90°. E_c and E_L are 180° out of phase, and so their vector sum is merely the difference between the two. The resultant vector is shown as the darker triangle in Fig. 7-6b.

The impedance of this circuit may be calculated by the formula

$$Z = \sqrt{R^2 + (X_L - X_c)^2}$$

By substituting the values given in the circuit of Fig. 7-5 into this equation, the impedance is found to be

$$Z = \sqrt{6^2 + (8 - 16)^2} = \sqrt{6^2 + (-8)^2} = \sqrt{100} = 10\,\Omega$$

With 30 V applied, the circuit current is found to be

$$I = \frac{E}{Z} = \frac{30}{10} = 3\text{ A}$$

Because the larger capacitive reactance completely cancels the inductive reactance, the circuit is said to be capacitive. The current leads the applied

Series Inductance–Capacitance–Resistance Circuit

(a)

(b)

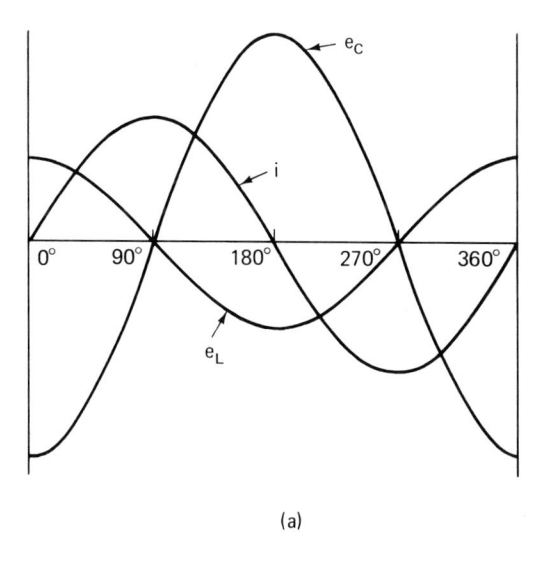

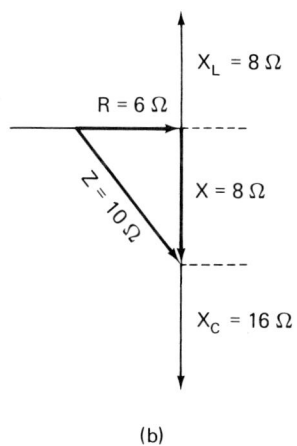

FIG. 7-6 Graph and vector representation of series LCR circuit shown in Fig. 7-5.

voltage by the phase angle θ, which is equal to

$$\cos\theta = \frac{R}{Z} = \frac{6}{10} = 0.60$$

$$\theta = 53°$$

The series LCR circuit illustrates the following important points:

1. The current in a series LCR circuit either leads or lags the applied voltage, depending on whether X_c is greater than or less than X_L.

2. A capacitive voltage drop in a series circuit always subtracts directly from an inductive voltage drop.
3. The voltage across a single reactive element in a series circuit can have a *greater effective value* than that of the applied voltage.

The impedance that a series *LCR* circuit offers to the ac voltage source depends upon the relative magnitudes of X_L, X_c, and R. If X_L is larger than X_c, the capacitive reactance is canceled out and the source sees only R and X_L in series. The reverse is true if X_c is larger than X_L. If X_L equaled X_c, they would cancel each other and the circuit's impedance would be equal to R.

7-4 THE *j* OPERATOR

Electrical engineers have developed a unique and simple notation to indicate when one electrical quantity is exactly 90° out of phase with another. By placing the letter *j* in front of a vector quantity, the implication is that it has gone through 90° of rotation. If the rotation is clockwise, a *minus* sign is placed in front of the *j* operator. For counterclockwise rotation a *plus j* is used.

A capacitively reactive voltage of 20 V would be written as $-j20$ V, implying that the voltage is lagging the current by 90°. If an inductive circuit has 2.3 A of current flowing through it, the prefix $-j$ would be used to indicate that the current was lagging 90° behind the voltage. A $-j$ in front of a quantity such as 83 Ω tells that it is a capacitive reactance. A plus *j* implies X_L.

Consider a simple series *RL* circuit where $R = 12\ \Omega$ and $X_L = 8\ \Omega$. Using the *j* operator, these values would be expressed as $12\ \Omega + j8\ \Omega$. A series *RC* circuit may have 67 *V* across the resistance and 92 *V* across the capacitance. This would be written as $67\text{ V} - j92\text{ V}$.

It is not possible to use the *j* operator in front of an impedance, because impedance is the resultant of a resistance and a reactance and will always be at some phase angle less than 90°.

7-5 POWER IN ALTERNATING-CURRENT CIRCUITS

In dc circuit analysis, the amount of power absorbed by the resistance of a circuit was determined by $P = I^2R$. In ac circuits, the determination of power is a more complicated process. Since both current and voltage vary with time, their product is also a function of time and is called the *instantaneous power*. In general, however, current and voltage in ac networks are out of phase by some angle θ.

In a purely capacitive circuit, such as in Fig. 7-7a, the power delivered to the capacitor on one half-cycle is *returned* to the generator on the other half-cycle. The power waveforms of Fig. 7-7b indicate this. During the first half-cycle *E* and *I* are both positive; hence the resultant power curve is posi-

Power in Alternating-Current Circuits

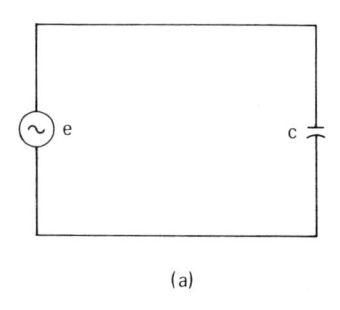

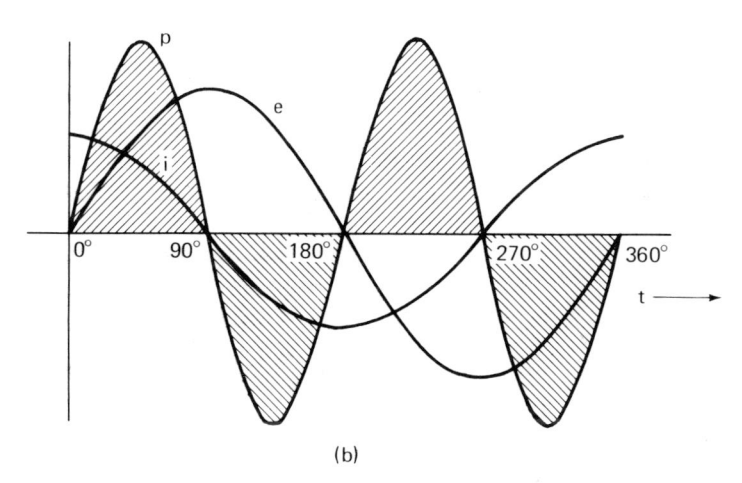

FIG. 7-7 (a) capacitive circuit; (b) power curve for purely capacitive circuit.

tive, implying power received from the generator. During the next half-cycle, E is positive but I is negative, and their product therefore is the negative power curve. Succeeding half-cycles repeat this pattern.

The inductive circuit presents the opposite situation of the capacitive circuit. The energy built up around the inductor in the magnetic field on one half-cycle is returned to the generator on the second half-cycle. The circuit and accompanying waveforms are seen in Figs. 7-8a and b.

The case of the resistive circuit appears in Figs. 7-9a and b. All the power delivered to the resistor is dissipated as heat. The power curves are all in the positive direction.

When a circuit contains resistance and capacitance, where $X_c = R$, the phase angle will be 45° with the current leading. By plotting the E, I, and P curves versus time, the waves shown in Fig. 7-10 result. Examination shows that the power curve is mostly positive. The total circuit power (positive areas minus negative areas) for the cycle is positive. The energy is dissipated in the circuit resistance.

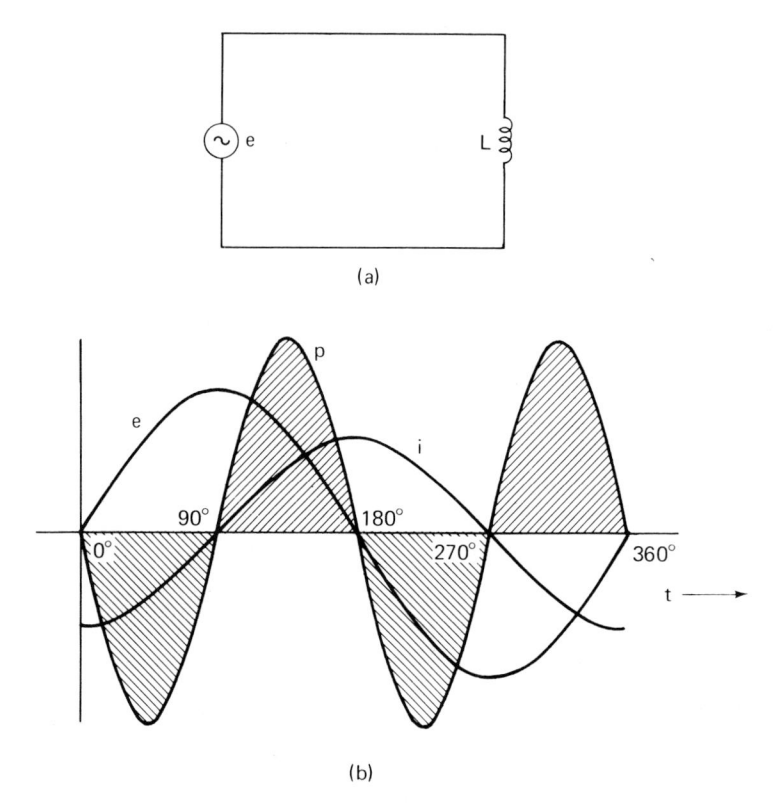

FIG. 7-8 (a) inductive circuit; (b) power curve for purely inductive circuit.

We may conclude that in any ac circuit containing reactive elements, the only power actually dissipated is the power absorbed by the resistance of the circuit. A reactive circuit, however, appears to consume large quantities of power. Thus, it is important to note that even though the generator receives back certain amounts of energy from the load, it must supply large amounts to the load. This power, which the generator must deliver (regardless of the return), is called the *apparent power* and is equal to the product of the effective value of the voltage and the current. Hence,

$$P(\text{apparent power}) = EI$$

When reactance is present, the apparent power must be multiplied by the ratio R/Z, which is the cosine of the phase angle θ. Therefore, true power is calculated by the formula

$$P_{\text{true}} = EI \cos \theta$$

An examination of the formula for true power reveals that if the phase angle θ equals 90°, its cosine is zero and the actual power absorbed by the

Power in Alternating-Current Circuits

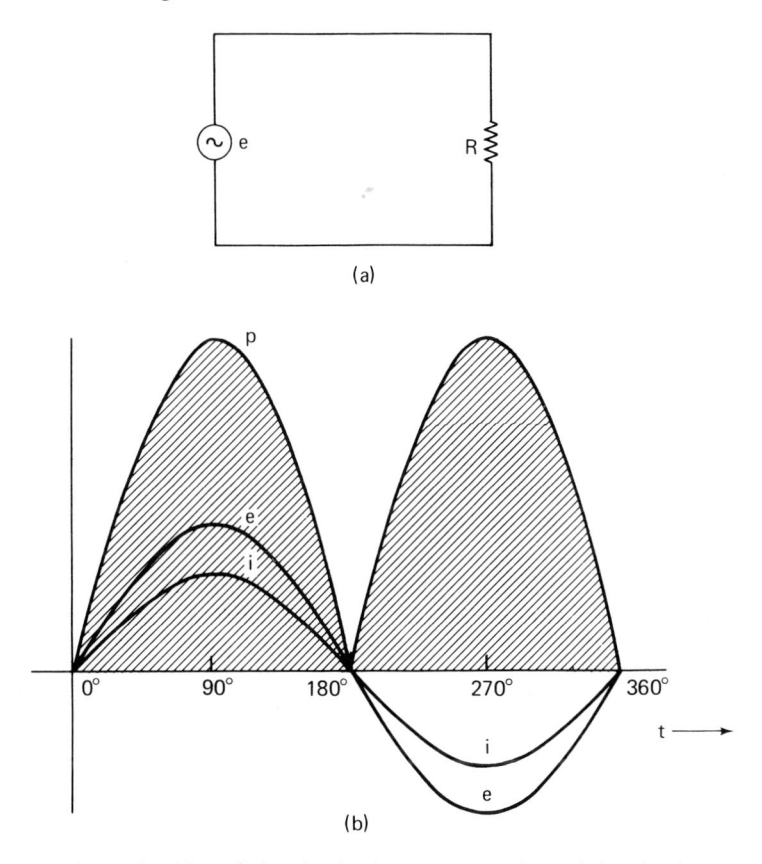

FIG. 7-9 (a) Resistive circuit; (b) power curve for resistive circuit.

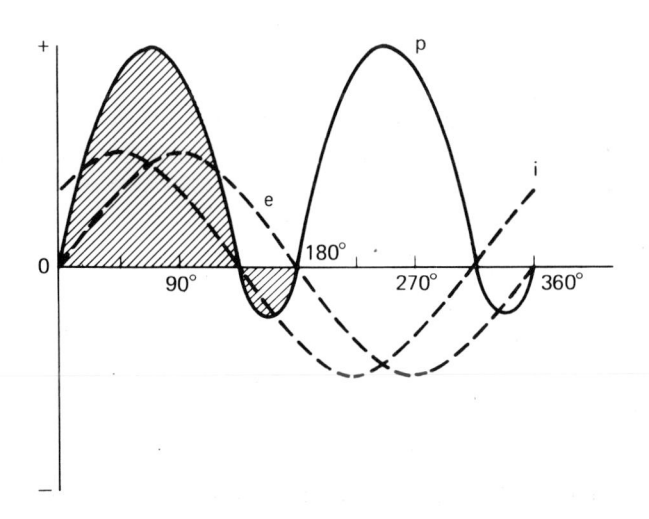

FIG. 7-10 Power curve of a typical RC circuit.

circuit is zero. Thus, a phase angle of 90° means that the circuit is *purely reactive* and returns as much power as it receives. If the phase angle is 0°, its cosine is 1, the circuit is purely resistive, and all the power produced by the source is absorbed by the load. The cosine θ, then, varies from 0 to 1 as the phase angle varies from 90 to 0°.

Example 7-2

Calculate the apparent power and true power of the circuit shown in Fig. 7-11a.

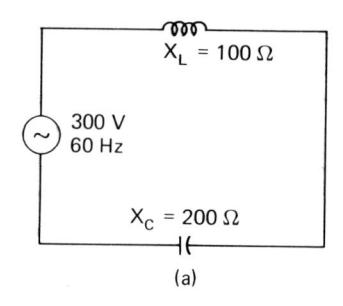

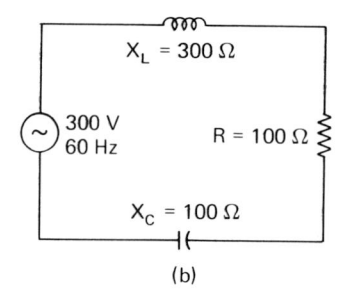

FIG. 7-11 Power relationships in ac circuits.

Solution

The effective reactance is 100 Ω capacitive. The circuit current is

$$I = \frac{E}{X_c} = \frac{300}{100} = 3 \text{ A}$$

The apparent power is

$$P_{app} = EI = 300 \times 3 = 900 \text{ W}$$

The phase angle is 90°; therefore, true power is

$$P_{true} = EI \cos \theta$$
$$= 300 \times 3 \times \cos \theta = 300 \times 3 \times 0 = 0 \text{ W}$$

Example 7-3

Calculate the apparent and true power of the circuit shown in Fig. 7-11b.

Solution

The impedance of the circuit is

$$Z = \sqrt{R^2 + (X_L - X_c)^2} = \sqrt{100^2 + 200^2} = 224 \, \Omega$$

The current flowing in the circuit is

$$I = \frac{E}{Z} = \frac{300}{224} = 1.34 \text{ A}$$

The phase angle of this effectively inductive circuit is

$$\cos \theta = \frac{R}{Z} = \frac{100}{224} = 0.446$$

$$\theta = 63.4°$$

The apparent power is

$$P_{app} = EI = 300 \times 1.34 = 401 \text{ W}$$

The true power is

$$P_{true} = EI \cos = 300 \times 1.34 \times 0.446$$
$$= 179 \text{ W}$$

Thus, 179 W of power is consumed in this circuit, but the generator must supply 401 W, 222 W of which are returned to the generator by the effective reactive element.

7-6 POWER FACTOR

In reactive ac circuits the relative amounts of apparent power and true power are an important consideration from the point of view of efficiency and circuit design. True power differs from apparent power by the cosine of θ. Thus, the cosine of θ determines the percentage of apparent power consumed as true power. The cosine of θ, then, is called the *power factor* of the circuit, and is expressed as

$$\text{power factor} = \cos \theta$$

From the previous discussion the power factor may also be expressed as

$$\text{power factor} = \frac{\text{true power}}{\text{apparent power}}$$

A power factor close to 1 is generally desired for all reactive ac circuits using appreciable power.

7-7 PARALLEL ALTERNATING-CURRENT CIRCUITS

The solution of parallel ac circuit problems is different than series ac circuits. You will recall that the *current* was *in phase* through each component of the series circuit, whereas the *voltages* across each were *out of phase*. The reverse is true for the parallel ac circuit. The voltage across each circuit element is in phase; the currents through the individual branches are out of phase. This presumes that the branches are composed of reactive as well as resistive components.

There are several methods that can be used to solve for the impedance of a parallel circuit. The two methods to be explained herein are, in the author's opinion, the easiest. The first method utilizes essentially the same approach used to find the equivalent resistance of several parallel resistors. We know that

$$\frac{1}{R_T} = \frac{1}{R_1} + \frac{1}{R_2} + \frac{1}{R_3} + \cdots$$

Because conductance, whose symbol is G, is the reciprocal of resistance, we may rewrite the equation as

$$G = \frac{1}{R_1} + \frac{1}{R_2} + \frac{1}{R_3} + \cdots$$

The answer is expressed in *siemens* (S), which is the new unit of conductance. The conductance formula may be used in parallel ac circuits containing resistance only. Only the total circuit conductance is determined, its reciprocal ($R = 1/G$) will give the circuit impedance. If any of the branches contain reactance, the formula cannot be used without some modification. This involves the use of two new terms. One is *susceptance*, which is the reciprocal of reactance, or $1/X$. It is measured in siemens and is represented by the symbol B. The second term is *admittance*, which is the reciprocal of impedance, or $1/Z$. It is likewise measured in siemens and is represented by the symbol Y.

These three formulas can be expressed as follows:

$$G = \frac{1}{R}, \quad B = \frac{1}{X}, \quad Y = \frac{1}{Z}$$

An illustration will serve to show how the total reactance of two parallel capacitors can be determined. Assume that one capacitor has 20-Ω reactance and the other 50 Ω. First, we must find the total susceptance as follows:

$$B_t = \frac{1}{X_c} + \frac{1}{X_c} \quad \text{or} \quad \frac{1}{20} + \frac{1}{50} \quad \text{or} \quad 0.05 + 0.02$$
$$= 0.07 \text{ S}$$

The reciprocal of susceptance is reactance; therefore,

$$X_t = \frac{1}{0.07} \text{ S} \quad \text{or} \quad 14.3 \ \Omega$$

Parallel Alternating-Current Circuits

The same procedure would be used if inductances were parallel rather than capacitances.

Let us next consider an inductive reactance of 8 Ω in parallel with a capacitive reactance of 14 Ω, as shown in Fig. 7-12. Because these are opposing reactances, it is necessary that one be subtracted from the other, as follows:

$$B_t = \frac{1}{X_L} - \frac{1}{X_c} \quad \text{or} \quad \frac{1}{8} - \frac{1}{14} \quad \text{or} \quad 0.125 - 0.0715$$
$$= 0.0535 \text{ S}$$

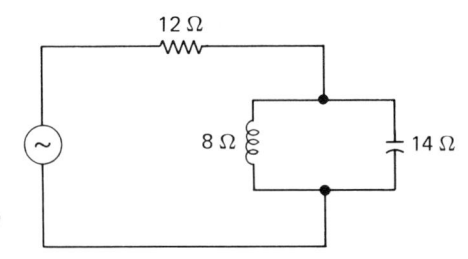

FIG. 7-12 A series resistor connected to two pure reactances in parallel.

The reactance of the parallel combination then becomes the reciprocal of B_t, or

$$X = \frac{1}{B_t} \quad \text{or} \quad \frac{1}{0.0535} \quad \text{or} \quad 18.7 \text{ Ω}$$

The inductive branch, having the lower reactance, will naturally have the greater amount of current flowing through it. Hence, the resulting reactance will be inductive, and the parallel combination will have a lagging power factor.

If we wish to find the impedance of the total circuit shown in Fig. 7-12, we proceed as though it was a 12-Ω resistance in series with an inductive reactance of 18.7 Ω, or

$$Z = \sqrt{12^2 + 18.7^2} \quad \text{or} \quad 22.2 \text{ Ω}$$

To solve for the impedance of a resistance and capacitance in parallel, we must use an adaptation of the conventional impedance formula for the series circuit. This formula indicates admittance and is

$$Y = \sqrt{G^2 + B^2}$$

The schematic diagram shown in Fig. 7-13 will serve to illustrate how

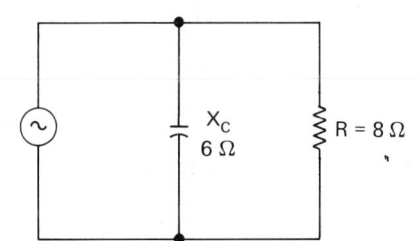

FIG. 7-13 Parallel *RC* circuit.

the impedance of this circuit can be calculated. First, it is necessary to determine the susceptance of the capacitive branch. This is

$$B = \frac{1}{X_c} \quad \text{or} \quad \frac{1}{6} \quad \text{or} \quad 0.167 \text{ S}$$

The conductance of the resistive branch is found by

$$G = \frac{1}{R} \quad \text{or} \quad \frac{1}{8} \quad \text{or} \quad 0.125 \text{ S}$$

By substituting the susceptance and conductance into the admittance formula, we have

$$Y = \sqrt{0.125^2 + 0.167^2} = \sqrt{0.0435} = 0.209 \text{ S}$$

The impedance of a circuit is the reciprocal of its admittance. Hence,

$$Z = \frac{1}{Y} \quad \text{or} \quad \frac{1}{0.209} \quad \text{or} \quad 4.78 \; \Omega$$

The vector diagram of the example appears in Fig. 7-14.

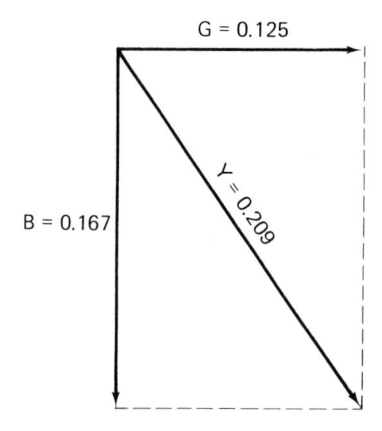

FIG. 7-14 Vector diagram of circuit shown in Fig. 7-13.

The phase angle or power factor of the circuit can be determined by the ratio of G over Y in Fig. 7-14. This gives a decimal fraction of 0.6, which is the cosine of the angle θ or approximately 53°, and represents a leading power factor.

The second method used to find the impedance of a parallel circuit involves the use of an *assumed voltage*. Any normal value of emf may be used, but experience indicates that some multiple of 10 is preferable, as it enables the use of the reciprocal scales on a slide rule. To illustrate this method, let us use the same circuit as in Fig. 7-13 and assume that 10 V ac is applied.

The first step in the solution is to calculate the amount of current that will flow through each branch with the assumed 10 V.

For the resistive branch the current is

$$I_R = \frac{10}{8} = 1.25 \text{ A}$$

Complex Circuits

For the capacitive branch the current is

$$I_C = \frac{10}{6} = 1.67 \text{ A}$$

These currents are in *quadrature* relationship (i.e., 90° apart) and must be added vectorally as follows:

$$I_T = \sqrt{I_R^2 + I_C^2} = \sqrt{1.25^2 + 1.67^2} = 2.08 \text{ A}$$

Now if this current is divided into the assumed voltage, the result will be the impedance of the circuit:

$$Z = \frac{10}{2.08} = 4.78 \text{ }\Omega$$

The simplicity of this method is apparent. It must be remembered that the reactive branches should be *pure reactances* with no resistance present in them, lest the reactive current not be 90° out of phase with the applied voltage.

The same approach is used in solving an *RL* parallel circuit. However, the vector diagram would be in the first quadrant.

In solving for the impedance of a parallel circuit containing *R*, *C*, and *L*, the procedures outlined above can be followed. Remember that B_c must be subtracted from B_L.

7-8 COMPLEX CIRCUITS

In the parallel circuits described in the previous section idealized conditions were assumed. That is, the reactances were pure and contained no resistance. From a practical standpoint this is not true. All inductors have some internal resistance. At high *radio frequencies* the current tends to travel *near the surface* of the conductors and causes the resistance to be higher than that at dc conditions. So far as capacitors are concerned, they also have losses at high frequencies, which are represented by a series resistor. In view of these practical considerations a typical parallel circuit would resemble the one shown in Fig. 7-15.

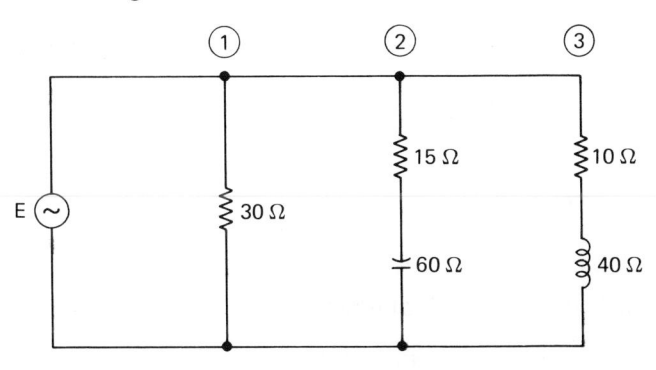

FIG. 7-15 Complex parallel *RCL* circuit.

In solving for the various parameters in this circuit the admittance formula can be used as before. Branch 1 consists of resistance only; hence, its conductance is

$$G = \frac{1}{R} = \frac{1}{30} = 0.033 \text{ S}$$

The solution of branches 2 and 3 is more difficult, as both reactance and resistance are present. To calculate the conductance component of circuits such as these, the following formula must be used:

$$G = \frac{R}{Z^2}$$

Inasmuch as we do not know the impedance of branch 2, we must first calculate it as follows:

$$Z = \sqrt{R^2 + X^2} = \sqrt{15^2 + 60^2} \cong 62 \text{ }\Omega$$

With this information it is possible to determine the conductance of branch 2.

$$G = \frac{R}{Z^2} = \frac{15}{62^2} = 0.00392 \text{ S}$$

To find the susceptance of this same branch, we use

$$B = \frac{X}{Z^2} = \frac{60}{62^2} = 0.0157 \text{ S}$$

The same approach must be taken to solve for G and B of branch 3. Therefore,

$$Z = \sqrt{R^2 + X^2} = \sqrt{10^2 + 40^2} = 41.2 \text{ }\Omega$$

The conductance of this branch is

$$G = \frac{R}{Z^2} = \frac{10}{41.2^2} = 0.00588 \text{ S}$$

The susceptance of this branch is

$$B = \frac{X}{Z^2} = \frac{40}{41.2^2} = 0.0235 \text{ S}$$

The admittance of the whole circuit may be determined from the basic formula

$$Y = \sqrt{G_T^2 + B_T^2}$$

where G_T = sum of individual conductances, including those in the reactive branches
  B_T = algebraic sum of the various susceptances

In solving for the impedance of the circuit in Fig. 7-15, it is necessary to use an expanded form of the admittance formula just given. This becomes

$$Y = \sqrt{(G_1 + G_2 + G_3)^2 + (\pm B_1 \pm B_2)^2}$$

where the inductive susceptance has plus value and the capacitance susceptance has negative value.

Substituting into the formula the several values previously calculated gives

$$Y = \sqrt{(0.033 + 0.00392 + 0.00588)^2 + (0.0235 - 0.0157)^2}$$
$$= \sqrt{(0.0428)^2 + (0.0078)^2} = \sqrt{0.00189} = 0.0435 \text{ S}$$

$$Z = \frac{1}{Y} = \frac{1}{0.0435} = 23 \ \Omega$$

To find the power factor of this circuit, we can use

$$\text{power factor} = \frac{P_{true}}{VA}$$

where VA = volt–amperes or apparent power
P_{true} = true power

The true power would be that consumed by the resistive elements in each branch. Unless an applied voltage were given, one could be assumed, as the power factor is independent of voltage. The cosine of the power factor will give the phase angle of the circuit.

Practice Problems

1. What is the impedance of a series circuit containing 30-Ω resistance and 40-Ω capacitive reactance?

2. A certain inductor has a resistance of 5 Ω and a reactance of 10 Ω. What is its impedance?

3. A series circuit contains a pure inductance having a reactance of 53.25 Ω, at a certain frequency, and a resistance of 40 Ω. If an alternating current of 1.5 A is flowing in the circuit, what voltage will appear across each circuit element? What is the value of the supply voltage?

4. What is the impedance of the circuit in problem 3?

5. Calculate the phase angle of the circuit in problem 3.

6. An ac series circuit is made up of two capacitors and two inductors whose reactances are $X_{c1} = 100 \ \Omega$, $X_{c2} = 50 \ \Omega$, $X_{L1} = 40 \ \Omega$, and $X_{L2} = 30 \ \Omega$. Calculate the net circuit reactance.

7. A series ac circuit has 40-Ω resistance, 90-Ω inductive reactance, and 60-Ω capacitive reactance. What is the circuit impedance?

8. What is the power factor of the circuit in problem 7? Is the current leading or lagging the applied voltage and by how many degrees?

9. What is the total conductance of a parallel circuit containing the following resistances: $R_1 = 22 \ \Omega$, $R_2 = 39 \ \Omega$, and $R_3 = 68 \ \Omega$? What is the equivalent resistance?

10. What is the susceptance of a parallel circuit containing the following reactances: $X_L = 10\ \Omega$ and $X_c = 27\ \Omega$? What is its reactance?
11. Find the impedance of a parallel combination of $R = 13\ \Omega$ and $X_c = 21\ \Omega$ (use admittance method).
12. Calculate the impedance of a parallel circuit having the following values: $R = 27\ \Omega$, $X_L = 30\ \Omega$, and $X_c = 60\ \Omega$ (use admittance method).

Commercial License Questions

Sections in which answers to questions are given appears in parentheses. A bracketed number following a question implies that it applies only to that element.

1. State Ohm's Law for ac circuits. (7–1)
2. What unit is used in expressing the ac impedance of a circuit? (7–1)
3. What is the impedance of a solenoid if its resistance is $5\ \Omega$ and 0.3 A flows through the winding when 110 V at 60 Hz is applied to the solenoid? (7–1)
4. Given a series circuit consisting of a resistance of $4\ \Omega$, an inductive reactance of $4\ \Omega$, and a capacitive reactance of $1\ \Omega$; the applied circuit alternating emf is 50 V. What is the voltage drop across the inductance? (7–3)
5. If a lamp rated at 100 W and 115 V is connected in series with an inductive reactance of $355\ \Omega$ and a capacitive reactance of $130\ \Omega$ across a voltage of 220 V, what is the current value through the lamp? [4] (7–3)
6. A potential of 110 V is applied to a series circuit containing an inductive reactance of $25\ \Omega$, a capacitive reactance of $10\ \Omega$, and a resistance of $15\ \Omega$. What is the phase relationship between the applied voltage and the current flowing in this circuit? [4] (7–3)
7. If an alternating current of 5 A flows in a series circuit composed of 12-Ω resistance, 15-Ω inductive reactance, and 40-Ω capacitive reactance, what is the voltage across the circuit? [4] (7–3)
8. A series circuit contains resistance, inductive reactance, and capacitive reactance. The resistance is $7\ \Omega$, the inductive reactance is $8\ \Omega$, and the capacitive reactance is unknown. What value must this condenser have in order that the total circuit impedance be $13\ \Omega$? [4] (7–3)
9. What is the meaning of *power factor*? (7–6)
10. What factors must be known to determine the power factor of an ac circuit? (7–6)
11. What effect does inductive reactance in an ac circuit have on the power factor of the circuit? (7–6)
12. What does the term *power factor* mean in reference to electric power circuits? [4] (7–6)
13. If an alternating voltage of 115 V is connected across a parallel circuit made up of a resistance of $30\ \Omega$, an inductive reactance of $17\ \Omega$, and a capacitive reac-

tance of 19 Ω, what is the total circuit current drain from the source? [4]

(7-7)

14. A parallel circuit is made up of five branches, three of the branches being pure resistances of 7, 11, and 14 Ω, respectively. The fourth branch has an inductive reactance value of 500 Ω. The fifth branch has a capacitive reactance of 900 Ω. What is the total impedance of this network? If a voltage is impressed across this parallel network, which branch will dissipate the greatest amount of heat? [4]

(7-7, 7-8)

8

Resonant Circuits

8-1 RESONANCE

Resonant circuits are extremely important in transmitters and receivers. Because of the principle of resonance, it is possible to adjust a transmitter so that it will transmit its RF energy at a *specific* frequency. If the resonant circuits were inadvertently detuned, the power output would be reduced and the transmitter might possibly be damaged. Receiving circuits must be tuned or resonated to the proper frequency so that the desired information can be received.

The principles of inductive and capacitive reactance have been studied in previous chapters. However, we have not considered what happens in a circuit when $X_L = X_c$. This causes a phenomenon known as *resonance*.

We have learned that X_L increases with frequency, whereas X_c decreases. The frequency at which both reactances are equal, called the *resonant frequency*, can be determined by solving for f in the equation

$$2\pi f L = \frac{1}{2\pi f C}$$

solving for f,
$$4\pi^2 f^2 LC = 1$$

$$f^2 = \frac{1}{4\pi^2 LC}$$

$$f = \frac{1}{2\pi\sqrt{LC}}$$

where f = frequency in hertz
L = inductance in henrys
C = capacitance in farads

Resonance

From this formula we can determine the frequency at which resonance occurs.

Example 8-1

What is the resonant frequency of a 2-mH inductance and an 80-pF capacitance?

Solution

$$f = \frac{1}{2\pi\sqrt{LC}} = \frac{1}{6.28\sqrt{2 \times 10^{-3} \times 8 \times 10^{-11}}}$$

$$= \frac{1}{6.28\sqrt{16 \times 10^{-14}}} = \frac{1}{6.28 \times 4 \times 10^{-7}}$$

$$= 398 \text{ kHz}$$

An examination of the resonance formula reveals that for any value of L and C there is only *one* definite resonant frequency. This frequency depends on the product of L and C alone, since all other factors are constant. Thus, various combinations of L and C may be used in a circuit to achieve resonance at a given frequency as long as the *product* of the two values is the same. A large value of L and a small value of C may resonate at the same frequency as a large value of C and a small value of L. For instance, in Example 8-1 an inductance of 0.5 mH and a capacitance of 320 pF resonates at the same frequency (398 kHz) as the 2-mH L and the 80-pF C used originally.

The larger the product of L and C the lower the resonant frequency, and vice versa. Thus, high-frequency tuned circuits generally use very small components, and low-frequency tuned circuits use relatively large coils and capacitors. Furthermore, if the value of either L or C is known, the value of the other component required for resonance may be computed from the formula.

Since

$$f^2 = \frac{1}{4\pi^2 LC}$$

$$L = \frac{1}{4\pi^2 f^2 C}$$

$$C = \frac{1}{4\pi^2 f^2 L}$$

Example 8-2

What value of capacitance must be shunted across a coil having an inductance of 56 mH in order that the circuit resonate at 5000 Hz?

Solution

$$C = \frac{1}{4\pi^2 f^2 L} = \frac{1}{39.4 \times 25 \times 10^6 \times 56 \times 10^{-3}}$$

$$= \frac{10^{-3}}{5.516 \times 10^4} = 18.2 \text{ nF}$$

8-2 SERIES RESONANT CIRCUITS

The condition of resonance can be achieved with *L* and *C* either in *series* or *parallel*. Each connection has distinctly different characteristics. The series connection will be discussed in this section.

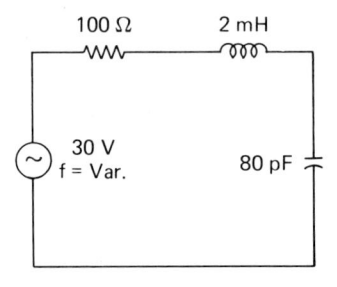

FIG. 8-1 Series resonant circuit.

Figure 8-1 shows a typical series resonant circuit whose resonant frequency is 398 kHz. The generator is capable of producing a variable frequency from approximately 100 to 800 kHz at a constant 30-V output. To understand this, or any series resonant circuit, let us begin by reviewing the characteristics of X_L and X_c. Figure 8-2 shows how X_L increases linearly, whereas X_c varies inversely with frequency. The particular frequency at which $X_L = X_c$ is called the resonant frequency. Below this frequency there is more X_c than X_L in the circuit. Therefore, the circuit looks *capacitive* to the generator. Above resonance, X_L is larger than X_c and the circuit looks *inductive* to

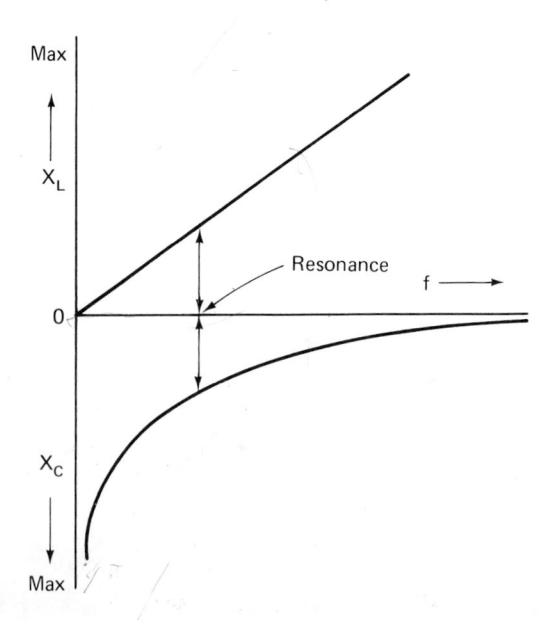

FIG. 8-2 Graph showing the variation of X_L and X_c with frequency.

Series Resonant Circuits

the generator. At resonance the generator sees only the resistance present, because X_L and X_c cancel.

To understand how the circuit's impedance and current vary with frequency, refer to Fig. 8-3. At frequencies considerably below resonance the circuit impedance is high and is determined principally by X_c as shown in the formula $Z = \sqrt{R^2 + (X_c - X_L)^2}$. As the generator frequency is increased, X_L begins to approach the value of X_c, and the total circuit reactance is diminished. At resonance the circuit impedance is $Z = \sqrt{R^2}$, or simply $Z = R$. Thus, the impedance is limited by the value of resistance present. The current flowing in the circuit produces a curve opposite to the impedance and is represented by the dotted curve I in Fig. 8-3.

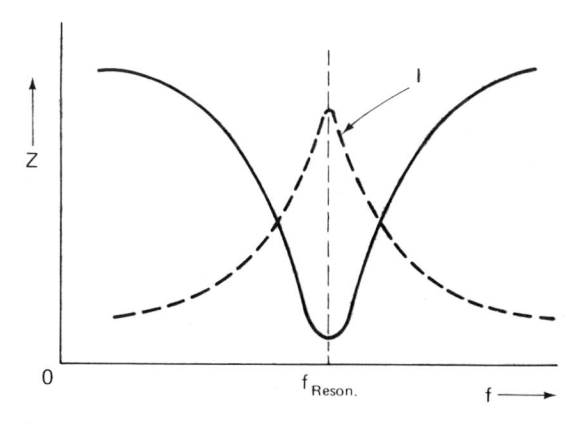

FIG. 8-3 Graph showing variation of Z and I with frequency for the series LC circuit.

Thus, series LCR circuit acts as a simple resistance at resonance. The total current flowing is limited only by the resistance. The voltages across each of the reactive elements, although equal and opposite, may be *very high* as determined by $E_L = IX_L$ and $E_c = IX_c$.

The amount of resistance present determines the *shape* of the curve. For example, in Fig. 8-4 the curve showing low resistance is very steep, whereas the one for high resistance is very much flattened out. The amout of resistance present does not change the resonant frequency. If the resistance of the circuit is too large, the circuit loses its ability as a frequency selector. This means that little *discrimination* in the amount of current flow (hence voltage) is made between frequencies at resonance and those not at resonance.

An example of a series resonant circuit appears in Fig. 8-5. This is a simplified schematic of a first RF stage of a radio receiver. Many voltages of different frequencies are induced into the antenna by the numerous radio stations. These voltages cause currents to flow in the primary of the transformer, which in turn induce voltages into the secondary. These voltages are considered to be in *series* with the inductance of the secondary. Hence, the secondary winding and the variable capacitor connected across it operate as a

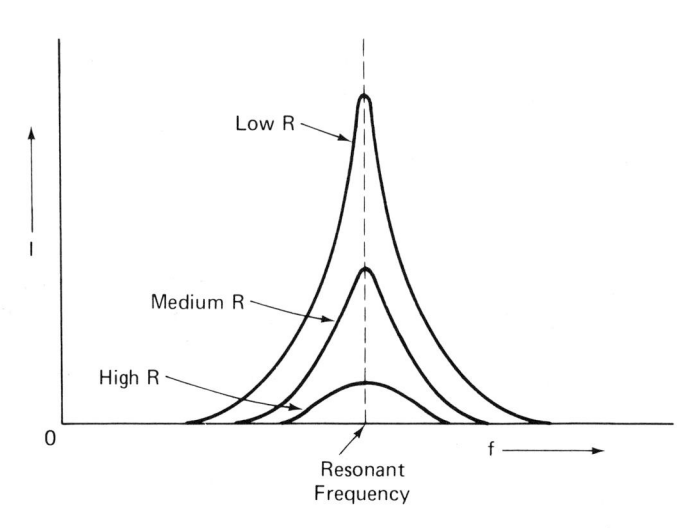

FIG. 8-4 Effect of resistance upon response curve of a series LC circuit.

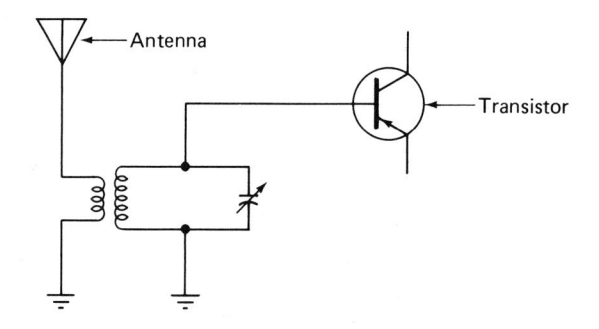

FIG. 8-5 Application of series resonant circuit.

series tuned circuit and not as a parallel resonant circuit, as it would appear to be. By varying the tuning capacitor, the secondary can be resonated to a particular frequency within the range determined by L and C. When resonanting at a certain frequency, $X_L = X_c$ and a large current will flow. This produces a relatively large reactive voltage, which is connected to the transistor where it is amplified.

At frequencies other than the resonant frequency the series circuit presents a high impedance; consequently, the current will be small. Therefore, little voltage will be fed to the transistor.

8-3 PARALLEL RESONANT CIRCUITS

A *parallel resonant* circuit is shown in Fig. 8-6. Unlike the series resonant circuit in which the induced or applied voltage is in series with the LC combination, the parallel resonant circuit has the applied emf *across* both L and C.

Parallel Resonant Circuits

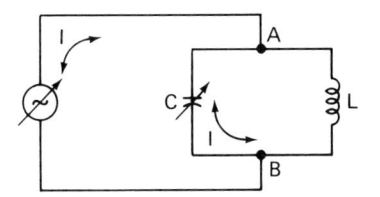

FIG. 8-6 Parallel resonant circuit.

The term *tank circuit* is frequently used in connection with resonant circuits, both series and parallel. The name is derived from the ability of an *LC* combination to *store energy*. This energy is in the form of an *electrostatic charge* when the circulating tank current is in the capacitor. When the current is in the inductor, the energy is in the *electromagnetic field*.

The vector diagram of a parallel resonant tank circuit is shown in Fig. 8-7. The applied voltage *E* appears across both *L* and *C* and is accordingly taken as the reference vector. Notice the I_c and I_L are 180° out of phase with each other and 90° out of phase with the applied emf.

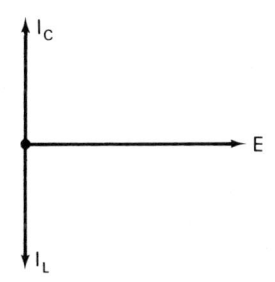

FIG. 8-7 Vector diagram of *E* and *I* relationships in an ideal parallel resonant tank circuit.

By Kirchhoff's law, the current flowing into junction *A* (Fig. 8-6) must equal the sum of the currents leaving *A*; therefore, the current *I* delivered by the generator is equal to $I_L + L_c$. The current through the inductance has the value E/X_L and lags *E* by 90°; the current through the capacitor has the value E/X_c and leads *E* by 90°. Thus, these two currents are 180° *out of phase with each other* and each current is 90° out of phase with the applied voltage. Partial or complete cancellation will take place in the line, depending on the amount of each of the branch currents. If the capacitive reactance X_C is *less* than the inductive reactance X_L, the circuit acts *capacitively*, since the current I_c will be greater than the current I_L.

If the inductive reactance X_L is *less* than the capacitive reactance X_c, the inductive current I_L will be *greater* than the capacitive current I_c, and the circuit will act inductively. However, when X_L is equal to X_c, the inductive and capacitive currents are equal, the branch currents cancel in the line, and the *line current in the ideal circuit is zero*. Thus, an ideal parallel resonant circuit presents *infinite impedance* to the line and acts as an open circuit.

The equation for parallel resonance is the same as for series resonance:

$$f = \frac{1}{2\pi\sqrt{LC}}$$

The frequency at which resonance occurs in a parallel *LC* circuit is sometimes called the *antiresonant* frequency to distinguish it from the resonant frequency of the series *LC* circuit.

Example 8-3

What is the resonant frequency of a parallel circuit having 150 μH and 160 pF?

Solution

$$f = \frac{1}{2\pi\sqrt{LC}} = \frac{1}{6.28\sqrt{1.5 \times 10^{-4} \times 1.6 \times 10^{-10}}}$$

$$= \frac{1}{6.28\sqrt{2.4 \times 10^{-14}}} = 1.026 \text{ MHz}$$

At frequencies considerably below resonance, X_L is much less than X_C, and the current in the inductive branch will be *nearly equal to the line current*. At frequencies much higher than resonance, X_C is much less than X_L, and the current in the capacitive branch is *almost equal* to the line current. At resonance the currents in the inductive and capacitive branches are equal, and the line current is nearly zero. Actually, just enough line current flows to make up for the losses in the tank due to resistance in the *L* and *C* branches. Off resonance the difference between the branch currents is the line current.

The impedance of a parallel *LC* circuit varies opposite to that of the series *LC* circuit. Graphically, this is shown in Fig. 8-8. At resonance the impedance is at a *maximum* value. This conforms with the fact that the line current is *minimum* at resonance. The *Z* and *I* of the circuit vary in opposite manner, as shown in the figure. Below resonance the circuit looks inductive

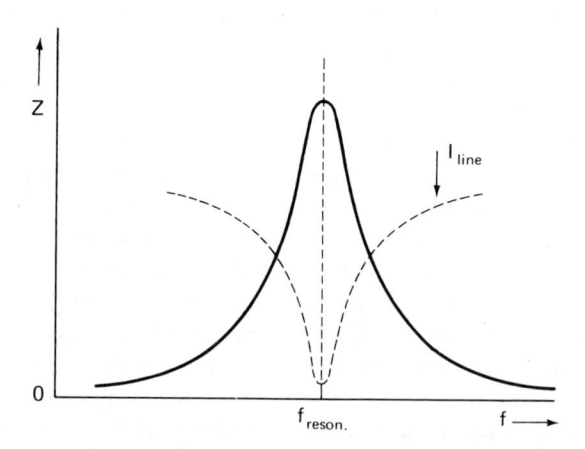

FIG. 8-8 Graph showing variation of *Z* and *I* with frequency for a parallel *LC* circuit.

Circuit Q

to the energy supplied to it; hence, most of the current is in the inductive branch. Above resonance the condition is reversed.

The amount of resistance present in the L and C branches determines the *height* and *steepness* of the impedance curve. If the resistance is increased, the height of the impedance curve is reduced and the curve flattens out. The current curve would be affected also in that it would tend to flatten out. However, the resonant frequency would remain unchanged.

The impedance of a parallel resonant circuit can be calculated by the formula

$$Z = \frac{X_L X_C}{X_C - X_L}$$

The following comparisons can be made between series and parallel resonant circuits:

Series	Parallel
Z is low	Z is high
I_{line} is maximum	I_{line} is minimum
E_C and E_L > line voltage	E_C and E_L = line voltage
Circuit is resistive	Circuit is resistive
	I_C and I_L > line current

8-4 CIRCUIT Q

The *quality* of a series resonant circuit, called Q, can be determined by comparing the voltage across one of the reactances to the line voltage. Thus;

$$Q = \frac{E_L}{E} \quad \text{or} \quad \frac{IX_L}{IR}$$

By canceling out the current I in both numerator and denominator, we find that Q can also be expressed as

$$Q = \frac{X_L}{R}$$

This equation indicates that Q varies *inversely* with circuit resistance. The curve labeled Low R in Fig. 8-4 would represent a circuit with a high Q. Circuits with medium and low Q's would be indicated by the curves marked Medium R and High R, respectively. Steep curves are synonymous with high Q circuits. They can provide good frequency discrimination. Thus, it will be seen that Q is a measure of the ability of a resonant circuit to *select* or *reject* a band of frequencies. The higher the Q of a series resonant circuit, the greater will be its value as a frequency selector—that is, the narrower will be the band of frequencies showing a voltage gain at resonance. Q is frequently referred to as a *figure of merit* of a tuned circuit.

In the series resonant circuit the quality or Q of the circuit is determined by the ratio of the voltage across either reactance as compared to the applied voltage. Since voltage is everywhere the same in the parallel resonant circuit, the Q is determined by the ratio of the current in the tank as compared to the line current. Then

$$Q = \frac{I_{tank}}{I_{line}}$$

The Q of a parallel resonant circuit is found to be the same as that of a series resonant circuit: the ratio of the inductive (or capacitive) reactance to the resistance of the circuit. If the Q of a circuit is known the total impedance may be determined as

$$Q = \frac{Z}{X_C}$$

Then
$$Z = QX_C \quad (\text{or } QX_L)$$

As with series resonance, the greater the resistance in the circuit, the lower the Q and, accordingly, the flatter and broader the resonance curve of either line current or circuit impedance.

At high radio frequencies, the current travels near the surface of the conductors and inductors. This results in a resistance value considerably higher than if only direct current were present. This is called *skin effect*, and substantially reduces the circuit Q. To reduce this effect, use larger wire or copper tubing; use Litz wire (a multistrand insulated wire); silver plate the conductors; and use fewer turns on the coil, but increase permeability by employing powdered-iron cores.

When a resistor is connected across a parallel tuned circuit, the Q will vary directly with the resistance. A large value of shunting resistance means a high Q, as in the formula

$$Q_p = \frac{R}{X_L}$$

The Q of a capacitor is normally very high because of the extremely high resistance of the dielectric. Any leakage implies that the dielectric resistance has reduced, and the Q will likewise be lowered. In some applications a shunt resistor is deliberately connected across the parallel LC combination to reduce the circuit Q.

8-5 BANDWIDTH

The *bandwidth* (BW) of a circuit is the total number of hertz above and below the resonant frequency that realize practically the same voltage gain as the resonant frequency itself. The width of this band is also called the *band pass* of the circuit. The effective limits of the band pass are taken to be the points on the resonance curve corresponding to 0.707 of the peak current or voltage, whichever is plotted. In Fig. 8-9 the shaded area represents the band of frequencies for which the current is greater than 0.707 of the peak

Bandwidth

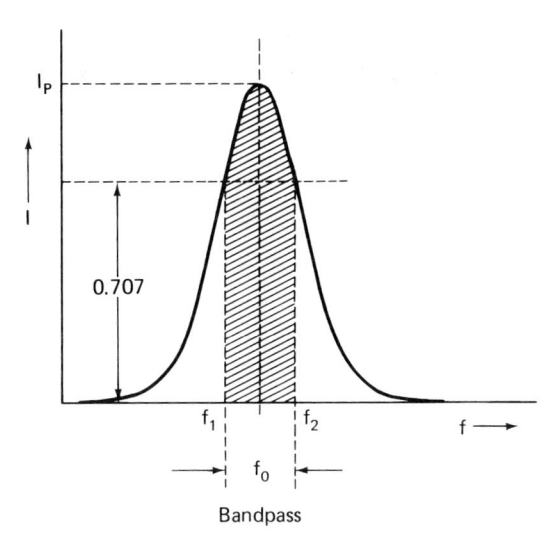

FIG. 8-9 Bandwidth of a series resonant circuit.

current. Note that one half the band lies above the resonant frequency (f_0 to f_2) and the other half lies below the resonant frequency (f_0 to f_1).

The width of the band pass may always be found from the resonance curve. However, since the Q of a circuit determines the overall width of the resonance curve, the band pass may also be found in terms of the resonant frequency and the Q of the circuit. Thus,

$$BW_{Hz} = \frac{f_0}{Q}$$

From this formula it can be seen that the higher the Q, the *smaller* the band pass.

In many communications receivers it is necessary to have narrow bandwidths (high Q circuits) so that only desired frequencies can be passed, while those outside the passband will be greatly attenuated. Example 8-4 illustrates the calculation of circuit Q and bandwidth.

Example 8-4

A parallel *LCR* circuit resonant at 100 kHz has an $X_L = X_C = 10 \text{ k}\Omega$ and a resistance of 200 Ω (combined R_L and R_C) and an applied voltage of 300 V.

Solution

The Q of the circuit is

$$Q = \frac{X_L}{R} = \frac{10,000}{200} = 50$$

The impedance to the source is
$$Z = QX_L = 50 \times 10{,}000 = 500 \text{ k}\Omega$$
The line current may be determined by
$$I = \frac{E}{Z} = \frac{300\text{V}}{500 \text{ k}\Omega} = 0.6 \text{ mA}$$
The current in the tank is
$$I_{\text{tank}} = Q(I_{\text{line}}) = 50 \times 0.6 = 30 \text{ mA}$$
The bandwidth of the circuit is
$$\text{BW} = \frac{f_0}{Q} = \frac{100\text{kHz}}{50} = 2\text{kHz}$$

On either side of resonance at the limits of the bandwidth, the impedance of the circuit is 0.707 of the maximum impedance of 500 kΩ or 353 kΩ. The power consumed by the circuit at resonance is equal to $E \times I$, since the impedance is a pure resistance.

If L is lowered and C increased, to maintain the same resonant frequency, then both X_L and X_C decrease, the Q of the circuit falls, and the bandwidth is increased. In may be said then that *a high L/C ratio is desirable if a narrow bandwidth is to be achieved.*

8-6 LOW PASS FILTERS

Low pass (LP) filters are most commonly used in power supplies to filter out the ripple frequency. Other uses of LP filters are as line noise filters, tone control circuits, crossover networks, and output filter of detector circuits. A typical, simple LP filter is shown schematically in Fig. 8-10. The inductance offers negligible opposition to low-frequency currents. As the input frequency increases, the reactance of the coil increases, making it more difficult for the higher frequencies to pass. The capacitor provides a low-impedance path back to the source to any high frequencies that may possibly get by the coil. If X_C is low compared to the load resistor, any high-frequency component of the input will thereby be *bypassed* around the load.

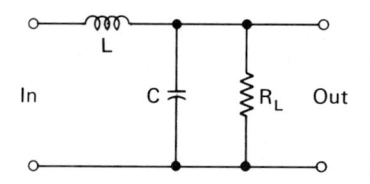

FIG. 8-10 Typical low pass filter.

The frequency response curve of a typical low pass filter appears in Fig. 8-11. Notice that all the low frequencies are passed, but that at some higher frequency the output begins to *roll off*. This is called the *cutoff frequency*. To determine the value of L and C required for a given cutoff fre-

High Pass Filters

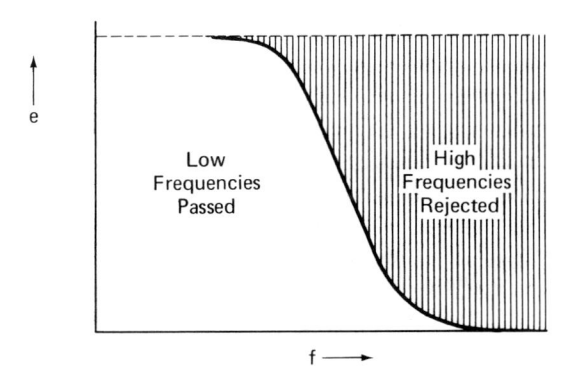

FIG. 8-11 Characteristic curve of low pass filter.

quency, the following formulas should be used:

$$L = \frac{R}{\pi f_c}, \quad C = \frac{1}{\pi f_c R}$$

where L = inductance in henrys
C = capacitance in farads
R = load or terminating resistance in ohms
f_c = cutoff frequency in hertz

If it is desired to find the cutoff frequency of a given value of L and C, the following formula may be used:

$$f_c = \frac{1}{\pi \sqrt{LC}}$$

For these formulae the *load* or *terminating resistor* should be approximately equal to the input or source resistance.

Example 8-5

Determine the cutoff frequency of a simple low pass filter where $L = 45$ mH and $C = 2$ μF.

Solution

$$f_c = \frac{1}{3.14\sqrt{45 \times 10^{-3} \times 2 \times 10^{-6}}} = \frac{1}{3.14\sqrt{9 \times 10^{-10}}}$$

$$= \frac{1}{3.14 \times 3 \times 10^{-5}} = 10.516 \text{ kHz}$$

8-7 HIGH PASS FILTERS

The difference between a simple high pass and low pass filter is that the reactive components have been *interchanged*. An example of a high pass filter is seen in Fig. 8-12. At low frequencies the capacitance provides a high

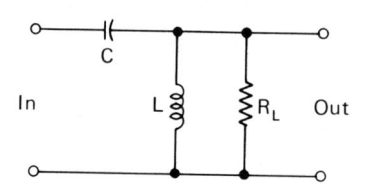

FIG. 8-12 High pass filter.

reactance, and these frequencies cannot pass. At higher frequencies X_C becomes small, and they can pass. Any low frequencies present are bypassed around the load by the inductance.

The frequency response characteristics of a high pass filter appear in Fig. 8-13. Below cutoff the low frequencies are *attenuated*. The values of L and C for a simple high pass filter can be determined by the following formulas:

$$L = \frac{R}{4\pi f_c}, \quad C = \frac{1}{4\pi f_c R}$$

High pass filters are used in tone control circuits, crossover networks, and between the antenna and receiver to keep out low-frequency interference.

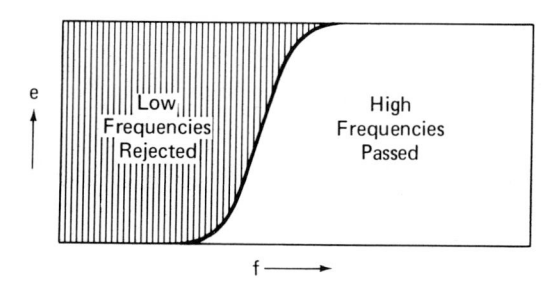

FIG. 8-13 Characteristic curve of high pass filter.

Simple low and high pass filters are sometimes referred to as *k derived* or *constant k*. This stems from the fact that the product of X_L and X_C is constant at all frequencies. To illustrate, at a certain frequency the X_L of an inductor may be 300 Ω while the X_C of the capacitor may be 100 Ω. The product of these two reactances is 30,000. If the frequency is doubled X_L becomes 600 and X_C is reduced to 50 Ω. The product of these two values is still 30,000. Hence, k is constant at 30,000, irrespective of frequency and assuming a given value of L and C.

Figures 8-14a and b illustrate more elaborate low pass and high pass filter circuits known as pi-type filters, so named from the resemblance of the schematic to the Greek letter π. The element nearest the input characterizes the filter. Thus, Fig. 8-14a shows a capacitor input pi-type low pass filter, and Fig. 8-14b an inductor input pi-type high pass filter. Figures 8-14c and d show inductor input low pass and capacitor input high pass filters, respectively.

Band Rejection and Band-Pass Filters

π-Type Low-Pass Filter
(a)

π-Type High-Pass Filter
(b)

Inductor Input
Low-Pass Filter
(c)

Capacitor Input
High-Pass Filter
(d)

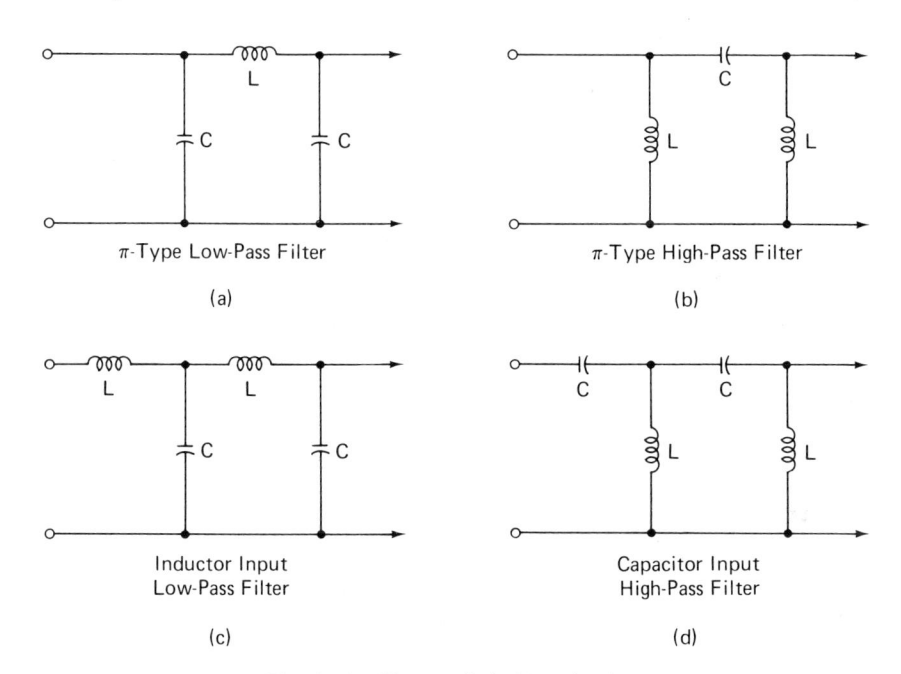

FIG. 8-14 Characteristic filter circuits.

8-8 BAND REJECTION AND BAND-PASS FILTERS

Figures 8-15a and b illustrate band-pass and band-rejection filters. In Fig. 8-15a the resonant band frequencies find the series resonant circuits S and S' low-impedance paths, and the parallel resonant circuit P a high-impedance path. Thus, the resonant band of frequencies is rejected or suppressed. All other frequencies find S and S' high-impedance paths and P a low-impedance path, and accordingly they pass from input to output with little opposition.

In Fig. 8-15b the resonant band frequencies find the parallel resonant circuits P and P' high-impedance paths, and the series resonant circuit S a low-impedance path. Thus, the resonant band is passed from input to output with little opposition. All other frequencies find P and P' low-impedance paths and S a high-impedance path, and accordingly these frequencies are rejected. Figures 8-16a and b show the response characteristics of these filters.

The band-pass filters considered above find wide application in electronics and communication circuits. By means of variable capacitors, the various resonant circuits may be adjusted to pass one band of frequencies and then readjusted to pass some other band, thus selecting first one broadcast station and then another to the exclusion of all others.

Band-Rejection Filter
(a)

Band-Pass Filter
(b)

FIG. 8-15 Band pass filter circuits.

(a)

(b)

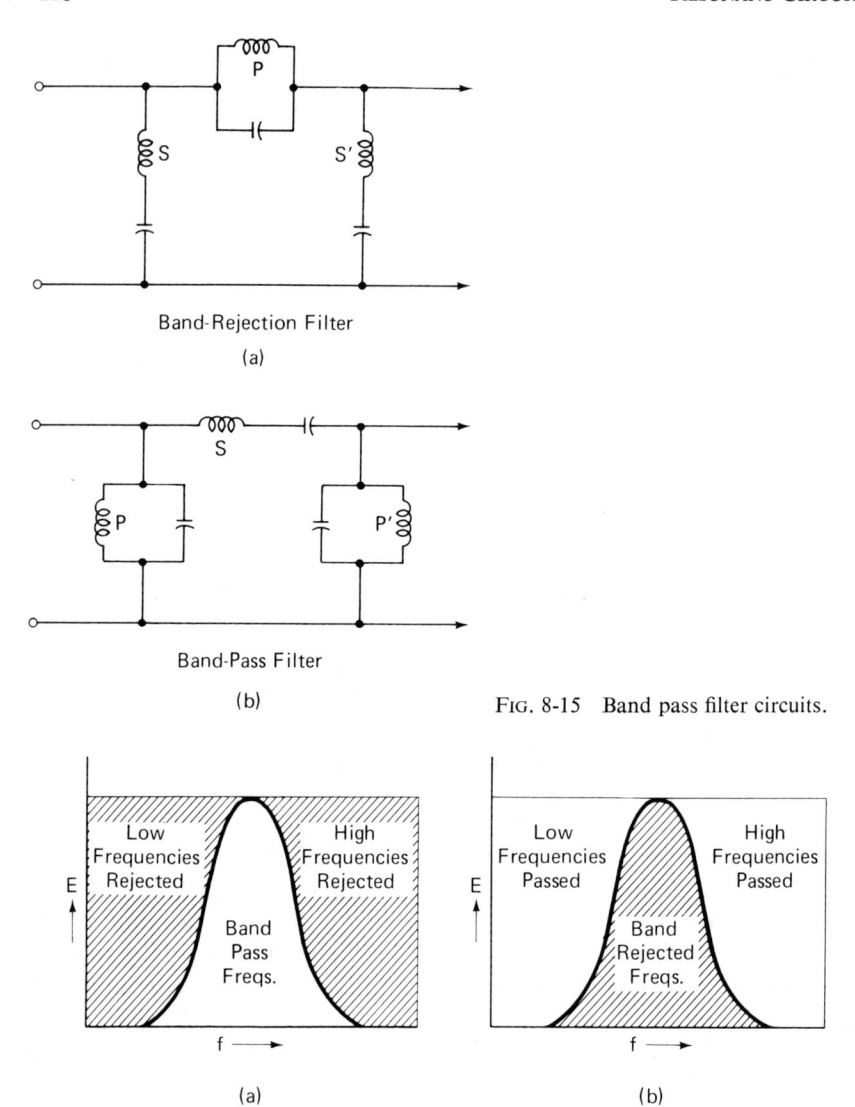

FIG. 8-16 Typical characteristics of (a) band pass filter, (b) band rejection filter.

Figure 8-17a shows a typical band-pass filter as used in the intermediate stages of a radio receiver. The heavy lines emphasize the filter; the functioning of the other elements shown need not be taken up at this point. Figure 8-17b is the same circuit drawn out as an equivalent filter. Parallel resonant circuit P offers high opposition and the series resonant circuit S little opposition; after passing through the transistor circuits, P' and S' offer impedance similar to P and S. Thus, this circuit is a band-pass filter.

Practice Problems

(a)

(b)

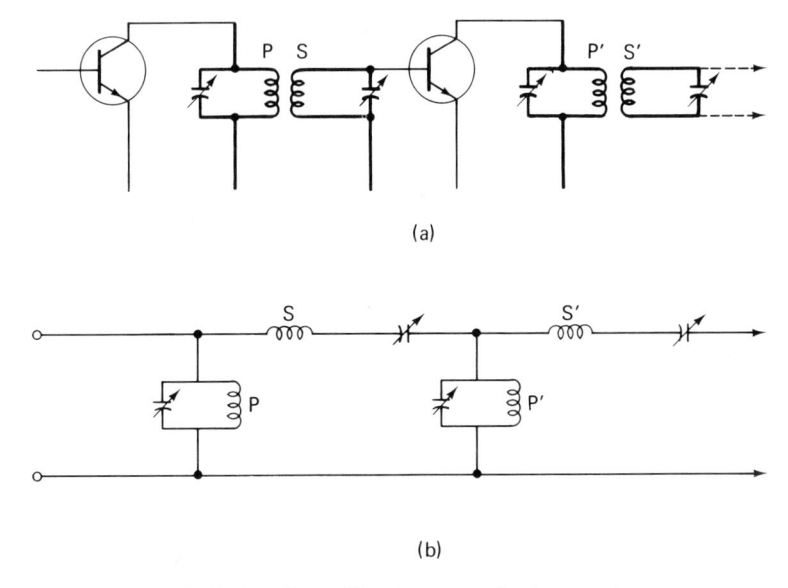

FIG. 8-17 Band pass filter in communications receiver.

There are numerous applications in which sharper cutoff is needed than is attainable with the constant-*k* filter; Figs. 8-17a and b are illustrations of such applications. Notice that tuned circuits, either series or parallel or both, are used. Filters of this type are called *m derived*. The *m* is considered as a ratio of the cutoff frequency to the frequency at which the output is reduced to zero, and may have any value between 0 and 1. The sharpness of cutoff increases as *m* approaches 0. When $m = 1$ the filter's characteristic is identical with the constant-*k* design. For most applications a value of $m = 0.6$ is a good compromise between sharp cutoff and what would be the output of a constant-*k* type filter.

Practice Problems

1. What is the resonant frequency of a series circuit having an inductance of 250 μH and a capacitance of 350 pF?
2. What value of inductance must be connected in series with a 250-pF capacitor to cause the circuit to resonate at 500 kHz?
3. What value of capacitance is required to resonate a 300-μH inductor at 600 kHz?
4. A television receiver has an absorption-type wave trap tuned to 19.75 MHz. If the wave-trap capacitor is 56 pF, what is the coil inductance?
5. Twenty-five volts, 600 kHz, is impressed across a series circuit consisting of 350 pF and 201 μH. At this frequency the effective resistance of the inductance

is 13 Ω. How much current flows in the circuit? How much power is dissipated in the circuit?

6. What voltages exist across the inductive and capacitive reactances in problem 5? What is the Q of the circuit?
7. Calculate the band pass of a parallel resonant circuit at 10.7 MHz if the circuit Q is 34.
8. Determine the Q of a tuned circuit if its resonant frequency is 45.75 MHz and the 0.707 points on the response curve are 4.5 MHz apart.
9. A parallel resonant circuit consists of 1.1 μH and 12 pF tuned to 44 MHz. Calculate the circuit impedance, assuming negligible resistance is present.
10. A low pass filter consists of an inductance of 240 mH and a capacitance of 0.06 μF. What is the cutoff frequency?

Commercial License Questions

Sections in which answers to question are given appear in parentheses.
A bracketed number following a question implies that it applies only to that element.

1. Given a series resonant circuit consisting of a resistance of 6.5 Ω, and equal inductive and capacitive reactances of 175 Ω, what is the voltage drop across the resistance, assuming the applied circuit potential is 260 V? (8–1, 8–2)
2. What is the value of total reactance in a series resonant circuit at the resonant frequency? (8–2)
3. State the formula for determining the resonant frequency of a circuit when the inductance and capacitance are known. (8–2)
4. If an ac series circuit has a resistance of 12 Ω, an inductive reactance of 7 Ω, and capacitive reactance of 7 Ω, at the resonant frequency, what will be the total impedance of twice the resonant frequency? [4] (8–2)
5. In a parallel circuit composed of an inductance of 150 μH and a capacitance of 160 pF, what is the resonant frequency? [4] (8–3)
6. What value of capacitance must be shunted across a coil having an inductance of 56 μH in order that the circuit resonate at 5000 kHz? [4] (8–3)
7. Under what conditions will the voltage drop across a parallel tuned circuit be a maximum? (8–3)
8. What is the value of reactance across the terminals of the capacitor of a parallel resonant circuit at the resonant frequency, and assuming zero resistance in both legs of the circuit? (8–3)
9. What is meant by the Q of an RF inductance coil? [4] (8–4)
10. What is a low pass filter? A high pass filter? [4] (8–6, 8–7)
11. Draw a diagram of a simple low pass filter. [4] (8–6)

9

Rectifiers

9-1 SEMICONDUCTOR MATERIALS

Diodes are widely used in electronics circuits in many different applications. Basically, they are *two-element* devices that permit current to flow in *only one direction* through them. Hence, if alternating current is applied, only *pulses* of one polarity will be permitted to pass through.

Two basic types of diodes are used: the *solid state* and the *vacuum tube*. The latter type is used mostly for replacement purposes in older equipment. Before studying solid-state diodes it is necessary to understand basic *semiconductors*, inasmuch as this is the material from which they (and transistors also) are made. Semiconductors are materials that lie midway between insulators and conductors in their ability to pass an electric current.

The material used for semiconductors is either *silicon* (Si) or *germanium* (Ge). In their pure (intrinsic) state these *elements* are insulators. To make them become semiconductors, it is necessary for the manufacturer to add extremely small, controlled amounts of other materials called *impurities*. The process is called *doping*.

The silicon atom has a charged nucleus of $+14$, the germanium atom has $+32$. In each case the positive charges in the nucleus are equalized by a similar number of electrons orbiting the nucleus. Hence, the atoms are neutral under normal conditions. The orbiting electrons possess mass and energy. Each electron in its relationship to its nucleus exhibits an energy value and functions at a definite and distinct energy level. This energy level is dictated by the electron's momentum and its proximity to the nucleus. Electrons that are closer to the nucleus are more tightly bound to the nucleus and require greater energy to break loose. Outer orbital electrons, frequently called *valence* electrons, are said to be stronger than those in the inner orbits because of their ability to break away from the parent atom. The outer orbit in which the valence electrons exist is called the *valence band* or *valence shell*. The electrons in these bands or shells are the ones that are involved in semiconductor operation.

Models of silicon and germanium atoms appear in Fig. 9-1 in two-dimensional form. In Fig. 9-1a a silicon atom with its 14 orbital electrons and +14 charge in the nucleus is shown. To the right is a simplified version of this atom with only the four outermost electrons shown and their corresponding plus charges in the nucleus. These electrons are called valence electrons. The

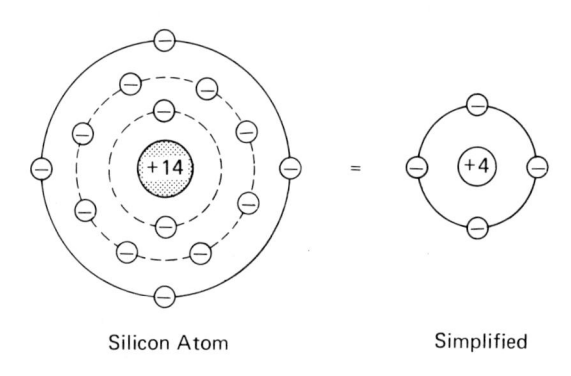

Silicon Atom   Simplified

(a)

Germanium Atom   Simplified

(b)

FIG. 9-1 Two-dimensional models of (a) silicon atom, (b) germanium atom.

Semiconductor Materials 123

germanium model is seen in Fig. 9-1b with its simplified version shown to the right.

When an electron is free from the valence band and moves into outer atomic space, it becomes a conduction electron and exists in the *conduction band*. Electrons possess the ability to move back and forth between valence and conduction bands.

Electrons rotate constantly in relatively fixed orbits about their nucleus. In a crystal such as silicon or germanium, the rotation of one valence electron in a given atom is coordinated with the rotation of one valence electron of an adjacent atom. This results in the formation of an *electron-pair bond or valence bond*. These are represented symbolically in Fig. 9-2 by the four pairs

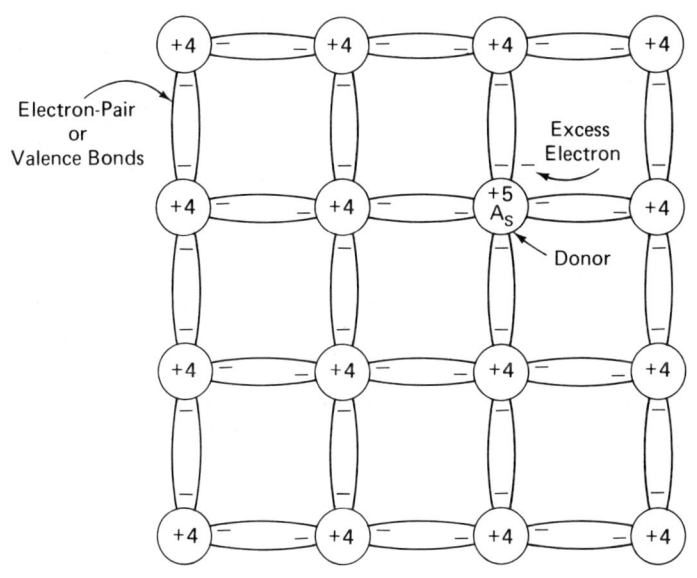

FIG. 9-2 The effect of an arsenic atom replacing a germanium atom.

of lines extending from each atom. The arrangement of cores and valence bands is referred to as a *crystal lattice*. This arrangement holds for both silicon and germanium crystals. With all outermost electrons in covalent bonds (ignore the circle with +5 for the moment), no free electrons exist to act as charge carriers throughout the crystal lattice. However, such perfect crystals cannot be manufactured, and all contain very small amounts of impurities.

To use the semiconductor material for diodes and transistors, it is necessary to add a very small amount of an element containing five electrons in the outermost orbit. Arsenic and antimony are typical of these and are called *pentavalent* elements. The arsenic or other pentavalent atom used takes the place of one of the silicon (or germanium) atoms. This *doping* process causes the pentavalent atoms to disperse uniformly throughout the lattice. A magnified part of a crystal containing one such atom (arsenic in this case) is shown in Fig. 9-2 by the circle with +5. The fifth electron is not part of any covalent bond and is simply held in place by the positive attractive power of the arsenic nucleus. This extra electron is only loosely bound to its nucleus. These are called *donor atoms*, inasmuch as they are donating a free electron to the crystal. This results in a crystal having excess negative charges, and is called *n-type* silicon or germanium. There electrons become the charge carriers in this type of semiconductor.

It is possible to add elements containing three electrons in their outermost orbits to the crystal. Examples of these would be indium, boron, and gallium and are sometimes called *trivalent* elements. One of these atoms can replace one atom in the crystal structure, as was the case with pentavalent elements. However, instead of having one excess electron, we find a deficiency of one electron in the electron-pair bond. The net result of this is to leave a *hole*, as indicated in Fig. 9-3. If an emf is applied, the electrons from other nearby bonds will be attracted to these holes, filling the gaps but creating a similar number of holes or *positive charges* in their former bonds. This action creates an equivalent movement of holes throughout the crystalline structure. We say that conduction is taking place by holes. These trivalent elements are known as *acceptors*, and the crystals are known as *p-type*.

Electrons are the *majority carriers* in *n*-type semiconductors. Because of manufacturing limitations, some holes are present in this type of semiconductor; they are known as *minority carriers*. These latter carriers are responsible for leakage currents. For *p*-type material, holes are the majority carriers. Again, because of imperfections electrons that are present are the minority

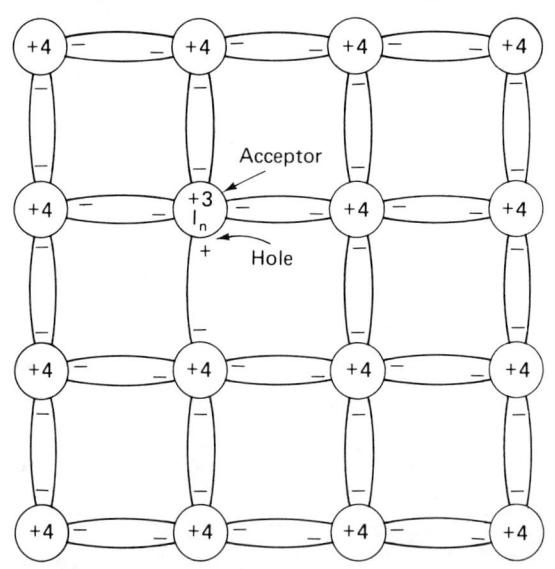

FIG. 9-3 The creation of a hole when a trivalent imurity, such as indium, is added to a germanium crystal.

carriers and are responsible for leakage currents in this type of semiconductor.

Holes in motion constitute an electrical current to the same extent that electrons in motion constitute an electrical current. There are differences, however, that must be kept in mind. The hole can exist only in a semiconductor material because it depends on a specific arrangement of electrons (electron-pair bonds). Holes do not exist in conductors such as copper and aluminum.

The hole is deflected by electric and magnetic fields in the same manner that electrons are deflected. Because the hole possesses a charge equal and opposite to that of the electron, the direction of deflection of the hole is opposite to that of the electron. Under the influence of an electric field, the electron moves toward the positive pole; the hole moves toward the negative pole.

9-2 *pn* JUNCTIONS

The *p*- and *n*-type materials shown separately in Fig. 9-4a are electrically neutral. By bringing the two types of materials together a semiconductor *rectifying junction* will be formed. This is not a mechanical joining, but one

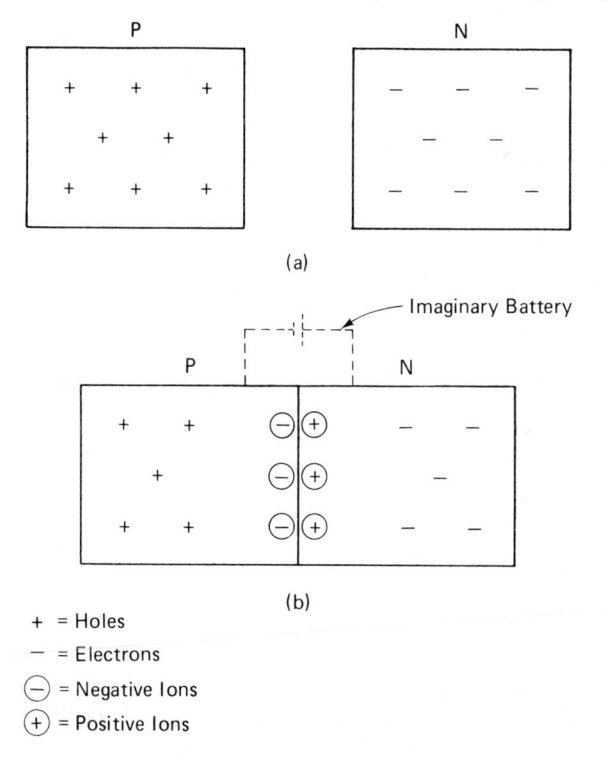

FIG. 9-4 (a) *p*- and *n*-materials. (b) Junction formed by combination of *p*- and *n*-materials.

in which the two materials are essentially melted together under exacting conditions to form a simple crystalline structure. At the moment the junction is formed there is a diffusion of charge carriers at the junction. Some electrons from the *n* region cross over and fill the holes in the *p*-type material near the junction.

A closer examination of this phenomenon is in order. When an electron from a donor atom leaves its parent atom and diffuses across the junction, its parent atom becomes a *positive ion*, because it now has one more positive charge in its nucleus than negative charges orbiting the nucleus. There is also a diffusion of holes across the junction into the *n*-type material. The atoms that permanently lose a hole become *negative ions*. This diffusion of carriers is confined to the immediate vicinity of the *pn* junction. This finite region is called the *depletion or space-charge region*, inasmuch as the impurity atoms therein have given up their charge carriers. The action just described results in an electrical charge being built up across the junction, which can be represented by a small *imaginary battery*, as shown in Fig. 9-4b. The battery symbol is used to merely illustrate the internal action, and the potential it represents is not directly measurable. The voltage differential of this equivalent battery is approximately 0.3 V for germanium and 0.7 V for silicon.

One may wonder why the diffusion action does not continue until sufficient holes in the *p* region and electrons in the *n* region drift across the junction and neutralize each other. The answer lies in the fact that once the barrier is formed additional electrons in the *n* region do not possess sufficient energy to overcome the charge of the negative ions in the *p* region. Likewise, holes in the *p* region do not have sufficient energy to overcome the repelling force of the positive ions at the junction in the *n* region. Hence, the diffusion of charge carriers stops very shortly after the junction is initially formed.

9-3 BIASING THE *pn* JUNCTION

When an external dc potential is connected across a *pn* junction, the amount of current flow is dictated by the polarity of the applied voltage and

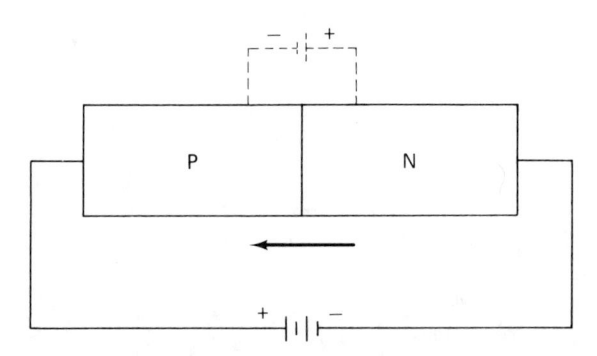

Fig. 9-5 Forward biased *pn* junction.

Biasing the pn Junction

the magnitude of its potential. Assume that the positive terminal of a battery is connected to the *p* material and the negative to the *n* material, as shown in Fig. 9-5. This results in a *forward-biased* junction, and relatively large amounts of current will flow, provided the battery emf is greater than the *barrier voltage*. The reason a large current flow results is that the positive potential of the battery repels the holes in the *p* region, forcing them toward the junction. In like manner the negative terminal of the battery repels the negative charges in the *n* region. These charges combine at the junction, resulting in a substantial current flow. Figure 9-6 shows the forward-bias curves for small-signal germanium and silicon *pn* junctions. For power diodes, such as used in large rectifiers, the current would be in amperes.

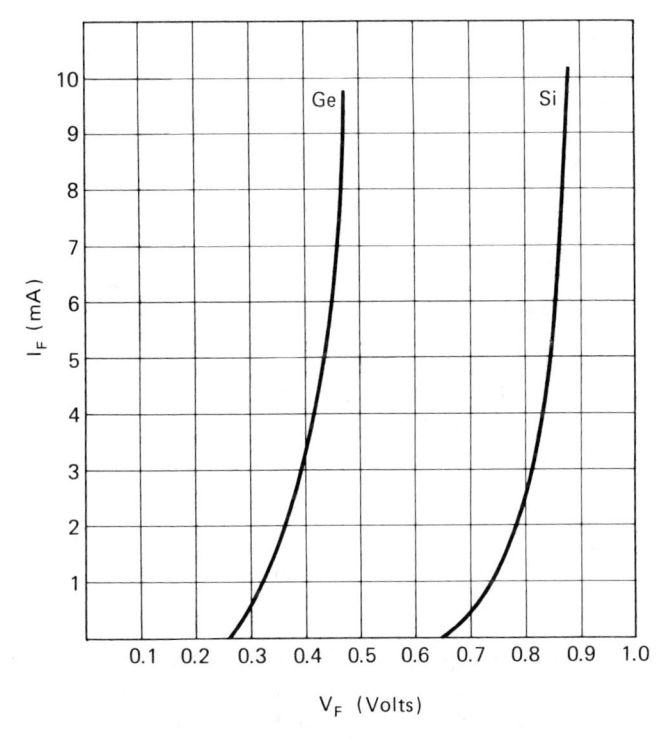

FIG. 9-6 Forward bias curves for germanium and silicon *pn* junctions.

Consider a *pn* junction that is *reverse biased*, as shown in Fig. 9-7. The current flowing in this case is very small and consists only of *leakage* between the *p*- and *n*-type materials due to the presence of minority carriers in both materials. The reason that very little current flows is that the majority carriers in both regions are attracted to the terminals of the battery, leaving the junction devoid of charge carriers.

128   Rectifiers

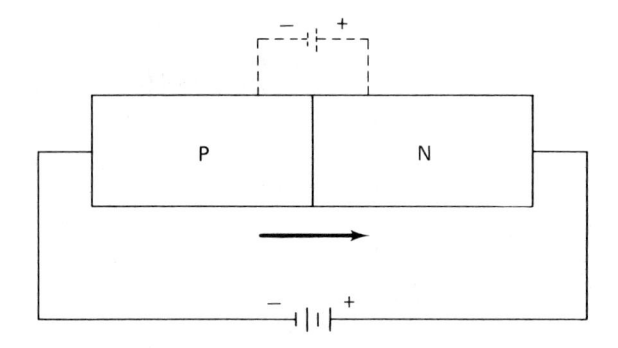

FIG. 9-7 Reverse biased *pn* junction.

The characteristics of the junction when reverse biased are shown in Fig. 9-8. Notice that even with a number of volts applied the reverse current is measured in microamperes. Diodes used in many power rectifier circuits are capable of withstanding several hundred volts with negligible leakage.

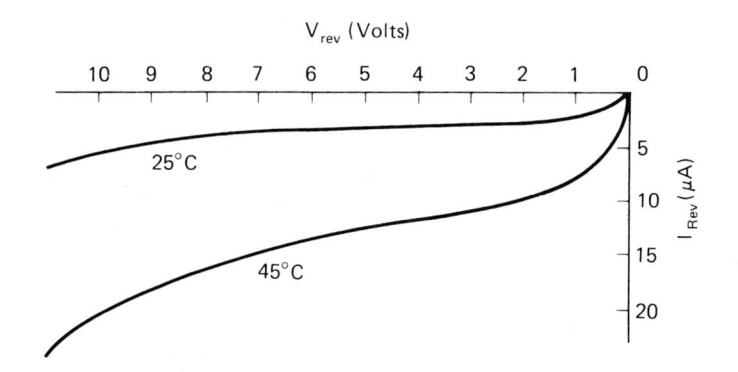

FIG. 9-8 Characteristics of *pn* junction under reverse bias conditions.

The amount of leakage current is also influenced by the *temperature* of the device. At 25°C the leakage is considerably less than at 45°C. Silicon diodes are capable of operating at much higher temperatures than germanium. For this reason they are more commonly used in rectifier cricuits.

Because the forward- and reverse-biased characteristics of a diode are so different, it is logical to expect a marked difference in the resistance of the junction under the two conditions. When forward biased, the resistance is only a *few ohms* whereas reverse bias should indicate a resistance of *many thousands of ohms*. This is sometimes referred to as the *front-to-back ratio* of resistance. Typical ratios might be 1 : 1000 or more. The symbol for a diode is shown in Fig. 9-9. The *p*-type material is referred to as the *anode* and the *n*-type as the *cathode*.

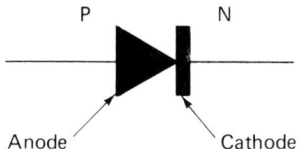

Fig. 9-9 Schematic symbol of a diode. Anode · Cathode

9-4 DIODE AS A RECTIFIER

The *pn* junction conducts readily when forward biased but not when reverse biased. This ability allows the *pn* junction to serve as a *rectifier* of ac voltage. If such a diode is connected in the circuit shown in Fig. 9-10a, rectifying action will result. When the generator voltage shown in Fig. 9-10b is applied to the circuit, current will flow only during the *positive halves* of each cycle. During the negative half-cycles, the diode is reverse biased and only leakage current flows, which normally is insignificant. The resulting current flowing through the resistor produces pulses of voltage as shown in Fig. 9-10c.

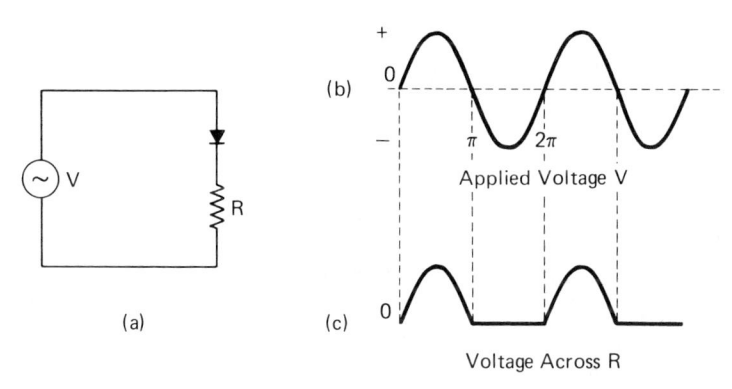

Fig. 9-10 Simple rectifier circuit and waveforms.

9-5 VACUUM-TUBE DIODE

Vacuum-tube diodes consist basically of two parts: cathode and anode. The function of the cathode is to furnish a supply of electrons, under proper conditions, and the anode collects them.

The simplest form of cathode is a tungsten wire called a *filament* heated to a very high temperature by passing an electric current through it. The power dissipated by the filament literally boils electrons from the surface of the filament. This is called *thermionic emission*.

A significant improvement in efficiency can be obtained by coating a tungsten cathode with a thin layer of *thorium*. When so constructed, the cathode is called a *thoriated tungsten cathode*. This kind of cathode requires less heating (lower operating temperatures) than a tungsten filament.

The most efficient of all is the *oxide-coated* cathode, which consists of a relatively thick layer of barium and strontium oxide on a nickel-alloy wire or ribbon. This type of cathode need only be heated to a temperature of 750°C to produce a large supply of electrons. Because of their high efficiency, oxide-coated cathodes are used extensively in low-power and mobile equipment in which power drain must be limited to a small value.

Thermionic cathodes are heated in one of two ways—directly or indirectly. The directly heated cathode is one in which the current used to supply the heat flows directly through the cathode emitting material. One type is illustrated in Fig. 9-11a.

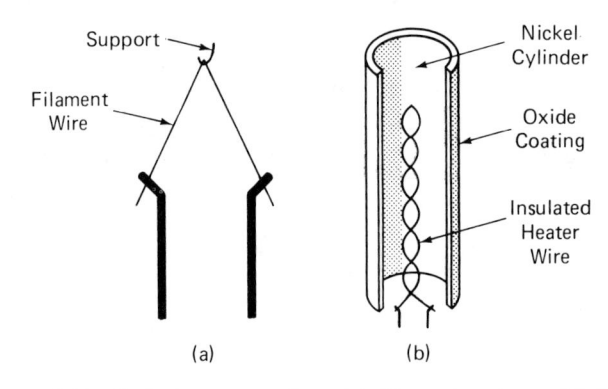

FIG. 9-11 Cathode construction: (a) filamentary type, (b) indirectly heated type.

As shown, a thin piece of wire called a *filament* is suspended on an insulated support. The wire can be of tungsten, thoriated tungsten, or oxide-coated nickel. A current is passed through the wire, causing it to be heated to incandescence. When hot, the filament (or coating) emits electrons and can be used as a cathode.

The use of alternating current as a source for directly heated cathodes should be avoided in vacuum tubes operating with weak signals. Due to the small mass of the filament wire, the filament temperature rises and falls in step with the ac heating current, causing periodic fluctuations in the number of emitted electrons. In weak signal circuits this can introduce *undesirable hum* into the signal.

A relatively constant rate of emission with a fluctuating heater current can be obtained by employing *indirectly heated cathodes*. The construction of an indirectly heated cathode is illustrated in Fig. 9-11b. In this type current does not flow through the emitting material.

The cathode consists of a thin nickel cylinder coated on the outside with barium and strontium oxides. A tungsten or tungsten-alloy wire called a *heater* is placed inside. This wire is used as a heating element only and does not supply any part of the emission in the tube. The cylinder is maintained

Vacuum-Tube-Diode Characteristics

at the correct temperature by the heat radiated from the heater. Ceramic insulating material is packed around the heater wire to electrically insulate it from the cathode and minimize ac hum. This also tends to keep the cathode at a relatively constant temperature, regardless of the 60-Hz variation in heater current. Most low-power tubes using indirectly heated cathodes require a warm-up time of 10 to 20 s.

The cathode is placed in the center of the tube structure and is surrounded by the plate. The plate is made of a material, such as carbonized nickel, nickel-plated steel, or molybdenum, that does not emit electrons readily and can radiate the heat generated during operation.

When assembled, the diode electrodes are placed inside a glass or metal envelope. To prevent the cathode emitting surfaces from becoming contaminated, and to allow the electrons freedom of motion without collisions with air molecules, the glass envelope is *highy evacuated* (hence the name vacuum tube).

To remove any gas left after evacuation, a small quantity of magnesium or barium, called a *getter*, is placed inside the tube. By heating this to a specified temperature, it fires and combines with the residual gas, forming a silvery deposit on the inner walls of the envelope. The elements within the tubes are brought out to base pins that mate with appropriate sockets to provide for easy removal and replacement. The schematic symbols of a vacuum diode are shown in Figs. 9-12a and b.

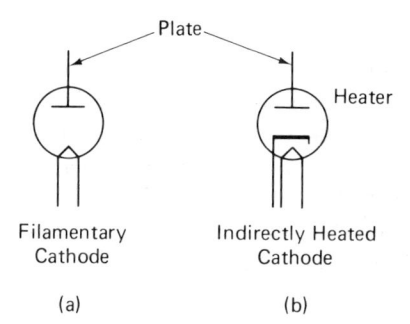

FIG. 9-12 Vacuum tube diode symbols: (a) filamentary cathode, (b) indirectly heated cathode.

9-6 VACUUM-TUBE-DIODE CHARACTERISTICS

To understand the characteristics of a vacuum diode, refer to the circuit shown in Fig. 9-13. The heater transformer provides the correct voltage to operate the cathode. As emission begins, electrons are emitted into the space surrounding the cathode. The majority of these have low velocities and do not travel far, but form a *space charge* around the cathode. If the plate supply voltage is zero, electrons from the space charge will not be attracted to the plate and no plate current flows. The plate circuit is *open* under these conditions.

If the plate is made positive, an electrostatic field is established between

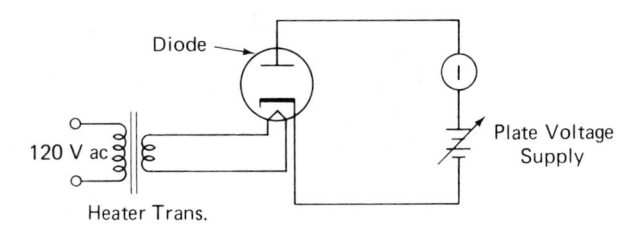

FIG. 9-13 Basic diode circuit.

plate and cathode. This will attract some electrons from the space charge and cause current to flow in the *plate circuit*. This is called *plate current* and will be indicated on the current meter. This is a simple *series circuit* in which the tube acts as a resistance. If the plate voltage is increased, the plate current increases as long as the space-charge reservoir of electrons exists.

As a result of the effects of the space charge, the resistance of the diode is *not constant* and is dependent on the amount of current through it. Since the resistance of a diode changes with changes in current, the volt–ampere characteristic of a diode is not a straight line. Hence, a diode is a *nonlinear device* wherein equal increases of source voltage do not produce equal increases of plate current.

9-7 VOLT–AMPERE CHARACTERISTIC

A very useful curve in determining the characteristics of a diode is the volt–ampere or $E_b I_b$ curve shown in Fig. 9-14. The curve consists of three distinct regions. The part OA, called the *heel* of the curve, is the *square-law region*. Here the current I_b increases approximately as the square of the applied emf, E_b. Part AB is the *linear* region where current varies directly with voltage. Part BC is the *saturation* region where an increase of E_b produces little increase of I_b. The tube's upper limit of conduction capabilities is reached in this region.

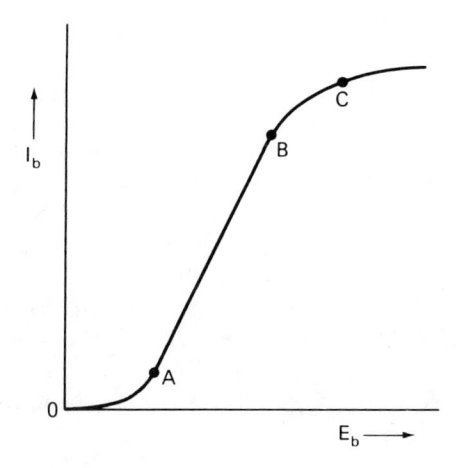

FIG. 9-14 Typical volt–ampere curve.

Plate Resistance

9-8 PLATE RESISTANCE

The tube offers opposition to the passage of dc current through it. This is called *dc plate resistance* and its designation is R_b.

To compute the dc plate resistance, the plate voltage and current must be known. The values can be obtained from the volt–ampere curve or from an actual circuit. Once values have been obtained, the dc plate resistance can be computed using Ohm's law:

$$R_b = \frac{E_b}{I_b}$$

where R_b = dc plate resistance in ohms
 E_b = potential between plate and cathode in volts
 I_b = plate current in amperes

The dc resistance of a diode is *not constant* throughout its operating current range. This can be verified by calculating the R_b at several different points along the volt–ampere curve shown in Fig. 9-15.

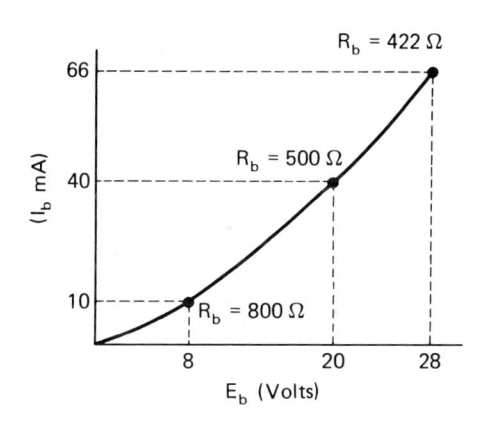

FIG. 9-15 E_b–I_b curve for a small current diode.

Example 9-1

Calculate the dc resistance of a diode, whose $E_b I_b$ curve is shown in Fig. 9-15, when $E_b = 8$ V and $I_b = 10$ mA.

Solution

$$R_b = \frac{E_b}{I_b} = \frac{8}{10 \times 10^{-3}} = 800 \ \Omega$$

Example 9-2

Calculate the dc plate resistance when $I_b = 40$ mA and
$$E_b = 20 \text{ V}.$$

Solution

$$R_b = \frac{20}{40 \times 10^{-3}} = 500 \ \Omega$$

Notice that the dc plate resistance *decreases* as the plate current *increases*.

When an ac voltage is applied to the plate of a tube, its internal resistance changes from the dc value. The term ac plate resistance is defined as the *opposition offered to the flow of alternating current*. The symbol used to designate ac plate resistance is r_p. The r_p of a diode is determined by using the formula

$$r_p = \frac{\Delta e_b}{\Delta i_b}$$

where r_p = ac plate resistance in ohms
 Δe_b = change in instantaneous voltage at the plate
 Δi_b = change in instantaneous current through the tube

Note that the ac plate resistance is computed using a small change (delta) in plate current and voltage. The values of current and voltage required can be obtained from the volt–ampere curve of from an actual circuit.

Example 9-3

Calculate the r_p of a diode if a change of plate voltage from 7 to 9 V causes the plate current to change from 8 to 11.8 mA. See Fig. 9-16.

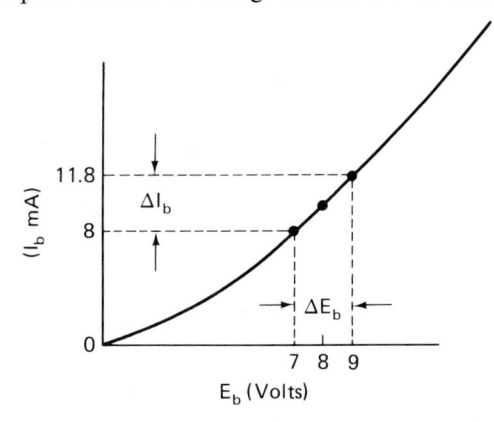

FIG. 9-16   Calculation of r_p.

Solution

$$r_p = \frac{\Delta e_b}{\Delta i_b} = \frac{9 - 7}{11.8 - 8} = \frac{2}{3.8 \times 10^{-3}}$$
$$= 526 \, \Omega$$

Notice that the ac resistance is less than the dc resistance at this approximate point (refer to Example 9-1).

In most applications the plate current of a diode occurs in *pulses*. During each pulse the plate of the tube heats up as a result of the current. Between pulses heat is radiated, allowing the plate to cool.

Zener Diodes 135

Due to the pulse-type nature of the plate-current waveform, the diode is given two current ratings, called the *maximum peak plate-current rating*, and the *maximum average plate-current rating*. The peak plate-current capabilities of a given tube depend on the emission available from the cathode. The maximum allowable average current depends on the amount of heat that can be safely dissipated by the plate.

Diodes also have a *peak inverse voltage rating*, abbreviated PIV. This rating determines the maximum negative voltage that may be applied to the plate with respect to the cathode and is a function of the physical spacing of those two elements. If the PIV rating is exceeded, an arc may occur between plate and cathode, causing damage to the tube.

9-9 ZENER DIODES

When a reverse bias is applied to a *pn* junction, a limited amount of reverse current will flow, usually measured in microamperes (Section 9-3). If the reverse voltage is increased, the reverse current remains essentially constant. However, at some specific reverse voltage (which varies for different types of diodes) a *sudden increase* in reverse current occurs. This phenomenon is called *voltage breakdown*. Under these conditions current flow is limited only by the impedance of the external circuit. The voltage at which this occurs is called the *Zener* or *avalanche voltage*. Diodes designed to operate in this region are called *Zener diodes*. Such a condition is indicated for a typical Zener diode in Fig. 9-17. If the reverse voltage is increased beyond this point, the reverse current will increase greatly. Notice that the Zener voltage decreases as the ambient temperature increases.

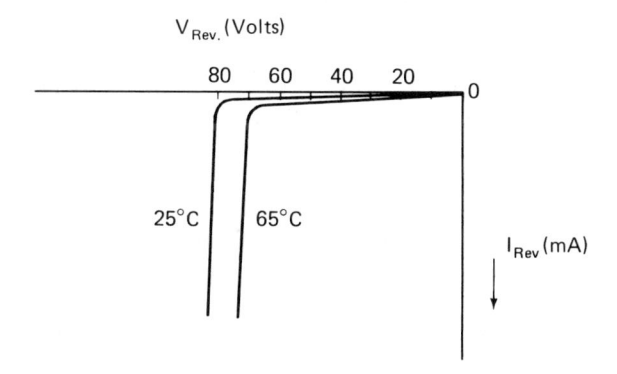

FIG. 9-17 Typical zener diode characteristic.

This characteristic of Zener diodes enables them to be used as *accurate voltage reference sources* in power supplies and other circuits. To illustrate, refer to the schematic of Fig. 9-18 wherein a variable dc voltage is connected across the resistor and Zener diode. The output of the variable supply must be at least 16 V in order to operate the diode in the Zener region. When the

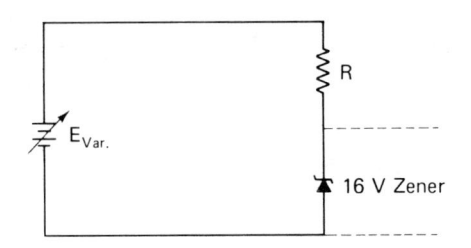

FIG. 9-18 Zener diode used as an accurate voltage reference source.

supply emf is greater than 16 V, the IR drop across R will be the difference between the Zener and supply voltage. Hence, if the supply were 24 V, the voltage drop across R would be $24 - 16 = 8$ V. There is a limitation to the circuit, and that is the power that the Zener and resistor can safely dissipate. If the supply voltage is large, then higher-wattage Zeners and resistors must be used in accordance with Watt's law. Note the symbol for the Zener diode in Fig. 9-18.

Practice Problems

1. A certain germanium diode has 5 mA of current flowing through it when 0.35 V is applied. What is its internal dc resistance?
2. What is the dc plate resistance of a vacuum-tube power diode if a plate voltage of 15 V causes 725 mA of plate current?
3. Calculate the current flowing through a small-signal diode if the instantaneous applied emf is 0.68 V and the internal resistance if 523 Ω.
4. A certain silicon diode exhibited the following characteristics when under test: $I_d = 63$ mA at $V_d = 0.71$ V; $I_d = 42$ mA at $V_d = 0.63$ V. Calculate the ac diode resistance.
5. A Zener diode passes 15 μA of current at 13.4 V_{dc}. When the voltage is increased 0.10 V, the current becomes 830 μA. What is the change of resistance of the Zener?

Commercial License Questions

Sections in which answers to questions are given appear in parentheses. A bracketed number following a question implies that it applies only to that element.

1. What is a semiconductor? (9–1)
2. What is n-type germanium? p-type? (9–1, 9–2)
3. Explain what is meant by forward and reverse biasing of a pn junction. (9–3)
4. What is meant by *thermionic emission*? (9–5)
5. What is a getter in a vacuum tube? (9–5)

6. What is the meaning of the term *plate saturation*? (9–7)
7. How is the dc plate resistance of a diode determined? (9–8)
8. What is the meaning of *ac plate rseistance*? (9–8)
9. What is a Zener diode? Describe the characteristics of a Zener diode. (9–9)

10

Transistor Fundamentals

10-1 BASIC TRANSISTOR

Transistors have several advantages as amplifying devices. They are very small, efficient, and have long life. No warm-up time is needed as with vacuum tubes.

Transistors are made by growing, alloying, or diffusing pieces of *p*-type and *n*-type materials together. A typical junction transistor (the kind to be considered in this chapter) is formed by joining two *pn* junctions together,

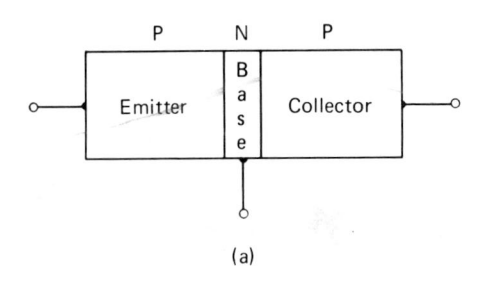

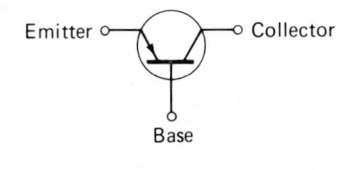

FIG. 10-1 (a) Functional diagram of a *pnp* transistor. (b) Schematic symbol.

138

npn Transistor

not in a mechanical sense but in a controlled fusion process. The result is a very thin layer of *n*-type material sandwiched between two layers of *p*-type material. Figure 10-1a is a pictorial of such an arrangement and is called a *pnp* transistor. One *p* section is called the *emitter* and the other the *collector*. The thin layer of *n*-material is the *base*. Electrical connections are made to each section so that they may be connected to an electrical circuit. The schematic symbol used to represent a *pnp* transistor is seen in Fig. 10-1b.

If a transistor were manufactured so that a *p*-type material were placed between two *n*-type sections, an *npn* transistor would result. Such an arrangement appears in Fig. 10-2a with its schematic symbol shown in Fig. 10-2b. Note that in this type of transistor the arrowhead on the emitter points *away* from the base—just the *opposite* to the *pnp* transistor.

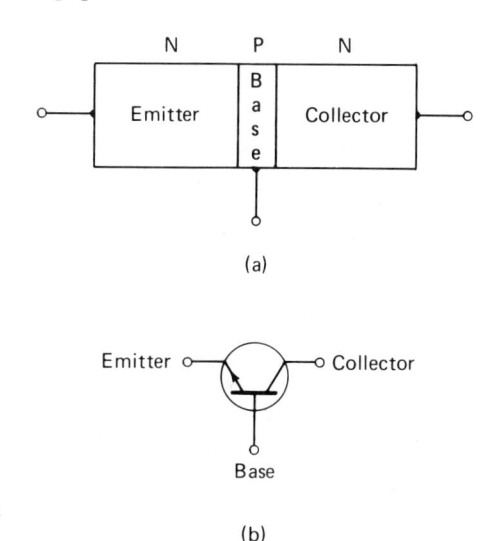

FIG. 10-2 (a) Functional view of an *npn* transistor. (b) Schematic symbol.

10-2 *npn* TRANSISTOR

A transistor is an electronically controlled device that *regulates* the current flowing through it. Current from a power source enters the emitter, passes through the very thin base region, and leaves via the collector. Current flow is always in this direction. This current can be made to vary in amplitude by varying the current flowing in the base circuit. It takes only a *small change* of base current to control a relatively large collector current. It is this ability that enables the transistor to amplify.

The action just described can be better understood by reference to Fig. 10-3, which shows an *npn* silicon transistor connected in a simple circuit. Two separate voltages are required to operate the circuit. For simplification these are represented by the two batteries V_{BB} and V_{CC}. The base–emitter junction *must* be forward biased to allow the transistor to operate, which means that

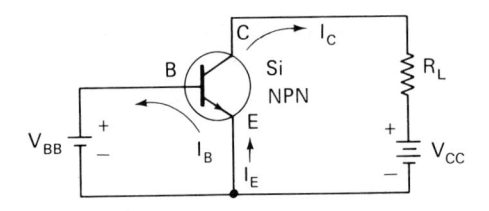

FIG. 10-3 Simple *npn* transistor circuit.

V_{BB} must deliver about 0.7 V. This reduces the base–emitter junction resistance to a low value and permits collector current, I_C, to flow. The magnitude of this current is dependent principally upon the amount of forward bias and the value of R_L. The amount of collector voltage supplied by V_{CC} does little to affect I_C, provided it is above a certain threshold value. The base–collector junction is *reverse biased* and, consequently, acts like a high resistance.

Assuming that R_L and V_{CC} are fixed, the collector current will vary directly with the amount of base current, I_B. This current is controlled by the base–emitter voltage, V_{BE}. As previously stated, if this voltage is sufficient to overcome the barrier voltage, I_B begins to flow. By varying V_{BE} only a few hundredths of a volt, the base current can be changed significantly.

The relationship between emitter current I_E, base current I_B, and collector current I_C is given by:

$$I_E = I_B + I_C$$

or

$$I_B = I_E - I_C$$

In most transistors the base current is very small compared to the emitter and collector currents. Therefore, $I_E \approx I_C$. There exists a very important relationship between the collector and base currents. The ratio of these two currents is called the *current gain* of the transistor and is represented by β (Greek letter beta). Expressed mathematically,

$$\beta = \frac{\Delta I_C}{\Delta I_B}, \quad V_{CE} \text{ held constant}$$

Example 10-1

Determine the beta of a transistor if the collector current changes from 6.5 to 8.3 mA when the base current is changed from 15 to 23 µA.

Solution

$$\beta = \frac{\Delta I_C}{\Delta I_B} = \frac{8.3 - 6.5}{23 - 15} = \frac{1.8 \text{ mA}}{8 \text{ µA}}$$
$$= 225$$

Beta is a number only and has no dimensions. Hence, we see that for every 1-µA change of base current the collector current is caused to change 225 µA.

Collector Family of Curves 141

10-3 *pnp* TRANSISTOR

A simplified *pnp* transistor circuit is shown in Fig. 10-4. Notice that the batteries are connected in reverse polarity compared to the *npn* configuration of Fig. 10-3. Except for this change, the circuit performance is the same. If a germanium transistor were used, the voltage source V_{BB} would have to be adjusted to about 0.25 V to overcome the barrier voltage and cause collector current to flow.

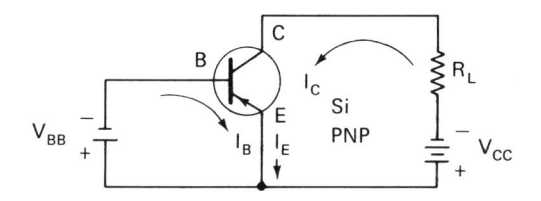

FIG. 10-4 Simple *pnp* transistor circuit.

A transistor is a *current-operated* device. This means that there must be base current flowing to have collector current. If no base current is flowing, the transistor is *cut off* and will not operate. On the other hand, if excessive base current is flowing for a particular transistor, it will be in *saturation*.

10-4 COLLECTOR FAMILY OF CURVES

Transistor manufacturers provide specifications for each device they produce. Included in this information is the collector family of curves, which can be used to predict the approximate performance of the transistor. A family of curves for a typical *pnp* transistor is shown in Fig. 10-5. The *X* axis

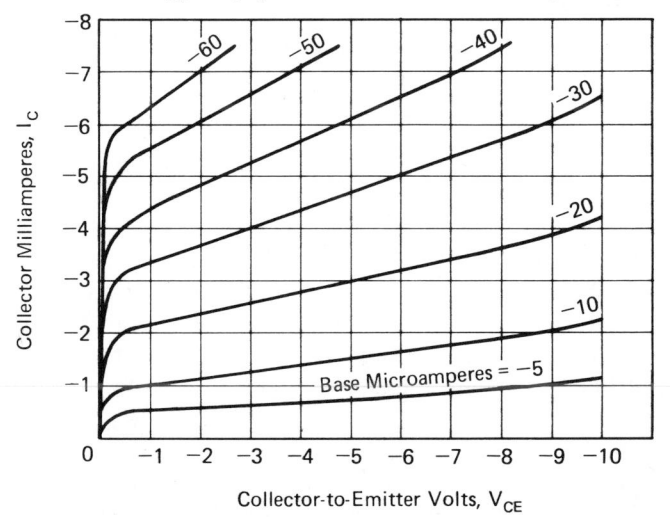

FIG. 10-5 Typical collector family of curves for a *pnp* transistor.

indicates collector–emitter voltage, V_{CE}, and the Y axis, collector current, I_C. The several curves indicate the amount of collect current that will flow for different values of base current.

No curve is shown for a base current of 0 μA because the device is cut off. If the V_{BE} is adjusted so that the base current is 5 μA, we can determine the amount of I_C for any value of V_{CE}. For example, if V_{CE} is -3 V, the collector current is approximately -0.7 mA. By increasing V_{CE} to -9 V, I_C increases to -1.0 mA. (The minus values of voltage and current indicate a *pnp* transistor.) Now if I_B is increased to -10 μA and V_{CE} is held at -3 V, I_C becomes approximately -1.3 mA. If at this value of collector voltage, I_B is increased to -20 μA, I_C is about -2.7 mA. Holding I_B constant but increasing V_{CE} causes I_C to increase. Hence, the collector current that will flow is determined by the base current and collector voltage.

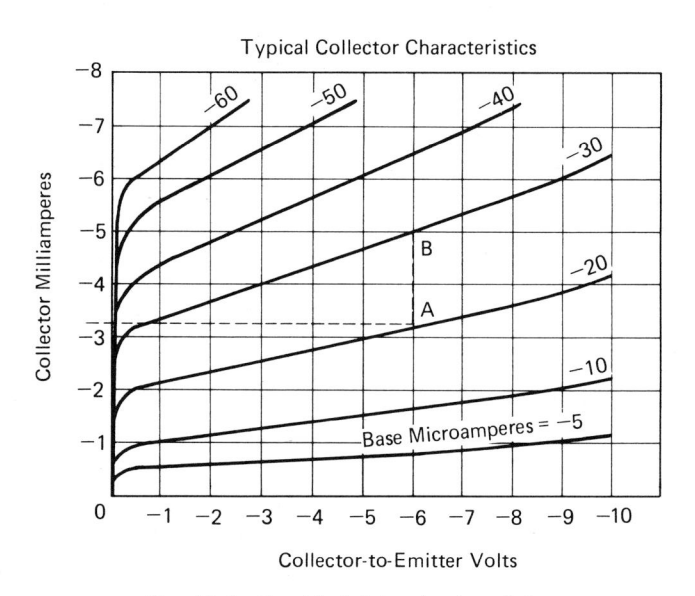

FIG. 10-6 Graphical determination of β.

The beta of a transistor can be graphically determined from its family of curves, as shown in Fig. 10-6. Assume that V_{CE} remains constant at 6 V. If the base current is changed from 20 to 30 μA (point A to B), the collector current varies from approximately 3.2 to 5.0 mA. This gives a ΔI_C of 1.8 mA for a ΔI_B of 10 μA. Applying these values to the formula for calculating beta,

$$\beta = \frac{\Delta I_C}{\Delta I_B} = \frac{1.8}{10} = 180$$

The betas of individual transistors, even though of the same type or number, may vary considerably when compared one to another.

10-5 LOAD LINES

A transistor must have a *load* to work into in order to develop a useful output. The effect of the load applied to the transistor can be predicted quite accurately in advance. This is accomplished by adding to the static collector family of characteristic curves a graphical representation of the load, known as the *load line*. An example of this appears in Fig. 10-7. This shows the distribution of V_{cc} between the load and the internal impedance of the transistor under differing conditions of collector current.

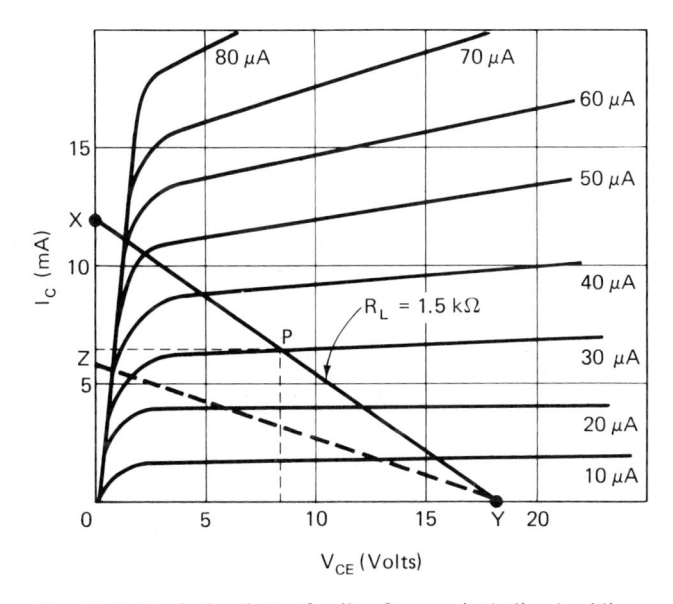

FIG. 10-7 Typical collector family of curves including load line.

The load line XY corresponds to a load R_L of 1.5 kΩ. The selection of 1.5 kΩ is arbitrary. It can be a higher or lower value, except that there are limits to the ohmic values of loads relative to the internal impedance of the transistor and that the fixed value of the collector supply voltage must be considered.

No connection exists between the load line XY and the static collector family. The *load line has no fixed association* with transistor constants or characteristics. It is simply a means of developing information concerning the behavior of a transistor used with a certain load. The load may be correct or incorrect for the device. An analysis of the load line will reveal if the load selected is correct or not.

A load line can easily be developed by making two assumptions: (1) the transistor is short circuited, and (2) the transistor is open.

Let us refer to the simple schematic in Fig. 10-8 to aid us in our analysis.

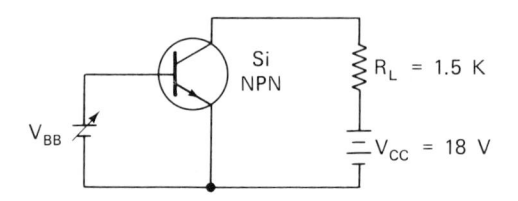

FIG. 10-8 Simple *npn* transistor circuit used to show development of load line.

If V_{BB} is adjusted so that the transistor goes into saturation (maximum conduction), it behaves as a short circuit (assumption 1). For practical purposes there will be no voltage drop across the transistor, because its internal resistance has been reduced nearly to zero. Therefore, V_{CC} is effectively connected across R_L. The current flow will be

$$I_{\text{sat}} = \frac{V_{CC}}{R_L} = \frac{18}{1.5} = 12 \text{ mA}$$

This is the saturation current that will flow and is marked X on the current axis in Fig. 10-7. This is one terminus of the load line.

If V_{BB} is readjusted so that the transistor is cut off or open (assumption 2), its internal resistance is very high compared to R_L. Consequently, the total supply voltage appears across the transistor, and no current flows. This condition is marked Y on the voltage axis (Fig. 10-7) and represents the other terminus of the load line. A straight line drawn between points X and Y is called a load line and is seen in Fig. 10-7. This is the load line for 1.5 kΩ.

If the load resistor in Fig. 10-8 were increased to 3000 Ω, it would be represented in Fig. 10-7 as the dotted line between points Z and Y. Notice that the saturation current is one half the value for the 1500-Ω resistor. Hence, the *slope* of the load line diminishes as the resistance increases.

Reference to Fig. 10-7 discloses that the load line intersects the collector-current curves at different points. To establish the distribution of voltages between the transistor and the load, an *operating point* must be selected. Assume that a base current of 30 μA is chosen. The load line crosses the collector-current curve for this bias at point P. A horizontal line drawn to the collector-current scale shows that the corresponding current is approximately 6.4 mA. A vertical line dropped from point P to the collector-voltage axis intersects it at approximately 8.4 V. Therefore, point P can be described as

$$I_b = 30 \text{ }\mu\text{A}$$
$$V_{CE} = 8.4 \text{ V}$$
$$I_c = 6.4 \text{ mA}$$

By projecting vertically downward from point P to the V_{CE} axis, we can determine the division of V_{CC} between the transistor and the load. For example, the voltage to the left of this intersection indicates the collector voltage

Biasing Techniques

V_c, which is 8.4 V. Similarly, the voltage to the right of this intersection is E_{R_L}, which is $18 - 8.4 = 9.6$ V. To verify this last statement, the voltage drop across the load resistor can be computed as

$$E_{R_L} = I_c R_L = 6.4 \text{ mA} \times 1.5 \text{ k}\Omega = 9.6 \text{ V}$$

From the standpoint of *linear operation* of an amplifier, this voltage distribution is desirable. By linear operation is meant the ability of an amplifier to amplify a signal with little or no distortion. An operating point should be chosen that gives approximately equal division of V_{CC} between the transistor and its load.

10-6 BIASING TECHNIQUES

To operate a transistor at a given *quiescent* point requires that it be properly biased. Several different biasing systems exist for transistorized circuits. The first system to be considered is called *simple bias*, and the circuit is seen in Fig. 10-9. For linear operation a bias point should be chosen that

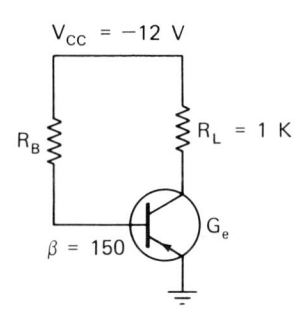

FIG. 10-9 Simple bias circuit.

will cause $V_{CE} \approx V_{R_L}$. With a V_{CC} of 12 V, this means about 6-V drop across each circuit element. The collector current will be

$$I_c = \frac{6 \text{ V}}{1 \text{ k}\Omega} = 6 \text{ mA}$$

The beta of the transistor is indicated as 150. This means that there will be 1 μA of base current for every 150 μA of collector current. Therefore,

$$I_B = \frac{I_c}{\beta} = \frac{6}{150} = 40 \text{ }\mu\text{A}$$

If we presume a germanium transistor, the barrier voltage, or V_{BE}, will be approximately 0.25 V. The *IR* drop across R_B will be

$$V_{R_B} = V_{CC} - V_{BE}$$
$$= 12 - 0.25 = 11.75 \text{ V}$$

With this information we may readily calculate the value of R_B as

$$R_B = \frac{V_{R_B}}{I_B} = \frac{11.75 \text{ V}}{40 \text{ }\mu\text{A}} = 282 \text{ k}\Omega$$

A modification of simple bias is the one shown in Fig. 10-10, which includes emitter resistor R_E. This circuit will provide more *stability* than the one in Fig. 10-9. This means that is operating point will tend to remain at the predetermined value in spite of temperature and other circuit variations. A method of determining the resistance values of the circuit will be given in the following example.

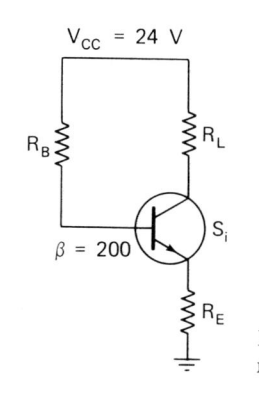

FIG. 10-10 Simple bias with emitter resistor.

Example 10-2

Determine the values of the three resistors in Fig. 10-10 if

$$I_C = 2 \text{ mA} \quad \text{and} \quad V_{CE} = V_{RL} = 10 \text{ V}$$

Solution

Since $V_{CE} + V_{RL} = 20$ V, it is apparent that $V_{RE} = 4$ V. The value of R_L is

$$R_L = \frac{10 \text{ V}}{2 \text{ mA}} = 5 \text{ k}\Omega$$

To determine R_B we must find the voltage drop across it. This is

$$V_{RB} = V_{CC} - (V_{BE} + V_{RE})$$
$$= 24 - (0.7 + 4) = 19.3 \text{ V}$$

The base current flowing through R_B is

$$I_B = \frac{I_c}{\beta} = \frac{2 \text{ mA}}{200} = 10 \text{ }\mu\text{A}$$

Therefore, the value of R_B is

$$R_B = \frac{19.3 \text{ V}}{10 \text{ }\mu\text{A}} = 1.93 \text{ M}\Omega$$

The value of R_E is

$$R_E = \frac{4 \text{ V}}{2 \text{ mA}} = 2 \text{ k}\Omega$$

Biasing Techniques

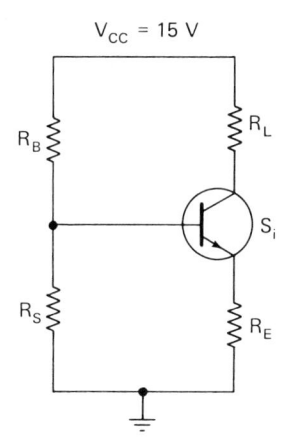

FIG. 10-11 Voltage divider bias.

A more stable bias circuit is shown in Fig. 10-11 and is called *voltage-divider bias*. An example of how the various resistor values can be calculated is shown in Example 10-3.

Example 10-3

Assume the following conditions for Fig. 10-11:

$$V_{RL} = V_{CE} = V_{RE} = 5 \text{ V},$$
$$\beta = 160 \quad \text{when} \quad I_c = 5 \text{ mA}$$

Solution

$$R_L = \frac{5 \text{ V}}{5 \text{ mA}} = 1 \text{ k}\Omega$$

$$I_B = \frac{I_c}{\beta} = \frac{5 \text{ mA}}{160} = 31.2 \text{ }\mu\text{A}$$

With voltage-divider bias it is desirable to have a bleeder current of approximately 10 times I_B flow through the divider network. This means that the current through R_s should be $10I_B$. Now the current through R_B must also be $10I_B$, *plus* the base current flowing in the transistor. Therefore, the *total* current through R_B is 11 times I_B. On the basis of the above information we can proceed to calculate the ohmic values of R_s and R_B.

$$V_{Rs} = V_{BE} + V_{RE} = 0.7 + 5 = 5.7 \text{ V}$$

Therefore,

$$R_s = \frac{V_{Rs}}{10I_B} = \frac{5.7 \text{ V}}{312 \text{ }\mu\text{A}} = 18.3 \text{ k}\Omega$$

The value of R_B is

$$R_B = \frac{V_{CC} - V_{Rs}}{11I_B} = \frac{9.3 \text{ V}}{343.2 \text{ }\mu\text{A}} = 27.1 \text{ k}\Omega$$

Another very stable bias circuit, called *simple bias with collector feedback*, appears in Fig. 10-12. The dc feedback feature results as follows: *Leakage currents* in the transistor will *increase* with temperature. Therefore, V_c (collector-to-ground voltage) will drop slightly and reduce the voltage drop across R_B. This, in turn, causes a small decrease in I_B, which results in less I_c. This reduces the IR drop across R_L and tends to restore V_c to approximately its original value.

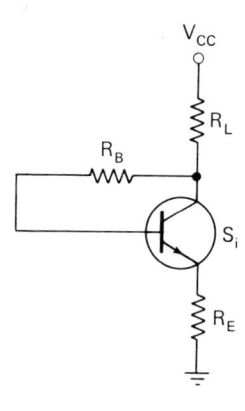

FIG. 10-12 Simple bias with collector feedback.

Example 10-4 indicates how the resistive values of the circuit can be determined.

Example 10-4

Referring to Fig. 10-12, assume that

$$\beta = 50 \quad \text{when} \quad I_c = 10 \text{ mA}$$
$$V_{CE} = V_{RL} = 10 \text{ V}$$
$$V_{BE} = 0.7 \text{ V} \quad \text{and} \quad R_E = 500 \text{ }\Omega$$

Solution

Solving for I_B,

$$I_B = \frac{I_c}{\beta} = \frac{10 \text{ mA}}{50} = 0.2 \text{ mA}$$

Therefore,

$$R_L = \frac{10 \text{ V}}{10.2 \text{ mA}} \approx 1 \text{ k}\Omega$$

(Note: $I_B + I_c$ flows through R_L.)

To solve for R_B it is necessary to find the IR drop across R_E. This is

$$V_{RE} = 10.2 \text{ mA} \times 500 \text{ }\Omega = 5.1 \text{ V}$$

From the above we determine that

$$V_{CC} = V_{RL} + V_{CE} + V_{RE} = 25.1 \text{ V}$$

The voltage drop across R_B is now determined as follows:

$$V_{R_B} = V_{CC} - (V_{R_L} + V_{R_E} + V_{BE})$$
$$= 25.1 - (10 + 5.1 + 0.7) = 9.3 \text{ V}$$

Therefore,
$$R_B = \frac{9.3 \text{ V}}{0.2 \text{ mA}} = 46.5 \text{ k}\Omega$$

10-7 TRANSISTOR CIRCUIT CONFIGURATIONS

The transistor, being essentially a three-terminal device, can be connected into a circuit in three basically different ways. Each configuration possesses certain distinguishing characteristics, which will be described. The circuit arrangement most commonly used is that of the *common-emitter* (CE) amplifier. This is the only one that has been shown in the several illustrations thus far and is the most commonly used circuit. The basic CE amplifier is shown in Fig. 10-13. The power supplies, shown in previous circuits, are not shown here because they do not determine the basic characteristics of the circuit. The circuit derives its name from the fact that the emitter connection is common to both input and output circuits.

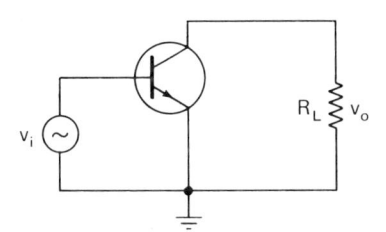

FIG. 10-13 Basic CE amplifier.

The signal to be amplified, designated v_i, is applied between base and emitter. It is this signal that is *superimposed* on the dc bias and causes the dc collector current to vary or be *modulated* at the signal rate. The output signal, v_o, appears across the load resistor R_L. The voltage gain, A_v, of the stage is equal to v_o/v_i. The CE amplifier has moderate input and output impedances, on the order of 1000 Ω for each. The CE amplifier can provide high voltage, current, and power gain.

The *common-base* amplifier, sometimes called *grounded base*, is the second basic configuration and is shown in simplified form in Fig. 10-14.

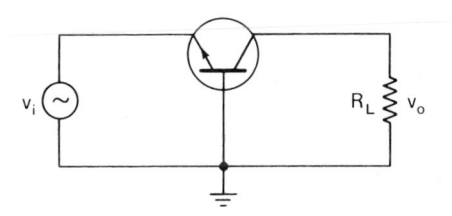

FIG. 10-14 Basic CB amplifier.

As is apparent, both from the name and schematic, the base is common to both input and output circuits. Characteristic of this arrangement is a low input impedance on the order of 50 Ω. Conversely, the output impedance is high, being typically 10 to 50 kΩ. This circuit configuration is used for impedance matching (i.e., low Z in to high Z out) and for certain high-frequency applications. Like the CE amplifier it can provide a good voltage gain.

The current gain of the CB amplifier is markedly different from the other configurations. It is called *alpha* and is represented by the Greek letter α. Alpha is the ratio of I_c/I_E and is slightly less than 1. Expressed mathematically,

$$\alpha = \frac{I_c}{I_E}$$

The final amplifier configuration is shown basically in Fig. 10-15 and is known as the *common-collector* or CC amplifier. The input signal is applied between base and collector and the output signal is taken between emitter and collector (ground).

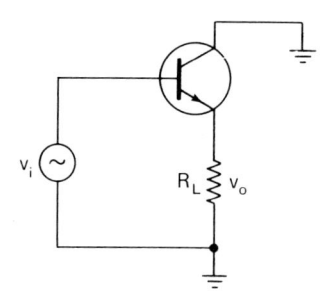

Fig. 10-15 Basic CC amplifier.

The CC amplifier has several distinguishing characteristics. One of these is its *high input impedance*, being on the order of 10 to 50 kΩ or more. The output impedance, on the other hand, is low, usually well below 1000 Ω. Therefore, this amplifier serves well as an impedance-matching device.

A second characteristic is its lack of voltage gain or amplification. The output is *always less* than the input with ratios of about 0.8 to 0.99 being typical.

Also characteristic of the CC amplifier is the high power and current gain it can provide. The current gain, A_i, is approximately equal to the transistor's beta.

The CC amplifier is more commonly known as an *emitter follower*. The reason is as follows: The collector current is modulated (caused to change or vary) by the input signal and is *in phase* with it. The collector current is essentially the same as the emitter current and also flows through load resistor R_L located in the emitter circuit. Consequently, the output voltage, v_o, is in phase with the input voltage. We say the output voltage *follows* (phase wise) the input voltage.

10-8 h PARAMETERS

Certain transistor charcteristics are defined by *hybrid parameters*, commonly referred to as *h parameters*. No detailed attempt will be made to show the derivation of these, but only to tell what they are. The reason for briefly describing them is that most manufacturers define some of their transistors' characteristics by *h* parameters, and therefore we should know what they are.

There are basically four *h* parameters for any transistor configuration. For the CE connection they are h_{ie}, h_{fe}, h_{oe}, and h_{re}. Each will be explained below.

The base–emitter junction of any transistor has a certain amount of resistance that the input signal sees, as mentioned in Section 10-7. This is defined as

$$h_{ie} = \text{input resistance}$$

where i = input
 e = CE configuration

If the manufacturer were indicating the input resistance for a CB or CC connection, the second subscript would be *b* or *c*, respectively. Mathematically, this parameter can be written as

$$h_{ie} = \frac{\Delta V_{BE}}{\Delta I_B}, \quad V_{CE} \text{ held constant}$$

The next *h* parameter to be considered is the *forward current transfer ratio* or h_{fe}. This is the same as beta, which has been discussed in Section 10-2. Mathematically, this is expressed as

$$h_{fe} = \frac{\Delta I_c}{\Delta I_B}, \quad V_{CE} \text{ held constant}$$

where f = forward current transfer ratio
 e = CE configuration

This is one of the most important of transistor characteristics and gives an indication of how much control the base current has over the collector current. If the measurement is made using different values of direct current in the base circuit, the parameter is written as h_{FE}, where the capital letter subscripts indicate dc values.

Another *h* parameter is called the *output conductance* and is represented as h_{oe}. Expressed mathematically,

$$h_{oe} = \frac{\Delta I_c}{\Delta V_{CE}}, \quad I_B \text{ held constant}$$

If changing V_{CE} has little effect on I_c, the transistor has a low conductance. Typical values of h_{oe} for the CE connection range from 10 to over 100 μS. Note that these values are equivalent to an output resistance (with input alternating current open circuited) from 100 to 10 kΩ, respectively.

The last h parameter is h_{re}, which is called the *reverse transfer voltage ratio* (measured with open input circuit). Expressed mathematically,

$$h_{re} = \frac{\Delta V_{BE}}{\Delta V_{CE}}, \quad I_B \text{ held constant}$$

There is not complete isolation between output and input circuits in a transistor. Some of the output voltage, a *very small* percentage, feeds back into the input circuit. This voltage is 180° out of phase with the input signal and partially cancels it. The reverse feedback voltage can *never* be equal to the input voltage; otherwise, there would be complete cancellation and, therefore, no output voltage. Typical values of h_{re} range from about 0.001 to 0.004. Transistor h parameters are used only for small-signal amplifiers and not for power amplifiers.

10-9 TRANSISTOR EQUIVALENT CIRCUIT

The equivalent circuit of a CE amplifier making use of h parameters is shown in Fig. 10-16. When voltage v_1 is applied to the input, a current i_1 flows through the input resistance h_{ie} of the transistor. This current causes an output current i_2 to flow, which is actually the input current multiplied by beta, or h_{fe}, and is represented by the current generator, labeled $h_{fe}i_1$. This output current is divided between the load resistor R_L (not shown) and the reciprocal of h_{oe} (a resistance). Voltage v_2 is the ac output of the transistor as a result of the output current flowing through these impedances.

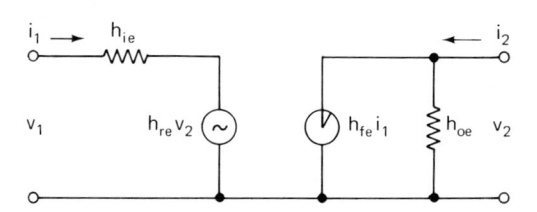

FIG. 10-16 Equivalent circuit of a CE amplifier showing h parameters.

Because there is not complete isolation between the input and output circuits of a transistor, a portion of v_2 is fed back into the input circuit. This voltage is 180° out of phase with the input voltage v_1, and is represented by the ac voltage generator $h_{re}v_2$.

10-10 LEAKAGE CURRENTS

Leakage currents are not confined to diodes alone, but apply also to transistors. These currents are *undesirable*. They add to the collector current of the transistor and can cause distortion of the amplified signal and, in some cases, overheating. Generally speaking, leakage currents are *directly related*

Other Transistor Types   153

to temperature so that if the device is not permitted to heat up beyond specifications, these currents are not too troublesome. Improved manufacturing techniques have reduced leakage currents to very small values.

The flow of currents I_E, I_B, and I_C in a *pnp* transistor connected as a CB amplifier is shown in Fig. 10-17. Notice that there is an additional current, I_{CBO}, that has not yet been discussed. This is a *leakage current* that flows because of the reverse bias voltage between the base–collector junction. It flows from the base to the collector. It is always present and can never be reduced to zero.

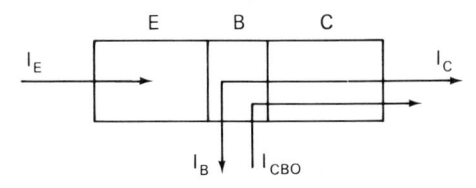

FIG. 10-17 Currents in a *pnp* transistor connected as a CB stage with normal applied voltages.

The I_{CBO} of a transistor can be measured by the circuit shown in Fig. 10-18. Because the emitter circuit is open, no I_C will flow. The only current that can flow is the leakage current I_{CBO}. The first two subscripts of this term designate the transistor elements between which the leakage is flowing. The second subscript also indicates the common terminal of the circuit configuration. The third subscript signifies the condition of the third lead in relation to the common terminal (emitter open).

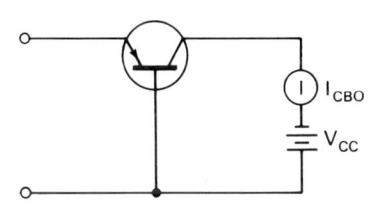

FIG. 10-18 Circuit for measuring I_{CBO} in a *pnp* transistor.

In a germanium transistor, I_{CBO}, may be expected to double for every 11°C increase in temperature, whereas silicon doubles approximately every 6°C. This would seem to indicate that germanium is better than silicon. This is not the case, because I_{CBO} in silicon devices is *much lower* than in germanium. If leakage currents are not kept under control, a condition known as *thermal runaway* occurs at higher temperatures and the transistor destroys itself.

10-11 OTHER TRANSISTOR TYPES

Several other types of transistors are in general use besides the standard *pnp* and *npn* types. A brief description of some of these and their symbols are presented.

The upper frequency limit of a transistor is controlled by the input and output capacitances of the device. An effective method of reducing these capacities is to restrict transistor action to a small portion of the semiconductor material. This creates the effect of a small-sized transistor with correspondingly low input and output capacitances. A device having these characteristics is called a *tetrode* transistor. Figure 10-19 shows the basic construction and schematic symbol for this device. By placing a negative bias on B_2 a portion of the base–emitter junction becomes reverse biased. Current flow between base and collector also occurs over a small portion of the base–collector junction. Hence, current flow through the device is restricted to a portion of the base region. Since the effective input and output capacitances involve only the active regions of these junctions, their capacities are substantially reduced.

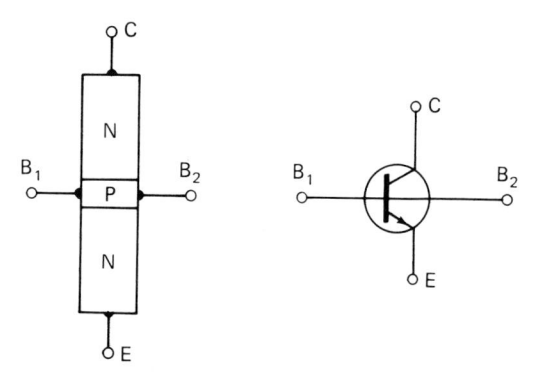

FIG. 10-19 Basic construction and symbol for a tetrode transistor.

The *unijunction transistor*, shown in Fig. 10-20, has found wide application in industry. The device consists of a silicon bar of n-channel semiconductor material with two base connections attached to the far ends. An emitter of p-type material is connected to the opposite side of the bar closer to the B_2 connection. A dc potential is applied between B_2 and B_1 with B_2 positive. This causes a *voltage gradient* to occur along the length of the n-type bar. When the voltage applied to the emitter is increased positively to the point where the emitter B_1 junction becomes forward biased, the resistance of this junction and also the resistance between the two base leads drops abruptly. This negative resistance effect is applied in numerous timing and oscillatory circuits.

The *field-effect transistor or FET* is a new generation of transistor that combines the advantages of both transistor and vacuum tube. Its characteristics are high input impedance, small size, low power, mechanical ruggedness, and a square-law transfer characteristic. Unlike the *bipolar* devices described in this chapter in which performance depends upon *two types of*

Other Transistor Types 155

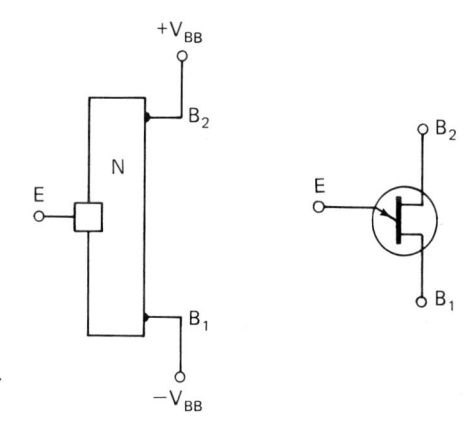

FIG. 10-20 Unijunction transistor pictorial drawing and schematic symbol.

charge carriers, the FET is a *unipolar* device whose operation is essentially a function of only one type of charge carrier. The junction FET was the first type developed of this kind and is still used. These use a reverse-biased semiconductor junction for the control electrode to vary the conductivity of the main channel semiconductor material.

The newest version of the FET is the *metal-oxide-semiconductor* known as the MOSFET, which uses a metal gate electrode separated from the semiconductor channel by an insulator. Like the *pn* junction, this insulated gate electrode can deplete the source-to-drain channel of active carriers when suitable bias voltages are applied. However, the insulated gate electrode can also enhance the conductivity of the channel by applying the proper potential to it. Figure 10-21 shows the basic symbols for the FET and MOSFET. Current flow is from the source (S) to the drain (D) wth the gate (G) controlling current flow. The substrate or base (B) in the MOSFET is usually connected to the source terminal. The direction of the arrow indicates whether the device is *n* or *p* channel.

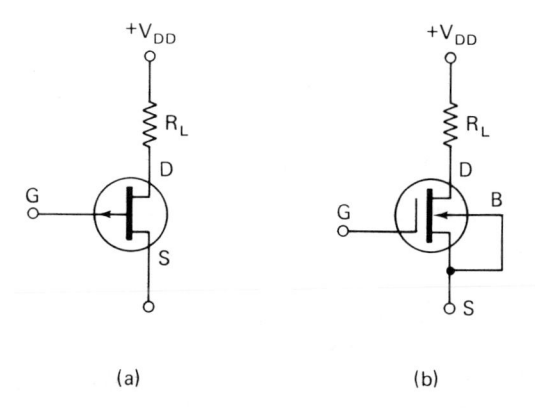

FIG. 10-21 Basic FET schematics: (a) junction type (*p*-channel) (b) MOSFET (*n*-channel depletion type).

Tunnel diodes are another type of semiconductor device used for very high frequency amplification and high-speed switching circuits. It is a very small *pn* junction having a very high concentration of *dopants* or impurities in both the *p*- and *n*-type materials. This high concentration makes the junction depletion region (space-charge region) so narrow that electrical charges can transfer across the junction by an action called *tunneling*. This causes a *negative resistance* region to occur on the characteristic curve of the device, which makes it possible to achieve amplification and RF energy generation.

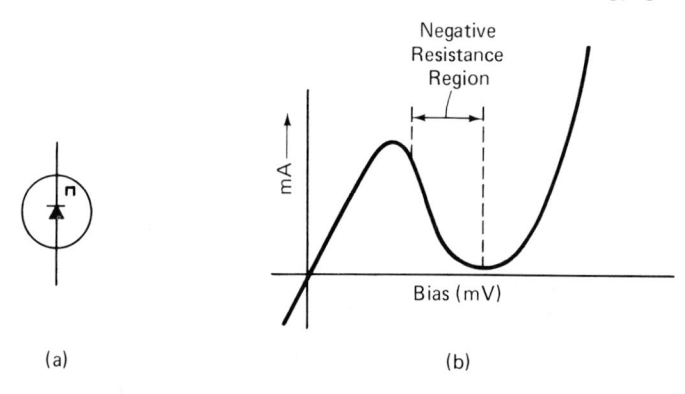

FIG. 10-22 (a) Tunnel diode symbol. (b) Characteristic curve of tunnel diode.

The symbol for a tunnel diode is shown in Fig. 10-22a. In Figure 10-22b can be seen the characteristic curve. By properly biasing the device so that its operation will be confined to the negative resistance region, it will provide amplification.

Practice Problems

1. A Ge transistor has 4.3 mA of collector current flowing when the base current is 15 μA. When the base current was increased to 25 μA, the collector current increased to 5.6 mA. What is the beta of the transistor?
2. What is the beta of a transistor whose collector-current change is 500 to 900 mA when the base current changes from 30 to 40 mA?
3. A certain transistor whose collector family of curves is represented in Fig. 10-7 is used in a circuit where $V_{CC} = 12$ V and $R_L = 1.2$ kΩ. If the base current is 20 μA, what is the quiescent collector voltage and current?
4. In problem 3, if an ac signal of 20 μA$_{p\text{-}p}$ is applied to the base of the transistor, what is the maximum and minimum excursions of collector voltage and current?

Commercial License Questions

5. In the simple bias circuit shown in Fig. 10-9, calculate the ohmic value of R_B under the following conditions: $V_{CC} = -16$ V, $R_L = 2$ kΩ, $Q_1 = $ Si transistor with $\beta = 80$ and $V_{CE} = 6$ V.

6. In the bias circuit shown in Fig. 10-10, calculate the ohmic value of R_B under the following conditions: $V_{CC} = 12$ V, $R_L = 1$ kΩ, $V_{R_C} = V_{CE}$, $R_E = 400$ Ω, and $Q_1 = $ Si transistor with $\beta = 120$.

7. In the bias circuit shown in Fig. 10-11, calculate the ohmic values of R_B and R_S, when $V_{CC} = 12$ V, $R_L = 1$ kΩ, $V_{R_L} = V_{CE} = 5$ V, $R_E = 400$ Ω, and $\beta = 100$.

8. In the bias circuit shown in Fig. 10-12, calculate the ohmic value of R_B under the following circuit parameters: $V_{R_L} = V_{CE}$, $V_{CC} = 16$ V, $R_L = 3.3$ kΩ, $R_E = 1.4$ kΩ, $I_c = 2$ mA, and $\beta = 200$.

9. What is the output conductance of a transistor whose collector current changes from 3.1 to 3.2 mA when the collector voltage is increased by 8 V and the base current remains constant?

10. Find the input resistance of a transistor whose base current changes from 35 to 40 μA when the base–emitter potential increases from 0.675 to 0.682 V and V_{CE} is constant.

11. Determine the reverse transfer voltage ratio of a Si transistor in which a change of 5 V of collector potential produces a base–emitter voltage change of 20 mV when the base current is held constant.

12. A certain Si transistor has a leakage current of 4.6 μA at 25°C. What is the approximate leakage current at 46°C?

Commercial License Questions

Sections in which answers to questions are given appear in parentheses. A bracketed number following a question implies that it applies only to that element.

1. Draw the schematic symbol for a *pnp* transistor. An *npn* transistor.
 (10–1, 10–2, 10–3)

2. Explain why the base–emitter junction of a properly biased transistor is usually a low impedance. (10–2)

3. Why is it that the output impedance of a transistor is a relatively high impedance? (10–2)

4. Explain what is meant by the *beta* of a transistor. (10–2)

5. What is meant by the *collector family* of curves of a transistor? What information can be learned from them? (10–4)

6. Discuss the purpose of load lines. (10–5)

7. Discuss the various circuits used to bias transistors. Draw the schematic diagram of each. (10–6)

8. Draw the schematic diagram of a simple CE amplifier showing proper supply potentials for the type of transistor used. (10–7)
9. What are typical values of input resistance for a CE amplifier stage? (10–7)
10. Explain what is meant by leakage currents in a transistor. How are they affected by temperature? (10–10)

11

Review of Vacuum-Tube Fundamentals

11-1 TRIODE

With the diminishing importance of vacuum tubes in most electronic circuits, except high-powered transmitters, this chapter will provide only a simple review of tube fundamentals.

The *triode* is the simplest of all vacuum tubes. It is a three-element device consisting of a *cathode*, *grid*, and *anode*. A cutaway view showing the physical relationships of these elements and their schematic symbol is shown in Fig. 11-1. Notice that the grid is wound in the form of a *helix* and surrounds the cathode. The function of the cathode is to emit electrons, and the grid controls their flow to the plate.

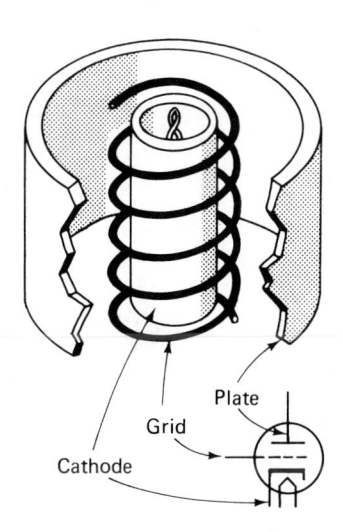

FIG. 11-1 Physical structure and schematic symbol of a triode vacuum tube.

Since the electrons comprising the plate current of a triode must flow between the grid wires, the potentials of both the *grid* and *plate* are effective in controlling plate current. The grid, however, being closer to the cathode, has more control over plate current than does the plate. It is this feature which produces amplification in a triode vacuum tube.

If the plate and grid voltages are adjusted to zero, all tube elements are at the same potential and no electric fields exist between them. The electrons propelled outward from the heated cathode form a *space charge* in the immediate vicinity of the cathode surface. For practical purposes no plate or grid current flows under these conditions.

As the grid potential is made increasingly negative with respect to the cathode, plate current becomes smaller. Eventually, a point is reached where the grid is sufficiently negative to prevent plate current flow and the tube is said to be *cut off*.

11-2 TUBE PARAMETERS

The ability of a small signal, applied to the grid of a tube, to control a relatively large plate current and thereby produce a large output voltage is a measure of the tube's amplifying capabilities. A term used to describe this is *amplification factor*. By definition this is the ratio between a small change in plate voltage to a small change in grid voltage, which results in the same change of plate current. Expressed as a formula,

$$\mu = \frac{\Delta e_b}{\Delta e_c}, \quad i_b \text{ constant}$$

where μ = amplification factor
Δe_b = small change of plate voltage
Δe_c = small change of grid voltage
i_b = plate current

The cutoff bias of a triode can be determined by dividing its μ into the plate voltage, as shown by the formula

$$E_{co} = \frac{E_p}{\mu}$$

where E_{co} = cutoff voltage
E_p = plate voltage

Plate current does not flow through the tube without encountering resistance. This can be calculated by Ohm's law, if only dc values are used, by the formula

$$R_{dc} = \frac{E_b}{I_b}$$

The ac plate resistance, designated r_p, is a more important parameter as it represents the tube's resistance under actual operating or *dynamic* conditions. It is calculated by the formula

$$r_p = \frac{\Delta e_b}{\Delta i_b}, \qquad e_c \text{ constant}$$

The ac resistance is usually less than the dc resistance for a given operating region on the family of curves.

Current flowing through the internal plate resistance of the tube creates heat ($P = i^2 r_p$). Tube manufacturers specify a *maximum plate dissipation* rating for every tube. This is the maximum power the tube can handle without damage to itself. This must not be exceeded as the tube temperature will rise beyond design limits and undesirable performance results. Also, tube life is substantially shortened. High-powered tubes usually require some kind of external cooling to keep temperatures within specified limits.

One of the most effective ways to determine the tube's performance is by measuring its *transconductance*—sometimes called *mutual conductance*. This parameter is the ratio of a change in plate current to the change in grid

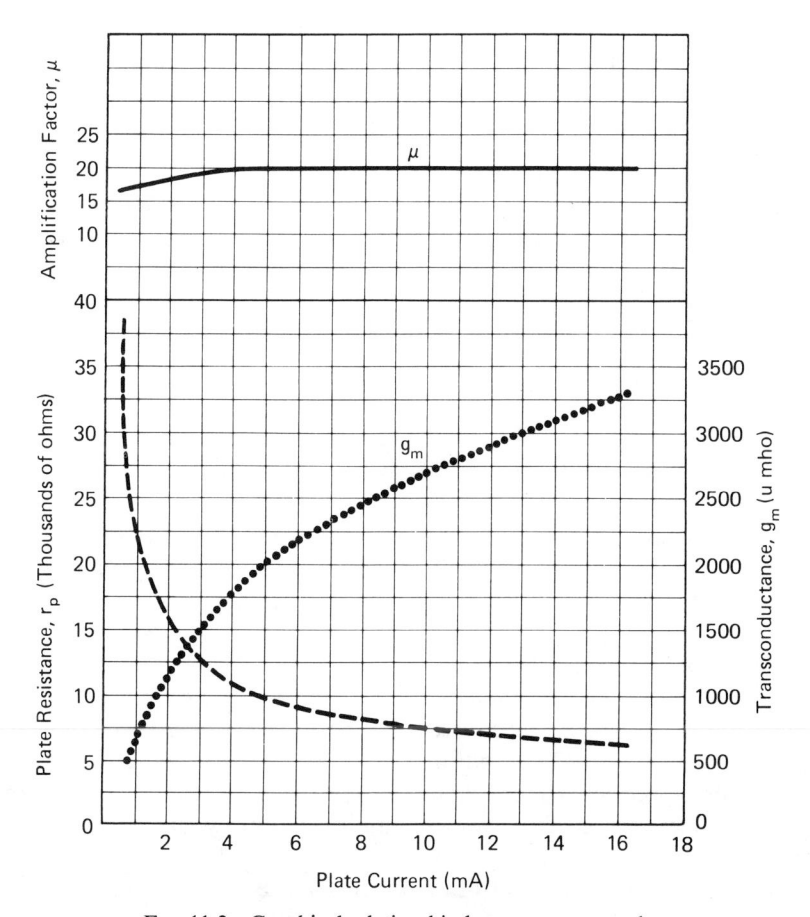

FIG. 11-2 Graphical relationship between μ, g_m, and r_p.

voltage that produced it, with the plate voltage held constant. This can be expressed mathematically as

$$g_m = \frac{\Delta i_b}{\Delta e_c}, \quad e_b \text{ constant}$$

where g_m is the transconductance measured in siemens.

11-3 RELATION BETWEEN μ, g_m, AND r_p

The three triode tube constants, μ, g_m, and r_p, are interrelated. The operating voltages applied to each type of tube determine the exact value of each constant, and for any one set of operating potentials each tube bears these three ratings. They vary in magnitude relative to each other in a definite manner. These data are illustrated in Fig. 11-2 for the 6J5 tube.

The relationship between these three tube parameters is given by the formula

$$\mu = g_m r_p$$

11-4 TETRODES

The relatively large values of interelectrode capacitances in the triode, particularly the grid to plate, impose a serious limitation on the tube as an amplifier at high frequencies. To reduce the grid-to-plate capacitance, a grid called a *screen grid* (G_2) is inserted between the control grid and plate of the tube, as shown in Fig. 11-3. Structurally, the screen grid is somewhat similar in appearance to the control grid.

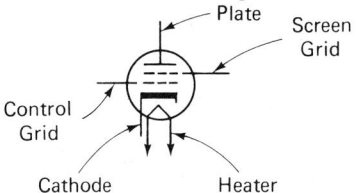

FIG. 11-3 Schematic symbol of a tetrode.

The primary purpose of the screen grid is to act as an *electrostatic shield* between these two tube elements. This effectively divides the grid–plate capacitance into two series capacitors, thereby greatly reducing the value of this capacitance.

To make the screen grid operate properly, it must be connected to either the cathode or chassis (electrostatically grounded). This cannot be a direct connection, however, since the screen grid must be at a positive dc potential for proper operation. This indicates that the external screen-to-cathode circuit must appear as a short circuit to alternating current, but as an open circuit to direct current. These two conditions are met by placing a capacitor between the screen grid and ground, as in Fig. 11-4. This is called a *screen bypass capacitor* (C_{sg}), and must have a capacitance large enough to present a very low reactance to the operating frequency of the circuit.

Pentodes

The ac plate resistance of a tetrode is very high compared to the triode. This is because of the effect the screen grid has on the plate current. The amplification factor is also much higher.

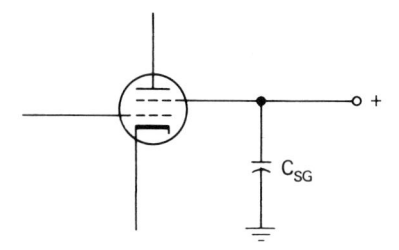

FIG. 11-4 Capacitor used to electrostatically ground the screen grid.

11-5 PENTODES

While the screen grid solved one problem (reduction of C_{gp}), it created another. The positive potential applied to the screen accelerates the electrons more than the plate would acting by itself. The result is that the electrons arriving at the plate have sufficient velocity to knock electrons out of the plate structure. This is called *secondary emission* and is undesirable. These electrons would, in large measure, be attracted to the positive screen, causing excessive screen-grid current. This, in turn, causes overheating of the screen grid and possible electron emission from this element. Under this confusing condition, the tube ceases to function properly.

To overcome the effects of secondary emission, a third grid, called a *suppressor grid*, is added between the plate and screen. This creates a five-element tube called a pentode, whose symbol is shown in Fig. 11-5.

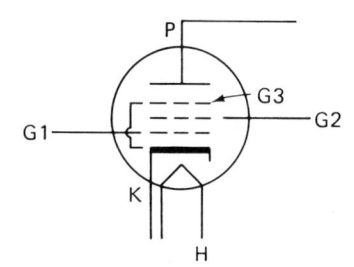

FIG. 11-5 Schematic symbol of a pentode.

A secondary advantage of the suppressor grid is a further reduction in interelectrode capacitance. This increases the available gain and extends the frequency range beyond that of the tetrode.

Connecting the suppressor grid to the cathode places it at zero potential with respect to the cathode, but at a negative potential with respect to the plate and screen grid. Being negative, the suppressor grid serves to repel or suppress secondary electrons from the plate. It also serves to slow down the primary electrons from the cathode as they approach the suppressor. These actions do not interfere with the flow of electrons from cathode to plate, but serve to prevent any interchange of secondary electrons between screen and plate.

Power-amplifier circuits normally use vacuum tubes that are specifically designed for power amplification. One such is the *beam power* tube. The plate characteristics are similar to those of a pentode, except that in the beam power tube the electrons are concentrated into sheets as they are attracted to the plate. The beams are formed by a set of *beam-forming plates* located inside the tube. The schematic symbols for beam power tubes are shown in Fig. 11-6.

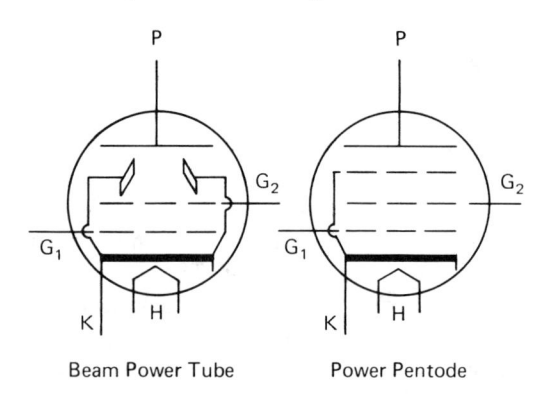

FIG. 11-6 Schematic symbols for beam power tubes.

11-6 MISCELLANY

Tubes that have more than three grids are commonly are referred to as *multigrid tubes*. For instance, if a grid is added to a pentode, a six-electrode tube results, known as a *hexode*. Other multigrid tubes are the *heptode*, which contains five grids, and the *octode*, which has six grids. In these tubes, as in the basic types, the grids are designated in numerical order starting with the control grid. Heptodes, also known as *pentagrid* tubes because of their five grids, are used mostly in frequency converters, or mixer circuits of superheterodyne receivers.

Special tube types are required for use in circuits operating at very high frequencies. For example, the interelectrode capacitances are very critical at high frequencies as they place a limit on the upper end of the frequency spectrum at which the tube can be used. In addition, the leads connecting the tube elements to the base pins have small but definite amounts of inductance that can look like appreciable reactances at high frequencies. Not to be overlooked is the *transit time* of the electrons in passing from cathode to plate. If this time interval becomes appreciable in respect to the applied frequency, the normal phase relationships of plate and grid voltages cease to exist. Special tubes have been developed for use at these high frequencies, such as the *lighthouse* and *acorn* tubes.

Occasionally, over a long period of time, a vacuum tube may lose part of its vacuum. When this happens, the tube is said to be *soft*. This is generally indicated by the presence of a soft bluish glow inside the tube (not visible in metal envelope tubes). Gassy, or soft tubes, tend to run hotter, are less efficient, and produce more internal noise.

Overheating in a tube is usually the result of any one of the following causes or any combination of them: excessive plate voltage or current, insufficient bias, overheated cathode, screen voltage too high, gas, or inadequate ventilation.

It is desirable to use an ac source of power for the filament supply of vacuum tubes because it is so readily available. Power transformers are almost always needed to step up the voltage so that after rectification and filtering the required plate supply voltage is available. It is easy for the transformer manufacturer to add a small step-down winding to the power transformer to furnish the required filament or heater voltage. If it were not, it would be necessary to have some other source of power, such as a battery or generator, and these have many limitations.

It is possible, in many cases, to reactivate the cathode or heater of a vacuum tube if it has an oxide coating such as thorium. This will partially restore the tube to its original emission. Reactivation is achieved by removing the tube from the circuit and applying a heater voltage about 25 per cent higher than normal. The additional heat produced "boils" some of the thorium out to the surface of the cathode to replace that which has disintegrated over a long period of usage. After operating the heater at this elevated temperature for several minutes, the voltage should be slowly reduced to normal. This completes the reactivation process.

Practice Problems

1. From the plate family of curves for the triode shown in Fig. 11-3 graphically determine the tube's μ at an operating point of -2-V bias and a plate voltage of 150 V.
2. Referring to problem 1, what is the approximate transconductance of the tube?
3. Referring to problem 1, what is the approximate dynamic plate resistance of the tube?
4. A certain tube has a plate current of 7.75 mA at zero bias and 3.8 mA at -4 V of grid bias. The dynamic plate resistance is 15.2 kΩ and the plate voltage remains constant. Calculate the μ of the tube.
5. The r_b of a certain tube is 0.15 MΩ when $e_b = 200$ V. The i_b changed from 12 to 19.5 mA when the grid bias was changed from -2 to -1 V. Calculate the transconductance of the tube if the plate voltage remains constant.
6. The i_b of a certain tube is 4.5 mA when $e_b = 90$ V. When the plate voltage is lowered to 40 V, $i_b = 0.9$ mA. Assuming no change in E_c, calculate the plate resistance of the tube.

Questions 7–9 refer to the following problem:

The data shown were recorded when a certain vacuum tube was tested:

E_b (V)	E_c (V)	I_b (mA)
250	−8	8
215	−6	8
250	−7	10

7. Calculate the amplification factor of the tube.
8. Calculate the transconductance of the tube.
9. Calculate the dynamic plate resistance.
10. A certain pentode operates from a supply voltage of 400 V. If $I_b = 3.2$ mA when $E_b = 250$ V and 0.8 mA of screen-grid current flows at a specified $E_{c2} = 100$ V, what is the value of the screen dropping resistor?

Commercial License Questions

Sections in which answers to questions are given appear in parentheses. A bracketed number following a question implies that it applies only to that element.

1. Describe the physical structure of a triode vacuum tube. (11-1)
2. What is the meaning of *electron emission*? (11-1)
3. What is *space charge* in a vacuum tube? (11-1)
4. What is the direction of electronic flow in the plate and grid circuits of vacuum-tube amplifiers? (11-1)
5. What is the meaning of the term *maximum plate dissipation*? (11-2)
6. What is the meaning of *mutual conductance* and *amplification factor* in reference to vacuum tubes? [4] (11-2)
7. What is the purpose of a screen grid in a vacuum tube? [4] (11-4)
8. Describe the physical structure of a tetrode vacuum tube. (11-4)
9. What is meant by *secondary emission* in a vacuum tube? [4] (11-5)
10. What is the primary purpose of a suppressor grid in a multielement vacuum tube? (11-5)
11. Describe the electrical characteristics of the pentode, tetrode, and triode. (11-1 to 11-5)
12. What are some possible causes of overheating vacuum-tube plates? (11-6)
13. What are the visible indications of a soft tube? (11-6)
14. What is meant by a *soft vacuum tube*? (11-6)
15. Why is it desirable to use an alternating current filament supply for vacuum tubes? (11-6)
16. What kind of vacuum tube responds to filament reactivation, and how is reactivation accomplished? (11-6)

12

Power Supplies

12-1 HALF-WAVE RECTIFIERS

Virtually all electronically operated equipment requires some form of power supply to furnish the necessary dc voltages for operation. Although this may come from batteries or generators, it customarily comes from power supplies.

Since a diode will pass current in only one direction, it is ideally suited for *converting* alternating current to direct current. If a sine wave of voltage is applied to a diode, it will conduct only during the positive alternation when the anode is positive relative to the cathode, as shown in Fig. 12-1a. Current flows during the entire period of time that the anode is positive. During the negative half-cycle (dotted polarity signs), the anode is negative and the diode is cut off. The circuit current is a series of positive pulses flowing in the same direction. The output is therefore pulsating direct current. This is shown in the waveform in Fig. 12-1b.

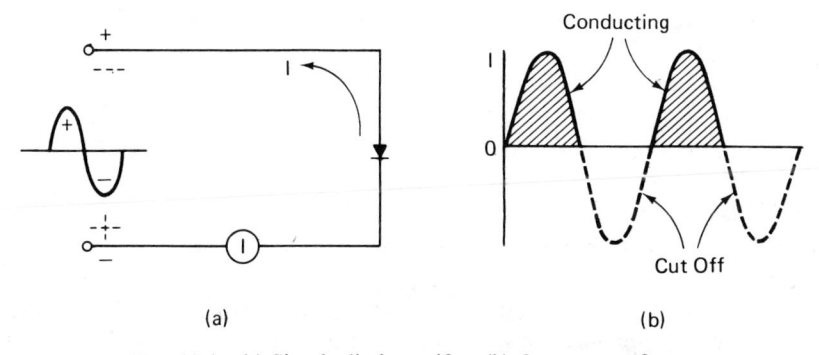

FIG. 12-1 (a) Simple diode rectifier. (b) Output waveform.

To utilize the diode as a rectifier, it is connected in series with the load through which the direct current is to flow. In many cases it is necessary to have a rectified voltage greater (or smaller) than the source voltage; therefore, the rectifier is often supplied power from a step-up (or step-down) transformer. A schematic diagram of a half-wave rectifier circuit is shown in Fig. 12-2. The rectified output is developed across load resistor R_L, which is connected in series with the diode and transformer secondary. The voltage across R_L depends upon the current flowing through it. The magnitude of this potential is nearly equal to the peak value of the supply voltage, as very little voltage drop occurs across the diode when conducting. Assuming the secondary emf of the transformer is 100 V_p, then about 99 V appears across R_L.

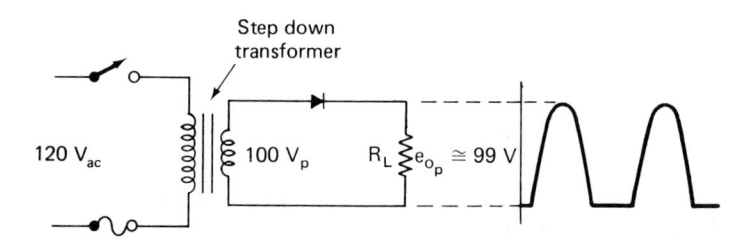

FIG. 12-2 Half-wave rectifier circuit and output waveform.

On the negative half-cycle the diode is reverse biased, and its internal resistance is very high compared to R_L. Consequently, the supply voltage appears across the diode and virtually no voltage is across R_L. The diode must be capable of withstanding this high *peak inverse voltage* without breaking down. This characteristic is one of the important specifications of any diode.

The pulsating dc voltage across R_L must be smoothed out or filtered before it can be used. This will be discussed in a subsequent section. A vacuum-tube rectifier could be used in place of the solid-state diode. In this case the transformer would need an additional winding to supply the filament or heater potential.

All rectifiers have some *internal resistance*. This varies according to the amount of current flowing through the device. For the vacuum-tube diode a typical value of plate resistance (the dynamic resistance between plate and cathode, designated r_p) is 500 Ω. For the amount of peak plate current flowing the drop across the diode's internal resistance (r_p) is approximately 21 V_p. This is about 5 per cent of the total supply voltage, which is not considered excessive. The power dissipated within the tube ($P = I^2 r_p$) produces heat, which materially contributes to the overall temperature rise of the device.

The solid-state diode has advantages over its vacuum-tube counterpart. Besides its much smaller size, it has considerably less internal resistance.

Basic Full-Wave Rectifier Circuit

For example, a silicon power diode handling approximately the same amount of current would have an internal resistance of about 20 Ω. This means that its *IR* drop will be something around 0.7 V, which provides for more rectifying efficiency. Another very significant factor is that the *dissipation loss* is much less. Hence, the solid-state diode does not tend to heat up nearly as much. When these diodes are used in high-current circuits, they generally have some form of stud mounting so they can be attached to an appropriate *heat sink* to radiate the heat generated.

12-2 BASIC FULL-WAVE RECTIFIER CIRCUIT

The half-wave rectifier utilizes the transformer during only one half the cycle and, therefore, for a given-sized transformer, less power can be developed than if the transformer were utilized on both halves of the cycle. This disadvantage limits the use of the half-wave rectifier to applications that require a small current drain.

A full-wave rectifier is a device that has two diodes arranged so that the load current flows in the same direction during each half-cycle of the ac supply.

A simple full-wave rectifier circuit is shown in Fig. 12-3a. Load resistor R_L is common to both diodes. During one half-cycle the polarity across the secondary winding will be as shown by the solid plus and minus signs. Under these conditions current flow will be from the lower end of W_1, through the load resistor and CR_1 back to the upper end of W_1, as indicated by the solid arrow. CR_2 cannot conduct, because W_2 is supplying a *reverse bias* to it.

On the alternate half-cycle the polarities are reversed, as shown by the dotted plus and minus signs. Now CR_2 will conduct and CR_1 will be idle. Current will flow from the upper end of W_2, through R_L and CR_2 to the bottom end of W_2, as shown by the dotted arrow. Notice that the rectified cur-

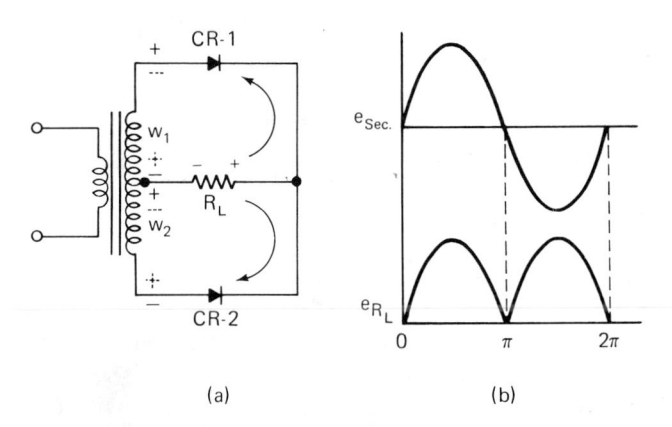

FIG. 12-3 (a) Simple full-wave rectifier circuit. (b) Voltage waveforms across secondary and load.

rent flowing through R_L is in the same direction when CR_2 conducts as when CR_1 is conducting. A pulse of current flows during each alternation of every input cycle. Since both alternations of the input cycle are used, the circuit is called a *full-wave rectifier*. The voltage waveform appearing across the load resistor is shown in Fig. 12-3b.

12-3 BRIDGE RECTIFIER

If four diodes are connected as shown in Fig. 12-4, the circuit is called a *bridge rectifier*. The input to the circuit is applied to diagonally opposite corners of the network, and the output is taken from the remaining two corners. During one half-cycle point A becomes positive with respect to point B by the amount of voltage induced into the secondary of the transformer. This forward biases diodes CR_1 and CR_3, causing a pulse of current flow through them and R_L, which is in series. The solid arrows indicate the direction of this rectified current.

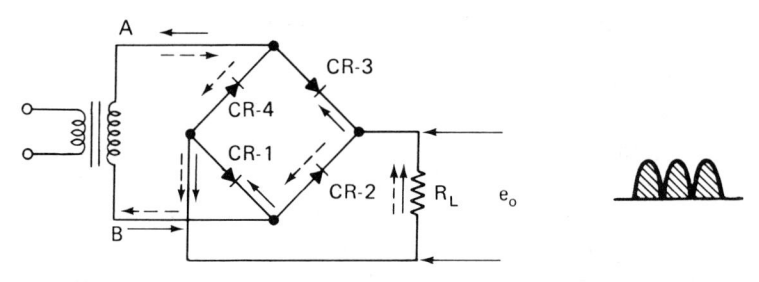

FIG. 12-4 The bridge rectifier.

One half-cycle later diodes CR_2 and CR_4 are forward biased, and current flows in the direction indicated by the dashed arrows. During this interval CR_1 and CR_3 are reverse biased. Notice that the rectified currents from both pairs of diodes always flow in the *same direction* through the load. The ripple frequency of the output voltage is *twice* the line frequency.

The bridge rectifier has several advantages over the full-wave rectifier shown in Fig. 12-3. These are

1. For a given transformer nearly twice the output voltage.
2. The secondary winding need not be center tapped.
3. Lower ratio of peak inverse voltage to average output voltage.

The basic vacuum-tube bridge rectifier operates on the same principles as described for the solid-state unit. A simplified circuit is seen in Fig. 12-5a. The solid and dotted arrows indicate the directions of rectified current flow for each half of the input cycle. In the actual circuit of Fig. 12-5b you will observe that *three* filament transformers are required for the four tubes.

Bridge Rectifier

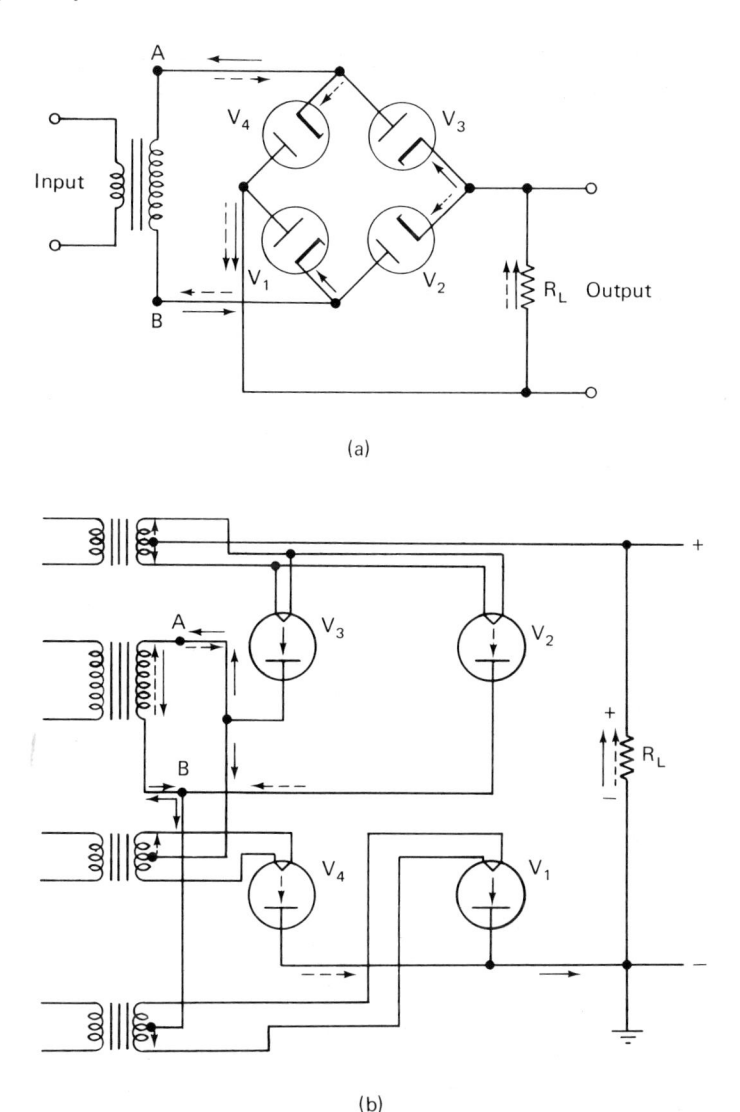

FIG. 12-5 (a) Simplified circuit of vacuum tube bridge rectifier.
(b) Actual circuit showing need of separate filament transformers.

Rectifiers V_2 and V_3 are operated at the same relative potential and therefore are connected to the same filament winding.

The filaments of V_1 and V_4, however, are returned to opposite ends of the high-voltage secondary of the power-supply transformer, and therefore operate at the full potential difference that exists across the load. Thus, if the filaments of V_1 and V_4 were supplied by a single filament transformer winding, the common connection would short circuit the load. Therefore, the filament windings of V_1 and V_4 must be insulated from each other to withstand the full output voltage across the load. If the lower end of the load, R_L, is grounded, filament windings for V_1 and V_4 operate alternately at the full difference of potential of the high-voltage secondary with respect to ground. Thus, V_1 and V_4 must be well insulated from ground also.

12-4 WAVEFORM ANALYSIS

If a series of current pulses like those obtained from a half-wave rectifier are applied to a load resistance, some average amount of power will be dissipated over a given period of time. For the output of a half-wave rectifier, such as shown in Fig. 12-6, this average is 0.318 of the peak value. It is the condition when the area designated A of one half-cycle equals area B. Thus, the pulses of current or voltage have the same effect on the load as a steady current or voltage having a value equal to 0.318 of the peak value of the pulses.

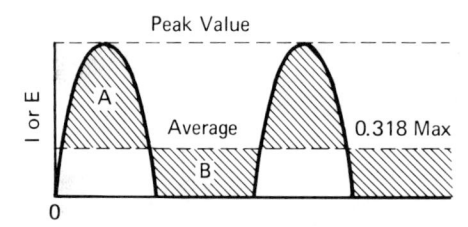

FIG. 12-6 Peak and average values for a half-wave rectifier.

The average value of current or voltage at the output of a full-wave rectifier is *twice* that of the half-wave circuit, or approximately 0.637 of the peak value. Expressed mathematically,

$$E_{avg} = 0.637 E_{max}$$

This is shown in Fig. 12-7.

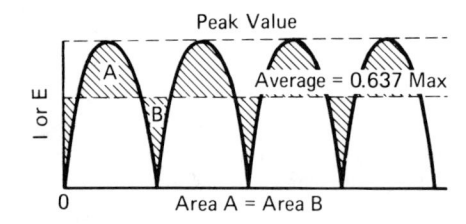

Fig. 12-7 Peak and average values for a full-wave rectifier.

Example 12-1

What is the average value of output voltage of a full-wave rectifier if the peak value is 163 V?

Solution

$$E_{avg} = 0.637 \times 163 = 103.8 \text{ V}$$

Capacitance Input Filter

Example 12-2

A power transformer delivers 300 V on either side of center tap to a full-wave rectifier. Neglecting the drop across the diodes, find the average load voltage.

Solution

First, convert the rms voltage to peak:

$$E_p = 1.414 E_{rms}$$
$$= 1.414 \times 300 = 424 \text{ V}$$

The average value then is

$$E_{avg} = 0.637 \times 424 = 270 \text{ V}$$

12-5 CAPACITANCE INPUT FILTER

The pulsating voltage and current delivered by the rectifier must be filtered to remove the ripple present. A simple method of smoothing out these fluctuations is to place a large capacitor across the load, as shown in Fig. 12-8a. As the output voltage rises, the capacitor begins to charge. It charges nearly as fast as the rate of rise of the voltage due to the limited reactance of the transformer secondary and low resistance of the diode.

When the rectifier output drops to zero, the voltage across the capacitor does not fall immediately. Instead, the energy stored in the capacitor is discharged through the load during the time that the rectifier is not conducting. If the load resistance is sufficiently high, the current drawn from the capacitor will be limited, due to the relatively long RC discharge time. Thus, the amplitude of the ripple is reduced. Figure 12-8b graphically shows this action. The heavy line shows the voltage variation across the load.

If a full-wave rectifier were used, the output pulses would be doubled, as shown in Fig. 12-7. Under these conditions the capacitor would not be able to discharge as much (as with the half-wave rectifier) before the next

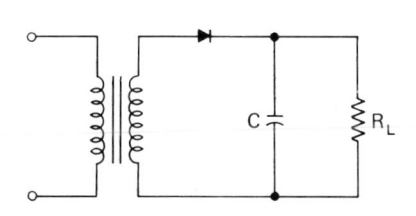

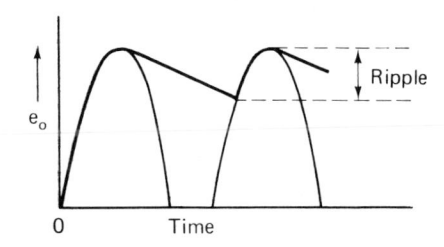

FIG. 12-8 (a) Half-wave rectifier with capacitor filter. (b) Output waveform with capacitor present.

pulse comes. Consequently, the amplitude of the ripple would be substantially reduced. This is a decided advantage of the full-wave rectifier.

After the capacitor has been charged (during the conducting period of the diode), the positive charge on the top plate of the capacitor holds the diode cathode positive, and the rectifier will not conduct again until a positive-going secondary voltage at the anode *exceeds* the positive voltage at the cathode. Hence, the diode does not begin to conduct until the secondary voltage of the transformer exceeds the charge on the capacitor. Diode current flows until slightly after the peak of the sine wave. This current flow is of short duration, but is sufficient to restore the charge on the filter capacitor. From this brief explanation we see that only short pulses of current actually flow through the diode—and the transformer. The diode must have a current rating sufficient to safely handle these pulses.

Capacitor input filtering, so called because the capacitor is the first (and only, in this case) filter element the diode "sees," has limitations. The amount of ripple is a function of the size of the capacitor and the value of load resistance. Large values of capacitance (usually several hundred microfarads) will reduce the amplitude of the ripple to a low value if the load resistance is fairly high (low load). If the load is relatively large (low resistance), the capacitor discharges more rapidly, resulting in increased ripple amplitude. Therefore, this type of filtering can only be used successfully when the current demands are low.

12-6 INDUCTANCE INPUT FILTER

In large current applications the *inductance* or *choke input* filter is used. A typical circuit is shown in Fig. 12-9. It is sometimes called an L-section filter because the L and C filter components form an inverted L.

Because an inductor resists changes in the magnitude of the current flowing through it, a large inductor placed in series with the rectifier output will absorb abrupt changes in the magnitude of the current supplied to the load. If the load current tries to fall, the magnetic field about the inductor begins to collapse and induces a voltage in the inductor, which will oppose the current change. Likewise, as the load current tries to increase, the mag-

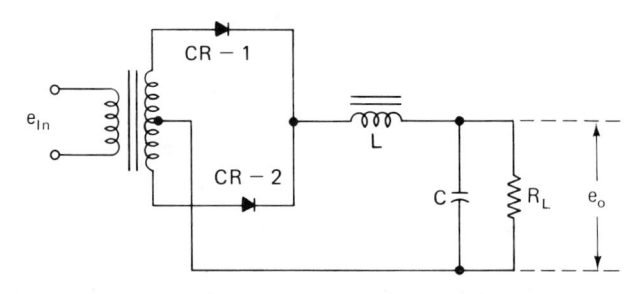

Fig. 12-9 Full-wave rectifier with L-section (choke input) filter.

Pi-Section Filter

netic field increases and produces a voltage in the inductor, which will again oppose the current change. Thus, the addition of the inductor prevents the current from building up or dying down quickly. If the inductor is made large enough, the load current becomes nearly constant.

The ripple amplitude across R_L in Fig. 12-9 will be less than if only the inductor were used, because of the additional filtering action of C. The large value of this electrolytic capacitor can also be thought of as bypassing any ripple frequency around R_L that may have passed through the choke. Typically, the reactance of C should be about one tenth the value of R_L to make it effective.

12-7 PI-SECTION FILTER

A typical *pi-section* filter is shown in Fig. 12-10. This filter arrangement is more effective than the ones previously discussed. It derives its name from its resemblance to the Greek letter π. This is also known as a *capacitor input* filter. Occasionally, when small loads are involved, the inductor is replaced by a resistor. Capacitor C_1 acts to filter the greatest portion of the ripple com-

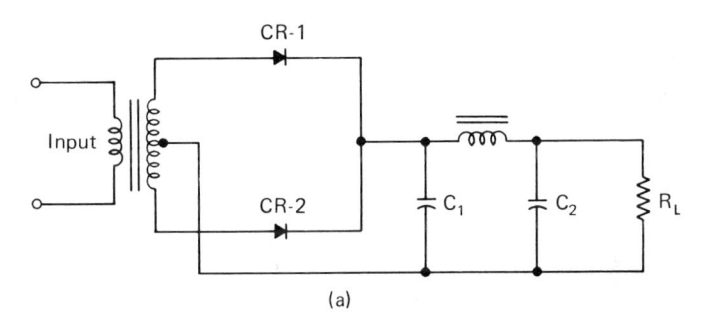

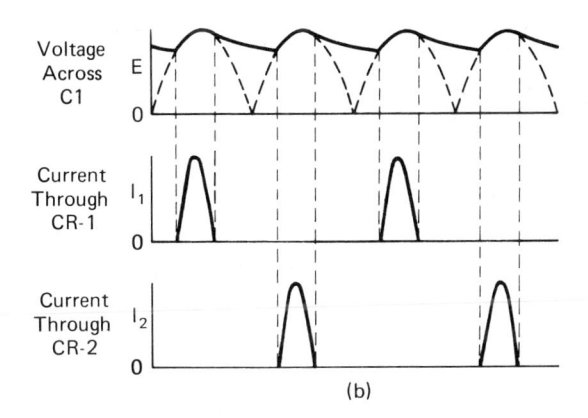

FIG. 12-10 (a) Full-wave rectifier with pi-section filter. (b) Voltage and current waveforms.

ponent. In all filters the major portion of the filtering action is accomplished by the input component.

The waveforms encountered in a typical pi filter are shown in Fig. 12-10b. Current flow through the diodes consists of a series of sharp, peaked pulses, because C_1 acts like a short circuit across the load while it is charging. Because of this high peak load on the rectifier tubes, the pi-section filter is used only in low-current equipments, such as radio recievers and small amplifiers. In power supplies designed to furnish relatively large currents, it is customary to add a small amount of resistance (1 to 10 Ω) in series with the diode(s) to *limit* the initial current flow and thereby prevent damaging the rectifiers. Output capacitor C_2 acts to bypass residual fluctuations existing after filtering by the input capacitor and series inductor.

12-8 MISCELLANEOUS FILTER CIRCUITS

Several variations exist in the arrangement of capacitors and inductors in filter circuits. Usually, the more filter elements, the better the filtering. Five possible combinations of filters are shown in Fig. 12-11.

Filters a and b are choke input types. The circuit in (b) is not customarily found, as there is a possibility that *hum* at the power-line frequency may be developed across it. The positioning of choke coils, relative to each other and to the power transformer, is important. Their stray magnetic fields may induce voltages into each other that are difficult to filter out. This problem can usually be eliminated by (1) using *fully shielded* transformers and chokes, and (2) by mounting them so that their magnetic fields are at right angles to each other.

Filters c, d, and e are all capacitor input types and will provide excellent filtering. In fact, some filters may provide more filtering than required for a particular installation. For example, very high gain amplifiers (those providing high amplification) need well-designed filters lest they amplify part of the ripple along with the signal. Low-gain and power amplifiers have less critical filtering requirements. Therefore, it is expedient to chose the proper type of filter circuit for the application.

The filter shown in Fig. 12-11d is unusual in that it uses a parallel resonant circuit in series with the output current. By selecting the proper values of L and C, the circuit can be designed to resonate at the ripple frequency. This affords a very high impedance and is effective in blocking ripple or hum from getting into associated circuits.

All filter circuits have capacitors across the output. This not only helps to improve the filtering, as previously explained, but causes the *output impedance* of the filter (and therefore the power supply) to be *very low*. Ideally, a power supply's output impedance would be zero. This shorts to ground any signal trying to get into another part of the circuit, via the common power supply, and thereby prevents oscillation.

Filter Chokes

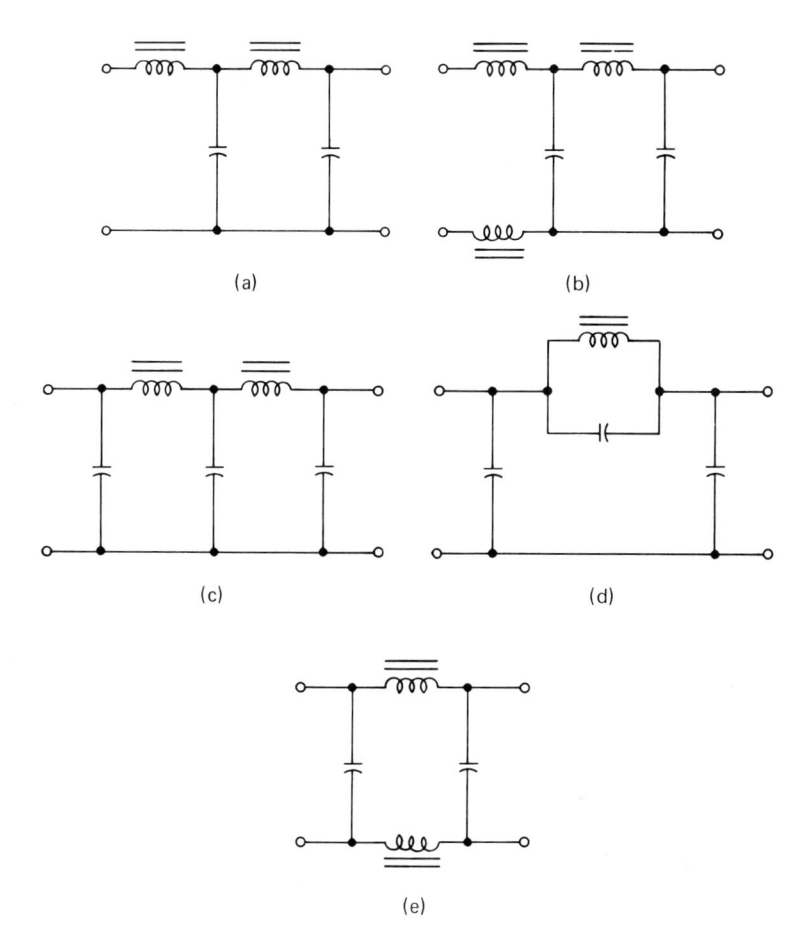

FIG. 12-11 Miscellaneous filter circuits for power supplies.

12-9 FILTER CHOKES

There are a number of factors to consider in selecting a filter choke. Some of the more important ones are value of inductance required, dc resistance and current-carrying capacity of the winding, insulation, losses, and distributed capacitance. A choke coil must present a high reactance to the ripple frequency. This means that the choke should consist of a large number of turns. While this may provide the necessary inductance, it might also create a winding having high dc resistance, resulting in excessive voltage losses across the coil. Of course, it is possible to use larger wire. However, for a given amount of inductance this will increase the physical size of the choke.

Choke coils are designed to provide a certain amount of inductance based on a given amount of dc current flowing through the winding. If the current falls below the specified value, the inductance increases. Conversely, if more current flows through the winding than it was designed for, saturation results with a substantial loss of inductance. This materially reduces the efficiency of the filter.

The core is *laminated* to reduce *eddy currents*. Soft iron is used, which has low hysteresis losses. The choke is frequently encased in a soft-iron shell to contain any stray magnetic fields.

It is good engineering practice to mount the filter choke and power transformer at the opposite end of the chassis from low-level signal circuits, thus reducing the possibility of induced hum. In some circuits where the signal levels are high, it is not necessary to use fully shielded choke coils. The distributed capacity should be low so that part of the ripple frequency (hum) will not be coupled across the choke.

Large reactive voltages may be built up across the choke; therefore, adequate insulation is needed. It is customary to have the insulation resistance able to withstand two or three times the normal values of reactive voltages appearing across the choke. For small-power applications this is about 1500 V_{dc}. For transmitter power supplies it will be much greater.

There are many electronics circuits whose current demands vary over a wide range. This means that the power supply must be capable of furnishing small and large values of current. To prevent the choke from saturating under these extreme conditions, the core is generally constructed with a very small *air gap*. This may be only a few thousandths of an inch, but it is sufficient to prevent the core from becoming saturated.

A *swinging choke* is a reactor so designed that its inductance will vary widely under changing load conditions. It is generally used as the first filter choke in such power supplies. The advantage of a swinging choke is that it acts to improve the voltage regulation of the supply. The amount of inductance required in the choke varies *inversely* with the load current. This means that the inductance must be high when the load is light, and vice versa. This quality is somewhat inherent in any iron-core choke, so that in some respects all seem to act as swinging chokes. However, by eliminating the air gap so that the iron core saturates when the load is large, it is possible to obtain a wide variation in inductance between low load and full load.

The low value of inductance for full load causes the filter to have the characteristics of a capacitor input filter, resulting in a tendency for the output voltage to rise. This compensates for the natural tendency of the supply voltage to drop with an increase in load. Thus, the regulation of the power supply is improved by using swinging chokes. A swinging choke might have an inductance as low as 5 H under full load and as high as 25 H at low loads.

12-10 VACUUM RECTIFIERS

The vacuum rectifier is used mostly for replacement purposes or in high-voltage power supplies for some large transmitters. These high-voltage rectifiers contain but one anode and cathode. The heater leads are brought out of the base, whereas the plate lead comes out of the top of the envelope.

A typical full-wave vacuum rectifier and filter are shown in Fig. 12-12.

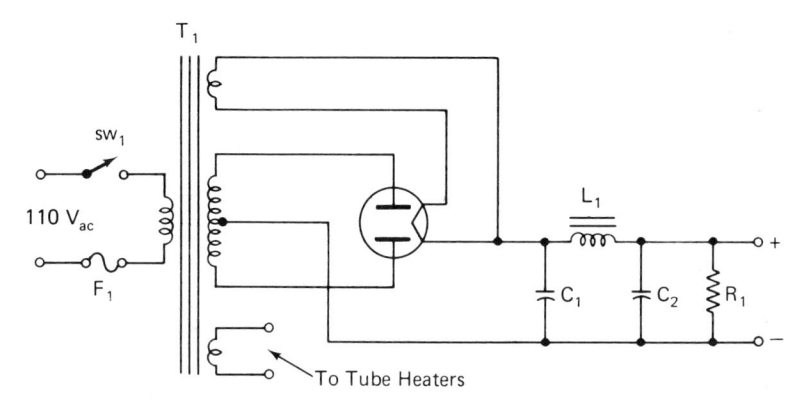

FIG. 12-12 Typical full-wave vacuum tube rectifier and filter.

The operation of this supply is essentially the same as the solid-state unit shown in Fig. 12-10. Transformer T_1 needs a separate winding to furnish filament power to the rectifier. The high-voltage secondary winding customarily develops about 600 V, which is connected to the rectifier plates. The rectifier output is delivered to the filter circuit. Typical values for the filter are $C_1 = 10$ to $20 \,\mu\text{F}$ at 450 V_{dc}; $L_1 = 5$–10 H; $C_2 = 20$ to $40 \,\mu\text{F}$ at 450 V_{dc}. Bleeder resistor R_1 is usually around 20 kΩ and serves to improve output regulation, as well as to discharge the filter capacitors when the supply is turned off. The lower secondary winding furnishes heater circuit power to other tubes in the equipment.

Vacuum rectifiers are characterized by the following:

1. More internal resistance than the solid-state diode and hence a larger voltage drop for a given load condition.
2. The voltage drop across the tube(s) is directly related to the current passing through it.
3. The power dissipated may be significant under high-current conditions. Added to this is the heat generated by the filament. The plate structure must be large enough to radiate much of this heat, or forced cooling may be required, or both.
4. For indirectly heated cathodes a warm-up time of 10 to 15 s is required before normal emission occurs. For filamentary type diodes, particularly low-power ones, conduction begins almost as soon as power is applied.
5. The rectifier works equally well into either a choke or capacitor input filter.

6. In high-power applications it is customary to have special circuits designed to bring the filament temperature up gradually when power is initially applied. This tends to relieve thermal stresses in the filament and prolong the life of the tube.
7. Occasionally a light-blue or purple glow can be seen between plate and filament, indicating the presence of gas. This results from the tube losing part of its vacuum. If the glow is faint, there may be little impairment of operation, although the tube should be frequently observed as there may be an incipient failure.

12-11 MERCURY-VAPOR RECTIFIERS

The relatively small amount of current-handling capability of the vacuum rectifier is overcome by the *mercury-vapor* rectifier. It can handle much more current, with greater efficiency, primarily because of the low voltage drop across the tube.

The space charge surrounding the cathode of the vacuum rectifier acts somewhat as a shield, impeding the flow of electrons between cathode and plate. This results in a large voltage drop. In the mercury-vapor rectifier a small amount of mercury is introduced into the tube envelope. Because of the *low pressure* within the tube, the mercury *vaporizes* as the tube reaches normal operating temperature. Electrons traveling toward the plate collide with many of the mercury-vapor atoms, creating positive ions. These are drawn toward the cathode, where they *neutralize* the space charge. Without the space charge the primary electrons are able to move to the plate with little opposition, thus resulting in a large plate current.

Ionization of mercury vapor occurs when the potential between plate and cathode is 10.4 V. Any additional increase in plate voltage will ionize more atoms—each in turn neutralizing a space electron—until the drop reaches 15 V. At that time the tube will no longer show a plate-voltage rise for a proportional rise in current.

Figure 12-13 shows the i_p-e_p relation during ionization. This shows that the voltage drop remains at a constant value of 15 V regardless of the current flowing through the tube, provided the rated tube current is not exceeded. On overload the voltage drop increases to some extent. When the tube drop exceeds 22 V, the filament may be damaged by excessive bombardment of positive ions.

Another important characteristic of the mercury-vapor tube is its *maximum inverse voltage rating*. This is the sparking voltage through the mercury vapor in a direction opposite to that of normal flow and is always less than it would be if the vapor were not present. A mercury-vapor tube always has a lower flashback voltage than a high-vacuum tube of similar construction. Nevertheless, mercury-vapor tubes with high inverse-voltage ratings have been developed. For example, mercury-vapor diodes having an

Mercury-Vapor Rectifiers

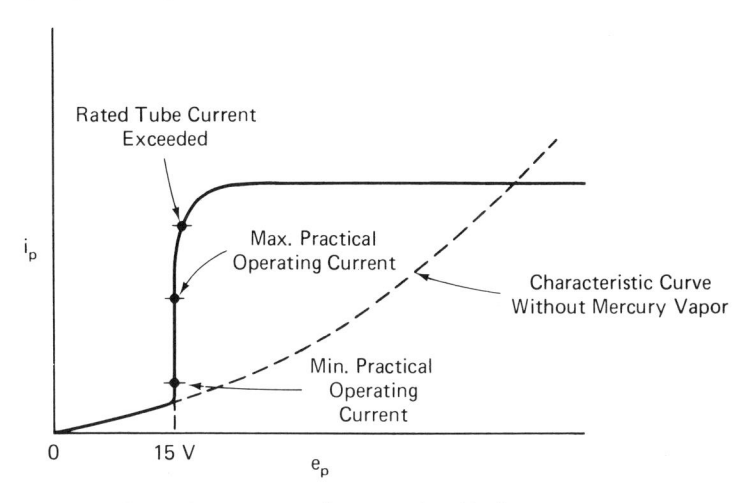

FIG. 12-13 Current vs. voltage relationship in a mercury-vapor rectifier.

output of 10 A and a maximum safe peak inverse voltage of 22,000 V are used extensively in broadcast transmitter power supplies.

In mercury-vapor rectifiers the vapor must reach its proper operating temperature before plate voltage is applied. If this precaution is not taken, the high voltage drop across the tube causes *secondary emission* from plate to cathode, and arc-back occurs. The cathode of the mercury-vapor tube must not only emit free electrons, but must also heat the surrounding space in order for the mercury-vapor temperature to be in the operating range of 20 to 60°C.

If the temperature becomes excessive, the gas pressure will likely increase to a point where the inverse ionization potential will fall below the peak voltage of the power transformer, causing flashback.

A typical full-wave mercury-vapor circuit is shown in Fig. 12-14. The circuit is essentially the same as any conventional full-wave rectifier. The small dot inside each rectifier tube implies the presence of gas. Switch S_1 must be closed first to apply filament power to the rectifiers. When the tubes have reached operating temperature (20 to 30 s), switch S_2 can be closed to apply plate voltage. This causes the tubes to give off a characteristic *bluish light*. The plate current that immediately begins to flow has a *steep leading edge* and represents the equivalent of a high-frequency pulse. If capacitor input were used, the very low reactance would essentially constitute a short circuit to such a high-frequency current. This would present a severe overload on the transformer–tube–capacitor circuit and cause *double ionization*. This phenomenon occurs when two electrons are knocked out of the mercury atoms. The resulting positive ions are attracted with twice the velocity toward the cathode, resulting in rapid deterioration of the cathode. Consequently, only choke input filtering can be used with mercury-vapor rectifiers.

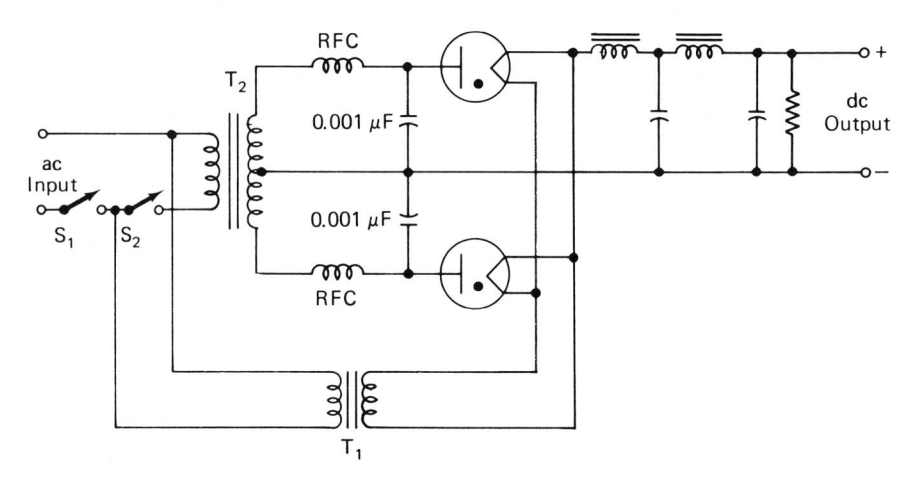

FIG. 12-14 Typical full-wave mercury rectifier with hash filter.

Although choke input greatly minimizes the peaks in plate current, there still remains a slight current surge at the moment of ionization. This results from the distributed capacitance in the circuit, mainly in the choke and power transformer. This, in conjunction with the circuit inductance, produces a *damped oscillatory wave* each time the rectifier passes a pulse of current. The signal produced by this action is similar to that of a spark transmitter, and radiation occurs over a wide part of the RF spectrum, producing interference in nearly all kinds of electronic equipment. An effective way to eliminate this is to place RF chokes in the plate circuit of each rectifier with mica bypass capacitors connected, as shown in Fig. 12-14. Typical values of these components are 10 to 20 mH for the chokes and 0.001 to 0.005 μF for the capacitors.

In practice it frequently is necessary to connect two mercury-vapor rectifiers in parallel to meet the current-handling requirements of the circuit. Although it has been stated that the *IR* drop across a tube is nominally 15 V, it might very well be that one tube will have 14.9 V while another has 15.1 V. Under these conditions the tube that strikes first will cause the same potential to be applied to the other tube (i.e., 14.9 V). The second tube will never ionize, and the first one will likely soon be burned out due to its taking the entire load.

To eliminate this, a resistor (typically 5 to 50 Ω) is connected in the plate lead of each rectifier. With these resistors in the circuit the voltage drop across the tube with the higher ionizing potential will be the 14.9 V across the other tube plus the *IR* drop across the resistor in series with the 14.9-V tube. This emf will inevitably be larger than 15.1 V, and hence this tube will start to conduct and assume its half of the load. For this reason these are known as *equalizing resistors*.

12-12 PEAK INVERSE VOLTAGE

The *maximum peak inverse voltage* is the highest instantaneous voltage that a diode can withstand in the nonconducting direction. For gas-filled and mercury-vapor tubes, it is the safe top value to prevent arc-back within the specified temperature range. Depending on design, the peak inverse voltage (PIV) may range from a few hundred to many thousands of volts.

The following example illustrates how PIV can be calculated for a full-wave rectifier circuit.

Example 12-3

Determine the PIV that the diodes in Fig. 12-15 are subject to under the conditions indicated.

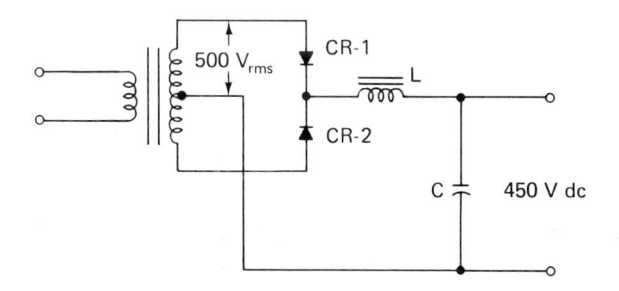

FIG. 12-15 Full-wave rectifier circuit used to illustrate PIV.

Solution

The peak value of the voltage on either side of center tap is $500 \times 1.414 = 707$ V. Capacitor C is charged to 450 V, which is the approximate cathode potential (neglecting the IR drop across L). Assume that diode CR_2 is nonconducting at the moment its anode is a maximum of 707 V *below* ground, while its cathode is 450 V above ground. Therefore, the potential difference between CR_2's anode and cathode is a maximum of $450 \text{ V} + 707 \text{ V} = 1157$ V. This represents the PIV the diode is subjected to. On the alternate half-cycle the conditions are reversed. A diode must be selected whose maximum PIV rating is greater than this in case flashback should occur.

Example 12-4

If mercury-vapor rectifier tubes were used, what PIV would they be subject to?

Solution

The peak voltage across the entire secondary is $1000 \text{ V} \times 1.414 = 1414$ V. Assuming that the top tube were conducting, it would have about a 15-V drop across it. The voltage across the bottom tube would have a maximum value of 1414 − 15 = 1399 V. Any tube selected for the circuit would necessarily have to have a PIV greater than this value.

Example 12-5

What is the maximum allowable voltage that a power transformer can deliver to a full-wave center-tapped rectifier circuit using diodes with a 6-kV PIV rating?

Solution

The maximum permissible voltage would be equal to the PIV rating of the diodes. Consequently, the effective value of the secondary voltage would be 6 kV × 0.707 = 4.24 kV. To allow for a margin of safety, the transformer should not deliver more than 90 percent of this value. This will accommodate line voltage surges and transients.

12-13 SERIES-CONNECTED FILTER CAPACITORS

Occasionally, the output voltage of a power supply is greater than the voltage rating of the available filter capacitors. The solution is to connect sufficient capacitors in series until the desired voltage rating is achieved. For example, if the supply delivers 800 V_{dc}, two electrolytic capacitors each having 450-WV_{dc} (working volts direct current) ratings could be series connected. This arrangement would provide a 100-V margin of safety. A typical configuration for such a circuit is shown in Fig. 12-16. The total capacity of a series combination such as C_1 and C_2 will be less than the smallest capacitor. If $C_1 = C_2 = 8\ \mu F$, then $C_T = 4\ \mu F$. Although the overall capacity is reduced, the voltage rating is increased.

Even though all capacitors have the same capacity, the voltage drop across each may vary considerably due to different *leakage currents*. The capacitor with the least leakage will have the greatest *IR* drop, and this may exceed the working voltage of the capacitor. By connecting an *equalizing resistor* across each capacitor, the total impedance of each device can be made

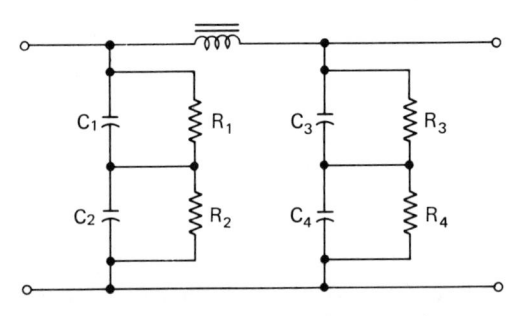

FIG. 12-16 Series-connected filter capacitors.

somewhat equal, thus equalizing the *IR* drops. It is not good engineering practice to series connect both electrolytic and paper capacitors, because their leakage resistances are so different.

A common failure in power supplies is shorted filter capacitors. This may result from voltage surges across the filter network or excessive leakage currents, causing the electrolytic capacitors to overheat. A shorted filter capacitor places an extreme load on the power transformer and rectifier, usually causing sudden failure. In vacuum and mercury-vapor rectifiers severe *internal sparking* and rapid disintegration of the filament or cathode structure occur. Simultaneously, the plate(s) become red hot. If the ac power is not immediately turned off, failure of the rectifier is inevitable. There is a good possibility that the power transformer may also be permanently damaged if it is not properly fused.

Before new rectifiers are installed it is imperative that the filter condensers be checked for possible short circuits.

12-14 BLEEDER RESISTORS AND VOLTAGE DIVIDERS

A resistor is frequently placed across the output terminals of a power supply (1) to *bleed off* the charge on the filter capacitor(s) when the rectifier is turned off, and (2) to apply a fixed load to the filter and thus *improve the voltage regulation* of the power supply. In the latter case, the resistor is designed to draw at least 10 per cent of the full-load current. Under these conditions the change of output voltage between no load and full load will be substantially reduced. In both conditions the resistor is called a *bleeder resistor*. If the bleeder resistor is tapped to provide several voltages, as shown in Figs. 12-17a and b, it is also called a *voltage divider*.

Customarily, one side of the voltage divider is grounded, as at *D* in Fig. 12-17a. Voltage measurements are usually made with reference to this point. However, in many applications it is necessary to have a voltage that is below ground. This can be accomplished by grounding any point, such as as *C* in Fig. 12-17b. In this case no part of the rectifier or filter can be grounded, as part of the voltage divider would be shorted out. Such a circuit can be used to furnish both plate and bias voltages from the same power supply.

In the voltage-divider circuits of Fig. 12-17 it has been assumed that no load was attached except across terminals *A* and *D*. As soon as a load is connected across parts of the divider, the voltage division changes. This is because the load forms a parallel circuit with part of the divider.

A simple voltage divider is shown in Fig. 12-18. The supply delivers a steady 500 V to load 2 and 400 V to load 1, through voltage divider R_1 and R_2.

Example 12-6

Find the ohmic values of R_1 and R_2 in Fig. 12-18 from the indicated currents and voltages.

186   Power Supplies

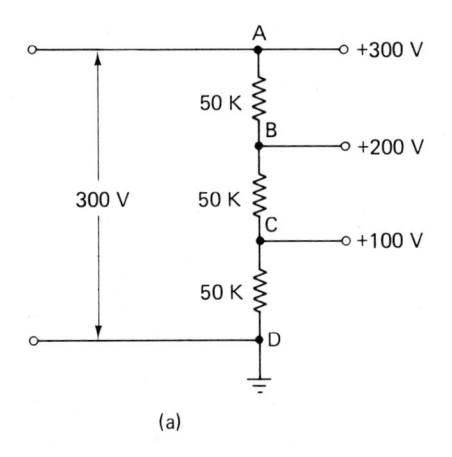

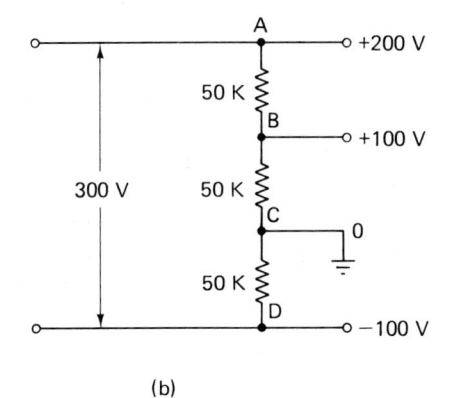

FIG. 12-17 Combination bleeder resistor and voltage divider.

(a)

(b)

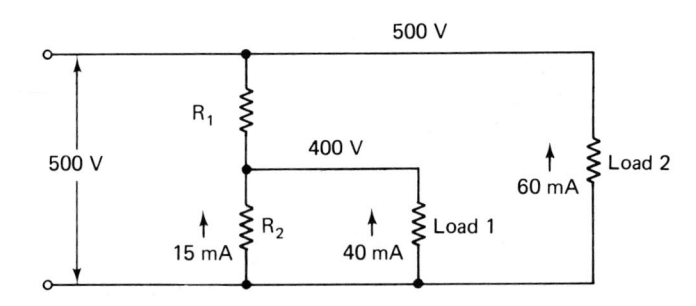

FIG. 12-18 Voltage divider network with two loads connected.

Solution

First, calculate R_2 as follows:

$$R_2 = \frac{400 \text{ V}}{15 \text{ mA}} = 26.67 \text{ k}\Omega$$

The current through R_1 equals the 15-mA bleeder current plus 40 mA

Voltage Regulation

from load 1. Therefore,

$$R_1 = \frac{500 \text{ V} - 400 \text{ V}}{55 \text{ mA}} = 1.82 \text{ k}\Omega$$

12-15 VOLTAGE REGULATION

Many electronic circuits require very stable operating voltages, even though their current demands change considerably. This problem can be solved by using a *voltage-regulator* circuit. Voltage-regulator circuits vary from simple to complex depending upon the *percentage of regulation* and power demanded.

All power sources have some internal resistance due to the resistance of the several components used. The current drawn from the supply produces an *IR* drop across this resistance, which subtracts from the terminal voltage of the supply. Consequently, when the load changes, the output voltage changes.

Suppose that a load requires a constant 18 V from a supply of 24 V. The simplest way to achieve this is to use an 18-V Zener diode, as shown in Fig. 12-19. The Zener current I_z flows through the diode and series resistor R. This current, plus the load current I_L, produces a 6-V drop across R. If I_L momentarily decreases, the drop across R attempts to diminish, thereby slightly increasing the voltage across the Zener. However, if V_z increases, even slightly, I_z increases and causes an increase in V_R. This restores the output voltage back to essentially the initial value. If I_L momentarily increased, the action would be reversed. Hence, the circuit is self-adjusting and delivers a regulated output voltage. The range of load changes over which the Zener is effective is limited.

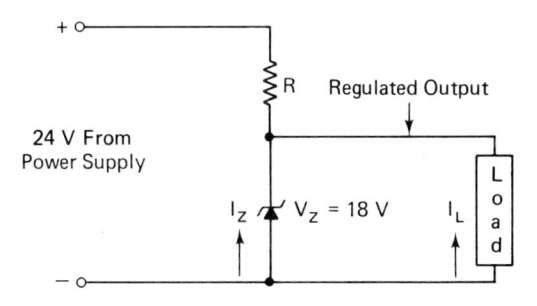

FIG. 12-19 Simple voltage regulator using a zener diode.

Another simple voltage regulator is the *gas-filled cold cathode glow tube*, sometimes called a VR tube, shown in Fig. 12-20. This tube does not contain a filament, but is capable of conducting electrons by *ionization* when sufficient voltage of the correct polarity is applied across its terminals. The released electrons cause the tube to glow, from whence it derives its name. Note the symbol for the tube in Fig. 12-20. The dot inside the circle implies the presence of gas.

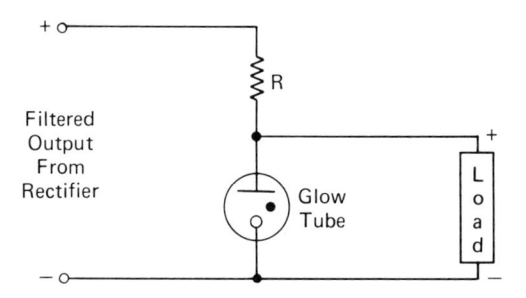

FIG. 12-20 Simple glow tube voltage regulator.

The circuit performs in about the same manner as the Zener regulator. If the load decreases, the terminal voltage will try to increase. This tendency for voltage rise causes increased ionization in the glow tube and thus reduces the resistance of the tube. The increase in current through the glow tube and the decrease in resistance of the tube are in such proportions that the voltage drop across the glow tube, and consequently across the load, remains constant. If the load had increased, the process would be reversed.

Voltage-regulator tubes are designed to regulate any of several low values of voltage (such as 75, 90, 105, or 150 V). Commercial names for these tubes are OA3 or VR75, OB3 or VR90, OC3 or VR105, and OD3 or VR150. The voltage across the tube should not exceed the specified value (except by a small margin required for initial ionization).

When a regulated voltage in excess of the maximum rating of one glow tube is required, two or more tubes may be connected in series. This arrangement permits several values of regulated voltage to be obtained from a single rectifier supply.

The regulation of a supply can be found by determining the ratio of the difference of voltages to the full-load voltage. Expressed as a formula,

$$\% \text{ regulation} = \frac{E_{nl} - E_{fl}}{E_{fl}} \times 100$$

where E_{nl} = no-load voltage
E_{fl} = full-load voltage

Example 12-7

What is the percentage of regulation of a power supply with a no-load voltage of 126.5 V and a full-load voltage of 115 V?

Solution

Substituting this information into the formula, we obtain

$$\% \text{ regulation} = \frac{126.5 - 115}{115} \times 100$$

$$= \frac{11.5}{115} \times 100 = 10\%$$

Example 12-8

If a power supply has an output voltage of 140 V at no load and the regulation at full load is 15 percent, what is the output voltage at full load?

Solution

Starting with the basic formula and transposing, we obtain

$$\% = \frac{E_{nl} - E_{fl}}{E_{fl}} \times 100$$

$$\% E_{fl} = 100 E_{nl} - 100 E_{fl}$$

$$\% E_{fl} + 100 E_{fl} = 100 E_{nl}$$

$$E_{fl} = \% + 100 = 100 E_{nl}$$

$$E_{fl} = \frac{100 E_{nl}}{\% + 100} = \frac{14{,}000}{115} = 121.7 \text{ V}$$

Power supplies having poor regulation have relatively high internal resistance. A well-designed, solid-state, 12-V power supply might have an internal impedance of 0.012 Ω and a voltage regulation of 0.01 per cent for a 100-mA load change.

12-16 REGULATED POWER SUPPLIES

The regulation of a dc power supply is accomplished by some type of feedback circuit that senses any change in the output voltage and develops a *control signal* to cancel this change. In the most common regulator circuits the dc output voltage is compared with a reference voltage. A differential emf results and is amplified and fed back to the base of the regulator transistor. In response to the feedback signal, the amount of conduction of this transistor is varied to regulate the output voltage. When the transistor is operated somewhere between cut off and saturation, the regulator circuit is referred to as a *linear voltage regulator*. If it operates in only the cutoff or saturation mode, the circuit is called a *switching regulator*.

All linear regulators may be classified as either *series* or *shunt*, depending upon the arrangement of the regulating transistor with respect to the load. As the name implies, in the series regulator the pass transistor (regulator) is in series with the load. Regulation is accomplished by varying the internal resistance of the pass transistor and, hence, the *IR* drop across it. In this way the dc voltage delivered to the load is maintained essentially constant.

In the shunt regulator the pass transistor is connected in parallel with the load, and a voltage-dropping resistor is connected in series with this parallel combination. If the load fluctuates, the current through the pass transistor is decreased or increased as required to maintain an essentially constant current through the series dropping resistor.

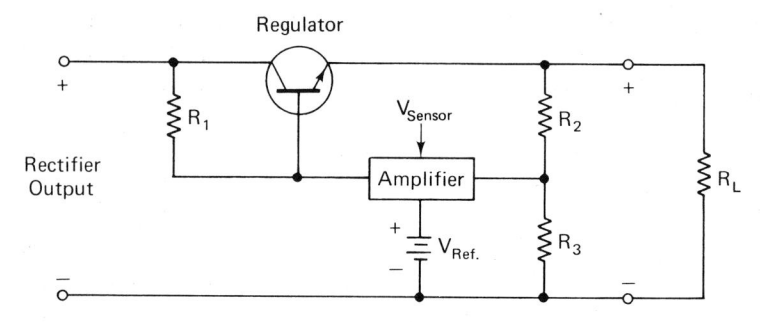

FIG. 12-21 Basic series voltage regulator.

Figure 12-21 shows a basic series regulator. The series pass transistor is usually operated as an *emitter follower*, and the error, or control signal, used to initiate the regulating action is applied to its base. The output of the regulator is not only delivered to the load but also to scaling resistors R_2 and R_3, which are integral with the regulator. The voltage appearing at the junction of R_2 and R_3 is applied to the input of the dc amplifier, where it is compared to a precise voltage reference source, indicated as V_{ref}. This is usually a Zener diode. This amplifier, which is part of the feedback loop between load and pass transistor, senses any change of output voltage as compared to the reference source. This error voltage is amplified and appears across R_1. The potential at the junction of R_1 and the amplifier output is applied to the base of the regulator. By controlling the base current of this transistor its *internal resistance* is regulated so as to maintain a constant potential across the load.

A simple, but effective, regulated power supply is shown in Fig. 12-22. It

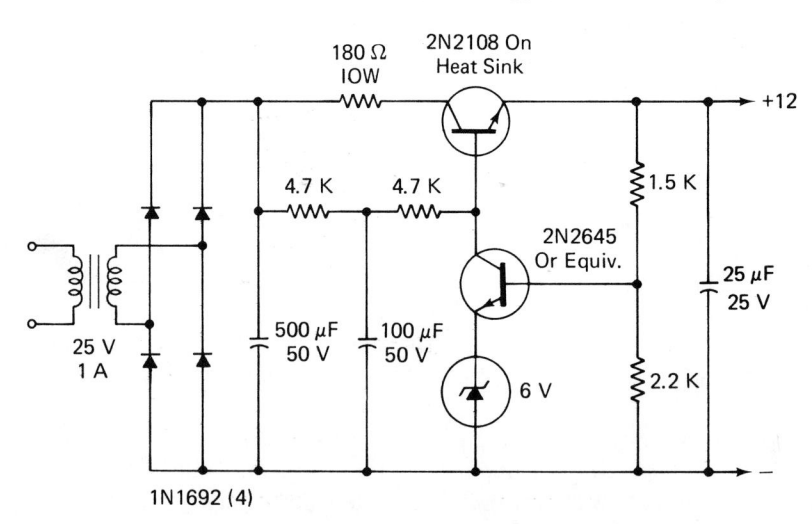

FIG. 12-22 General-purpose regulated dc supply.

Regulated Power Supplies

is designed to put out 12 V at 100 mA. The 180-Ω resistor provides *short-circuit protection*, limiting the output current to less than 200 mA. The 100-μF capacitor and 4.7-kΩ resistors provide an effective filter for the base current to the 2N2108 series regulating transistor, reducing the output ripple to less than 80 μV under full-load conditions. A portion of the regulated output voltage is connected to the base of the 2N2645 dc amplifier by the 1.5 and 2.2-kΩ scaling resistors. The voltage is compared to the 6-V reference Zener diode. Any tendency of the output voltage to change produces an error voltage that is amplified and fed to the base of the 2N2108 regulator. Its internal impedance is changed accordingly, and the output voltage is maintained at the assigned value.

The output voltage of the step-down transformer is connected to a bridge rectifier consisting of four 1N1692 diodes. The 500-μF capacitor across the rectifier output provides bulk filtering. The 25 μF provides some additional filtering and helps to maintain the output impedance of the supply at approximately 0.65 Ω. For line voltage variations of ± 10 per cent the output voltage regulation is better than ± 0.3 per cent.

A typical vacuum-tube VR circuit is shown in Fig. 12-23. Tube V_1 acts as the pass element, whose resistance is varied by a change of bias. When the ouput voltage tends to decrease, the grid voltage of V_2 becomes less positive. As the cathode voltage of V_2 is maintained at a constant potential with respect to ground by voltage regulator V_3, and positive with respect to the grid, a drop in grid voltage will increase the bias and decrease the plate current of V_2. This causes the plate voltage of V_2 to increase. Because the plate is tied directly to the grid of V_1, its bias will decrease. This results in a drop of plate-cathode potential of V_1, tending to restore the original voltage across the

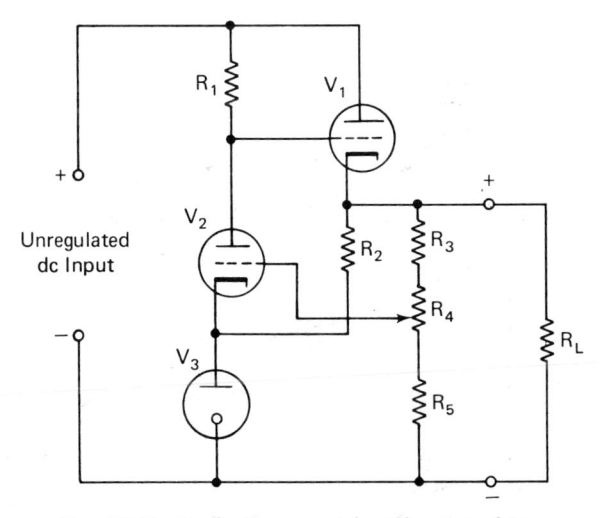

FIG. 12-23 Typical vacuum-tube voltage regulator.

load R_L. The output voltage can be adjusted to a different value by repositioning potentiometer R_4.

12-17 VOLTAGE-DOUBLER CIRCUITS

Higher voltages may be developed by rectifier-multiplier circuits than are available by conventional rectifiers using the same ac power-line source. The currents available under these conditions are relatively low. These circuits rely on the alternate charging of two capacitors. It is possible to obtain ac voltages under no-load conditions of about twice the peak value of the ac input. Under loaded conditions the dc output is approximately equal to the peak value of the ac input.

A typical *voltage-doubler circuit* is shown in Fig. 12-24. Examination of the circuit shows that it is essentially a combination of two half-wave

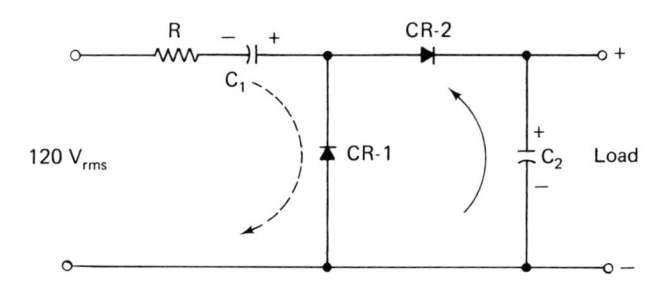

FIG. 12-24 Typical voltage-doubler circuit.

rectifier circuits. Consider that the input is 120 V_{rms} and that the instantaneous polarity is such that the upper terminal is negative and the bottom positive. Current will flow through CR_1 in the direction of the dotted arrow and charge C_1 to about 170 V_p (120 $V_{rms} \times 1.414 \approx 170\ V_p$). When the ac input reverses, the output capacitor C_2 is charged through CR_2. The voltage available to charge this capacitor is the peak value of the ac input plus the dc charge on C_1, which are in series at this moment. Hence, neglecting minor voltage drops around the circuit, C_2 charges to about 340 V_p (no load).

The peak inverse voltage on the diodes is about equal to twice the value of the peak input voltage. Typical values of capacitors range from 40 μF to over 100 μF each. Their working voltage ratings must be at least equal to the peak value of voltage appearing across them. Larger capacitors provide better voltage regulation and higher output voltages.

12-18 VIBRATOR POWER SUPPLIES

Vibrator power supplies are used for transforming low dc voltages, such as obtained from storage batteries, to the high voltages required to operate the plate and screen grid circuits of vacuum tubes. Because dc voltage

Vibrator Power Supplies

cannot be directly stepped up in a transformer, some method must be used to interrupt it and supply it to the primary of a step-up power transformer. Such a device is called a *vibrator*, because it contains a vibrating reed or contact that supplies pulses of direct current to the primary winding. The transformer cannot readily distinguish between pulses of direct (properly timed) and alternating current. Consequently, it steps the voltage up in the conventional manner. Vibrator power supplies are basically of two types: *nonsynchronous* and *synchronous*.

A conventional nonsynchronous vibrator power supply is shown in Fig. 12-25a. The vibrator, usually a plug-in device, is shown within the dotted rectangle. When power switch SW is closed, battery voltage is applied to electromagnet M via the transformer primary and the vibrating reed contact. This causes a pulse of current to flow through the primary before the vibrating reed contact is broken by the pull of magnet M. At this moment the current in the primary drops to zero. The buildup and collapse of the current pulse induces an ac voltage into the transformer secondary, where it is rectified by CR_1 and then filtered by the pi filter shown. Once the vibrating reed contact is broken, the spring action of the reed (like a door bell or buzzer) returns it to its original position. The magnet M is once again energized and the cycle is repeated. In this manner the reed sets up vibrations on the order of 135 Hz.

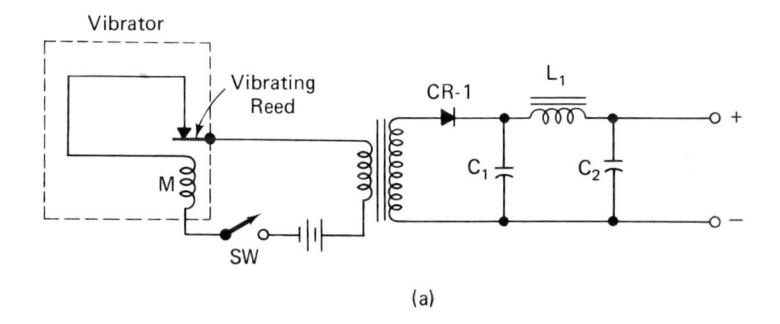

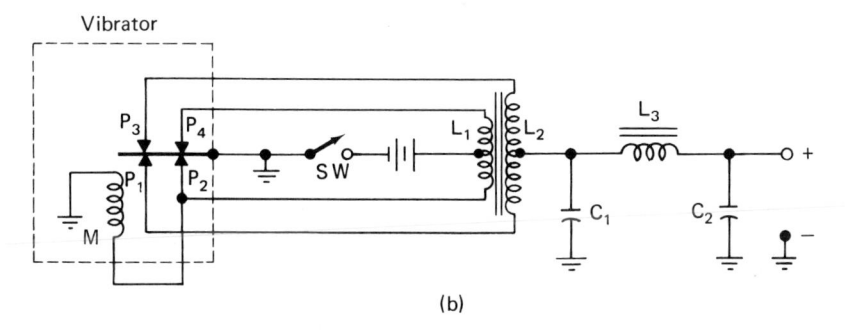

FIG. 12-25 Vibrator power supplies: (a) non-synchronous, (b) synchronous.

The synchronous vibrator is characterized by *dual sets of contacts* on the vibrator and no external rectifier. Such a circuit appears in Fig. 12-25b. When switch SW is turned on, direct current is applied to electromagnet M via the lower half of L_1. The magnet pulls the reed down, contacting points P_1 and P_2. This allows a pulse of direct current to flow through the lower half of L_1. The lower end of L_2 is grounded through P_1 and the vibrating reed. Simultaneously, the grounding of P_2, through the reed, removes the current flowing through the vibrator magnet M. With the magnet deenergized the vibrating reed swings in the opposite direction, establishing contact with P_3 and P_4, grounding them. This causes a pulse of direct current to flow through the upper half of primary L_1.

The pulses of direct current alternately flowing through each half of L_1 induce a high voltage in the secondary L_2. Because alternate ends of transformer winding L_2 are grounded in synchronism with the switching of pulses in the primary L_1, the direction of the induced voltage in the secondary is *unidirectional* or direct current. Hence, no rectifier is required with this kind of circuit. Necessary filtering is provided by the pi filter.

Normally, vibrator power supplies are capable of delivering relatively small amounts of power. Excessive loading may cause severe arcing at the contacts with a substantially reduced service life. Supplies of this type generally create *switching transients* that can be very troublesome, due to the noise or interference caused in receivers, high-gain amplifiers, and so forth. To eliminate this, it is necessary to install an RF choke in series with the hot lead of the battery, properly bypassed to ground. Fully shielding the entire vibrator supply will likely also reduce the "noise."

12-19 THREE-PHASE POWER SUPPLIES

It is customary for commercial transmitting stations to use three-phase power to operate the rectifier system. One of the principal reasons is that *polyphase rectifiers* produce a more nearly constant output than full-wave single-phase systems. Even in half-wave rectifier systems the ripple frequency is three times that of the power line. This makes for easier filtering, with smaller filter components for a given percentage of ripple. A number of circuit combinations are possible using polyphase systems. They may be conveniently subdivided into systems giving three and six-phase output waves, that is, ripple frequencies of three and six times the supply, respectively. The latter category makes the filtering requirements even less stringent.

A modest three-phase half-wave rectifier circuit is shown in Fig. 12-26. The transformer T_2 is Δ-to-Y (delta-to-wye) connected, which provides the optimum step-up voltage ratio betwen primary and secondary windings for a given turns ratio. Switch SW_1 is the filament power switch connecting single-phase power to the primary of the step-down filament transformer T_1. It should be turned on sufficiently ahead of high voltage power switch SW_2 to

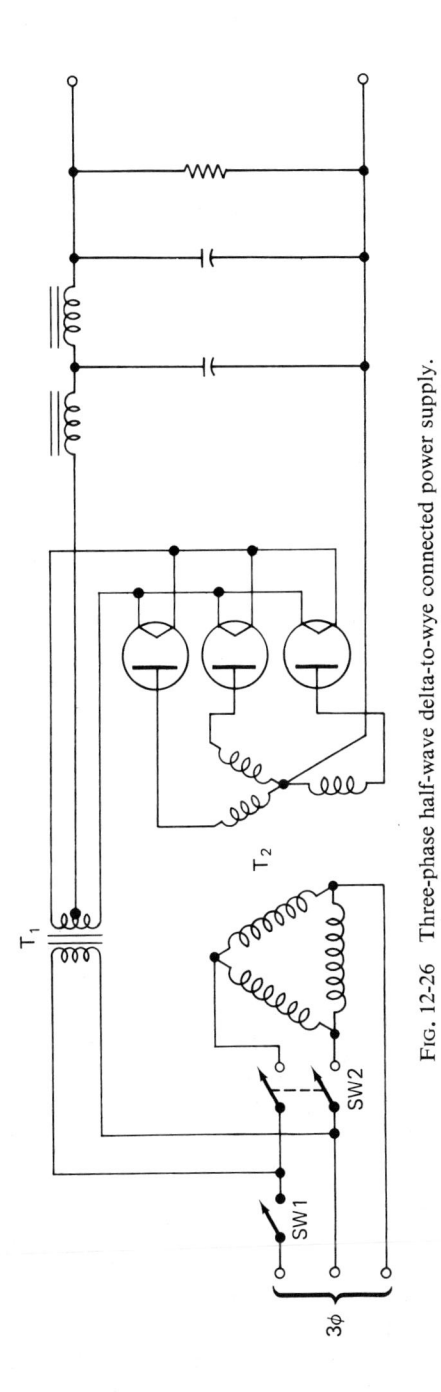

FIG. 12-26 Three-phase half-wave delta-to-wye connected power supply.

permit the rectifier tubes to come up to operating temperature. The ripple frequency in this circuit is three times the power-line frequency.

Practice Problems

1. The secondary winding of a power transformer supplies 350 V_{rms} on either side of the center tap to a full-wave rectifier. Assuming negligible IR drops across the rectifiers, what is the average output voltage?
2. What must the rms input voltage be to a half-wave rectifier if it must deliver a dc component of 100 V?
3. What is the ripple frequency of a half-wave rectifier fed from a 400-Hz aircraft alternator? What would the ripple frequency be for a full-wave rectifier operating under the same input conditions?
4. A small power transformer develops a total of 500 V across the center-tapped secondary. Neglecting any voltage drop in the rectifiers, what is the approximate dc output voltage if choke input filtering is used?
5. A power transformer supplying 325 V on each side of the center tap is connected to solid-state rectifiers and a capacitor input filter system. Calculate the approximate dc output voltage.
6. A power supply must deliver 275 V_{dc} to its load. If choke input filtering and solid-state rectifiers are employed, what must the output voltage of the power transformer be?
7. A transistorized power supply delivers 11.75 V_{dc} under load. If the no-load voltage is 12.2 V, what is the percentage of regulation?
8. A power supply delivers 36 V_{dc} under no-load conditions. If its regulation is 1.3 per cent, what is the full-load voltage?

Commercial License Questions

Sections in which answers to questions are given appear in parentheses. A bracketed number following a question implies that it applies only to that element.

1. List the main advantages of a full-wave rectifier as compared to a half-wave rectifier. (12–1)
2. What is the predominant ripple frequency in the output of a single-phase full-wave rectifier when the primary source of power is 110 V at 60 Hz? [4]
(12–2)
3. Draw a diagram of a bridge rectifier giving full-wave rectification without a center-tapped transformer. Indicate polarity of output terminals. [4]
(12–3)
4. What is the principal function of the filter in a power supply? (12–5)
5. Explain the operation of a vacuum-tube-rectifier power supply and filter.
(12–5, 12–10)

Commercial License Questions

6. What are the characteristics of a choke input filter system as compared to a condenser input system? (12–5, 12–6)

7. How may a condenser be added to a choke input filter system to increase the full-load voltage? [4] (12–7)

8. If the reluctance of an iron-cored choke is increased by increasing the air gap of the magnetic path, in what other way does this affect the properties of the choke? (12–9)

9. Why is it desirable to have low-resistance filter chokes? (12–9)

10. Why is it not advisable to operate a filter choke in excess of its rated current value? [4] (12–9)

11. What does a blue haze in the space between the filament and plate of a high-vacuum rectifier tube indicate? (12–10)

12. What is the effect upon a filter choke of a large value of dc flow? (12–10)

13. What is the principal function of a swinging choke in a filter system? (12–10)

14. What are the primary advantages of a mercury-vapor rectifier as compared to the thermionic high-vacuum rectifier? (12–10, 12–11)

15. What are the primary advantages of a high-vacuum rectifier as compared to the hot cathode mercury-vapor rectifier? (12–10, 12–11)

16. What are the primary characteristics of a gas-filled rectifier tube? (12–11)

17. What is the value of voltage drop across the elements of a mercury-vapor rectifier tube under normal conducting conditions? [4] (12–11)

18. What factors permit high conduction currents in a hot cathode type of mercury-vapor rectifier tube? (12–11)

19. Why is a time-delay relay arranged to apply the high voltage to the anodes of mercury-vapor rectifier tubes some time after the application of filament voltage? [4] (12–11)

20. Why is it important to maintain the operating temperature of a mercury-vapor tube within specified limits? [4] (12–11)

21. When mercury-vapor tubes are connected in parallel in a rectifier system, why are small resistors sometimes placed in series with the plate leads of the tubes? [4] (12–11)

22. What is meant by the *inverse peak voltage rating* of a rectifier tube? [4] (12–12)

23. What is meant by *arc-back* or *flashback* in a rectifier tube? [4] (12–12)

24. How is the inverse peak voltage to which the tubes of a full-wave rectifier will be subjected determined from the known secondary voltages of the power transformer? Explain. [4] (12–12)

25. May two condensers of 500 V operating voltage, one an electrolytic and the other a paper condenser, be used successfully in series across a potential of 1,000 V? Explain your answer. (12–13)

26. When filter condensers are connected in series, resistors of high value are often connected across the terminals of the individual condensers. What is the purpose of this arrangement? (12–13)

27. If the plate, or plates, of a rectifier tube suddenly become red hot, what might be the cause, and how could remedies be effected? (12–13)

28. If a high-vacuum-type, high-voltage rectifier tube should suddenly show severe internal sparking and then fail to operate, what elements of the rectifier-filter system should be checked for possible failure before installing a new rectifier tube? (12–13)

29. A rectifier-filter power supply is designed to furnish 500 V at 60 mA to one circuit and 400 V at 40 mA to another circuit. The bleeder current in the voltage divider is to be 15 mA. What value of resistance should be placed between the 500- and 400-V taps of the voltage divider? [4] (12–14)

30. What is the purpose(s) of a bleeder resistor as used in connection with power supplies? (12–14)

31. What effect does the resistance of filter chokes have on the regulation of a power supply in which they are used? (12–15)

32. If a power supply has an output voltage of 140 V at no load and the regulation at full load is 15 per cent, what is the output voltage at full load? [4] (12–15)

33. What is the percentage regulation of a power supply with a no-load voltage output of 126.5 V and a full-load voltage output of 115 V? (12–15)

34. What is the definition of *voltage regulation* as applied to power supplies? (12–15)

35. If a power supply has a regulation of 11 per cent when the output voltage at full load is 240 V, what is the output voltage at no load? [4] (12–15)

36. Draw a diagram of a voltage-doubling power supply using two half-wave rectifiers. [4] (12–17)

37. Draw a diagram of a synchronous-vibrator power supply. A nonsynchronous-vibrator power supply. (12–18)

38. Describe the principle of operation of a synchronous type of mechanical rectifier. (12–18)

39. In what circuits of a radio station are three-phase circuits sometimes employed? (12–19)

13

Audio Amplifiers

13-1 CLASSIFICATION OF AMPLIFIERS

The very small voltage from microphones, phono pickups, tape reproducer heads, and so forth, must be amplified before they can be useful. Amplifiers designed for this purpose are called *voltage amplifiers*.

All voltage amplifiers are not the same. For example, one designed for *audio frequencies* (AF) has different components than one for *radio frequencies* (RF). Hence, one way to classify amplifiers is by the band of frequencies they are designed to amplify, such as AF, RF, VHF, microwaves, and so on. These bands are identified in Table 4-1.

Amplifiers are also classified as A, AB, B, and C. This has reference to the location of the operating point, which is determined by how much bias is used. Only the first three types can be used for audio amplification. The class C amplifier is used in the output stage of transmitters because of its high efficiency. The first three types will be discussed in subsequent paragraphs, the latter is covered in Chapter 14.

13-2 CLASS A AMPLIFIERS

The operational characteristics of the several classes of amplifiers can be best understood by referring to their *dynamic transfer characteristic curves*. To illustrate, refer to Fig. 13-1 which shows such a curve for a typical transistor. This is a plot of base versus collector currents for a given value of load resistor and supply voltage. The slope of the curve is determined principally by the value of load used. The curve portrays more exactly the performance of the device under dynamic conditions than does the load line. Dynamic transfer curves are also used with vacuum tubes and field-effect transistors.

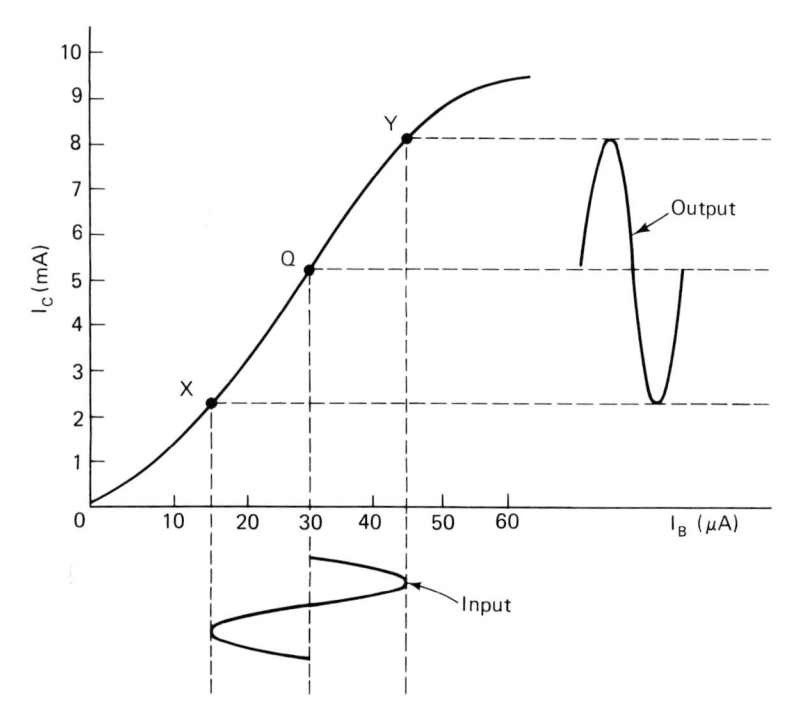

FIG. 13-1 Dynamic transfer characteristic curve showing class A (linear) operation.

In Fig. 13-1 the transistor is biased with 30 μA of base current, which establishes an operating point Q in about the center of the *linear part* of the dynamic curve. The peak values of input current cause the collector current to vary between X and Y. The resultant output current is a replica of the input signal, due to the linearity of the curve. This is referred to as *class A* operation. Class A operation may also be defined as that condition in which the bias and input signal are adjusted so that output current flows for 360° of the input cycle.

In vacuum-tube amplifiers grid current does not flow in class A operation. To show this, the subscript 1 may be added to the letter or letters of the class identification. The subscript 2 indicates that grid current flows during some parts of the input cycle. Thus, if the grid is not driven positive at any time in the class A cycle, no grid current will flow, and the amplifier is designated class A_1. If no subscript is shown with the letter A, it is assumed that no grid current flows.

Class A amplifiers are characterized by *minimum distortion, low power output* (relative to class B and C amplifiers), *high voltage amplification*, and relatively *low plate efficiency*. The latter is a ratio of the ac power output to the dc input. Expressed mathematically.

Class B Amplifiers 201

$$\text{plate efficiency} = \frac{P_o}{P_i} \times 100$$

where P_o = ac power, in watts, delivered to load
  P_i = dc input power ($E_b I_b$)

The same basic formula is used for transistors and FETs except that the input power is the product of the collector or drain voltage and the current through the device. Efficiencies of class A amplifiers vary from about 15 to 30 per cent.

13-3 CLASS B AMPLIFIERS

Increased efficiency for a given transistor or tube is obtainable when it is operated in class B. However, distortion products are somewhat higher than with class A. To operate in class B, the bias must be adjusted so that the operating point is *at or near cutoff*. Therefore, when no signal is present, the output current is essentially zero. Figure 13-2 shows the operating point of a vacuum tube biased class B. Only the positive halves of the signal cause plate current to flow. Notice that the output current waveform is similar to that of a half-wave rectifier. This severely distorted current pulse is rich in *harmonic*

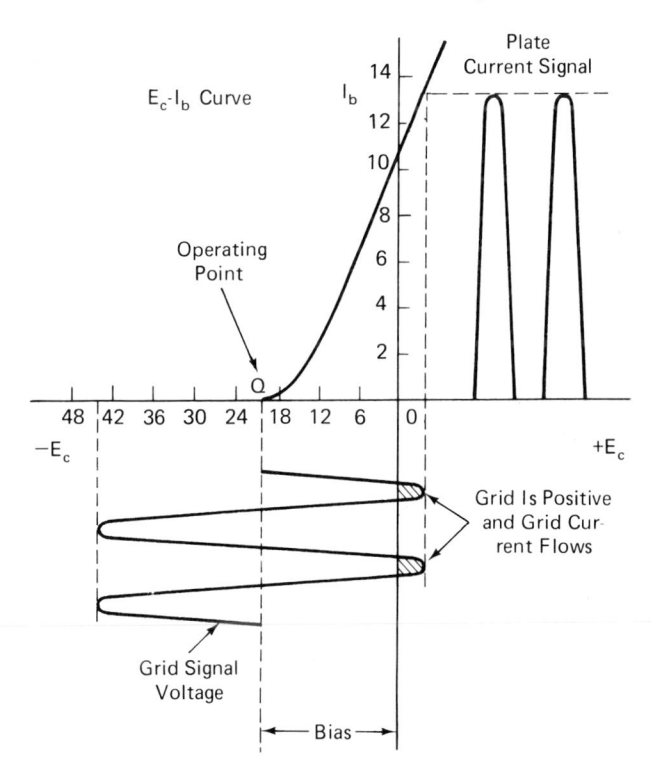

FIG. 13-2 Biasing and driving requirements for class B operation.

energy generated by the non-linear operation of the tube. Because of this, a single tube operated class B is unsuitable for audio amplification. However, two tubes can be combined in a single stage such that the majority of the distortion is eliminated. This is called a *push–pull amplifier*.

To obtain maximum output power from a class B amplifier, a *large driving signal* must be applied to the grid. This is generally several times greater than required for class A when using the same tube. At the peaks of the positive half-cycles the grid is driven slightly positive with respect to the cathode, and grid current flows as represented by the shaded areas in Fig. 13-2.

Unless the signal source used to drive the class B amplifier has a low internal impedance (good regulation), the grid current will cause clipping of the positive peaks of the grid signal, thus increasing distortion.

The characteristics of class B amplifiers are *medium output power*, *medium plate efficiency* of 40 to 60 per cent, *large bias voltage*, and *large driving signal* requirements.

To determine the approximate cutoff voltage for a triode, the following formula can be used:

$$E_{co} = \frac{E_b}{\mu}$$

Example 13-1

What is the cutoff voltage of a triode operating with 300-V plate potential and $\mu = 30$?

Solution

$$E_{co} = \frac{300}{30} = 10 \text{ V}$$

13-4 CLASS AB OPERATION

When a compromise is needed between the low distortion of class A and the higher efficiency of class B, *class AB operation* is used. In this case the bias and input signals are adjusted so that output current flows for appreciably more than half the input cycle, but for less than the entire cycle. This condition is indicated in Fig. 13-3.

If the input signal drives the grid positive with respect to the cathode, grid current will flow during the positive peaks and the amplifier is designated as a class AB_2 amplifier. In this case the operating point would have to be shifted a little closer toward cutoff than is shown for the class AB_1 operation in Fig. 13-3. Although a class AB_2 amplifier delivers slightly more power to its load, the class AB_1 amplifier has the advantage of presenting to its drive tube a *constant impedance*. In contrast with this effect, the amplifier that draws grid current over a portion of its input cycle presents a changing im-

Class C Amplifiers

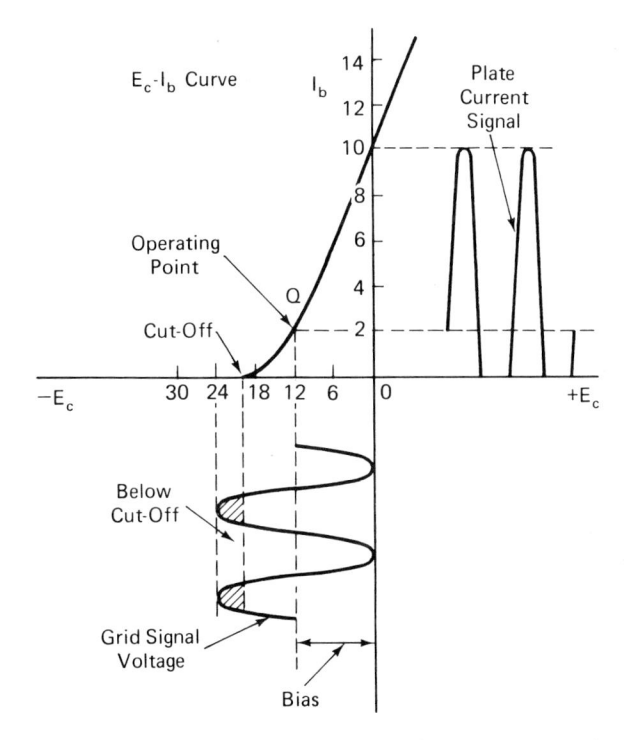

FIG. 13-3 Biasing and signal requirements for class AB operation.

pedance to its driver at the point where grid current starts to flow. Thus, when grid current flows, the impedance falls to a relatively low value. The driver that supplies this kind of load must be designed to supply undistorted power to the load during these periodic intervals of low impedance.

It is not possible to operate a single tube or transistor in class AB because of the distortion that would result. By using two of the devices in push–pull, most of the distortion can be eliminated.

13-5 CLASS C AMPLIFIERS

Class C amplifiers are biased from *two to five times cutoff;* consequently, output current flows for less than half (160° or less) of each cycle of the input signal. Because of the brief period during which conduction occurs, the average current is small compared to the peak current, and efficiencies of 70 to 80 per cent are obtainable.

Class C amplifiers are not used as audio amplifiers because of the extreme distortion they produce, but are used as RF power amplifiers in transmitters, where tank circuits are used in the output to restore the missing portions of each cycle.

The conditions for class C operation are illustrated in Fig. 13-4. Notice that the operating point is far below cutoff, and that an extremely large grid

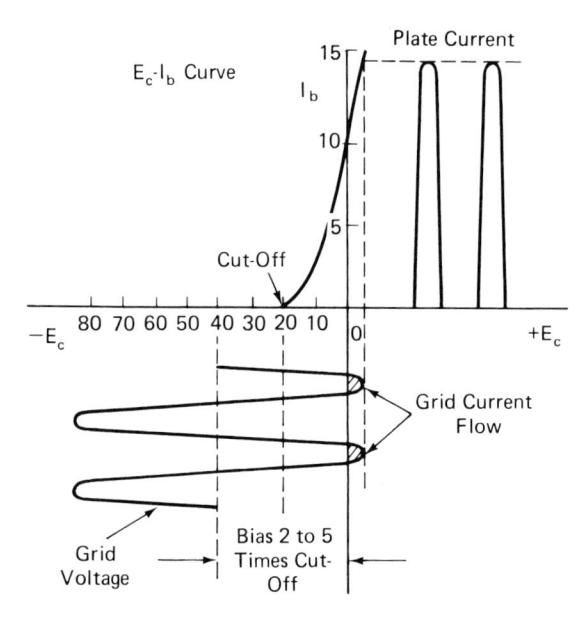

FIG. 13-4 Biasing and driving requirements for class C operation.

signal is required to bring the tube into conduction. To drive the tube to full output, the grid signal must be large enough to drive the grid substantially positive at the crest of the positive alternation. Class C RF power amplifiers in which the control grid is driven from 50 to several hundred volts positive are not uncommon.

13-6 BIASING

The techniques for biasing transistors have been shown in Section 10-6. This section will show the basic biasing methods for the vacuum tube and FET. There are two basic methods of biasing these devices: *fixed and self-bias*. Two examples of fixed bias are shown in Fig. 13-5. In Fig. 13-5a battery E_{cc} supplies the negative voltage for the grid, which will be below ground, by the emf of the battery. With the cathode at ground the bias on the tube then equals E_{cc}. There is no *IR* drop across grid resistor R_c because no grid current flows. Its primary function is to allow the charge that would build up in the coupling capacitor to leak back to ground. If this resistor were omitted, the grid would *block* or become negative in the case of vacuum tubes and FETs. No current would flow through it. It would be cut off. The tube can be biased at any desired operating point by selecting the right value of E_{cc}.

A method of providing fixed bias for a JFET (junction field-effect transistor) is seen in Fig. 13-5b. The similarity between the two circuits in this figure is obvious. Bias is supplied by V_{GG} through the gate return-to-ground resistor R_g, whose function is the same as for the vacuum-tube circuit.

Biasing 205

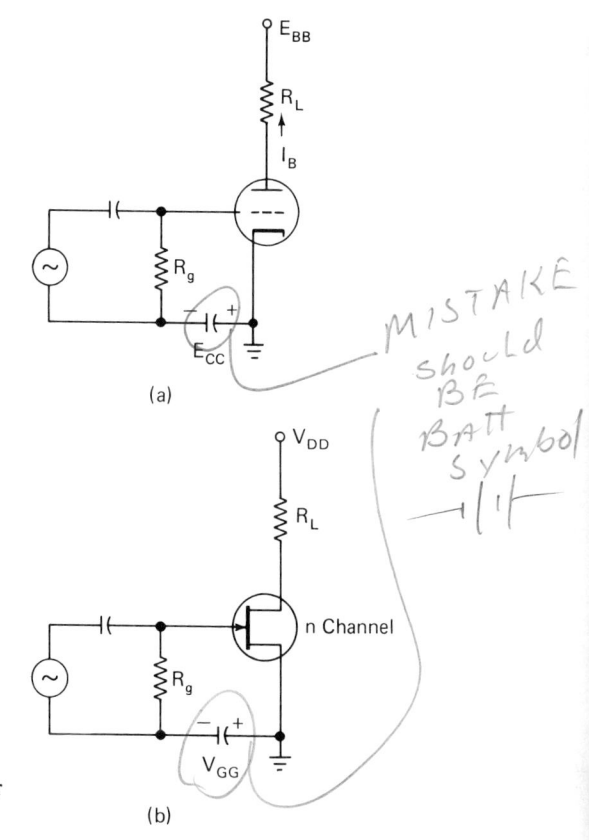

FIG. 13-5 Fixed bias arrangements for (a) vacuum tube, (b) JFET.

The gate–source junction is reverse biased by V_{GG}, and therefore no current flows through R_g. Power supplies are sometimes used as sources of fixed bias. This form of bias is limited mostly to class C amplifiers, although some class B amplifiers use it.

A more practical method of biasing is called self-bias, which eliminates the need for bias batteries. Such an arrangement is shown in Figs. 13-6a and b. In the tube example resistor R_K is called the *cathode biasing* resistor. In the case of the JFET it is called the *source biasing* resistor. Their functions are the same in either case, so this discussion will be confined to the vacuum tube.

Cathode resistor R_K is in series with V_1 and load resistor R_L. The plate current flowing in the circuit must also pass through R_K, creating an IR drop across it. This raises the cathode *above ground* by an amount equal to this voltage. Inasmuch as the grid is at ground potential, the bias on the tube is determined by the voltage across R_K. The ohmic value of R_K can be determined for any class A operating condition by knowing the quiescent current and bias voltage desired.

Example 13-2

What value of R_K is necessary to bias a tube if an analysis of the load line indicates an I_b quiescent of 6.5 mA and $-E_{cc} = 3$ V is required?

Solution

$$R_K = \frac{3 \text{ V}}{6.5 \text{ mA}} \simeq 460 \, \Omega$$

When a signal is applied to the grid, the plate current varies accordingly, and a small voltage variation develops across R_K. The polarity of this voltage is such as to *oppose* the change in plate current. For example, if the input signal makes the grid less negative, I_B increases. This produces a larger bias voltage across it, which tends to reduce plate current. Thus, the varying cathode voltage prevents the input signal from producing the full change in plate current, causing a reduction in gain. This process, whereby a signal causes itself to be partially canceled, is called *degeneration or negative feed-*

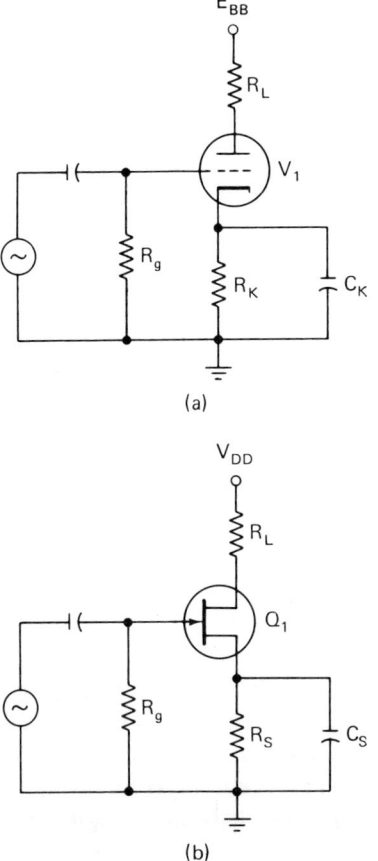

FIG. 13-6 Self biasing used for (a) vacuum tube, (b) JFET.

Coupling Circuits

back. Although degenerative feed back causes a loss in amplification, it reduces the distortion introduced into the signal and is often used for this purpose.

The degeneration effect of the signal voltage across R_K can be eliminated by placing a bypass capacitor across it, as shown in Figs. 13-6a and b. If the correct-sized capacitor is used, two things are accomplished: the IR drop across R_K charges C_K on the positive peaks of the input signal. The capacitor partially discharges before the next peak comes, thereby holding the potential constant. The second purpose is to bypass the signal around R_K so that it is not attenuated.

The value of the cathode bypass capacitor can be computed by assuming that its reactance is one tenth the resistance of R_K at the *lowest frequency* to be amplified. By rearranging the equation for capacitive reactance, a formula can be derived for computing the value of C_K.

$$C_K = \frac{1}{\omega X_c}$$

Since X_c should equal one tenth R_K,

$$C_K = \frac{10^6}{\omega(R_K \times 0.1)}$$

where C_K = value of C_K in microfarads
R_K = value of the cathode resistor in ohms
$\omega = 2\pi f$, where f is the lowest frequency to be amplified

Example 13-3

What value capacitor is required to properly bypass source resistor R_S (Fig. 13-6b) if its ohmic value is 500 Ω and the lowest frequency to be amplified is 80 Hz?

Solution

$$C_s = \frac{10^6}{6.28 \times 80 \times 50} = 39.8 \ \mu F$$

The closest commercially available value is 40 μF.

13-7 COUPLING CIRCUITS

When the gain provided by a single amplifier stage is insufficient, two or more stages may be *cascaded* (connected in series) to give the required amplification. These stages can be coupled in any one of several basic methods: *direct coupling*, *inductive or transformer coupling*, *resistance–capacitance coupling (RC)*, and *impedance coupling*. The most widely used method is *RC* because it is inexpensive and provides good frequency response.

A typical *RC* coupling network is shown in Fig. 13-7a. The amplified

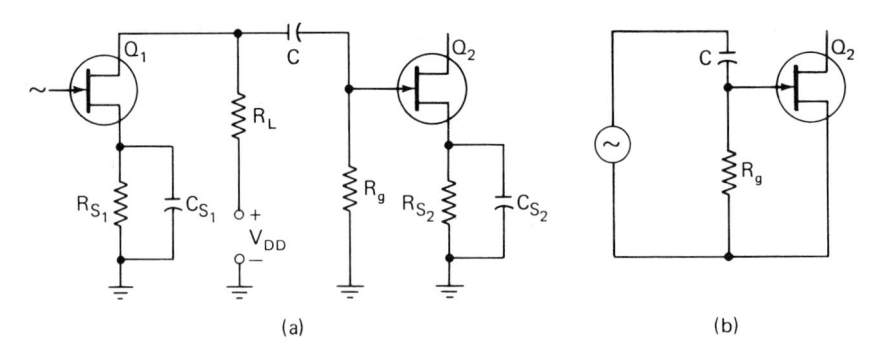

FIG. 13-7 (a) *RC* coupling between two FET's. (b) Equivalent circuit.

signal voltage from Q_1 appears across R_L. Because the power supply's impedance is very low, the bottom end of R_L can be considered at ground potential so far as the signal is concerned. Therefore, the alternating current developed across R_L is impressed across the series network of CR_g. The equivalent circuit appears in Fig. 13-7b. If the coupling components are properly selected, the reactance of C will be small compared to the ohmic value of gate return resistor R_g. Hence, practically all the signal developed across R_L will appear across R_g. It is this voltage that is applied to the input of Q_2 (gate–source terminals). Notice in the equivalent circuit that the source of Q_2 is grounded. This is because the reactance of C_{s2} is low, compared to R_{s2}, and effectively grounds the source at the signal frequencies.

Example 13-4 will indicate how the proper value of coupling capacitor can be determined for a given *RC* network.

Example 13-4

What value of coupling capacitor is needed in the circuit of Fig. 13-7a if $R_g = 1$ MΩ and the lowest frequency to be amplified is 30 Hz?

Solution

In a well-designed circuit the reactance of C should be about one tenth the value of R_g at the lowest frequency. Therefore,

$$C_{\mu F} = \frac{10^6}{6.28 \times 30 \times 10^5} = \frac{1}{18.84} \simeq 0.053 \ \mu F$$

In practice a 0.05 μF would be used.

Coupling capacitors are also called *blocking capacitors*. This is because they block the direct current present at the drain (plate or collector for other devices) from being applied to the input of the next stage. If this happened, the bias of the stage would be upset and severe distortion would result. A common failure in amplifiers is leaky coupling capacitors, where some of the direct current in the preceding stage appears across R_g and changes the operating point of the stage.

Impedance coupling can be used between two stages, as seen in Fig. 13-8. The conventional load resistor is replaced by an inductance L. To obtain as much amplification as possible, particularly at the lower frequencies, the inductance is made as large as practicable. To avoid undesirable magnetic coupling, a closed-shell type of inductor is used. Because of the low dc resistance of the inductor, less dc voltage appears across it. Thus, the tube can operate at a higher plate voltage, or a lower E_{bb} may be used.

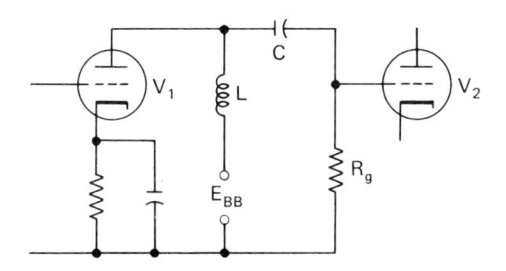

FIG. 13-8   Impedance coupling.

Amplification is not uniform as it is with RC coupling because the load impedance, Z_L, varies with frequency. Since the output voltage appears across Z_L, the voltage gain increases with frequency up to the point where the *shunting capacitance* limits it. The shunting capacitance includes not only the interelectrode and distributed wiring capacitances found in RC-coupled amplifiers, but also the distributed capacitance associated with the turns of the inductor. The distributed capacitance between the turns of the coil greatly increases the capacitance to ground, and plays a major part in the limiting of the use of this coupling at higher frequencies.

Transformer coupling can be used between stages and has several advantages. The voltage amplification of the stage may exceed that of the tube if the transformer has a step-up turns ratio. Direct current isolation of the grid of the next tube is provided without the need for a blocking capacitor. This type of coupling is also used to couple a *high-impedance source to a low-impedance load*, or vice versa, by choosing a suitable turns ratio. An example of transformer coupling is shown in Fig. 13-9.

Transformer coupling has the disadvantages of greater cost, greater space requirements, the necessity for greater shielding, and the possibility of poorer frequency response at the higher and lower frequencies.

In a *direct-coupled* amplifier the plate of one tube is connected directly to the grid of the next tube. This arrangement presents a problem of voltage distribution. Since the plate of a tube must have a positive voltage with respect to its cathode, and the grid of the next tube must have a negative voltage with respect to its cathode, the two cathodes cannot operate at the

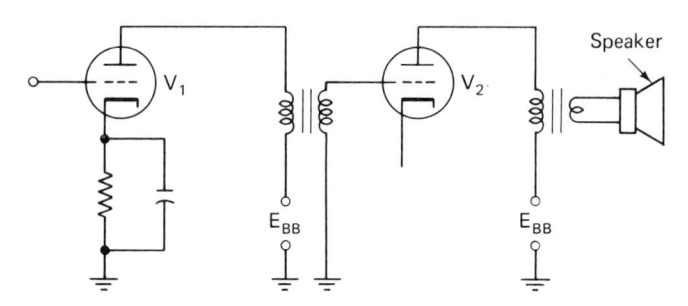

FIG. 13-9 Transformer coupling.

same potential. Proper voltage distribution is obtained by a voltage divider, as shown in Fig. 13-10. In this amplifier the plate of V_1 is connected directly to the grid of V_2. The grid of V_1 is returned to point A through R_{g_1}. The cathode of V_1 is returned to point B and the grid bias for V_1 is developed by the voltage drop between points A and B of the voltage divider. The plate of V_1 is connected through its plate load resistor, R_L, to point D on the divider. R_L also serves as the grid resistor for V_2.

Since the plate current from V_1 flows through R_L, a certain amount of supply voltage appears across R_L. This voltage drop must be allowed for in choosing point D on the divider. Point D is so located that approximately half the available voltage is applied to the plate of V_1. The plate of V_2 is connected through a suitable output load, R, to point E, the most positive point on the divider. Since the voltage drop across R_L may place too high a negative bias on the grid of V_2, it may be necessary to connect the cathode of V_2 at point C, which is negative with respect to point D, to lower the bias on the grid of V_2. Point C, together with the value of R, determines the proper voltage for V_2.

The entire circuit is a *complex resistance network* that must be adjusted carefully to obtain the proper plate and grid voltages for both tubes. If

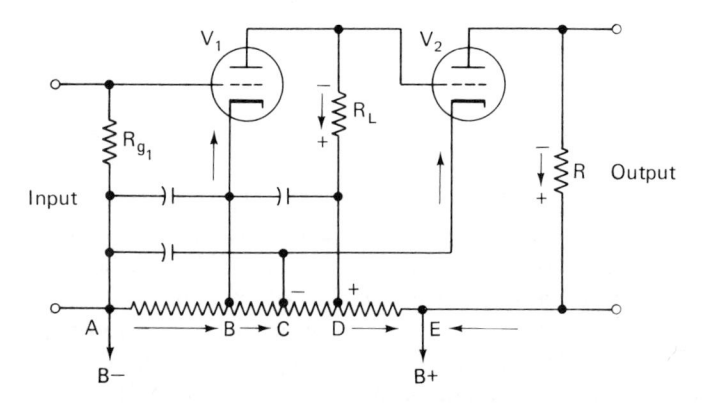

FIG. 13-10 Direct coupled amplifier.

Distortion 211

more than two stages are used in this type of amplifier, it is difficult to achieve stable operation. Any small changes in the voltages of the first tube will be amplified, and will thus make it difficult to maintain proper bias on the final tube connected into the circuit. Because of the instability thus encountered, direct-coupled amplifiers are practically always limited to two stages. Furthermore, the power supply must be twice that required for one stage. This is known as a *Loftin–White* amplifier.

13-8 DISTORTION

Normally, distortion of the output waveform of an amplifier is undesirable. However, because of some nonlinearity in all active devices, there is always a certain amount of distortion produced. Fortunately, this can be kept to a minimum by good circuit design and improved active devices. The types of distortion that are produced fall into three categories: *amplitude*, *frequency*, and *phase distortion*.

Amplitude or nonlinear distortion occurs whenever the signal operates over a *nonlinear section* of an amplifier's characteristic curve. This can be caused by improper bias or too large an input signal. Severe amplitude distortion is produced if peak clipping occurs as a result of driving the tube into cutoff or saturation.

Frequency distortion results when some frequency components of a complex signal are amplified more than others. This occurs because of reactances present in the circuits. Inasmuch as these vary with frequency, some discrimination will occur, principally at the low and high ends of the passband. Frequency distortion can be kept to a minimum by ensuring that the *bandwidth* of the amplifier is wide enough to include all significant frequency components of the signal.

Phase distortion exists when the phase relationships between the frequency components of the output signal are not the same as in the input signal. Phase shift results when an amplifier becomes highly reactive at frequencies well above and below midband. At these frequencies the phase shift becomes nearly 90°. Thus, if high-, medium- and low-frequency signals are simultaneously applied to the input of an amplifier, the midrange frequency will be amplified with little or no phase shift, the higher frequencies will lag, and the lower frequencies will lead with respect to their original phases. Since the three signals have been shifted in time, relative to each other, they no longer produce the same complex waveform when added together in the output of the amplifier.

13-9 AMPLIFIER GAIN

Amplifier stages can provide *voltage*, *current*, or *power gain* depending on circuit design and the type of active device used. The first few stages of an amplifier usually provide voltage gain, the output stages provide the power gain.

For triode vacuum tubes the voltage gain can be calculated as

$$A_v = \frac{\mu R_L}{r_p + R_L}$$

This indicates that the voltage gain, A_v, is determined in large measure by the value of R_L. However, if R_L becomes excessively large (i.e., more than 10 times r_p), the tube's internal plate resistance begins to increase faster than R_L and the gain will decrease.

Example 13-5

Calculate the voltage gain of a triode whose circuit parameters are $\mu = 20$, $r_p = 7.7$ kΩ, and $R_L = 25$ kΩ.

Solution

$$A_v = \frac{20 \times 25 \text{ k}\Omega}{7.7 \text{ k}\Omega + 25 \text{ k}\Omega} = \frac{500 \text{ k}\Omega}{32.7 \text{ k}\Omega} = 15.3 \text{ k}\Omega$$

Typical values of R_L for triodes is about 5 to 7 times r_p. In special cases it may vary from 2 to as high as 10 times r_p.

A different condition exists in the case of a pentode. The r_p of a typical pentode is on the order of 1 MΩ. If the same rule for the triode were to apply, then R_L should be about 5 to 7 times r_p. This would mean that the total series resistance of r_p plus R_L would be about 6 MΩ. This would force the tube to operate nearly at cutoff, and severe distortion would result. A practical approach to this problem is to use a value of R_L that is 10 to 25 per cent of r_p. These values have been determined empirically. The voltage gain at midfrequency becomes

$$A_v = g_m R_L$$

For a more exact calculation, the shunting effect of the input resistance of the following stage should be taken into consideration. The parallel effect of R_L and R_g can be called R_{eq} for equivalent resistance. The gain formula for the pentode then becomes

$$A_v = g_m R_{eq}$$

Example 13-6

Calculate the gain of the pentode section of a 6AU8 if $g_m = 8000$ μS and $R_{eq} = 47$ kΩ.

Solution

$$A_v = 8 \times 10^{-3} \text{ S} \times 47 \times 10^3 \text{ } \Omega = 536$$

Comparison of Transistor Configurations 213

The formula shown in the preceding paragraph is a general one that can apply to all active devices.

The power gain of a transistor stage is calculated by finding the product of the current and voltage gains. Expressed mathematically,

$$A_p = A_i A_v$$

Example 13-7

What is the power gain of a transistor if its current gain is 38.4 and voltage gain is 476?

Solution

$$A_p = A_i A_v = 38.4 \times 476 = 18{,}278$$

13-10 COMPARISON OF TRANSISTOR CONFIGURATIONS

A comparison of the three transistor amplifier configurations is shown in Table 13-1. These apply to small signal amplifiers and not power devices. The input and output resistances may vary over broader limits than shown, depending upon such things as the input signal level, quiescent current level through the device, and the kind of transistor used.

TABLE 13-1 COMPARISON OF TRANSISTOR CONFIGURATIONS

Item	CB *Amplifier*	CE *amplifier*	CC *Amplifier*
Input resistance	30–150 Ω	500–1500 Ω	20–500 kΩ
Output resistance	300–500 kΩ	30–50 kΩ	50–1000 Ω
Voltage gain	500–1500	300–1000	Less than 1
Current gain	Less than 1	25–50	25–50
Power gain	20–30 dB	25–40 dB	10–20 dB

13-11 PARAPHASE AMPLIFIERS

Paraphase amplifiers (phase splitters) produce, from a single input waveform, two output waveforms that have exactly equal but opposite instantaneous polarities. This type of circuit is used to drive such circuits as push–pull amplifiers.

The simplest method of obtaining two signals with equal magnitudes and opposite polarities is by the use of a transformer with a center-tapped secondary, as shown in Fig. 13-11a. When no center tap is available, two resistors of equal value may be connected across the secondary to achieve the same results, as shown in Fig. 13-11b. The dots by the primary and secondary windings indicate like instantaneous polarities.

The transformer phase splitter has several disadvantages, among which are its size and cost. It also has limited application because of distortion and losses inherent in transformers. For example, the loss in voltage through

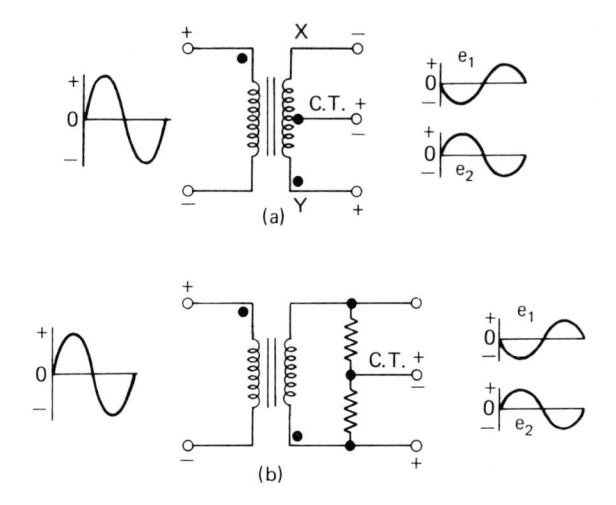

FIG. 13-11 Transformer phase splitters: (a) with center-tapped secondary, (b) without center tap.

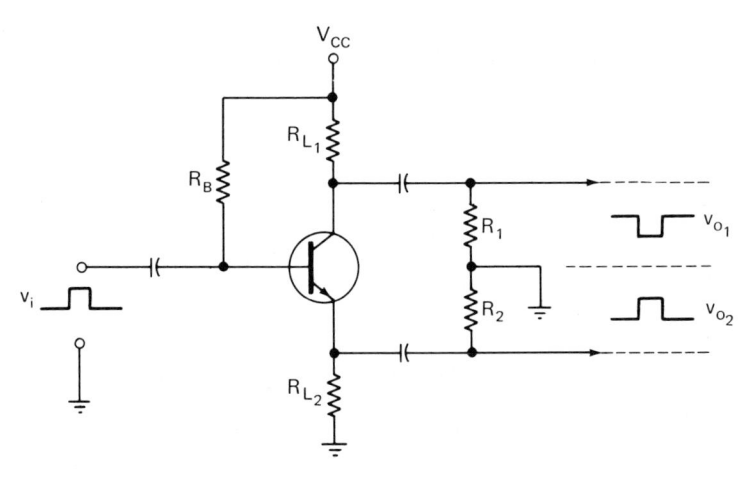

FIG. 13-12 Single transistor paraphase amplifier.

leakage reactance is greater for higher frequencies than it is for lower frequencies. The shunting capacitance effect and hysteresis losses also increase with frequency. Since in many circuits harmonics must be transmitted unattenuated and undistorted, the transformer phase inverter is generally replaced with a solid-state or vacuum-tube circuit that performs phase inversion without the use of transformers.

The simplest type of phase splitter is shown in Fig. 13-12. For convenience, a positive square-wave input signal is shown. Because of the 180° phase inversion across the transistor (the same is true for vacuum tubes and

FETs when connected in this manner), the signal at the collector swings negative. A characteristic of this circuit is that half the load, R_{L1}, is in the collector circuit while the other half, R_{L2}, is in the emitter circuit. Because i_E and i_c are in phase, the signal at the emitter will increase in amplitude in a positive direction (in phase with the input signal), while the collector signal swings in the negative direction. For balanced output, R_{L1} must equal R_{L2}. Likewise, R_1 and R_2 must be equal—but not necessarily the same ohmic value as R_{L1} or R_{L2}. This type of circuit is frequently called a *split-load phase inverter*.

The voltage gain of this circuit is always less than 1 because of the amount of *degeneration* resulting from R_{L2} being unbypassed. At higher frequencies the output of the two phases may not be equal, due to the varying *distributed and interelement capacitances* in the circuit.

Other phase-inverter circuits are available using two active devices that can provide considerable voltage gain. To maintain both outputs at the same level, it is necessary to take part of the output of the first stage and divide it down so that it is equal in amplitude to the input of this stage. This signal is then coupled to the second stage of the phase inverter, where it is amplified. If both active devices are *matched*, the two output signals will not only bear a 180° relationship to each other but will have the same amplitudes.

This type of circuit has some disadvantages. One is that it can only be perfectly balanced over a relatively narrow range of frequencies. This is because of the phase shift introduced at the low- and high-frequency ends of the band as a result of reactances present. A second disadvantage is that the output of the second device has more *amplitude distortion* than that from the first stage. This is because the distortion produced by the first stage is fed to the second, which amplifies and distorts it an additional amount because of its own nonlinear operating characteristics.

13-12 PUSH–PULL POWER AMPLIFIERS

The primary function of a power amplifier is to *efficiently* deliver power to some kind of load, such as a loudspeaker or antenna. The power amplifier may be operated as a *single-ended* or as a *push–pull stage*. A single-ended amplifier stage may consist of only one active device or of two or more connected in parallel. Such a parallel arrangement provides greater power output for a given input than do single active devices of the same kind. Single-ended power amplifiers can only be operated class A.

A typical push–pull audio amplifier using vacuum tubes is shown in Fig. 13-13. This configuration will provide a greater power output for a given amount of distortion than will comparable tubes used in parallel. Push–pull amplifiers can be biased for either class A, AB, or B operation. More power output is obtainable in class B than class A or AB. Audio power amplifiers are *never* operated class C due to the extreme distortion that would result.

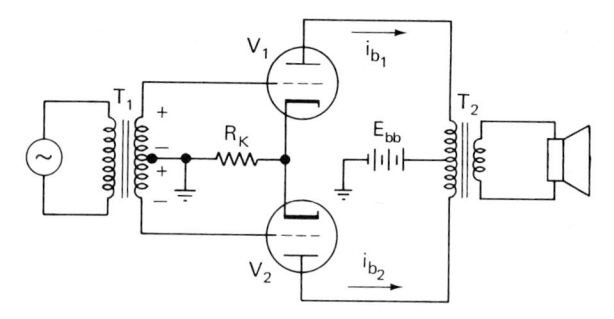

FIG. 13-13 Simple push–pull amplifier.

The following are requirements for push–pull operation:

1. The input signals must have equal amplitudes and opposite polarity.
2. The circuit must be balanced so that the signals will be amplified equally.
3. The active devices must be matched.

An analysis of the push–pull circuit shown in Fig. 13-13 is as follows: since equal plate currents flow through each half of the primary in opposite directions, the resulting magnetic fields are equal in intensity but are opposing each other. Thus, the magnetizing effect of the direct currents on the iron core is *canceled*, and there is no dc core saturation of the output transformer.

Bias for the push–pull amplifier may be provided by a bias tap in the power supply (fixed bias) or, as shown, by a common cathode resistor. The signals applied to the tubes will have the instantaneous polarities shown; that is, the grid of V_1 is positive with respect to the center tap at the instant the grid of V_2 is negative with respect to the center tap. Plate current increases through V_1 (i_{b_1}) and decreases through V_2 (i_{b_2}). The proportion of increase and decrease through each tube is equal. Assume that i_{b_1} flowing through the top half of the primary produces a counterclockwise field. Then i_{b_2} flowing up through the lower half of the primary will produce a clockwise field. This is so because the entire primary is wound in the same direction, but the current flows in opposite directions through each half. Thus, if an expanding counterclockwise field induces a positive voltage in the secondary (caused by an increase in i_{b_1}), the collapsing clockwise field caused by the decreasing i_{b_2} will also induce a positive voltage. This will occur for one half-cycle of the inputs, with conditions reversing during the other half-cycle. This makes the outputs of the two tubes *additive* at all times. If both fields expand or collapse equally at the same time, there would be no voltage induced in the secondary. This explains the necessity of two input signals of opposite polarity.

Some distortion of the signal is produced by each tube as they are alternately driven out of cutoff. This is due to the nonlinearity of the $E_G I_p$ curve near the cutoff region. These distortion products contain harmonics of

Transistor Amplifier

the fundamental, that is, second, third, fourth, fifth, and so on. The first three or four are, or would be, the troublesome ones, except that the *even* harmonics cancel in the primary winding of the output transformer. The power represented by the odd harmonics is relatively low and causes little undesirable effects. While the signal currents of the fundamental frequency in the primary of the output transformer are going in opposite directions at any instant of time, and therefore are *additive*, the even harmonics are going in the *same direction* (i.e., increasing or decreasing at the same instant) and *cancel*.

If one tube in a class A push–pull amplifier should burn out, the amplifier will still operate, but at about one half the normal output. If class B operation is used, only half the input signal will be reproduced. Therefore, severe distortion results.

The power output of an amplifier can be easily calculated if the voltage across the load is known, as shown by the following example:

Example 13-8

What is the power output of an audio amplifier if the rms voltage across a 500-Ω load is 40 V?

Solution

$$P_o = \frac{e^2}{R_L} = \frac{40^2}{500} = 3.2 \text{ W}$$

Another advantage of push–pull operation is that there is no hum in the output caused by power-supply ripple. It cancels out in the primary of the output transformer.

13-13 TRANSISTOR AMPLIFIER

A simple three-stage transistor amplifier employing a class B output stage appears in Fig. 13-14. One characteristic of the circuit is that the first stage, a *pnp* transistor, is directly coupled to an *npn* stage. The advantages of this are simplicity of design and excellent frequency response. Because no coupling capacitor is used, the response of this part of the overall circuit is from direct current to whatever the frequency cutoff of the transistors might be. The frequency response of the entire amplifier is limited only by the characteristics of the driver and output transformers. A disadvantage of direct coupling, such as between Q_1 and Q_2, is that great care must be exercised in selecting suitable transistors and in designing the bias network. The collector potential of Q_1 must be properly established (by selecting the proper values for voltage-divider bias resistors R_1 and R_2) so that the quiescent point for Q_2's operation will provide linear operation. If for any reason the beta of Q_1 should change over a period of time, the bias of Q_2 will be seriously affected.

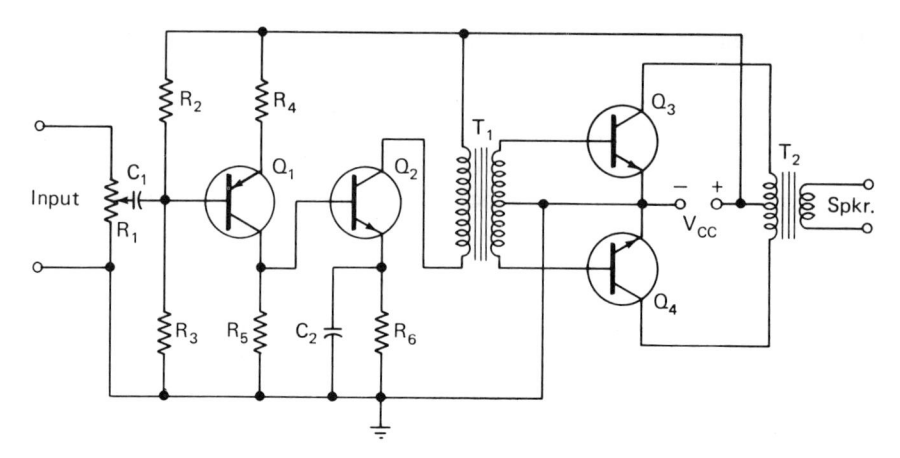

FIG. 13-14 Basic three-stage class B AF amplifier using direct coupling between the first two stages.

The push–pull output stage is biased class B. No base current flows, and hence no collector current, in the absence of a signal.

13-14 COMPLEMENTARY AMPLIFIERS

The output transformer can be eliminated in class B transistorized amplifiers if *complementary* transistors are used. These are two transistors, an *npn* and *pnp*, that have identical electrical characteristics. With transistors of this type it is possible to design a series-output type of power amplifier that does not require push–pull drive. Hence, no phase inversion is necessary and the driver circuit is substantially simplified. An example of this configuration is shown in Fig. 13-15. Transistor Q_1 is the driver. The signal developed

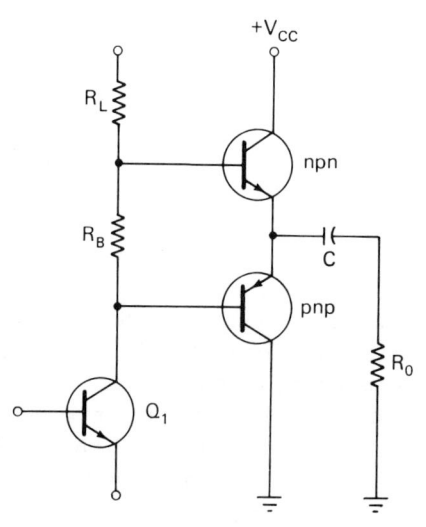

FIG. 13-15 Basic complementary amplifier.

Inverse Feedback

across its load resistor, R_L, is applied to the bases of the complementary output transistors. When the signal at the collector of Q_1 is on the positive half-cycle, the *npn* transistor is driven on while the *pnp* transistor is driven past cutoff. On the other half-cycle the condition reverses. Hence, a push–pull type output is achieved.

A small amount of forward bias is required for the complementary transistors to eliminate *crossover distortion*. This is caused by each transistor as it is driven out of cutoff. Bias resistor R_B provides just enough voltage to operate these transistors above the "heel" of their dynamic transfer curves (class AB). In practice, a forward-biased diode is used in place of R_B. Its small IR drop is sufficient to maintain the quiescent current at a reasonable value with variations in junction temperatures. It is usually mounted on the *heat sink* that the complementary transistors are on, so that it *tracks* with the V_{BE} of these transistors.

13-15 INVERSE FEEDBACK

The performance of an amplifier can be substantially improved by the use of *inverse* or *negative feedback*. This involves taking a small percentage of the output and feeding it back out of phase to a preceding stage. The overall gain will be *reduced* by this action, but several desirable characteristics result. The total harmonic distortion produced by the amplifier is reduced, and the frequency response increased. Also the *gain* of the amplifier is stabilized. Because of these important advantages, most amplifiers use some inverse feedback.

To understand how the gain of an amplifier is influenced by negative feedback, refer to Fig. 13-16. A portion of the output voltage is taken from the potentiometer and fed back, in series opposition with v_s, to the input of the amplifier. The Greek letter β (beta) is used to signify the amount of feedback voltage. This is a decimal fraction and is always less than 1. Hence, the feedback voltage is designated as βv_o. For example, suppose the v_o of an amplifier is 10 V and that one tenth of this is fed back to the input. Then

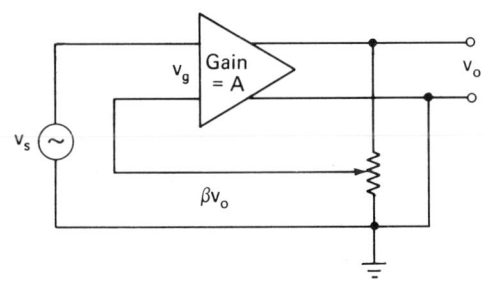

$$\beta = \tfrac{1}{10} = 0.1$$

FIG. 13-16 Negative feedback using series injection.

Negative feedback may be obtained in a number of ways. It may involve one or two stages, and in rare instances more than two stages. Also, it may use *voltage feedback*, *current feedback*, or a combination of voltage and current (compound) feedback.

An example of how negative feedback may be obtained across a single stage of amplification is illustated in Fig. 13-17. The feedback network consists of C_1, R_1, and R_2. The prime function of C_1 is to block the dc drain voltage from the gate circuit, yet present a very low impedance to all signals passed through the amplifier. By inspection it can be seen that R_1 and R_2 form a voltage divider or scaling network. The portion of v_o appearing at the gate of the FET is the ratio of R_2 to $R_1 + R_2$. This is *voltage feedback*.

Another method of obtaining negative feedback is seen in Fig. 13-18. This is called *current feedback*. Here the source resistor bypass capacitor has been omitted. The degenerative action may be analyzed as follows: assume that the input signal swings the gate voltage in a positive direction. The increase in drain current causes an increase in the voltage drop across R_s.

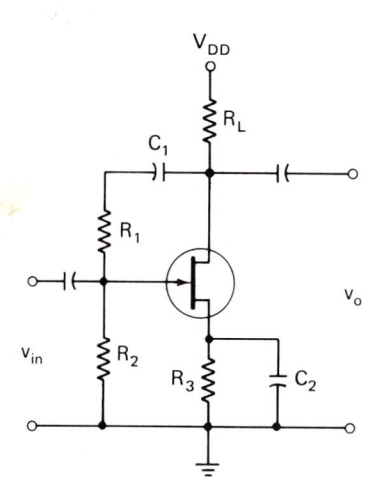

Fig. 13-17 FET amplifier stage using inverse feedback.

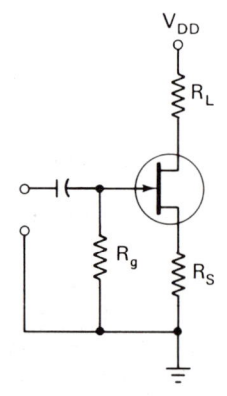

Fig. 13-18 Degenerative amplifier using current feedback.

Cathode and Emitter Followers

Since R_s is not bypassed, drain circuit signal currents flowing through it will add to the bias produced by the no-signal component. The gate-to-source voltage on the positive half-cycle is equal to the difference in the input and the drop across R_s. The magnitude of the gate voltage swing in a positive direction is not as great as it would be without feedback, because the drop across R_s is increased.

On the negative half-cycle the input signal swings the gate voltage in a negative direction and drain current decreases. The decrease in current through R_s causes a decrease in voltage across it. During this half-cycle, the gate-to-source voltage is equal to the sum of the input voltage and the drop across R_s. The magnitude of the negative swing of gate voltage is less than it would be without feedback, because the drop across R_s is less. The reason this is called current feedback is that the amount of current flowing through R_s determines the percentage of feedback.

If proper phase relations are established, negative feedback involving more than one stage may be used. Figure 13-19 shows a two-stage negative-feedback amplifier using voltage feedback. In this case, special attention must be paid to the phase relations throughout the circuit.

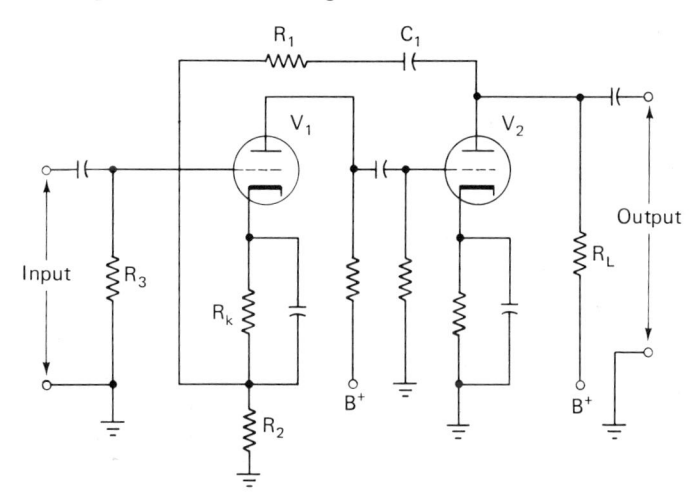

FIG. 13-19 Degenerative 2-stage amplifier employing voltage feedback.

13-16 CATHODE AND EMITTER FOLLOWERS

There are occasions when it is necessary to match a high-impedance source to a low-impedance load such as a transmission line. This can be accomplished by using the cathode-follower circuit shown in Fig. 13-20a. Although a transformer could possibly be used, this arrangement has the advantage of better frequency response and is usually less expensive. Notice that the load is in the cathode circuit. Because the signal current flowing

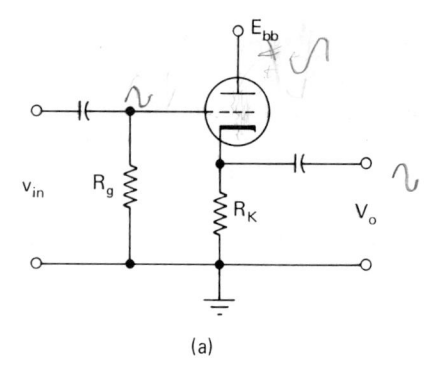

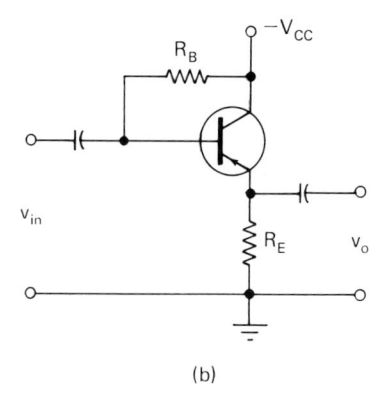

FIG. 13-20 (a) Cathode follower. (b) emitter follower or common collector amplifier.

through R_k is in phase with v_i, the output, v_o is following the phase of the input signal. Hence, the circuit is called a cathode follower. The circuit operates as a class A degenerative amplifier having a broad frequency response. The voltage gain of the stage is less than 1, although it can provide considerable power gain. The cathode biasing resistor must be unbypassed in this circuit.

The transistorized equivalent of the cathode follower is the *emitter follower* shown in Fig. 13-20b. This circuit possesses all the characteristics of cathode-follower stages. The ouptut impedance in both circuits is almost always considerably less than 1000 Ω.

Practice Problems

1. What is the approximate cutoff voltage of a 5893 low-power transmitting triode with a $\mu = 27$ and 350-V plate potential?
2. A 3–250 A2/250 TL medium-power transmitting triode is operated at 2500-V_{dc} plate potential. Find the mu of the tube if the cutoff voltage is −178 V.
3. What is the ac power output of the tube in problem 2 if it is operating under the following circuit parameters: $E_b = 2500$ V, $I_b = 225$ mA, and the efficiency is 71 per cent?

Commercial License Questions

4. A 6CU6 self-biased beam power amplifier is operated in a circuit having the following parameters: $E_b = 250$ V, $I_b = 55$ mA, $E_{c2} = 150$ V, $I_{c2} = 2.1$ mA, and a bias voltage of -22.5 V. What ohmic value of cathode biasing resistor is required?

5. If the circuit referred to in problem 4 is designed to pass frequencies as low as 30 Hz, what value of capacitor is required to properly bypass the cathode biasing resistor?

6. A triode operates in a circuit having the following parameters: $r_p = 7700$ Ω, $R_L = 20$ kΩ, $\mu = 20$, $g_m = 2600$, and an input signal of 150 mV$_{p-p}$. Calculate the gain of the stage.

7. How much gain will a tube provide if the following conditions exists: plate load resistor is 34 kΩ, amplification factor is 70, plate resistance is 53 kΩ, and the plate voltage is 275 V?

8. A certain pentode has a transconductance of 2600 μS, a load resistance of 220 kΩ, an I_b of 1.7 mA, and is connected to a stage whose grid resistor is 500 kΩ. What is the gain of the stage?

9. What is the voltage gain of an amplifier having a transconductance of 1300 μS and a load resistance of 27 kΩ?

10. The final power amplifier tube of a transmitter has a plate voltage of 2150 V_{dc} at 3.42 A. If the power output is 5 kW, what is the stage's efficiency?

Commercial License Questions

Sections in which answers to questions are given appear in parentheses. A bracketed number following a question implies that it applies only to that element.

1. What are the frequency ranges included in the following frequency subdivisions: MF (medium frequency), HF (high frequency), VHF (very high frequency), UHF (ultrahigh frequency), and SHF (superhigh frequency)?
(13–1, Table 4–1)
2. Describe what is meant by a *class A amplifier*. (13–1)
3. What is meant by the *load* on a vacuum tube? (13–2)
4. Draw a graph indicating how the plate current in a vacuum tube varies with plate voltage, grid bias remaining constant. (13–2, 13–3, 13–4)
5. Explain the operation of a triode vacuum tube as an amplifier. (13–2)
6. Does a properly operated class A audio amplifier produce serious modification of the input waveform? (13–2)
7. Describe the characteristics of a vacuum tube operating as a class A amplifier. (13–2)
8. What are the characteristics of a class A audio amplifier? (13–2)
9. What will be the effect of incorrect grid bias in a class A audio amplifier? (13–2, 13–5)

10. When a signal is impressed on the grid of a properly adjusted and operated class A audio-frequency amplifier, what change in average value of plate current will take place? (13–2)

11. Does dc grid current normally flow in a class A amplifier employing one tube? (13–2, 13–5)

12. Describe the characteristics of a vacuum tube operating as a class B amplifier. (13–3)

13. During what portion of the excitation voltage cycle does plate current flow when a tube is used as a class B amplifier? (13–3)

14. What is the approximate efficiency of a class A vacuum-tube amplifier? class B? class C? (13–2, 13–3, 13–4)

15. Why does a class B audio-frequency amplifier stage require considerably greater driving power than a class A amplifier? (13–3)

16. Discuss the input circuit requirements for a class B audio-frequency amplifier grid circuit. (13–3)

17. Why are tubes, operated as class C amplifiers, not suited for AF amplification? (13–4)

18. During what approximate portion of the excitation voltage cycle does plate current flow when a tube is used as a class C amplifier? (13–4)

19. Draw a diagram of a resistance load connected in the plate circuit of a vacuum tube and indicate the direction of electronic flow in this load. (13–5)

20. Is the dc bias normally positive or negative in a class A amplifier? (13–5)

21. What is the purpose of a bias voltage on the grid of an AF amplifier tube? (13–5)

22. What are the factors that determine the bias voltage for the grid of a vacuum tube? (13–5)

23. Explain how you would determine the value of cathode bias resistance necessary to provide correct grid bias for any particular amplifier. (13–5)

24. What is the purpose of bypass condensers connected across an AF amplifier cathode bias resistor? (13–5)

25. Draw a diagram showing a method of obtaining grid bias to an indirectly heated cathode-type vacuum tube by use of a resistance in the cathode circuit of the tube. (13–5)

26. What factors may cause low plate current in a vacuum-tube amplifier?

27. What is meant by a *blocked grid*? (13–5)

28. Draw a simple schematic circuit showing a method of resistance coupling between two triode vacuum tubes in an AF amplifier. (13–6)

29. What would be the effect of a short-circuited coupling condenser in a conventional resistance-coupled audio amplifier? (13–6)

30. If the value of capacitance of a coupling condenser in a resistance-coupled audio amplifier is increased, what effect may be noted? (13–6)

Commercial License Questions 225

31. Draw a simple schematic diagram of a method of impedance coupling between two vacuum tubes in an AF amplifier. (13-6)
32. Draw a simple schematic diagram showing a method of transformer coupling between two triode vacuum tubes in an AF amplifier. (13-6)
33. Draw a diagram illustrating direct or Loftin–White coupling between two stages in AF amplification. (13-6)
34. Draw a simple schematic diagram of a triode vacuum-tube AF amplifier inductively coupled to a loudspeaker. (13-6)
35. Draw a simple schematic circuit showing a method of coupling a high-impedance loudspeaker to an AF amplifier tube without flow of tube plate current through the speaker windings, and without the use of a transformer. (13-6)
36. What circuit and vacuum-tube factors influence the voltage gain of a triode AF amplifier stage? (13-8)
37. Under what circumstances will the gain per stage be equal to the voltage amplification factor of the vacuum tube employed? (13-8)
38. What is the most desirable factor in the choice of a vacuum tube to be used as a voltage amplifier? (13-8)
39. What is the stage amplification obtained with a single triode operating with the following constants: plate voltage 250, plate current 20 mA, plate impedance 5000 Ω, load impedance 10,000 Ω, grid bias 4.5 V, and amplification factor 24? [4] (13-8)
40. What are the advantages of using two tubes in push–pull as compared to the use of the same tubes in parallel in an AF amplifier? (13-11)
41. What are the advantages of push–pull operation compared to single-tube operation in amplifiers? (13-11)
42. What will occur if one tube is removed from a push–pull class A audio-frequency amplifier stage? [4] (13-11)
43. What is the power output of an audio amplifier if the voltage across the load resistance of 500 Ω is 40 V? [4] (13-11)
44. Draw a simple schematic circuit diagram of a two-stage audio amplifier using transistors. (13-12)
45. Why is degenerative feedback sometimes used in an audio amplifier? [4] (13-14)
46. What is the purpose of deliberately introduced degenerative feedback in audio amplifiers? [4] (13-14)
47. Draw a diagram of an audio amplifier with inverse feedback. [4] (13-14)
48. In a low-level amplifier using degenerative feedback, at a nominal midfrequency, what is the phase relationship between the feedback voltage and the input voltage? [4] (13-14)

14

Radio-Frequency Amplifiers

14-1 RADIO–FREQUENCY TRANSFORMER COUPLING

The amplifiers studied thus far are basically *untuned*. When an amplifier is designed for use at radio frequencies, it is usually connected to *tuned circuits*. By amplifying only a certain band of frequencies the amplifier can be made to perform better, that is, provide higher gain, usually with a better signal-to-noise ratio. Examples of tuned amplifiers are found in the RF and IF (intermediate frequency) amplifiers in superheterodyne receivers and in all kinds of transmitter circuits.

One of the most conventional methods of coupling two RF stages is by the *double-tuned* transformer. This is used when a single frequency or a narrow band of frequencies is to be amplified. The resonant conditions in the network result in a voltage gain characteristic that is very selective. In these stages *selectivity* is of primary importance. In some practical applications a manual adjustment changes the degree of coupling between the primary and secondary windings of the transformer. This permits a maximum transfer of energy with a specified amount of selectivity. Occasionally, a double-tuned transformer may be *overcoupled* to produce a widespread characteristic.

An example of double-tuned or RF transformer coupling is shown in Fig. 14-1. The primary winding of the IF transformer is permeability tuned to resonate at 455 kHz with C_1 connected in parallel. The secondary is also permeability tuned to the same frequency. The relatively low impedance of the input circuit of Q_2 is matched to the IF transformer secondary by tapping down on the coil. The dotted rectangle around the transformer indicates a light-weight metallic can or shield. This prevents stray magnetic fields from inducing unwanted signals into the windings and also prevents their radiating

Radio–Frequency Transformer Coupling

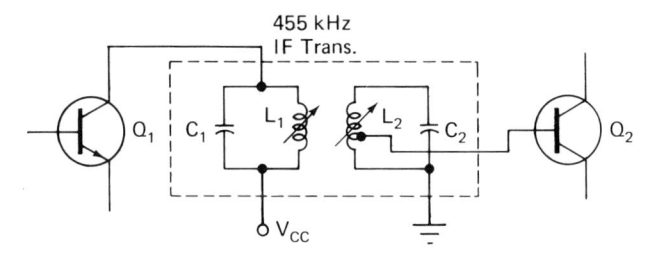

FIG. 14-1 Double-tuned amplifier.

energy into adjacent components. A small opening at each end of the can permits adjusting the powdered iron slug in each winding to determine the correct amount of inductance for resonance.

The double-tuned transformer-coupled amplifier has a passband characteristic that depends in part on the *degree of coupling* and on the *circuit Q*.

Under proper operating conditions, essentially uniform amplification of a relatively narrow band of frequencies may be achieved, and amplification of frequencies outside this band may be sharply reduced.

Since the slope of the response curve is not perfectly vertical, the circuit cannot completely *discriminate* against frequencies just outside the desired channel without also attenuating to some extent the frequencies at the upper and lower limits of the passband. However, double-tuned amplifiers approach an ideal band-pass charactistic much more closely than single-tuned amplifiers, which have rounded response curves.

A better understanding of the gain at resonance, as well as the response throughout the passband, may be gained from the curves of secondary current versus frequency, as shown in Fig. 14-2a.

When the coefficient of coupling is *low* (less than critical), the response frequency and the *passband are very narrow*. As the coupling is increased to the critical value, maximum current flows in the secondary, and the output voltage across the secondary is also at its maximum. The passband is still relatively narrow and would attenuate the sideband frequencies farthest removed from the resonant frequency.

If the coupling is increased until the optimum value is reached, the gain is still relatively high; but the passband has been increased and the response is essentially uniform.

As the coupling is again increased, the *humps* at F_1 and F_2 are well defined, and the gain at resonance is considerably reduced. Although the passband is now much wider, the gain throughout the band is not sufficiently uniform.

The two humps in the curve are due to the reactance that is coupled into the primary on each side of resonance as the coupling is increased. Below resonance this reactance is inductive, and above resonance it is capacitive.

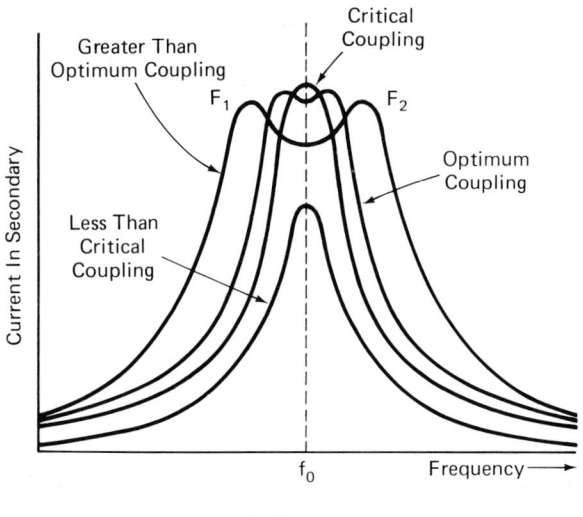

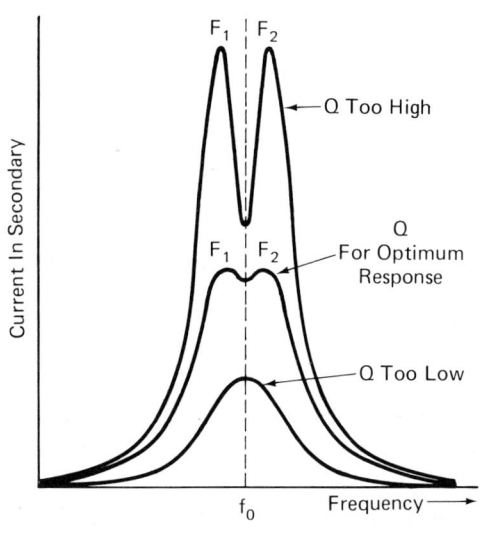

FIG. 14-2 Response curves for a double-tuned, transformer-coupled amplifier.

For the same frequency the coupled reactance has the opposite sign to that of the primary, and the impedance of the primary is therefore reduced. Accordingly, there is an *increase in primary current* at frequencies slightly off resonance, and a corresponding *increase in secondary induced voltage and current* at these frequencies.

Class C Amplifiers

The *width* of the passband may be as important as the response within the band. The bandwidth may be determined by the following formula:

$$BW = \frac{f_0}{Q}$$

where f_0 is the resonant frequency.

The frequencies at the two humps, F_1 and F_2, which define the practical lower and upper limits of the passband, are determined by the following equations:

$$F_1 = \frac{f_0}{\sqrt{1+k}}$$

$$F_2 = \frac{f_0}{\sqrt{1-k}}$$

where k is the coefficient of coupling.

Figure 14-2b shows the effects of varying the Q while maintaining a constant coefficient of coupling. Actually, the desired response curve could be achieved by the proper manipulation of both k and Q, because they are interrelated.

14-2 CLASS C AMPLIFIERS

Class C amplifiers are used principally in RF power applications because of their *increased efficiency*. A comparison of class A, B, and C amplifiers is shown in Fig. 14-3. The angle of current flow in a class A amplifier is 360°, due to its being biased in the center of the linear portion of the dynamic curve. The angle of current flow in the class B amplifier is about 180°, because it is biased at or near cutoff. The angle of current flowing in a class C amplifier is less than 180°; typically, it varies from 120 to 160°.

The *percentage of efficiency* is determined by the class of operation. The dc power supplied to an amplifier is always greater than the ac power it delivers to the load. Some power is used up by the device in converting part of the dc input power into ac energy for the load. The power used by the device is the product of its voltage drop and current. Since the angle of current flow is less than 180°, the *average* current is less than in class A or B for a given operation.

To better understand class C operation, refer to the schematic and waveforms of Fig. 14-4. From the indicated values, the cutoff bias is

$$E_{co} = \frac{-E_{bb}}{\mu} = \frac{-1000 \text{ V}}{20} = -50 \text{ V}$$

Using the value of three times cutoff and E_{co} of -50 V, the necessary bias is -150 V, which is supplied by E_{cc}. The voltage actually applied to the grid consists of the bias plus the driving or exciting voltage. If the peak

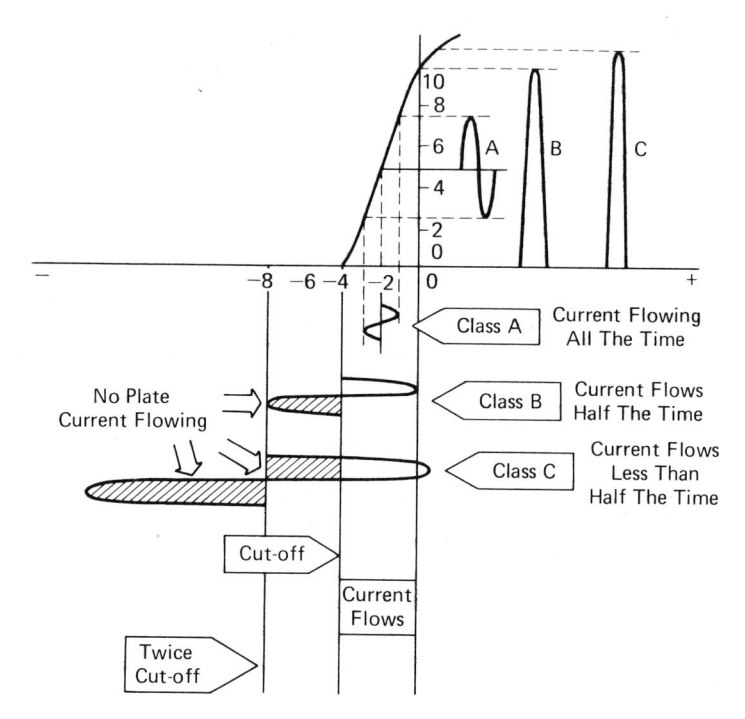

FIG. 14-3 Operation of Class A, B, and C amplifiers.

amplitude of the *exciting* voltage is 180 V when the grid end of the RF input tank circuit is positive, the peak positive grid-to-cathode voltage is 180 − 150 = +30 V. On the other half-cycle, the peak negative grid-to-cathode voltage is −330 V.

When the grid voltage is above cutoff, plate current flows. When the grid becomes positive with respect to the cathode, grid current will flow. If grid leak bias were used, this flow of grid current would produce the necessary bias for class C operation. The duration of grid current flow, measured in degrees, is labeled θ_g. If the distorted plate-current waveform is broken down into its associated harmonics, it is found to consist primarily of the *fundamental, second,* and *third harmonics*.

Since the parallel resonant circuit offers maximum impedance at the resonant frequency, a large voltage drop will be developed across the tank circuit during each pulse of plate current. The voltage will be maximum when the current reaches its maximum value. The instantaneous plate voltage is the algebraic sum of the $B+$ voltage and the ac voltage across the tuned circuit. When plate current starts flowing, the voltage at the plate end of the tuned circuit goes negative and subtracts from the $B+$ voltage, causing the plate voltage to decrease in value toward zero.

Class C Amplifiers

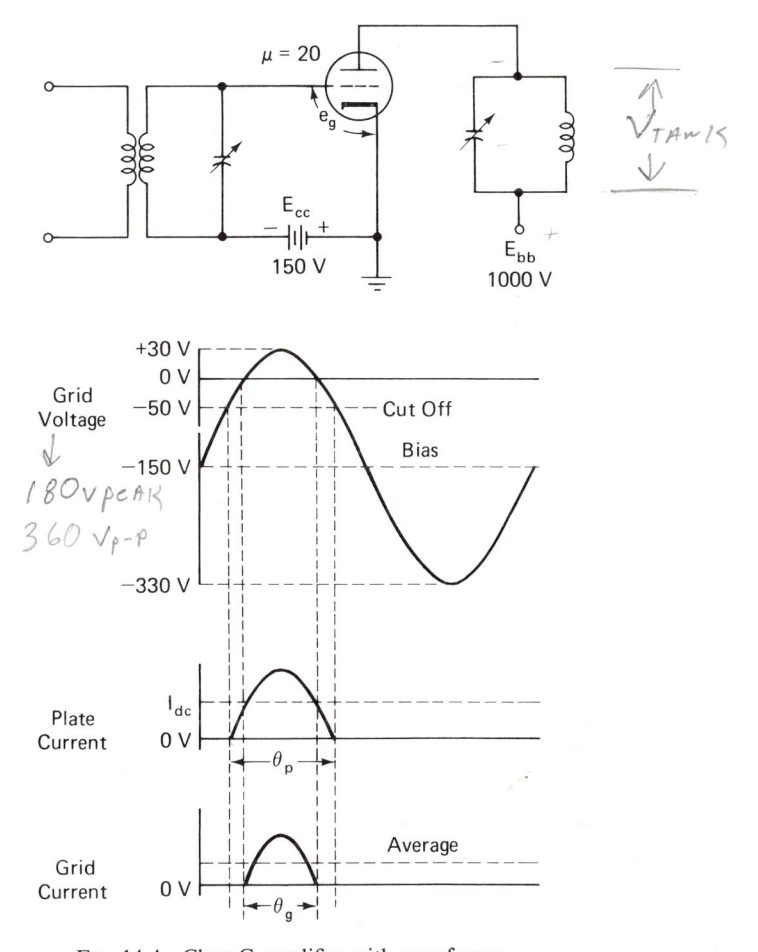

FIG. 14-4 Class C amplifier with waveforms.

When the tube is driven into cutoff, plate current ceases and the field about the inductor *collapses*, causing plate voltage to rise toward $B+$. The collapsing magnetic field of the inductor maintains current in the same direction through the capacitive branch of the tank circuit, charging the capacitor in the opposite direction. When the field has completely collapsed, the capacitor charge is maximum with a polarity *opposite* to the initial charge. When the voltage across the tuned circuit goes positive at the plate end, it adds to the $B+$ voltage, causing the instantaneous plate voltage to be nearly *double* the value of E_{bb}. Once fully charged, the capacitor begins to discharge, since there is no potential difference across it to sustain the charge. Its discharge path will be through the inductor. This action is known as the *flywheel effect*. When the grid signal again reaches a value sufficient to bring the tube out of cutoff, the tank circuit receives another burst of energy due to tube conduction. These large current pulses sustain the flywheel effect of the tuned tank.

14-3 GRID CURRENT LOADING

The grid circuit, while drawing current, causes some adverse effects that must be considered. Figure 14-5 shows the input circuit to an amplifier operated class C. The driving device will be considered to be a generator that possesses a large value of internal resistance (R_i). The tube is biased by the negative potential E_{cc} supplied to the grid, with the cathode returned to ground. No grid current flows until the input signal e_i rises sufficiently to equal and effectively remove the biasing voltage E_{cc}. Any additional rise of e_i drives the grid positive with respect to the cathode, and grid current flows.

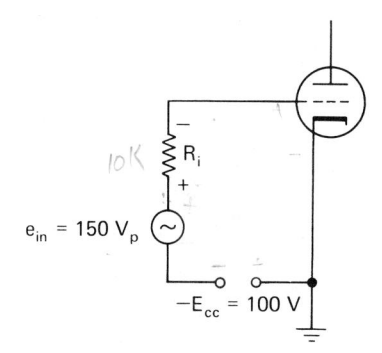

Fig. 14-5 Grid current loading.

When grid current flows, the grid-cathode resistance drops from its normally infinite value to something on the order of 1000 Ω. A typical value for the internal impedance of the generator is 10 kΩ. Consequently, most of the *IR* drop (i.e., $i_g R_i$) will be across the generator's *internal impedance*, and will be of a polarity that *opposes* the positive signal applied to the grid. For example, suppose that the fixed bias voltage $E_{cc} = -100$ V and that the peak value of grid current, i_g, is 10 mA. The *IR* drop across R_i will have a peak value of 100 V with the polarity indicated in Fig. 14-5. With a 150-V_p input signal, as indicated, the instantaneous voltage between grid and cathode will be 50 V_p. This causes a *flattening* of one half the plate-current waveform (positive half) and distortion results—called *peak clipping*. To reduce peak clipping as much as possible, a driving source with as low an output impedance as possible should be used.

Since there is grid current flowing during class C operation and practical generators possess some value of internal resistance, there will be an I^2R loss in the grid circuit. The power dissipated in the grid circuit is called the *driving* or *exciting power*. Its value is usually expressed as an average power. However, the instantaneous power dissipated at the grid will be the product

Series and Shunt Feed 233

of the instantaneous voltage and current. Part of the exciting or driving power will be dissipated in the form of heat at the tube grid, and part of it will be dissipated by the internal impedance of the bias source, or, if grid-leak bias is used, across the grid-leak resistor.

Example 14-1

What is the value of bias voltage in a class C amplifier operating under the following conditions: $E_b = 1000$ V, $I_b = 150$ mA, $I_G = 10$ mA, and with a 5-kΩ grid-leak resistor?

Solution

$E = IR$; $E = 0.01 \times 5000 = 50$ V. (*Note:* The plate voltage and current are not needed for the solution of this problem.)

The bias voltage can be increased or decreased by (1) changing the *amount of drive* to the amplifier, (2) varying the *coupling* between the driver stage and the amplifier, or (3) varying the *value of the grid-leak resistance*.

There is a disadvantage to using grid-leak bias alone for class C amplifiers. Consider what would happen if the excitation to the amplifier were to fail. The high value of plate voltage and current that would flow through the tube would cause *excessive plate dissipation* and the tube would likely be destroyed. Therefore, it is customary with class C amplifiers to use some external bias supply or to place a resistance in series with the cathode circuit in addition to the grid-leak bias. In this way the stage will not be left without some bias in the event that drive power fails. The voltage produced across the cathode-biasing resistor, although insufficient to bias the tube in class C operation, will at least *provide some bias* to prevent the plate current from reaching proportions that will destroy the tube. It is necessary to place an RF bypass capacitor across this resistor so that the cathode can be maintained at ground potential.

Many transmitters use an *overload relay* in their plate circuits. This is designed to open up the plate circuit if excessive current flows. If a tetrode or pentode is being used, the overload relay is designed to also open up the screen-grid circuit.

14-4 SERIES AND SHUNT FEED

Radio-frequency amplifiers may be either *series* or *shunt fed*. An example of these two methods of plate-current feeding is shown in Fig. 14-6. The series-fed circuit, although having slightly fewer parts, has the disadvantage of having the plate current flow through the tank inductor. This lowers the Q of the tuned circuit and can slightly reduce the output voltage. The shunt-fed circuit in Fig. 14-6b isolates the plate current from the tank circuit. The RF voltage built up across the *RFC* is capacitively coupled, by

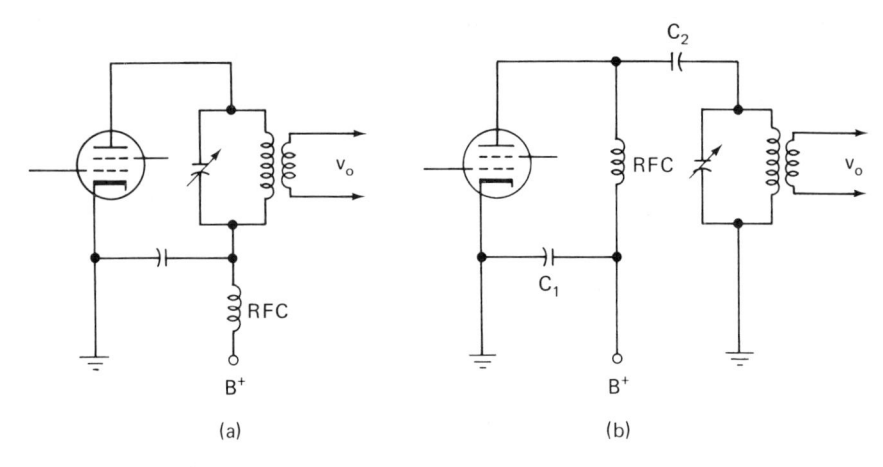

FIG. 14-6 RF amplifiers with (a) series-feed, (b) shunt-feed.

C_2, to the tank circuit. Capacitor C_2 also blocks the dc voltage from the tuned circuit, enabling one side of the tuning capacitor to be grounded. In both circuits capacitor C_1 serves to place the bottom side of the tank at RF ground potential. In either circuit the plate current will dip to a minimum value when the tank circuits are tuned to resonance. Either method of feed may be used with single-ended, push–pull, or parallel output stages.

14-5 INTERSTAGE COUPLING TECHNIQUES

A number of schemes have been devised for coupling two stages. Inductive coupling is frequently used, because the tuning of the primary and secondary windings can provide additional *selectivity* and *sensitivity*. However, as previously explained, if the coils are too close, overcoupling may result. Placing the windings farther apart will reduce this problem, but may occupy more space than is available. One method of eliminating this problem is the use of *impedance* or *capacitive coupling* between RF stages, such as shown in Fig. 14-7, which shows a method for coupling two tetrode stages together. The amount of coupling can be controlled in two ways: (1) by adjustment of the variable coupling capacitor, or (2) by varying the tap on the inductance of the tuned circuit. Remember that the bottom side of the tank is at RF ground potential, so that by placing the tap closer to the top of the tuned circuit more RF energy can be capacitively coupled to the next stage. The *RFC* in the grid circuit of the second tetrode prevents bypassing the coupled signal to ground, while at the same time permitting grid current to flow through resistor *R* to provide the necessary grid-leak bias. The simplicity of this coupling is readily apparent, but it does have one severe drawback in that harmonics or undesirable frequencies present in the tank circuit would be coupled to the next stage.

A very popular way of coupling RF energy between stages is by means

Interstage Coupling Techniques

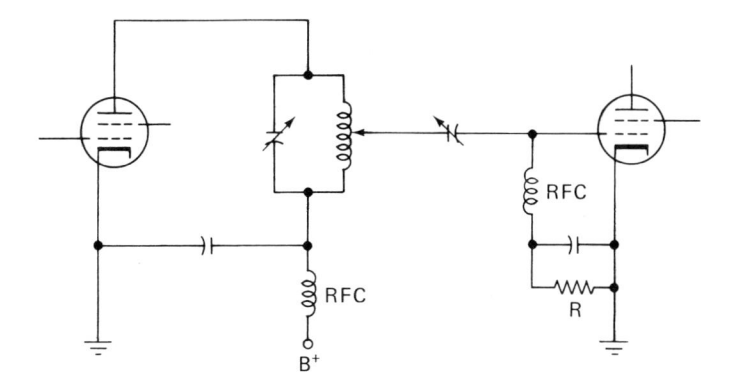

FIG. 14-7 Impedance or capacitive coupling between RF stages.

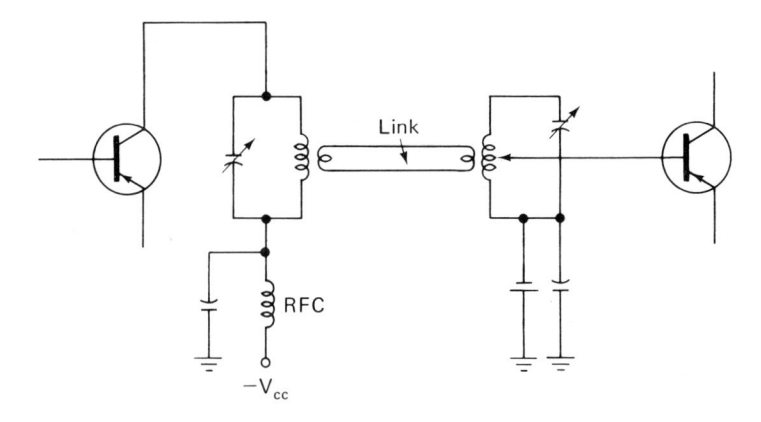

FIG. 14-8 Link coupling between tuned RF stages.

of a *link* or low-impedance transmission line. One of the principal advantages is that two tuned circuits may be separated by relatively large distances. The schematic diagram of a typical link-coupling circuit appears in Fig. 14-8. The link circuit consists of a very few turns of wire that are coupled to the low-impedance end of the tank circuit. The link is connected to a length of low-impedance transmission line and terminated at the other end by the same number of turns, which is again coupled to the low-impedance point of the input tank circuit. The low-impedance end of the tank is that end to which the RF bypass capacitor is connected. In push–pull circuits the low-impedance point would be the center tap and the link would have to be close to this point. Because of the low-impedance nature of the link or transmission line, great flexibility is available as far as mechanical construction is concerned. That is, the circuits being coupled need not be close together. The left-hand end of the link essentially operates as an RF *step-down transformer*, whereas the right side is an RF *step-up transformer*. The interelement capacitances of one active device have practially no effect on the impedance ratio of the other tank circuit.

14-6 NEUTRALIZATION—GENERAL

When triodes are used as RF amplifiers, the grid-plate capacity, although only a few picofarads, allows some of the amplified energy in the plate circuit to feed back to the grid circuit. If this voltage has the same frequency and phase as the grid voltage, oscillations will occur. Although this type of internal feedback is frequently employed in oscillator circuits, it is undersirable in amplifier applications because it causes *distortion, spurious radiations*, and interference to nearby radio receivers.

It is possible to eliminate these oscillations in triodes by a process called *neutralization*. This involves a network that feeds back to the grid a voltage which is at all times *equal but opposite in phase* to the voltage fed back to the grid through the plate-grid capacitance. Since the feedback voltage through C_{gp} is canceled out by voltage fed back through the external circuit, oscillations cannot take place. The use of well-shielded tetrodes or pentodes usually makes neutralization unnecessary, because the plate and grid are shielded from each other by the screen grid and its associated RF bypass capacitor, which holds the screen at RF ground potential. However, the overall efficiency of these tubes is not as great as that of triodes, since there is a screen-grid power loss.

There are several well-known neutralization systems in use. Two of these, the *plate* or *Hazeltine neutralization* system, and the *grid* or *Rice system*, have the advantage of being useful over a wide frequency range and derive their names from the part of the circuit in which the feedback voltage is developed.

14-7 PLATE NEUTRALIZATION (HAZELTINE)

Plate neutralization is the most frequently used and is shown in Fig. 14-9. This is a typical transformer-coupled RF amplifier to which have been added a *neutralizing inductance* L_2 closely coupled to L_1 and a *neutralizing capacitor* C_N. L_2 is connected in such a manner that the polarity of voltage at point B of this autotransformer is in phase opposition to the voltage at point A. The center of this autotransformer (point D) is placed at RF ground through the low-reactance bypass capacitor C_2. C_{gp} is the grid-to-plate internal capacitance, represented in the schematic as a capacitor external to the tube. C_N is the neutralizing capacitor through which the neutralizing signal is coupled to cancel the effects of C_{gp}.

The operation of the plate-neutralization circuit can best be understood with the aid of the equivalent circuit of Fig. 14-10. Point D of the plate tank circuit is effectively the same point as the bottom of the grid tank through the low reactance of C_2 (Fig. 14-9). The low reactance of the coupling capa-

Plate Neutralization (Hazeltine)

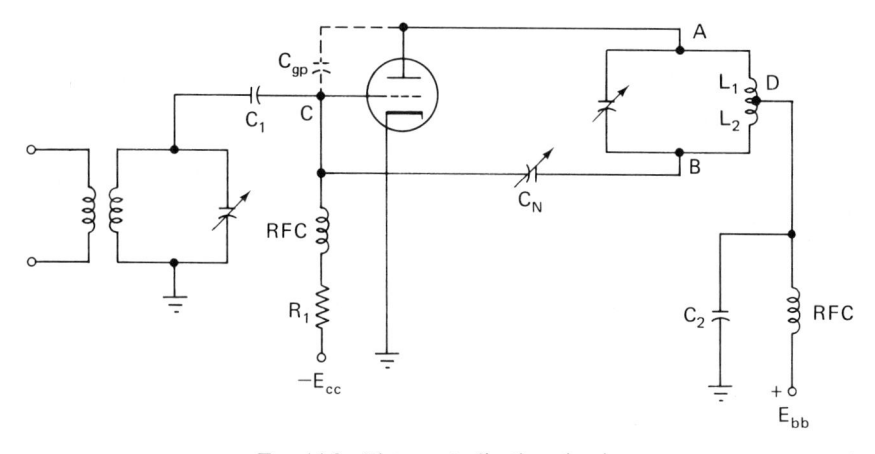

FIG. 14-9 Plate neutralization circuit.

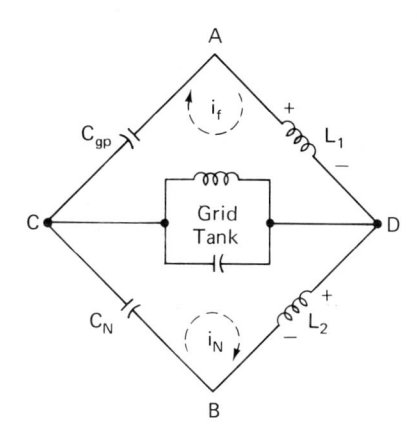

FIG. 14-10 Equivalent circuit of plate neutralization.

citor C_1 makes the top of the grid tank effectively the same as point C. Therefore, the plate-neutralization circuit resolves down to a *bridge arrangement* consisting of the grid tank circuit, L_1 and L_2, and C_{gp} and C_N (Fig. 14-10).

When the potential across the plate tank appears as in Fig. 14-10 (positive at the top and negative at the bottom), currents will be caused to flow in the directions indicated by the dotted arrows. When the plate tank voltage reverses its polarity, the direction of current flow will be opposite to the direction indicated. Assuming the potentials generated by L_1 and L_2 are equal and the values of capacitances C_{gp} and C_N are equal, the two currents will also be equal. When these currents are equal, there will be *no difference in potential* between points C and D and, therefore, no resultant current will flow through the grid tank circuit. No current flow indicates that no energy has been fed back from the output to the input; thus, no regeneration or degeneration can take place and the circuit is considered to be neutralized. If the neutralizing capacitor C_N were smaller in value than C_{gp}, its higher reactance would cause i_f to be the predominant current, resulting in a feedback current through the input tank circuit, which is *regenerative*. This regenerative signal would result in circuit oscillation. If the neutralizing capacitor were larger in value than C_{gp}, its lower reactance would cause i_n to be the predominant current. This current through the grid input tank would develop a feedback voltage that is *degenerative*. This degenerative signal will result in reduced output from the stage. Therefore, proper neutralization of an RF amplifier stage is realized when the feedback voltage through the neutralizing capacitor (C_N) cancels the feedback voltage through C_{gp}, resulting in no energy transfer from the output circuit to the input circuit.

14-8 GRID NEUTRALIZATION (RICE)

Another circuit that provides a means of neutralizing the grid-plate capacitance is the grid-neutralization circuit shown in Fig. 14-11. It differs from plate neutralization in that the split-tank circuit which provides the neutralizing voltage is located in the grid circuit. Except for this difference, the operation is somewhat the same as for plate neutralization.

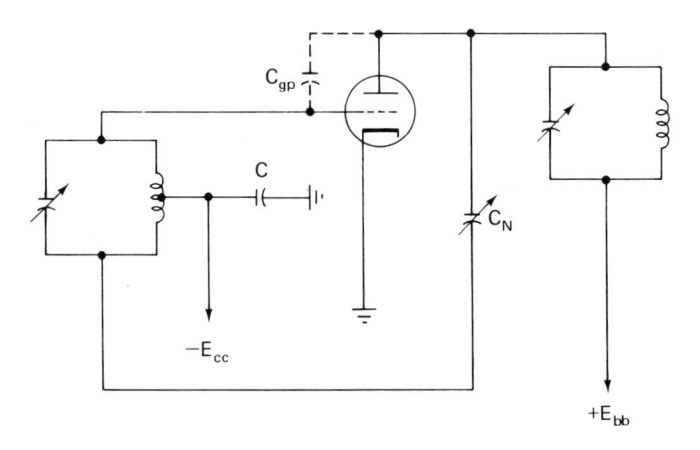

FIG. 14-11 Grid neutralization circuit.

The neutralizing capacitor C_N is connected between the plate and bottom end of the grid tank circuit. The low reactance of capacitor C places the center tap connection at RF ground. Initially, before the application of an input signal, C_{gp} and C_N will charge to the dc value of the plate potential. Their charge path is from $-E_{cc}$ through the respective capacitors to the $B+$ supply. When a signal is applied and the top of the grid tank circuit is positive, the plate current increases and the plate potential decreases. As the plate voltage decreases, C_{gp} and C_N will commence to discharge through the grid tank. If C_N equals C_{gp}, their discharge currents through the grid tank inductance will set up *opposing fields* and effectively neutralize each

Neutralization of Push–Pull Stages 239

other. If the feedback current through C_{gp} is greater, the resultant field will develop a feedback voltage across the grid tank that would be regenerative.

14-9 COIL NEUTRALIZATION

In *coil neutralization* (Fig. 14-12) the grid-plate capacitance can be nullified by paralleling it with an inductor having the same value of reactance. Since the two reactances are equal and opposite, a *parallel resonant* circuit exists between the grid and the plate. Therefore, there is no transfer of energy through the circuit from plate to grid due to the high impedance of the circuit. C_1 is a blocking capacitor that prevents the plate supply voltage from being felt on the control grid of the tube.

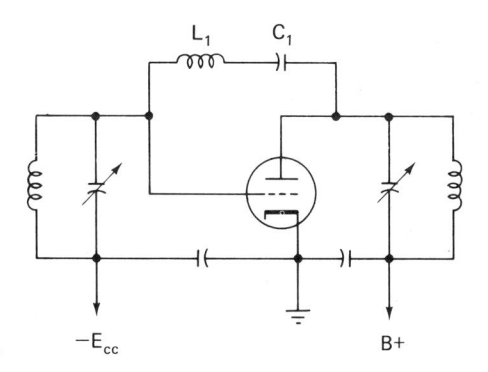

FIG. 14-12 Coil neutralization.

14-10 NEUTRALIZATION OF PUSH–PULL STAGES

An inspection of the circuit shown in Fig. 14-13 reveals that the circuit is a combination of both plate and grid neutralization. The two grid inputs and the two plate outputs (respectively) are *out of phase with each*

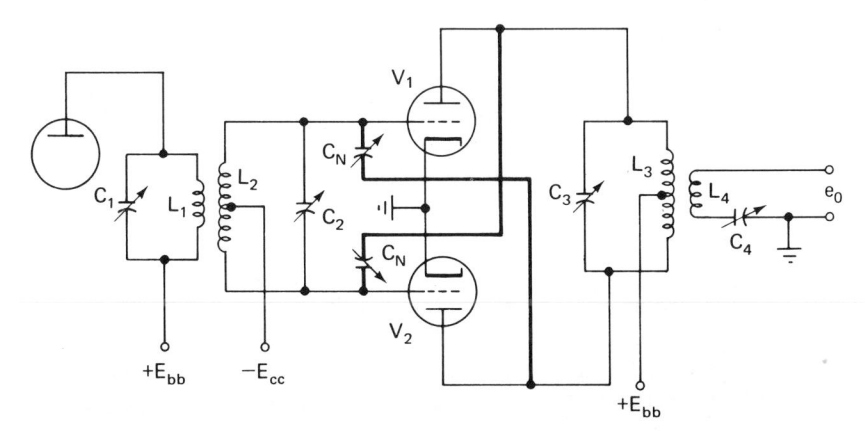

FIG. 14-13 Neutralization circuit for push–pull operation using triodes.

other. Hence, we already have the ingredients for grid or plate neutralization. All that is necessary is to feed the output of one tube through its appropriate neutralizing capacitor back to the input of the other tube. Such a circuit is also called *cross neutralization*. It is essential to adjust both neutralizing capacitors at the same time to maintain equal capacitance values for each. Circuits of this kind are inherently stable, assuming that they are evenly balanced electrically. When the stage is properly neutralized, it tends to remain so over a relatively wide band of frequencies.

14-11 NEUTRALIZATION PROCEDURES

The procedures for neutralizing are almost independent of the type of neutralizing circuit used. At the start of neutralization the *plate voltage is removed* from the stage to be neutralized so that any signal present in the plate circuit is due to the *interelectrode capacity* coupling between the grid and plate. Then the master oscillator and those amplifier stages that precede the unneutralized stage are tuned. This will provide a strong signal to the grid of the unneutralized stage. The next step depends on the indicator used, but it always results in the adjustment of the neutralizing capacitor (C_N) until there is a minimum amount of energy transferred to the plate circuit. Assume that an oscilloscope and small pickup coil connected to the vertical input are used to check for the presence of oscillation. Place the pickup coil near the plate tank. C_N is then adjusted so that no RF voltage appears on the scope when the plate tank is tuned to resonance. Under these circumstances the RF current divides equally through C_{gp} and C_N. The resulting RF currents in the plate tank flow in opposite directions and cancel the tank inductive effect so that no resonant buildup occurs between the coil and capacitors. A neon glow lamp, a loop of wire attached to the filament connections of a flashlight bulb, or a sensitive RF galvanometer may be used if an oscilloscope is not available.

If there is a milliammeter in the amplifier grid circuit, the adjustment of C_N may be made by observing the grid meter as the plate tank is tuned through resonance, with no plate voltage applied. When there is an unbalance between C_{gp} and C_N, the plate becomes alternately positive and negative as the plate tank approaches resonance. On positive swings, plate current flows. As the plate tank circuit is tuned to the resonant frequency, some of the electrons that were going to the grid now go to the plate, thereby causing a dip in the grid current. However, if C_N is adjusted to neutralize the amplifier stage, the RF current from the input stage divides equally and flows in opposite directions in the two halves of the plate tank coil, thus canceling the inductive effect of the coil and preventing the buildup of resonance in the tank. There is no rise in tank current and voltage, and the triode plate remains at zero potential. Therefore, with C_N properly adjusted no dip in grid current occurs as the plate tank is tuned through the resonant frequency.

Parallel Operation of Radio–Frequency Amplifiers

In some transmitter circuits it is more convenient to *turn off the filament voltage* on the amplifier stage instead of removing plate voltage. If this is done, the process of neutralizing the amplifier is carried out in the same way, except that no current flows in the amplifier grid circuit. The absence of radio frequencies in the amplifier plate tank is evidence of the correct adjustment of C_N.

Once a neutralizing capacitor is adjusted for a particular tube, it will require only occasional checks. However, if the *tube is changed* for a new one, the neutralizing capacitor will need adjustment, since the new tube may have a slightly different value of C_{gp}.

14-12 PARALLEL OPERATION OF RADIO–FREQUENCY AMPLIFIERS

It is possible to parallel the output of similar-type tubes when an RF power output is needed greater than a single tube is capable of providing. Such an arrangement is shown in Fig. 14-14, where twice the RF output is obtainable over that provided by a single tube. The circuit has some characteristics different from that of single-ended operation. For example, the plate impedance will be *one half* that of the single tube. This will likely require lowering the value of inductance and raising the capacitance of the tuned circuit to provide a better match between tube and load. The fact that approximately twice the plate current will flow may very well require that the size of the inductance in the plate circuit be larger in order to handle the increased plate current. *Twice the driving power is required* for the grid circuit, inasmuch as grid current will be flowing in two tubes rather than one. The value of the grid-leak resistor R will have to be chosen accordingly.

The mechanical construction of such a circuit is very straightforward. However, certain problems may appear in connection with the operation.

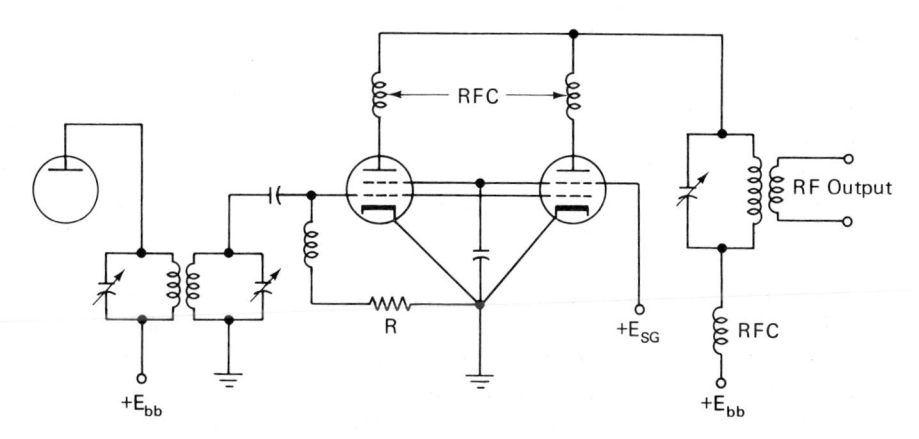

Fig. 14-14 RF power amplifier using tetrodes in parallel.

For example, there may be more susceptibility to *parasitic oscillations*, which almost always requires the inclusion of an *RFC* in the plate lead of each output stage *mounted as close as possible* to the tube. It may also be necessary, although it is not shown, to install *RFC*s in the screen circuit of each stage. Again, they should be installed as close as possible to the tube. Observe in the schematic that all the ground leads are brought to a *common point*. This is typical for high-frequency operation. All leads must be kept as short as possible. Lead dress for these circuits is very critical. Longer leads tend to increase the possibility of parasitic oscillation. The interelectrode capacitances of both tubes are in parallel with each other, thus doubling their value for a single tube. These higher values may *limit the performance* of parallel operation at higher RF frequencies.

14-13 PUSH–PULL OPERATION OF RADIO–FREQUENCY AMPLIFIERS

The operation of RF amplifiers in push–pull has some advantages over parallel operation. Because *even-order harmonics are canceled* in the output, it is possible to obtain more power output from two given tubes operated push–pull than in parallel. In order for even-order harmonics to be canceled, it is necessary that the drive supplied to each tube be of the same *amplitude*. The grid and plate tanks must be accurately center tapped, and the coupling to the output must be taken equally from both halves of the tank circuit. Needless to say, the two tubes must be matched in their electrical characteristics.

Another advantage of push–pull operation is the fact that the interelectrode capacities which are effectively connected across the tank circuit are made up of the capacitances of the tubes acting in series. The result is only half the capacitance across the tube circuit that would be present if a single tube were being used. This means that the same tubes can be used for higher-frequency operation than if they were parallel connected. A typical push–pull circuit showing the use of triodes appears in Fig. 14-13.

14-14 FREQUENCY MULTIPLIERS

The stability of an oscillator is a function of its frequency. The higher it is designed to operate, the more susceptible it is to *frequency drift*. The relationship is *not linear;* that is, the frequency stability may be several times worse at double the frequency. As a consequence, it is customary for high-frequency operation of a transmitter to operate the oscillator at a lower frequency and *double*, or *multiply*, the frequency to that desired. This multiplication can only be at some whole number such as 2, 3, 4, or 5 and not 1.5, 2.5, 3.5, and so on. The transmitter is said to be operating at a *harmonic* of the oscillator frequency.

Frequency Multipliers

Harmonics are multiples of fundamental frequency. If a certain fundamental frequency is 400 Hz, the second harmonic would be 800 Hz, and the third harmonic would be 1.2 kHz.

Example 14-2

What is the seventh harmonic of 360 kHz?

Solution

360 kHz × 7 = 2.520 MHz.

A pure sine wave has *no harmonic* content. However, generated frequencies are very seldom pure. They usually contain harmonics of the fundamental frequency. For example, if the master oscillator (operating class C) generates a 1-MHz frequency, the wave is *rich in harmonics*. Therefore, if a signal rich in harmonics is connected to the grid of a tuned amplifier, the plate tank circuit may be tuned to any one of the harmonics present in the original grid signal. The process by which the input frequency to the grid is converted to a higher one in the plate by tuning to a harmonic of the fundamental is called *frequency multiplication*. If the second harmonic is selected, the stage is called a *frequency doubler;* if the third is used, the circuit is referred to as a frequency *tripler*, and so forth.

A typical frequency-doubler circuit is shown schematically in Fig. 14-15. The evidence that it is a doubler, and not a tuned amplifier or oscillator (covered in Chapter 16), is the notation that the plate tank circuit is tuned to the second harmonic. Also, the grid bias supply is indicated as being about 10 times cutoff. The fact that a *triode vacuum tube is used* and *no neutralization* circuit is evident is another indication that the plate tank must be tuned to a harmonic and not the fundamental frequency.

The frequency-doubler circuit is operated class C with the plate tank

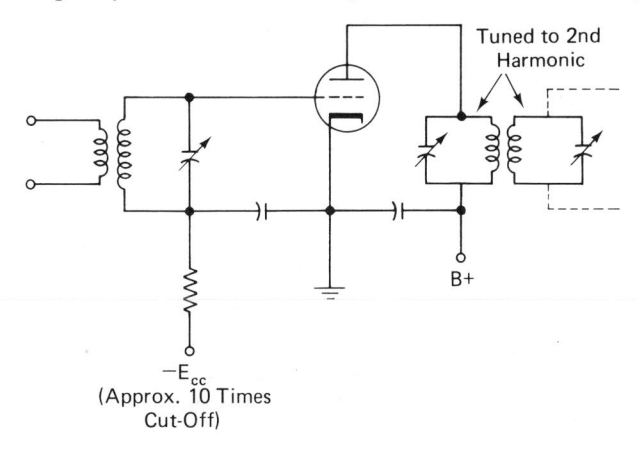

FIG. 14-15 Simple schematic diagram of frequency doubler.

resonant to double the grid signal frequency. A combination of fixed and grid-leak bias is incorporated, fixed bias for no signal tube protection and grid-leak bias for *amplitude stability*. When the signal applied to the grid rises above the cutoff value of the tube, there will be a pulse of current at the same frequency as the input signal. Since the pulse of plate current contains appreciable energy at the second harmonic and the resonant frequency of the plate tank circuit is determined by the values of L and C, the pulse of tube current excites the plate tank circuit and causes it to resonate at a frequency twice the grid signal frequency. When the tube goes into cutoff, the energy supplied to the plate tank circuit is sufficient to continue oscillations between current pulses. The reason the tuned circuit continues to oscillate is that the pulses of current always arrive at the same time during alternate cycles of the doubled frequency, thus energizing the plate tank circuit at the right time.

The circuit in Fig. 14-15 may be operated as a frequency *tripler* by tuning the plate tank circuit to the third harmonic of the input signal. The pulses of plate current arrive at the tuned circuit during every third cycle of output voltage, and deliver enough energy to the tuned circuit to sustain oscillations during those cycles when no current flows.

To increase the efficiency of the stage when it is operating as a frequency doubler, the angle of plate current flow must be reduced. Decreasing the angle of plate current flow may be accomplished by increasing the bias. The shorter the length of current pulses, the higher plate circuit efficiency will be when generating a particular harmonic; however, the grid exciting voltage must be *increased* if appreciable power output at this harmonic is to be realized. The power output of a frequency multiplier *varies inversely* with the extent of frequency multiplication due to decreasing angle of plate current. Values for θ_p representing a practical compromise between high efficiency and high power output are given in Table 14-1.

TABLE 14-1

Harmonic	θ_p (degrees)	Percentage of Power Output
2	90–120	65
3	80–120	40
4	70– 90	30
5	60– 70	25

In every case it is necessary to increase the operating bias and the grid driving signal as the frequency multiplication increases in order not to cause overheating. The flywheel effect in the plate tank supplies the missing cycles of grid drive, and the output is approximately an undamped wave having sine waveform.

Three important conditions must prevail to obtain frequency multi-

Grounded-Grid Amplifiers 245

plication: (1) high grid driving voltage, (2) high grid bias, and (3) plate tank circuit tuned to the desired harmonic.

Certain amplifier circuits are suited to the generation of even harmonics and others to the generation of odd harmonics. Push–pull amplifiers permit only odd-harmonic frequency multiplication—third, fifth, seventh, and so forth.

If the grids of two triodes are connected in push–pull and the plates in parallel, *even-order* harmonics can be produced. The grid signals are 180° out of phase. Thus, pulsating plate current flows first in one tube and then in the other. Because the plates are connected in parallel, *two* output pulses occur for each input cycle at the grids. This type of doubler is capable of greater output and higher plate efficiency than the single tube type.

14-15 GROUNDED-GRID AMPLIFIERS

The *grounded-grid amplifier* is widely used for VHF and UHF applications in which more conventional amplifier circuits fail to work properly. As the name implies, the circuit is characterized by the grid being maintained at ground potential, which makes the grid act as a shield between plate and cathode, reducing the plate-cathode capacity to a very low value. With triode tubes designed for this type of operation, an RF amplifier can be built that is free from the type of feedback that causes oscillation.

Characteristic of the grounded-grid amplifier is the fact that the input signal is applied between the cathode and grid and the output is taken between the plate and grid. Hence, the grid serves as a *common element*. A study of the circuit reveals that the ac component of the plate current has to flow through the signal source, which means that the signal source is in series with the plate current. Consequently, some of the power in the load is supplied by the signal source. In transmitting applications this feed-through power is on the order of 10 per cent of the total power output.

A typical grounded-grid amplifier circuit appears in Fig. 14-16. Because of the cathode-heater capacity, it is necessary that the heater circuit be maintained *above RF ground potential*. To accomplish this, *RFC* coils are added in series with each heater lead. The bottom ends are bypassed to ground to prevent any of the RF energy from getting back into the heater supply. Observe that the RF input is connected to the cathode.

Because the input and output are in series with each other, the gain of the stage equals $\mu + 1$. In this type of circuit there is *no phase reversal* between input and output signals. If any feedback should take place, it would have to be between the plate and cathode circuit (C_{pk}), which in this circuit is very small. Hence, the circuit is *extremely good* for high-frequency operation and *no neutralization* is required. The input impedance of the grounded-grid amplifier consists of a capacitance in parallel with an equivalent resistance, representing the power furnished by the driving source of the

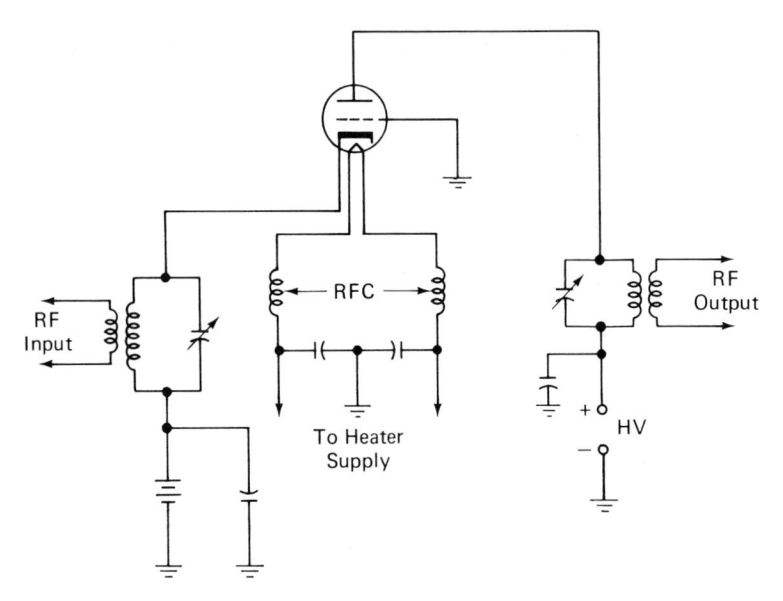

FIG. 14-16 Typical grounded-grid amplifier.

grid and to the load. This resistance is on the order of a few hundred ohms. The output impedance, neglecting the interelectrode capacitance, is equal to the plate impedance of the tube.

14-16 VERY HIGH AND ULTRAHIGH-FREQUENCY AMPLIFIERS

Circuits designed for operation in the UHF and VHF bands require special tube types, special parts, and refined construction techniques. Tubes must be chosen that have low interelectrode capacitances and *short transit times* (the time required for electrons to pass from the grid to the plate). Special tubes, such as *lighthouse* and acorn types, are required for satisfactory performance at these high frequencies. Capacitors must be high-quality mica or the equivalent, having short leads. Generally, inductors are fashioned from wire that is larger than for conventional work, because at these high frequencies currents tend to travel at the surface of the wire.

Parts must be placed relative to each other such that unwanted voltages will not be induced from one to the other. The *shielding* of components becomes extremely critical. Often the circuit must be *compartmentized* and the leads leading from one section to another must pass through *bulkhead feedthrough capacitors*. Any good conductor, such as aluminum, copper, or brass, may be used as shielding material to prevent stray magnetic fields from interfering with circuit operation. The placement of parts must be as close together as possible to minimize lead lengths. Wiring should go as directly as possible to each component *without bending the leads*. In some cases silver-plated wire is used to reduce skin effect.

14-17 COUPLING THE RADIO-FREQUENCY AMPLIFIER TO THE LOAD

To accomplish any useful work, the RF output of the amplifier must be connected to a load. This may be a tuned or untuned circuit driving the following stage or an antenna. The amount of energy transferred is dependent upon several factors, such as the closeness of coupling, the Q of the circuit(s), how close the impedances are matched, and, of course, the power available. If loose coupling exists, little power will be transferred and improper tuning of the tank will result. For example, as the LC circuit is tuned through resonance, the plate current will likely dip to too low a value for efficient operation. This may cause *excessive* screen current.

Increasing the coupling will draw more energy from the tank circuit. This in turn causes an increase in plate current, and the dip at resonance will not be as great—perhaps 50 per cent of what the plate current is when the tank is tuned off resonance.

If the coupling is *too tight*, there may be *no noticeable dip* as the tank is tuned through resonance. This implies excessive coupling. The question arises as to what is the correct value? The only way to be sure is to refer to the maintenance manual for the transmitter and follow the procedures outlined therein. If such is not available, refer to a tube manual, or better still, a specification sheet for the device. In the absence of any of this information the following procedure may be adopted. Adjust the coupling so that there is about a 50 to 60 per cent dip in plate current from the off-resonance value. This is a starting point. Increase the coupling slightly and redip the plate current. *It is necessary to retune the tank every time the coupling is changed* because of the reflected reactance back into the tank from the load. The grid current should show some increase as the tuning procedure continues. The screen-grid current should also increase as the proper resonant condition is achieved, although this may be difficult to monitor as screen-grid current meters are not usually installed in RF amplifiers. Continue increasing the coupling in increments and tuning the plate tank until reaching a point where the dip in plate current is about 25 per cent less than the off-resonance value. For example, if the maximum value of plate current is 300 mA when the tank is detuned, a satisfactory reading would be about 225 mA when properly loaded and tuned.

The procedure described does not indicate how much power is being delivered to the load. A typical circuit showing how power to the load can be calculated appears in Fig. 14-17. The power supplied to the load is

$$P = I_{RF}^2 R_L$$

During the tuning process it would have been observed that the RF power output increased as the dc power supplied to the power amplifier increased. A degree of coupling will be reached when the RF power delivered

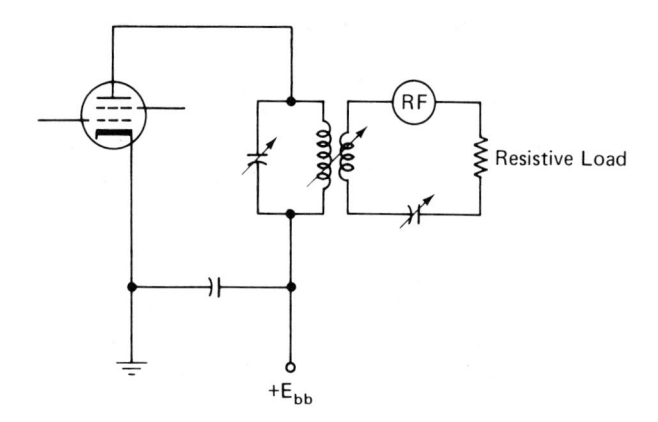

FIG. 14-17 Simplified circuit showing RF amplifier connected to a load.

to the load will not increase even though the dc power supplied to the amplifier increases as the coupling is increased. This indicates the point of *optimum coupling*.

It is usually possible to make slight improvements in loading by decreasing the capacitance and increasing the inductance, or vice versa, and going through the tuning procedure again. When maximum RF power is delivered to the load for the minimum amount of dc power supplied, optimum performance has been achieved. At this point the tube's plate impedance is approximately equal to the tank impedance. The ratio is 1 : 1. *This does not imply that the plate efficiency is maximized;* only that maximum power is being delivered to the load. Plate efficiency, which is another factor, can be determined by the following formula:

$$\text{efficiency } (\%) = \frac{P_{RF}}{P_{dc}} \times 100$$

This is simply the ratio of RF output to dc input. Some of the dc input is dissipated in the plate resistance of the tube, some as heat in the tank, and a small amount in the wiring. The balance of the input power is converted to RF energy and is delivered to the tank. For example, if a power amplifier has an operating efficiency of 75 per cent and the dc input is 800 W, the RF output is 600 W.

From the foregoing it should be evident that a power amplifier stage should not be adjusted to achieve maximum efficiency as the sole consideration. Remember that for maximum transfer of power the load impedance must match the tube's impedance, and when the load impedance goes beyond the maximum power transfer point, the efficiency rises but the RF power output diminishes.

Troubleshooting Procedures 249

14-18 TROUBLESHOOTING PROCEDURES

Effective, rapid troubleshooting is only possible by having a thorough knowledge of the functioning of all circuit components. One must be familiar with the normal performance of a stage(s) so that abnormalities may be quickly recognized. Most power RF amplifiers have meters in the plate and grid circuits to monitor current and voltage readings. Malperformance will show up with meter readings that are not normal. For example, if the grid current meter reads low, the most likely cause of trouble is *insufficient drive*. This in turn may be due to a weak driver stage or improper coupling. Naturally, with inadequate drive the final plate current will be *higher* than normal. Other causes of low grid current may be low filament voltage, weak tube (emission falling off), or too much bias.

Other evidences of malfunctioning include smoke, excessive heat, discoloration of components due to being overheated, arcing or the presence of unwanted gases inside tubes, and unusual smells. Resistors, transformers, and certain other components customarily give off an acrid odor if they are overheated. By quickly turning the equipment off under such conditions, more serious troubles may be avoided. In any event, the circuit must be checked to localize the trouble. Once the trouble has been identified and corrected, power may be reapplied.

Another possible cause of trouble is the RF choke. In a shunt-fed plate circuit a shorted choke would allow the RF output to be short circuited

TABLE 14-2

Meter	Symptoms	Possible Trouble
DC grid current	Zero reading	1. No grid drive 2. No RF in plate of previous stage 3. Bad tube
	Low reading	1. Weak grid drive 2. Mistuned plate circuit of previous stage 3. Bad tube
DC grid current	High reading	1. Open screen dropping resistor (no screen voltage) 2. No plate voltage
DC plate current	Zero reading	1. Bad tube 2. No grid drive (fixed bias, below cutoff) 3. Open screen dropping resistor (fixed bias, below cutoff) 4. No plate voltage 5. Open filament return to ground
	Low reading	1. Weak tube 2. Open grid-leak resistor
	High reading	1. No grid drive (no fixed bias) 2. Mistuned plate circuit

through the power supply, or power supply bypass capacitor. The amplifier stage would cease to function, the plate current meter might be burned out by the RF current, and a heavy dc plate current would flow as a result of the detuning of the tank circuit.

A troubleshooting chart is shown in Table 14-2 for an RF power amplifier, which lists abnormal meter readings and their possible causes. Any one of the items in the column of possible troubles can cause the abnormal reading.

Commercial License Questions

Sections in which answers to questions are given appear in parentheses. A bracketed number following a question implies that it applies only to that element.

1. Draw a diagram illustrating inductive coupling between two tuned RF circuits. (14–1)
2. What effect does a loading resistance have on a tuned RF circuit? [4] (14–1)
3. What is the principal advantage of a class C amplifier? [4] (14–2)
4. Draw a grid voltage–plate current characteristic curve of a vacuum tube and indicate the operating points for class A, B, and C amplifier operation. (14–2)
5. Describe the characteristics of a vacuum tube operating as a class C amplifier. (14–2)
6. Compare the design and operating characteristics of class A, B, and C amplifiers. (14–2)
7. Why is the plate circuit efficiency of an RF amplifier tube operating as class C higher than that of the same tube operated as class B? If the statement above is false, explain your reasons for such a conclusion. (14–2)
8. What are the advantages of using a resistor in series with the cathode of a class C radio-frequency amplifier tube to provide bias? (14–3)
9. In adjusting the plate tank circuit of an RF amplifier, should minimum or maximum plate current indicate resonance? [4] (14–4)
10. What is the purpose of an RF choke? (14–4)
11. Indicate by a simple diagram the series-fed plate circuit of an RF amplifier. [4] (14–4)
12. Indicate, by a simple diagram, the shunt-fed plate circuit of an RF amplifier. [4] (14–4)
13. Draw a simple schematic diagram showing a method of coupling between two tetrode vacuum tubes in a tuned RF amplifier. (14–5)
14. Draw a diagram illustrating capacitive coupling between two tuned RF circuits. (14–5)

Commercial License Questions   251

15. Describe what is meant by *link coupling* and for what purpose(s) is it used?
(14–5)
16. What is the purpose of neutralizing an RF amplifier stage? [4]   (14–6)
17. Why must some RF amplifiers be neutralized?   (14–6)
18. Why does a screen-grid tube normally require no neutralization when used as an RF amplifier?   (14–6)
19. Does a pentode vacuum tube usually require neutralization when used as an RF amplifier?   (14–6)
20. What is the principal advantage of a tetrode over a triode as an RF amplifier?
(14–6)
21. Draw a simple schematic diagram of a system of neutralizing the grid-plate capacitance of a single electron tube employed as an RF amplifier.   (14–7)
22. If, upon tuning the plate circuit of a triode RF amplifier, the grid current undergoes variations, what defect is indicated? [4]   (14–7)
23. Explain the process of neutralizing a triode RF amplifier.   (14–7 to 14–11)
24. Explain the purposes and methods of neutralization in RF amplifiers.
(14–7 to 14–11)
25. Draw a simple schematic diagram showing a method of coupling between two triode vacuum tubes in a tuned RF amplifier, and a method of neutralizing to prevent oscillation.   (14–8)
26. Why is it necessary or advisable to remove the plate voltage from the tube being neutralized? [4]   (14–11)
27. What tests will determine if an RF power amplifier stage is properly neutralized?
(14–11)
28. What instruments or devices may be used to adjust and determine that an amplifier stage is properly neutralized?   (14–11)
29. How may the distortion effects caused by class B operation of an RF amplifier be minimized?   (14–13)
30. How may the generation of even harmonic energy in a RF amplifier stage be minimized?   (14–13)
31. For what purpose is a doubler amplifier stage used?   (14–14)
32. What are the characteristics of a frequency-doubler stage?   (14–14)
33. What class of amplifiers is appropriate to use in an RF doubler stage?
(14–14)
34. Draw a circuit of a frequency doubler and explain its operation.   (14–14)
35. Draw a simple schematic circuit of an RF doubler stage, indicating any pertinent points which will distinguish this circuit as that of a frequency doubler.
(14–14)
36. What is meant by a *harmonic*?   (14–14)
37. What is the seventh harmonic of 360 kHz?   (14–14)

38. Under what circumstances is neutralization of a triode RF amplifier not required? [4] (14–15)
39. Why are grounded-grid amplifiers sometimes used at very high frequencies? [4] (14–15)
40. Draw a diagram of a grounded-grid amplifier. [4] (14–15)
41. In a class C radio-frequency amplifier, what ratio of load impedance to dynamic plate impedance will give the greatest plate efficiency? [4] (14–17)
42. What would be the result of a short circuit of the plate RF choke coil in an RF amplifier? (14–18)
43. What material is used in shields to prevent stray magnetic fields in the vicinity of RF circuits? [4] (14–1)

15

Measuring Instruments

15-1 BASIC DIRECT–CURRENT METER MOVEMENT

There are many different kinds of test equipment. Knowing which ones to use is an important part of an operator's or technician's responsibility. Usually there are several instruments commercially available to do a particular job, but some will be better than others. The better ones are usually the most expensive.

The meter movement most commonly used is the *moving coil* or *galvanometer* (invented by D'Arsonval), preferred because of its accuracy, ruggedness, and linearity. It works on the principle of *magnetic attraction* and *repulsion*. Suppose that a small bar magnet is mounted on a shaft that is free to rotate between two poles of a horseshoe magnet. The bar magnet will turn until its north pole is lined up with the south pole of the horseshoe magnet.

Suppose that a small spring is attached to the bar magnet in such a fashion that the spring will have no tension when the north poles of the two magnets are lined up. However, because of the repelling force of the like poles, the bar magnet will rotate until its turning force is balanced by the force of the spring, as shown in Fig. 15-1.

If the bar magnet were replaced with a small electromagnet, we would have a galvanometer. The strength of the field around the electromagnet would be a function of the shape, size, and number of turns in the coil together with the amount of current flowing through it. By causing a small electric current to flow through the coil, the magnetic forces between the electromagnet and the horseshoe magnet cause the coil to turn until the *magnetic turning force* is balanced by the *force due to tension* in the spring.

The actual magnetic circuit is as shown in Fig. 15-2. The soft-iron pole

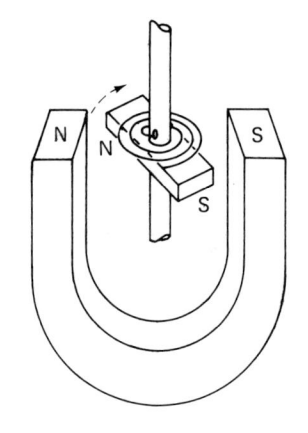

FIG. 15-1 Strength of magnetic field affects meter torque.

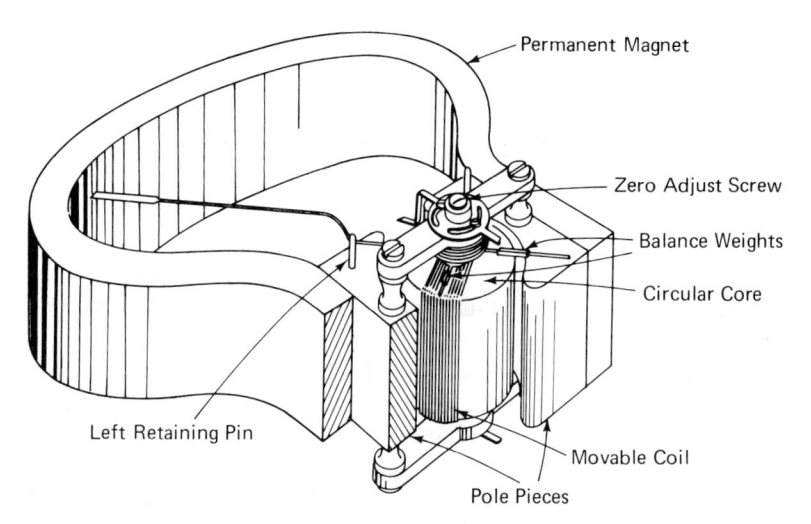

FIG. 15-2 Cut-away view of a moving-coil or D'Arsonval meter movement.

pieces and circular core create a strong, uniform magnetic field in the space between them.

A moving coil consisting of several turns of fine wire is wound on a rectangular aluminum frame. This is referred to as a *bobbin*. Because the coil must be light and able to swing freely, only a limited number of turns of fine wire can be placed on it. Because of the meter design and sensitivity, only a very little current is necessary to produce a full-scale deflection of the needle.

The pointer is attached to the coil assembly and swings with it. Tiny counterweights are used to balance the assembly on its pivots. The two ends of the coil are connected to two springs, one on each side of the bobbin. The other ends of these springs go to the external circuit from which current is fed to the coil. These are wound in *opposite directions* to compensate for temperature changes.

If no *damping action* were provided for the meter movement, it would swing rather quickly toward maximum, when a current was passed through the coil, and overshoot its mark. It would then swing in the opposite direction past the required mark and continue to oscillate back and forth until it came to rest at the exact reading.

Damping is a form of *braking action* on the coil. When current flows through it, a magnetic field induces a current in the aluminum frame. This frame acts like a coil having only one turn of wire. The induced current sets up a magnetic field, which opposes that of the coil. The braking action that results slows the movement of the coil. When the coil has deflected to the correct position, its magnetic field becomes static and no additional voltage is induced in the frame.

Additional damping is also provided for in the following manner. When current first begins to flow through the coil, a *counter emf is generated* within it that bucks the original current. This acts to cut down the flow of current. The counter emf gradually decreases, and the current through the coil gradually increases toward its normal value until the coil comes to rest.

Every meter coil has some dc internal resistance. Generally, this is determined by the size of wire and number of turns on the bobbin. As a general rule, instruments that are designed to read large amounts of current have low internal resistance; those designed to operate from very minute currents have a larger internal resistance.

Meter sensitivity may be defined as the amount of current required to provide full-scale deflection. When more turns are added to a coil, a stronger magnetic field results, and a smaller amount of current is necessary for full-scale deflection. Sensitivities of meters vary from $5\mu A$ to approximately 50 mA. The smaller the amount of current necessary for full-scale deflection, the greater the sensitivity. Sensitivity is a characteristic of the design of the meter movement itself and cannot be altered.

Meters are *delicate instruments* and must be handled carefully. If they are exposed to strong magnetic fields, the magnet may be partially degaussed, rendering future readings inaccurate. Meters are designed to be mounted on steel, aluminum, or Bakelite panels. If mounted on improper types of panels, the magnetic field of the instrument may be affected and erroneous readings result.

When a current-reading meter is inserted in a circuit, the operation may be slightly upset, owing to the internal resistance of the meter. The actual current flowing may be a little less than that which was calculated due to the meter resistance.

The meter should always be read from a position at *right angles* to the meter face. Since the meter divisions are small and the meter pointer is raised above the scale, reading the needle position from an angle will result in an inaccurate reading, often of as much as an entire scale division or more. This type of incorrect reading is called *parallax*. Most meters are slightly inaccurate due to the meter construction, and adding error from a parallax reading may result in a very inaccurate reading.

15-2 DIRECT–CURRENT AMMETER CONNECTIONS AND SHUNTS

The schematic symbol for a current meter is a circle enclosing the abbreviation A, mA, or μA to indicate the basic current range. The symbol for a voltmeter is likewise a circle enclosing the letter V. Plus and minus signs are frequently used to indicate meter polarity.

In the series circuit of Fig. 15-3 the current flow is 100 mA as calculated by Ohm's law. The current meter will also indicate the same current. It does not matter in which part of the series circuit the meter is installed, because the current is the same in all parts of the circuit. Observe the meter polarity. The meter terminal closest to the negative side of the battery (going around the circuit) is always negative, and vice versa.

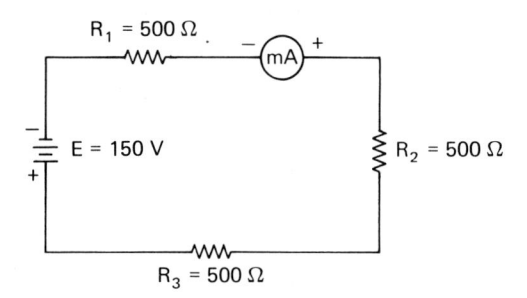

FIG. 15-3 Connection of a current reading meter in a series circuit.

Current meters must always be connected in *series* with the circuit components in which it is desired to measure the current flow. If they should be accidently connected in parallel with one of the circuit elements, the *IR* drop across the element would likely cause so much current flow through the meter as to burn it out.

A dc meter usually has its *terminals* marked for polarity, either with plus and minus signs or red for plus and black for minus. Current flow must be into the negative side through the meter movement and out the positive side of the meter in order for the needle to read up scale.

When the full circuit current to be measured is more than the meter can handle, a *shunt* must be used. This is a *precision resistor* connected across the meter for the purpose of shunting or bypassing a specific fraction of the current around the meter movement, as shown in Fig. 15-4.

A meter with an internal shunt has the scale calibrated to take into account the current through the shunt; therefore, the scale reads the total

Direct-Current Ammeter Connections and Shunts 257

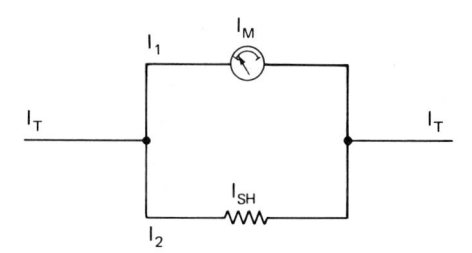

FIG. 15-4 Shunt resistor connected across a current meter.

circuit current. The meter resistance is actually lower with the shunt, thereby reducing the effect of the meter in the circuit. The shunt is usually not shown in the schematic symbol for a current meter.

The value of shunt resistance necessary to extend the range of a current meter can be calculated by the formula

$$R_{sh} = \frac{R_m I_m}{I_{sh}}$$

where R_{sh} = shunt resistance
R_m = meter movement resistance
I_m = meter movement current for full-scale deflection
I_{sh} = shunt current

Example 15-1

It is desired to extend the range of a 0 to 1-mA meter having 27-Ω internal resistance to read a total of 10 mA. What value of shunt is required?

Solution

Substituting the given values into the formula gives

$$R_{sh} = \frac{27 \times 0.001}{0.009} = 3 \, \Omega$$

Most current-reading meters used in industry have *several ranges* available for measuring current. A *switching arrangement* is used to change ranges, as illustrated by the simple schematic of Fig. 15-5.

Assume that it is desired to extend the range of a basic 0 to 10 milliammeter to permit measurements from 0 to 10 mA, 0 to 100 mA, 0 to 1 A, and 0 to 10 A. No shunt is required for the 0- to 10-mA range, since the meter movement is designed to handle this amount of current.

For the 0- to 100-mA range,

$$R_{sh} = \frac{9 \times 0.001}{0.09} = 1 \, \Omega$$

For the 0- to 1-A range,

$$R_{sh} = \frac{9 \times 0.001}{0.99} = 0.091 \, \Omega$$

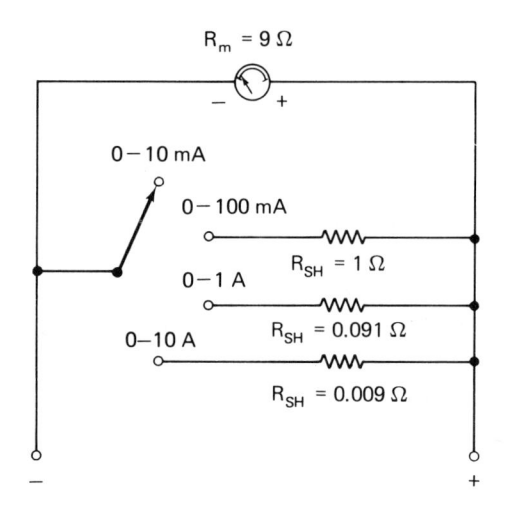

FIG. 15-5 Multirange ammeter using separate shunts and switch.

For the 0- to 10-A range, 10 A is used to compute the shunt instead if the actual current of 9.99 A because there is practically no numerical difference in the answer using either figure.

A disadvantage of this circuit is that the *total current* flowing in the circuit would *momentarily* flow, or try to flow, through the meter itself during the short interval of time required for the selector switch to move from the one position to the other. This would most likely burn out the meter. To prevent this, it is necessary to use a *shorting switch* to avoid damage to the meter. When switching from one shunt to another, the new shunt must be connected before contact is broken with the shunt in use. This can be done with the switching arrangement illustrated in Fig. 15-6.

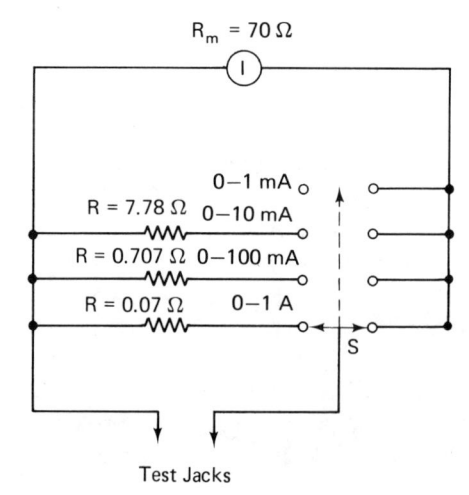

FIG. 15-6 Multirange milliammeter showing suitable switching arrangements.

15-3 DIRECT–CURRENT VOLTMETERS

Voltmeters are used to measure potential differences. They are basically sensitive current-reading meter movements with *series multiplying* resistors.

To find the series resistance required to measure any given voltage, the full-scale deflection, internal resistance of the meter, and voltage range to be measured must be known. The necessary series resistor can then be calculated by means of Ohm's law. For example, assume that it is desired to use a 0- to 1-mA movement, having 50-Ω internal resistance, to measure 10 V at full-scale deflection. Refer to Fig. 15-7. The maximum permissible

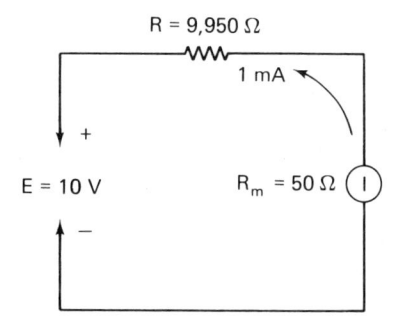

FIG. 15-7 Voltmeter made by adding series-multiplying resistor to milliammeter.

current through the movement is 1 mA. Since the total voltage (10 V) and the current (1 mA) are known, the total meter circuit resistance must be

$$R = \frac{E}{I} = \frac{10 \text{ V}}{1 \text{ mA}} = 10 \text{ k}\Omega$$

Actually, the value of the multiplying resistor would be 10,000 Ω minus the meter resistance of 50 Ω, or 9950 Ω.

When the series resistance is very large compared with the internal resistance of the meter, it is usually not necessary to subtract the meter resistance from the total value of the series resistance required. In this example the 50-Ω meter resistance is a very small percentage (less than 1 per cent) of the total 10,000-Ω circuit resistance. Hence, the error produced will be small. From a practical standpoint the basic meter movement would likely only have an accuracy of 2 to as high as 5 per cent unless high-accuracy laboratory standards were used. In more sensitive meter circuits the internal meter resistance may run as high as 2000 Ω.

Another method for finding multiplier resistance is based on the *ohms-per-volt* rating of the meter. For example, a given meter movement has a current sensitivity of 1 mA. It is desired to have the meter read 100 V full scale. Current sensitivity can be translated into ohms per volt by calculating the amount of resistance that must be placed in series with the meter to measure 1 V. The resistance required is

$$R = \frac{E}{I} = \frac{1\text{ V}}{1\text{ mA}} = 1000\text{ }\Omega$$

Thus, to measure a 1-V potential with full-scale deflection, 1000 Ω must be placed in series with the movement. Therefore, to read 100 V full scale, it is necessary to put 100 kΩ in series. Hence, we may refer to a 1-mA meter movement as having a 1000-Ω/V sensitivity.

The ohms-per-volt rating is another way of expressing meter sensitivity. The less current required for full-scale deflection, the more resistance must be placed in series with the movement to measure each volt and, therefore, the greater the sensitivity. When the ohms-per-volt rating is given, it is necessary only to multiply this figure by the desired range in volts to get the required value of multiplier resistance.

Example 15-2

It is desired to increase the range of a 0- to 50-μA movement, having 20,000-Ω/V sensitivity, to read 100 V. What value of multiplier resistance must be used?

Solution

$100 \times 20,000 = 2,000,000$-Ω or 2-MΩ resistance.

Suppose that it is desired to have a switching arrangement with four ranges and separate multipliers, using a 0- to 2-mA, 18-Ω meter movement. The ranges required are 5, 50, 250, and 500 V. The schematic diagram for this appears in Fig. 15-8. First, find the meter sensitivity:

$$\text{sensitivity} = \frac{1}{2\text{ mA}} = 500\text{ }\Omega/\text{V}$$

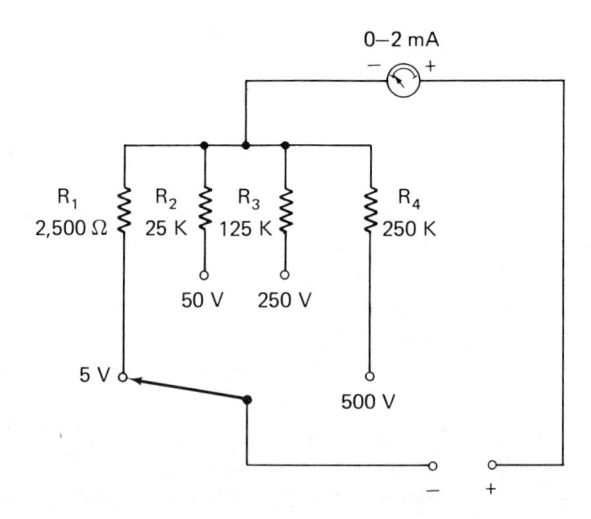

FIG. 15-8 Multirange voltmeter (separate multipliers).

Direct–Current Voltmeters

The circuit resistance for the 5-V range is therefore 2500 Ω. (The meter resistance should be subtracted from this.) R_2 for the 50-V range is 25 kΩ; R_3 for the 250-V range is 125 kΩ; R_4 for the 500-V range is 250 kΩ. At these higher values of multiplying resistors the meter's internal resistance can be neglected, because it is less than one tenth of 1 per cent of the total circuit resistance.

Figure 15-9 shows a *multirange voltmeter* using series multipliers. In this example a 0- to 50-μA meter movement having an internal resistance of 2 kΩ is used. It is necessary to calculate the value of the multiplier resistance for the lowest range first, or R_1.

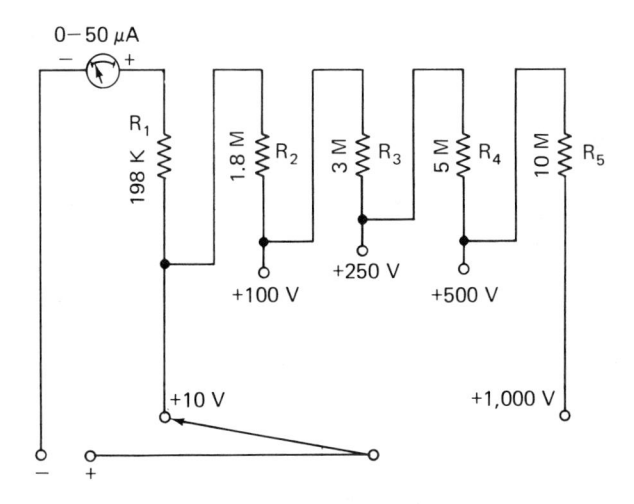

FIG. 15-9 Multirange voltmeter (series multipliers).

In this position the circuit resistance must be

$$R = \frac{E}{I} = \frac{10 \text{ V}}{50 \text{ μA}} = 200 \text{ kΩ}$$

From this value we subtract the 2-kΩ meter resistance to get the value of 198 kΩ for R_1.

Next, find the total multiplier resistance for the 100-V range. R_1 will then be subtracted from this to give the value for R_2. The calculations for this are

$$R = \frac{100 \text{ V}}{50 \text{ μA}} = 2 \text{ MΩ}$$

Subtracting R_1 plus R_m from this value gives 1800 kΩ for the value of R_2. Hence, when the selector switch is in the 100-V position, a total of 2 MΩ is in the circuit, which limits the current to 50 μA.

For the 250-V range, the multiplier resistance necessary is 2.5 times that required for the 100-V position. We may simply multiply the total resistance required for the 100-V position by 2.5. This gives an overall value required of 2 MΩ times 2.5, or 5 MΩ. From this we would have to subtract the sum of R_1, R_2, and the meter resistance. Thus, 5 MΩ minus (198 kΩ + 1800 kΩ + 2 kΩ) equals 3 MΩ for the value of R_3.

The same approach would be used to calculate the values of R_4 and R_5, which turn out to be 5 and 10 MΩ, respectively. From these examples we find the voltmeter has a sensitivity of 20,000 Ω/V.

15-4 OHMMETERS

The *ohmmeter* is an instrument that measures resistance. It is also used to locate *shorted* or *open* circuits and to check circuit continuity. It consists of a sensitive current meter, a source of low-voltage direct current, and some form of current-limiting resistor. The meter usually is a conventional dc moving-coil type, and a battery supplies the necessary dc voltage.

Resistance values vary from fractions of an ohm to megohms. Most ohmmeters are constructed to cover a number of ranges from very low to very high resistance by connecting various values of current-limiting resistors in the ohmmeter circuit. The different ranges then can be selected as required by means of a selector switch.

In the basic ohmmeter circuit of Fig. 15-10 a 4.5-V battery, a variable resistor R_A, and a fixed resistor R_B are connected in series with a milliammeter. P_1 and P_2 represent test leads that are connected across the resistance

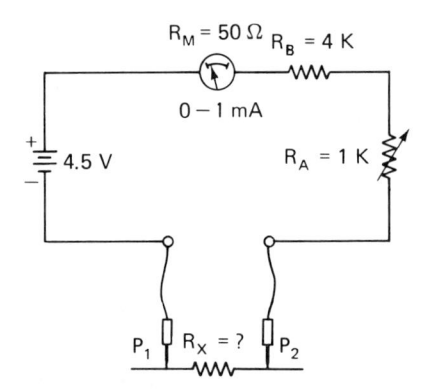

FIG. 15-10 Basic circuit of series-type ohmmeter.

to be measured, R_x. The fixed resistor R_B limits the flow of current and is placed in the circuit to prevent damage to the meter. If no limiting resistor is placed in the circuit and the variable resistor is adjusted to a low value, the current becomes excessive. Resistor R_A adjusts the series resistance so that 1 mA flows when the test leads are shorted. This occurs when the total series resistance is 4.5 kΩ. In the circuit illustrated the internal resistance of the meter is 50 Ω, the resistance of R_B is 4000 Ω, and R_A is adjusted to 450 Ω, making up the required series resistance of 4.5 kΩ.

A knob on the front of the panel marked 0 Ω controls resistor R_A

and is adjusted to provide full-scale deflection of the meter needle with the leads shorted. This position corresponds to zero resistance. Always zero the meter before using it. In the *series-type ohmmeter* full-scale deflection of the meter pointer indicates the *lowest resistance;* the opposite end of the scale represents the highest resistance. Resistor R_A can be readjusted to compensate for an aging battery.

To read the value of an unknown resistor R_x, connect it to the test leads. This reduces the current flowing in the circuit. If the value of R_x is equal to the combined resistance of $R_m + R_B + R_A$, the total circuit resistance is 9 kΩ. The current in the circuit is now 0.5 mA and the meter scale is calibrated to read 4500 Ω. If the value of the unknown resistor R_x is twice the internal resistance of the ohmmeter, the total circuit resistance is tripled, and the current is reduced to one third the full-scale deflection value. The meter pointer deflects to one third full scale and corresponds to an R_x of 9000 Ω.

Care must be taken to avoid connecting the ohmmeter across circuits in which a voltage exists, since such a connection can result in damage to the meter.

15-5 VOLT–OHM–MILLIAMMETERS

Voltage, resistance, and current measurements or a combination of these are often required in checking a circuit to determine normal operation or to localize a defective component. Usually, a combination meter such as a *volt–ohm–milliammeter*, abbreviated VOM, is used rather than individual meters. The VOM is frequently called a *multimeter* because of its ability to measure voltage, resistance, and milliamperes of current. It consists of a sensitive dc meter with switches, battery, and appropriate terminals. By selecting the proper terminals and placing the selector in the desired position, the meter can be made to read voltage, current, or resistance. The meter face generally has several voltage and current scales plus a resistance scale. Most VOMs have built-in rectifiers to permit reading of ac voltages. These may be read on separate ac scales on the meter face.

15-6 DIRECT–CURRENT VACUUM-TUBE VOLTMETERS

The 20,000-Ω/V voltmeter has an advantage over the 1000-Ω/V type in that it takes less current. Nevertheless, some power is required for its operation, and this comes from the circuit in which the voltage measurements are being made. The advantage of the *vacuum-tube voltmeter* (VTVM) is that it can amplify a signal without taking power from the source of voltage that it is connected to. The high input impedance of the vacuum tube makes it desirable for use in a voltmeter circuit. Because of this, the VTVM can be used to measure small voltages in high-impedance circuits. The input impedance is approximately 11 MΩ. The VTVM requires a source of power for its operation. Because of the high gain available, it is *susceptible to RF pickup* when being used around operating transmitters unless they are well shielded.

The VTVM circuit shown in Fig. 15-11 is basically a balanced bridge when no voltage is being measured. Plate load resistors R_1 and R_3 are equal, as are cathode biasing resistors R_4 and R_5. Therefore, there is no potential difference across meter M and it reads zero.

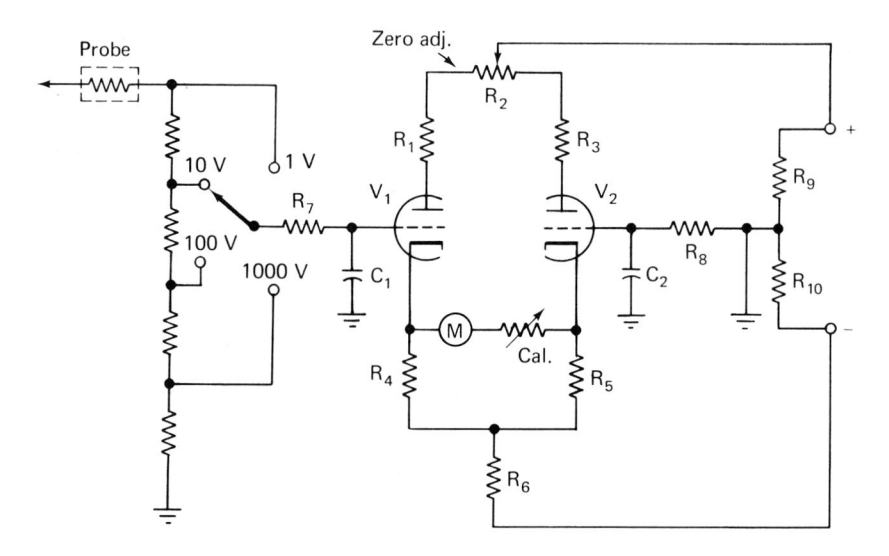

FIG. 15-11 Vacuum-tube voltmeter circuit.

When an unknown positive voltage is applied to the grid of V_1, more current flows through that tube than the other, and the circuit is unbalanced. This creates a potential difference between the cathodes of V_1 and V_2, causing the meter to indicate. By properly calibrating the meter scale, the amount of needle deflection is made directly proportional to the value of the applied voltage. Resistor R_6, common to the cathodes of both tube sections, is the feedback resistor that stabilizes the system and makes the readings linear. Resistor R_7 and capacitor C_1 form a filter for any ac component that may be present. Resistor R_7 is balanced by R_8 connected to the grid of the second tube.

The dynamic transfer characteristic of a vacuum tube is *not linear* over a very broad range. This limits the input voltage to 3 V *or less* in the average commercial instrument. To read higher values of voltage, a scaling network is used. The 1-MΩ resistance in the probe tends to minimize capacitive loading effects when measuring dc voltages in RF circuits.

The exact values of resistors to be used in the circuit depend upon the value of supply voltage and the sensitivity of the meter. Resistors R_6, R_8,

Alternating–Current Vacuum-Tube Voltmeters 265

and R_{10} should be adjusted so that the voltmeter circuit can be brought to balance and give full-scale deflection with 3 V or less applied to the left-hand grid. A *meter reversing switch* would normally be incorporated in a practical circuit so that the meter will still read up scale when negative voltages are measured with respect to ground. In some circuits the meter is connected between the plates of the dual triode.

A voltmeter circuit designed around a matched pair of field-effect transistors appears in Fig. 15-12. Basically, the vacuum tubes have been replaced with FETs. The input impedance is as high or higher than that of vacuum tubes; consequently, they lend themselves very nicely for use in high-impedance voltmeter circuits. The advantage of the circuit is that power can be supplied from a small battery, which should provide operation for a prolonged period of time. The very small current requirements of this circuit enable a voltmeter of this type to be completely portable. The FETs must be *matched pairs* and have pinch-off voltages that are within 10 per cent of each other.

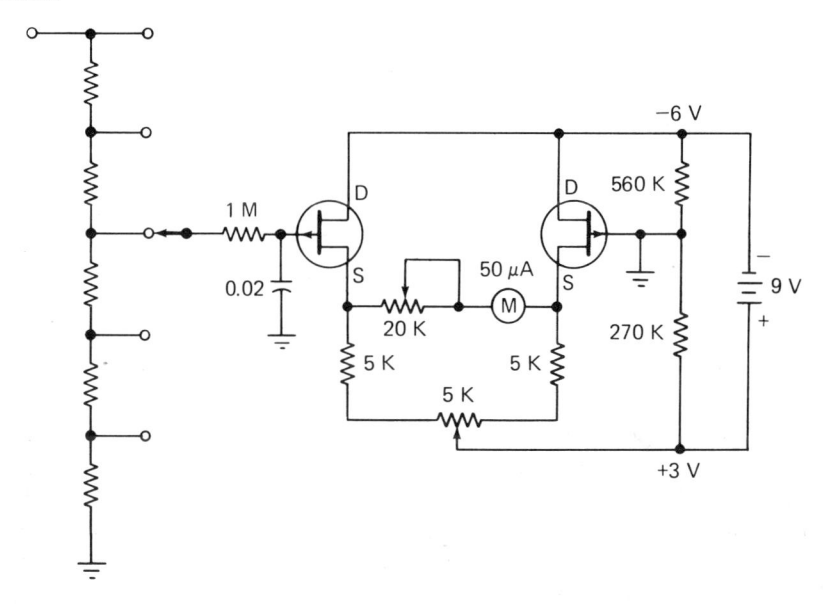

Fig. 15-12 FET voltmeter.

15-7 ALTERNATING–CURRENT VACUUM-TUBE VOLTMETERS

When ac voltages are to be measured, it is necessary that a *diode* be used to rectify the alternating current before it is applied to the amplifier. A dual diode is located in the voltmeter probe and is connected to the voltmeter by an appropriate cable. The voltage to be measured is applied to one half of CR_1 (Fig. 15-13) of the dual diode. The other half of CR_1 is connected to the grid of the second triode section V_2 through a divider network.

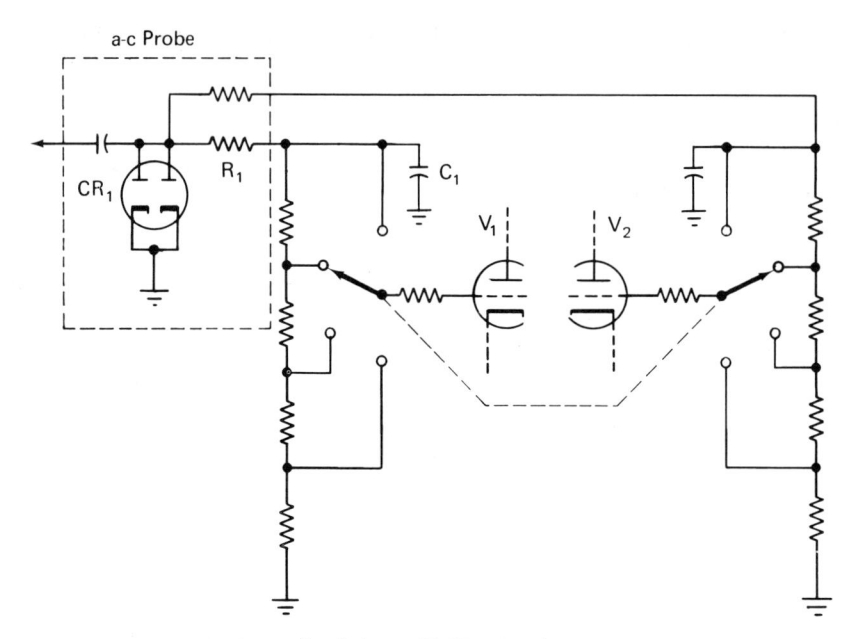

FIG. 15-13 Input circuit for ac VTVM showing contact-potential-canceling diode.

With no voltage applied to rectifier diode CR_1, its plate is approximately 1 V negative relative to ground. This voltage, referred to as *contact potential*, is the result of electrons leaving the heated cathode with sufficient velocity to reach the plate of the tube. This flow of electrons causes a voltage drop across the plate load of about 1 V and appears at the grid of V_1, unbalancing the amplifier. To compensate for this and restore the amplifier to perfect balance, an equivalent potential is applied to the grid of V_2 by connecting the other diode of CR_1 (the right-hand diode in the schematic drawing) to the grid of V_2.

The correct amount of contact potential to counteract the effect of the voltage on the grid of V_1 is obtained by adjusting the zero adjust potentiometer (not shown on this schematic) until the amplifier circuit is balanced. Diode CR_1 functions as a half-wave rectifier, and the rectified current flowing through R_1 and the scaling network in the grid circuit of V_1 drive the grid of V_1 negative. This negative voltage unbalances the circuit and causes a current to flow through the meter that is proportional to the applied voltage.

Capacitor C_1 filters the rectified voltage. The capacitor located in the ac probe serves to block any dc voltage present in the circuit from being measured. Placing the rectifier in the ac probe permits the use of the basic dc amplifier circuit for all ac measurements. When the applied ac voltage is rectified in the probe, a wider frequency range is available, since all cable capacitance and inductance have no effect on dc voltages.

Alternating–Current Rectifier Meters 267

The rectified output of CR_1 in the ac probe is proportional to the *peak amplitude* of the ac wave rather than the average value. Because the negative and positive peaks of a *complex wave* may not have equal amplitudes, a different reading may be obtained when the voltmeter terminals are reversed. This turnover effect is inherent in all peak-indicating devices, but is not necessarily a disadvantage. The fact that the readings are not the same when the voltmeter readings are reversed is an indication that the waveform under measurement is *asymmetrical*.

The scale calibration of the ac VTVM is based usually on the *effective value* of a sine wave. The rms reading can be converted to a peak value by multiplying by 1.414. When reading low values of ac voltage, it is customary to find a special ac scale located on many VTVMs. This is because of the nonlinearity of the diode when small signals are applied. For convenience, many VTVMs have a special scale(s) that indicates peak-to-peak value. Thus, an ac voltage can be read either in rms or peak-to-peak values.

Alternating-current VTVMs are *very sensitive* on the low voltage ranges. If the probe should be touched or placed near a strong field, the meter will indicate this pickup. The indication may be a rather violent swing of the needle across the face of the instrument. Generally speaking, the probe should never be touched when on low ranges. Zero should be set with the probe shorted to the common lead.

15-8 ALTERNATING–CURRENT RECTIFIER METERS

It is possible to use a simple ac rectifier in conjunction with a dc meter to read ac voltages. The bridge rectifier circuit shown in Fig. 15-14 is the most frequently used type. The meter would normally indicate the average value of the rectified alternating current. However, the manufacturer calibrates the scale in rms values.

In many instances copper oxide or selenium rectifiers are used. Because of the relatively high distributed capacitance across each of the rectifier elements, the accuracy is *seriously impaired* at higher frequencies.

Meters of this type are usually not used above the *audio frequencies* because of the errors that are introduced. Like any other ac voltmeter, the scale calibration is based upon the use of *sinusoidal waves*. If nonsinusoidal

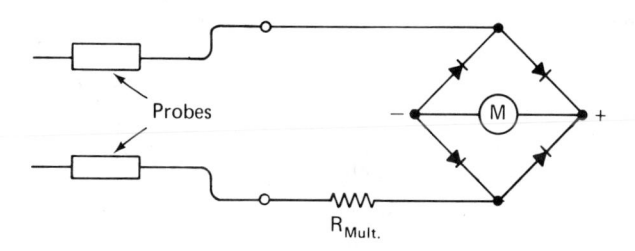

FIG. 15-14 Typical ac rectifier meter.

or complex waves are to be measured, the scale readings will not be accurate.

15-9 PEAK-READING VOLTMETERS

Any ac VTVM may be used to read peak values of voltages, provided the voltages are sinusoidal and the meter reading is multiplied by 1.414. Also, the frequency of the measured voltage must be within the operating range of the meter. *Peak-reading* vacuum-tube voltmeters are available. In these instruments the grid bias is usually adjusted to appreciably greater than cutoff, and the rectified current is determined primarily by the *peaks* of the positive cycles.

For audio and broadcast work it is possible to use a circuit very similar to that shown in Fig. 15-14. If a capacitor is connected across the meter, the rectified ac pulses charge the capacitor to approximately the peak value of the voltage. Because of the high resistance of the meter circuit, the capacitor does not appreciably discharge between successive pulses of voltage and the meter tends to read the peak value of voltage. Most ac VTVMs actually respond to the peak value of the input signal because of the capacitance across the probe although their scales are calibrated in effective values.

15-10 ELECTRODYNAMOMETERS

The *electrodynamometer-type* meter differs from the galvanometer in that no permanent magnet is used. Instead, *two fixed coils* are used to produce the magnetic field. Additionally, *two movable coils* are also used. The movable coils are designed to rotate inside the larger fixed coils. The pointer is attached to the movable coil assembly. The two coil assemblies are connected in *series*.

The construction of an electrodynamometer is shown in Fig. 15-15. The two movable coils are positioned coaxially and are connected in series with each other. The unit is pivot mounted between the fixed coils. The central shaft is restrained by spiral springs that hold the pointer at zero when no current is flowing through the coil. These also serve as conductors for supplying the current to the movable coils. Because these springs are very small, the meter is not capable of carrying a very heavy current.

When current flows through the two coil assemblies, their magnetic fields tend to line up, causing the moving coil to turn against the action of the hair springs attached to the movable coil. As an increase in current will increase both the magnetic field of the fixed coils and the moving coils, the swing of the indicating needle will be *proportional to the square of the current*. Hence, the meter scale cannot be linearly calibrated.

Because of the series connections of the coils, *no change* in the direction of the torque will result with a change in current direction. Consequently,

Wattmeters

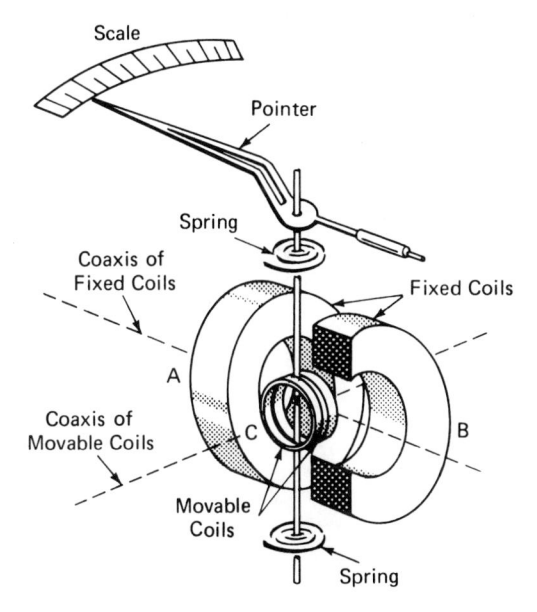

FIG. 15-15 Construction of an electrodynamometer.

electrodynamometers may be used with either direct or alternating current. The instrument is characterized by a lack of sensitivity, nonlinear scales, and poor accuracy as compared to other meters.

The electrodynamometer is mostly used in ac voltmeters. The meter is *mechanically damped* by means of aluminum vanes that move in enclosed air chambers.

15-11 WATTMETERS

Wattmeters are used to measure electrical power. They are similar to electrodynamometers in that they have a pair of fixed coils, known as *current coils*, and a movable coil, known as a *potential coil*. A drawing of this type of meter is shown in Fig. 15-16a. The current coils are stationary and are connected in series with the load; the potential coil is connected *across the line*. The interaction of the magnetic fields of these coils produces the actuating force. The force acting on the movable coil at any instant is proportional to the product of the instantaneous values of line current and voltage.

The potential coil has a high resistance connected in series with it for the purpose of making it as *purely resistive as possible*. As a result, current in the potential circuit is practically in phase with line voltage. Therefore, when voltage is impressed on the potential circuit, current is proportional to and in phase with the line voltage. The schematic diagram of a wattmeter is shown in Fig. 15-16b.

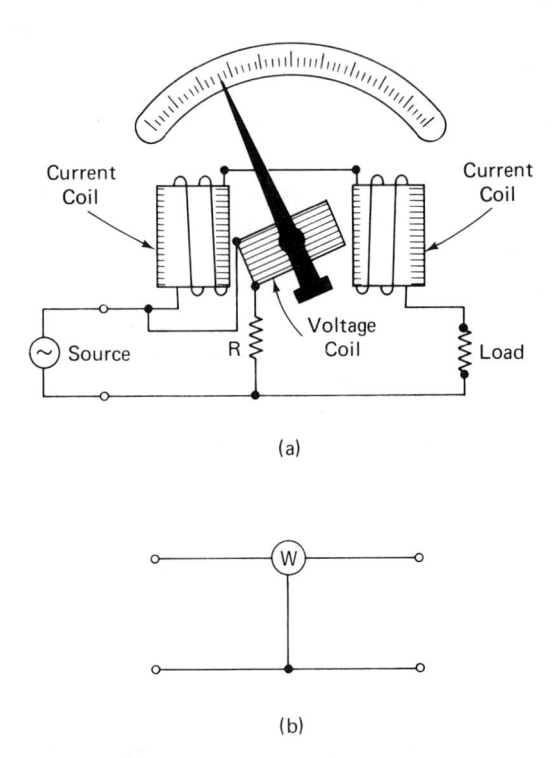

FIG. 15-16 (a) Simplified electrodynamometer wattmeter circuit. (b) Schematic symbol for wattmeter.

15-12 WATTHOUR METERS

A watthour meter is an instrument designed to record the amount of electrical power consumed over a *period of time*. Its principle of operation is essentially the same as that of the wattmeter. However, instead of an indicating needle the current and voltage coils produce a torque that rotates the armature of an *electric motor*. As the motor rotates, it drives a series of indicator hands, which indicate the amount of power being consumed. The electric power meters installed in houses and buildings by the local utility companies are examples of watthour meters. There are usually four (sometimes five) small dials geared together so that each is reduction geared to 10 times the preceding indicator. The first dial reads kilowatt hours, the second tens of kilowatt hours, the third hundreds of kilowatt hours, and so on. The electric bill is calculated on the number of kilowatt hours used in a given period of time. (Usually 1 month multiplied by the rate per kilowatt hour.)

15-13 RADIO–FREQUENCY AMMETERS

It is necessary to use *thermocouple ammeters* when measuring RF currents. The conventional ammeter cannot be used, as the RF energy would

Radio–Frequency Ammeters

cause a large reactive voltage to appear across the meter coil, burning it out. In the thermocouple meter a small dc voltage is developed across the open ends of a *junction of dissimilar metals* that is proportional to the *square* of the current flowing through it. This voltage also depends on the material of which the wires are made and on the difference in temperature between the heated junction and the open ends.

The junction is heated electrically by the flow of alternating or direct current through a heater element. The maximum current that may be measured depends on the current rating of the heater, the heat that the thermocouple can stand without being damaged, and on the current rating of the meter used with the thermocouple.

A simplified schematic diagram of the thermocouple is shown in Fig. 15-17. The input current flows through the heater strip, via the terminal blocks, and heats the thermocouple. The open ends of the thermocouple are connected to the center of two copper *compensating strips*. These strips radiate heat so that the open ends of the wires (thermocouple) will be much cooler than the junction end, thus permitting a higher voltage to be developed. The compensating strips are thermally and electrically insulated from the terminal blocks.

Thermocouple ammeters are calibrated for specific frequencies, since a

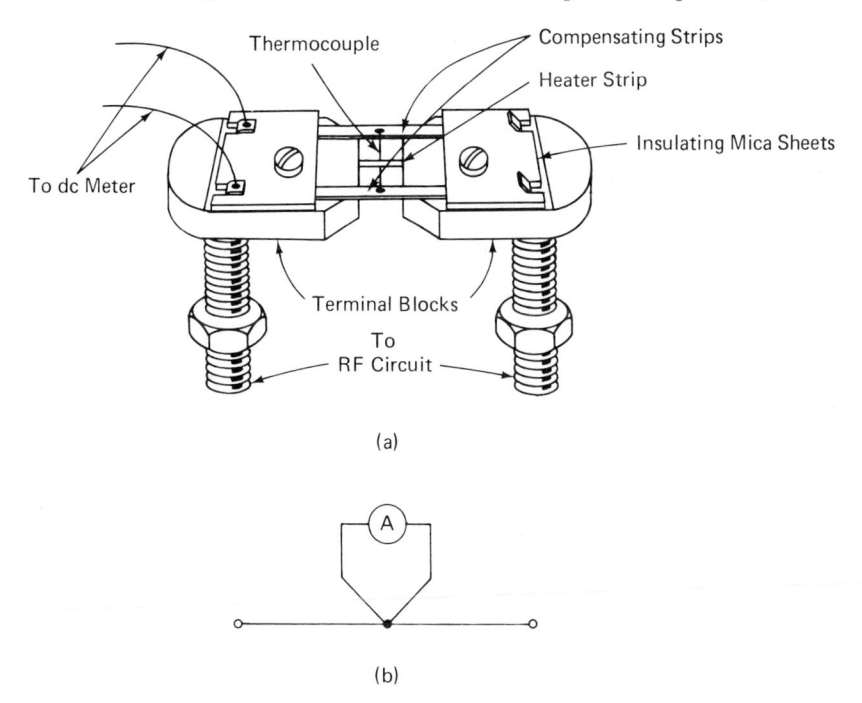

FIG. 15-17 (a) Internal view of thermocouple ammeter. (b) Schematic symbol of thermocouple ammeter.

frequency increase causes more skin effect (the tendency of an RF current to travel at the surface of the wire). This effectively increases the resistance of the heater wire. Meter deflection, therefore, changes with frequency, although the amount of current does not change. Heaters made of a short section of very thin wall tubing reduce the amount of error caused by skin effect. When thermocouple meters are required to operate over a considerable frequency range, *correction data charts* are provided by the manufacturers.

Deflection is proportional to the amount of heat in the heater wire and this, in turn, is proportional to the *square of the current* passing through it. The thermocouple ammeter, therefore, has a *square-law scale*, such as shown in Fig. 15-18. Since the bottom of a square-law scale is crowded, readings in that area are inaccurate. Thermal meters should be selected so that the expected value of current is at least half the full-scale range. For example, if the estimated current is 7 A, the meter range should not exceed 10 A.

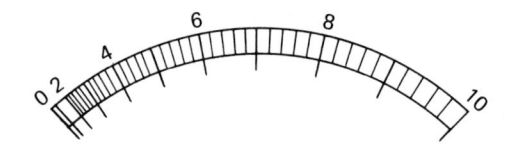

FIG. 15-18   Square-law scale.

15-14 VOLUME–UNIT METERS

The *VU or volume-unit meter* is used in audio equipment to indicate the input power to an amplifier or transmission line. It is a copper oxide rectifier-type ac voltmeter with a standardized speed of pointer movement, speed of return, and calibration. The VU meter pointer has a *rapid rise* and *slow fall*, making it easy to follow audio peaks and modulation envelope values. The meter is marked to indicate the percentage of modulation.

The VU meter gives an indication of volume level as so many VU above or below a zero reference level of 1 mW. It also indicates the amount of power-providing peaks that will modulate a transmitter 100 per cent.

The meter gives correct VU readings across only a 600-Ω load, and zero VU is indicated when 0.775 V is applied across a 600-Ω load. This is a power level of 0.001 W. A change of 1 VU is equal to a change of 1 decibel (dB). Therefore, any number of volume units equals the same number of decibels. When 0 dB means that 1 mW is being delivered to 600 Ω, then decibels and volume units are identical. A VU meter always can be used as a decibel meter, but a decibel meter can be used as a VU meter only when there are steady audio outputs. When the audio level is varying, the decibel meter and the VU meter read different instantaneous values.

Oscilloscopes

15-15 OSCILLOSCOPES

The *cathode-ray oscilloscope* is the most versatile and frequently used instrument. This results from its ability to measure peak-to-peak voltages, frequency, waveshapes, phase relationships, and distortion.

Operationally, an oscilloscope consists of a number of interconnected electronic circuits that control the performance of a *cathode-ray tube* (CRT). This is a special electron tube in which a stream of electrons, emitted by a heated cathode, are focused to form a narrow beam and accelerated to high velocity. The beam strikes the inside face of the tube, which is coated with a *fluorescent* material, causing a visible spot to appear. A cross-sectional view of a CRT is seen in Fig. 15-19.

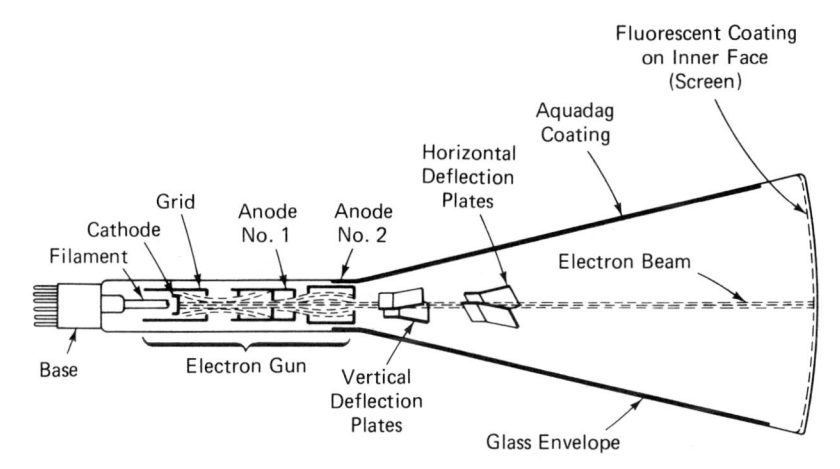

Fig. 15-19 Basic parts of the cathode-ray tube showing the electron optical system.

The *electron-gun* part of the tube consists of a cathode, control grid, and two anodes. The grid consists of a cylinder surrounding the cathode with a small opening in the end closest to the fluorescent screen. The electrostatic field between grid and cathode (bias voltage) causes the electrons to converge and pass through the grid opening. Thus, the grid and its potential act as an electrostatic lens system concentrating the electrons into a beam. An *intensity control* (brightness) determines the bias voltage and, consequently, the number of electrons in the beam striking the screen.

The electron beam passes through two anodes on its way to the screen. These are cylinders that have metal diaphragms with holes in them. Anode 1 is the *focusing anode*. By connecting the proper voltage to this element (usually a few hundred volts), the beam can be further focused. A second, larger-diameter cylinder, called the *accelerating anode*, is charged with a much higher potential (several thousand volts).

In electrostatic deflection systems, such as shown in Fig. 15-19, the beam must pass between two pairs of deflection plates. The first pair is perpendicular to the second. If no electrostatic field exists between the plates of either pair, the beam will not be deflected. A voltage applied to one set of plates will cause the beam to bend toward the plate that has a positive potential and away from the plate having a negative potential. The amount of deflection is proportional to the magnitude of voltage applied. The second pair of plates deflects the beam in the same manner, but in a plane perpendicular to the first. Because the mass of an electron is so small, the deflection of the beam can be considered instantaneous.

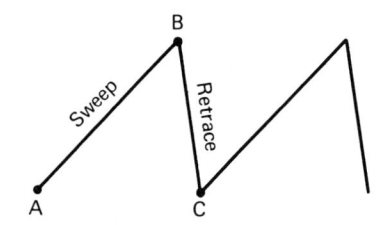

FIG. 15-20 Sawtooth waveform.

By applying proper voltages to the horizontal and vertical deflection plates, any waveform can be made to appear on the screen. In practice, a *sawtooth voltage* having a very linear rate of rise (Fig. 15-20) is generated in a circuit called the *time base generator*. It is amplified by the horizontal amplifier and applied to the horizontal deflection plates. As the voltage rises from *A* to *B*, it causes the electron beam to sweep from left to right across the face of the CRT. The voltage then drops quickly from *B* to *C* to commence the next sweep cycle. In more expensive *scopes* the retrace is blanked out by appropriate circuitry and does not cause any indication on the screen. The time base generator can be adjusted, by controls on the front panel of the instrument, to provide a wide range of sweep frequencies. Because of the *persistence of fluorescence* of the phosphors on the screen and the *persistence of vision* of the human eye, the rapidly sweeping electron beam produces a horizontal straight line across the screen.

The voltage or waveform to be observed is connected to the vertical amplifier, where it is amplified sufficiently and then connected to the vertical deflection plates. If the frequency of the sweep is synchronized with the voltage waveform to be observed, it will appear *stationary* on the screen. This permits the operator to closely examine the waveform for amplitude, distortion, and so forth.

Because electrons that strike the screen are traveling at very high speeds, they may cause other electrons to be knocked out of the phosphorescent material, producing *secondary emission*. These would form a *space charge* around the screen and interfere with normal tube action. To prevent this, part of the inside of the CRT is coated with *Aquadag*, a conducting material, which extends to about $\frac{1}{2}$ in. from the screen. Since it is at a positive

Absorption Frequency Meters 275

potential with respect to the cathode, the Aquadag removes the electrons caused by secondary emission and serves the added purpose of providing a shield for the electron beam.

15-16 ABSORPTION FREQUENCY METERS

The simplest method of measuring a frequency is to use a resonant circuit tunable over the desired frequency range and having its tuning dial calibrated in frequency. Such a circuit operates on the principle that when it is brought close to an oscillating tank it will *absorb maximum energy* when it is tuned to the resonant frequency of the tank. For this reason it is called an *absorption frequency meter*.

Because of the interaction between the two coupled circuits, some detuning occurs. This can be kept to a minimum if the absorption frequency meter, sometimes called a *wavemeter*, is loosely coupled to the output coil of the device whose frequency is being measured. It is not usually possible to obtain a very high degree of accuracy with the instrument, because the Q of the tuned circuit cannot be high enough to avoid uncertainty as to the exact dial setting. Nevertheless, it is a very useful instrument. It has the advantage of being inexpensive, requiring no power supply, and being compact.

A typical absorption-frequency-meter circuit appears in Fig. 15-21. It is shown loosely coupled to the output tank circuit of an RF power amplifier. The external coil is designed to be a *plug-in* type so that various frequency bands can be covered by different coils. The capacitor is connected to an *accurately calibrated vernier dial*, with the calibration in terms of some arbitrary unit. The frequency or wavelength is then determined by means of calibration curves or charts that relate the dial setting (and the particular coil being used) to either frequency or wavelength.

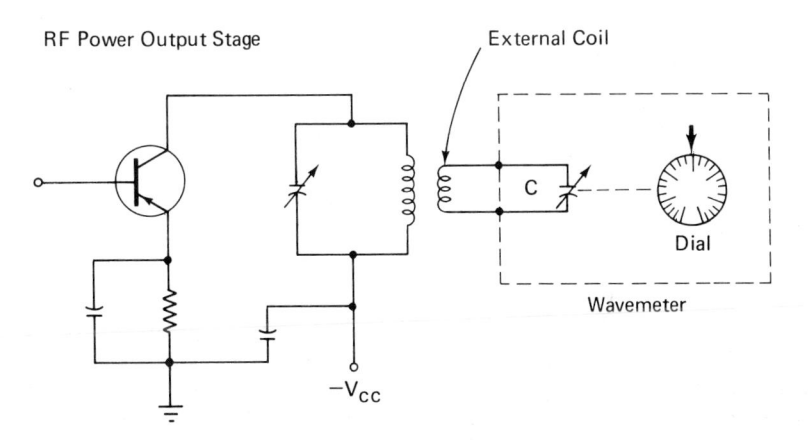

FIG. 15-21 Absorption frequency meter shown coupled to a typical RF power output stage.

When the absorption wavemeter is used to check the frequency of a transmitter, the plate (or collector) current of the stage under test can provide the necessary resonance indication. When the wavemeter is loosely coupled to the tank circuit, the plate (or collector) current will give a slight upward movement as the meter is tuned through resonance. For greatest accuracy the *loosest possible coupling* should be used.

Caution! The LC tank of the circuit being measured may have high direct current as well as RF voltages, and unless extreme caution is exercised the operator may be injured. It is preferable that the wavemeter be loosely coupled to the cold end of the LC tank so that hand capacity will have little detuning effect on the resonant circuits.

Some variations of the basic wavemeter circuit are represented in Fig. 15-22. In Fig. 15-22a a sensitive RF indicator, such as a thermocouple meter, is used to indicate resonance. In b a small flashlight bulb is used. Figure 15-22c is a modification of a in that a small silicon or germanium diode is employed to rectify the RF so that essentially direct current flows through the meter. The meter should have relatively high resistance and sensitivity so as not to cause excessive loading of the tuned circuit. Capacitor C serves to bypass any unrectified RF current around the meter and thereby prevents possible damage. Circuit d operates on the principle that the voltage buildup

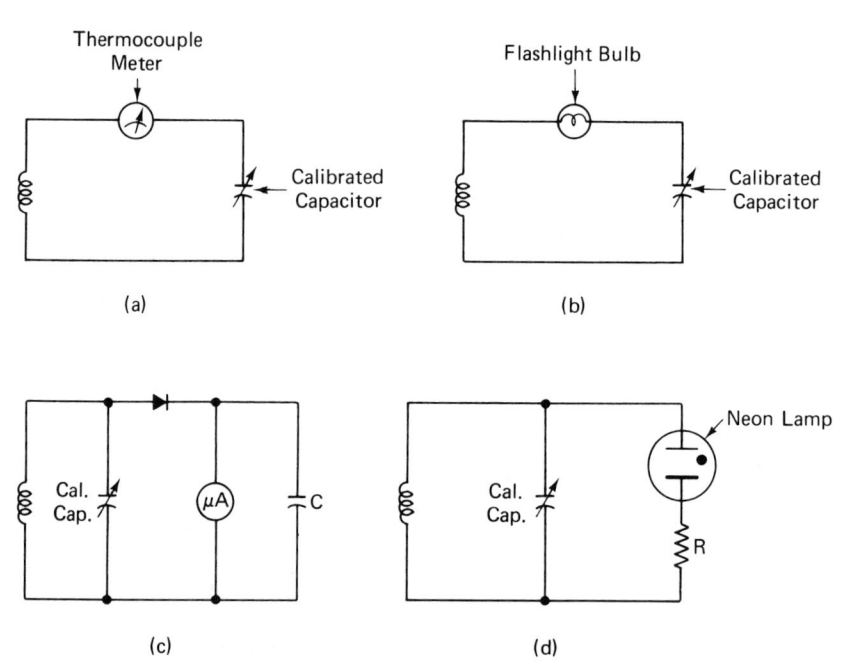

FIG. 15-22 Four variations of the basic absorption frequency meter.

Absorption Frequency Meters

across the parallel tuned circuit is sufficient to light the neon lamp. A high-ohmic-value, low-wattage resistor R serves to limit the ionized current through the lamp and simultaneously to prevent appreciably lowering the Q of the wavemeter tuned circuit. A better (but more complex) arrangement would be to use a diode-type VTVM connected across a portion of the inductor. The advantage of this type of indicator is that its input impedance is very high, and it therefore permits a higher Q to be developed in the wavemeter circuit.

Certain precautions should be observed when using an absorption-type frequency meter to measure the frequency of a self-excited oscillator. Even with loose coupling, power is taken from the circuit being measured. If the coupling is excessive, an impedance will be reflected back into the measured circuit, *changing its basic operating frequency* and resulting in erroneous readings.

If the wavelength of a particular frequency is desired, it can be calculated by the formula

$$\lambda = \frac{300 \times 10^3}{f_{kHz}}$$

where λ is in meters (m).

Example 15-3

What is the wavelength of a 560-kHz signal?

Solution

$$\lambda = \frac{300 \times 10^3}{560} = 0.535 \times 10^3 = 535 \text{ m}$$

If the amount of error in hertz per second at a given frequency is known, it is possible to determine the error at some other frequency. Consider the following example:

Example 15-4

If a wavemeter having an error proportional to the frequency is accurate to 20 Hz when set at 1000 kHz, what is the error when set at 1250 kHz?

Solution

Since the error is proportional to frequency, a ratio may be set up:

$$\frac{20}{1000} = \frac{x}{1250}$$

$$x = \frac{25{,}000}{1000} = 25 \text{ Hz}$$

15-17 GRID–DIP METERS

Another instrument that has found widespread use as a frequency measuring device is the *grid-dip meter*. It is interesting to note that this instrument can determine the resonant frequency of a tuned circuit even though the circuit is not energized. Power does not have to be applied to the *LC* tank whose resonant frequency is to be determined. This is an advantage, as the operator is not exposed to the high dc and RF voltages customarily encountered in transmitting equipment. Grid-dip meters can also be used in receiving circuits where such hazards are not so common.

The grid-dip meter consists of an active device used in an oscillatory circuit, such as the Hartley or Colpitts. A sensitive current-indicating meter is connected in the input circuit. When the circuit is oscillating, grid current flows. The circuit is designed so that different inductors can be plugged in and resonanted by the tuning capacitor. This coil serves as a *tuning probe*. This permits tuning over a wide band of frequencies. When the probe of the oscillating tank is coupled to any other *LC* circuit tuned to the same frequency, energy will be drawn out of the grid-dip meter's oscillating tank. This reduces the grid excitation to the oscillator and causes a reduced reading on the sensitive milliammeter. If the grid-dip meter is coupled to a tuned circuit whose resonant frequency is unknown, the grid-current meter will dip as the instrument is tuned through the resonant frequency of the circuit in question.

As in the case of the absorption wavemeter, a minimum amount of coupling, consistent with satisfactory operation, should be used between the grid-dip meter and the tank whose frequency is to be determined.

Because several plug-in coils are used to cover different frequency bands, it is necessary to have a *calibration scale* for each coil.

The schematic diagram of a transistorized grid-dip meter is shown in Fig. 15-23. The transistor is used in a common-base oscillator stage powered

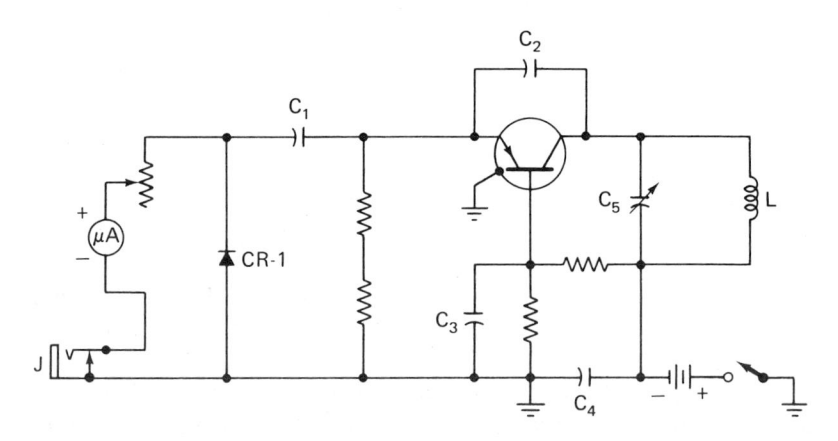

FIG. 15-23 Transistorized "grid-dip" meter.

Primary Frequency Standards

by a small battery. The oscillatory circuit consists of inductor L and capacitor C_5. Feedback to sustain oscillations in the resonant circuit is coupled by capacitor C_2 from the collector to the emitter. Some of this RF voltage in the base–emitter circuit is coupled by C_1 to diode CR_1. The rectifier output appears on the dc microammeter. When power is absorbed from the LC tank, the amount of feedback voltage is reduced and the meter reads a decreased current.

15-18 PRIMARY FREQUENCY STANDARDS

How can TV, FM, and broadcast stations know when their transmitters are exactly on frequency? The same question could be asked of the broadcast industry and radio amateurs. The answer can be found at the National Bureau of Standards (NBS), a branch of the U. S. Department of Commerce. Part of its function is to transmit several exceedingly accurate frequencies that can be received throughout most of the Western Hemisphere. Companies as well as individuals can readily check the calibration of their equipment against these known standards.

The National Bureau of Standards operates two radio transmitting stations; WWV at Fort Collins, Colorado, and WWVH at Puunene, Maui, Hawaii, for broadcasting standard radio frequencies of great accuracy. Station WWV transmits on 2.5, 5, 10, 15, 20, and 25 MHz. Its broadcasts are continuous night and day, except for a silent period of approximately 4 min each hour commencing at 45 min past each hour.

Station WWVH broadcasts on standard radio frequencies of 5, 10, and 15 MHz. The broadcast is interrupted for approximately 4 min each hour (silent period) beginning at 15 min after each hour.

Two additional stations are operated by the Bureau: WWVB and WWVL, both located at Fort Collins, Colorado. The former transmits on the standard frequency of 60 kHz and the latter on 20 kHz. Transmission is continuous in both cases.

The frequencies transmitted by WWV are held stable to 5 parts in 10^{11} at all times. Deviations are normally less than 1 part in 10^{11} from day to day. Incremental frequency adjustments not exceeding 1 part in 10^{11} are made at WWV as necessary.

Standard audio frequencies of 440 and 600 Hz are broadcast on each radio carrier frequency at WWV and WWVH. These are transmitted alternately at 5-min intervals starting with 600 Hz on the hour.

The frequency of 440 Hz represents A above middle C on the music scale and is the standard of the music industry in the United States as well as many other countries.

The RF signals of both stations are also modulated by pulses at 1 Hz. Intervals of 1 min are marked by the omission of the 59th pulse of each min. The next minute commences with two pulses spaced by 0.1s. The first of these two pulses marks the beginning of the minute. The 2-, 3-, and 5-min intervals are synchronized with the seconds' pulses and are marked by the beginning or ending of the periods when the audio frequencies are not transmitted.

Universal Time (UT) (referenced to the zero meridian at Greenwich, England) is announced in International Morse Code each 5 min from WWV and WWVH. This provides a ready reference to correct time where a timepiece may be in error by a few minutes. The 0- to 24-hour (h) system is used starting with 0000 at midnight at longitude zero. The first two figures give the hour, and the last two figures give the number of minutes past the hour when the tone returns. For example, at 1655 UT the four figures 1-6-5-5 are broadcast in code. The time announcement refers to the end of an announcement interval, that is, to the time when the audio frequencies are resumed.

At station WWV a voice announcement of Eastern Standard Time is given during the last half of every fifth minute during the hour. At 10:35 A.M., EST, for instance, the voice announcement given in English is: National Bureau of Standards, WWV, when the tone returns Eastern Standard Time will be 10 hours, 35 minutes.

At WWVH a similar voice announcement of Hawaiian Standard Time occurs during the first half of every fifth minute during the hour.

A forecast of *radio propagation conditions* is broadcast in International Morse Code during the last half of every fifth minute of each hour on each of the standard frequencies from WWV. This tells users the condition of the ionosphere at the regular time of issue and the radio quality to be expected during the next 6 h. The NBS forecasts are based on data obtained from a worldwide network of geophysical and solar observatories. The forecast announcements from WWV refer to propagation along paths in the North Atlantic area, such as Washington, D.C., to London, or New York City to Berlin.

The forecast announcement is broadcast as a letter and a number. The letter portion identifies the radio quality at the time the forecast is made. The letters denoting quality are N, U, and W, signifying, respectively, that radio propagation conditions are either normal, unsettled, or disturbed. The number is the forecast of radio propagation quality during the 6 h after the forecast is issued. Radio quality is based on a 1-to-9 scale defined as follows:

Disturbed Grades (W)	*Unsettled Grade (U)*	*Normal Grades (N)*
1. Useless	5. Fair	6. Fair to good
2. Very poor		7. Good
3. Poor		8. Very good
4. Poor to fair		9. Excellent

It is possible to check the accuracy of an oscillator (in any equipment

Secondary Frequency Standards 281

such as a transmitter or receiver) by comparing it to one of the NBS frequencies. For example, a local oscillator supposedly operating on 10 MHz needs to be checked for accuracy. Begin by tuning a receiver in on this NBS carrier frequency. Next adjust the local oscillator so that its output when coupled to the receiver is about the same amplitude. Any difference between the two frequencies will appear in the output as a *beat note* that wavers in intensity a certain number of times per second. This varying signal represents the difference between the local oscillator's and the NBS frequencies. By adjusting the local oscillator so that no beat note (zero beat) is heard, its accuracy will be equal to that of the NBS standard or 1 part in 10 million (at least until the local oscillator drifts off).

15-19 SECONDARY FREQUENCY STANDARDS

A convenient instrument to use in checking frequencies is the *secondary frequency standard*. It is simple to use and can provide accurate frequency checkpoints throughout a wide band of the RF spectrum. Basically, the instrument consists of an oscillator, usually 100 kHz, that *generates a large number of harmonics*. Since the harmonics are multiples of 100 kHz, some of them can be compared directly with the standard frequencies transmitted by WWV.

A typical circuit is shown in Fig. 15-24. The oscillator frequency is determined by crystal Y_1, which resonates at 100 kHz. No *LC* tank is in the collector circuit, only an RF choke, which presents a high impedance to the harmonics of the crystal.

To calibrate the secondary frequency standard, tune in a WWV frequency on a receiver and couple the output of the secondary standard to the receiver's antenna by means of a short length of coaxial cable. If the receiver

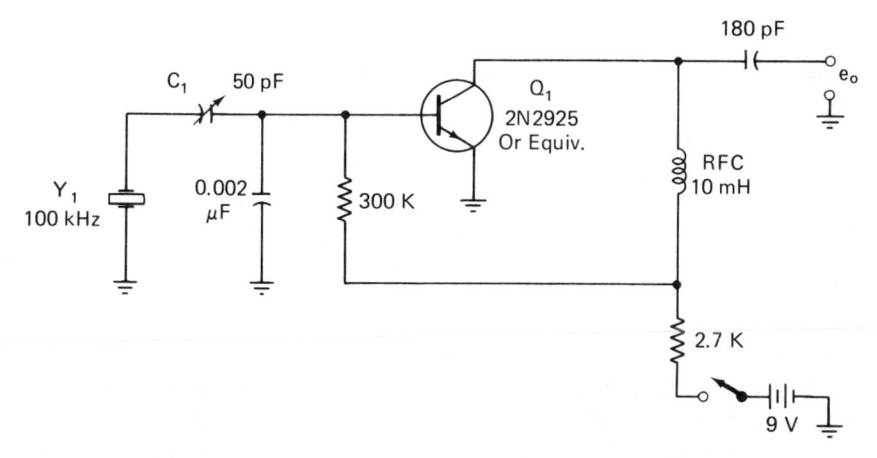

FIG. 15-24 Typical circuit of a 100 kHz secondary frequency standard.

has a *beat frequency oscillator* (BFO), it should be turned off. Wait until no tone modulation is present and turn on the 100-kHz oscillator. Adjust the oscillator trimmer capacitor C_1 until the oscillator's harmonic is in zero beat with WWV. For best results the signal from WWV and the 100-kHz oscillator should be about the same strength.

The block diagram of a more sophisticated secondary frequency standard appears in Fig. 15-25. Notice that a 100-kHz crystal-controlled oscillator is used, but only to synchronize the 100-kHz *multivibrator* (see Chapter 16). The multivibrator circuit is designed to free run at a frequency slightly less than 100 kHz. When the output of the crystal oscillator is fed into the multivibrator (via the buffer amplifier), its frequency will increase to the point where it will *synchronize* with the oscillator.

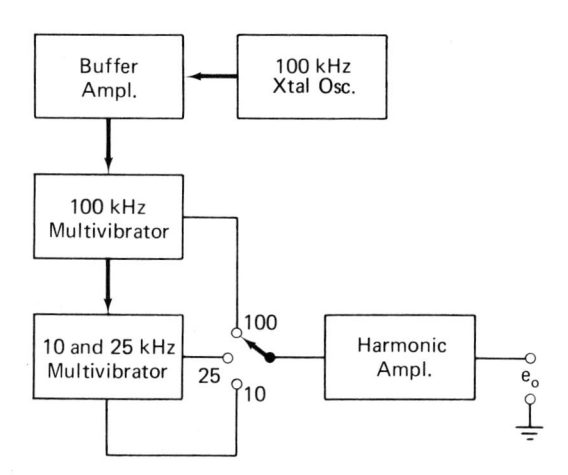

FIG. 15-25 Block diagram of typical secondary frequency standard.

The multivibrator generates a *symmetrical square wave rich in harmonics* whose upper limit substantially exceeds those produced by the circuit of Fig. 15-24. The harmonic amplifier has wide band characteristics and serves to amplify the hamonics fed into it.

The output of the 100-kHz multivibrator is also used to sync the 10- and 25-kHz multivibrators. These circuits also produce square waves containing high-order harmonics. By connecting either of these outputs through appropriate switching to the harmonic amplifier, an output is provided that will give check points every 25 or 10 kHz on a receiver dial.

Although the 100-kHz secondary standard does not make possible the exact measurement of a particular frequency, it is easy to determine whether or not the signal to be measured is in a particular 100-kHz segment. Suppose that the unknown signal falls between 27,400 and 27,500 kHz as indicated by the marker signal in the receiver. Obviously, the unknown signal lies somewhere within this band. If a more accurate measurement is necessary,

Heterodyne Frequency Meters 283

a reasonably good estimate can be made by counting the number of dial divisions between the two 100-kHz checkpoints and dividing the number into 100. This will give the number of kilo-hertz per dial division. With a secondary standard of the type, shown in Fig. 15-25, the selector switch could be placed in either the 25- or 10-kHz positions, thus narrowing the band that the unknown signal falls in.

15-20 HETERODYNE FREQUENCY METERS

A versatile instrument for measuring unknown RF signals, as well as generating signals to check the alignment of receivers and other types of RF signal-generating equipment, is the *heterodyne frequency meter*. It contains a very stable crystal oscillator that is used to calibrate the frequency of the variable oscillator. The crystal oscillator produces a number of harmonics, permitting calibration of the meter at a number of frequencies. These points of calibration are called crystal checkpoints, and the frequencies at which they occur are given in a calibration book used to determine the frequency of the dial setting. The variable oscillator is designed to provide *straight line frequency dial calibrations* and minimal drift.

The heterodyne meter consists of a frequency-calibrated oscillator that beats, or heterodynes, against the frequency to be measured. The pickup antenna is loosely coupled to the device under test. The calibrated oscillator is then tuned so that the difference in frequency between the oscillator and the unknown frequency is in the AF range. This difference frequency is known as a *beat frequency*, and when it is detected and amplified, it can be heard in the earphones. If the dial setting of the calibrated oscillator is tuned to the same frequency as the device under test, the difference frequency is zero or zero-beat frequency. When zero-beat frequency is achieved, the position of the pointer on the dial setting represents the unknown frequency.

A basic heterodyne-frequency-meter circuit is shown in Fig. 15-26. The circuit of V_1 is the variable-frequency oscillator whose output beats against the unknown frequency entering by way of the pickup antenna. The circuit containing V_2 (mixer stage) mixes these signals and serves also as a pentagrid converter when the variable oscillator is being calibrated. During calibration the crystal oscillator is in the control-grid circuit of V_2, and the output of V_2 is fed to an AF amplifier V_3, which drives the headphones.

The circuit of V_1 forms an electron-coupled oscillator that has good stability under varying load conditions. Switch S_1 permits the oscillator to operate on two different frequency ranges. C_4 is a trimmer capacitor that corrects for frequency deviation when the variable oscillator is being calibrated.

With switch S_2 in the off position, the control grid of V_2 is grounded, and V_2 becomes a mixer stage. It mixes the output of V_1, which is coupled through capacitor C_8, and the unknown frequency, which is coupled through

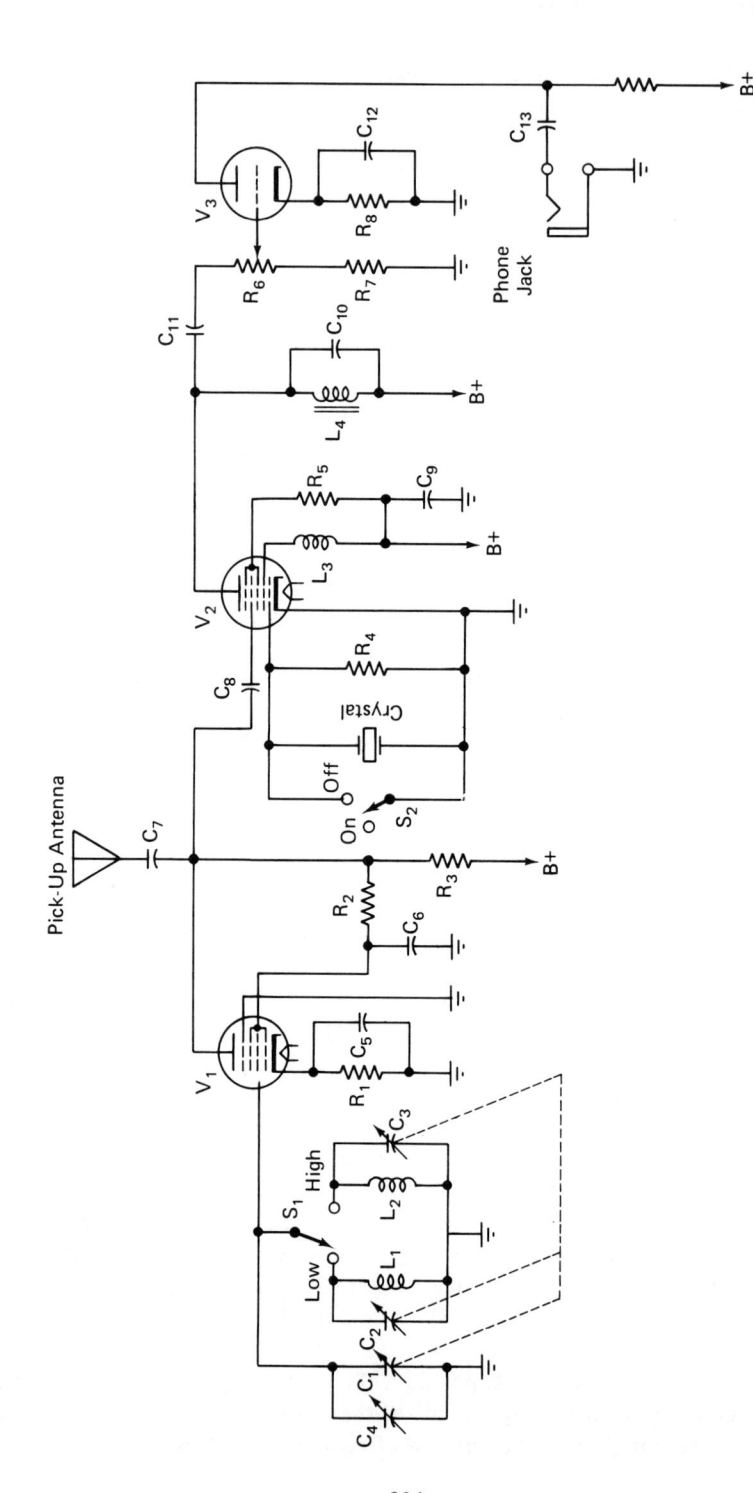

FIG. 15-26 Basic circuit of a heterodyne frequency meter.

Heterodyne Frequency Meters

C_7. When it is desired to calibrate the variable oscillator, S_2 is placed in the on position, and V_2 becomes a pentagrid converter with the cathode and the two grids directly above it, constituting a crystal oscillator with the remaining electrodes forming a mixer stage. The output of V_2 is coupled through C_{11} to V_3, and the output of this tube is fed to the headphones.

The front panel of a typical heterodyne frequency meter is shown in Fig. 15-27. When the lock screw is loosened, the tuning knob can be rotated to obtain various dial settings. The dial is calibrated in hundredths, units, and tenths for accurate readings. Rotating the dial changes the frequency of the variable-frequency oscillator. When calibrating, the crystal off–on switch is turned to the on position. The corrector control varies capacitor C_4 (see Fig. 15-26) and compensates for any deviation of the variable-frequency oscillator. The Freq Band switch selects the frequency range in which the meter is to be operated (typically 125 to 2000 kHz in the LF band and 2 to 20 MHz in the HF band).

The dial setting in Fig. 15-27 can be read in the following manner: The thin line marked on the window of the hundreds dial indicates the approximate reading of the dial. Since it is between 3800 and 3900, the ultimate dial reading must lie between these numbers. The reading on the units dial is read directly below the arrow on the tenths vernier. To obtain the reading that lies between 76 and 77, the tenths vernier must be read. The tenths value

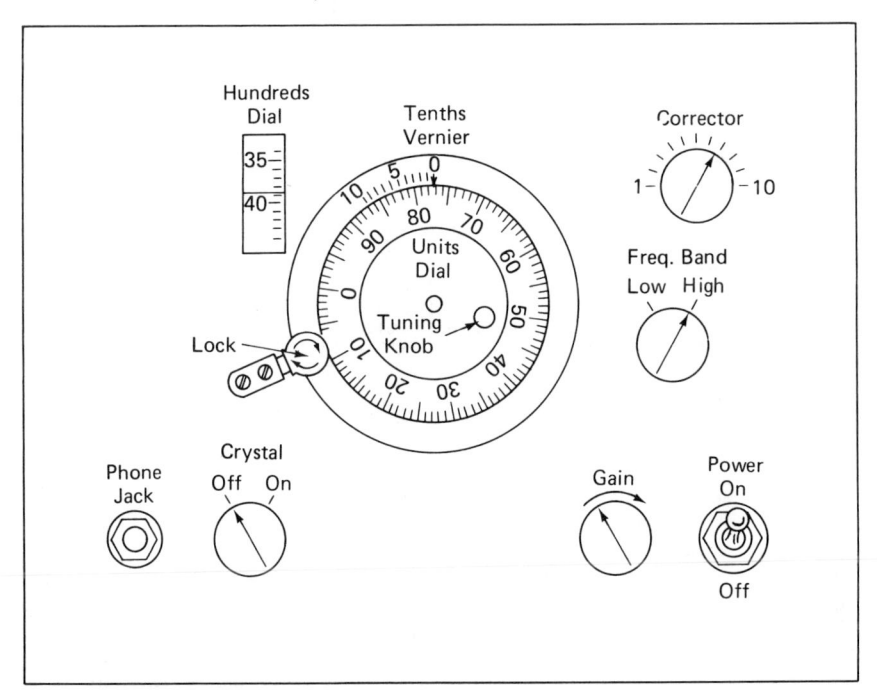

FIG. 15-27 Typical control panel of a heterodyne frequency meter.

is obtained from the dial by finding the line on its scale that coincides most closely with a line on the unit dial. The value of 0.7 on the tenths dial corresponds with 83 on the unit dial; therefore, the final reading is 3800 plus 76 plus 0.7, or 3876.7. The frequency reading that corresponds to this number must be obtained from a calibration book, which is included with each frequency meter.

A typical page of a calibration book is shown in Fig. 15-28. The page is divided into dial and frequency columns, and each dial setting corresponds to four different frequencies. The frequency column immediately following the dial setting represents the fundamental frequency of the variable oscillator.

| Frequency | 3600– 3650
7200– 7300
14400–14600
18000–18250 | Dial | 3761.8–3880.5 | Page 60 |

Dial	Frequency				Dial	Frequency			
					3821.1	3625	7250	14500	18125
					3823.4	3626	7252	14504	18130
3761.8	3600	7200	14400	18000	3825.8	3627	7254	14508	18135
3764.2	3601	7202	14404	18005	3828.1	3628	7256	14512	18140
3766.6	3602	7204	14408	18010	3830.5	3629	7258	14516	18145
3768.9	3603	7206	14412	18015					
3771.3	3604	7208	14416	18020	3832.8	3630	7260	14520	18150
					3835.2	3631	7262	14524	18155
3773.7	3605	7210	14420	18025	3837.6	3632	7264	14528	18160
3776.1	3606	7212	14424	18030	3840.0	3633	7266	14532	18165
3778.5	3607	7214	14428	18035	3842.4	3634	7268	14536	18170
3780.8	3608	7216	14432	18040					
3783.2	3609	7218	14436	18045	3844.8	3635	7270	14540	18175
					3847.1	3636	7272	14544	18180
3785.6	3610	7220	14440	18050	3849.5	3637	7274	14548	18185
3788.0	3611	7222	14444	18055	3851.9	3638	7276	14552	18190
3790.3	3612	7224	14448	18060	3854.3	3639	7278	14556	18195
3792.7	3613	7226	14452	18065					
3795.1	3614	7228	14456	18070	3856.7	3640	7280	14560	18200
					3859.1	3641	7282	14564	18205
3797.5	3615	7230	14460	18075	3861.5	3642	7284	14568	18210
3799.8	3616	7232	14464	18080	3863.8	3643	7286	14572	18215
3802.2	3617	7234	14468	18085	3866.2	3644	7288	14576	18220
3804.6	3618	7236	14472	18090					
3806.9	3619	7238	14476	18095	3868.6	3645	7290	14580	18225
					3871.0	3646	7292	14584	18230
3809.3	3620	7240	14480	18100	3873.4	3647	7294	14588	18235
3811.7	3621	7242	14484	18105	3875.7	3648	7296	14592	18240
3814.0	3622	7244	14488	18110	3878.1	3649	7298	14596	18245
3816.4	3623	7246	14492	18115					
3818.7	3624	7248	14496	18120	3880.5	3650	7300	14600	18250

Nearest Crystal Check Point – 3500, 7000, 14,000, 17,500–3528.2

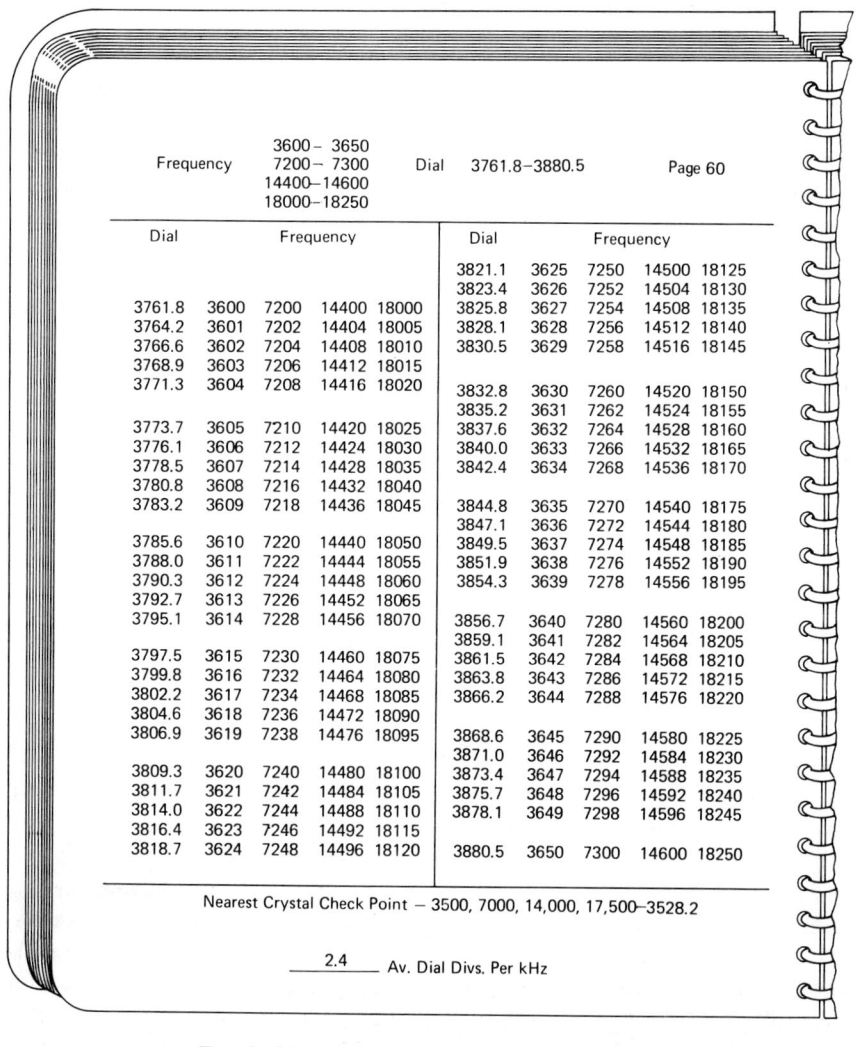

_____2.4_____ Av. Dial Divs. Per kHz

FIG. 15-28   Page of a typical calibration book.

Heterodyne Frequency Meters

For example, a dial reading of 3785.6 indicates that the fundamental frequency of the variable oscillator is 3610 kHz. The other three frequency columns correspond to the harmonics generated by the variable oscillator. Consider an oscillator under test whose frequency output is approximately 14,000 kHz. When determining its exact frequency output by means of a frequency meter, a dial reading of 3768.9 is obtained. Looking in the third frequency column in the calibration book, the exact oscillator frequency is seen to be 14,412 kHz.

At the bottom of each page are the words *nearest crystal checkpoint*. The numbers immediately following these words are the crystal checkpoints nearest the desired frequency and represent harmonics of the crystal oscillator.

The observed dial setting may fall between two values listed in the calibration book. To aid in the calculation of the frequency corresponding to an intermediate dial setting, the following method, called *interpolating*, is used. The dial reading shown in Fig. 15-27 is 3876.7. The calibration book (Fig. 15-28) shows that this reading lies between the dial settings of 3875.7 and 3878.1.

$$
\begin{array}{ccc}
 & \text{dial setting} & \\
 & 3875.7 & \\
\text{diff} = 2.4 & 3876.7 & \text{diff} = 1.4 \\
 & 3878.1 &
\end{array}
$$

The fundamental frequencies of these two dial settings are 3648 and 3649 kHz.

$$
\begin{array}{ccc}
 & \text{frequency} & \\
 & 3648 & \\
\text{diff} = 1 \text{ kHz} & \text{unknown freq.} & \\
 & 3649 & \text{diff} = x \text{ kHz}
\end{array}
$$

Therefore, a proportion is set up:

$$\frac{1 \text{ kHz}}{x \text{ kHz}} = \frac{2.4}{1.4}$$

$$x = \frac{1.4}{2.4} = 0.583 \text{ kHz}$$

The unknown frequency is 3649 minus 0.583, or 3648.417 kHz. The last two significant figures can be discarded for all practical purposes.

Frequency meters are used to check unknown frequencies and also to check or set the frequency of a receiver or a transmitter. To tune a transmitter to a desired frequency, find the page in the calibration book that shows the dial setting of the frequency to which the transmitter is to be tuned; with the crystal oscillator turned on, the frequency-meter dial is then set to the nearest crystal checkpoint. The corrector control is then adjusted to obtain zero beat in the headphones, the crystal oscillator is turned off, and the frequency-meter dial is set to the desired frequency. The transmitter is then loosely coupled to the frequency meter and tuned until zero beat is obtained. An identical procedure is used when tuning a receiver or other equipment.

Consider the following example of interpolation to determine an unknown frequency. If a heterodyne frequency meter having a straight-line relation between frequency and dial reading has a dial reading of 31.7 for a frequency of 1390 kHz, and a dial reading of 44.5 for a frequency of 1400 kHz, what is the frequency of the ninth harmonic of the frequency corresponding to a scale reading of 41.2? The solution is

$$44.5 = 1400 \text{ kHz}$$
$$31.7 = 1390 \text{ kHz}$$
$$\text{diff is} \quad \overline{12.8 = 10 \text{ kHz}}$$

Hence, 12.8 dial divisions represents 10 kHz. To find the frequency corresponding to a dial reading of 41.2, we must next find the scale difference between 41.2 and 31.7, which is 9.5. By setting up a simple proportion we can determine how many kilohertz above 1390 kHz (dial reading = 31.7) the dial reading of 41.2 corresponds to.

$$\frac{9.5}{12.8} = \frac{x}{10 \text{ kHz}}$$
$$12.8x = 9.5 \times 10$$
$$x = \frac{95}{12.8} = 7.42 \text{ kHz}$$

Therefore, the frequency corresponding to a dial reading of 41.2 is 7.42 kHz above 1390 kHz, or 1397.42 kHz. The ninth harmonic of this frequency is 12,576.78 kHz.

If a heterodyne frequency meter, having a calibrated range of 1000 to 5000 kHz, is used to measure the frequency of a transmitter operating on approximately 500 kHz by measurement of the second harmonic of this transmitter, and the indicated measurement was 1008 kHz, what is the actual frequency of the transmitter output? Since the second harmonic is 1008 kHz, it should be apparent that the fundamental or operating frequency of the transmitter will be one half of 1008 kHz, or 504 kHz.

Consider the following example, which is illustrative of the linear proportionality between dial error and frequency. If a frequency meter having an overall error proportional to the frequency is accurate to 10 Hz when set at 600 kHz, what is its error in cycles when set at 1110 kHz? Because the error is proportional to frequency, it can be found by setting up a proportion:

$$\frac{600}{1110} = \frac{10}{x}$$
$$x = \frac{10 \times 1110}{600} = 18.5 \text{ Hz}$$

Carrier-Shift Detector

There are several precautions that must be observed before using a heterodyne frequency meter. For example, if the operating potentials are not correct, particularly in the oscillator stages, the *frequencies may shift*. It is necessary that the unit be allowed to *warm up* for about $\frac{1}{2}$h before using. If this is not done, the oscillator frequency will drift from its assigned value. Before using, the frequency meter should be *calibrated* by zero beating against the crystal oscillator. If any of the components should become loose or vibrate in any way, variations in zero beat will be detected in the output.

Whenever it becomes necessary to change a tube in a heterodyne frequency meter, it is essential that the unit be left on for a period of time to allow for some "aging" of the tube. The unit may then be calibrated and used in the normal manner.

15-21 CARRIER-SHIFT DETECTOR

It is customary for a broadcast station to have a *carrier-shift detector*. This instrument is designed to indicate any shift in the carrier, either upward or downward, when modulation takes place. The meter has a center zero scale with approximately 10 divisions on either side. Each division represents 1 per cent carrier shift. The modulation monitor generally includes the carrier-shift indicator. Actually, the oscillator does not move from its assigned frequency, even though some carrier shift may occur with modulation.

A carrier-shift detector is shown in Fig. 15-29. The inductance L is coupled to the tank of the power output stage it is desired to monitor. The

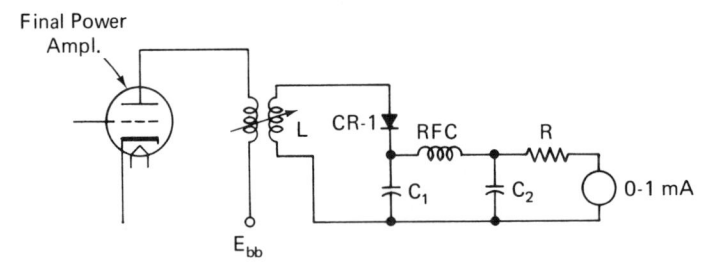

FIG. 15-29 Simple carrier-shift detector.

degree of coupling is adjusted until the meter indicates zero. This corresponds to one half full-scale (zero is the center) reading. Diode CR_1 provides half-wave rectification of the modulated carrier wave. The pi filter, made up of C_1, *RFC*, and C_2 keeps radio frequency out of the meter. Resistor R is chosen to compensate for different meter sensitivities that might be used and also serve as part of the diode load. The meter is observed as the transmitter is modulated. If no carrier shift or overmodulation occurs, the needle will remain centered on zero. If the needle swings up scale (to the right of zero), a positive carrier shift is indicated, and vice versa.

15-22 FREQUENCY MEASUREMENTS WITH LECHER WIRES

At very high frequencies it is possible to determine the wavelength of an unknown frequency be measuring the distance between the standing waves on a resonant line. Such a transmission line is called *Lecher wires*, when used for this purpose. The line can be made of two lengths of No. 12 bare copper wires, or small diameter copper tubing, stretched tightly to eliminate movement, held about 1 in. apart and supported by insulators only at the far ends.

The fundamental principle involved is illustrated in Fig. 15-30. As the position of the sensitive current meter is moved, starting near the coupling loop, a series of sharply defined positions will be found where the current

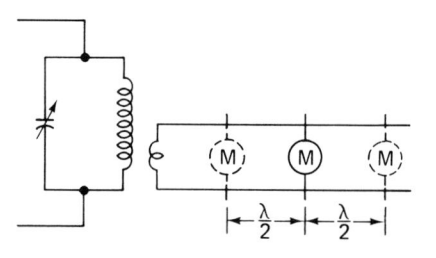

FIG. 15-30 Lecher-wire arrangement for measuring wave length at very high frequencies.

through the thermocouple meter is a minimum. *Minima*, rather than maxima, are used, since they are *sharper*. These positions of resonance are *almost exactly one half-wave length apart*. The frequency can then be calculated as

$$f = \frac{300{,}000{,}000}{\lambda}$$

where λ is one wavelength (twice the distance between adjacent minima as measured).

f is in hertz.

Measurements of wavelength made with a Lecher wire system may reach 0.1 per cent accuracy if care is used.

Practice Problems

1. Calculate the value of shunt resistance necessary to extend the range of a 0- to 1-mA meter movement, having 40-Ω internal resistance, to read a total of 50 mA.
2. A certain 50-μA meter having 2000 Ω internal resistance is to be used in a circuit to read a total of 100 mA. What value of shunt resistance is needed?
3. What value of series multiplying resistor must be connected to a 0- to 1-mA meter having 50-Ω internal resistance to enable it to read a maximum of 10 V_{dc}?
4. It is desired to use a 50-μA meter having 2-kΩ resistance as a dc voltmeter to read 250 V. What value of multiplying resistor is required?

Commercial License Questions

5. Two resistors, one 50 kΩ and one 100 kΩ, are series connected across a 150-V_{dc} source. If a voltmeter having 100-kΩ resistance is connected across the 100-kΩ resistor, what voltage will it read? What is the actual voltage across the 100-kΩ resistor before the meter is connected?

6. Referring to the basic ohmmeter circuit shown in Fig. 15-10, if a 3.0-V battery is used and $R_B = 2.5$ kΩ, what resistance value must R_A be adjusted to in order for the meter to read full scale? If an unknown resistance connected between the probes causes the meter to read 0.4 mA, what is its value?

7. What is the wavelength of a 1650-kHz signal?

8. A wavemeter, having an error proportional to frequency, is accurate to 45 Hz when set at 5 MHz. What is its error when set at 6.25 MHz?

Commercial License Questions

Sections in which answers to questions are given appear in parentheses. A bracketed number following a question implies that it applies only to that element.

1. Describe the construction and characteristics of a D'Arsonval-type meter.
   (15–1)
2. What is the purpose of a shunt as used with an ammeter?   (15–2)
3. If two ammeters are connected in parallel, how may the total current through the two meters be determined?   (15–2)
4. If two ammeters are connected in series, how may the total current through the two meters be determined?   (15–2)
5. How may a dc milliammeter, in an emergency, be used to indicate voltage?
   (15–3)
6. What is the purpose of a multiplier resistance used with a voltmeter?
   (15–3)
7. What is the ohms per volt of a voltmeter constructed of a 0 to 1 dc milliammeter and a suitable resistor which makes the full-scale reading of the meter 500 V? [4]   (15–3)
8. If a 0 to 1 dc milliammeter is to be converted into a voltmeter with full-scale calibration of 100 V, what value of series resistance should be connected in series with the milliammeter?   (15–3)
9. If two voltmeters are connected in series, how would you be able to determine the total drop across both instruments?   (15–3)
10. Draw a diagram of an ohmmeter and explain its principle of operation.
    (15–4)
11. What single instrument may be used to measure electrical resistance? Electrical power? Electrical current? Electromotive force?   (15–5)
12. What type of meter is suitable for measuring the AVC voltage in a standard broadcast receiver? [4]   (15–6)

13. What type of voltmeter absorbs no power from the circuit under test? [4]
    (15–6, 15–7)

14. Why are copper oxide rectifiers, associated with dc voltmeters for the purpose of measuring alternating current, not suitable for the measurement of voltages at radio frequencies? (15–8)

15. What type of meter is suitable for measuring peak ac voltage? [4] (15–9)

16. Describe the construction and characteristics of a dynamometer-type indicating instrument. (15–10)

17. Does an ac ammeter indicate peak, average, or effective values of current?
    (15–10)

18. What instrument measures electical energy? (15–11)

19. What instrument measures electric power? (15–12)

20. A current-squared meter has a scale divided into 50 equal divisions. When 45 mA flows through the meter, the deflection is 45 divisions. What is the current flowing through the meter when the scale deflection is 25 divisions? [4]
    (15–13)

21. What type of meters may be used to measure RF currents? (15–13)

22. Describe the construction and characteristics of a thermocouple type of meter; of a wattmeter. (15–13)

23. What are cathode rays? (15–15)

24. Draw a diagram of an absorption-type wavemeter and explain its principle of operation. (15–16)

25. What precautions should be observed in using an absorption-type frequency meter to measure the frequency of a self-excited oscillator? Explain your reasons. (15–16)

26. What is the formula for determining the wavelength when the frequency, in kilocycles, is known? (15–16)

27. With measuring equipment that is widely available is it possible to measure a frequency of 10,000,000 Hz to within 1 Hz of the exact frequency?
    (15–18, 15–19)

28. What is the meaning of *zero beat* as used in connection with frequency-measuring equipment? (15–18, 15–19)

29. What is the device called that is used to derive a standard frequency of 10 kHz from a standard-frequency oscillator operating on 100 kHz? [4] (15–19)

30. Describe the technique used in frequency measurements employing a 100-kHz oscillator, a 10-kHz multivibrator, a heterodyne frequency meter of known accuracy, a suitable receiver, and standard frequency transmission. [4]
    (15–19)

31. In frequency measurements using the heterodyne zero-beat method, what is the best ratio of signal emf to calibrated heterodyne oscillator emf? [4]
    (15–20)

Commercial License Questions 293

32. If a heterodyne frequency meter having a calibrated range of 1000 to 5000 *kHz* is used to measure the frequency of a transmitter operating on approximately 500 kHz by measurement of the second harmonic of this transmitter, and the indicated measurement is 1008 kHz, what is the actual frequency of the transmitter output? (15–20)

33. If a frequency meter having an overall error proportional to the frequency is accurate to 10 Hz when set at 600 kHz, what is its error in cycles when set at 1110 kHz? (15–20)

34. What procedure should be adopted if it is found necessary to replace a tube in a heterodyne frequency meter? [4] (15–20)

35. What precautions should be taken before using a heterodyne type of frequency meter? (15–20)

36. If a heterodyne frequency meter having a straight-line relation between frequency and dial reading has a dial reading of 31.7 for a frequency of 1390 kHz and a dial reading of 44.5 for a frequency of 1400 kHz, what is the frequency of the ninth harmonic of the frequency corresponding to a scale reading of 41.2? [4] (15–20)

37. Draw a schematic diagram of test equipment that may be used to detect carrier shift of a radiotelephone transmitter output. [4] (15–21)

38. What is the reason why certain broadcast-station frequency monitors must receive their energy from an unmodulated stage of the transmitter? [4] (15–21)

39. What is the purpose of using a frequency standard or service independent of the transmitter frequency monitor or control? [4] (15–21)

40. Discuss the properties and use of Lecher wires. (15–22)

16

Oscillators

16-1 REQUIREMENTS FOR OSCILLATION

The function of an *oscillator* is to generate alternating voltages. The frequency of these voltages may be from several hertz up to the gigahertz region, depending upon the type of oscillator used. Some are designed to operate at a *fixed frequency*, others may be *variable*. Generally, variable-frequency oscillators are designed to operate over specific bands of frequencies, such as audio, broadcast, UHF, VHF, or microwaves. It is not practical to design an oscillator that could operate over the entire electromagnetic spectrum. Oscillators employ transistors, vacuum tubes, or special tubes to generate the high-frequency energy.

Essentially, an oscillator is an amplifier that derives its *input signal from its own output terminals*. The input signal must be of the *correct phase* and *amplitude* to sustain oscillation. This is called *positive* or *regenerative* feedback. Because power must be supplied by the oscillator to some external circuit, it is essential that the amplifying device be capable of supplying this loss of power. Four requirements are necessary to produce continuous oscillations. These are

1. A tuned circuit.
2. An amplifying device.
3. Positive or regenerative feedback.
4. A power supply.

The tuned circuit determines the frequency at which oscillations take place. This is usually an *LC* combination that resonates at the desired frequency, which can be calculated by the formula

$$f_0 = \frac{1}{2\pi\sqrt{LC}}$$

The amplifying device must be capable of operating in the frequency spectrum desired. This may pose serious problems for very high frequency oscillators, as the amplifier may not function properly at the higher frequencies.

In the early days of radio communication *spark-gap oscillators* were used. The circuit consisted of a spark gap in series with an *LC* combination, such as shown in Fig. 16-1. The high voltage developed across the secondary

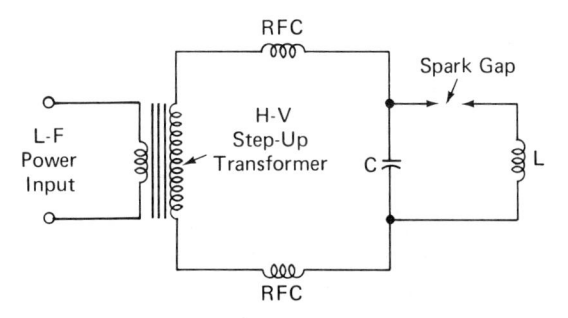

FIG. 16-1 Spark gap oscillator.

of the transformer charged the capacitor. When the charging emf reached a critical potential, the air between the points of the gap *ionized*. This provided a low-resistance path across the gap, and allowed the capacitor to discharge into the inductor. This started the oscillatory action that generated a *highly damped* wave. The *RFC*s keep the high-frequency energy in the tank from getting into the secondary winding.

The disadvantage of this circuit is that radiation occurred over a *very wide portion* of the *RF* spectrum. This caused serious interference in receiving circuits that were tuned to considerably removed frequencies. Many ocean-going vessels still maintain these old transmitters strictly for *emergency purposes*. If a coded distress message should ever be sent by such a transmitter, virtually all vessels within receiving distance would be bound to receive it, regardless of the frequency(s) their receivers were tuned to.

16-2 TWO-STAGE AMPLIFIER USING REGENERATIVE FEEDBACK

A regenerative feedback circuit is shown in Fig. 16-2. Feedback occurs as follows: stages Q_1 and Q_2 are impedance coupled amplifiers. Assume that stage Q_1 is properly biased and receives a positive-going pulse at its base. Because of the beta of the transistor, the signal will be amplified and appear as a negative-going pulse in the collector circuit. The amplified signal appears across the collector load impedance L_1 and is capacitively coupled to the base

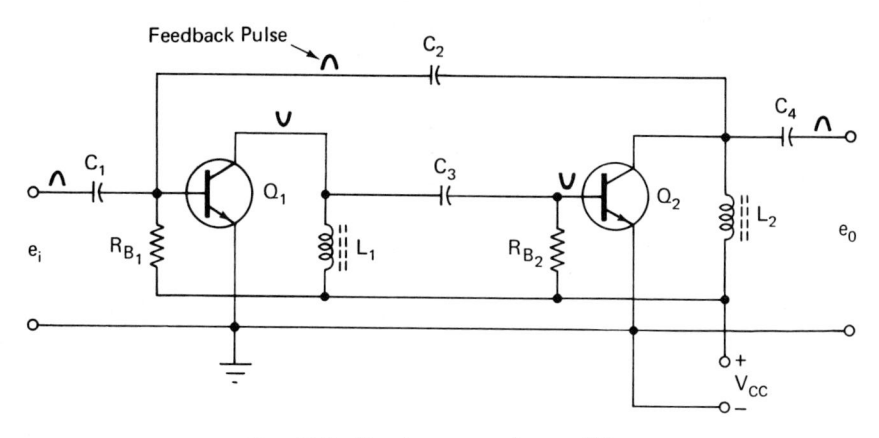

FIG. 16-2 Simple regenerative amplifier.

of transistor Q_2 via capacitor C_3. The negative-going pulse at the base of Q_2 is further amplified and appears across the collector load impedance L_2 as a positive-going pulse. Capacitor C_4 couples this pulse to the output circuit. A portion of this signal is fed back to the base of Q_1 via feedback capacitor C_2. Observe that the feedback pulse is in phase with the initial signal supplied to the base of Q_1.

16-3 ARMSTRONG OSCILLATOR

The Armstrong oscillator circuit, named after its inventor, consists of a tuned circuit in the input of the amplifier that is supplied energy from a *tickler coil* in the output circuit. This type of oscillator is shown in Fig. 16-3. Oscillations begin spontaneously; hence, no external trigger source is required. In Fig. 16-3a when the switch is closed, the battery supplies forward bias to the base–emitter junction through resistor R_B. The collector current that begins to flow causes a magnetic field to be built up around the tickler coil, which induces a voltage into the tank circuit connected to the base of the transistor. A portion of the voltage developed across the tuned circuit is capacitively coupled via C_1 to the base of the transistor. Because the impedance of the base–emitter junction of a common-emitter-type amplifier is relatively low, it is necessary that capacitor C_1 be connected to a tap on the tank inductance to match the transistor's impedance. This prevents *seriously loading* the tank circuit and reducing its Q. The stability is improved by lowering the tap, but the output power suffers.

If the leads connecting to the tickler coil should be reversed, the phasing will not be proper to sustain oscillation. The amount of voltage induced into the tank circuit is not only a function of the magnitude of the collector current, but the amount of *coupling* between the tickler coil and the tank. In some designs it is possible to vary the position of the tickler coil relative to the tank

Armstrong Oscillator 297

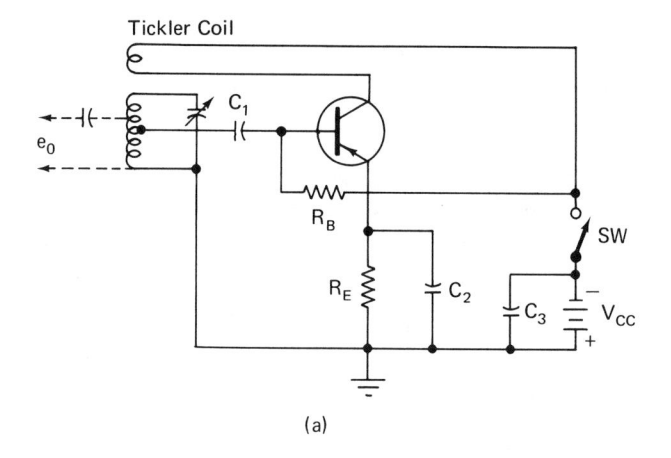

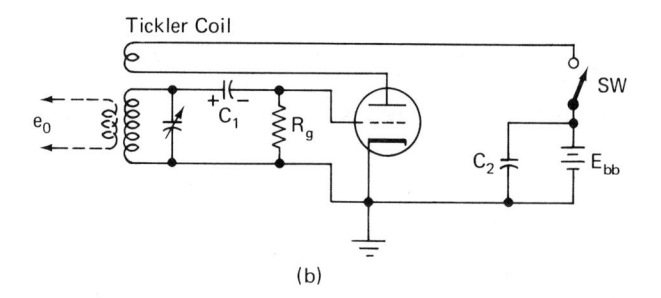

FIG. 16-3 Armstrong oscillator circuits using (a) pnp transistor, (b) triode vacuum tube.

inductance. This controls the amplitude of the signal developed in the tank, and hence that supplied to the base of the transistor. The emitter resistor R_E and its associated capacitor C_2 provide a *stabilizing circuit to prevent thermal runaway*.

The operation of the triode Armstrong oscillator circuit is essentially the same as that of the transistorized version. Assuming that the cathode is sufficiently heated for normal operation, plate current begins to flow the moment the switch is closed. This current induces a magnetic field around the tickler coil, which induces a voltage into the tank circuit. This induced voltage causes the tank to oscillate. These oscillations are amplified by the triode and appear across the tickler coil, providing the regenerative feedback necessary for oscillation. C_1 and R_G provide proper grid bias by means of grid-leak action.

Various methods of coupling the output to the next stage are available. These include *capacitive-*, *transformer-*, and *impedance-coupling* networks. The choice of coupling depends on the specific application of the oscillator.

The output, in any case, usually is taken from the oscillating *LC* circuit. This has the effect of loading down the circuit, and is equivalent to increasing *R* in the tank. For this reason, the load *increases the losses*, increasing the amount of regenerative feedback required. Too great a load *damps the oscillations* in the tank circuit, causing the oscillations to die out.

This oscillator can be designed to operate in both the AF and RF ranges. Too high a frequency of oscillations, however, introduces undesirable effects as a result of the high interelectrode capacitance of the triode, reducing the output available. For this reason, low μ triodes commonly are used.

16-4 OSCILLATOR BIASING

The performance of an oscillator is determined in part by the type of biasing used. If fixed bias were used, the oscillator would not likely be self-starting. Therefore, some method of self-biasing is employed. *Grid-leak biasing* provides the most satisfactory method. Not only will this guarantee the oscillator to be self-starting, but if properly designed will also ensure the correct amount of bias voltage. A typical circuit employing a series type of grid-leak biasing is shown in Fig. 16-4. Compare this arrangement to the shunt type shown in Fig. 16-3b. Performance is essentially the same with either arrangement.

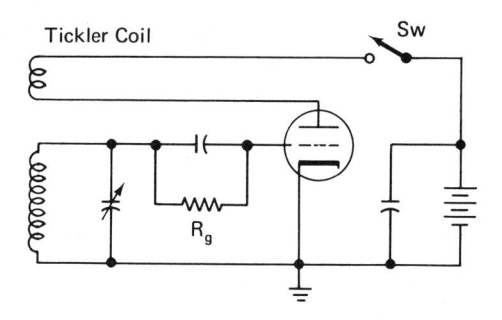

FIG. 16-4 Series type of grid leak biasing.

For an explanation of how grid-leak biasing is achieved, refer to the schematic shown in Fig. 16-4. Assume that the cathode is operating at normal temperature. When the switch is closed, plate current immediately begins to flow. At this instant there is no bias, and the plate current rises toward saturation. A magnetic field is quickly built up around the tickler coil in the plate circuit, inducing a voltage into the tank. The polarity of this voltage is such that the signal coupled to the grid will be going positive, which tends to further increase plate current. This regenerative action continues until the plate is driven into *saturation*, at which time the field surrounding the tickler coil can no longer expand. Induction into the *LC* circuit will now cease, and because of the oscillatory nature of the tank current, an opposite voltage begins to appear across the tank, driving the grid negative.

Let us examine the action that takes place during the moment that the grid is driven positive. Some of the electrons traveling toward the plate are attracted to the grid and accumulate on the grid capacitor. If no grid-leak resistor, R_g, were present, the charge building up on the right side of the capacitor would quickly be large enough to completely block the tube, and oscillation would cease. To prevent this, a *grid-leak resistor* is used, which permits the electrons to leak off and return back to the cathode via the tank coil.

The value of the coupling capacitor will vary from as high as 1000 pF to as low as 50 pF depending upon whether the oscillator is designed to operate at low or high frequencies. A typical value would be 100 pF. Typical values of grid-leak resistors range from 25 to 100 kΩ.

With the proper combination of grid-leak resistance and capacitance, the electrons that are trapped on the grid and the plate of the capacitor form a highly negative charge which *will not leak off completely*, even during the negative half of the tank circuit oscillation. In other works, the time constant of the *RC* combination has to be long, relative to the frequency of oscillation. With proper design, sufficient bias is generated to cause the tube to operate in class C, thus providing efficiencies from 50 to 70 per cent. If too large a capacitor or grid-leak resistor is employed, excessive bias voltage will be developed. This will cause the plate current flowing to be reduced to such a value that *little if any power* can be drawn from the oscillator. If the time constant of the grid-leak circuit is too large, *low-frequency RC oscillations* may occur simultaneously with the RF energy being generated. This creates *parasitic oscillations*, which are undesirable.

16-5 TUNED–PLATE TUNED–GRID OSCILLATOR

A common oscillator circuit is one that employs tuned circuits in both the input and output, as shown in Fig. 16-5a. It is known as a *tuned-plate tuned-grid* (TPTG) oscillator and uses capacitive coupling between output and input circuits. The capacity is that existing between *grid and plate* and is represented as C_{gp} in Fig. 16-5a. Oscillators of this type almost always employ triode vacuum tubes, because of the larger C_{gp} involved.

The grid circuit, $L_1 C_1$, is tuned to the resonant frequency desired. When the first surge of current starts this circuit oscillating, the oscillations appear at the grid and are amplified. The amplified RF energy appearing across the plate tank circuit is divided across a *capacitive voltage-divider* network comprising capacitor C_3, C_{gc}, and C_{gp}. C_3, the plate bypass capacitor, is normally very large in comparison to these other capacitors and presents a very low reactance. Therefore, the major portion of the plate tank voltage appears

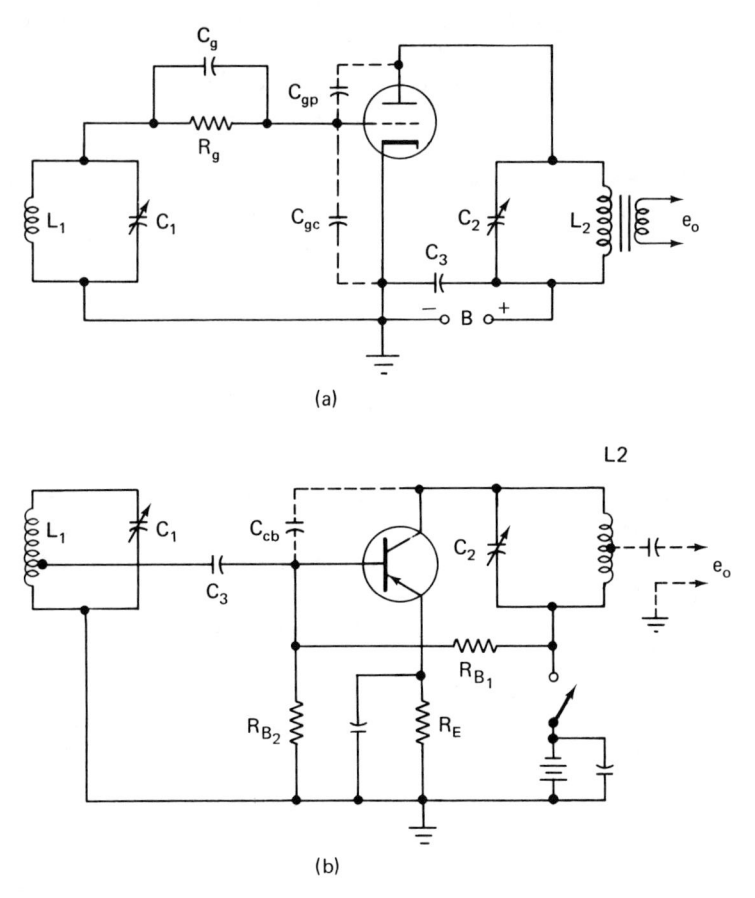

FIG. 16-5 (a) Vacuum tube tuned-plate–tuned-grid oscillator; (b) transistorized version of tuned-collector–tuned-base.

across C_{gc} and C_{gp}. The feedback path, then, is through the grid-plate capacitance to the grid tank circuit. If the plate tank is tuned to the same frequency as the grid tank, the phase of the feedback is *not proper* to sustain oscillations. For this reason the plate circuit is made *inductive* at the frequency of oscillations of the grid circuit to make the feedback regenerative. This is done by tuning the plate circuit to a slightly higher frequency. Because there are two tuned circuits with this type of oscillator, two frequencies of operation are possible. Normally, the tank circuit having the *higher Q* will dictate the oscillating frequency. When low-frequency oscillations are desired, it is frequently necessary to add a small capacitor, on the order of 10 to 50 pF, in parallel with the tube's grid-plate capacity, in order to feed back sufficient energy from the plate to the grid circuit.

The transistorized version of this oscillator appears in Fig. 16-5b. In principle, the operation is the same as previously explained. Because transis-

Hartley Oscillator 301

tors are normally biased at cutoff, it is necessary to *forward bias* the base–emitter junction to get collector current to flow. The voltage-divider network, R_{B1} and R_{B2}, provides the necessary voltage at their junction to forward bias the transistor. The collector–base capacity appearing across the reverse-biased junction provides the capacitive path for energy in the collector tank to be fed into the base tank circuit. To effect a better impedance match and also to avoid lowering excessively the Q of the $L_1 C_1$ tank circuit, capacitor C_3 is connected to a tap on L_1. The oscillator output can be coupled to a following stage by inductive or capacitive means, such as those shown.

16-6 HARTLEY OSCILLATOR

The Hartley oscillator is a modification of the Armstrong circuit. The customary tickler coil has become an integral part of the tank inductance L_1 and L_2, shown in Fig. 16-6a. In the series-fed arrangement of this oscillator

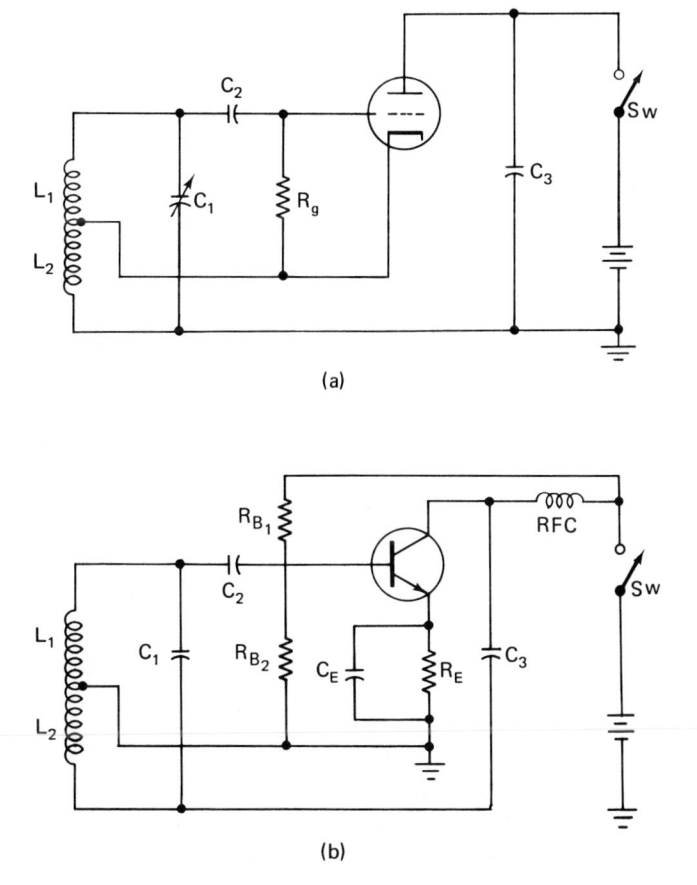

FIG. 16-6 Hartley oscillator circuits: (a) series-fed vacuum tube version, (b) shunt-fed transistorized version.

(Fig. 16-6a) the cathode is connected to a tap on the tank inductance. When the power switch is closed, plate current flows from the negative terminal of the battery through L_2 into the cathode of the tube, and from thence to the plate, completing the circuit. The sudden increase in plate current induces voltage into the upper half of the tank, L_1, and the polarity of the voltage is such that the signal coupled through capacitor C_2 drives the grid positive, thus initiating oscillatory action. Grid bias is provided by capacitor C_2 and resistor R_g. In this arrangement the tuned circuit contains the dc component of plate current in addition to the ac signal. The Hartley oscillator will also operate if no mutual inductance exists between L_1 and L_2 in the tank, because of the coupling effect of the tank capacitor C_1.

Better oscillator performance is usually achieved when current is kept out of the tank circuit. Such an arrangement is called *shunt feeding* and is indicated in the transistorized version of the Hartley oscillator shown in Fig. 16-6b. The presence of dc current in a tank circuit reduces the circuit Q. Radio-frequency energy developed across the *RFC* is capacitively coupled to L_2 via capacitor C_3.

The *tapped tank inductance* is characteristic of all Hartley oscillators. To guarantee the circuit to be self-starting when the switch is closed, the scaling resistors R_{B1} and R_{B2} forward bias the base–emitter junction to cause collector current to flow. Any tendency for the oscillator to shift from its operating point is compensated for by the emitter swamping resistor R_E.

16-7 COLPITTS OSCILLATOR

The Colpitts oscillators shown in Figs. 16-7a and b are similar to the shunt-fed Hartley oscillators with the exception that the Colpitts oscillator uses a *split tank capacitor* as part of the feedback circuit instead of a split tank inductor. The frequency of oscillation is determined by the values of L and C_1 and C_2. The connection between the two capacitors is employed as the *center tap* of the circuit. These two capacitors are in series; consequently, the grid-end and plate-end capacitances are larger than would normally be used for a tuning capacitor used in an equivalent LC circuit of the Hartley oscillator. The grid-plate capacitance of the tube is in shunt with C_1 and the plate-cathode capacitance is in shunt with C_2. Because of the large capacitive values of these tank capacitors, the interelectrode capacitance of the tube has little effect upon the circuit performance. Like the Hartley, the center tap of the tank circuit is *not exactly* in the center. If the plate-end tuning capacitor has a larger capacitance, a greater voltage drop will appear across the grid capacitor, resulting in improved frequency stability. However, if the grid-end tuning capacitor should be larger, then the greater voltage will appear across the plate-end, resulting in higher power output. The functioning of the grid-leak combination and the plate bypass capacitors are the same as in other oscilla-

Colpitts Oscillator

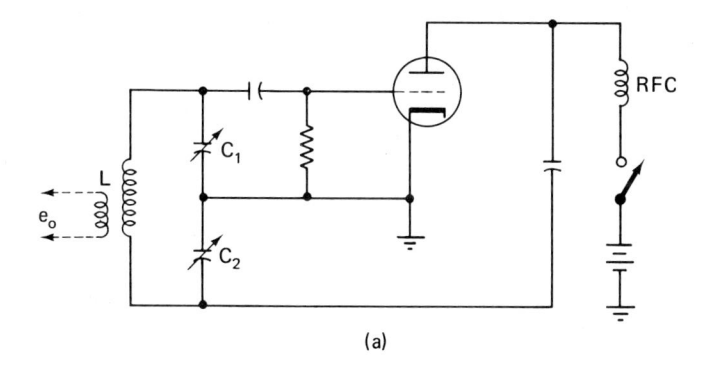

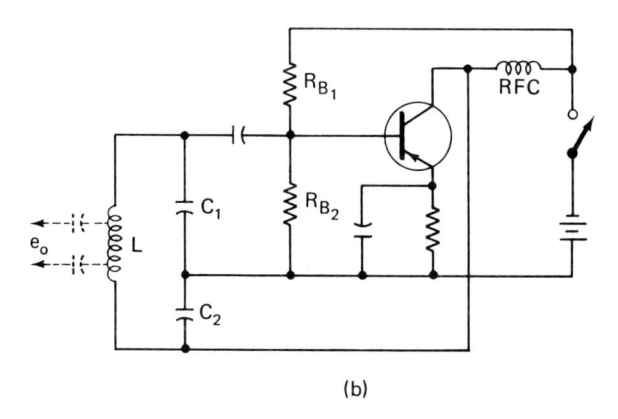

FIG. 16-7 Colpitts oscillator circuits: (a) using a triode vacuum tube, (b) using a pnp transistor.

tor circuits. A transistorized version of the Colpitts oscillator is shown in Fig. 16-7b.

Two methods of drawing power from the oscillator circuit are shown. In Fig. 16-7a a small *link*, or inductance, is magnetically coupled to the tank circuit, in Fig. 16-7b the tank coil is tapped by two capacitors that draw the required amount of power.

An ultrahigh-frequency form of the Colpitts oscillator is known as the *ultra-audion* and is shown schematically in Fig. 16-8. The grid-cathode and plate-cathode interelectrode capacitances, which make the operation of the ultra-audion oscillator similar to that of the Colpitts, are indicated by dotted lines in the figure. Shunt feed is employed and the *RFC* prevents the ac component of the plate voltage from entering the power supply. Capacitor C_3 provides a low reactive path for *RF* current and blocks direct current from the tank. The voltage drop across C_{gc} is appreciable at the frequency employed

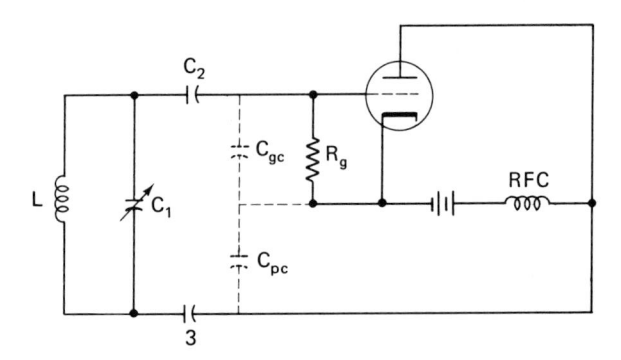

FIG. 16-8 Ultra-audion oscillator.

and provides the grid excitation. The total tank capacitance is made up of C_1 in parallel with the series combination of C_2, C_{gc}, C_{pc}, and C_3. Capacitors C_2 and C_3 are relatively large so that they will offer negligible reactance to *RF* current.

If two oscillator circuits are located near each other so that they are capacitively or magnetically coupled and are operating on adjacent frequencies, they have a tendency to *synchronize or lock together* and operate on the same frequency.

16-8 ELECTRON-COUPLED OSCILLATORS

If the output of an oscillator is coupled directly to a power amplifier, *undesirable loading effects* occur. There can be distortion of the output waveform or even a stopping of oscillation. In addition, the frequency of the oscillation will not be stable. A *buffer amplifier*, therefore, is used to couple the oscillations to the power amplifier. An ordinary voltage amplifier can serve as a buffer, since it draws little power from the oscillator.

By using a *multielectrode* tube, the oscillator and buffer stages can be replaced by one circuit that performs both functions. Such a circuit is called an *electron-coupled oscillator* and is shown in Fig. 16-9. The cathode, the control grid, and the screen grid perform the function of the triode in a Hartley oscillator. The cathode connection taps the split-inductance tank.

In the electron-coupled oscillator the screen grid collects only that portion of the current needed for feedback. The output portion of the current passes through the screen grid to the pentode plate, where it is collected and passed through the output tank circuit consisting of C_3 and L_3. Capacitors C_2 and C_4 serve to bypass oscillations around the power supply.

The only connection between the oscillator and the output circuit is the electron stream itself. This isolates the oscillator from the load and has all the advantages of a separate oscillator and buffer.

Crystal-Controlled Oscillators 305

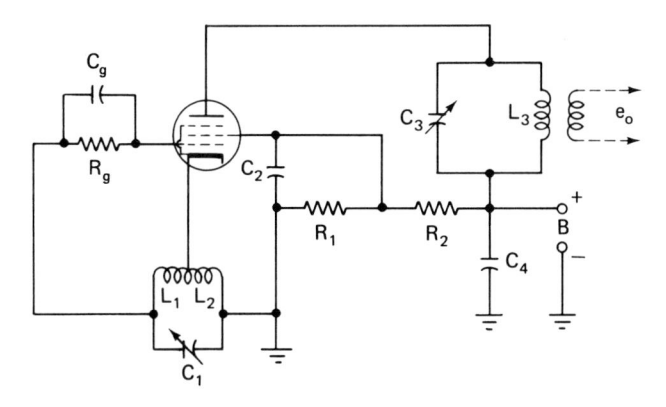

FIG. 16-9 Electron-coupled oscillator.

16-9 CRYSTAL-CONTROLLED OSCILLATORS

Oscillators having the highest frequency stability are those which are controlled by *crystals*. The crystals are usually quartz (although Rochelle and tourmaline crystals have the same basic characteristics) and exhibit characteristics known as the *piezoelectric* effect. When a piezoelectric material is compressed or stretched in certain directions, an electric charge appears on the surfaces that are perpendicular to the axis of strain. Conversely, when a piezoelectric material is placed between two metallic surfaces and a potential difference is applied, a mechanical strain is set up within the crystal. When the potential is removed, the mechanical strain is relieved, and the crystal snaps back to its original position and, in doing so, generates a small voltage. If the potential is reversed, the strain is in the opposite direction. If the potential is again removed, the crystal returns to its original position and generates a reverse potential.

Quartz is the most commonly used crystal material because it is mechanically rigid, inexpensive, and has a low temperature coefficient. Very small slabs are cut from the quartz and then polished. The final dimensions, particularly the *thickness*, determine the frequency at which the crystal resonates. Only a small emf need be applied to the crystal to make it oscillate due to its *very high Q*. The voltage it generates is much larger than that required to sustain oscillation.

A crystal-controlled oscillator employing a triode tube appears in Fig. 16-10. The equivalent circuit of crystal Y is shown to the left in the drawing. It is apparent, from the equivalent circuit, that when the crystal is oscillating it behaves as a *series resonant* circuit. The inductance L represents the electrical equivalent of the crystal mass that is effective in causing mechanical vibration. R is the electrical equivalent of the internal resistance due to friction, and C_2 is the capacity effect of the metal crystal holders. The series capacitor

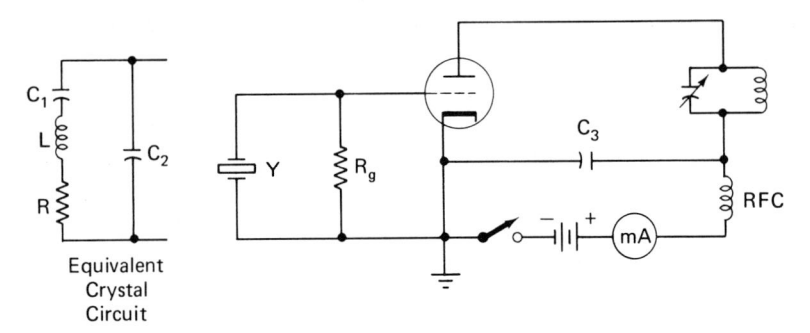

FIG. 16-10 Crystal-controlled oscillator and equivalent circuit of the crystal.

C_1 is the reciprocal of the crystal stiffness, that is, compliance, which is the equivalent of capacitance in the electrical system. The circuit oscillates by virtue of the feedback taking place through grid-plate capacitance of the tube. Grid-leak bias is established across resistor R_g. The grid capacitor is usually omitted because the crystal blocks any dc flow to ground. To make the circuit operate properly, the plate tank must be tuned *slightly higher* than the frequency of the crystal. This ensures that the feedback voltage will be of the proper phase relationship to sustain oscillations. Note the similarity of this circuit to the conventional TPTG oscillator.

Proper tuning can be determined by observing the action of the dc milliammeter in the plate circuit. If plate current decreases as the tank capacitor is tuned, it is an indication that the circuit is oscillating and developing grid bias. As the oscillator operates more strongly, the plate current further decreases. Crystal oscillators exhibit certain unique characteristics when tuned. For example, as the plate tank circuit is decreased in frequency, the plate current gradually decreases to a minimum value and then pops up to a maximum. At the point where the plate current suddenly increases to its maximum value, the oscillations cease. Although the minimum plate-current readings suggest *strongest oscillations* of the circuit, they *do not indicate optimum operating conditions*. By tuning the plate tank circuit so that the plate current drops to a value about three quarters of the way down from maximum, the most satisfactory performance will be obtained. Even though the output is reduced, the operation is much more stable.

Slight changes in loading will not cause the oscillator to cease functioning. It is possible to vary the oscillator's frequency by a few hertz by placing a *small variable capacitor* across the crystal. If the capacitor is made too large, the crystal will likely not oscillate.

If the oscillator tube is replaced, the plate tank will likely have to be slightly retuned, due to the difference in interelectrode capacities of the tubes. Another factor affecting the frequency is the pressure on the crystal by the plates. Slight changes in this pressure will cause the oscillations to vary a few

Crystals and Their Characteristics

hertz. For circuits employing low-frequency crystals it may be necessary to add a small capacitance between the grid and plate to increase the feedback. Too large an external capacitor will cause the crystal to *fracture*.

A typical transistorized crystal-controlled oscillator appears in Fig. 16-11. Biasing resistors R_1 and R_2 ensure proper forward bias to guarantee the oscillator to be self-starting. The collector is tapped down on the tank inductance for impedance matching.

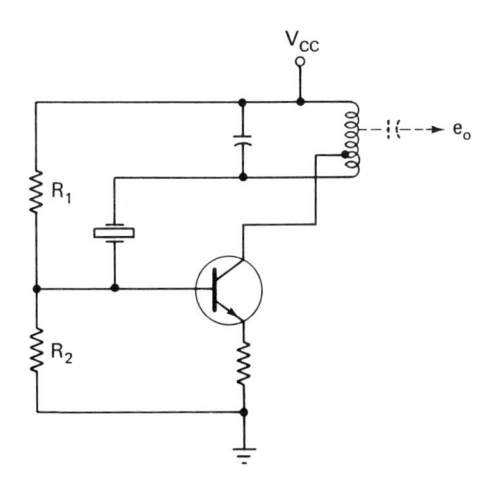

FIG. 16-11 Typical transistorized crystal-controlled oscillator.

16-10 CRYSTALS AND THEIR CHARACTERISTICS

Crystals, although rarely symmetrical in shape, have the general form of a *hexagonal prism*, usually containing hexagonal pyramids at each end. The cross section of a symmetrical crystal is shown in Fig. 16-12a. The major axes are designated as X, Y, and Z. The angle at which the small plates are cut from the crystal determines its electrical characteristics. When a plate is cut parallel to the Z axis with its faces perpendicular to the X axis, it is called an X cut. This is the type shown in Fig. 16-12b. To obtain a Y cut, it is necessary to cut the plate perpendicular to the Y axis, as shown in Fig. 16-12c. By rotating the plane of the cut around one or more axes, a number of different cuts, such as the AT, BT, CT, DT, GT, and NT, are obtained. If, for example, the Y-cut plate shown in Fig. 16-12c is rotated clockwise, from the Z axis, an AT cut results.

It is possible to obtain crystal frequencies from as low as several kilohertz to as high as 30 MHz by proper selection of the type of cut, dimensions of the plate, and mode of vibration. Operation at frequencies much higher than this is not practical, because the plate becomes very thin and fragile. It is possible to extend the upper frequency range by finishing the plates so that they can be excited at the *third* or *fifth harmonic* of their fundamental

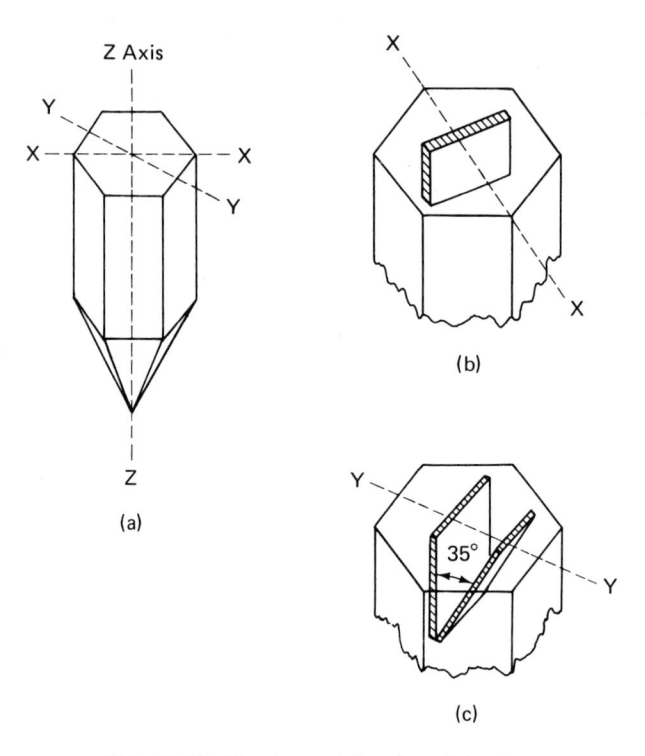

FIG. 16-12 Quartz crystal and crystal cuts.

frequency. This results in what is known as *overtone crystals*, which will operate up to 100 MHz.

Different cuts exhibit different *temperature coefficients* and characteristics. The meaning of temperature coefficient can be explained in the following way: If a change in ambient temperature produces a relatively small deviation in oscillator frequency, the crystal is said to have a *low temperature coefficient*. Conversely, if a change in ambient temperature produces a large deviation in oscillator frequency, the crystal is said to have a *high temperature coefficient*. If the oscillator frequency is not changed by any variation in ambient temperature, the crystal is said to have *zero temperature coefficient*. If the crystal frequency increases with an increase in temperature, the crystal has a positive temperature coefficient, and vice versa.

Some crystals have negative as well as positive temperature coefficients depending upon the temperature to which they are subjected. For example, about 70°F the coefficient may be positive, and negative at temperatures above 140°F. Between these two limits there is a crossover point at about 110°F at which the crystal exhibits, for all practical purposes, a zero temperature coefficient. X-cut crystals have temperature coefficients that vary from about -10 to -25 Hz/°C/MHz. Y-cut crystals have a range of about -25

to $+100\,\text{Hz}/°\text{C}/\text{MHz}$. This cut of crystal is characterized by having two modes or frequencies of operation that usually are close together. Proper care must be used in tuning an oscillator using this type of cut to ensure the correct frequency. The other cuts previously described have temperature-frequency characteristics superior to those of either the X or Y cuts. For example, a GT cut has a -1- to a $+1$-$\text{Hz}/°\text{C}/\text{MHz}$ range. Hence, we see the *GT cut* has almost a zero temperature coefficient over a wide temperature range. However, this particular cut is useful only to a few hundred kilohertz.

The following example illustrates how temperature coefficient information is used to determine a crystal oscillator's frequency:

Example 16-1

A 600-kHz, X-cut crystal, calibrated at 50°C and having a temperature coefficient of -20 parts per million (ppm) per degree centigrade, will oscillate at what frequency when its temperature is 60°C?

Solution

The -20 ppm/°C implies that the fundamental frequency of the crystal is reduced 20 Hz/°C/MHz. The crystal frequency expressed in megahertz is 0.6 MHz. Therefore, the above expression can be read as -20 Hz/°C/MHz $\times$ 0.6. Inasmuch as the crystal temperature increased 10°C, the frequency change will be $-20 \times 10 \times 0.6 = -120$ Hz. Consequently, the crystal oscillates at 120 Hz below what it did at 50°C, or 599.88 kHz. Had the crystal possessed a positive coefficient, the 120 Hz would have been added rather than subtracted.

The crystal must be mounted in some sort of holder that will provide electrical connection to both surfaces and at the same time prevent contaminants from interfering with the oscillatory action. Several types of holders are in common usage. One consists of a small Bakelite box, or other insulating material, that contains two metal plates and a spring. The crystal is placed between the two plates and the combination held securely in position by a spring (refer to Fig. 16-13).

A second type of holder uses metal plates with the corners raised several thousandths of an inch, which clamp the crystal at its corners. The small air gap thus provided permits more ease of vibration, particularly at higher frequencies.

Another method involves the use of crystals with the electrodes directly plated on them. By a sputtering or evaporation process a thin film of metal, such as gold, silver, or platinum, can be deposited on the two major surfaces of the crystal. Electrical connection is usually made by small wires soldered to the metal film.

If crystals are removed from their holders, they should be handled with clean, lint-free cloth and only by their edges, as any dirt or oil, such as might

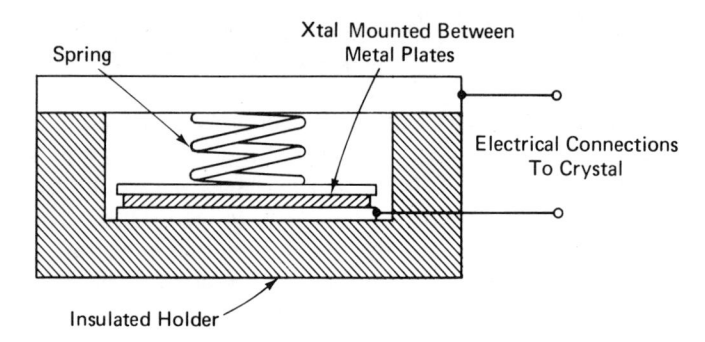

FIG. 16-13 Typical crystal holder.

be on bare hands, will likely prevent the crystal from oscillating. Crystals may be cleaned, if necessary, by using soap or detergent and warm water. Carbon tetrachloride, or similar cleaning slovents, can likewise be used, but only with adequate ventilation as their fumes may be toxic.

Crystals are inherently delicate and should be handled carefully. If dropped or subjected to excessive voltage, they may fracture. If the break should occur near an edge, it is sometimes possible to grind the edge smooth and restore operation, but the frequency will almost inevitably be higher. If grinding is necessary, it should be done on a flat piece of plate glass, or similar surface, using a very fine carborundum powder mixed with water. Lapping a crystal surface with this compound will *raise* its resonant frequency. It is possible to lower the frequency by several hundred hertz by marking an × from corner to corner on both surfaces of the crystal, using lead or solder.

There is a category of crystals known as *overtone crystals* that operates on an odd multiple of their fundamental frequency. For example, a 24-MHz third-overtone crystal is actually ground for operation on 8 MHz. A 40-MHz fifth-overtone crystal is also ground for operation at 8 MHz. Actually, overtone crystals do not operate at an exact odd multiple of the fundamental frequency. Manufacturers of these crystals usually recommend circuits that will guarantee, within a certain tolerance, the overtone frequency of their crystals. Generally speaking, any significant departure from the prescribed circuit will not produce the exact desired frequency.

16-11 CRYSTAL OVENS

To guarantee that a crystal will operate at the precise frequency, it is necessary that it be maintained at a very uniform temperature. This is accomplished by mounting the crystal in an *oven*, which is a small, well-insulated compartment containing the crystal, a heater element, and some form of transducer to sense changes of temperature in the oven. The temperature within the oven is usually well above ambient, typical values being from 110 to 150°F. Because it takes considerable time for the oven to reach operating

Crystal Ovens

temperature, *it is customary to have the oven circuit in operation at all times*, thus ensuring the crystal being at the proper frequency whenever it is turned on.

A number of different circuits are available for controlling the oven temperature. One of the simplest is illustrated in Fig. 16-14. A *bimetallic*

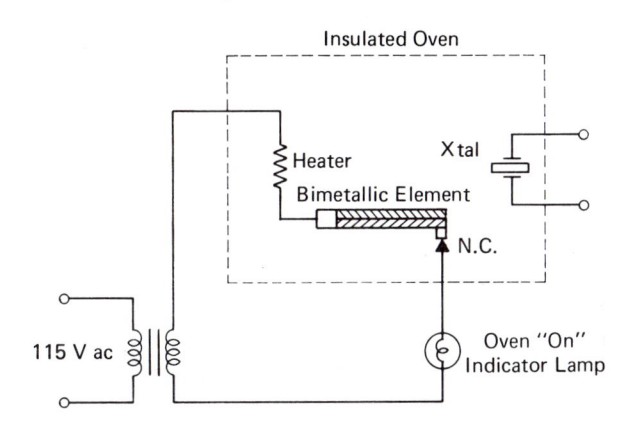

FIG. 16-14 Temperature-controlled oven using a bimetallic switch or sensing element.

element is used to sense temperature. This is normally closed below ambient temperatures. The secondary circuit of the power transformer is completed through the heater, bimetallic switch, and an indicator lamp. Until the oven comes up to temperature, the bimetallic switch is closed and the heater causes a temperature rise within the oven. The oven on indicator lamp, usually mounted on a control panel, indicates that the heater circuit is functioning properly. When the oven comes up to temperature, the bimetallic switch opens the circuit. When the temperature drops slightly, the switch closes again and current is supplied to the heater. Proper temperature within the oven is maintained by the bimetallic element cycling the heat on and off, as required.

An alternative method of controlling the oven temperature appears in Fig. 16-15. The temperature-sensing element is a *thermistor*, indicated by the resistor with a dotted circuit around it with the letter T alongside. A thermistor is a semiconductor device having a *negative temperature coefficient*. They are available in a wide range of temperature and resistance values. In selecting a thermistor it is important to choose one whose resistance versus temperature characteristics satisfy the temperature requirements of the oven. The circuit operation is as follows: When the oven is below temperature, the thermistor resistance is high. Bias resistor R_B in series with the thermistor forms a voltage-divider network. The junction of these two resistive elements is connected to the base of the transistor. Under these circumstances the voltage across the thermistor will be high and a forward bias will be provided

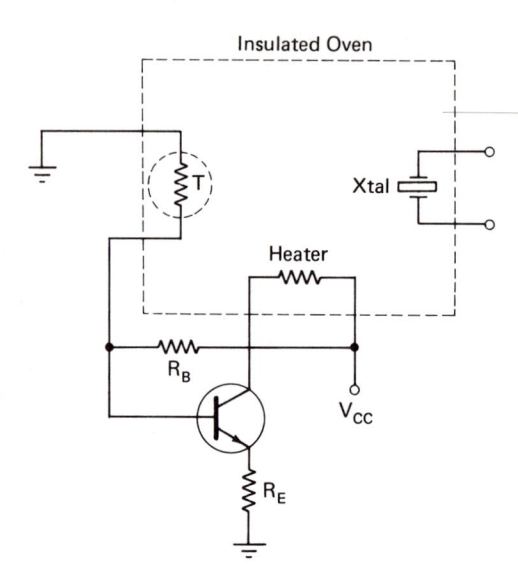

FIG. 16-15 Basic thermistor and transistor control circuit for crystal oven.

for the base–emitter junction of the transistor. This causes a large collector current to flow, which passes through the heater element. A power transistor would be required in this circuit. When the oven reaches the required temperature, the thermistor resistance will have decreased to a point where the voltage drop across it will be insufficient to forward bias the transistor, and collector current will cease or drop to a low value. As the resistance of the thermistor changes with oven temperature, the transistor is cycled on and off, thus controlling the heat applied to the oven.

A third method of controlling oven temperature appears in Fig. 16-16. The control element consists of a *mercury thermometer* whose column has been elongated in the region in which temperature control is desired. The net result of this design is that the mercury column will be caused to move a *large* amount for a small change in temperature, thus increasing the sensitivity in the specified operating range. Circuit operation is as follows: biasing resistors R_{B1} and R_{B2} are connected across V_{CC}. The ratio of voltages across these resistors is such that the transistor is biased on when the oven is below temperature. The collector current thus flowing closes the relay, and power is supplied to the heater element. As the oven temperature increases, the mercury column rises until it establishes contact with the upper electrode, which has been fused inside the column. The temperature at which the mercury column establishes contact with the upper connection corresponds to the desired oven temperature. At this point bias resistor R_{B2} is short circuited by the mercury column and the bias on the transistor is reduced to the point where collector current is insufficient to close the relay. This opens the heater

Other Crystal-Controlled Oscillator Circuits 313

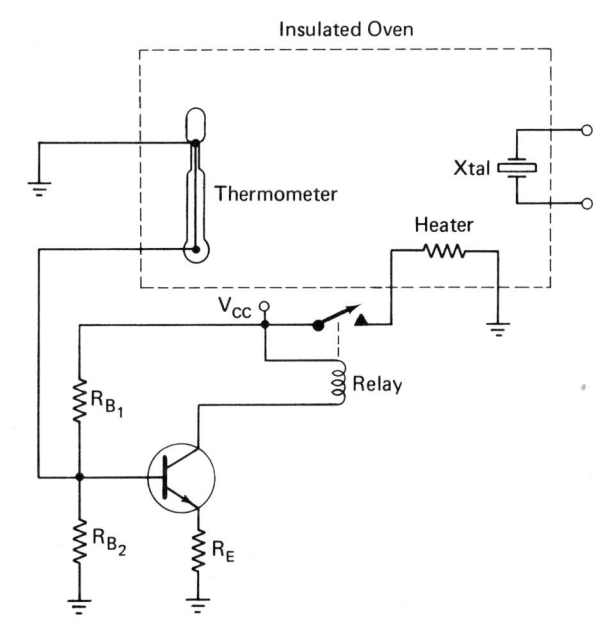

FIG. 16-16 Typical oven control circuit using mercury thermometer as control element.

circuit. When the temperature drops, the mercury column likewise drops, removing the short across R_{B2}. Proper bias is reestablished, and the collector current recloses the relay, supplying power to the heater.

16-12 OTHER CRYSTAL-CONTROLLED OSCILLATOR CIRCUITS

A popular circuit, known as a Pierce oscillator, is shown in Fig. 16-17. It has no tuned circuit. The crystal is excited by a portion of the energy in the plate circuit being fed back through the crystal to the grid. Because of the extremely high Q of the crystal, typical values running from 10,000 to over 100,000, the feedback signal to the grid is that of the crystal frequency. One limitation of the circuit is that the crystal is subject to a large percentage of

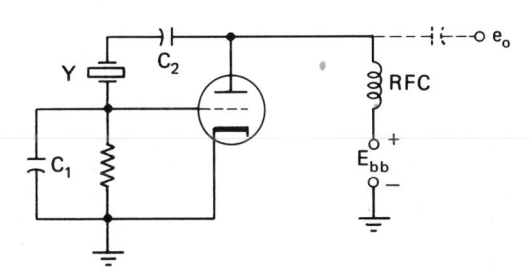

FIG. 16-17 Pierce oscillator circuit.

the RF output of the tube. Fracturing of the crystal is possible. To preclude this possibility, it is customary to operate the plate voltage at a relatively low value. The function of C_2 is to block the direct current present at the plate from the crystal.

An adaptation of the Pierce oscillator using a pentode is shown in Fig. 16-18. The cathode, control, and screen grid function as a triode with the RF

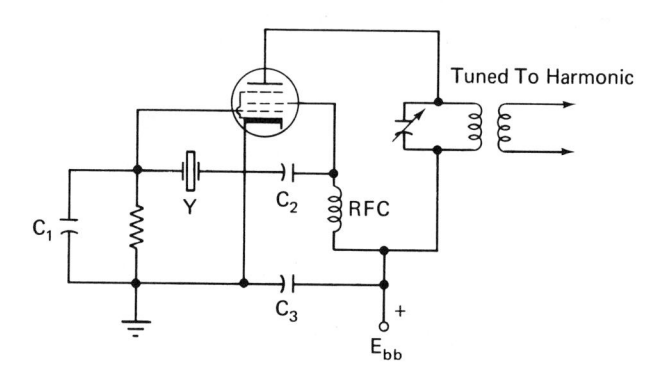

Fig. 16-18 Typical Pierce oscillator designed for operation on a multiple of the crystal.

voltage developed across the *RFC* being coupled, via C_2, to the crystal. The circuit provides a degree of isolation between the plate circuit and the crystal. An advantage of this circuit is that the plate tank can be tuned to a multiple of the crystal fundamental, providing an output that is two, three, or even four times that of the crystal.

In any oscillator circuit it is imperative that the supply voltage be *well regulated*. Any tendency for the supply voltage to shift with changing loads will cause the oscillator frequency to vary. It is not uncommon to have a separate power source available for the oscillator in transmitter circuits.

16-13 AUDIO OSCILLATORS

Amateurs and other experimenters are often interested in constructing a simple audio oscillator for code practice or other use. A typical circuit

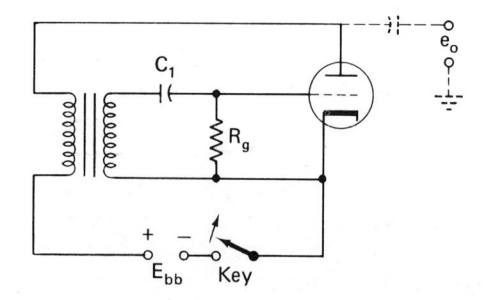

Fig. 16-19 Simple audio oscillator.

Phase-Shift Oscillator 315

appears in Fig. 16-19. Essentially, it is an Armstrong circuit, as indicated by the fact that plate current must flow through the primary winding of a transformer that serves as a tickler coil. The induced secondary voltage provides the grid excitation. The frequency is determined by the inductance of the transformer windings together with the value of C_1 and R_g. Output can be derived through a capacitor connected to the plate, as shown by the dotted capacitor. For code practice a telegraph key can be inserted in the lead between $-E_{bb}$ and the cathode of the tube.

16-14 PHASE-SHIFT OSCILLATOR

The transistor *phase-shift oscillator* is normally used in the CE configuration, as shown in Fig. 16-20, but with the proper phase-shifting networks

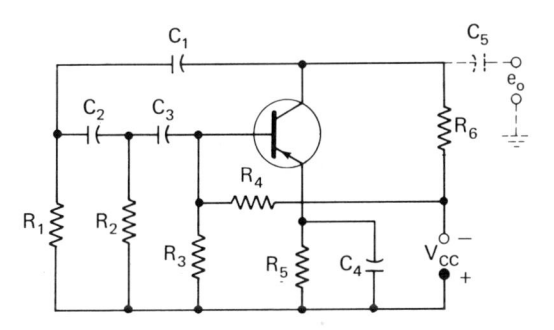

FIG. 16-20 Transistor phase-shift oscillator.

it can be used in other configurations. In the CE circuit the necessary 180° phase shift is obtained by using a three-section *RC* network. These sections are C_1, R_1; C_2, R_2; and C_3, R_3. Each section provides a 60° phase shift. At least three *RC* sections are required to provide the 180° phase shift.

By increasing the number of phase-shift sections comprising the network, the losses of the total network can be decreased. Since the loss per section is decreased as the amount of phase shift (per section) is reduced, many oscillators employ networks consisting of four, five, and six sections. Assuming that the values of *R* and *C* are equal for each section, the individual sections are designed to produce phase shifts per section of 45, 36, and 30°, respectively.

The *RC* phase-shift oscillator is normally fixed in frequency, but can be made variable by providing ganged variable capacitors or resistors in the phase-shift network. An increase in the value of either *R* or *C* will produce a decrease in the output frequency.

In Fig. 16-20 resistors R_3 and R_4 establish forward bias for the transistor. Resistor R_5 is the emitter swamping resistor, and provides stabilization for the circuit. Resistor R_6 is the collector load resistance across which the output signal is developed. Capacitor C_5 is the output coupling capacitor.

Oscillations are started by any random noise in the power source or the transistor when power is first applied. A change in the base current results in an amplified change in collector current, which is shifted in phase 180°. The output signal developed across the collector load resistance, R_6, is returned to the transistor base as an input signal inverted 180° by the action of the feedback and phase-shift network, making the circuit regenerative. The output waveform is sinusoidal.

If the components comprising the phase-shift network should change value, the frequency of oscillation will change to the frequency at which a phase shift of 180° will occur to sustain oscillations.

16-15 HIGH-FREQUENCY OSCILLATORS

As the frequency of an oscillator is raised, the RF circuit losses increase because of

1. Increasing skin effect.
2. Greater capacitance charging currents.
3. Dielectric loss in the glass parts of the tube.
4. Eddy-current losses in the adjacent conductors.
5. Energy loss by direct radiation from the circuit.

The net effect of all these causes the loading on the tuned circuit to be increased so that the Q of the tank circuit is decreased, resulting in poor efficiency.

Skin effect causes the current to flow near the surface of the conductor. Consequently, the higher the frequency, the thinner will be the layer in which current flows. The I^2R loss therefore increases with frequency.

To reduce this undesirable effect, conductors of large diameter are used so that the distribution of current throughout the cross-sectional area will be more uniform, even though the depth of penetration is small. Hence, using large leads for UHF circuits not only reduces the lead inductance but also the lead resistance. The conductors are often plated with silver to further reduce the skin effect. If corrosion is a problem, gold plating is often used.

Because the reactance of the interelement and distributed capacitance become small for UHF, their charging currents increase. These currents contribute nothing to the power output. Because of skin effect, these currents travel along the surface of the conductors and may cause excessive *localized heating* at the junctions of the electrodes and the envelope, which may result in cracked seals and failure. Hence, UHF oscillators are designed to have the maximum possible inductance. As a result, the amount of capacitance necessary to resonate with the inductance may be just that of the tube or transistor alone. Special tubes or transistors are employed that have very low capacitances. In addition, special oscillatory circuits using push–pull connections,

High-Frequency Oscillators

such as shown in Fig. 16-21, are used to further reduce the effective capacitance. A study of the circuit reveals that the interelectrode capacitances are in series; consequently, the effective capacitance shunting the resonant circuit is reduced to one half that for a single tube.

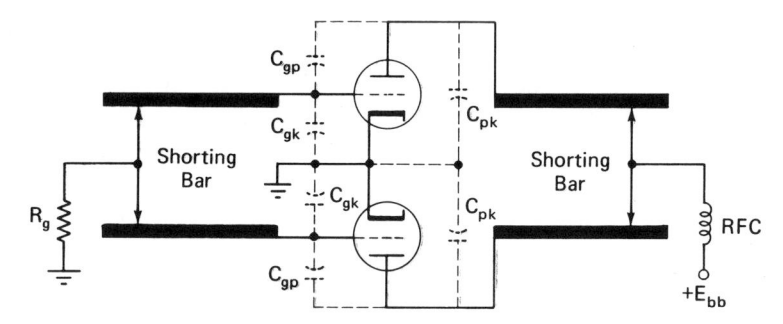

Fig. 16-21 Parallel-rod transmission-line oscillator.

Losses resulting from eddy currents and radiation in adjacent conductors may be substantially reduced if the parallel conductors are closely spaced, so that the field around one conductor neutralizes the field around the other. However, this tends to increase the RF resistance of the conductors if they are too close together. Also a serious limitation is placed on the maximum RF voltage that can exist between the conductors without breakdown. These losses can almost entirely be eliminated by using *coaxial lines* instead of open wire lines.

Dielectric losses are objectionable not only because of the reduction of output and efficiency, but because they may cause failure of the seals where the leads of the tubes or transistors are brought out through the headers. In vacuum-tube circuits dielectric losses are minimized by *eliminating the tube base* and providing connections directly to the load, and by bringing the conductors through the glass at points on the conductors where the RF voltage is minimum.

It is important that the circuits have the highest possible Q when not loaded. Consequently, tuned circuits associated with these oscillators are usually *resonant sections of transmission lines*, rather than coil and capacitor combinations. The Q of a quarter-wave short-circuited section of transmission line can be made much higher than that of a conventional tank circuit, because it is more feasible to make a tuned line using conductors of larger diameter than is possible with the conventional inductor. Additionally, tuned transmission lines are used as circuit elements in UHF oscillators, because the tube or transistor leads may act as extensions of the transmission line. Hence, the *interelement capacitances and lead inductances* are incorporated as part of the tuned circuit.

In the oscillator shown in Fig. 16-21, the resonant tanks in both grid and plate circuits consist of the large parallel rods. Inductance exists between the parallel rods. The capacitance is provided by that existing between the parallel rods and the tube interelectrode capacitances. Both grid and plate circuits are adjusted to resonance by moving the *shorting bars* back and forth until they are at a quarter-wavelength from the tube end of the rod. This provides the resonant condition necessary for oscillation. Because the losses in this type of circuit are relatively low, the Q is high. The parallel rods and shorting bars are sometimes called *Lecher lines*. The grid bar may be varied for maximum stability, whereas the plate bar is adjusted for maximum output.

Coupling to a resonant line may be accomplished by a *hairpin loop* about one quarter of the length of the line.

For even higher frequencies *coaxial tanks* may be used in both input and output circuits, such as shown in Fig. 16-22. The coaxial tanks consist of

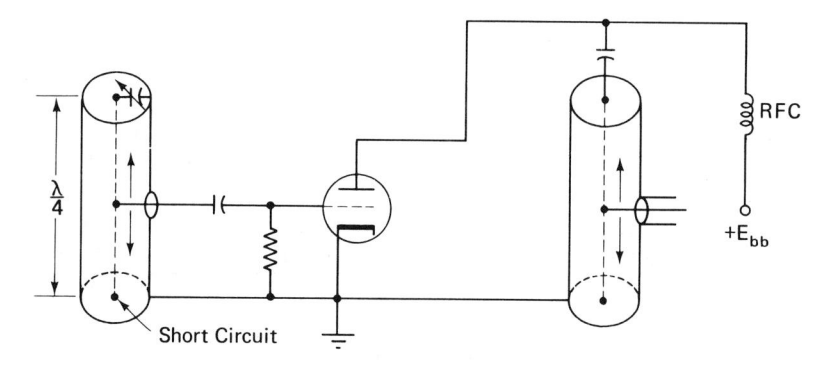

FIG. 16-22 TPTG oscillator using quarter-wave coaxial tanks.

copper tubes with a conductor running up the center and designed to be one quarter-wavelength long at the required frequency. One end of the tank is closed off and provides a short circuit. The copper tube behaves as an inductor having a single turn. The capacity present between the inner surface of the tube and the central conductor running up the center provides that necessary for resonance. A small trimmer capacitor connected across the open end of the tank may be used for fine adjustment of the resonant frequency. Concentric tanks are well adapted for use at these higher frequencies because of the complete shielding and high Q they offer. Although the line may be resonant, if its effective electrical length is near any multiple of a quarter-wave, the Q of the circuit is decreased by making the line longer than a quarter-wave. There are many other types of ultrahigh-frequency oscillators in use, such as the *cavity resonators* used in microwave and radar circuits.

16-16 DYNATRON OSCILLATORS

The oscillator circuits previously described have relied upon capacitive or inductive feedback to sustain oscillation. *Dynatron* oscillators operate

Dynatron Oscillators

upon the principle of *negative resistance*. A circuit is said to exhibit negative resistance if an increased voltage results in a decrease of current flow, which is the reverse of what was learned in Ohm's law. Certain tubes and diodes can be made to operate in such a mode as to exhibit negative resistance. This effect is present in screen-grid tubes when an increase in plate voltage is accompanied by a decrease in plate current, provided the plate voltage does not exceed that of the screen. This negative-resistance effect depends upon *secondary emission* from the plate. Under these conditions the characteristic shown in Fig. 16-23b results. It is apparent that there is an appreciable range in which a positive increase in plate voltage causes a negative increment in plate current. If the plate potential is raised beyond the value indicated at point C in Fig. 16-23b, the plate voltage begins to regain control of the secondary electrons, and secondary emission ceases.

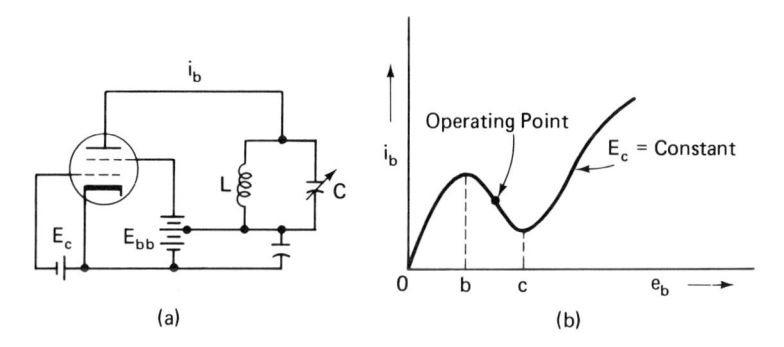

FIG. 16-23 Basic dynatron oscillator circuit.

If an oscillatory circuit is connected across this negative resistance, oscillations will develop, provided the absolute value of negative resistance is less than or equal to the equivalent resistance of the tuned circuit. The amplitude of oscillation may be varied by means of the control-grid voltage, which determines the slope of the current–voltage characteristic in the negative-resistance range. For optimum operation an operating point should be chosen in the center of the linear region of the negative-resistance characteristic, and the amplitude of oscillation should be kept small. In addition to an excellent waveform, the dynatron oscillator possesses good frequency stability and a reasonably simple circuit. Its chief disadvantage lies in its dependence upon secondary emission, which is a property of a tube that is extremely variable with age, even within the same tube types. The essential parts of a dynatron oscillator are shown in Fig. 16-23a.

Another device exhibiting negative resistance is the *tunnel diode*. These are small *pn*-junction devices having a very high concentration of impurities. In both the *n*- and *p*-type semiconductor materials this high impurity content makes the junction depletion region so narrow that electrical charges can transfer across the junction by a quantum or chemical reaction referred to as *tunneling*. The tunneling effect produces a negative-resistance region on its characteristic curve. For proper operation the diode must be biased in the center of the negative-resistance region. Tunnel diodes can be used in oscillator circuits operating at frequencies as high as 5 GHz. These oscillators are inexpensive, require only a fraction of a volt for dc bias, and are rugged and reliable in severe environments. A tunnel-diode oscillator is shown in Fig. 16-24. The circuit values shown are designed to provide oscillations at approximately 100 kHz.

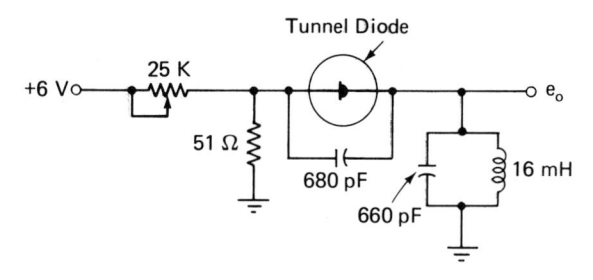

FIG. 16-24 Basic tunnel diode oscillator.

16-17 RESISTANCE–CAPACITANCE OSCILLATORS

The oscillatory circuits thus far studied have produced sinusoidal outputs. Voltages having *sawtooth waveforms* are widely used in television, radar, and other circuits. Generally speaking, in these applications the sawtooth wave is used to deflect an electron beam across the face of a cathode-ray tube. A relatively simple but effective circuit for developing a sawtooth wave is the neon-tube sawtooth generator shown in Fig. 16-25a. The circuit consists of a series RC combination connected across a supply source with a neon tube connected across the capacitor. The circuit operation is as follows: capacitor C is charged through resistor R until its potential reaches a sufficiently high value to ionize the gas in the tube. Until this time the tube has a high impedance. At the ionization potential its impedance drops to a very low value, discharging C rapidly. When the emf across C drops below the deionizing potential, the tube stops conducting, and its initial high impedance is reestablished. The capacitor stops discharging because now the voltage across C is less than that required to ionize the tube. The capacitor again begins to charge.

For a given supply voltage, the frequency of the sawtooth voltage depends upon the RC time constant and is controlled by adjusting R. A study of Fig. 16-25b reveals that the output voltage varies between the *ionizing* and *deionizing* potential of the tube. The full value of E_{bb} is not applied across C, because the firing potential is a lower value, and the difference appears across R. Likewise, C does not completely discharge, because when the deionizing potential is reached C stops discharging. The capacitor voltage follows a

Unijunction Transistor Oscillator 321

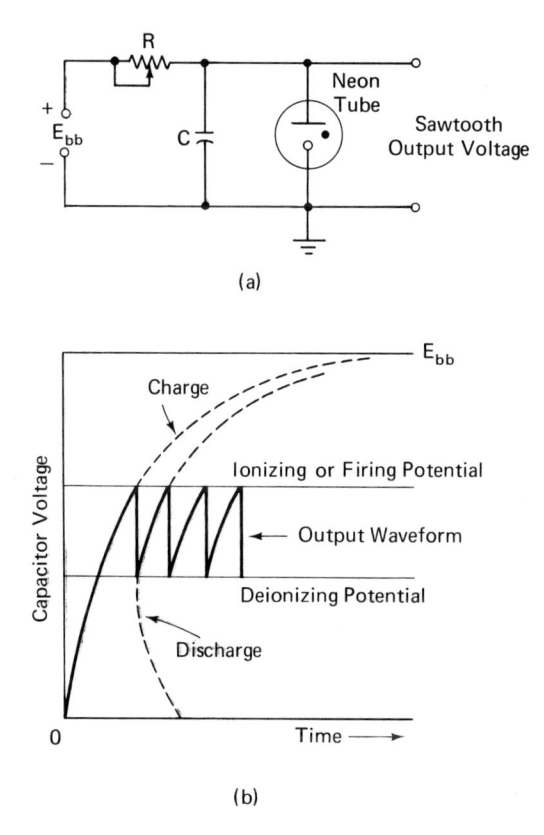

FIG. 16-25 (a) Neon tube sawtooth generator; and (b) output waveform.

normal RC charging curve between these two limits. The discharge follows a similar curve, except that the discharge time is only a small fraction of the charge time, because the resistance of the discharge path is so small. If E_{bb} is relatively high and the ionizing and deionizing potentials of the neon tube are relatively low, the charge across the capacitor will be quite linear.

16-18 UNIJUNCTION TRANSISTOR OSCILLATOR

A very popular timing circuit is the one shown in Fig. 16-26. It is a form of relaxation oscillator inasmuch as it employs an RC time constant whose output voltage drives the *unijunction transistor*. The unijunction transistor is characterized by two base connections, designated B_1 and B_2, connected to a silicon bar, usually of *n*-type material. A single rectifying contact, called the emitter, is made between base 1 and base 2. With no emitter current flowing, the silicon bar acts like a simple voltage divider with approximately 5000 to 10,000 Ω of resistance between B_1 and B_2. If the voltage on the emitter

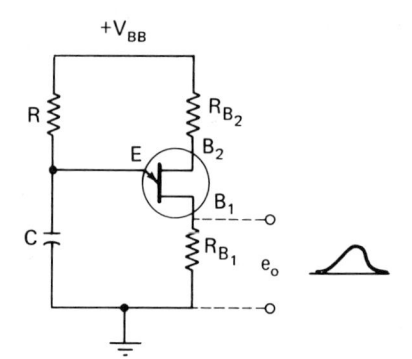

FIG. 16-26 Basic unijunction transistor oscillator circuit.

(with respect to ground) is less than the voltage gradient between B_1 and B_2 at the point where the emitter makes contact with the silicon bar, the emitter–B_1 junction will be reverse biased.

When power is initially applied to the circuit, capacitor C begins to charge through resistor R. As the voltage across C rises exponentially, it reaches some critical voltage at which the emitter begins to inject holes into the silicon bar. The emitter–B_1 junction of the device now conducts heavily, and its internal resistance drops to a very low value. The charge across C is effectively delivered to resistor R_{B_1}. A pulse of voltage appears across this resistor and is shown to the right of the drawing. Depending upon circuit parameters, the duration of this pulse is on the order of 30 to 50 microseconds (μs). The *pulse repetition rate* (prr) is determined by the RC time constant of the network, which is supplying the potential to the emitter. Once the capacitor has discharged to some low value, the cycle repeats. The output waveform consists of a series of pulses.

16-19 MULTIVIBRATORS

Multivibrators are examples of relaxation oscillators that use an RC time constant for determination of output waveform and frequency.

Multivibrators are classified as either *free running* or *driven* (triggered). The free-running oscillator is one in which the oscillations begin once power is applied. The triggered oscillator is controlled by a *synchronizing* or *triggering* external signal.

The free-running (astable) multivibrator is essentially a nonsinusoidal two-stage oscillator in which one stage conducts while the other is cut off, until a point is reached at which the stages reverse their conditions. The output is a *square wave*.

The basic collector-coupled transistor multivibrator of Fig. 16-27 is a two-stage, RC coupled, CE amplifier with the output of the first stage coupled to the input of the second stage, and the output of the second stage coupled to the input of the first stage. Since the signal in the collector circuit of a CE amplifier is reversed in phase with respect to the input of that stage, a portion

Multivibrators

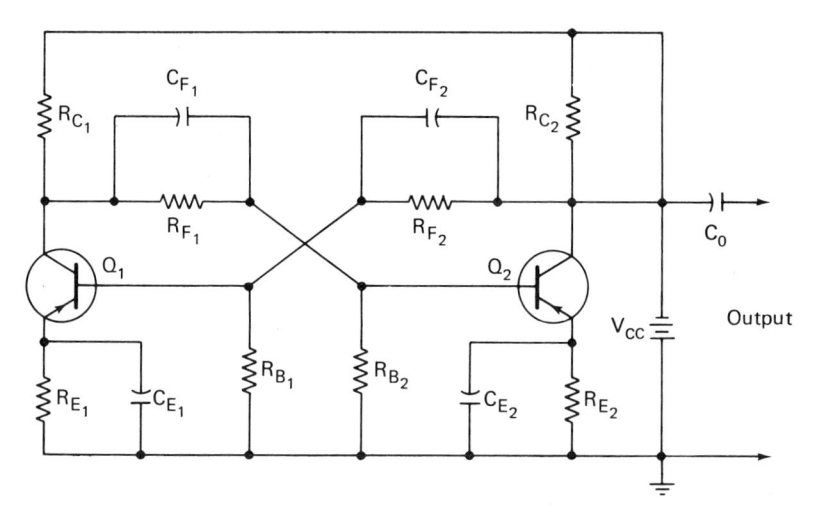

FIG. 16-27 Transistor multivibrator.

of the output of each stage is fed to the other stage in phase with the signal on the base electrode. This regenerative feedback with amplification is required for oscillation.

Because of circuit variations, one transistor will conduct before or more heavily than the other. Assume this to be Q_1. Its collector current rises rapidly, causing its collector voltage to decrease (become more positive). This emf is fed to the base of Q_2, driving it toward cutoff and causing its collector voltage to rise toward V_{cc} (negative). This potential is coupled to the base of Q_1 via R_{F2} and causes Q_1 to go into saturation. This action happens so quickly that capacitor C_{F1} does not get a chance to discharge, and the increased positive voltage at the collector of Q_1 appears across R_{B2}.

Capacitor C_{F1} now begins to discharge, and more of the previously increased positive voltage at the collector of Q_1 appears across C_{F1} and less across R_{B2}. This decreases the reverse bias on the base of Q_2 until a time is reached when forward bias is reestablished across Q_2. Now Q_2 begins to conduct. Its collector becomes less negative or more positive. This voltage coupled through capacitor C_{F2} to the base of transistor Q_1 drives it more positive and causes a decrease in current flow through transistor Q_1. The resulting increased negative voltage at the collector of transistor Q_1 is coupled through capacitor C_{F1} and appears across resistor R_{B2}. The collector current of transistor Q_2 therefore increases. This process continues rapidly until transistor Q_1 is cut off. Transistor Q_1 remains cut off (and transistor Q_2 conducts) until capacitor C_{F2} discharges through resistor R_{F2} enough to decrease the reverse bias on the base of transistor Q_1. The cycle then repeats itself.

The output of a multivibrator is a square wave whose frequency is determined by the values of R and C in the circuit. The output may be obtained from either collector.

16-20 PARASITIC OSCILLATIONS

It is not uncommon for a circuit to operate at *spurious* frequencies, usually higher than that at which it was designed to function. They are called *parasitic oscillations*, not alone because they are unwanted, but because they derive power from the normal circuit. The net result is that the output power is reduced and the output waveform contains *high-frequency oscillations* that are very undesirable. They may occur in almost any type of circuit, such as amplifiers, power supplies, oscillators, transmitters and receivers. Parasitic oscillations usually result from *poor construction practices*, such as excessively long leads, improper component layout, and so forth. Lead inductances and stray capacitances (including distributive capacitances) combine to form the resonant condition.

The presence of parasitic oscillations is generally manifest in *erratic operation*. An example might be a transmitter that is operating, but erratically. The plates of the tubes show *excessive red color*, plate current is abnormal, and perhaps some capacitors may arc over for no apparent reason.

The Q of parasitic circuits is usually very high. An effective way to suppress them is to install a *parasitic choke* in either the plate or the grid lead of the circuit involved. These should *be installed as close as possible* to the tube (to the base and collector leads if transistors are used). A parasitic choke may be constructed by winding approximately six turns of wire around a 25- to 50-Ω, 1-W carbon resistor. In some instances the simple addition of a resistor, usually 50 to 200 Ω, in the plate, screen, or grid leads will add sufficient resistance to reduce the Q of the circuit to the point where oscillations will not be sustained. The use of parasitic chokes in a typical oscillator circuit is shown in Fig. 16-28.

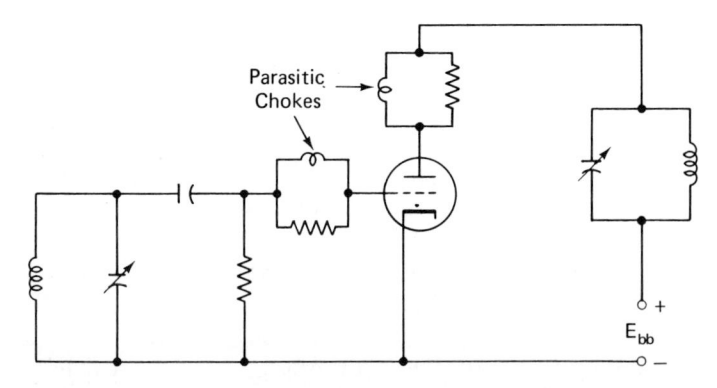

FIG. 16-28 Use of parasitic chokes to suppress parasitic oscillations.

A person might be unaware of the presence of parasitic oscillations in a circuit except for disturbances in nearby receivers or TV sets. It is not uncommon to experience low-frequency parasitic oscillations in amplifiers. Their presence is detected by the audible nature of the oscillation, which is usually a "putt-putt-putt" that is referred to as *motor boating*. This type of undesirable oscillation can be eliminated by using *decoupling circuits* between stages of the amplifier.

16-21 METHODS USED TO DETECT OSCILLATION

There are a number of reliable methods by which the presence of oscillations may be detected. These are

1. RECEIVER   A radio receiver in the immediate vicinity of the circuit suspected will produce a whistle as the oscillator is tuned across the receiver's frequency. This is the result of the oscillator's frequency beating with the local oscillator of the receiver.

2. OSCILLOSCOPE   This is one of the best instruments for testing for the presence of oscillations. Several turns of insulated wire may be fashioned in the form of a loop and connected to the vertical amplifier of the scope. The loop is placed near the oscillating tank circuit. The presence of RF energy will be shown on the screen. If the scope does not have good high-frequency characteristics, it may be necessary to feed the RF directly to the vertical deflection plates of the scope.

3. NEON LAMP   A neon lamp provides an excellent indication of the presence of oscillation, provided the circuit is generating sufficient voltage to ionize the lamp. All that is necessary is to touch the lamp against the grid lead or plate lead of a low-power oscillating circuit. This method may be dangerous in high-power circuits.

4. RADIO-FREQUENCY INDICATOR   Another simple but effective method is to connect a flashlight lamp in series with a loop of insulated wire. The lamp will glow when the loop is coupled closely to the oscillator tank coil, provided the oscillator produces 1 W or more of power. An alternative method is to use a 0- to 1-mA meter (with a silicon or germanium diode connected across it) connected to the pickup loop. The presence of RF will give an indication on the meter. If a sensitive RF thermal galvanometer is available, it may be used instead of the lamp.

5. BIAS VOLTAGE   If an oscillator is operating correctly, there will be proper bias voltages and currents flowing. If the oscillator is stopped from oscillating (i.e., by shorting the capacitor in the tank circuit), a change of bias voltage and current will be observed.

6. PLATE CURRENT   When an oscillator is operating properly, its output current will dip to a relatively low value due to the presence of bias voltage being developed. Nonoscillation is generally indicated by high values in collector or plate current.

7. LEAD PENCIL A simple but dangerous method it to use a soft lead pencil. When the lead is touched to the oscillating tank circuit, it will produce a spark if it is oscillating and generating more than a few watts of energy. This method is not recommended with circuits generating more than 50 W of RF power.

16-22 OSCILLATOR STABILIZATION

The most important function of an oscillator is to generate a *highly stable frequency*. There are a number of factors that must be considered to achieve this objective. The principal ones are

1. Low output power. Reduce oscillator output coupling so that collector or plate currents will be low and the tank Q maintained at a high value.
2. Well-regulated supply voltage. A separate power supply should be used for the oscillator stage or a very well-regulated supply if used with several stages.
3. Buffer stage. A buffer stage should be used between the oscillator and the output stage of the transmitter to prevent changes in the output circuit constants from reflecting back to the oscillator and causing frequency shifts.
4. Temperature control. Oscillator stages, particularly crystal-controlled stages, should be located within a temperature-controlled oven.
5. Mechanical construction. All components in the oscillator stage must be rigidly mounted, as any vibration will tend to affect the frequency of the oscillator.
6. Shielding. Proper shielding of all components is important; otherwise, hand capacitance, humidity, and other factors will adversely affect the performance of the resonant circuits.
7. Reduced collector or plate current. By maintaining these currents at relatively low values, little heat will be generated within the oscillator. Consequently, component values will not change appreciably.
8. Use of high C/L ratios. By keeping the capacitance in the tank circuit high, external adjustments will have less effect on oscillator frequency.
9. Grid-leak condensers and resistors. The correct choice of grid-leak capacitor and resistor is essential to good stability.
10. Impedance matching. By tapping the tank coils down particularly for transistorized circuits, better impedance matches can be obtained, resulting in higher circuit Q's.

Commercial License Questions

Sections in which answers to questions are given appear in parentheses. A bracketed number following a question implies that it applies only to that element.

1. Describe how a vacuum tube oscillates in a circuit. (16-1, 16-2)
2. Draw a simple schematic diagram showing a tuned-grid Armstrong-type triode oscillator with series-fed plate. Indicate power-supply polarity.
(16-3)
3. Explain how grid bias voltage is developed by the grid leak in an oscillator.
(16-4)
4. Draw a simple schematic diagram showing a tuned-plate tuned-grid triode oscillator with shunt-fed plate. Indicate polarity of supply voltage. (16-5)
5. Draw a simple schematic diagram showing a tuned-plate tuned-grid oscillator with series-fed plate. Indicate polarity of supply voltage. (16-5)
6. By what means is feedback coupling obtained in a tuned-grid tuned-plate type of oscillator? (16-5)
7. Draw a simple schematic diagram showing a Hartley triode oscillator with shunt-fed plate. Indicate power supply polarity. (16-6)
8. What are the differences between Colpitts and Hartley oscillators?
(16-6, 16-7)
9. Draw a simple schematic diagram showing a Colpitts-type triode oscillator with shunt-fed plate. Indicate power-supply polarity. (16-7)
10. Why is a high ratio of capacity to inductance employed in the grid circuit of some oscillators? (16-7)
11. Do oscillators operating on adjacent frequencies have a tendency to synchronize oscillation or drift apart in frequency? (16-7)
12. List the characteristics of an electron-coupled type of oscillator. (16-8)
13. Draw a simple schematic diagram of an electron-coupled oscillator, indicating the circuit element necessary to identify this form of oscillatory circuit.
(16-8)
14. Draw a simple schematic diagram of an electron-coupled oscillator, indicating power-supply polarities where necessary. (16-8)
15. Draw a diagram and describe the electrical characteristics of an electron-coupled oscillator circuit. [4] (16-8)
16. Draw a simple schematic diagram of a quartz-crystal-controlled oscillator, indicating the circuit elements necessary to identify this form of oscillatory circuit. (16-9)
17. Draw a simple schematic diagram of a crystal-controlled vacuum-tube oscillator. Indicate power-supply polarity. (16-9)

18. Why is a separate source of plate power desirable for a crystal oscillator stage in a radio transmitter? (16–9)

19. What are the principal advantages of crystal control over tuned-circuit oscillators? (16–9)

20. What is the function of a quartz crystal in a radio transmitter? (16–9)

21. For maximum stability, should the tuned circuit of a crystal oscillator be tuned to exact crystal frequency? [4] (16–9)

22. What precautions should be taken to ensure that a crystal oscillator will function at one frequency only? [4] (16–9, 16–10)

23. What may result if a high degree of coupling exists between the plate and grid circuits of a crystal-controlled oscillator? (16–9, 16–10)

24. What crystalline substance is widely used in crystal oscillators? (16–10)

25. Why is the crystal in some oscillators operated at constant temperature? (16–10, 16–11)

26. What is meant by negative temperature coefficient of a quartz crystal when used in an oscillator? (16–10)

27. A 600-kHz X-cut crystal, calibrated at 50°C and having a temperature coefficient of —20 ppm/°C will oscillate at what frequency when its temperature is 60°C? [4] (16–10)

28. What is the approximate range of temperature coefficients to be encountered with X-cut quartz crystals? (16–10)

29. Is it necessary or desirable that the surfaces of a quartz crystal be clean? If so, what cleaning agents may be used that will not adversely affect the operation of the crystal? (16–10)

30. What will result if a dc potential is applied between the two parallel surfaces of a quartz crystal? (16–10)

31. What does the expression *low temperature coefficient* mean as applied to a quartz crystal? (16–10)

32. What does the expression *positive temperature coefficient* mean as applied to a quartz crystal? (16–10)

33. What is the purpose of maintaining the temperature of a quartz crystal as constant as possible? (16–11)

34. Why are quartz crystals in some cases operated in temperature-controlled ovens? [4] (16–11)

35. What are the advantages of mercury thermostats as compared to bimetallic thermostats? (16–11)

36. Draw a simple schematic diagram of a crystal-controlled vacuum-tube oscillator using a pentode-type tube. Indicate power-supply polarity where necessary. (16–12)

37. Draw a diagram of a one-tube audio oscillator using an iron-core choke. (16–13)

Commercial License Questions 329

38. List the characteristics of a dynatron-type oscillator. (16–16)
39. Upon what characteristic of an electron tube does a dynatron-type oscillator depend? (16–16)
40. Draw a simple schematic diagram of a dynatron-type oscillator, indicating the circuit elements necessary to identify this form of oscillatory circuit.
 (16–16)
41. What is a multivibrator and what are its uses? (16–19)
42. Draw a simple schematic diagram of a multivibrator oscillatory circuit.
 (16–19)
43. What determines the fundamental operating frequency range of a multivibrator oscillator? (16–19)
44. Define *parasitic oscillations* (16–20)
45. What may be the result of parasitic oscillations? (16–20)

17

Basic Transmitters

17-1 FUNDAMENTAL CONCEPTS

A transmitter is a device for converting intelligence, such as voice or code, into electrical impulses for transmission either on closed lines or through space from a radiating antenna. Transmitters take many forms, have varying levels of power, and employ numerous methods of sending the desired information or energy component from one point to another.

The function of a radio transmitter is to supply power to an antenna at a definite radio frequency and to convey intelligence by means of the radiated signal. Radio transmitters radiate waves of two general types:

1. *Continuous wave* (CW), which has a waveform like that of the RF current in the tuned tank circuit of a power output stage. The peaks of all the waves are equal and evenly spaced along the time axis. The waveform is sinusoidal.
2. *Modulated wave.* The amplitude may be modulated by means of a signal of constant frequency or by means of speech or music and is called *amplitude modulation* (AM). If the frequency of the wave is varied with time, it is called *frequency modulation* (FM).

Continuous wave is used principally for radio telegraphy. The advantages of this kind of transmission are narrow bandwidth and a high degree of intelligibility, even under severe noise conditions.

Transmitters are designed for either *fixed-frequency* (i.e., broadcast, police, etc.) or *variable-frequency* operation (i.e., aeronautical, amateur, etc.). They may also be classified according to power output or the general band of frequencies they operate in (see Table 4-1).

Generally speaking, lower frequencies are used for long-distance point-to-point communication. They have a tendency to follow the curvature of the earth. Very high frequencies are normally good for *line-of-sight* communication. They are not so adversely affected by various forms of static. Construction techniques, the type of amplifying devices used, and the circuit components used may vary tremendously for low-frequency transmitters as compared to UHF gear. Nonetheless, the basic principles of operation are essentially the same.

All transmitters are designed to operate on a specific frequency. The informtion to be transmitted is sent out on, or carried by, this frequency. Hence, it is referred to as the *carrier frequency*. It is this particular frequency that a receiver must be tuned to in order to recieve the information that is being transmitted. Every frequency has a specific *wavelength*. The wavelength *varies inversely* with frequency and can be calculated by the formula

$$\lambda = \frac{3 \times 10^8}{f_{Hz}}$$

where λ = wavelength in meters
3×10^8 = velocity of radio waves through free space in meters/sec

The wavelength of a frequency is the distance between adjacent peaks.

One major problem to be considered in designing any transmitter is the *stability* of the oscillator that controls the carrier frequency. If it is allowed to drift beyond FCC specifications, all receivers tuned to its frequency will occasionally have to be retuned in order to receive an undistorted signal. Oscillators that are self-excited are more susceptible to drift than are those that are crystal controlled. This is true even though the oscillator may be mounted inside a temperature-controlled oven.

17-2 SINGLE-STAGE TRANSMITTER

The simplest type of transmitter is a single-stage unit transmitting a CW signal. There are four essential components of such a transmitter:

1. A generator of RF oscillations.
2. A means of amplifying these oscillations (in a single-stage transmitter this would also be the oscillator tube).
3. A method of turning the RF output on and off (keying) in accordance with the information to be transmitted.
4. An antenna to radiate the keyed output of the transmitter.

A typical schematic diagram of a single-stage transmitter employing these essential components appears in Fig. 17-1. This is a simple tuned collector oscillator with a crystal in the feedback circuit. The crystal frequency is amplified by the transistor, and the output is coupled to an antenna by means

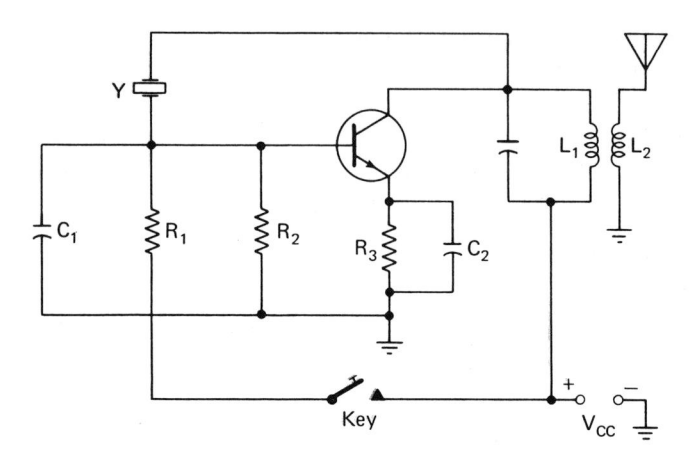

FIG. 17-1 Simple single-stage transmitter.

of a small antenna-coupling coil. The RF energy fed to the antenna is radiated into space.

Before this simple oscillator-transmitter circuit will actually function, it is necessary that the key be closed. This supplies forward bias to the base-emitter junction via voltage-divider resistors R_1 and R_2. The transistor is biased for class C operation by the combined bias network C_1 and R_1 and the emitter-bias network C_2 and R_3.

The oscillator signal developed across L_1 is inductively coupled by L_2 to the load, which is the antenna. The impedance transformation provided by L_1 and L_2 adequately matches the relatively high collector impedance to the lower antenna impedance.

The oscillator in a transmitter changes frequency when it is being *warmed up* and when its *load varies*. Most of the frequency drift is due to changes in the physical size of the components with variations in temperature. Placing the frequency-determining components of the oscillator in a temperature-controlled oven minimizes this drift. The stability of an oscillator can be seriously affected if its coupling to the antenna, or other load, is too tight. Any variation in the antenna characteristics, such as height and swaying wires, will reflect back into the oscillator and cause instability. Vibration also affects oscillator stability. In some transmitters the oscillator is suspended on springs and snubbed by rubber cushions to isolate vibrations.

17-3 MASTER OSCILLATOR POWER AMPLIFIER

The single-stage transmitter has a number of shortcomings. These are *very low* power output, *poor frequency stability*, and little rejection of oscillator harmonic output. These and other difficulties can be largely overcome, if not eliminated, by using at least one stage following the oscillator. Such a circuit is shown in Fig. 17-2 and is known as a *master oscillator power amplifier*. (MOPA).

Master Oscillator Power Amplifier 333

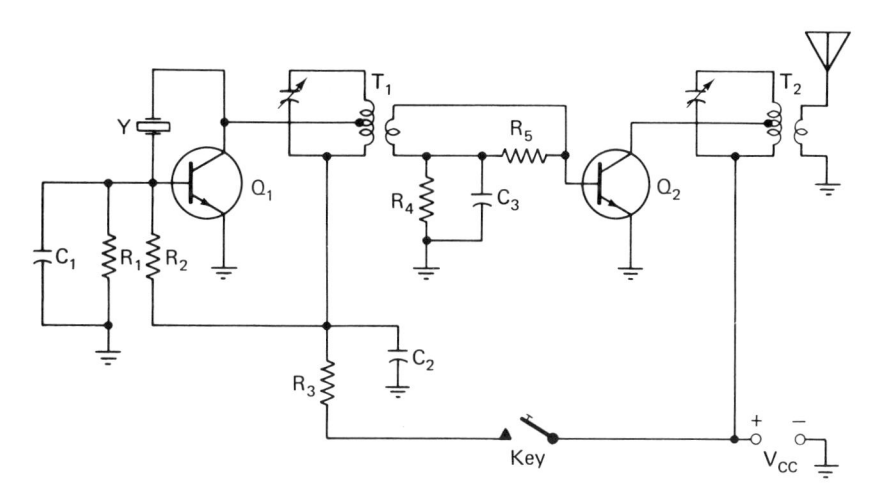

FIG. 17-2 An MOPA type transmitter.

In the MOPA any changes in antenna loading, which might be caused simply by the swaying of the antenna in the wind, would not be reflected back to the oscillator circuit to the same extent existing in a simple oscillator transmitter. Also, the fact that more tuned circuits are required provides a greater degree of rejection to oscillator harmonics.

In Fig. 17-2 transistor Q_1 serves as a modified Pierce oscillator. Capacitor C_1, usually a silver mica of approximately 1000 pF, controls the amount of feedback. Keying the transmitter is accomplished by supplying V_{cc}, via the key, to the collector of Q_1, and to the base by biasing resistors R_1 and R_2. The function of R_3 and C_2 is to form a shaping network to give a *click-free* CW signal. Typical values for this combination are 100 Ω and 10μF.

Coil assembly T_1 is designed for a good impedance match between the collector of Q_1 and the base of Q_2. Resistor R_4 and capacitor C_3 are connected between the cold end of the secondary of T_1 and ground. R_4 is the base-leak resistor and has a typical value of between 10 and 100 Ω, which permits Q_2 to be driven farther into the class C bias region than would be possible without it, adding to the efficiency of the stage. Capacitor C_3 effectively places the lower end of this coil at ground potential. The function of R_5 is to provide some loading to the secondary winding of T_1. Its presence adds stability to the circuit by providing a somewhat constant load on the oscillator. Transistor Q_2 provides additional amplification and isolation between the oscillator and antenna.

Master oscillator power amplifier transmitters may consist of more than two stages. Several *buffer stages* may exist between these two stages. An example of a MOPA transmitter is shown in block diagram form in Fig. 17-3. The master oscillator operates on 2017.5 kHz. The desired output frequency is 8070 kHz. To achieve this, the oscillator is fed into a *doubler stage*, whose output is 4035 kHz. The output of this stage is fed into a second frequency

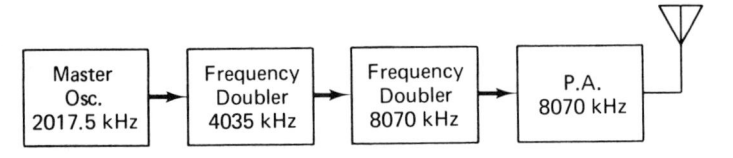

```
Master          Frequency        Frequency
 Osc.    -->    Doubler    -->   Doubler    -->   P.A.
2017.5 kHz      4035 kHz         8070 kHz         8070 kHz
```

FIG. 17-3 Block diagram of MOPA transmitter.

doubler that is tuned to the second harmonic of 4035 kHz, thus providing the desired output frequency. It is proper to refer to the two doubler frequency stages as buffer amplifiers.

17-4 BREAK-IN OPERATION

It is customary to use the same antenna for transmitting and receiving. This requires some method of quickly shifting the antenna from the transmitter to the receiver the moment transmission ceases. A *keying relay* can be used to accomplish this shift. It is a fast-acting switch having double-pole double-throw action. One set of poles is used to key the transmitter; the other connects the antenna to either the transmitter or the receiver. When the key is down, the transmitter is turned on and the antenna connected to the transmitter. When the key is up, the transmitter is turned off and the antenna connected to the receiver, permitting incoming signals to be heard. This is known as *simplex* operation.

The advantage of this system is that the operator at the receiving station can break in on the transmitting operator if necessary by merely holding his key down for several seconds. The transmitting operator can hear the signal from the receiving station whenever his key is up. He would hear a steady tone between the dots and dashes he is sending, and stop transmitting. The distant operator can then explain the reason for his *breaking in.*

The type of material used for relay contacts is important. Silver is frequently used because of its low resistance. However, if any appreciable amount of current is to be interrupted, the heat produced by the arcing will cause *pitting* and *burning* of the contacts. Under these conditions relay contacts should be made of *nickel* or *tungsten,* which have much higher melting points. Even then pitting will occur, but over a longer period of time. Periodically burnishing the contacts with a small, flat file will remove the built-up metal. Relays are generally designed to provide a slight *wiping action* as they open and close. This tends to keep the contact area clean and polished, thus providing reliable operation over a longer period of time.

17-5 BUFFER AMPLIFIERS

The buffer amplifier is usually a *low-gain* stage having low Q circuits and draws no grid or base current from the oscillator. Hence, it presents a *high-impedance load* to the oscillator and does not affect its Q. Therefore,

Power Amplifiers 335

changes in tuning of subsequent stages have little or no effect upon the oscillator's stability. In a buffer stage using vacuum tubes, usually a *tetrode* or *pentode* is used rather than a triode in order to reduce capacitive coupling between the oscillator and succeeding stages. Because of the low gain of the buffer, it is usually unnecessary to neutralize it. The use of low-Q circuits tends to afford a wide passband characteristic.

17-6 POWER AMPLIFIERS

The power amplifier increases the magnitude of the RF current and voltage. A typical circuit is shown in Fig. 17-4a. The amplification factor is 20

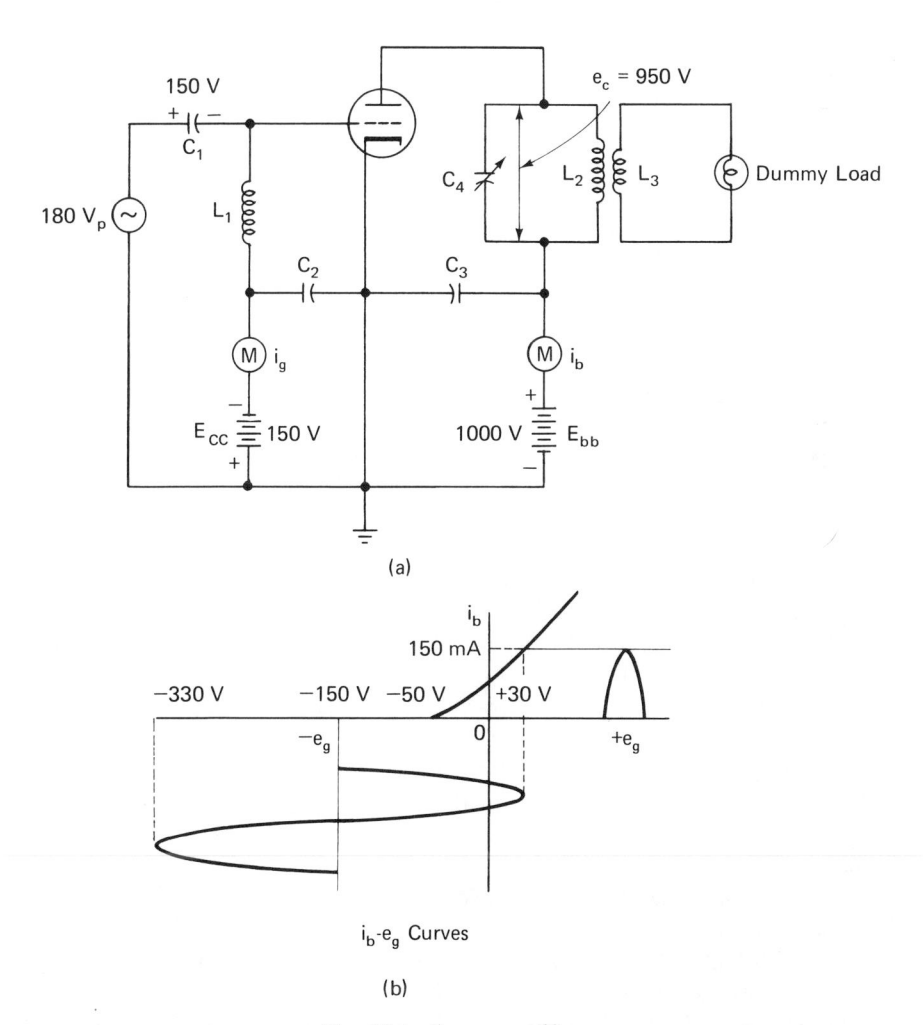

Fig. 17-4 Power amplifier.

and the plate supply voltage is 1000 V. The cutoff bias, e_{co}, is therefore

$$e_{co} = \frac{-e_p}{\mu} = \frac{-1000}{20} = -50 \text{ V}$$

The operating bias, e_o, is approximately three times the cutoff, and therefore

$$e_o = 3(-50) = -150 \text{ V}$$

The maximum value of the RF input signal is 180 V. Thus, when the grid end of the RF input is positive, the peak positive grid-to-cathode voltage is $180 - 150 = +30$ V. On the other half-cycle the peak negative grid-to-cathode voltage is $(-180) + (-150) = -330$ V.

When the grid voltage is above cutoff, plate current flows, and at the instant the grid voltage is $+30$ V, the plate current is 150 mA (Fig. 17-4b). Tuning capacitor C_4 charges up to nearly the full value of the supply voltage, or 950 V. During this charging process, the lower capacitor plate is positive and the upper plate is negative. Thus, the instantaneous plate-to-ground voltage is $1000 - 950 = 50$ V. This value is called e_{min} and represents the lowest value of plate-to-cathode voltage in the entire cycle.

The relations between plate voltage, plate current, grid excitation voltage, and resonant plate tank circuit voltage and current are shown in Fig. 17-5. The flywheel effect in the plate tank circuit causes the capacitor to periodically reverse its polarity and continue the ac cycle within the tank when the grid voltage is below cutoff and no energy is being supplied from the power supply.

The plate tank circuit in Fig. 17-4 may have an *artificial load* applied to it for the purpose of tuning the amplifier prior to coupling the antenna to it. An incandescent lamp of approximately the same power rating as the amplifier can be used as a dummy load if no regular artificial antenna is available, although it may produce some radiation.

The plate supply voltage is reduced, and tuning capacitor C_4 is adjusted for resonance, as indicated by the *dip* in the plate milliammeter (line current is minimum at resonance). The sharp decrease in plate current is accompanied by a corresponding increase in tank current. As resonance is approached, grid current *increases* as plate current *decreases*. The load on the tank circuit may be increased by moving inductor L_3 closer to L_2 and increasing the plate supply voltage to the normal value.

The increased load on the amplifier increases the current through the lamp and decreases the current in the plate tank circuit. The decrease in current in the resonant tank is accompanied by a decrease in voltage, e_c, across the tank. Thus, in Fig. 17-5, e_{min} becomes larger ($e_{min} = E_{bb} - e_c$), and plate current increases with the load. For a given cathode emission, plate current increases and grid current decreases as the load on the tank increases.

It is desirable to use an ac filament supply for transmitting tubes because it is almost always readily available. Furthermore, it is a relatively simple matter to provide the correct voltage by means of step-down transformers.

Power Amplifiers 337

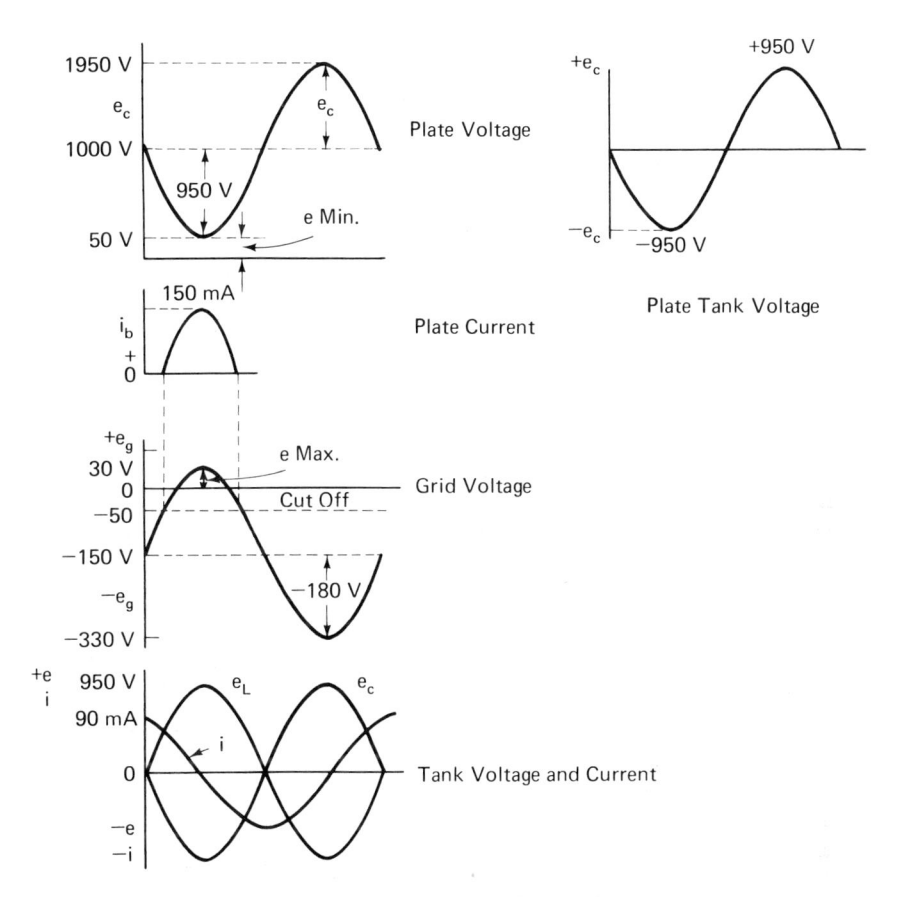

FIG. 17-5 Power amplifier current and voltage relationships.

When alternating current is used for the filament supply, it is usually necessary to provide a filament center tap on the step-down transformer to prevent *hum voltages* from modulating the signal. If the grid and plate return leads were connected to one side of the transformer winding, the other side of the filament would vary at the power-supply frequency. This would cause the bias voltage to change at the supply frequency. This malpractice would cause unwanted modulation of the carrier frequency. Center tapping the grid and plate return lines effectively cancels out the equal and opposite voltage changes on either side of the filament. Hence, the bias remains essentially constant and no hum appears. See Fig. 17-6 for a typical circuit.

Some high-power vacuum tubes use filament supplies operating from dc sources. Under these circumstances it is advisable to *periodically reverse* the polarity of the filament potential in order to lengthen the life of the filament. This is because there will be a greater potential difference between the positive side of the filament and the grid than between the negative side of the

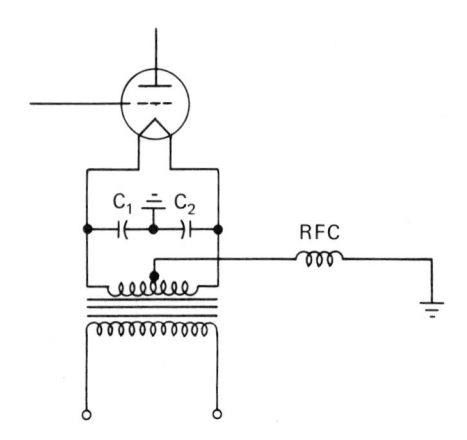

FIG. 17-6 Typical circuit showing center tapping of filament transformer RF power amplifier.

filament and the grid. Hence, the effective bias is larger on the positive side than on the negative side. Therefore, more current will be drawn from the negative side of the filament. Additionally, the plate to filament voltage is greater on the negative side, which increases the emission from this side of the filament.

In high-power transmitting tubes, *directly heated cathodes* are preferred. The reason for this is that directly heated cathodes, constructed from thoriated tungsten, have the ability to withstand higher differences in potential than barium- or strontium-coated indirectly heated cathodes. When a few thousand volts is connected between the plate and barium- or strontium-coated cathodes, the electrical stress on the coating may be sufficiently high to physically pull the coating from the cathode surface. With high plate voltages, the life expectancy of these tubes is very short. Therefore, directly heated cathode power tubes possess a longer life than their indirectly heated counterparts operated under the same conditions.

17-7 DUMMY ANTENNAS

It is very undesirable to have a transmitter connected to an antenna while it is being adjusted. Many spurious frequencies would be radiated, causing interference over a wide area. This can be eliminated by using an artificial or *dummy antenna*, which will absorb the power during these testing procedures. Ideally, a dummy antenna will possess characteristics very similar to the antenna that the transmitter is to be connected to. The circuit of a typical dummy antenna appears in Fig. 17-7. The capacitor should have a reactance equal to that of the antenna. The resistive element should have an ohmic value equal to the resistance exhibited by the antenna when connected to the transmitter. This must be a noninductive resistor and capable of dissipating power equal to that which is normally delivered to the antenna. The ammeter should be of the RF type to indicate the amount of current being delivered to the load.

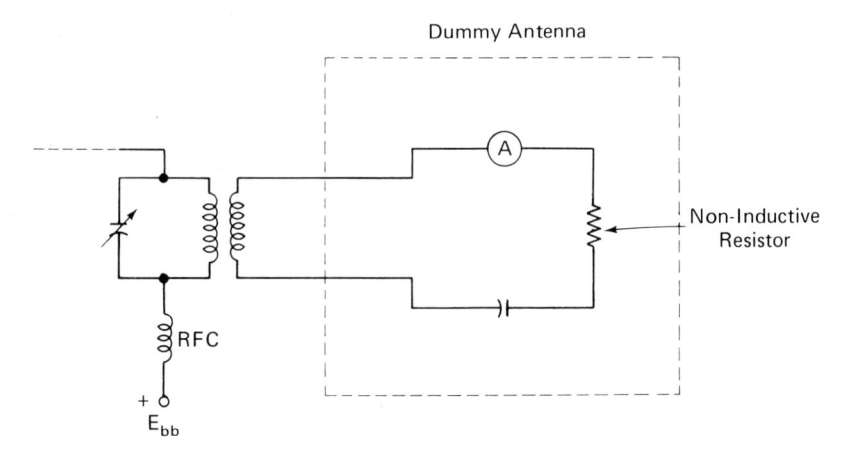

FIG. 17-7 A typical dummy-antenna circuit.

The RF power being delivered to the antenna can be calculated by Watt's law, where power equals I^2R. It is not uncommon to use an *incandescent* lamp in series with a capacitor as a dummy load. This works quite well for low-powered transmitters. The lamp resistance serves as a noninductive resistor. An approximation of the amount of power delivered to the lamp can be determined by its brilliance relative to normal operation.

Initial tuning of a transmitter should be accomplished with the plate and screen supplies *reduced* to about 25 to 50 per cent of normal value. This reduces the probability of damage to the amplifier. Also, it eliminates the radiation of spurious signals during adjustment procedures. Once the adjustments have been made and all tuning procedures accomplished, the dummy antenna may then be removed and the transmitter connected to its normal antenna.

17-8 BIAS METHODS

Because of the power output requirements of transmitters, class C radio-frequency amplifiers are used most often, and these normally employ grid-leak bias.

A grid-leak bias circuit is shown in Fig. 17-8a. The triode is operating as a class C amplifier with a peak driving voltage of 180 V, a cutoff bias of -50 V, and an operating bias of -150 V. The grid is assumed to be conducting. The input signal is developed across the *RFC* because of its high reactance. The positive peak value of the driving signal is such that the grid is driven 30 V positive with respect to the cathode. This causes grid current to flow, which charges capacitor C_1. Some of the charge on C_1 leaks off through the *RFC* and grid-leak resistor R to ground, producing an *IR* drop across R. In this illustration, R is assumed to be 15 kΩ and the grid-leak current 10 mA; consequently, the grid bias voltage is 150 V_{dc}. This voltage charges

(a)
Peak Positive Grid-to-Cathode Voltage

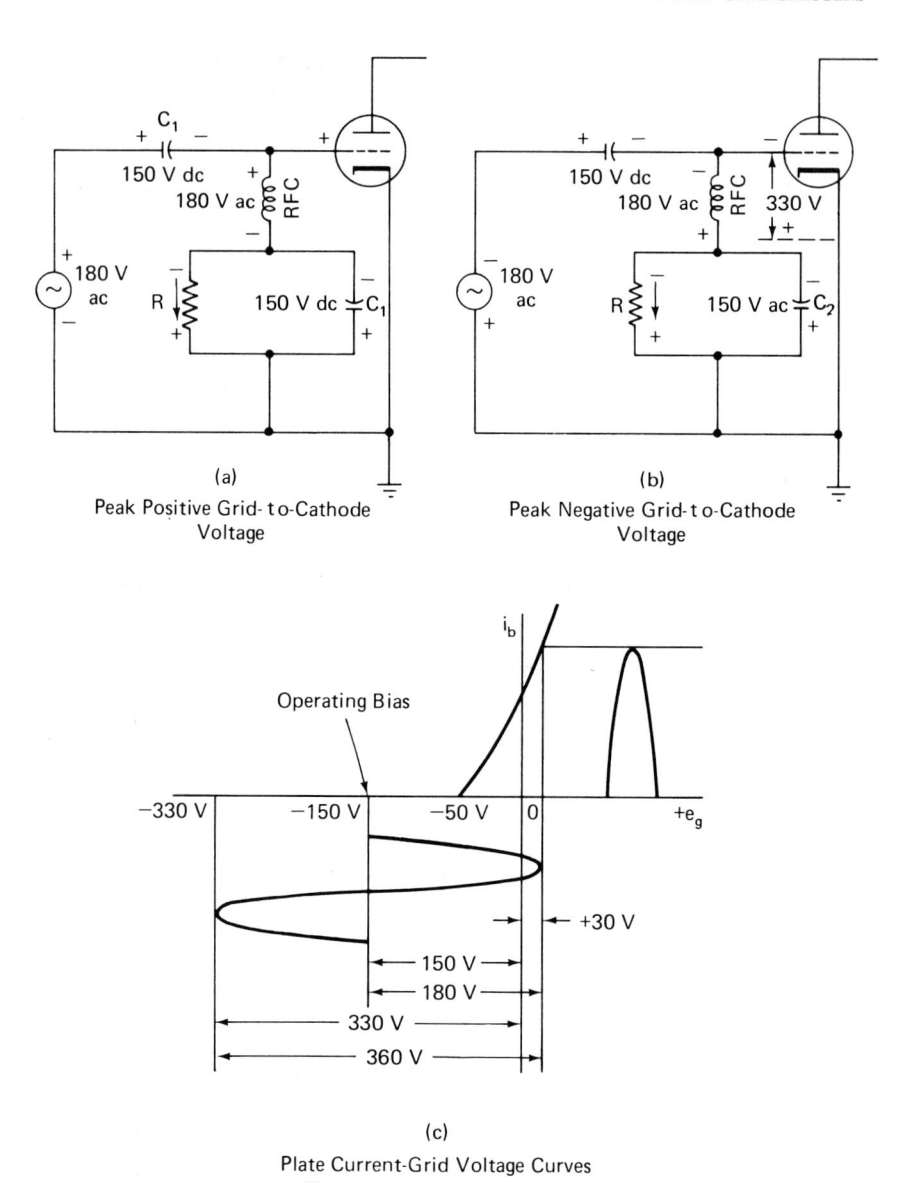

(b)
Peak Negative Grid-to-Cathode Voltage

(c)
Plate Current-Grid Voltage Curves

FIG. 17-8 Analysis of grid-leak bias.

C_2, whose capacity is large enough to maintain the potential across R until the succeeding cycle of grid excitation voltage arrives. C_2 also serves as an RF bypass across R placing the bottom end of the *RFC* at ground. If the charge built up on C_1 were not permitted to leak off, its charge would continue to build up until the tube was blocked.

Reference to Fig. 17-8c shows the input signal waveform and the corre-

Bias Methods

sponding plate current pulses. Observe that during a small portion of the input cycle the grid is driven positive. The grid current that flows for this brief period charges C_1.

During the negative swing of the input cycle, the signal voltage appearing across the *RFC* adds to the bias voltage across C_2. This presents a peak of -330 V between grid and cathode, as indicated in Fig. 17-8b. A graphical representation of this condition appears in Fig. 17-8c.

Grid-leak bias has the desirable characteristic of adjusting its value *automatically* when the amplitude of the grid driving voltage varies in magnitude. For example, an increase in driving voltage increases the operating bias, which checks the increase in grid current; or if the grid driving voltage decreases, the decrease in grid current is checked by a shift of the operating bias in a positive direction. The correction is automatic in either case, because it is the flow of grid current through the grid resistor that produces the operating bias. Thus, the grid current is maintained at the proper value automatically over an appreciable range of input voltage.

Removing the driving voltage or lowering its amplitude below the value that drives the grid positive causes a loss in grid current and operating bias. Plate current then becomes *dangerously high* and the tube may be damaged.

Separate bias may be used to prevent excessive plate current when grid bias is removed. Figure 17-9 shows a circuit that uses *protective bias* in series

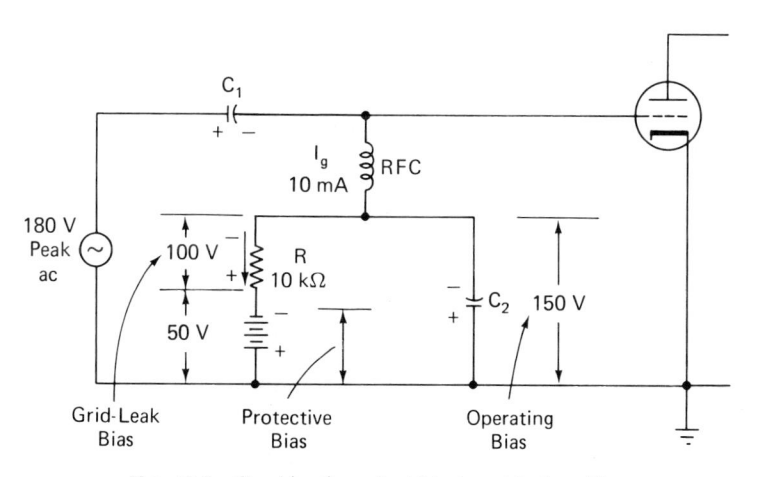

Fig. 17-9 Combination of grid-leak and battery bias.

with grid-leak bias. In this example, the grid bias is developed as a result of a presumed flow of 10 mA of grid current through a 10-kΩ resistor. This voltage in series with the 50-V battery provides a bias of 150 V_{dc} for this stage. In the absence of a driving signal, there still remains the 50-V battery potential on the grid, which is assumed to be the cutoff voltage.

17-9 KEYING REQUIREMENTS

There is more involved in keying a transmitter than simply turning it on and off with a switch or key. If the transmitter is suddenly turned on and its output is permitted to go from zero to maximum instantaneously, *harmonics* or *key clicks* will be generated for many kilohertz on either side of the carrier frequency. Conversely, if the output drops from maximum to zero instantaneously, harmonics will again be generated.

To comply with FCC regulations regarding the avoidance of key clicks, the transmitter output must be *shaped* to provide proper rise and decay times for the carrier. The sideband energy will decrease if longer rise and decay times are used. Some compromise is necessary between the extremes of zero rise and decay times and that of a long rise and fall time. Figure 17-10a

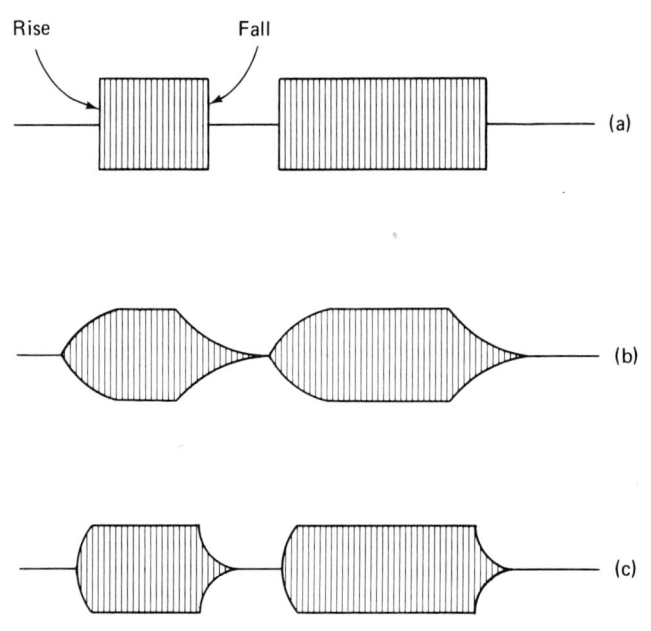

FIG. 17-10 Typical oscilloscope displays of coded transmission: (a) with key clicks, (b) "soft" signal, (c) ideal signal with practically no clicks.

represents a keyed carrier having zero rise and fall times. This would present a *chirping signal* when monitored in a receiver. With improper keying circuits, signals may be transmitted having an envelope similar to that shown in Fig. 17-10b. This is what is known as a *soft signal* and is not nearly as easy to copy. With proper keying circuits a waveform similar to that shown in Fig. 17-10c will be accomplished. Thus, the signal will have practically no key clicks and will be easy to copy.

Key-Click Filters 343

The FCC requirements demand that "the frequency of the emitted wave shall be as constant as the state of the art permits." Therefore, there must be no appreciable change in the transmitter frequency while keying occurs. If the oscillator is not adequately *buffered* from the keyed circuits, a fast frequency change will occur during the transmission cycle, producing chirps. This is the result of a varying load on the oscillator.

If the transmitter output is not reduced to zero when the key is up, some radiation will still occur, referred to as *backwave*. This makes it more difficult for an operator to copy the information being transmitted. If the backwave is 40 dB or more below that when the key is down, the signal-to-noise ratio is sufficient to not cause any objectionable performance to the copying operator. Usually, this condition is not discernable beyond a short distance from the transmitting station. If the circuit to be keyed is carrying dc or ac currents, a small spark will occur every time the key makes or breaks contact. This generates a *highly damped wave or click*, which extends over an extremely broad range of frequencies. The click is synchronous with the keying of the transmitter and consequently has no effect on the transmitted output signal. If one is monitoring his own transmission by means of a receiver, it is necessary that theses clicks be eliminated in order to listen critically to the transmitted signal.

17-10 KEY-CLICK FILTERS

The presence of key clicks can be eliminated by properly designed *key-click filters*. Two possible arrangements of filters appear in Fig. 17-11. The

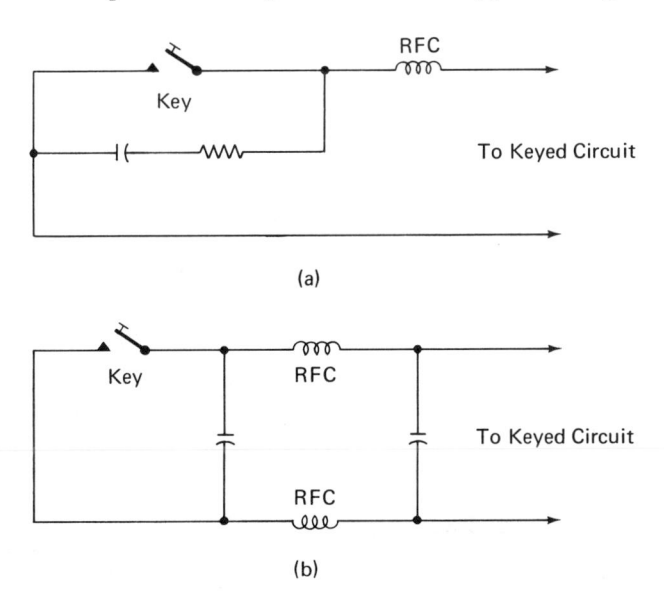

FIG. 17-11 Key-click filters: (a) simple filter, (b) improved filter.

simplest circuit involves connecting a capacitor and series resistor across the key, such as shown in Fig. 17-11a. Each time the key opens, the spark that would normally occur across the contacts charges the capacitor. When the key is closed, the capacitor discharges. Without the series resistor (usually 10 to 100 Ω) there would likely be a small spark each time the key closed. The resistor should be connected between the capacitor and the "hot" side of the circuit. The *RFC* presents a high impedance to any small spark that may possibly appear, isolating it from other transmitter circuits.

A better circuit is shown in Fig. 17-11b. The added capacitor and *RFC* should suppress even the most troublesome clicks. Typical values of C in both drawings vary from 0.001 to 0.01 μF. The *RFC*s should be somewhere between 0.5 and 2.5 mH with a current-carrying capacity adequate for the current involved. The filter should be mounted as close to the key as possible. In many transmitter circuits the telegraph key is used to operate a *keying relay* located within the transmitter. If this is the case, it is necessary that the key-click filter be mounted as close to the keying relay as possible.

17-11 OSCILLATOR KEYING

A transmitter can be keyed in either its oscillator or amplifier stages. There are advantages and disadvantages to each method. Oscillator keying has an advantage in that a low-level stage is keyed, and only relatively small currents or voltages need by interrupted. However, it is almost impossible to avoid chirps, particularly at higher frequencies. By choosing proper resistor and capacitor values, the oscillator output may be shaped to minimize chirps. This involves changing the oscillator's operating conditions so that the output does not rise immediately to full value. Consequently, the drive to the following stage is not constant. This implies that the load reflected back to the oscillator is varying. Attempting to operate the oscillator from no voltage to full voltage and under a varying load presents stringent demands. Actually, no oscillator has been designed that has no frequency change under these conditions. In spite of these difficulties, acceptable standards can be achieved if the oscillator is keyed on the lower frequency bands.

A basic keyed oscillator circuit is shown in Fig. 17-12. With the key connected as shown the grid circuit is closed at all times and the negative side of the plate circuit is keyed. This technique is called *plate keying*. When the key is open, no plate current flows and the circuit does not oscillate. An alternative method can be used in which the key is placed in the cathode circuit at the point marked X. With the key up the cathode circuit is open, and neither grid nor plate current can flow. This is called *cathode keying*.

17-12 BLOCKED-GRID KEYING

Blocked-grid keying is a satisfactory method of achieving coded transmissions. One circuit for accomplishing this appears in Fig. 17-13, where the

Blocked-Grid Keying 345

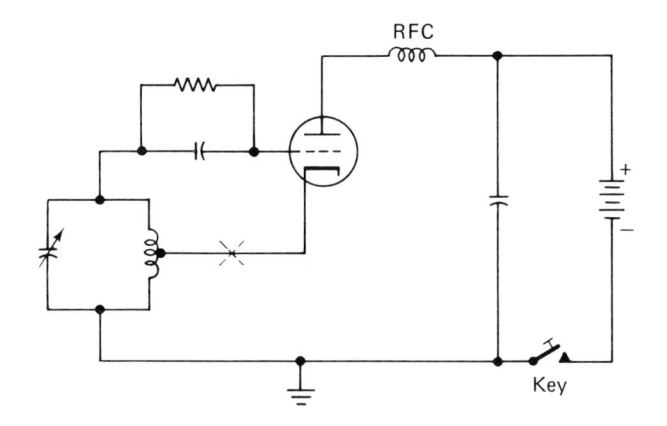

FIG. 17-12 Basic oscillator keying circuit.

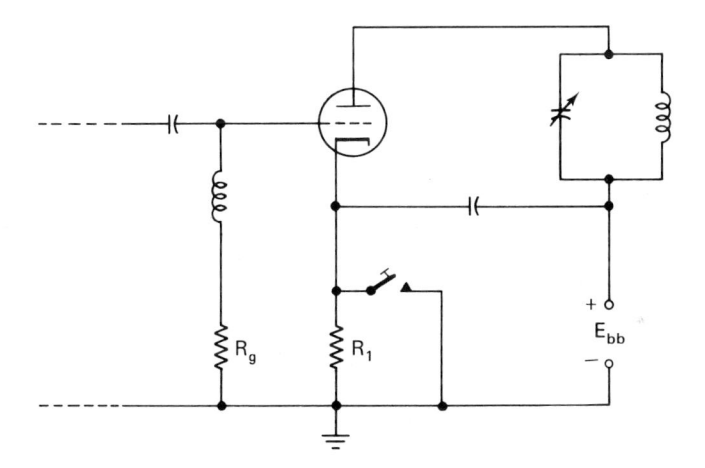

FIG. 17-13 Blocked-grid keying with key across cathode resistor.

key shorts cathode resistor R_1, allowing normal plate current to flow. With the key open reduced plate current flows through resistor R_1. If R_1 has a high enough value, the bias developed is sufficient to effectively cut off plate current. Complete cutoff is not possible, because the bias voltage developed across R_1 depends on the flow of some plate current. However, the blocking is sufficient for practical keying. This method of keying is applied to the buffer stage in a transmitter.

The block-grid keying method shown in Fig. 17-14 affords complete cutoff of plate current and is one of the best methods for keying amplifier stages in CW transmitters. In the voltage divider, with the key open, 667 V are developed across the 200-kΩ resistor and 333 V are developed across the 100-kΩ resistor. The grid bias is −100 plus −333, or −433 V. Because this is below cutoff, no plate current flows and plate voltage is 667 V. With the

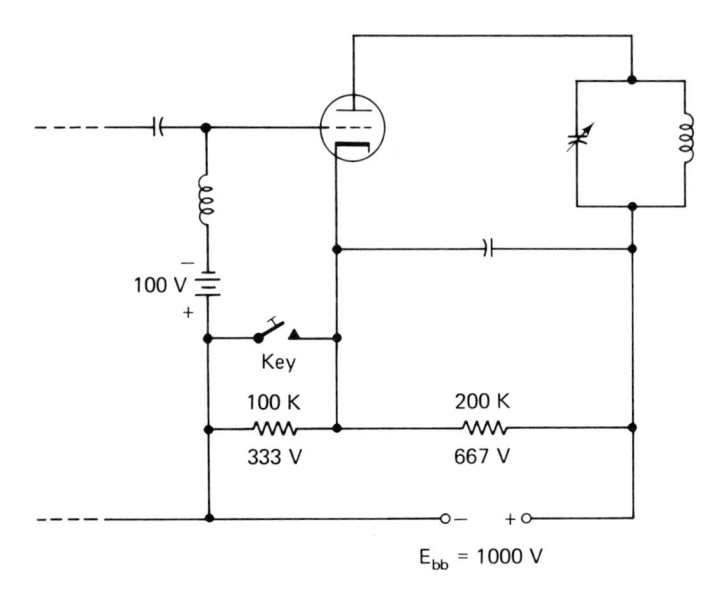

FIG. 17-14 Blocked-grid keying with key across grid resistor.

key closed, the 100-kΩ resistor is shorted out, and the voltage across the 200-kΩ resistor is increased to 1000 V. Thus, the plate voltage becomes 1000 V at the same time the grid bias becomes −100 V. Grid bias is now above cutoff and the triode conducts.

When greater frequency stability is required, the oscillator should remain in operation continuously while the transmitter is in use. The oscillator circuit must be *carefully shielded* to prevent radiation and interference to the operator while he is receiving.

17-13 VACUUM-TUBE KEYING

Another method of keying a transmitter is by *vacuum-tube keying*. This technique provides very fast operation and high-speed transmission. It operates on the principle of dropping the supply voltages to the keyed stage to a low value and simultaneously adds bias to the point where there is no output. The *keyer tube*, usually a low μ triode, is connected in series with the cathode circuit of the stage to be keyed. By controlling the bias of the keyer tube, it can be made to saturate or cut off and thereby control the flow of plate current of the RF stage.

An example of vacuum-tube keying is shown in Fig. 17-15. The series arrangement of the keyer tube and RF stage is clearly evident. A source of negative voltage is required for the key tube. This may be obtained from the bias supply of the transmitter or from a separate bias source. When the key is up, the *full negative voltage* is connected between grid and cathode of the keyer tube, biasing it beyond cutoff. Its internal resistance becomes very high,

Cathode Keying

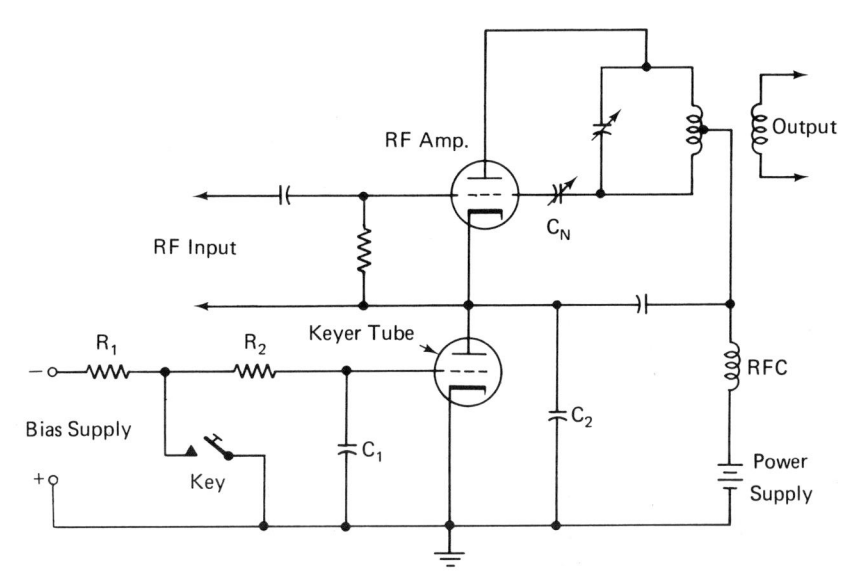

FIG. 17-15 Typical vacuum-tube keying circuit.

and nearly all the power-supply voltage appears across this tube. With virtually no voltage applied to the RF amplifier, its output drops to zero. Concurrently, capacitor C_1 has charged to the full value of the bias supply.

When the key is closed, the bias voltage is removed from the keyer tube and the total bias supply is connected across R_1, which prevents the bias supply from being shorted. Capacitor C_1 discharges through R_2 and the key. The values of R_2 and C_1 determine the rate of discharge and hence the time required to reduce the grid bias to zero. This is one of the principal factors in controlling the shape of the leading edge of the RF output pulse. With no bias on the keyer tube, its internal resistance drops to a low value and it conducts heavily. This restores normal operating voltage and current to the RF amplifier.

The rate of charge and discharge of C_1 is determined by R_1 and R_2. The value of the charge on C_1 is also a function of the bias supply voltage. These parameters largely determine the *softness* of the keying.

If the total RF amplifier current is greater than can be handled by a single keyer tube, then two or more of the same type can be parallel connected.

17-14 CATHODE KEYING

One of the simplest keying methods is to key the cathode circuit. By interrupting the circuit at this point (see Fig. 17-16, point A), complete control of plate and grid current can be achieved. It would be necessary to bypass the key with a capacitor, as shown by the dotted lines, in order to maintain the

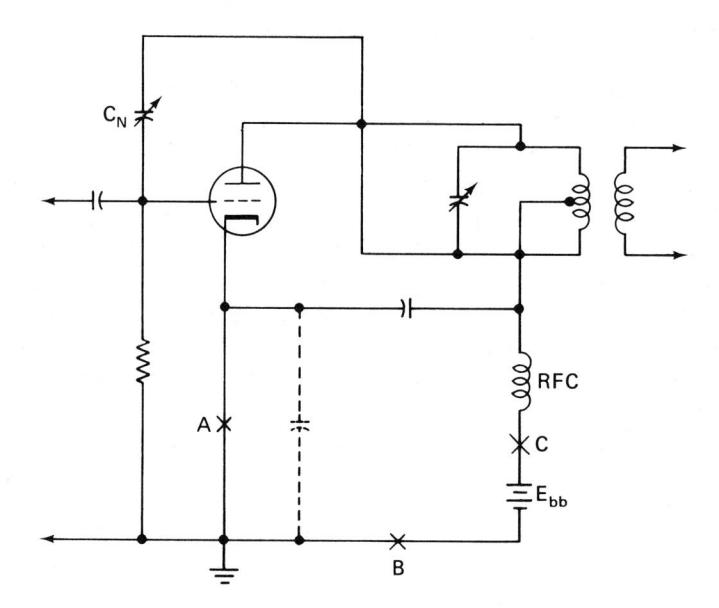

FIG. 17-16 Cathode-keying is accomplished at point *A*. Plate-circuit keying may be at either *B* or *C*.

cathode at RF ground potential. Arcing would also be minimized by this capacitor. It may be more practical to use a keying relay at point *A* in order to safely carry the current.

If the same basic circuit were to have the current interrupted at points *B* or *C* it would be known as *plate keying*. It is just as feasible to key at *B* because of the low potential involved, although grid current is not interrupted. If keying should be attempted at point *C*, it would be desirable to use a relay because of the high voltage involved. The danger of using a hand key at this point, with its attendant hazard to the operator, should be obvious. These simple methods of keying are prone to produce clicks.

An amplifier can be keyed in any way that will reduce the output to zero. For example, the *screen-grid voltage* may be keyed. Simply reducing the screen voltage to zero is usually not enough, as some plate current will continue to flow. Generally, it is necessary to connect the grid to some negative value to reduce the plate current to zero during key-up conditions.

Output stages that are not neutralized, such as tetrode and pentode stages, will generally leak a little and create some backwave, irrespective of how keying is accomplished. By keying two stages it is usually possible to eliminate this problem. This may present another problem in that the keyed stages are not far enough removed, or buffered, from the oscillator to prevent pulling.

17-15 FREQUENCY-SHIFT KEYING (F1)

Coded signals may be difficult to receive if there is a poor signal-to-noise ratio. This situation becomes untenable in commercial radiotelegraphy and radioteletype (abbreviated RTTY) where reliability is paramount. A system has been developed that overcomes most of these difficulties. It is called *frequency-shift keying*. The carrier frequency is not turned on and off, but rather shifted a given amount. This alternative provides for one frequency to be used for *mark intervals* and the other for the *spacing intervals*. This is a form of frequency modulation and consequently possesses inherent discrimination against noise that is weaker than the signal. It is this characteristic that is helpful in reducing errors when operating automatic printing equipment.

Although FCC regulations permit any value of frequency shift up to 900 Hz, general practice is to use 850 Hz. This relatively small shift permits the receiver to receive both signals at one dial setting, and thus remain in a quieted condition. The receiver is designed to respond to these frequency changes rather than amplitude changes, and therefore provides an improvement in reception.

Frequency-shift keying can be accomplished in several ways. Two simple methods are shown in Fig. 17-17. In (a) the resonant frequency of the

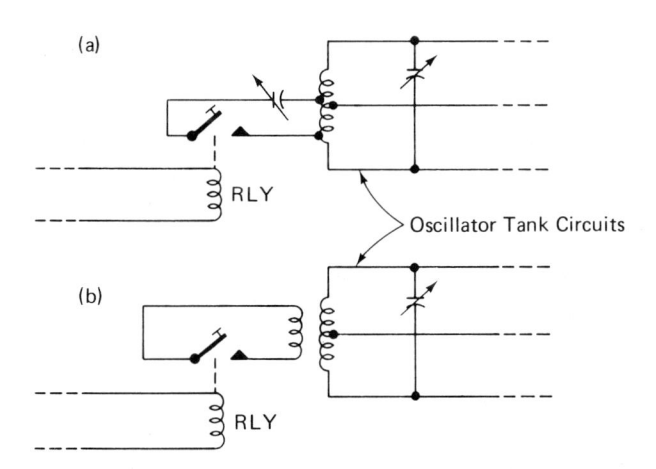

Fig. 17-17 Two methods of frequency-shift keying changing the resonant tank frequency by adding capacitance as in (a), and changing the inductance by the use of a shorted loop as in (b).

oscillator tank is shifted downward by placing a capacitor across a part of the inductance. The size of the capacitor and the number of turns it is connected across determine the frequency shift. In Fig. 17-17b the shift is brought about by coupling a shorted loop to the *LC* tank. The amount of shift in this case is a function of the coupling and the number of turns on the loop. This has the effect of reducing the inductance and raising the frequency.

It is not practical to replace the relays shown in Fig. 17-17 with a telegraph key. There are two basic reasons for this. First, the key would likely be located several feet or more from the transmitter whose oscillator is being keyed, and the *unpredictable* inductance and capacitance of the leads would result in erratic operation. Second, depending upon the type and power of the oscillator circuit being used, some dc and RF voltage may be present at the key and present a hazard to the operator.

Relays are not without problems, however. Because of the desirability of keeping leads short, the relay must be mounted close to the oscillator. But the opening and closing of the contacts may set up *vibrations* in parts of the oscillator circuit and result in some frequency modulation of the carrier. Therefore, it is necessary that all mechanical parts of the circuit be very secure to eliminate this unwanted condition.

In commercial applications more sophisticated methods are employed to accomplish the frequency shift. The block diagram of one such method is shown in Fig. 17-18. The frequency used is the sum (or difference) frequency

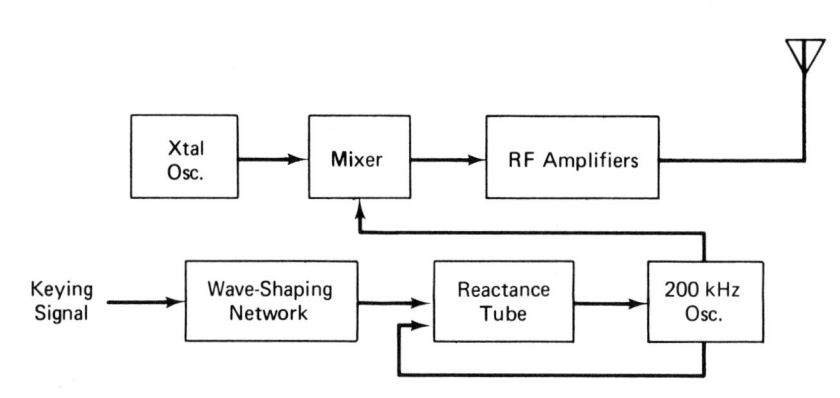

FIG. 17 18 Block diagram of a typical frequency-shift keying system.

obtained by combining the output of the crystal oscillator with the output of the 200-kHz oscillator. This latter oscillator is controlled by the *reactance tube*, which in turn is controlled by the keying signal. The entire system is electronic—no relays are needed, and the limitations of the circuit shown in Fig. 17-17 are overcome.

17-16 MODULATED CONTINUOUS WAVE SIGNALS (A2)

High-frequency transmissions, particularly over long distances, are subject to considerable fading. Fading is a phenomenon whereby a frequency increases and decreases in strength at the receiving station. This results from the fact that the signal may take any one of *several paths* or a combination of them. This is the result of the signal being reflected back and forth between the earth and the *ionosphere*. If these signals arrive in phase, they reinforce each other, but if they are out of phase, they cancel. Because the ionospheric conditions fluctuate, fading is common. This condition can be overcome, at least in part, by modulating the carrier with audio. As we shall learn in Chapter 18, the modulation process causes the carrier frequency to vary in amplitude in accordance with the modulating frequency. Inherent in this process is the automatic generation of signal frequencies, called *sidebands*, on either side of the carrier. The modulating frequency(s) determines how far the sidebands extend on either side of the carrier. Typically, a frequency somewhere between 500 and 1000 Hz is used to modulate the carrier. Let us assume that a 1-kHz frequency or tone is used. This will produce two sidebands, one above the carrier by 1 kHz and another below by the same amount.

The carrier and sideband frequencies will likely travel slightly different paths, so that even if one fades the others may be received. By keying a tone-modulated signal such as this, any good CW receiver will provide a readable signal, even though the carrier or one sideband has temporarily faded. For this reason it is advisable that all distress messages be made using A2 transmission.

17-17 VARIABLE-FREQUENCY-OSCILLATOR OPERATION

Outside of radio amateur activity there is little use of the *variable-frequency oscillator* (VFO) transmitter. All commercial, police, military, and so forth, operation is crystal controlled. Nevertheless, there is sufficient amateur use of the VFO transmitter to warrant discussion.

Variable-frequency oscillators must be well designed and preferably temperature controlled. The elimination of mechanical vibration is particularly important. It is desirable to use tuning capacitors having thick, small plates. The coil should be rigidly mounted to avoid the slightest movement. Leads should be short. Use flexible wire where possible as it has less tendency to vibrate. The entire oscillator assembly should be shock mounted to isolate all mechanical vibration.

A VFO should be checked frequently and thoroughly before it is placed on the air. Because the stages following the oscillator may affect the frequency, final tests should be made with the complete transmitter in operation. Most VFOs will appear stable and show signals of good quality when operating without a load.

17-18 REDUCING HARMONIC RADIATION

Harmonic radiation is troublesome to the operator and can be very annoying to neighbors. Only the best of construction and design techniques can reduce this type of distortion to minimum standards. This must follow through with *each stage* of the transmitter.

All active devices are inherently nonlinear—some more than others. This means that even though a pure sine wave may be generated in an oscillator it will suffer some distortion as it passes through subsequent amplifier stages. It can be shown mathematically that any wave which deviates from a pure sinusoid contains harmonics. The amount of deviation is related to the harmonic content.

The carrier should be as free from harmonic content as the state of the art permits. Harmonics will frequently mix with other RF signals and generate *spurious* responses over a broad portion of the spectrum.

Effective harmonic suppression has three separate phases:

1. Reducing the amplitude of harmonics generated within the transmitter. This is a matter of proper circuit design and operating conditions.
2. Preventing harmonics from reaching the antenna.
3. Preventing stray radiation from the transmitter and its associated wiring. This demands adequate shielding and filtering of all circuits and leads from which radiation may take place.

Items 1 and 2 will be discussed in the following paragraphs. Item 3 will be discussed in Section 17-19.

Radio-frequency power amplifiers need to be driven hard to maintain high efficiency, but this produces harmonics. Some compromise between best efficiency and minimum harmonic generation must be achieved. For example, frequency-multiplier stages are prone to generate a lot of harmonics. Consequently, the compromise is to operate them at *low power levels* using plate voltages not exceeding 250 to 300 V. When the frequency has been multiplied to the required value, it is desirable to use as few stages as possible in building up the final output power level and to use tubes that require a *minimum* of driving power. Power amplifiers are likely to have rather large amplitude harmonic currents flowing in both their grid and plate circuits. Their effect can be minimized by properly bypassing these circuits to the cathode. The capacitive reactance of the capacitors in both grid and plate circuits is usually much lower to these high harmonic frequencies than is the effective reactances of the tank coils. Consequently, most of the harmonic currents go through the capacitors in these circuits.

The *length of the leads* forming these paths is very important, since their inductance will resonate with the tube capacitances at some frequency usually in the VHF range. It is almost impossible to eliminate resonances of this type. However, by keeping the leads short, the resonant frequency can be kept high enough to likely fall between the VHF and UHF TV bands and thus reduce television enterference (TVI). The inductance of leads running from the tube to the tank capacitor may be reduced also by using *flat strip* instead of wire conductors. It is also better to use the *chassis* as the return from the blocking capacitor or tuned circuit to cathode, since chassis paths generally present less inductance than almost any other form of connection.

Link coupling between the driver stage and power amplifier usually provides for a more optimum placing of components than if capacitive coupling is used. This generally permits shorter lead lengths and, consequently, a more favorable condition for bypassing harmonics. Link coupling also reduces the coupling between the driver and amplifier at harmonic frequencies, thus preventing driver harmonics from being amplified.

The amount of harmonic content in RF waves is affected by the amount of grid bias and current flowing in the input circuit. Generally speaking, the *harmonic content increases as drive increases*, although this is not necessarily true of all the harmonics. Third- and higher-order harmonics usually go through fluctuations in amplitude as the drive is increased. Under some conditions a high value of grid current will minimize one harmonic as compared with another. If one particular harmonic is troublesome, it might be that changing the drive to the stage will eliminate or at least minimize it. If a certain harmonic is particularly troublesome, it may be necessary to insert a *harmonic trap* between the plate lead and the tank circuit. It will present a very high impedance to the harmonic current without appreciably affecting the basic tank circuit. The L/C ratio is not critical. Usually, a high C circuit will have less effect on the performance of the plate tank at normal operating frequencies. The harmonic trap itself will radiate unless the transmitter is well shielded. The trap should be mechanically mounted so that there is little coupling to the plate tank conductor.

The interference produced by a transmitter as a result of radiating spurious signals is dependent upon power level and the distance between transmitter and receiver. In radio amateur work one of the major problems relates to television interference (TVI). This is particularly troublesome if the TV signal is weak, or if the TV receiver and amateur transmitter are close together and the amateur station is operated at high power.

The third and final step in minimizing harmonic generation or TVI is to keep the spurious energy from passing from the final stage over the transmission lines to the antenna. It is futile to attempt this if the radiation from the transmitter and its associated connecting wire has not been reduced to the point where no TVI is produced in a nearby TV receiver when the transmitter is connected to a dummy load. If interference is still seen, it is almost certain that harmonics will be radiated or coupled to the antenna system regardless of what preventative measures are taken. With *inductively coupled* output systems some spurious energy will be transferred. These harmonics may be greatly reduced by providing adequate selectivity between the final tank and the transmission line. Twenty to thirty decibels of reduction of the second harmonic and more reduction for higher harmonics may be achieved by an appropriate antenna coupler.

Interference problems must be solved one step at a time and in logical order. The following pieces of test equipment are required in checking for and eliminating interference: grid-dip meter, dummy antenna, and an accurately calibrated field strength meter.

17-19 SHIELDING

Nearly all spurious radiation from a transmitter can be eliminated by *proper shielding*. However, to be effective, the shield must *completely enclose* the entire circuit and parts, and must have no opening that will permit RF energy to escape. Ordinary metal boxes and cabinets are not likely to provide good shielding, since they may have such openings as louvers, lids, and large holes for bringing in leads. These openings will provide excellent paths for RF energy to escape. An essential requisite to shielding is that all joints make good electrical connection along their entire length. Small openings or holes to accommodate power leads usually present little trouble. Larger holes, such as used to mount meters, can be *very troublesome*. If larger openings are required or if ventilation is a problem, it is possible to use wire screen, as this makes effective shielding. The screen must make good electrical connection where the individual wires cross over each other. Perforated aluminum paneling such as found in most hardware stores can provide effective shielding provided the holes are small. There must be good electrical connection or bonding between the edges of the screen or perforated aluminum paneling and the other parts of the metal-enclosed chassis.

Where joints are formed such as at corners or where metal screen overlaps a part of the metal cabinet, at least $\frac{1}{2}$ in. of overlap should be provided. Fastening should be with sheet metal or other type screws closely spaced to guarantee tight connections around the entire joint. The metal contact areas should be *cleaned* before joining together.

The effectiveness of shielding may be nullified unless the leads connecting to external power supplies and other circuits are also properly shielded. Such conductors form excellent paths for RF to escape. Shielded wires should be used for such purposes, and the shield should be maintained from the point where the lead connects to the RF circuit or tube through the point where it leaves the chassis to the other chassis, such as the power supply. The braid should be grounded to the chassis at *both ends* and at frequent intervals along the way. If meters are to be panel mounted in the RF unit, they should be enclosed in metal covers or shields with the connections to the meter being made with shielded wire. Each end of the lead of the wire should be bypassed with a small ceramic disc capacitor of about 500 to 1000 pF.

The other end of each capacitor should be connected to the metal braid. An alternative way of mounting the meter would be to mount it behind the panel with panel opening covered by a piece of wire screen carefully bonded to the panel all around the hole.

Shielding is even more important at high frequencies. For example, the intermediate-frequency (IF) coils in receivers are almost always encased in a metal shield. This eliminates coupling, both electromagnetic and electrostatic, between other coils and components. This also decreases or prevents regeneration, oscillation, parasitics, effects of hand or body capacitance, degeneration, and other factors that can cause circuit instability. These shields may consist of any metal that is a suitable conductor of electricity. Because of cost factors, weight, and ease of forming, aluminum is most frequently used.

In UHF applications it is customary to *compartmentize* individual stages. The thin metal shields, often plated to provide excellent electrical conductivity, used in these applications are mechanically bonded to a common chassis. The necessary connecting leads between stages usually pass through these shields by means of *bulkhead feed-through capacitors*, thus eliminating undesirable frequencies. These partitions provide excellent electrostatic shielding. They also afford good electromagnetic shielding at these higher frequencies. The stray magnetic fields induce currents into the shields, which in turn create an opposing magnetic field in accordance with Lenz's law.

Good construction techniques dictate that no high voltage leads or "hot" RF parts are exposed to personnel. All adjustable controls, particularly for transmitters, are brought to the front panel by means of *insulated shafts*. This makes it unnecessary to reach inside the unit where the operator may be exposed to high dc and RF voltages. In fact, it is customary to provide *interlock switches* on this kind of equipment so that all power is turned off wherever any door or panel is opened. In very large transmitters it is common to have *shorting bars* mechanically placed across large filter capacitors to discharge them. Any metal part of a transmitter that is exposed should be bonded to the main frame or chassis and grounded. This prevents the accumulation of static charges and protects the operating personnel against shocks.

Harmonics present in the tank circuit of the power amplifier may be capacitively coupled to the antenna. To eliminate this, a *Faraday shield* may be installed between the final tank and the antenna. This consists of a number of closely spaced parallel wires connected together only at the bottom and grounded. This provides an excellent electrostatic shield and substantially reduces the capacitive coupling between output tank and antenna.

Transmitters are protected against overloads by *fuses* and *circuit breakers*. Fuses are usually located so that they can be replaced from outside the equipment. Circuit breakers are usually of the electromagnetic overload type and trip whenever excessive current is drawn. They, too, are mounted so that they may be reset from outside the equipment. Their contacts are usually made of tungsten or silver because of their relatively low resistance.

17-20 TYPICAL CONTINUOUS WAVE TRANSMITTER

Many possible circuit arrangements exist for CW transmitters. One such is shown in Fig. 17-19, which is a transistorized unit designed for 50-MHz operation. The crystal oscillator generates the 50-MHz low-level signal to excite the buffer amplifier Q_2. Forward bias for Q_1 is provided by R_1 and R_2. A small part of the RF energy present in the tank circuit associated with Q_1 is fed back through the interelement capacitance of this transistor to excite the crystal and sustain oscillations. Keying is accomplished by interrupting the V_{cc} supply to Q_1.

The 50-MHz output of the oscillator is coupled by L_3 to the base of Q_2, which serves as a *predriver stage*, or low-level amplifier. The step-down transformer action of L_2 and L_3 matches the collector impedance of Q_1 to the low-impedance base circuit of Q_2. The tank circuit in the collector of Q_2 is tuned to resonate at 50 MHz. To match the relatively low impedance of the base circuit of Q_3, the tank inductor L_4 is tapped down. This also serves to keep the Q of this tank circuit as high as possible. Q_3 is a silicon power transistor, which develops the necessary power to drive the output stage.

Driving power for the output stage is coupled from the collector of Q_3 through a band-pass filter made up of L_7, C_{13}, C_{12}, and L_8 to the base of Q_4. Additional filter circuits in the collector of the power amplifier provide the necessary rejection of spurious and harmonic frequencies. The output from this latter filter section is coupled through an appropriate length of 50-Ω coaxial cable to the antenna. Capacitors C_{16} and C_{18} are adjusted to provide optimum impedance match between the transmitter and antenna. Up to 40 W of RF power can be delivered to the antenna depending upon the type of transistor used as the power amplifier. Decoupling between stages is provided by RF bypass capacitors C_6, C_{10}, and C_{14}, in conjunction with bulkhead feed-through capacitors C_7, C_{11}, and C_{15}, and inductors L_5, L_6, and L_9.

17-21 TUNING A CONTINUOUS WAVE TRANSMITTER

All radio transmitters must be properly tuned to ensure efficient operation on the assigned frequency. They should be *tuned on continuous wave* even if modulated continuous wave or voice modulation is to be used.

Before attempting any tuning the high voltage supply to the final amplifier should be *disconnected*. Power to all other stages should be turned on and at normal values. If the transmitter uses vacuum tubes, they must be turned on long enough to allow the equipment to properly warm up. If the oscillator is located in an oven, it must be brought up to required temperature.

If a VFO is used, the oscillator tuning dial is set to the desired fre-

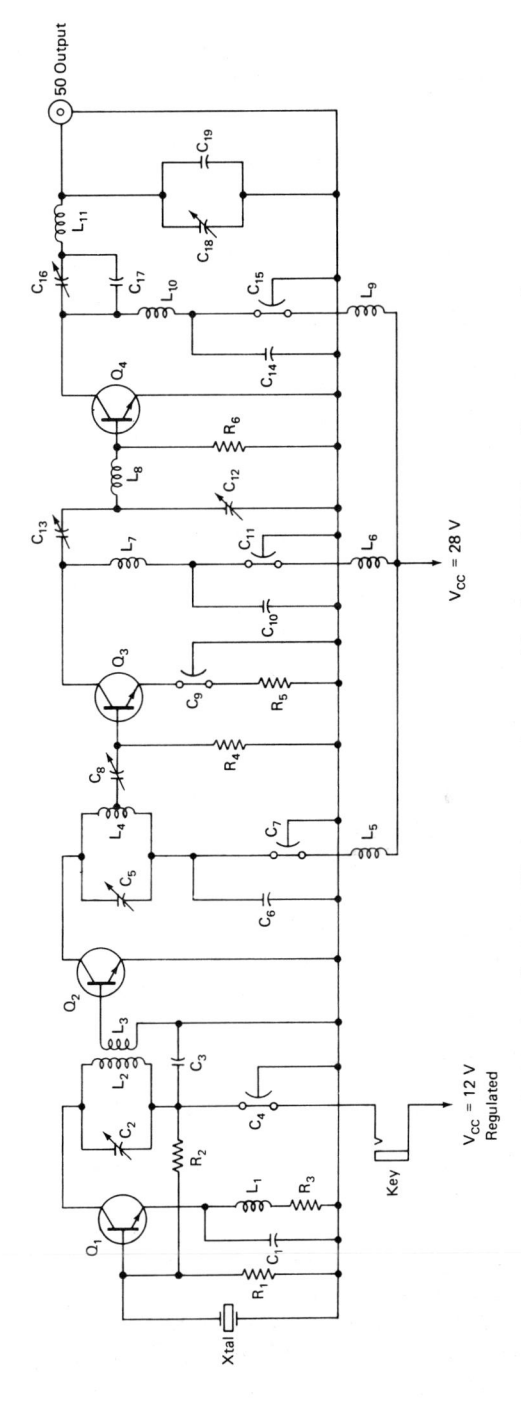

FIG. 17-19 A typical 50-MHz CW transmitter capable of up to 40W depending on types of transistors used.

quency. The key is depressed and the oscillator output is connected to the various buffer and doubler stages (if frequency multipliers are used). The plate or collector currents to these stages should be monitored on an appropriate current-reading meter. This may be a single meter for each stage or one meter to read the combined currents of the several stages. The buffer and doubler stages are tuned by observing the plate- or collector-current readings or the grid- or base-current meter in the input of the final stage. The plate- or collector-current readings should continue to drop in value as the stages are tuned until a dip is reached. During this tuning sequence the grid- or base-current meter in the final stage should show an *increase*.

The high voltage supply is now turned on, but at a *reduced level*. The output tank circuit is now tuned for resonance as observed by a dip in the plate- or collector-current meter. Next, the antenna is tuned to resonance. This is indicated by a rise in plate or collector current. The antenna coupling is now increased. This slightly detunes the final, and redipping is necessary. Now the full supply voltage is applied to the power amplifier and the antenna coupling adjusted until the plate or collector current reaches the specified value. Some adjustment of the antenna circuit may be necessary subsequent to the tuning procedure.

17-22 LOCATING TROUBLES

Many troubles can occur in a transmitter. Some of the more common ones and their indications will be presented. Some are very insidious and hard to locate; others are very apparent. Typical of the latter category are such things as overheated transformers or filter chokes, burned insulation, arcing, discolored resistors due to overheating, leaking oil-filled capacitors, gassy or burned-out tubes, blown fuses, clattering of an overload circuit breaker, or failure of the keying relay to close when the key is depressed.

The meters normally installed in various circuits of a transmitter can be of tremendous help in diagnosing and locating troubles. Some typical indications are listed. Although reference is to vacuum-tube circuits, they also apply to transistor circuits.

OSCILLATOR

Increased plate-current reading. This generally implies that the stage is not oscillating. This may be caused by a fractured crystal, shorted tube, or detuned tank circuit.

Decreased plate-current reading. Weak tube, reduced supply voltages, or reduced coupling to the following stage are likely causes.

Zero plate-current reading. Some obvious causes would be burned-out tube, loss of filament or plate supply voltage, broken leads, blown fuse, or open meter.

Doo key

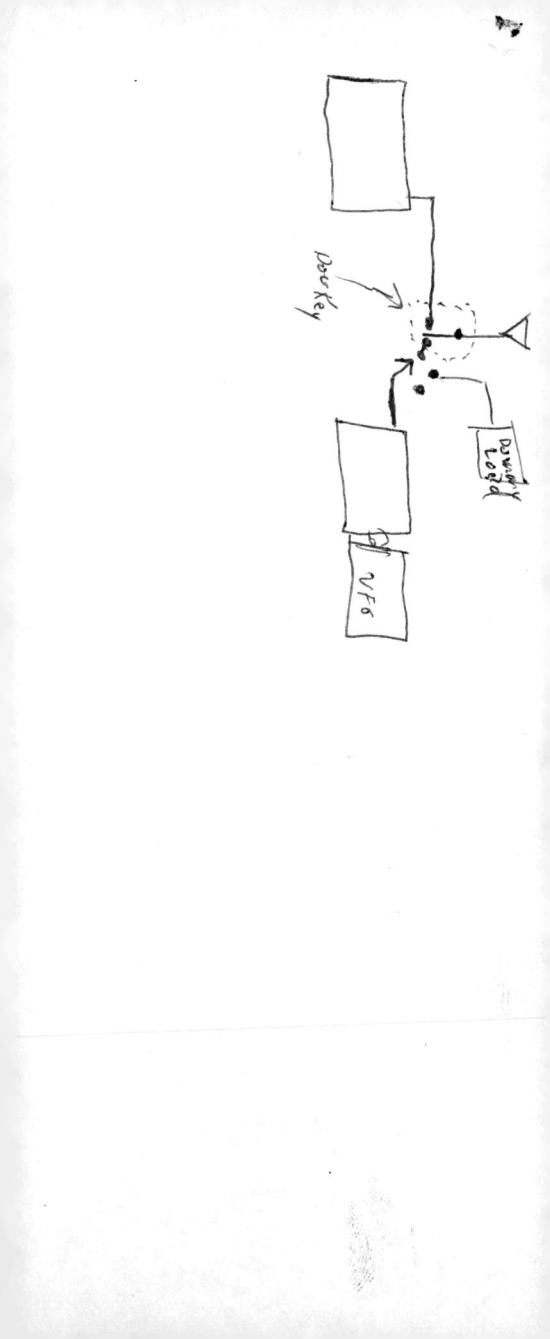

Doorway Lead

VFO

RADIO-FREQUENCY AMPLIFIERS

Increased grid-current reading. This may result from an increased drive from a preceding stage, the loss of bias, shorted grid-leak resistor, or reduced plate voltage.

Decreased grid-current reading. Reduced drive from preceding stage, low emission from tube, low filament voltage, detuned driving circuit, or increased plate current. Grid current varies *inversely* with plate current. Consequently, anything that causes an increase in plate current reduces grid current.

Zero grid-current reading. Lack of filament voltage, burned-out tube, open grid circuit, no drive from preceding stage if fixed bias is used, or an open meter circuit.

Increased plate-current reading. Detuned plate circuit, loss of RF drive if grid-leak bias is used, gassy tube, increased coupling to following stage, loss of fixed bias supply (if used), or too high screen-grid voltage.

Decreased plate-current reading. Low emission, low supply voltages (plate, screen, or filament), reduced coupling to plate circuit load, excessive bias, or open antenna circuit.

Zero plate-current reading. Burned-out amplifier tube, blown fuse or tripped circuit breaker, no plate- or screen-grid voltage, open circuit, shorted plate bypass capacitor, or a failure of the power supply.

ANTENNA

Increased antenna-current reading. The most likely cause of this indication would be an increase in the output power of the power amplifier due to an increase in power supply voltage to the final. This is an unlikely event.

Decreased antenna-current reading. This is the most likely indication to be observed and results from such things as reduced coupling to the final tank circuit or increased coupling. Detuning of the final plate tank circuit, reduced supply voltages, weak tubes, and insufficient drive to the RF amplifiers are other causes. In fact, anything that causes a reduction in the power developed by the final amplifier will result in a decreased antenna-current reading.

Zero antenna-current reading. This indication would generally mean that the power amplifier stage is not delivering RF power. A broken antenna connection may also produce such an indication. Another possibility is a burned-out RF ammeter.

Commercial License Questions

Sections in which answers to questions are given appear in parentheses. A bracketed number following a question implies that it applies only to that element.

1. What are the lowest radio frequencies useful in radio communication? (17–1)
2. What radio frequencies are useful for long-distance communications requiring continuous operation? (17–1)
3. What frequencies have substantially straight-line propagation characteristics analogous to that of light waves and unaffected by the ionosphere? (17–1)
4. What is meant by *carrier frequency*? (17–1)
5. Define the following types of emission: A_0, A_1, A_2, A_3, A_4, A_5. (See Appendix D)
6. What are the disadvantages of using a self-excited-oscillator type of transmitter? (17–2)
7. Describe the effect of a swinging antenna on the output of a simple oscillator. (17–2, 17–3)
8. What is the effect of too much coupling between the output of an oscillator and an antenna? (17–2)
9. Draw a simple schematic diagram showing a method of coupling the RF output of the final power amplifier stage of a transmitter to an antenna. (17–2, 17–3)
10. What is the purpose of a buffer amplifier stage in a transmitter? (17–3, 17–5)
11. What are the advantages of a master oscillator power amplifier type of transmitter as compared to a simple oscillator transmitter? (17–3)
12. What is meant by *break-in operation* at a radiotelegraph station, and how is it accomplished? (17–4)
13. What materials are used for relay contacts? (17–4)
14. What is meant by self-wiping contacts as used in connection with relays? (17–4)
15. What is the principal advantage of a tetrode over a triode as an RF amplifier? (17–5)
16. What operating conditions are favorable to harmonic generation in an RF doubler amplifier? (17–5)
17. Discuss the following with respect to harmonic attenuation properties in a transmitter: link coupling, tuned circuits, degree of coupling, bias voltage, decoupling circuits. (17–5, 17–6, 17–8, 17–20)
18. What is the purpose of an RF amplifier in a transmitter? (17–6)
19. Why is it advisable to reverse periodically the polarity of the filament potential of high-power vacuum tubes when a dc filament supply is used? (17–6)
20. What is the purpose of a dummy antenna? (17–7)
21. What is the primary purpose of a grid leak in a vacuum-tube transmitter? (17–8)

Commercial License Questions 361

22. Suppose that a class C amplifier, using grid-leak bias only, lost its RF grid excitation. What would happen? (17–8)
23. Draw a diagram of a key-click filter. (17–10)
24. What are the various points in a radiotelegraph transmitter where keying can be accomplished? (17–10 through 17–14)
25. Why might an RF amplifier tube have excessive plate current? (17–8, 17–22)
26. Why should all exposed metal parts of a transmitter be grounded? (17–18, 17–19)
27. Explain the operation of an overload relay. (17–19, 17–22)
28. What is the purpose of a Faraday screen between the final tank inductance of a transmitter and the antenna inductance? (17–19)
29. What is the purpose of an electrostatic shield? (17–19)
30. Describe three methods for reducing the RF harmonic emission of a radiotelephone transmitter. (17–18, 17–19)
31. In the adjustment of a radiotelephone transmitter, what precautions should be observed? (17–21)
32. Describe the order in which circuits should be adjusted in placing a transmitter in operation. (17–21)
33. What effect on the plate current of the final amplifier will be observed as the antenna circuit is brought into resonance? (17–21)
34. If the plate current of the final RF amplifier in a transmitter suddenly increased and radiation decreased although the antenna circuit was in good order, what would be the possible causes? (17–22)
35. In a class C radio-frequency amplifier stage of a transmitter, if plate current continues to flow and RF energy is still present in the antenna circuit after grid excitation is removed, what defect would be indicated? (17–22)
36. What would be the possible indications that a vacuum tube in a transmitter had subnormal filament emission? (17–22)
37. What are some possible indications of a defective transmitting vacuum tube? (17–22)
38. A MOPA-type transmitter has been operating normally. Suddenly the antenna ammeter reads zero, although all filaments are burning and plate and grid meters are indicating normal voltages and currents. What would be the possible causes? (17–22)

18

Amplitude Modulation

18-1 BASIC MODULATION CONCEPTS

A sine wave conveys very little information because it continuously repeats itself. The carrier of a transmitter is such a wave and will not produce any sound in the loudspeaker of an ordinary receiver. If it is to carry more information than by code, some part of the wave must be made to vary in accordance with the information to be transmitted. One way of doing this is to cause the *amplitude* of the carrier to vary with the information to be transmitted. This is called *amplitude modulation* (AM). An AM wave is defined as one whose envelope contains a component similar to the waveform of the signal to be transmitted. Also, the amplitude of the carrier is varied by the *strength* of the transmitted signal.

An example of an RF carrier that is being amplitude modulated is shown in Fig. 18-1. The unmodulated carrier is seen in a, the modulating signal in c, and the resultant modulated carrier in b. Suppose that the carrier frequency is 1 MHz (approximately the center of the broadcast band) and a 1-kHz tone is to be transmitted. This results in 1000 RF cycles to every audio cycle. The RF cycles will be made to vary in amplitude with the AF modulating signal, as shown in Fig. 18-1b. When this kind of wave is received at a station, it is *demodulated or detected*, amplified, and connected to a loudspeaker, which reproduces the original tone.

When two different frequencies are present simultaneously in a *linear circuit*, there is no interaction between them. The resultant current or voltage will be the sum of the instantaneous values of the two at every instant. This results because there can only be one value of current or voltage at any part of the circuit at any moment of time. Hence, a linear circuit (one in which Ohm's law holds) cannot be used to effect modulation. This holds true

Basic Modulation Concepts 363

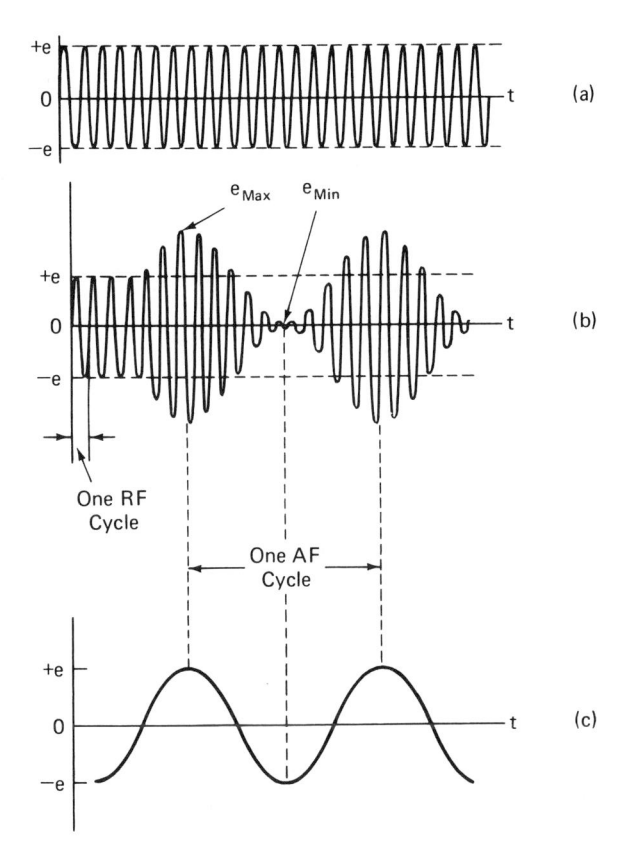

FIG. 18-1 (a) unmodulated CW carrier; (b) modulated RF carrier; (c) audio frequency sine wave producing the modulation.

for any active device operating only over the *linear part* of its dynamic transfer characteristic curve.

If the operating point is shifted to the lower end of the curve, *nonlinear* operation results. If two different frequencies are applied to such an amplifier, they will interact with each other and new frequency components will be generated. The output signal will contain frequencies *that were not present* in the input signal. Two of these newly generated frequencies are of great significance in amplitude modulation. One has a frequency equal to the *sum* of the two original frequencies; the other is equal to the *difference* between the original two. This process of developing sum and difference frequencies is called *heterodyne action* or *mixing*, and is inherent in amplitude modulation.

The new frequencies are called *beat frequencies*. In modulation parlance, the more specific term for the new sum frequency is *upper side frequency* and the new difference frequency is called the *lower side frequency*.

In practice, the modulation signal usually consists of a number of frequencies, as found in speech or music. These may properly be referred to as bands of frequencies. In these instances the side frequencies are grouped into the upper and the lower sideband.

The relationship of the carrier, audio, and sideband frequencies is illustrated in Fig. 18-2. It is assumed that the carrier is 100 per cent modulated, as shown in Fig. 18-1b, and operates on a frequency of 1000 kHz at

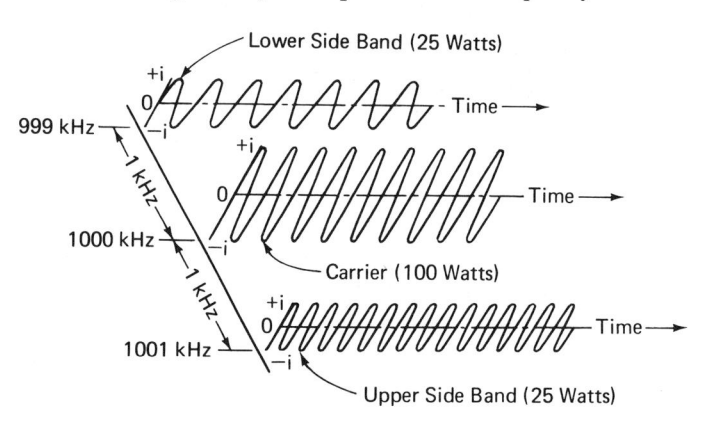

FIG. 18-2 Carrier wave and associated sideband frequencies.

100 W. The audio modulating frequency is a 1-kHz tone at 50 W. Each sideband is displaced 1 kHz from the carrier frequency, as shown in Fig. 18-2. The power in each sideband is *one half the audio modulating power* of 50 W, or 25 W. Therefore, *one sixth* the total transmitter output of 150 W is present in each sideband.

Note that the amplitude of each of the three frequencies (when considered alone) is constant. However, because these frequencies appear simultaneously at the output, they add to form the modulation envelope shown in Fig. 18-1b. The carrier and the sidebands are not merely a mathematical abstraction; they may be separated from one another by suitable filters and used individually.

In an AM wave *only the sidebands contain the intelligence* to be transmitted; the audio frequency itself is *not* transmitted. Because the sidebands contain the modulating frequencies, power in them should be as high as possible.

The modulation envelope rises and falls in amplitude as the continual phase shift between the carrier and sidebands causes these signals to first aid and then oppose one another. These cyclical variations in amplitude of the envelope have the same frequency as the modulating voltage. It is this characteristic variation in amplitude which often gives rise to the descriptive, but oversimplified and erroneous, concept of the audio intelligence riding or being superimposed on the carrier.

18-2 BANDWIDTH OF AMPLITUDE-MODULATED WAVE

When a carrier is modulated, the sideband frequencies *use up a portion of the frequency spectrum*. This is referred to as the *bandwidth* of the signal and is a function of the frequencies contained in the modulating signal. For example, when a 100-kHz carrier is modulated by a 5-kHz audio tone, sideband frequencies are created at 95 and 105 kHz. This signal requires 10 kHz of space in the spectrum. If the same 100-kHz carrier is modulated by a 10-kHz audio tone, sideband frequencies will appear at 90 and 110 kHz, and the signal will have a bandwidth of 20 kHz. Some broadcast stations may cause interference on adjacent channels if they are modulated with 10-kHz audio signals. From these examples it can be seen that at any instant the bandwidth of an AM wave is two times the highest modulating frequency. Thus, if a 400-kHz carrier is modulated with 3, 5, and 8 kHz simultaneously, sideband frequencies will appear at 392, 395, 397, 403, 405, and 408 kHz. This signal extends from 392 to 408 kHz and has a bandwidth of 16 kHz, twice the highest modulating frequency of 8 kHz.

Musical instruments produce complex sound waves containing a great number of frequencies. To transmit music with a high degree of fidelity, modulating frequencies up to 15 kHz must be included. This requires a bandwidth of at least 30 kHz. If only voice frequencies are to be transmitted, the bandwidth requirements are less. Intelligible voice communications can be carried out as long as the communications system retains audio frequencies up to several thousand hertz per second.

18-3 BASIC MODULATION SYSTEM

There are several ways in which the AF signal can be combined with the carrier. If the audio signal is applied to the grid of one of the RF amplifiers, it is known as *grid modulation*. The audio signal is superimposed on the bias of the stage and modulates the RF carrier. If the modulating signal is applied to the plate of the tube, it is called *plate modulation*. Thus, the common methods of modulation are *plate, control-grid, screen-grid, suppressor-grid,* and *cathode modulation*. A modulation method also exists in which the modulating signal is applied to both screen grid and plate.

A block diagram of an AM radiotelephone transmitter is shown in Fig. 18-3. The RF section generates the high-frequency carrier, amplifies it, and connects it to the antenna. The AF part of the transmitter includes three basic sections: *speech amplifier, driver,* and *modulator*. The speech amplifier builds up the weak microphone signals to a value sufficient to provide proper *excitation* to the driver. The driver consists of a power amplifier to convert the AF signal into a relatively large voltage and appreciable current to drive the modulator. The modulator stage is usually a class B amplifier designed to provide considerable AF power output. The output of this stage is coupled by means of a *modulation transformer* to the RF power amplifier in such a

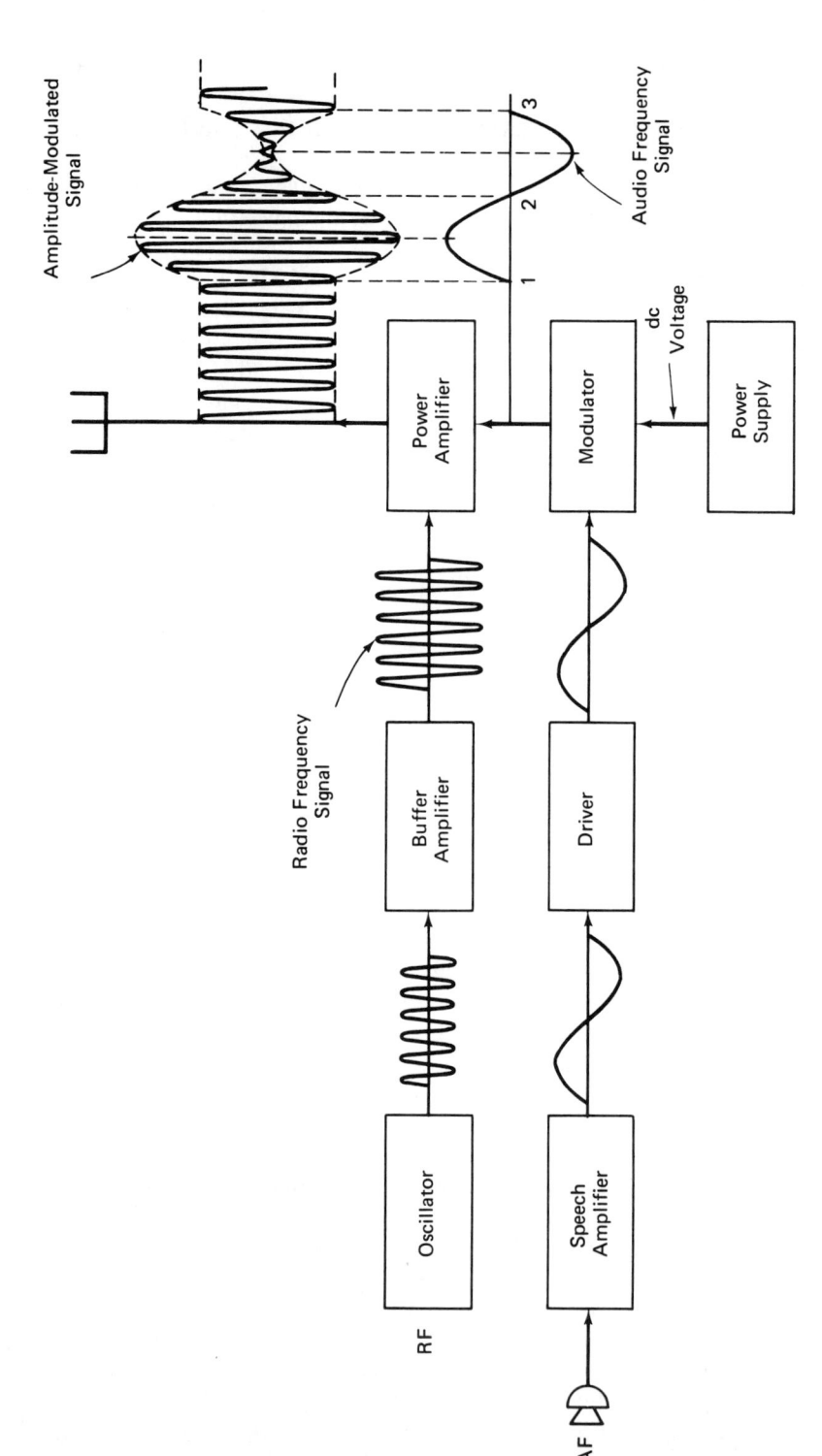

Fig. 18-3 Block diagram of an AM radiotelephone transmitter.

Plate Modulation 367

way as to alternately add to and subtract from the plate voltage of the RF amplifier.

This action causes the amplitude of the RF field at the antenna to increase during the time the audio cycle is going positive. During the negative half of the cycle, it subtracts from the plate supply voltage to the final, thus reducing the radiated RF power.

The waveforms in Fig. 18-3 indicate this action. The power supply furnishing the dc power to the modulator also supplies the RF power amplifier. The dc power to this latter stage flows through the secondary winding of the modulation transformer. Consequently, the AF output of the modulator is *superimposed* on the dc supply to the RF power amplifier. This modulates the RF carrier at the AF rate.

18-4 PLATE MODULATION

The most common method of obtaining an AM wave consists in plate modulating a class C amplifier. A typical circuit arrangement illustrating this is shown in Fig. 18-4. The modulator is a class B amplifier connected to modu-

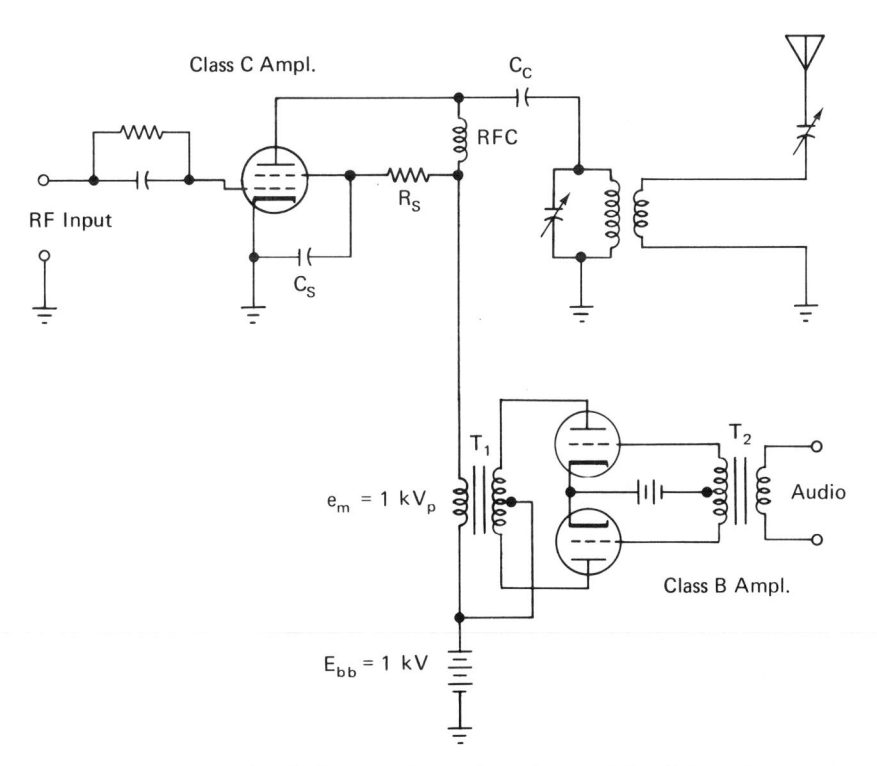

FIG. 18-4 Circuit diagram of a typical plate modulated class C amplifier.

lation transformer T_1. The load impedance for the secondary of this transformer is the plate circuit of the class C amplifier.

The *RFC* must offer a high impedance to RF voltages, but a low impedance to modulating frequencies. This allows the AF to pass through to the plate circuit of the final, but keeps RF out of the modulation transformer. The coupling capacitor, C_c, must be large enough to permit the RF output of the final to pass on to the plate tank circuit, but small enough to be a high impedance at modulating frequencies.

When a plate-modulated class C radio-frequency amplifier uses a tetrode, it is important that the *screen voltage* be modulated simultaneously with the *plate voltage*. This causes the total space current to increase at the positive peak of the modulating voltage when the plate current has to be greatest, and vice versa. By varying the total space current in accordance with the plate requirements, the screen currents, and hence the screen circuit power loss, are minimized. If the screen voltage is not modulated, the space current tends to be constant throughout the modulation cycle, making modulation difficult, if not impossible. The modulating voltages are applied to the screen grid through a series resistor R_s, as shown in Fig. 18-4. Screen bypass capacitor C_s must be small enough to not effectively bypass the high-

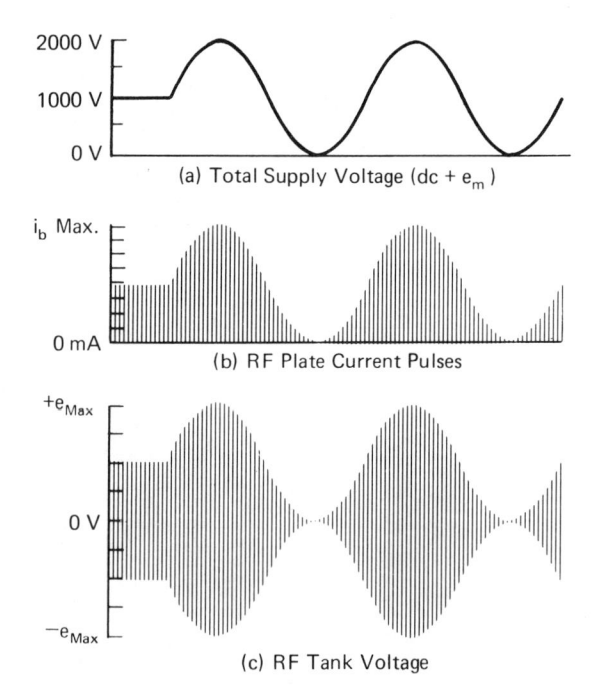

Fig. 18-5 Current and voltage waveforms in a class C plate modulated RF power amplifier.

Percentage of Modulation 369

est modulating frequency. Simultaneously, C_s must be sufficiently large to effectively bypass the carrier frequency.

The current and voltage waveforms in a class C plate-modulated RF amplifier are shown in Fig. 18-5. Part a of this illustration shows the effects of superimposing the modulating voltage e_m on the supply voltage to the final power amplifier (FPA). This shows that the voltage varies between 0 and 2 kV in accordance with the modulating signal.

When the modulating voltage is applied to the plate of the FPA, the plate current pulses vary as shown in Fig. 18-5b. The instantaneous products of the voltage and current shown in Figs. 18-5a and b produce a waveform in the plate tank, as seen in Fig. 18-5c. This represents 100 per cent modulation.

18-5 PERCENTAGE OF MODULATION

A 100 per cent modulated carrier envelope is shown in Fig. 18-6a; the carrier and sidebands are shown in b. Since for 100 per cent modulation the peak audio modulating voltage is approximately equal to the peak RF volt-

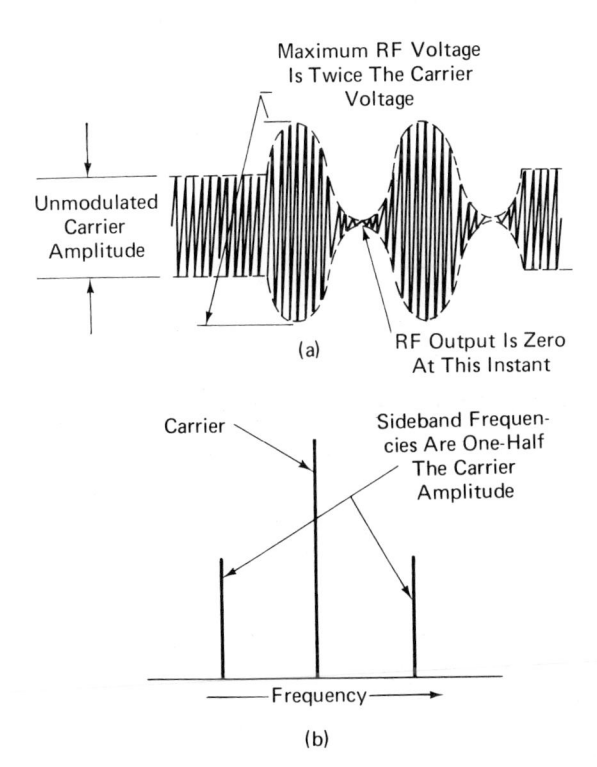

FIG. 18-6 Conditions for 100% modulation.

age, the combined sideband voltage is equal to the carrier voltage. Because the sideband voltage is divided between two sideband frequencies, at 100 per cent modulation each side frequency has an amplitude equal to one half the amplitude of the carrier.

If the audio modulating voltage is increased beyond the amount required to produce 100 per cent modulation, the negative peak of the modulating signal becomes larger in amplitude than the dc plate supply voltage to the FPA. This causes the final plate voltage to be negative for a short period of time near the negative peak of the modulating signal. During this interval, no RF energy is developed across the plate tank circuit, and the output voltage remains at zero, as shown in Fig. 18-7.

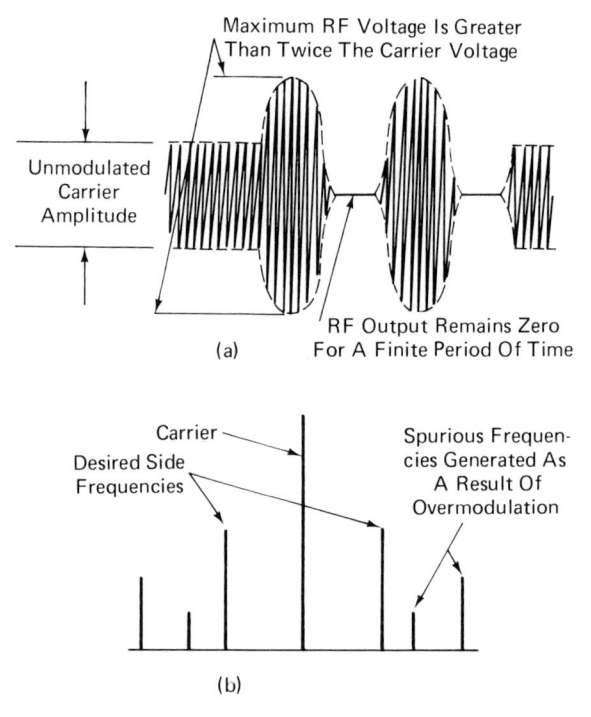

FIG. 18-7 Distortion and interference produced because of overmodulation.

If a receiver is tuned to a frequency near, but somewhat outside, the channel on which the transmitter is operating, overmodulation is found to generate unwanted sideband frequencies that appear for a considerable distance above and below the desired channel. This effect is sometimes called *splatter*. These unwanted or *spurious* frequencies, shown in Fig. 18-7b, cause interference to other stations operating on adjacent channels. Hence, *overmodulation* and its attendant distortion and interference are to be avoided.

Example 18-1

A modulated carrier has a peak amplitude of 100 V when the unmodulated carrier amplitude is 200 V. Determine the percentage of modulation.

Solution

$$\%M = \frac{e_{max}}{e_{car}} \times 100$$

$$\%M = \frac{100}{200} \times 100 = 50\%$$

18-6 SIDEBAND POWER

A transmitter is normally rated according to the amount of *unmodulated carrier power* it delivers to an antenna. This is called the station's *authorized power*. The sideband power generated as a result of modulation exists in addition to the unmodulated carrier power. At 100 per cent modulation the sideband power is 50 per cent of the unmodulated carrier power. If the same carrier is modulated 50 per cent, the sideband power *decreases* to one fourth the amount obtained at 100 per cent modulation. This illustrates the importance of a high percentage of modulation.

When a carrier is modulated 100 per cent, the total radiated power is 1.5 times the unmodulated carrier. A 50 per cent modulated carrier is shown in Fig. 18-8.

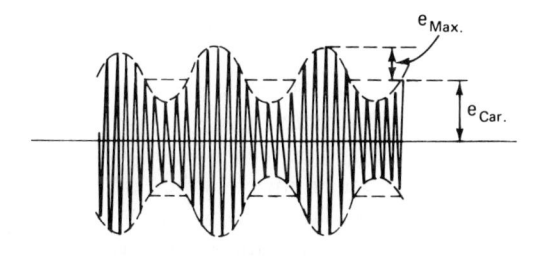

FIG. 18-8 RF carrier showing 50% modulation.

18-7 MODULATOR POWER REQUIREMENTS

The power output of a plate-modulated stage is contained in the carrier and sidebands; the carrier power must come from the FPA and the *sideband power from the modulator*. An illustration showing the power distribution for a 400-W, plate-modulated transmitter is shown in the block diagram of Fig. 18-9. Assume that the transmitter is modulated 100 per cent and that the FPA and modulator stages have efficiencies of 80 and 40 per cent, respectively.

If the modulating signal is reduced to zero, only the carrier frequency

AMPLITUDE MODULATION

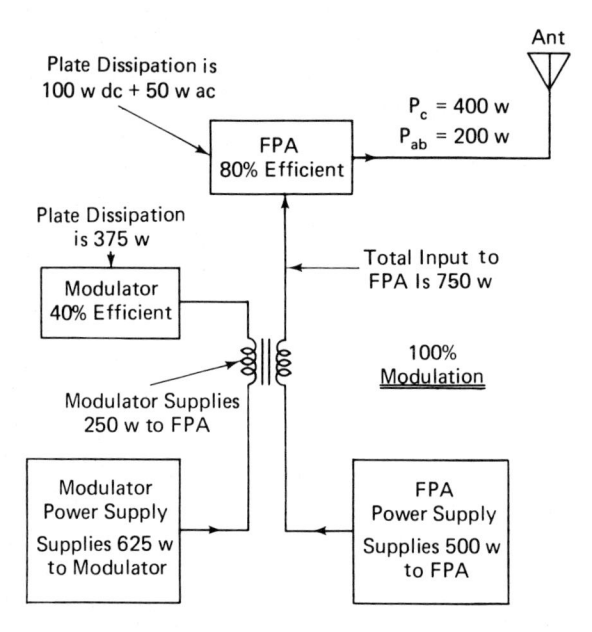

```
Plate Dissipation is                               Ant
100 w dc + 50 w ac
                      ┌─────────────┐   P_c = 400 w
                      │    FPA      │   P_ab = 200 w
                      │ 80% Efficient│
                      └─────────────┘
Plate Dissipation           ↑
    is 375 w                │
       ↓                    │
┌─────────────┐             │  Total Input to
│  Modulator  │─────┐       │  FPA Is 750 w
│ 40% Efficient│    │       │
└─────────────┘    ))||((
                   ))||((   100%
                   ))||((   Modulation
Modulator Supplies  │
   250 w to FPA     │
┌─────────────┐     │     ┌─────────────┐
│  Modulator  │     │     │    FPA      │
│Power Supply │     │     │Power Supply │
│Supplies 625 w│────┘     │Supplies 500 w│
│ to Modulator│           │   to FPA    │
└─────────────┘           └─────────────┘
```

FIG. 18-9 Power distribution in a plate-modulated stage.

will exist in the output of the FPA. Since this amplifier operates class C and has an efficiency of 80 per cent, some power is dissipated at the plate of the tube, and the power supply must deliver more than 400 W of carrier power. The power drawn from the FPA power supply is

$$P_i = \frac{P_o}{\text{efficiency}} = \frac{400}{0.80} = 500 \text{ W}$$

The difference between the 500-W dc power input and the 400-W RF power output represents the *plate dissipation* of the FPA.

When the transmitter is 100 per cent modulated, the two sideband frequencies contain one half as much power as the carrier. Since the sidebands occur as a result of modulation, the power which they contain is supplied by the modulator. The sideband power at the antenna is 200 W. This represents 80 per cent of the audio power applied to the final power amplifier plate circuit by the modulator. The exact amount of power the modulator must supply for 100 per cent modulation can be calculated as

$$P_i = \frac{P_o}{\text{efficiency}} = \frac{200}{0.80} = 250 \text{ W}$$

Therefore, the modulator must supply 250 W to the plate of the FPA in order to produce 100 per cent modulation. Of this 250 W only 200 reach the antenna, with 50 W being lost at the plate of the FPA as plate dissipation.

Unlike the final, the modulator likely operates class AB or B. In the illustration the modulator has an efficiency of 40 per cent. Therefore, only

Modulator Power Requirements

250 W of audio power is obtained from the 625 W of dc power supplied to the plate of the modulator. Modulator plate dissipation represents the difference between 625 W of input power and 250 W of audio output, or a 375-W loss. Of the 625 W of input power to the modulator, only 200 W are converted into useful sideband energy. This increase in power will naturally cause the antenna current to rise. *The change in antenna current can therefore be used as a measure of the modulation level.*

At the crest of the modulation cycle, the amplitude of the modulated wave rises to *double* the unmodulated value (100 per cent modulation). The instantaneous peak power in this case will be *four times* the unmodulated value. Therefore, it is necessary to reduce the ratings of tubes and semiconductors when they are used as modulators, compared to their service as class C radio-frequency power amplifiers.

The preceding explanations have been predicated on the use of a *sinusoid* as the modulating waveform. When *complex waves* are used, no such simple relationships exist. Complex waves, such as speech or music, do not as a rule contain as much average power as a sine wave. Ordinary speech waveforms have about *one half* as much average power as a sine wave for the same peak amplitude in both waveforms. Thus, for the same modulation percentage, the sideband power with ordinary speech will average only about half the power with sine-wave modulation, since it is the peak envelope amplitude, not the average power, that determines the percentage of modulation.

When choosing a modulation transformer, its power-handling capabilities and its impedance-matching characteristics must be considered. For example, a class C amplifier with a plate voltage of 1000 V and a plate current of 150 mA is to be modulated by a class A amplifier with a plate voltage of 2000 V, plate current of 200 mA, and plate impedance of 15 kΩ. What is the proper turns ratio for the coupling or modulation transformer?

The plate impedance of the class C amplifier is

$$Z_p = \frac{E_b}{I_b} = \frac{1000}{0.15} = 6667 \; \Omega$$

The amplifier should work into an impedance approximately *twice* that of the plate resistance of the stage. Therefore, with a modulator plate resistance of 15 kΩ, the load should ideally be 30 kΩ. Therefore, the impedance ratio between the modulated class C amplifier and the modulator is

$$Z_{ratio} = \frac{30,000}{6667} = 4.5 : 1 \quad \text{step down}$$

The turns ratio of the modulation transformer will be equal to the square root of the impedance ratio, or

$$\text{turns ratio} = \sqrt{4.5} = 2.12 : 1 \quad \text{step down}$$

18-8 MODULATION LEVEL

The modulating signal can be introduced into any active element of a tube or transistor. In addition, the modulating signal can also be applied to any of the RF stages in the transmitter.

A modulation circuit can be classified into one of two categories according to the level of the carrier wave at the point in the system where the modulation is applied. The FCC defines *high-level modulation* as "modulation produced in the plate circuit of the last radio stage of the system." This same document defines *low-level modulation* as "modulation produced in an earlier stage than the final." The regulation does not define what classification is involved when the modulating signal is applied to some element in the FPA other than the plate.

The popularity of plate modulation comes from its high plate circuit efficiency, good linearity, and simplicity of adjustment. The only disadvantage is the large amount of audio power required from the modulator.

Broadcast stations are permitted certain percentages of distortion for various levels of modulation. For example, the equipment must be capable of at least *95 per cent modulation*. When the percentage of modulation is between 85 and 95 per cent, the total harmonic content of the output power must not exceed 7.5 per cent. When the modulation is less than 85 per cent, the harmonic content must not exceed 5 per cent. In operation the modulation percentage must be maintained as high as possible, *consistent with good quality of transmission*, and in no case less than 85 per cent on positive peaks and no more than 100 per cent on negative peaks.

18-9 PLATE AND SCREEN-GRID MODULATION

The plate current of a screen-grid tube is almost independent of plate voltage, but very much dependent on screen-grid voltage. Thus, the modulating voltage is normally applied to the screen and plate circuit simultaneously when plate modulating a screen-grid tube.

A combination plate- and screen-grid-modulated amplifier is shown in Fig. 18-10. Notice that the modulating voltage across the secondary of the modulation transformer T_1 is in series with the dc supply to both the plate and screen circuits.

To prevent screen degeneration at the carrier frequency, capacitor C_2 is connected between the screen grid and cathode of the tube. The value of this capacitor is chosen so that its reactance approaches a short circuit at radio frequencies, but appears as an open circuit to the AM frequencies.

Capacitor C_3 serves a purpose similar to that of capacitor C_2. It must prevent RF from developing across the secondary of T_1 while having little if any effect on the audio.

Control-Grid Modulation

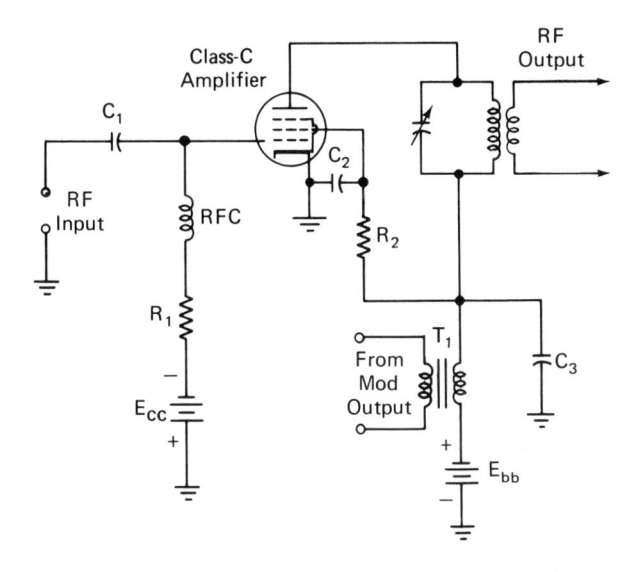

FIG. 18-10 Circuit employing a combination of plate- and screen-grid modulation.

18-10 CONTROL-GRID MODULATION

A disadvantage of plate modulation is the large amount of power required of the modulator. By applying the modulating voltage to the grid circuit of the RF amplifier, the modulator power requirements can be reduced considerably.

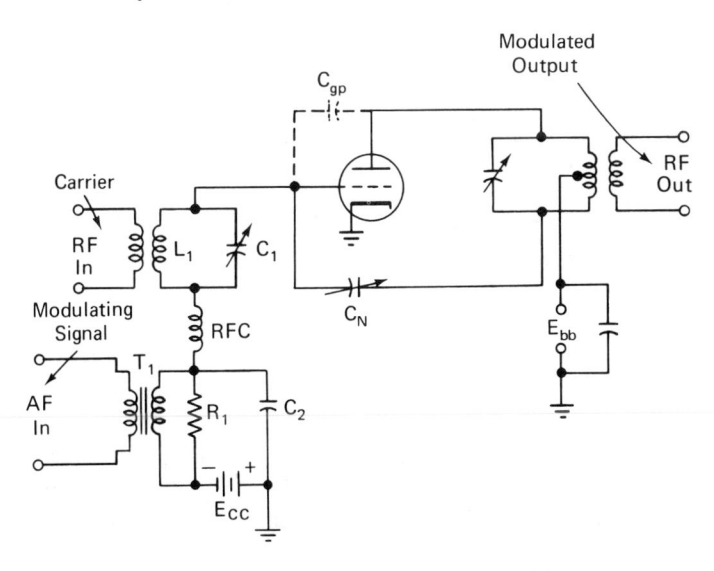

FIG. 18-11 Typical grid-modulation circuit.

A circuit for producing control-grid modulation is shown in Fig. 18-11. The circuit resembles that of a normal class C amplifier, except for the inclusion of modulation transformer T_1 in series with the grid bias supply E_{cc}. During periods of 100 per cent modulation, the most positive peaks of the grid signal will drive the grid positive and draw grid current. This places a heavier load on both the modulator and the grid bias supply during the time interval that grid current is drawn. To prevent fluctuations in bias as a result of grid current, the bias supply must be *well regulated*.

To minimize the loading effect that grid current has on the modulator, the modulator stage should have a *low internal impedance*. This can be accomplished by using negative feedback loops in the modulator amplifier circuits. Additional stability can be gained by connecting a resistor, R_1, across the secondary of the modulation transformer. This places a constant load on the modulator and makes the grid current a small portion of the total load on the modulator.

The operation of a grid modulator circuit can be better understood by referring to Fig. 18-12. This shows the dynamic transfer curve for the modu-

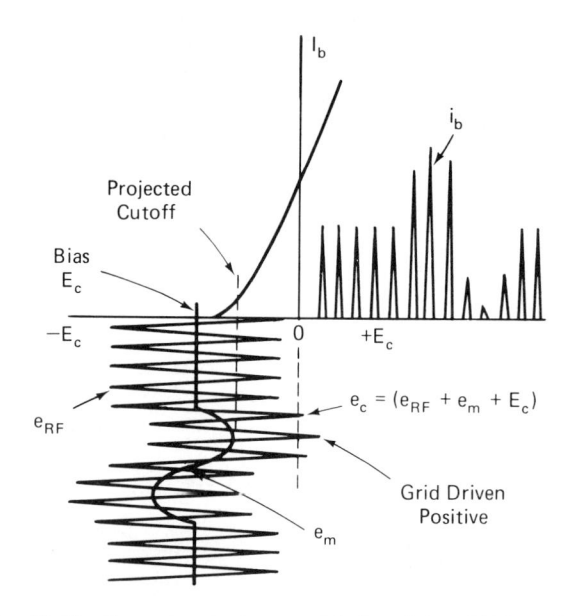

FIG. 18-12 Curve showing how modulation of the grid bias voltage produces amplitude modulate plate current pulses.

lated RF amplifier. The bias voltage is adjusted such that the characteristics of the circuit are essentially those of a class C amplifier. If the grid bias is properly adjusted, there will be little evidence of any variation of plate current when modulation takes place. If distortion is present, this is not true.

Figure 18-12 shows that the positive peaks of the carrier wave shift

Suppressor-Grid Modulation

upward on the curve during the positive alternation of the modulating signal. To allow for this, the amplitude of the carrier applied to the grid is smaller than would be the case for a plate-modulated amplifier. When the carrier is unmodulated, the peak amplitude of the plate current pulses is only about *one half* the amplitude they have at the peak of the modulating signal. This means that the carrier power obtainable is considerably less than the amount the tube can produce as a conventional class C amplifier. The small amplitude of carrier input also results in decreased efficiency.

When compared with plate modulation, the grid-modulated class C amplifier has the advantage of requiring only a small amount of modulating power. However, it has the disadvantage of *much lower plate efficiency*. Also, the grid-modulated amplifier tends to be more *nonlinear* in its characteristic; that is it has higher amplitude distortion. It is also more critical in adjustment than is the plate-modulated amplifier. For wideband applications, however, notably in TV transmitters, the difficulties in generating the large amount of video modulating power needed for plate modulation make grid modulation standard practice.

18-11 SUPPRESSOR-GRID MODULATION

Amplitude modulation may be obtained in a pentode class C amplifier by applying to the *suppressor grid* a modulating voltage superimposed upon a suitable negative bias. Such an arrangement is shown in Fig. 18-13. This results in a relationship between RF voltage and modulating signal similar to that encountered in a grid-modulated amplifier. The plate efficiency and the

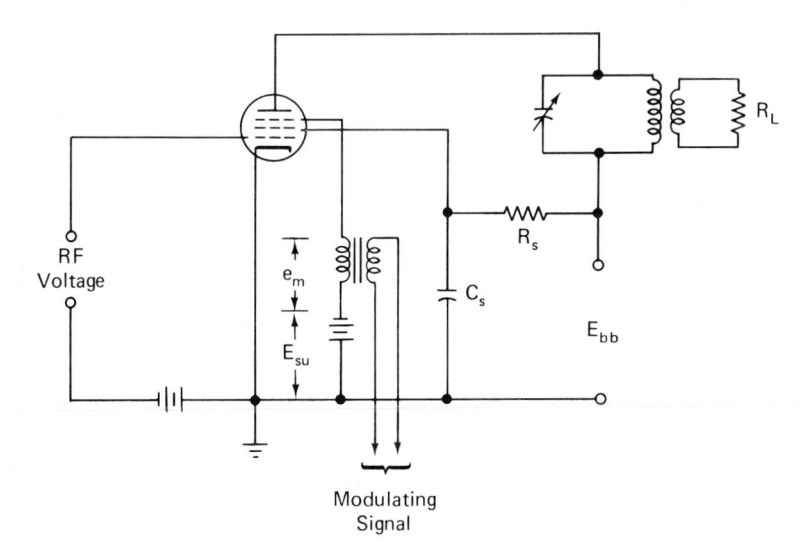

FIG. 18-13 Circuit of suppressor-grid-modulated class C amplifier.

modulating power requirements are about the same for control-grid and suppressor-grid modulation.

There is a tendency in suppressor-grid modulators for the average screen current to be excessively high. This results because a virtual cathode exists in the tube throughout the modulation cycle, except at the positive modulation peaks. To overcome this limitation, the screen-grid voltage can be obtained from the plate supply source through a suitable dropping resistor, such as R_s shown in Fig. 18-13. When the screen-grid current increases, the screen voltage drops thereby limiting the screen losses by simultaneously reducing the screen voltage and current. Capacitor C_s must be large enough to bypass the carrier frequency but not have any effect on the highest modulation frequencies.

18-12 HEISING MODULATION

A unique method of accomplishing plate modulation is known as *Heising modulation*. With this circuit the plate of the modulated stage and the modulator stage are series connected and fed through a common AF inductor. This is known as a *constant-current* system, since the total current through the plate choke remains substantially constant during the modulation cycle.

A Heising modulation circuit is shown in Fig. 18-14. The plate inductor common to both modulator and RF amplifier is L and has an inductance that is very high, on the order of 30 to 200 H. To understand the operation of the system, consider first that no modulating signal is present at the grid of the modulator tube, but that there is RF input to the grid of the FPA. Under this quiescent condition a certain total amount of current will flow through L. When an audio signal is applied to the grid of the modulator tube, fluctuations in its plate current will result. These current variations are in phase with the signal on the grid. Therefore, when the grid of the modulator is driven in the positive direction, the modulator plate current increases. Inductor L tends to oppose this increase by developing a reactive voltage across its terminals. The net result is that the total dc voltage available at the plate of the RF amplifier is reduced by an amount equal to the counter emf developed across the choke. This causes a reduction of plate current in the RF amplifier by an amount that is proportional to the increased modulator plate current. Consequently, the total current flowing through the inductor remains essentially constant.

When the excitation to the modulator grid is swinging in the negative direction, its plate current decreases. The inherent nature of the choke coil tends to oppose this by developing a voltage across its terminals, which aids the supply voltage. This increases the plate potential of the RF amplifier by an amount equal to the reactive voltage across the modulator plate inductor. This increased potential causes the RF amplifier plate current to increase by an amount approximately equal to the decreased modulator plate current.

Heising Modulation

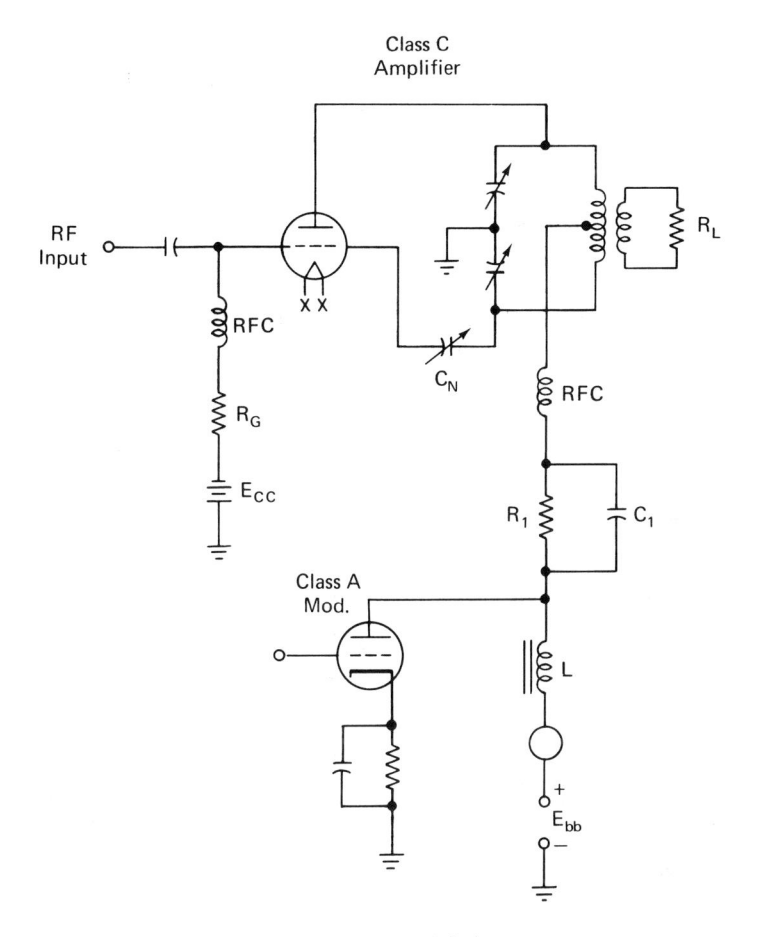

FIG. 18-14 Heising modulation system.

Consequently, the total current through the inductor L remains essentially constant.

In the circuit thus far described, the highest percentage of modulation obtainable would be about 80 per cent. To achieve 100 per cent modulation, the modulating voltage must have a peak value equal to the dc plate potential of the FPA. However, a class A modulator (the type used in this system) cannot develop an output voltage of more than about 80 per cent of the value of its dc plate voltage without generating considerable distortion. To increase the modulation capability to 100 per cent, it is necessary to use a series dropping resistor, R_1, in the plate circuit of the RF amplifier. The IR drop across this resistor will reduce the dc plate voltage on the RF amplifier, but not on the modulator plate. If the power supply delivers 800 V and the modulator can develop 600 V peak, the series resistor must be of sufficient value to provide an IR drop of at least 200 V for 100 per cent modulation.

Resistor R_1 not only reduces the value of the dc plate potential on the FPA, but also causes some drop of modulating voltage supplied to the plate of the FPA. The function of C_1 is to provide an AF bypass around this resistor so as not to attenuate them.

18-13 COLLECTOR MODULATION

The *collector-modulated circuit* shown in Fig. 18-15 is the solid-state equivalent of plate modulation. The collector circuit of Q_2, the FPA, is shunt fed via RFC_2. Transformer T_1 is the modulation matching transformer.

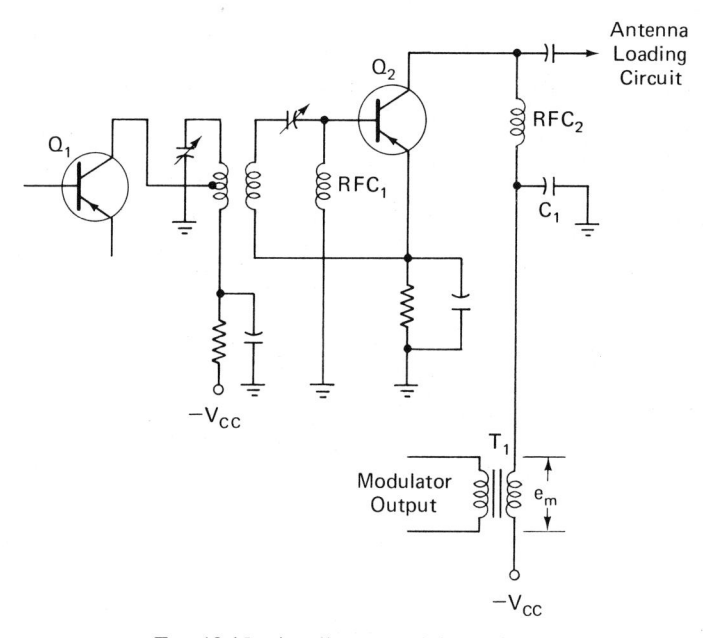

FIG. 18-15 A collector-modulated circuit.

It would be assumed that for 100 per cent modulation the modulating voltage, e_m, would be equal to V_{cc}. However, this condition is not realized in the circuit of Fig. 18-15. Because the transistor saturates during part of the cycle, its output voltage does not rise in proportion to the increase in collector voltage. To achieve high levels of modulation, it is essential that the *driving transistor* Q_1 be modulated, as well as the FPA. A typical circuit that accomplishes this is shown in Fig. 18-16. Observe that a portion of the modulating voltage is tapped down on the secondary of T_1, and supplied, in series with V_{cc}, to the collector of Q_1. Collector-modulation circuits provide high efficiency, low distortion, and ease of adjustment.

Checking Percentage of Modulation 381

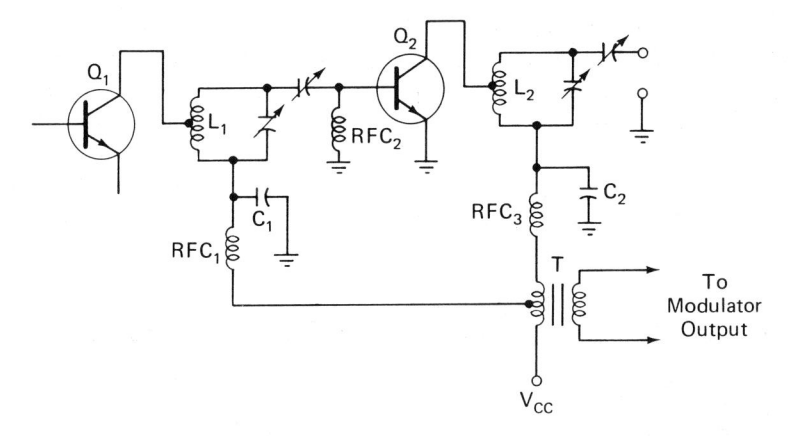

FIG. 18-16 Typical collector-modulation circuit to produce high-level modulation.

18-14 BASE MODULATION

Base modulation possesses the same basic characteristics as grid modulation. The partial schematic of a shunt-fed base-modulation circuit is indicated in Fig. 18-17. Modulation transformer *T* not only furnishes the required

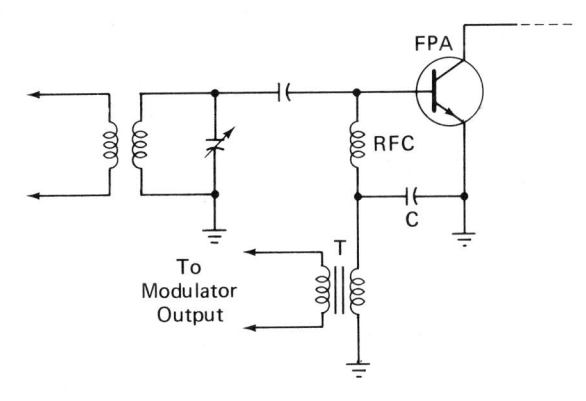

FIG. 18-17 Base modulation circuit.

modulating voltage, but also provides an impedance match between modulator and the base circuit of the FPA. Compared to collector modulation, this circuit has lower collector efficiency, reduced power output for the same transistor type, poorer linearity, and is more difficult to adjust.

18-15 CHECKING PERCENTAGE OF MODULATION

One of the best ways to measure the degree of modulation and check for linearity is by use of the oscilloscope. The scope can be used in several ways, but only the two most common ones will be discussed. One involves the use of *sinusoidal patterns* and the other *trapezoidal patterns* on the scope screen.

Figure 18-18a shows how to check modulation percentage by the use

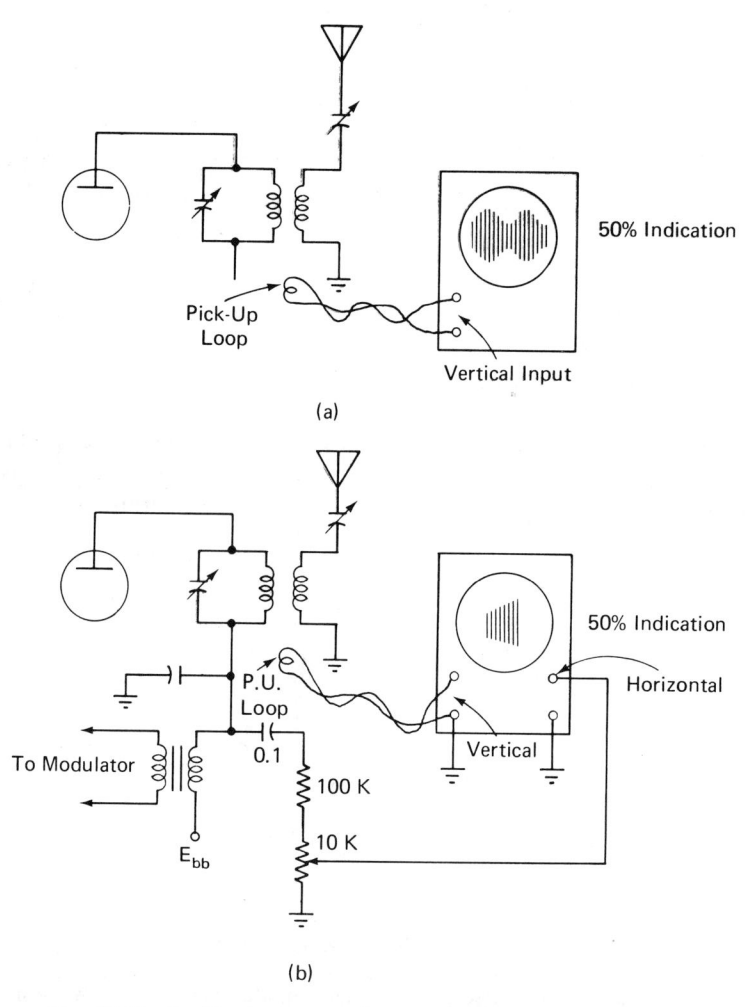

FIG. 18-18 Oscilloscope connections for monitoring modulation: (a) sinusoidal, (b) trapezoidal.

of a sinusoidal scope display. All that is needed is a small loop or coil consisting of two or three turns of wire that are connected to the vertical input. The loop is loosely coupled to the tank of the FPA. *Great care* must be exercised when placing the pickup coil near the tank because of the *dangerous dc and RF potentials* likely to be encountered. The horizontal sweep frequency should be set to the modulating frequency (or some submultiple of it) to

Checking Percentage of Modulation

permit the modulated wave to be seen on the screen as a stationary pattern. A typical scope display for approximately 50 per cent modulation is shown in Fig. 18-18a.

Differing degrees of modulation produce various-shaped patterns, as shown in Fig. 18-19 in the right-hand column. When no modulation

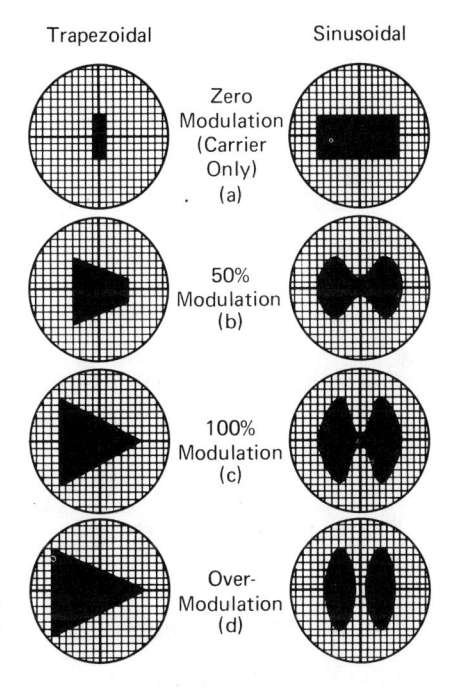

FIG. 18-19 Four possible oscilloscope displays for trapezoidal and sinusoidal patterns.

is present, the display is as shown in a. When the carrier is modulated at 50 or 100 per cent, the pattern takes on the shapes shown in the right-hand column, b and c. Overmodulation is shown in d.

If the transmitter is being modulated by a sine wave, any deviation from a sinusoid, as shown on the CRT, represents *distortion* being produced in the modulation system. The distortion can be located by checking back, stage by stage, until the defective stage is found.

A more accurate analysis of the precentage of modulation can be made from a trapezoidal pattern displayed on the scope. The circuit necessary to produce such a pattern is represented in Fig. 18-18b. The scope is set for *external sweep*, and the modulating voltage provides the sweep when connected to the horizontal input terminals. One of the easiest ways to obtain the sweep voltage is to connect a capacitor to the FPA side of the modulation transformer. The other side of the capacitor connects to any suitable signal-divider network, such as indicated. The capacitor must have a working voltage high enough to withstand the peaks of modulation. It should be obvious that *this connection to the modulation transformer should not be made until* the power supplies for the transmitter are turned off because of the dangerously high voltages involved. The potentiometer is used to adjust the phase of the sweep voltage. Proper phasing is achieved when the CRT display has no three-dimensional effects.

The trapezoidal displays for 0, 50, and 100 per cent modulation and overmodulation are shown in Fig. 18-19, left-hand column. The degree of modulation can be found by the following equation, where A and B are as shown in Fig. 18-20.

$$\%M = \frac{B - A}{B + A} \times 100$$

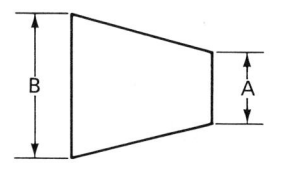

FIG. 18-20 Trapezoidal pattern showing how the percentage of modulation can be determined.

For example, if A is one half the length of B, and B is assumed to have a length of 10, the percentage of modulation is

$$\% M = \frac{10 - 5}{10 + 5} \times 100 = 33\%$$

The trapezoidal patterns have a distinct advantage over sinusoidal patterns when checking for linearity of modulation. The trapezoid will have *straight sides when the linearity is good and curved sides when the linearity is poor.* The sides may curve inward or outward.

18-16 LINEAR RADIO-FREQUENCY AMPLIFIERS

Linear operation is achieved by operating over the straight portions of the $E_g I_p$ curve. Under these conditions, little if any distortion will result. The only amplifiers that produce undistorted output are class A, AB, and B. A class C amplifier is nonlinear because it is operated over the entire portion of the $E_g I_p$ curve.

In high-level modulation systems, where the FPA feeds directly into the antenna, linear amplifiers are not needed. Any distortion of the plate current pulses will be smoothed out by the flywheel effect of the tank circuit. When the modulated stage is not the final amplifier, all the following RF amplifier stages *must be linear;* otherwise, the modulated envelope will be distorted. This means that only class A, AB, or B amplifiers can be used. Because of efficiency, only class AB or B amplifiers would be used. The block diagram for a typical low-level modulated transmitter using a linear amplifier following the modulated stage is shown in Fig. 18-21.

The circuit of a modulated class C pentode RF amplifier followed by a linear, push–pull, class B amplifier is shown in Fig. 18-22. The push–pull arrangement allows for two current pulses to be supplied to the tank circuit

Linear Radio-Frequency Amplifiers 385

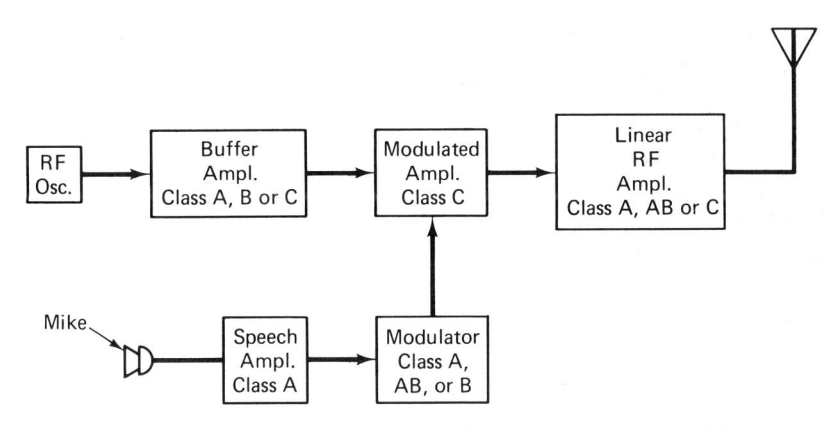

Fig. 18-21 Block diagram of a low-level modulation system followed by a linear RF amplifier.

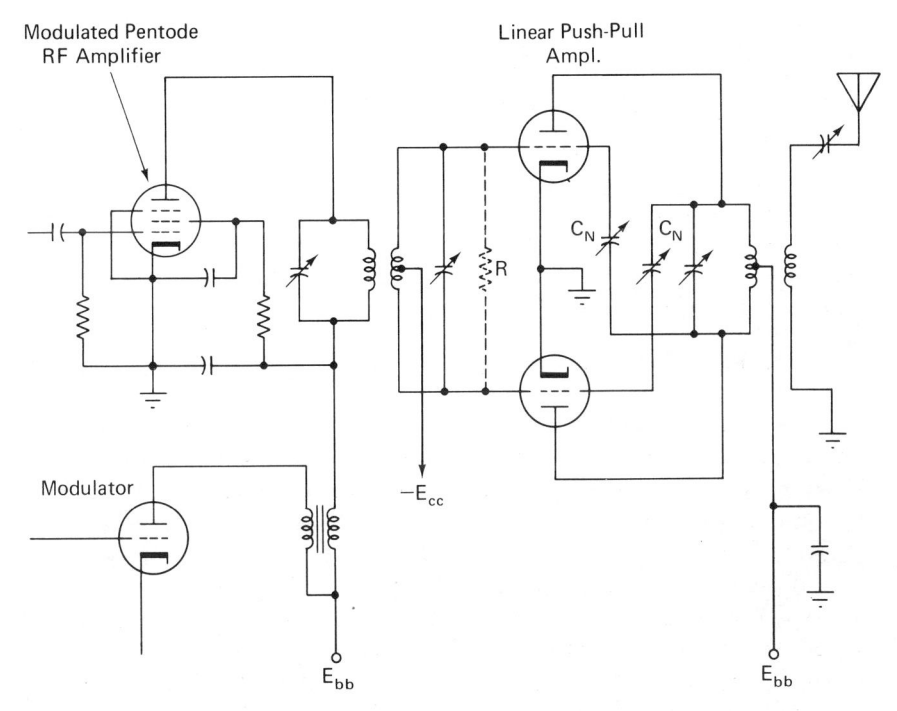

Fig. 18-22 Low-level modulation system using a pentode-modulated RF amplifier and a linear push–pull amplifier.

for each cycle of modulated input. This provides for stronger oscillations in the push–pull tank than could be achieved with single-ended operation using the same tube type. Neutralization is accomplished by the two capacitors designated C_N in the schematic.

To produce modulation without distortion, the modulated stage must work into a *constant load*. To ensure this condition, it is common practice to place a noninductive resistor across the grid tank of the push–pull class B, linear amplifier. This is shown as a dotted resistor R in the diagram. Its value is chosen so as to reflect the desired impedance into the plate tank circuit of the modulated stage.

The bias setting of the linear, class B, amplifier is critical. The unmodulated carrier input must not drive the tubes into saturation. If it should, then when modulation is present the tubes are driven into saturation and the positive modulation peaks will be flattened, although the negative peaks will be produced in the output. This results in severe distortion. An indication of this can be seen on a plate-current meter, which will *decrease* when modulation is present. This causes the antenna current to also go down. Under proper operating conditions *there should be very little, if any, change in plate current*, whether the transmitter is modulated or not.

A linear amplifier possesses some of the characteristics of a grid-modulated stage; that is, the amplitude of the plate current pulses vary directly as the excitation on the grid. Doubling the amplitude of the excitation voltage doubles the amplitude of the plate current pulses. Because the RF power developed in the load is equal to the square of the current, it is apparent that there is a fourfold increase in power.

18-17 CARRIER SHIFT

A symmetrically modulated carrier has equal positive and negative pulses. Consequently, the average carrier voltage remains constant, whether modulation is present or not. A simple circuit to detect *carrier shift* was shown in Section 15-21.

If the modulated carrier swings more in the positive direction than in the negative, the result is a shifting of the average carrier voltage in the positive direction. This is known as *positive carrier shift*. If the opposite situation prevails, it is known as *negative carrier shift* or downward modulation.

Distortion is produced with carrier shift. A small amount, perhaps up to 5 per cent, is scarcely detectable to the ear. Excessive negative shift produces the same effect as though the modulation were low.

Negative or positive carrier shift *does not* denote a change of carrier frequency, although added harmonic frequencies are produced. It only implies that there is an average increase or decrease in carrier voltage. This produces a corresponding change on the dc plate-current meter of the FPA and affects the output power.

There are many circuit or component malfunctions that will cause a deviation on the carrier-shift indicator. Some of these are listed for high- and low-level modulated stages.

Plate modulation—decrease in antenna current:

1. Insufficient bias on the modulated RF amplifier.
2. Excessive overloading of the class C modulated RF amplifier.

Antenna Ammeter Indications

3. Incorrect load impedance for the class C modulated RF amplifier.
4. Insufficient excitation to the modulated stage.
5. Defective tube.
6. Poor regulation of the power supply.

Plate modulation—increase in antenna current: This is a normal situation. However, an upward "kick" of current may occur, which is an abnormal condition. Possible causes are

1. Overmodulation.
2. Incomplete neutralization of the modulated amplifier.
3. Parasitic oscillation in the modulated amplifier.

Grid modulation—decrease in antenna current:

1. Excessive RF drive to the grid of the modulated amplifier.
2. Distortion in the speech amplifier or modulator.
3. Insufficient bias on the grid of the modulated stage.
4. Too much resistance in the grid bias power supply.
5. Poor regulation of the plate circuit power supply for the modulated RF amplifier.
6. Inadequate loading of the plate circuit of the modulated stage.
7. Defective tube.

Grid modulation—increase in antenna current:

1. Distortion in the speech amplifier or modulator.
2. Overmodulation.
3. Excessive grid bias in the modulated amplifier.
4. Incomplete neutralization of the modulated amplifier.

18-18 ANTENNA AMMETER INDICATIONS

An RF ammeter connected in series with the antenna will give an increased indication when the carrier is modulated.

Consider that a certain transmitter delivers 1000 W of unmodulated RF power into an antenna whose impedance is a constant 10 Ω. What percentage of antenna-current increase should be expected between unmodulated conditions and 100 per cent sinusoidal modulation? Let us first determine the antenna-current reading without modulation using the power formula:

$$i = \sqrt{\frac{P}{Z}} = \sqrt{\frac{1000}{10}} = 10 \text{ A}$$

With 100 per cent sinusoidal modulation the sideband power equals one half the carrier power. This means that the total power delivered to the antenna now becomes 1500 W. The antenna current that is now flowing can be easily determined as follows:

$$i = \sqrt{\frac{P}{Z}} = \sqrt{\frac{1500}{10}} = 12.25 \text{ A}$$

The rise in antenna current is 2.25 A or

$$\frac{12.25}{10} \times 100 = 22.5\,\%$$

From this example we see that if an RF carrier is modulated 100 per cent by a pure sine wave the antenna current can be expected to increase 22.5%.

Suppose that the same RF carrier is modulated only 50 per cent. What increase in antenna current can be expected? The sidebands contain only one fourth as much power when 50 per cent modulated as compared to 100 per cent modulation. Therefore, the sideband power is 125 W and the total antenna power is 1125 W. The antenna current is

$$i = \sqrt{\frac{1125}{10}} = 10.6 \text{ A}$$

This is an increase of

$$\frac{10.6}{10} = 6\,\%$$

These conditions have assumed sinusoidal modulation. Speech and musical frequencies are complex waves, and when they are used to modulate the carrier, the increase in antenna current *may hardly be perceptible*, even though peaks of modulation may reach 100 per cent. The only time the antenna-current meter could register a 22.5 per cent increase with voice modulation would be when there is considerable overmodulation.

18-19 CHARACTERISTICS OF RADIOTELEPHONE TRANSMITTERS

A typical class C modulated RF amplifier and class B modulator are shown in Fig. 18-23. The voltages shown are illustrative only; exact values would have to be determined from the manufacturer's specifications and data sheets for the types of tubes used. As an aid in analyzing the performance of the circuit, several current-indicating meters have been shown. It is assumed that all filaments in the transmitter are on, but that plate voltage to the final and modulator output are off. Assume that the oscillator tank, buffer amplifiers, frequency multipliers, and so on, that may be used are all properly tuned, and the antenna is loosely coupled to the FPA tank.

Meter M_1 indicates the grid current in the FPA. As the grid tank is tuned to resonance, the meter reading will *peak* at some particular value. If the current is not the correct amount, the coupling to the preceding stage may be increased or decreased as required.

The power supply for the plate circuit of the FPA is *set to a low voltage* and then turned on. The plate tank is brought into resonance as indicated by

Characteristics of Radiotelephone Transmitters

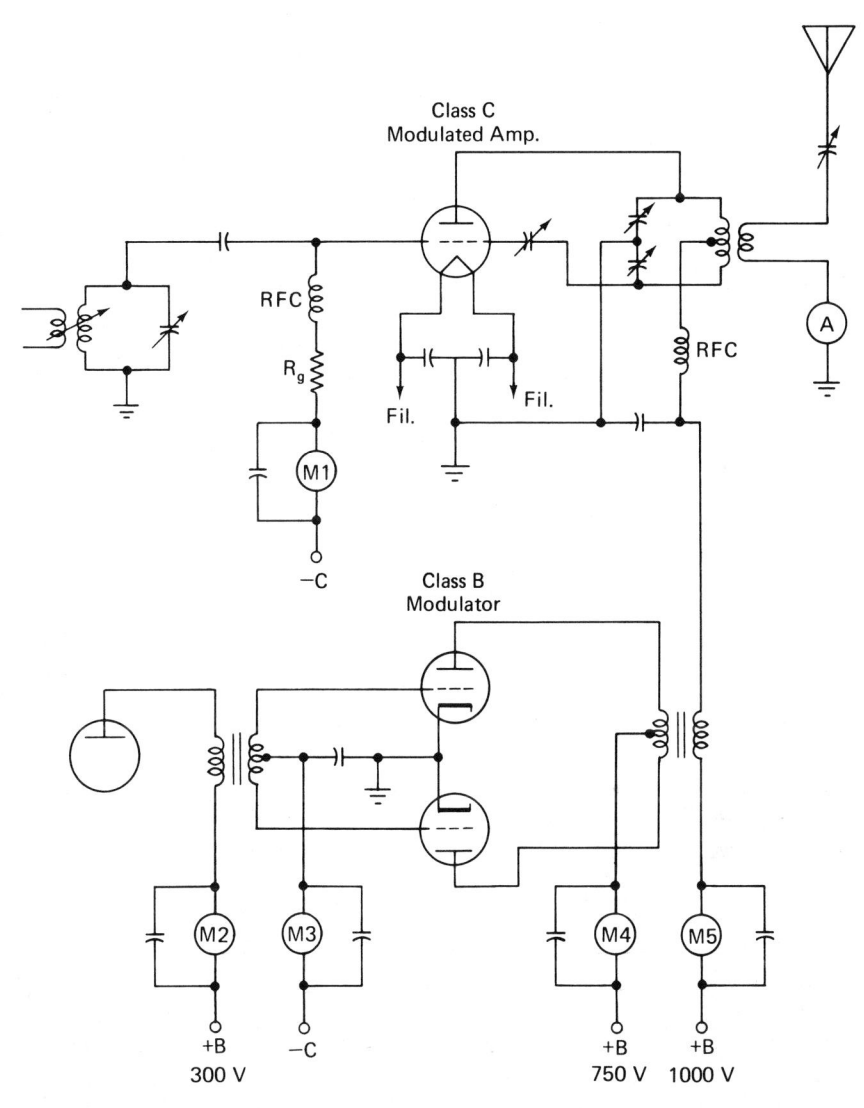

FIG. 18-23 Schematic diagram of class B push–pull modulator and modulated RF amplifier.

a *dip* in meter M_5. Next, the antenna coupling is increased and resonated, as indicated by a peak reading on antenna ammeter A. Meter M_5 will also register an increased reading under these conditions. The plate voltage to the modulated amplifier can now be raised to the proper value, and the antenna coupling increased until the specified amount of current is indicated on ammeter A. This may necessitate a slight retuning of the FPA plate tank and antenna circuits to optimize performance.

The audio level control of the speech amplifier (not shown) should be increased just to the point where M_2 begins to increase when the amplifier is fed with a sine wave. At this level, distortion is beginning. Ideally, the level control should be backed off slightly from this setting so that the amplifier is not overdriven. This position represents the highest setting of the level control, without distortion, for sine-wave input.

The push-pull class B modulator should next be checked for proper operation. With normal drive there will be grid current flow, which can be read on M_3. If the modulator is biased for AB_2 operation, some grid current will flow, but the amount should be less than for class B. It will be necessary to refer to the manufacturer's data sheet to determine normal values of grid current.

With the level control set to zero (no excitation to the modulator) meter M_4 will only read the *quiescent* plate current. This should be very low due to class B operation.

To check overall performance, the gain or level control should be turned up partway. The RF ammeter should rise along with meters M_4 and M_5. At this low level of excitation there is little likelihood of grid current flow in the modulator, so M_3 will probably remain at zero. When the level control is brought up to a high value, all meters should increase except M_2.

The *average plate current* of the modulated class C amplifier should remain constant, whether modulated or not. If any change occurs, it indicates the presence of *distortion* of the modulation envelope and attendant carrier shift. Experience indicates that a 1 to 2 per cent drop in current of M_5 is normal at high levels of modulation. This may also be expected for grid modulation, as well as for linear amplifiers.

At this point the signal generator may be removed and a microphone connected to the speech amplifier input. There should be a *slight rise* in antenna current as the microphone is spoken into. The percentage of modulation and linearity can now be checked with an oscilloscope.

Occasionally, a modulation transformer will develop a short between turns of either the primary or secondary windings. At high levels of modulation the reactive voltage across these windings may become very large. Any defect in the insulation between turns will permit arcing and cause a short, resulting in heavy current flow in the shorted part. Whether the short occurs in the primary or secondary, the result is a reduction of impedance in the primary and an increase of current beyond normal values as read on M_4. The percentage of modulation falls off, negative carrier shift results, and the antenna current will not rise to normal values. Additional evidence of shorted turns will be overheating of the transformer, audible humming, and, possibly, smoke. The plates of the modulator will overheat. The replacement of the transformer is necessary before normal operation can be restored.

Suppose that the peak output of a modulator is decreased from 1000 to 10 W; what would this be equal to in decibels?

$$\text{dB gain or loss} = 10 \log_{10} \frac{P_1}{P_2}$$
$$= 10 \log_{10} \frac{1000}{10} \; 10 \log 100$$
$$= 10 \times 2 = 20 \text{ dB}$$

If the first speech amplifier tube in a radiotelephone transmitter were overexcited, but the percentage modulation capabilities of the transmitter were not exceeded, the modulation envelope would be distorted. The same would apply to any stage of the speech amplifier, provided it were properly biased.

18-20 PRINCIPLES OF SINGLE-SIDEBAND TRANSMISSION

Since both sidebands contain the same information, only one theoretically has to be transmitted. Such a system has been developed and is in common use. It is known as *single sideband*, and is abbreviated SSB. Actually, the carrier need not be transmitted if the receiver has appropriate circuitry. This is known as *single-sideband suppressed carrier*, or SSSC. Because the carrier is not transmitted, there is considerable increase in operating efficiency. Transmitters designed to operate in this manner require appreciably less power input and are considerably smaller in size and weight. This is an advantage in aircraft applications.

There are other advantages that accrue from SSB operation. For example, the *bandwidth requirements are cut in half*. This is significant when considering the overcrowding in the 2- to 30-MHZ band. At the receiving end the reduced bandwidth also means less noise and better signal-to-noise ratios. Eliminating the carrier also eliminates the heterodyne interference that so often spoils communications in congested bands. With only one sideband being transmitted there is less tendency for the received signal to be subjected to the normal fading of signals, as experienced with conventional AM transmissions and reception.

As might be expected, there are some limitations of SSB operation. For example, in AM receivers no specific circuitry is required to maintain local oscillator stability, yet the results are quite satisfactory. On the other hand, with SSB *frequency stability is extremely important*. Because the carrier is not transmitted, it must be generated locally within the receiver. For optimum fidelity this locally generated carrier must be identical to the original carrier in the transmitter, both in *frequency* and *phase*. A frequency change of a relatively few hertz will be noticeable. Consequently, transmitter stability in the order of 0.5 ppm or better is desirable for SSB operation. At the receiving end high stability is also required. Well-designed receivers either use AFC (automatic frequency control) or precision frequency control circuitry to maintain stability. If these features are not available, the operator must constantly monitor the signal to maintain communications.

A graphic representation of the carrier and the sidebands produced by voice modulation (A3) is shown in Fig. 18-24a. By appropriate filtering, SSB operation may be achieved as shown in Fig. 18-24b, where the lower

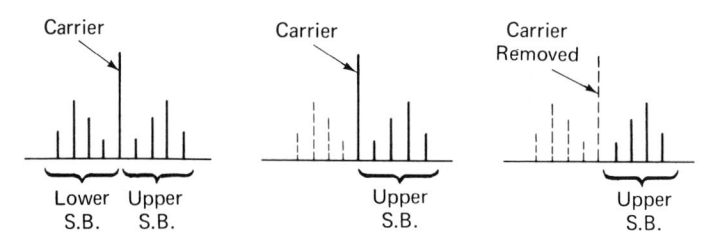

FIG. 18-24 (a) Graphic representation of carrier and sidebands produced by voice modulation; (b) upper sideband and carrier; (c) upper sideband with suppressed carrier.

sideband has been removed (A3H). By appropriate circuitry and filtering, the carrier may be suppressed and the lower sideband removed, as shown in Fig. 18-24c. This is known as A3J transmission.

One method of generating an SSB signal is shown in the block diagram of Fig. 18-25. The crystal oscillator generates a low-frequency RF signal that is fed into the balanced modulator, which eliminates the carrier frequency while permitting sidebands to be generated. The design of the balanced modulator is such that there would be more output when more audio signal were present. When audio is applied, the balance is upset, and one side of the modulator will conduct more than the other. In this modulation process *sum and difference* frequencies are generated. The modulator is not balanced for the sidebands (they are not canceled out) and, therefore, will appear in the output. By suitable filtering the appropriate sideband is passed. Although either sideband can be transmitted, standardization favors transmission only of the upper sideband.

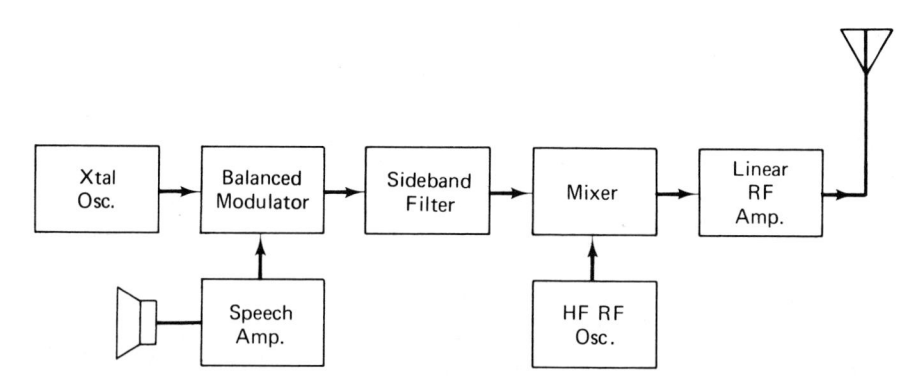

FIG. 18-25 Block diagram of filter-type SSB transmitter using a balanced modulator.

Balanced Modulator Circuits

Because voice frequencies are used for modulation, it should be realized that the sidebands will not be very far apart. For example, at an audio frequency of 50 Hz the upper and lower side frequencies would be plus and minus 50 Hz, or only 100 Hz apart. It is difficult to obtain the necessary selectivity with conventional filters to discriminate against these low frequencies. Consequently, voice frequencies below 100 Hz are usually *suppressed*. The amount of separation of the sidebands is also a function of the carrier frequency. By using carrier frequencies on the order of 50 to 100 kHz, the two side frequencies will be farther apart (percentage wise) than if a carrier frequency of perhaps 2 MHz is used. The demands on the filter system are less stringent under these conditions.

With the carrier and one sideband being used there still remains the problem of raising the sideband to the desired RF value. It is not possible to use frequency multipliers, because the bandwidth would be increased and the original modulating frequency altered. The most satisfactory method of raising the frequency to the required value is by means of *heterodyning*, or *frequency translation*. By judiciously selecting the right high-frequency RF oscillator frequency to feed into the mixer, the two side frequencies produced will not only appear at the proper place in the desired band, but will be separated far enough so that the selectivity of the tuned circuit in the linear RF amplifier will be able to discriminate against the undesired beat frequency, permitting only the correct one to be amplified and fed on to the antenna.

18-21 BALANCED MODULATOR CIRCUITS

The schematic of a simple bridge-type modulator is shown in Fig. 18-26. The basic principle of any balanced modulator is to feed the carrier into the modulator in such a way that it does not appear in the output, but that the sidebands will. To accomplish this, it is necessary to introduce the

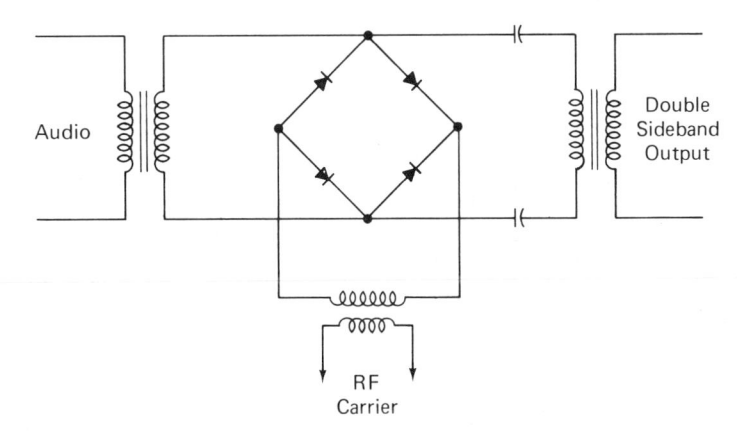

FIG. 18-26 Bridge type of balanced modulator.

audio in a push–pull manner, the RF drive in parallel, and to connect the output in push–pull.

A typical vacuum-tube balanced modulator appears in Fig. 18-27. As

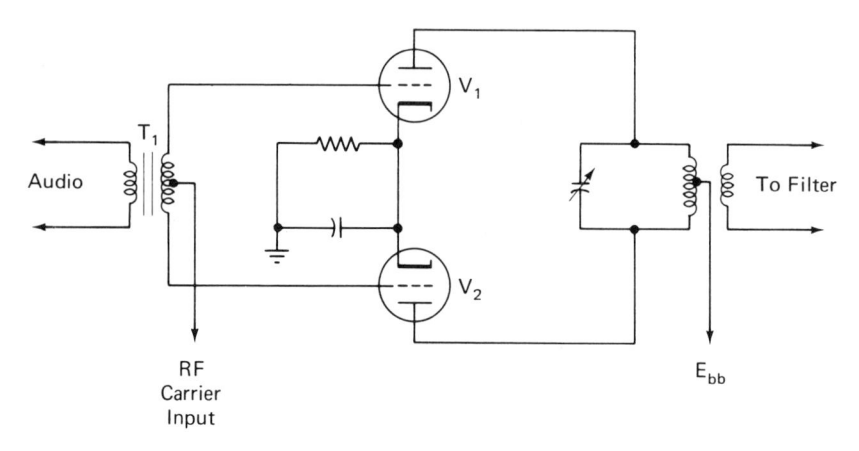

FIG. 18-27 Vacuum-tube type of balanced modulator.

far as the audio signal is concerned, the circuit appears as a push–pull amplifier. Unlike the conventional push–pull amplifier, however, is the presence of a tank circuit in the output. The tank presents little impedance to the audio frequencies; consequently, no audio is developed in the output. However, let us consider the RF and see what is taking place. This is not a push–pull input. The RF is applied to each grid in series with one half the secondary of T_1. Therefore, the grid signals on V_1 and V_2 are in phase; consequently, the RF components in the plate circuits are also in phase. Because these currents flow in opposite directions through the push–pull tank circuit, their effects will cancel, and no RF output at the carrier frequency is available.

To better understand the operation of the circuit, it is necessary to consider the combined action of audio and RF. The tubes, which must be *well matched*, are biased in the nonlinear region of the characteristic curve so that modulation products will be produced. Because of the push–pull input of the audio signal, the bias of the tubes will vary at an AF rate. Therefore, the sideband components are increasing in one tube and decreasing in the other. These effects are additive in the plate tank circuit, which is resonant to these frequencies. Hence, an output is developed at the sum and difference frequencies. The result is an output containing only the side frequencies. The output of the plate tank circuit is inductively coupled to the filter where the lower sideband is removed.

18-22 SIDEBAND FILTERS

We have previously mentioned the closeness of the two sidebands in the output of the balanced modulator. To separate them (filter out the lower sideband), it is necessary that *multisection high-Q filters* be used. One frequently used circuit takes advantage of the high Q available with quartz crystals. By using a *crystal lattice filter* the necessary response characteristics may be obtained. Crystal filters provide excellent results, particularly at frequencies around 100 kHz. However, they are costly and relatively bulky when compared to other types of components. They are also more susceptible to shock and vibration and, consequently, not always suitable for mobile applications.

The *electromechanical* filter has found wide use in this type of application. The RF input is fed to a *magnetostrictive transducer* that excites the *mechanical resonators*. The energy is passed through the mechanical resonator to another magnetostrictive transducer located in the output. The resonators may be in the form of rods, plates, or discs, depending upon the manufacturer and the design. Electromechanical filters are available to cover a multiplicity of frequency ranges. They have sharp cutoff characteristics, low insertion loss, good stability, and may be constructed to withstand considerable shock and vibration.

18-23 PHASE-SHIFT METHOD OF GENERATING SINGLE SIDEBANDS

There is another method for generating SSB operation without the need of filters. The block diagram for such a system is shown in Fig. 18-28. This is known as a *phase-shift method* of generating single-sideband suppressed carrier. The unwanted sideband is canceled or suppressed by producing and combining two components that are 180° out of phase with respect to each other. One inherent advantage of this system is that the cancellation process

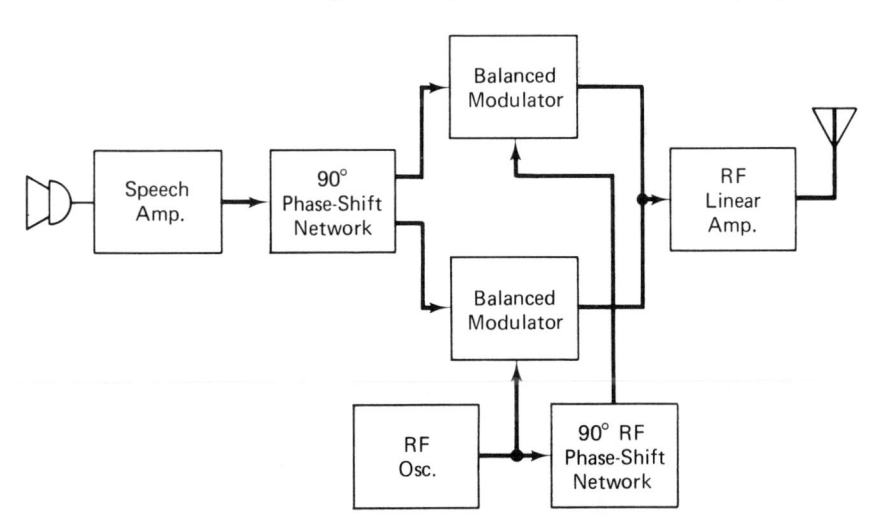

Fig. 18-28 Phase-shift method of generating SSB suppressed-carrier signal.

is not predicated upon how much separation exists between the upper and lower sidebands. This feature makes it possible to use crystal oscillator frequencies in the band that operation is desired in. The modulator may be operating at the final carrier frequency, and frequency translation or heterodyning is unnecessary.

Observe that two balanced modulators are used. The modulating signal and carrier are fed to both modulators. As in the system previously described, the carrier will be suppressed, and the output of the balanced modulators will contain only the upper and lower sidebands. However, because of the 90° phase shifts in the input signal applied to these stages, one sideband will be in phase opposition to the other and will therefore cancel. The other sideband is additive, producing a single-sideband output. If the RF oscillator has been discretely chosen, the desired sideband will be at the final transmitting frequency and may be connected directly to the linear amplifier.

Commercial License Questions

Sections in which answers to questions are given appear in parentheses. A bracketed number following a question implies that it applies only to that element.

1. During 100 per cent modulation, what percentage of the average output power is in the sidebands? [4] (18–1, 18–6)
2. What is the ratio of modulator power output to modulated amplifier plate power input for 100 per cent amplitude modulation? (18–1, 18–6, 18–7)
3. What is the effect of 10,000-Hz modulation of a standard broadcast station on adjacent channel reception? [4] (18–2, 18–5)
4. If a 1500-kHz radio wave is modulated by a 2000-Hz sine-wave tone, what frequencies are contained in the modulated wave? (18–2)
5. Discuss the characteristics of a modulated class C amplifier. [4] (18–3, 18–4)
6. What is the term for the last AF amplifier stage that modulates the RF stage? [4] (18–3)
7. Draw a simple schematic diagram showing a method of coupling a modulator tube to an RF power amplifier tube to produce plate modulation of the amplified RF energy. (18–4)
8. Draw a simple schematic diagram of a class B audio high-level modulation system, including the modulated RF stage. [4] (18–4)
9. Why is a high percentage of modulation desirable in AM transmitters? (18–5, 18–6)
10. What undesirable effects result from overmodulation in a broadcast transmitter? [4] (18–5)
11. What are some of the possible results of overmodulation? (18–5)

Commercial License Questions 397

12. If you decrease the percentage of modulation from 100 to 50 per cent, by what percentage have you decreased the power in the sidebands? [4]
(18–6, 18–6)

13. A certain transmitter has an output of 100 W. The efficiency of the final, modulated amplifier stage is 50 per cent. Assuming that the modulator has an efficiency of 66 per cent, what plate input to the modulator is necessary for 100 per cent modulation of this transmitter? Assume that the modulator output is sinusoidal. [4]   (18–7)

14. How is the load on a modulator that modulates the plate circuit of a class C RF stage determined? [4]   (18–7)

15. Given a class C amplifier with a plate voltage of 1000 V and a plate current of 150 mA which is to be modulated by a class A amplifier with a plate voltage of 2000 V, plate current of 200 mA, and a plate impedance of 15,000 Ω. What is the proper turns ratio for the coupling transformer? [4]   (18–7)

16. What is meant by *high-level modulation*? [4]   (18–7)

17. What is meant by *low-level modulation*? [4]   (18–8)

18. Define high-level and low-level modulation. [4]   (18–8)

19. When the transmitter of a standard broadcast station is operated at 85 per cent modulation, what is the maximum permissible combined audio harmonic output? [4]   (18–8)

20. Draw a simple schematic diagram showing the proper method of obtaining dc screen-grid voltage from the plate supply in the case of a modulated pentode, class C amplifier.   (18–9)

21. Draw a simple schematic diagram showing a method of coupling a modulator tube to an RF power amplifier tube to produce grid modulation of the amplified radio frequency energy.   (18–10)

22. Draw a simple schematic diagram of a grid bias modulation system, including the modulated RF stage.   (18–10)

23. What is meant by *grid modulation*? By *plate modulation*?   (18–10)

24. In a properly adjusted grid bias modulated RF amplifier, under what circumstances will the plate current vary as read on a dc meter? [4]   (18–10)

25. How would loss of RF excitation affect a class C modulated amplifier when using a grid-leak bias only?   (18–10)

26. Draw a simple schematic diagram showing a method of suppressor-grid modulation of a pentode-type vacuum tube.   (18–11)

27. Draw a simple schematic diagram showing a Heising modulation system capable of producing 100 per cent modulation. Indicate power-supply polarity where necessary.   (18–12)

28. What pattern on a cathode-ray oscilloscope indicates overmodulation of a standard broadcast station? [4]   (18–15)

29. Draw a simple sketch of the trapezoidal pattern on a cathode-ray oscilloscope screen indicating low percentage modulation without distortion. [4]   (18–15)

30. Draw a diagram of a carrier wave envelope when modulated 50 per cent by a a sinusoidal wave. Indicate on the diagram the dimensions from which the percentage of modulation is determined.   (18–15)

31. Under what class of amplification are the vacuum tubes in a linear amplifier stage, following a modulated stage, operated?   (18–16)

32. Draw a diagram of a low-level plate-modulated pentode RF amplifier coupled to a push–pull linear RF amplifier and explain the principles of operation.   (18–16)

33. If a final RF amplifier operated as class B linear were excited to saturation with no modulation, what would be the effects when it was undergoing modulation?   (18–16)

34. Doubling the excitation voltage of a class B linear RF amplifier gives what increase in RF power output?   (18–16)

35. Name four causes of distortion in a modulated amplifier stage output. [4]   (18–16, 18–17)

36. What might be the cause of variations in plate current of a class B type of modulator?   (18–16, 18–19)

37. What do variations in the final amplifier plate current of the transmitter employing low-level modulation usually indicate? [4]   (18–16)

38. What may be the cause of a decrease in antenna current during modulation in a class B linear RF amplifier? [4]   (18–16)

39. What are causes of downward fluctuation of the antenna current of an AM transmitter when the transmitter is modulated? [4]   (18–16, 18–17)

40. What might be the cause of variation in plate current of a class B type of modulator?   (18–16)

41. In a modulated amplifier, under what circumstances will the plate current vary as read on a dc meter? [4]   (18–16)

42. What is the effect of carrier shift in a plate-modulated class C amplifier?   (18–17)

43. What might be the cause of a decrease in antenna current of a high-level AM radiotelephone transmitter when modulation is applied?   (18–17)

44. Should the plate current of a modulated class C amplifier stage vary or remain constant under modulation conditions? Why?   (18–17)

45. What may cause upward fluctuation of the antenna current of an AM transmitter when the transmitter is modulated?   (18–17)

46. In a class C radio-frequency amplifier stage feeding an antenna system, if there is a positive shift in carrier amplitude under modulation conditions, what may be the trouble? [4]   (18–17)

Commercial License Questions 399

47. What may cause unsymmetrical modulation of a standard broadcast transmitter? [4] (18–17)

48. In a modulated class C radio-frequency amplifier, what is the effect of insufficient excitation? (18–17)

49. What percentage of antenna current increase should be expected between unmodulated conditions and 100 per cent sinusoidal modulation? (18–18)

50. What percentage increase in average output power is obtained under 100 per cent sinusoidal modulation as compared with average unmodulated carrier power? [4] (18–18)

51. Draw a diagram of a complete class B modulation system, including the modulated RF amplifier stage. Indicate points where the various voltages will be connected. [4] (18–19)

52. What would be the effect of a shorted turn in a class B modulation transformer? In a class A modulation transformer? (18–19)

53. If the first speech amplifier tube of a radiotelephone transmitter were overexcited, but the percentage modulation capabilities of the transmitter were not exceeded, what would be the effect upon the output of the transmitter?
(18–19)

54. If tests indicate that the positive modulation peaks are greater than the negative peaks in a transmitter employing a class B audio modulator, what steps should be taken to determine the cause? [4] (18–19)

55. If the power output of a modulator is decreased from 1000 to 10 W, how is the power reduction expressed in decibels? [4] (18–19)

56. During 100 per cent modulation, what percentage of the output power is contained in the sidebands for (a) A3 emission, (b) A3H emission, (c) A3J emission?
(18–20)

57. Explain the principles involved in a single-sideband suppressed-carrier transmitter. (18–20)

58. How do the bandwidth and power requirements of a SSSC emission compare with that of full carrier and sidebands? (18–20)

59. Draw the block diagram of a SSSC transmitter (filter type). (18–20)

60. Explain the principle of operation of a vacuum-tube type of balanced modulator.
(18–21)

61. Draw the block diagram of a phase-shift method of generating a SSSC signal.
(18–23)

19

Frequency Modulation

19-1 REASONS FOR FREQUENCY MODULATION

A disadvantage of AM is the presence of *interference* and *static* (noise), which may be experienced over the entire spectrum. If the amplitude of the transmitter signal is increased, the amplitude of the noise also increases. With frequency modulation (FM) the amplitude of the RF modulated wave remains *constant* while its *frequency is varied* at a rate determined by the frequency of the information to be transmitted.

There is some amplitude modulation of the FM carrier due to noise. However, in the receiver the positive and negative peaks of the carrier are *clipped* (removed), and the noise is removed. Because the transmitted information is a function of the *rate* at which the carrier is deviated from its center frequency and not dependent upon variations in carrier amplitude, there is no degradation of the signal when clipping or limiting occurs.

Frequency modulation requires a *greater bandwidth* than AM and, consequently, must operate higher in the RF spectrum.

The most familiar application of FM is in broadcasting. These stations, operating in the frequency spectrum of 88 to 108 MHz, are noted for their fidelity and relative freedom from noise. Another use of FM is for mobile stations, such as police, taxicabs, fire communication, emergency services, and business applications. This latter group all use voice frequencies for modulation. A third application is with television, where the voice frequencies are transmitted via FM. (*Note:* The video information is transmitted via AM.) A fourth application is for amateur use, where *narrowband FM* is mostly used. However, there are several bands where wideband FM may be used. All these are voice modulated.

The types of FM transmission authorized by the FCC are

Fundamental Principles 401

F0: Absence of any modulation.

F1: A carrier that is shifted in frequency (FSK) according to some code, such as Morse or radioteletype.

F2: A keyed carrier that is being modulated at some audio frequency.

F3: FM transmission using voice or music as the modulating frequencies.

F4: Facsimile transmission.

F5: A carrier shifted in frequency in accordance with picture information. Experimental TV—not used on commercial TV.

F6: Four-frequency, diplex system with 400-Hz spacing between frequencies. Channels are not synchronized. Bandwidth is 2.05 kHz (complex telegraphy).

P0: Absence of any modulation intended to carry information in pulsed-type emissions.

19-2 FUNDAMENTAL PRINCIPLES

Frequency modulation differs in a number of ways with respect to AM. In FM the amplitude of the carrier remains constant. The modulation applied to the carrier causes *changes in frequency* of the transmitted signal. The strength of the AF signal determines the *shift* or *deviation* in frequency.

To understand how an FM signal is generated, refer to the circuit in Fig. 19-1. This is a shunt-fed Hartley oscillator whose frequency is determined by *LC*. Observe that a condenser microphone is connected across capacitor *C*. The principle of the condenser microphone is that its *capacity changes* as sound waves strike its diaphragm. The changes are very small. However, for

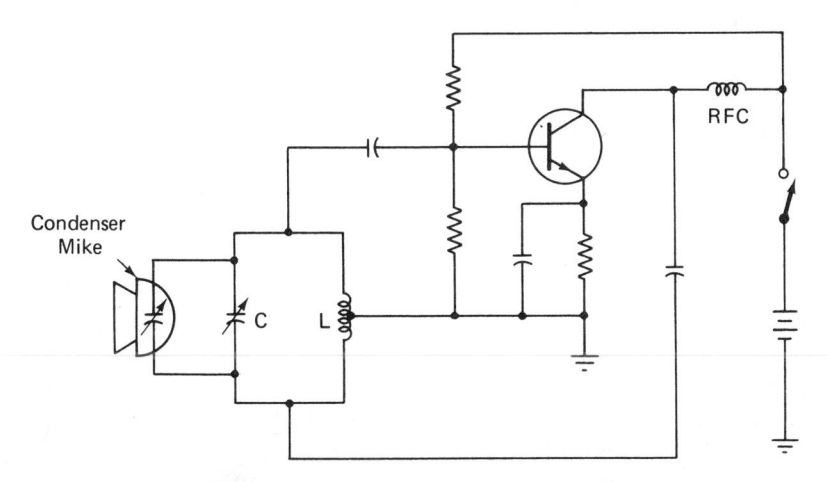

FIG. 19-1 Simple oscillator circuit which is frequency-modulated by means of a condenser microphone.

this illustration we shall assume that the capacity changes are large. With the circuit oscillating but the microphone idle, the circuit is adjusted to the desired operating frequency by means of capacitor C. This is known as the *center frequency*. This may also be called the mean or average frequency if we consider the possible deviations on both sides. If the microphone is actuated by sound waves, as represented in Fig. 19-2, the diaphragm will vibrate, thus

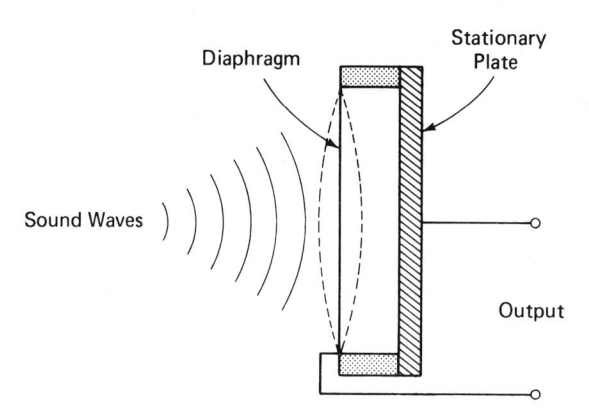

FIG. 19-2 Sound waves striking diaphragm of microphone cause the capacity to change in accordance with the sound vibrations.

varying the space between it and the stationary plate. If the diaphragm moves closer to the stationary plate, the capacity is increased. Conversely, if the diaphragm moves outward, the capacity will diminish. The stronger the audio signal, the greater will be the change in capacity.

Since the microphone is in parallel with the tuned circuit (Fig. 19-1), it is apparent that changes in microphone capacity will result in changes of the resting or mean frequency of the oscillator. A whisper may change the frequency by ± 2 kHz. A louder sound may cause a frequency change of ± 40 kHz. An amplitude change equivalent to a shout might cause the frequency to change by ± 75 kHz. *It has been established that a change of frequency of ± 75 kHz will be equivalent to 100 per cent modulation of an amplitude modulated wave.* Thus when the carrier is deviated or changed 75 kHz either side of its resting frequency, we have 100 per cent modulation.

What happens when the frequency of the sound changes? It varies the *rate* at which the frequency of the oscillator is changed. The higher the frequency, the greater the number of times per second the oscillator frequency will be changed. Thus, if a 1-kHz tone of a certain intensity changes the oscillator frequency by ± 30 kHz at a rate of 1000 times/s, an AF tone of 4 kHz, of the same intensity will provide the same oscillator deviation 4000 times/s. Similarly, a 100-Hz tone of the same intensity will change the frequency of the oscillator by ± 30 kHz 100 times/s.

To further illustrate the principles of FM, refer to Fig. 19-3. An AF

Modulation Index 403

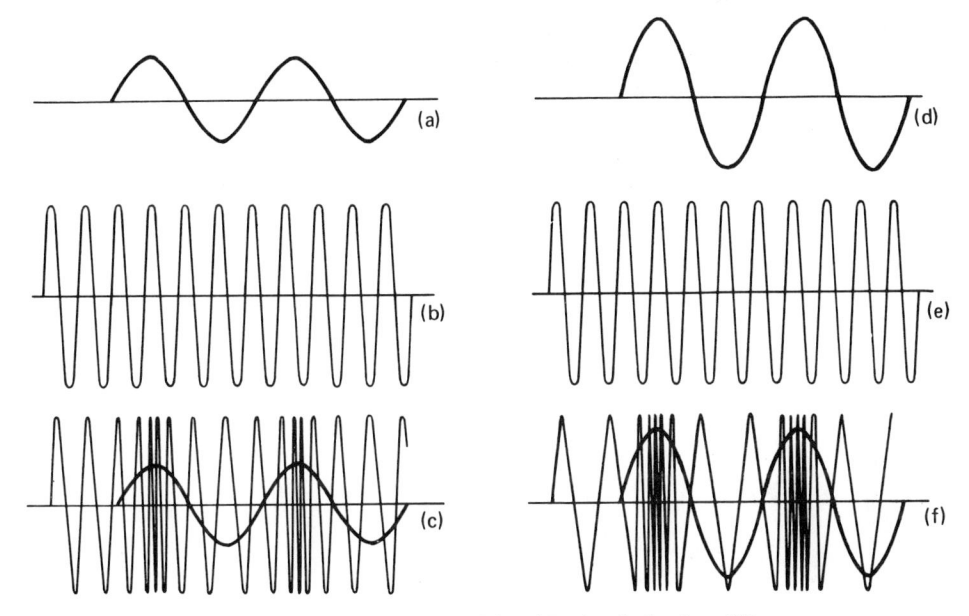

FIG. 19-3 FM waves produced by AF signals having different amplitudes.

signal and an RF signal are shown separately in a and b. When they are combined in the modulation process, the resulting signal is the FM wave shown in c. As the amplitude of the AF signal increases in the positive direction, the modulated wave is *compressed*. When the AF signal swings in the negative direction, the carrier frequency is *spread out*. These variations in the spacing of the modulated wave represent *instantaneous changes in frequency*.

When the modulating signal is increased in amplitude, as in Fig. 19-3d, the changes in the spacing of the wave are proportionately greater, as shown in Fig. 19-3f. Therefore, the frequency of the modulated wave is *directly proportional* to the amplitude of the modulating signal. When the audio voltage reaches its peak in the positive direction, the carrier has been shifted to its highest value above center frequency. When the modulating voltage reaches its negative peak, the frequency of the carrier is shifted down to its lowest value below the center carrier frequency. Maximum frequency deviation therefore takes place at the *peaks* of the audio signal. The sine waves superimposed on the carrier shown in Figs. 19-3c and f have been shown only to help understand the relationship between the AF modulation signal and the modulated carrier.

19-3 MODULATION INDEX

The limits of frequency shift on either side of the carrier are known as the *frequency-deviation* limits. The ratio between the maximum frequency deviation and the maximum frequency of the modulating signal is called the *modulation index*.

$$\text{modulation index} = \frac{\text{maximum frequency deviation}}{\text{maximum frequency of modulation signal}}$$

Several examples will help to illustrate this. In FM broadcasting 15 kHz is the highest required AF signal to be transmitted, and the widest frequency deviation is 75 kHz. Hence, the modulation index is

$$M = \frac{75 \text{ kHz}}{15 \text{ kHz}} = 5$$

If a 10-kHz tone were used to 100 per cent modulate the carrier, the modulation index would be

$$M = \frac{75}{10} = 7.5$$

With 50 Hz as the lowest required frequency to be transmitted, the modulation index becomes as high as

$$M = \frac{75 \text{ kHz}}{50 \text{ Hz}} = 1500$$

The *deviation ratio* of an FM signal is the ratio of the maximum allowable deviation (75 kHz) to the highest modulating frequency (15 kHz). Thus,

$$\text{deviation ratio} = \frac{75 \text{ kHz}}{15 \text{ kHz}} = 5$$

It should be observed that the deviation ratio and modulation index have the same value only when maximum deviation results from the highest modulating frequency.

19-4 PERCENTAGE OF MODULATION

In the FM broadcasting industry 100 per cent modulation has a different meaning than with AM. The AF signal varies only the frequency of the oscillator. Therefore, the transmitter operates at *maximum efficiency* continuously and supplies a constant input power to the antenna, *regardless* of the degree of modulation. A modulation of 100 per cent simply means that the carrier is deviated in frequency by the full permissible amount. For example, an 88-MHz FM station has 100 per cent modulation when its audio signal deviated the carrier 75 kHz above and below 88 MHz. For 50 per cent modulation the frequency would be deviated 37.5 kHz above and below the resting frequency.

Consider that an FM broadcast transmitter is modulated 40 per cent by a 5-kHz signal. What will be the frequency swing of the transmitter if the percentage of modulation is doubled? The frequency swing will be 80 per cent of 75 kHz, or 60 kHz.

19-5 SIDEBANDS AND BANDWIDTH

It is possible to obtain a frequency deviation many times the frequency of the modulating signal itself. In broadcast equipment the deviation frequency may be as high as 100 kHz or more, even though the modulation frequency is but a few kilohertz. Therefore, the *sidebands* generated by FM are not restricted to the sum and difference between the highest modulating frequency and the carrier, as in AM. In FM, *many sidebands are generated*. The *first pair* of sidebands are those of the *carrier frequency plus and minus the modulating frequency*, and a pair of sidebands will appear also at *each multiple* of the modulating frequency. As a result, an FM signal occupies a greater bandwidth than does an AM signal. The total number of sidebands present of significant amplitude (more than 1 per cent of the amplitude of the unmodulated carrier) depends on the modulation index.

If the frequency deviation is kept constant, the number of sidebands *increases* as the modulating frequency *decreases*, and the total bandwidth occupied decreases as the modulating frequency decreases. The total bandwidth, however, can never be less than the bandwidth set by the peak-to-peak deviation alone, no matter how low the frequency of the modulating signal becomes. If the amplitude of the modulating signal increases and its frequency remains constant, the deviation increases and the modulation index increases. This means that *more energy* goes into the outer sidebands and, correspondingly, more of them increase to significant amplitude. The result is an increase in the number of useful sidebands as well as an increase in bandwidth.

The position of the sideband pairs for a single sinusoidal modulating wave depends only on the frequency of the modulating wave. The amplitude of the sidebands depends on the modulation index. The modulation index, in turn, depends on the amplitude of the modulating signal, because frequency deviation is proportional to signal amplitude. For a given modulation index and sinusoidal modulating frequency, the sideband pairs appear on either side of the carrier frequency, as shown in Fig. 19-4. All the sideband components taken together form the frequency spectrum of the FM wave. For example, if the modulating frequency is 15 kHz and the frequency deviation is 75 kHz, the modulation index will be $\frac{75}{15}$ or 5, and the frequency components beyond the *eighth pair* of sidebands will be less than 1 per cent of the unmodulated carrier amplitude, and considered negligible.

The number of significant sidebands produced by the modulation process is determined by Bessel functions. The mathematics involved is beyond the scope of this book.

When the transmitted signal is unmodulated, there is a certain constant power in the carrier. When modulation is applied, *power is taken from the carrier and forced into the sidebands;* therefore, the carrier amplitude, or center-frequency component, is reduced. The maximum power is carried in the fourth sideband, which is 4 times 15, or 60 kHz, away from the carrier

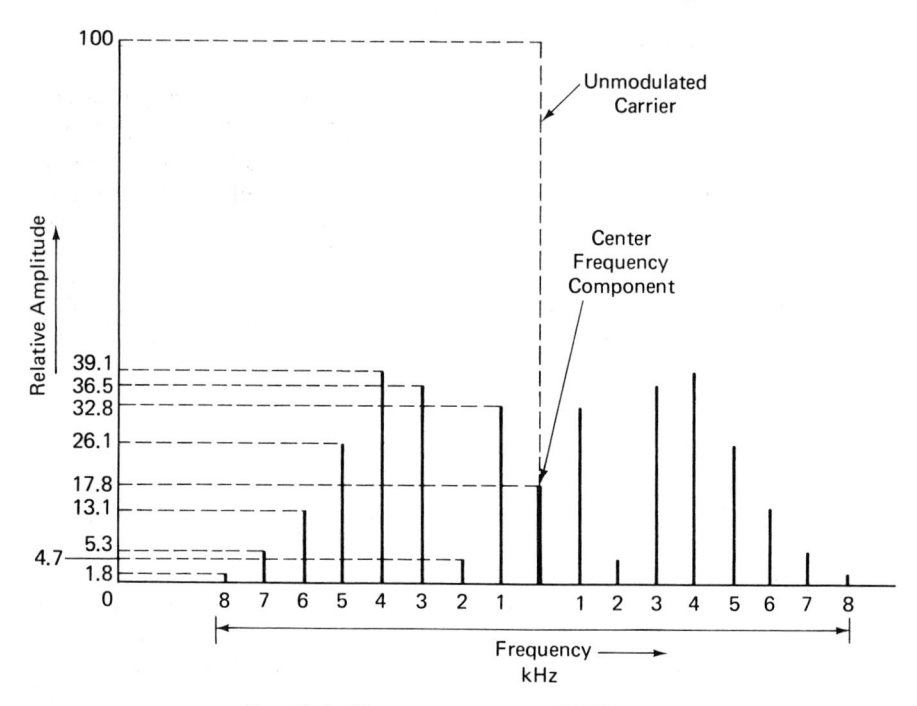

FIG. 19-4 Frequency spectrum of FM wave.

frequency (see Fig. 19-4). Since no information is in the carrier, reducing its amplitude increases the efficiency of operation. For some value of modulation index and modulating frequency, carrier amplitude falls to zero and all the power is contained in the sidebands.

The effective sidebands must be at least as far from the carrier as the frequency deviation limits. Because of this, it is necessary to provide a channel or bandwidth that will handle the highest sideband component, plus a *guard band* that will accommodate any sidebands that extend beyond these limits. Since the modulating signal cannot always be specified and may vary over wide limits, it is more convenient to assign the channel in terms of deviation limits, and then to set aside some additional frequency space for guard bands on either side of the deviation limits. Figure 19-5 indicates the manner in which the FM broadcast band channel limits have been established. For 100 per cent modulation the carrier will be deviated 75 kHz either side of center frequency. Additionally, there is a 25-*kHz guard band* beyond the limits of the maximum deviation. Thus, the *total bandwidth* is 200 kHz.

The FM band extends from 88 to 108 MHz. Inasmuch as each station uses 200 kHz of bandwidth (including guard bands), a total of 100 FM stations can be accommodated. The FCC does not assign stations to adjacent channels in any one area to prevent interference or overlapping of some of

Preemphasis and Deemphasis 407

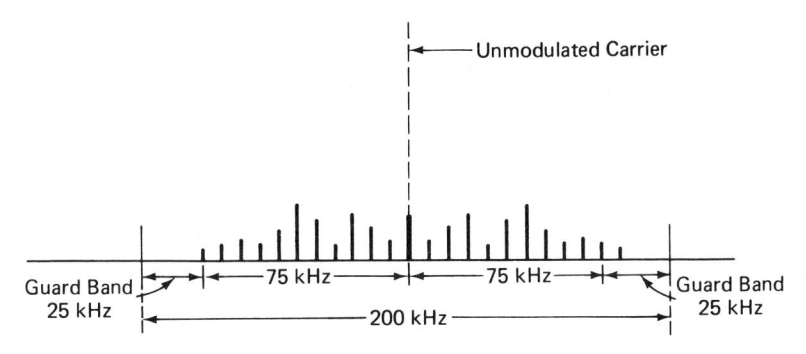

FIG. 19-5 Bandwidth of FM channel.

the farthest sidebands. As a consequence, the maximum number of channels that could be assigned to any one local area is 50.

The question is sometimes asked, "Why can't FM be used in the standard broadcast band so that the same wide AF response can be enjoyed as with FM?" The answer lies in the fact that the broadcast band is approximately 1100 kHz wide (550 to 1650 kHz). The 200-kHz bandwidth required for FM would accommodate only five and one half FM channels! This would result in a tremendous reduction of commercial broadcast stations that the public now can receive.

19-6 PREEMPHASIS AND DEEMPHASIS

As signals pass through the transmitter, the receiver, and the space between them, some unwanted noise and distortion are superimposed on the desired speech. This is distributed uniformly throughout the audible spectrum. Therefore, the ratio of the signal to the unwanted noise *decreases* in the higher frequencies, because the speech amplitudes in this range do not have the intensity that the lower frequencies have. Moreover, the distortion increases in the high-frequency portion of the spectrum. The high frequencies make the greatest contribution to intelligibility of speech waves, since the consonants, which form the majority of speech sounds, have their peak energy in this part of the audio band.

To avoid degrading the reproduction of higher frequencies, a certain amount of *added amplification* (preemphasis) is provided for these frequencies. The result of this process does not sound unnatural when received and deemphasized at the receiver. This combination (preemphasis and deemphasis) provides a more uniform signal-to-noise ratio throughout the audio range. A transmitter using preemphasis has a *wider sideband spectrum* for speech than one without it.

The fact that preemphasis results in a greater bandwidth for a given deviation always must be taken into account. However, the possibility of overmodulation is not likely, since the high-frequency components of the signal originally are weak, and the preemphasis merely brings them up to the level of the low tones. It does not cause overmodulation of an FM transmitter, although the deviation limits set for the particular unit will be increased.

The preemphasis characteristics of the speech amplifier of an FM transmitter are such that the output remains relatively constant from 50 to about 500 Hz. The output *increases gradually* from 500 to 2000 Hz. From 2 to 15 kHz the output increases approximately eight times, or about 18 dB.

The characteristics of preemphasis and deemphasis normally are achieved by simple electrical combinations of resistance, capacitance, and inductance connected to give the desired relationship between the input and output voltages of the network. The characteristics of speech are complicated, and therefore the networks chosen represent a compromise between duplicating the exact loss of high frequencies and using as few parts as possible. In general, preemphasis and deemphasis circuits are very simple *CR* or *LR* combinations.

19-7 REACTANCE-TUBE MODULATORS

It is possible to control the frequency of an oscillator by controlling the amount of reactance, capacitive or inductive, that is present in the circuit. This is accomplished by a circuit known as a *reactance modulator*, which uses the characteristics of a vacuum tube to control the reactance in the tank circuit of the oscillator. By *simulating* a capacitance or an inductance across its output terminals, it is said to inject reactance into the tank circuit. The simulated reactance is controlled by the audio signal. The reactance-tube modulator frequency modulates the oscillator in accordance with the audio signal.

Figure 19-6 shows a reactance-tube modulator circuit with C_L and R_L forming the plate load. These circuit elements are across the oscillator tank C_L. With no AF present the only signal appearing at the grid of the reactance modulator is the RF voltage appearing across R_L. This is amplified and applied across the plate load $C_L R_L$. The reactance of C_L is made very large in respect to the resistance of R_L, and therefore the capacitive reactance, X_{C_L}, determines the resultant current flow, causing it to lead the voltage across it by approximately 90°. This current flowing through R_L results in a voltage drop across R_L, which also leads the applied voltage by 90°.

This RF voltage fed back to the grid of the reactance tube causes an RF variation of plate current, which is coupled to the tank circuit of the oscillator through C_c. However, the current in the oscillator tank is in phase with the RF voltage, since the circuit is operating at resonance, while this additional current resulting from the same voltage leads it by 90°. The additional current supplied by the reactance tube acts as if it were caused by a capacitor.

Applying an audio signal at some single frequency to the grid of the

Solid-State Frequency-Modulated Modulator 409

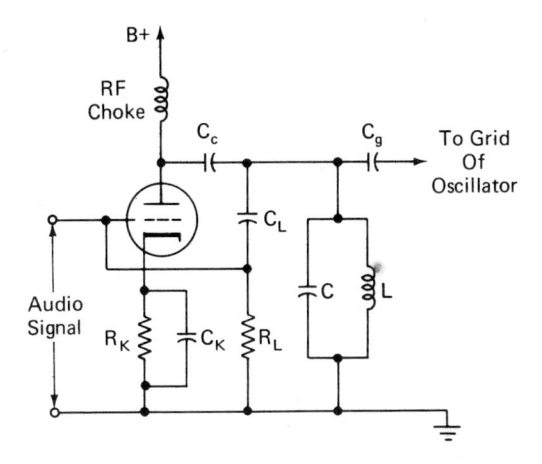

FIG. 19-6 Typical reactance-tube modulator circuit.

reactance tube causes *two voltages* to be present at the grid—an AF and an RF voltage. The RF voltage is responsible for the reactive plate current flow, and the audio signal changes the amount of plate current flowing in accordance with its amplitude. Changing the amount of plate current coupled to the tank circuit means that its reactive effect is varied, and results in the injection of a changing reactance into the oscillator tank. This changing reactance adds a changing capacitance to the oscillator tank. Oscillator frequency is varied accordingly, and the result is an FM signal at the oscillator output.

19-8 SOLID-STATE FREQUENCY-MODULATED MODULATOR

It is possible to design an FM oscillator using voltage-sensitive semiconductor devices called *varicaps* or *varactors*. They are small, reverse-biased diodes characterized by fast response, reliability, and stability under shock and vibration. Essentially, they are silicon diodes that exhibit a variable capacitive effect when subjected to a variable voltage. The reverse bias creates a *depletion or barrier region* free of mobile charges, which acts as a *dielectric*. This combination of dielectric contained between two conducting areas has capacitive properties. The capacity is approximately equal to the reciprocal of the square root of the reverse-biased voltage, as shown by

$$C \cong \frac{1}{\sqrt{V}}$$

A typical circuit using a varicap to frequency modulate an oscillator is represented in Fig. 19-7. The varicap, CR_1, is in series with tuning capacitor C. A voltage divider consisting of R_1 and R_2 establishes the correct negative bias on the varicap so that its capacity will be in the proper range with respect to capacitor C. The AF signal developed across the secondary winding of modulation transformer T is superimposed on the negative bias voltage of

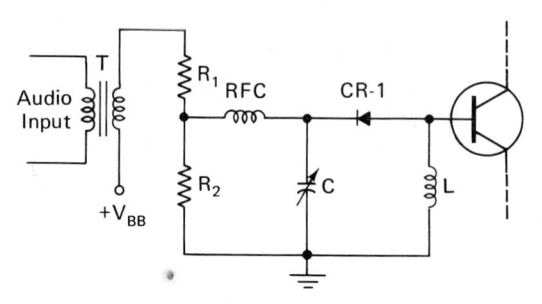

FIG. 19-7 Varicap used to frequency modulate an oscillator.

CR_1. The fluctuating ac voltage alternately increases and decreases the capacity of the varicap, which in turn affects the total capacity across the tank inductor. The net result is that the oscillator is caused to vary above and below its center frequency. The stronger the AF modulating voltage, the greater will be the oscillator swing. The RF choke isolates the high-frequency oscillator energy from the reverse-biased network, while allowing the AF to pass.

19-9 PHASE MODULATION

Besides changing the amplitude or frequency of the carrier, its *phase* can be changed to produce a signal bearing intelligence. This is called *phase modulation*. When frequency modulation is used, the phase of the carrier wave is indirectly affected. Similarly, when phase modulation is used, the carrier frequency is affected. Familiarity with both frequency and phase modulation is necessary for an understanding of either.

Two cycles of a carrier are represented by curve *A* in Fig. 19-8. At time

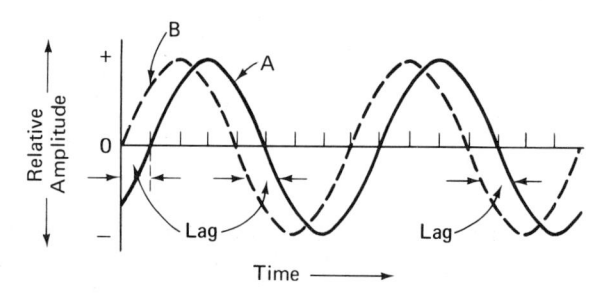

FIG. 19-8 Determining relative phase between two waves of the same frequency.

zero, curve *A* has some negative value. If another curve, *B*, of the same frequency is drawn having zero amplitude at zero time, it can be used as a reference in describing curve *A*.

Curve *B* starts at zero and swings in the positive direction. Curve *A* starts at some negative value and also swings in the positive direction, not reaching zero until a fraction of a cycle after curve *B* has passed through zero. This fraction of a cycle is the amount by which *A* is said to lag *B*. Because the two curves have the same frequency, *A* will always lag *B* by the same amount. If the positions of the two curves are reversed, then *A* is said to lead *B*. The amount by which *A* leads or lags the reference is called its phase. Since the reference given is arbitrary, the phase is relative.

The phase angle between the two waves shown in Fig. 19-8 is approximately 90°. The phase angle can be measured in either *degrees* or in *radians*. Since 360° equals 2π rad, this angle can be expressed as $\pi/2$ radians.

In phase modulation the relative phase of the carrier is made to vary in accordance with the intelligence to be transmitted. The carrier phase angle, therefore, is no longer fixed. The *amplitude* and the *average frequency* of the carrier are held constant while the phase at any instant is being varied with the modulating signal.

The effect of a positive swing of the modulating signal is to increase the phase angle of the carrier with respect to its unmodulated value. A negative modulating signal does the opposite. A *vector* could be used to represent this change by rotating it, first clockwise, then counterclockwise, with respect to its reference position.

The phase at any instant is determined by the *amplitude* of the modulating signal. The rate of change in modulating-signal amplitude depends on two factors—the modulation amplitude and the modulation frequency. If the amplitude is increased, the phase deviation is increased. The carrier vector must move through a *greater angle* in the same period of time, increasing its speed, and thereby increasing the carrier frequency shift. If the modulation frequency is increased, the carrier must move within the phase-deviation limits at a *faster rate*, increasing its speed and thereby increasing the carrier frequency shift. When the modulating-signal amplitude or frequency is decreased, the carrier frequency shift is decreased also. The faster the amplitude is changing, the greater the resultant shift in carrier frequency; the slower the change in amplitude, the smaller the frequency shift.

A phase-modulation system is shown in the block diagram in Fig. 19-9. The transmitter oscillator is crystal controlled. This frequency is then amplified. The audio signal is applied to the RF by means of a combining network. The output of the combining network is fed into a series of class C amplifiers that provide the necessary frequency multiplication. The output of these multipliers is fed to the FPA.

19-10 OTHER METHODS OF PRODUCING PHASE MODULATION

E. H. Armstrong developed an FM system based on the principles of phase modulation. The carrier oscillator and the audio signal (suitably corrected) are fed into a *balanced modulator*. Here, the carrier is suppressed, but upper and lower AM sidebands are developed. Meanwhile, part of the crystal

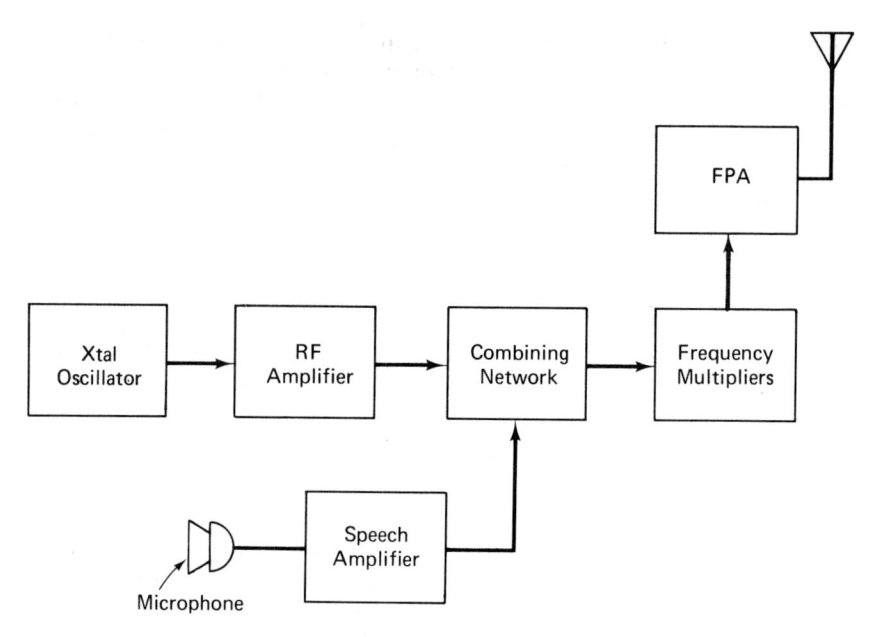

FIG. 19-9 Block diagram of phase-modulated transmitter.

oscillator output is connected also to a phase-shifting (*RC*) network to provide a 90° phase shift. The outputs of this phase-shift circuit and the balanced modulator are fed to a *linear network*, sometimes called a *combining circuit*, whose output is an FM signal.

The limitation of this technique is the very low amount of deviation obtainable. Typical values are 10 to 25 Hz. To meet the ±75 kHz-deviation required for 100 per cent modulation, *frequency multiplication* is required. Because this will also multiply the oscillator frequency, it needs to be low. A total multiplication of 3000 to 5000 times is essential to produce the ±75- kHz deviation. This requires oscillators to operate in the 15- to 30-kHz range, which is too low for crystal oscillators.

In practice, an oscillator frequency of about 200 kHz is used. Its output is fed to a series of frequency multipliers until about one third of the final operating frequency is reached. This is then reduced by frequency conversion, using a second crystal oscillator. The frequency of this latter oscillator is selected so that the converter output, when multiplied by a second group of frequency multipliers, will provide the desired carrier frequency. This is then fed to the FPA.

The advantage of the Armstrong system over direct FM systems is the excellent frequency stability provided by crystal oscillators. Many circuits are required for this system.

Another indirect FM system was developed many years ago around a special tube called a *phasitron*. It produced a shift, or deviation, of approximately 175 Hz from a starting crystal-controlled oscillator frequency of about 200 kHz. A series of frequency multipliers was needed to raise the deviation to 75 kHz. The number of multipliers was less than required for the Armstrong system. The manufacturer has discontinued production of these tubes except for replacement purposes.

19-11 BLOCK DIAGRAM OF A FREQUENCY-MODULATED TRANSMITTER

The operation of a typical direct FM transmitter can be understood with the aid of a block diagram such as shown in Fig. 19-10. The direct system

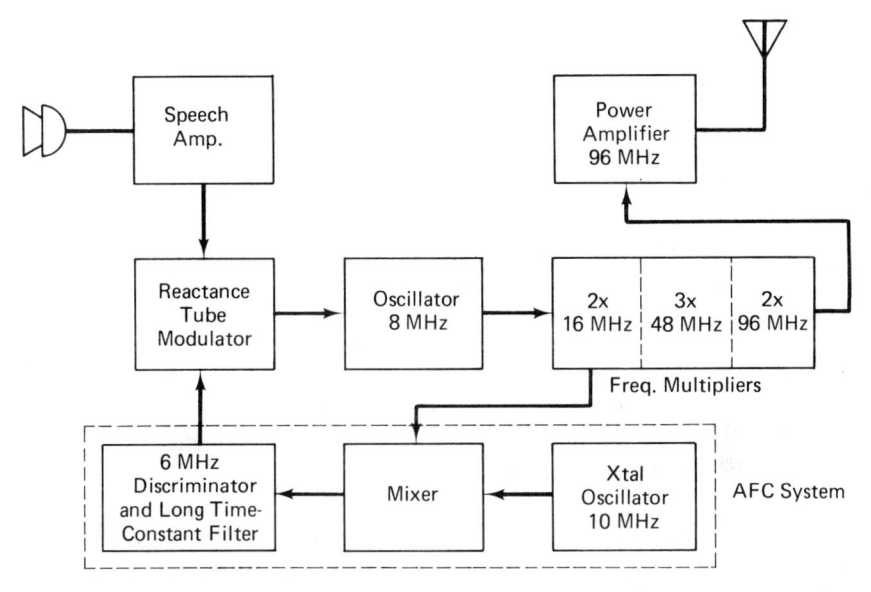

FIG. 19-10 Block diagram of an FM transmitter using reactance tube modulator and an AFC system.

of FM is also known as the *Crosby method*. The speech amplifier drives the reactance-tube modulator, which controls the deviation of the 8-MHz oscillator. A series of three frequency multipliers is used. The same frequency multipliers increase the original oscillator deviation to the required ± 75 kHz for 100 per cent modulation.

The final amplifier consists of one or more stages to raise the output power to that necessary to meet the station's license requirements. The antenna is generally a multielement, high-gain, horizontally polarized unit, mounted high atop a building or mountain to provide the maximum service area possible.

With any reactance-tube modulator–oscillator combination any voltage change in any part of the circuit will result in a shift of oscillator frequency. Some means must be provided to ensure the oscillator returning to its center, or resting frequency, in the absence of modulation. An effective way for accomplishing this is to use an *automatic frequency control* (AFC) circuit. Such an arrangement appears in Fig. 19-10 within the dotted rectangle, labeled AFC system.

A brief explanation of how the AFC system operates will enable us to understand how the oscillator can be kept on frequency. A more detailed explanation of an AFC system is given in Chapter 21. The block diagram shows that part of the output of the first doubler (16 MHz) is fed to the mixer. The mixer also has another input from a crystal-controlled oscillator, in this example 10 MHz. These two frequencies are heterodyned to produce a 6-MHz *difference frequency*. This is fed into a discriminator circuit (described in detail in Chapter 21), which is tuned to 6 MHz. The discriminator produces no output voltage when its input is exactly 6 MHz. Suppose, however, that the 8-MHz oscillator drifts upward from its center or resting frequency when no modulation is applied. The output of the first frequency doubler will now be slightly greater than 16 MHz. The mixer output will also increase, and the discriminator will develop an output voltage. This voltage is fed to the reactance-tube modulator, changing its reactance and shifting the oscillator frequency *downward* to its correct value. At this point there will be no output from the discriminator, and the system stabilizes at the correct frequency. If for any reason the 8-MHz oscillator should drift downward, the polarity of the output voltage from the discriminator would be reversed. The reactance of the modulator is changed, and the oscillator is returned to the correct frequency again. Automatic-frequency-control systems are widely used in FM receivers and numerous industrial control systems.

In practice, the oscillator frequency is alternately above and below its center frequency as a result of the speech amplifier controlling the reactance-tube modulator. This means that the output of the first frequency multiplier will be varying about the value indicated. Consequently, there will be an ac output from the discriminator which is constantly trying to swing the oscillator in a direction opposite to that demanded by the reactance-tube modulator. To overcome this limitation, a *long time constant filter* (*RC* network) is added to the output of the discriminator. This filters out the ac and lets only *quasi dc voltages* pass, which represent relatively long-term changes of oscillator frequency.

The FCC's regulations require a center-frequency stability of ± 2 kHz for broadcast stations. This requirement is best satisfied by some form of automatic frequency control.

19-12 PUBLIC SAFETY RADIO SERVICE

A very large number of FM transmitters are found in the public-safety section of the community. This broad field includes local and state police, fire departments, forestry–conservation, highway maintenance, and local government. To this group should be added the communications facilities of taxicabs and other segments of private industry wherein salesmen and service personnel can be in contact with their offices. Voice communication is all that is required, and consequently the AF bandwidth can be confined to about 200 Hz to 3 kHz. This is adequate for speech transmission.

It is unnecessary to have a bandwidth of 200 kHz as with FM broadcast stations. For public-safety service only 20 kHz is required. Likewise, only ± 15-kHz deviation is necessary. In fact, most of the public safety radio service transmitters in use today deviate only ± 5 kHz. This results in what is called *narrowband FM* and provides more FM channels.

By reducing the deviation to ± 5 kHz and limiting the audio to 3 kHz, a deviation ratio of 1.67 can be achieved. Phase modulators are generally used in these applications, as they give increased frequency deviation and little additional distortion. Multiplying the output of the phase modulator about 12 times provides the required ± 5-kHz carrier deviation. At the receiver the reduced deviation results in less audio output. This is partially compensated by reduced noise resulting from narrow receiver bandwidth.

Public Safety Radio Services operating in the 25- to 470-MHz spectrum are permitted a bandwidth of 20 kHz. The frequency deviation is ± 5 kHz. This is known as 20 F3 emission, where the 20 represents the bandwidth in kilohertz. In the 470- to 950-MHz part of the RF spectrum the authorized bandwidth is 40 kHz and the frequency deviation is ± 15 kHz. This is called 40 F3 emission. When amplitude modulation is used for telephony, the modulation percentage should be sufficient to provide efficient communication and shall be normally maintained above 70 per cent on peaks, but shall not exceed 100 per cent on negative peaks. When phase or frequency modulation is used for telephony, the deviation arising from modulation shall not exceed ± 15 kHz from the unmodulated carrier. The requirements specified in this and the preceding paragraph do not apply to transmitters authorized to operate as mobile stations with a maximum plate input to the final RF stage of 3 W or less.

Transmitters are *type accepted* (prior approval by the FCC to the manufacturer for particular models of equipment) and range from 1 to 60 W for mobile units. Base station transmitters range from 60 W to 2 kW maximum. Both transmitting and receiving antennas are *vertically polarized*.

The licensee of each station shall employ a suitable procedure to determine that the carrier frequency of each transmitter falls within the limits prescribed in Table 19-1. This determination shall be made, and the results thereof entered in the station records, in accordance with the following:

1. When the transmitter is initially installed.
2. When any change is made in the transmitter that may affect the carrier frequency or the stability thereof.

TABLE 19-1

| | All Fixed Stations (% of carrier) | All Mobile Stations | |
Frequency Band (MHz)		3 W or Less (% of carrier)	Over 3 W (% of carrier)
Below 25	0.01	0.02	0.01
25–50	0.002	0.005	0.002
50–1000 (except for 450–470)	0.0005	0.005	0.0005
450–470	0.00025	0.0005	0.0005
Above 1000	*	*	*

*Specified in the station authorization.

3. At intervals not to exceed 6 months for transmitters employing crystal-controlled oscillators.
4. At intervals not to exceed 1 month for transmitters not employing crystal-controlled oscillators.

The licensee of each station shall employ a suitable procedure to determine that the plate power input to the final RF stage of each base station or fixed station transmitter, authorized to operate with a plate power to the FPA in excess of 3 W, does not exceed the maximum figure specified in the current station authorization. This determination shall be made, and the results thereof entered in the station records, in accordance with items 1, 2, and 3 of the preceding paragraph.

In addition to the preceding requirements, the licensee of each station must employ suitable procedures to determine that the modulation of each transmitter, with an input to the FPA of more than 3 W, does not exceed the limits specified. These measurements shall be made at the time intervals indicated in the preceding paragraph. All measurements required by the FCC for base and fixed stations may be made, at the option of the licensee, by any qualified engineering measurement company. In this case the required record entries shall show the name and address of the company as well as the name of the person making the measurements.

In the case of mobile installations the measurements called for in the preceding paragraphs may be made at a test or service bench, provided the measurements are made under *load conditions equivalent to actual operating conditions*. Subsequent to these tests, and after the installation, the transmitter must be given a routine test to determine that it is capable of being satisfactorily received by an appropriate receiver.

Before any radio station can be established, whether in the Public Safety Radio Service or otherwise, it is necessary to submit an *application for a station authorization* to the FCC. These forms are available from any FCC field office or from the Commission's office in Washington, D.C. When the station authorization is received, the installation may commence. When the station is completed, the engineer-in-charge of the local field office must be notified before tests are made. In an emergency it is possible to obtain telephone authorization to operate a new station, such as a mobile unit, for a period of less than 10 days.

It is necessary that some *control point* be established for each communication system that must be under the supervision of the licensee. Monitoring facilities are to be located at this point, and persons responsible for the maintenance of the system are stationed there. For this kind of operation the mobiles as well as the base station may transmit and receive on the same frequency or use separate ones. Station identification can be by the *assigned call* or by *other codes*, provided FCC approval has been obtained. Station identification must be made at the *end of each transmission* or at least once every 30 min.

19-13 STEREO MULTIPLEX

The *binaural* nature of man's hearing has prompted engineers to develop a system that permits the simultaneous transmission of speech or music from two separate sources, such as both sides of an orchestra. When these signals are reproduced over two separate loudspeakers, *spatial effect* is produced that gives three-dimensional qualities to the sound. This provides considerable realism to the transmitted information and is called *stereophonic sound*. With this system it is essential that the information picked up by the left microphone be transmitted over a separate channel and be reproduced by the left loudspeaker. The same must apply to the right channel.

To transmit stereophonic sound over a broadcast station, some method must be provided whereby both channels may be transmitted independently without interfering with each other. This has been accomplished by the use of an *FM multiplex system*. The two separate channels are appropriately combined to frequency modulate the carrier of the transmitter. One requirement of an FM stereo multiplex system is that it be *compatible* with a conventional monaural system. The bandwidth necessary to transmit the two stereophonic channels requires the use of FM.

To understand this system, let us begin by analyzing the process that occurs as far as *monophonic* reception is concerned. The amplified output of both the left and right microphones is combined and fed into the FM modulator at the transmitter, as shown in the block diagram of Fig. 19-11. This constitutes what is known as the $L + R$ signal. The output of both microphones is combined to provide a point source of sound. If only one microphone had been used, the sound nearest the microphone would be reproduced much louder than that coming from farther away. The listener would have

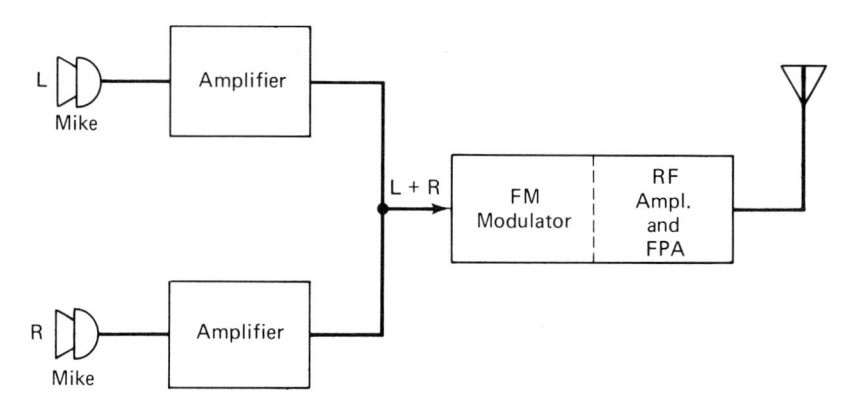

FIG. 19-11 The combination of the L and R channels to provide monophonic transmission.

the feeling that he was sitting closer to one side of the orchestra than the other.

In a typical FM stereo multiplex system the output of the right and left microphone amplifiers, after preemphasis, is fed into a *matrix*, as shown in Fig. 19-12. The matrix contains an *addition* circuit $(L + R)$ and a *subtraction* circuit $(L - R)$. In the matrix the R signal is reversed in polarity and then combined with the L signal. This produces the $L - R$ signal component.

The $L + R$ and $L - R$ outputs of the matrix are fed to their respective amplifiers for additional amplification. The $L + R$ amplifier output is fed to the reactance-tube modulator of the transmitter. The $L - R$ amplifier output is fed to a balanced modulator.

Next, consider the function of the balanced modulator. Observe that a 19-kHz crystal oscillator is used. This is referred to as the *pilot* oscillator. This is fed to the reactance-tube modulator and a doubler. This 38 kHz, referred to as a *subcarrier*, is modulated by the signal coming from the $L - R$ amplifier. The result of this amplitude modulation is the generation of *upper* and *lower sidebands* with a *suppressed carrier*. The output of the modulator, referred to as DSSC (double-sideband suppressed carrier) on the block diagram, is fed to the reactance-tube modulator. Hence, the reactance-tube modulator is being modulated by several signals. This is why it is called a *multiplex system*.

Let us consider the relationships of the several frequencies fed to the reactance-tube modulator by examining the diagram shown in Fig. 19-13. The frequency base of this diagram represents a bandwidth of approximately 75 kHz, which is the carrier deviation of an FM broadcast station when 100 *per cent* modulated. The $L + R$ signal occupies the *lower part* of this spectrum, from 50 Hz to 15 kHz. The pilot subcarrier signal, used at the receiver for *reinser-*

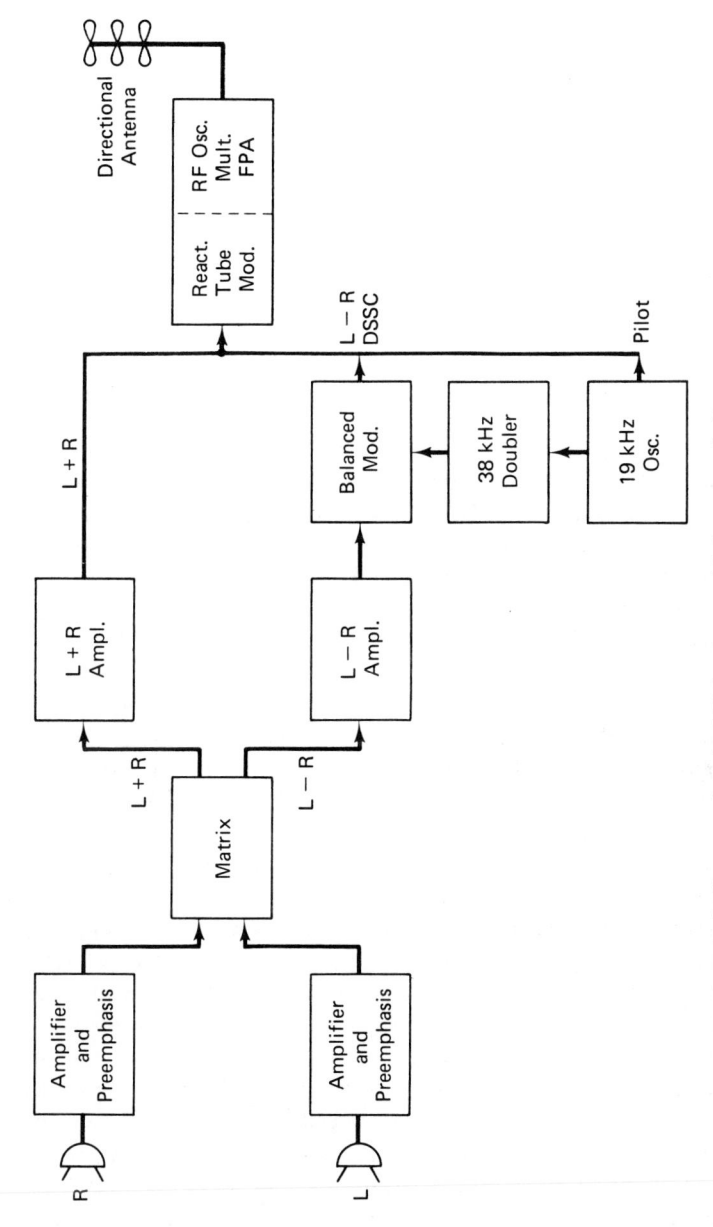

Fig. 19-12 Block diagram of typical FM stereo multiplex transmitting system.

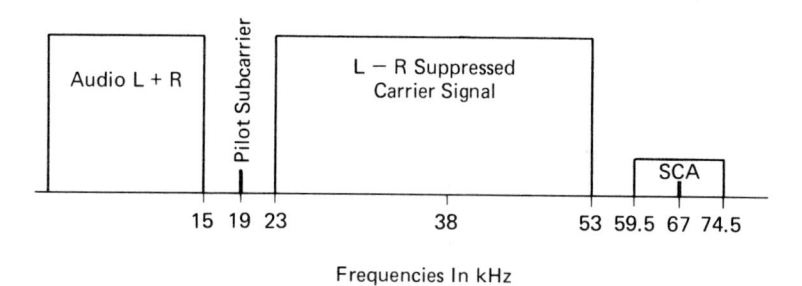

FIG. 19-13 Modulation and subcarrier frequency distribution.

tion of the suppressed carrier, is centered on 19 kHz. The 38-kHz subcarrier, which is fed to the balanced modulator, has AM sidebands. These sidebands extend 15 kHz above and below the 38-kHz subcarrier when the station is transmitting the maximum allowable AF range of signals. Hence, the lower sideband extends from 23 to 38 kHz and the upper sideband from 38 to 53 kHz. Very selective filters are employed at the transmitter to prevent any undesirable beat frequencies from being generated between the 19-kHz pilot signal and audio frequencies above 15 kHz from the $L + R$ channel, and frequencies below 23 kHz from the modulation effect of the lower sideband $L - R$ channel. Theoretically, there should be no *crosstalk* between the stereophonic and monaural channels; however, experience indicates there is always a little. The 19-kHz pilot subcarrier is above the audible spectrum and is not reproduced in the FM receiver. According to FCC regulations, this crosstalk must be held down to a level of 40 dB or more between channels.

There is a small box in the spectrum between 59.5 and 74.5 kHz, as shown in Fig. 19-13. The FCC has made provisions for the broadcasting of a special channel, which can be transmitted to organizations that desire to subscribe to background music. This is known as *Subsidiary Communication Authorization* (SCA). This musical background is transmitted on a carrier centered at 67 kHz. The modulation of this carrier by audio frequencies up to 7.5 kHz produces the upper and lower sidebands indicated in the drawing. This type of multiplex transmission can be applied to any FM broadcast station. Maximum SCA modulation cannot exceed 10 per cent of the total stereophonic signals to be transmitted. There must be at least 60 dB of separation between SCA and the main carrier at 100 per cent modulation. The transmitting station is required to keep a log of SCA transmissions. This log will contain the time the subcarrier generator is turned on and off and the time that modulation is applied to and removed from the carrier.

The following tabulation indicates the various percentages of carrier modulation for the different types of signals transmitted.

Station and Type of Signal	Carrier Modulation (%)
Monaural station without SCA	
Main channel	100
Monaural station with SCA	
Main channel	90
SCA	10
Stereo station without SCA	
Main channel	45
Subchannel	45
Pilot subcarrier	10
Stereo station with SCA	
Main channel	40.5
Subchannel	40.5
Pilot subcarrier	9
SCA	10

To better understand how the L and R signals can be combined to provide stereo information, let us briefly consider the action that takes place in a stereo receiver. It contains a matrix section, also, that does the adding and subtracting necessary to reconstruct the original L and R signals for stereo reproduction. When the signal components are added in the receiver, the following action takes place. Signal 1 equals $(L + R) + (L - R)$, which when added provides $2L$. When the signals are combined out of phase in the receiver, the following results: signal 2 equals $(L + R) - (L - R)$ equals $2R$. It is the matrix section of the receiver that provides this action.

Practice Problems

1. What is the modulation index of a standard FM broadcast station when the modulating frequency is 5 kHz?
2. Referring to problem 1, if the modulating signal is 25 kHz, what is the modulation index?
3. What will be total frequency swing of a standard FM broadcast station that is (a) 10 per cent modulated? (b) 90 per cent modulated?
4. What is the modulation index of a narrowband FM transmitter (Public Safety Radio Service) when modulated with a 2.5-kHz signal?

Commercial License Questions

Sections in which answers to questions are given appear in parentheses. A bracketed number following a question implies that it applies only to that element.

1. What are the merits of an FM communication system compared to an AM communication system? (19–1)

2. What type of modulation is largely contained in static and lightning radio waves? (19–1)
3. Explain what is meant by the following types of emission: F0, F1, F2, F3, F4, and P0 emission? (19–1)
4. What characteristics of an audio tone determine the percentage of modulation of an FM broadcast transmitter? [4] (19–2)
5. What determines the rate of frequency swing of an FM broadcast transmitter? [4] (19–2)
6. What is the meaning of the term *center frequency* in reference to FM broadcasting? [4] (19–2)
7. What frequency swing is defined as 100 per cent modulation for an FM broadcast station? [4] (19–2)
8. Draw a diagram of a means of modulation of an FM broadcast station. (19–2, 19–7, 19–8, 19–9)
9. What is the AF range that an FM broadcast station is required to be capable of transmitting? [4] (19–3)
10. In an FM radio communication system what is the meaning of *modulation index*? Of *deviation ratio*? What values of deviation ratio are used in an FM radio communication system? (19–3)
11. Define amplifier gain, percentage deviation, stage amplification, and percentage of modulation. Explain how each is determined. [4] (19–4)
12. An FM broadcast transmitter is modulated 40 per cent by a 5-kHz tone. When the percentage of modulation is doubled, what is the frequency swing of the transmitter? [4] (19–4)
13. An FM broadcast transmitter is modulated 50 per cent by a 7-kHz tone. When the frequency of the test tone is change to 5 kHz and the percentage of modulation is unchanged, what is the transmitter frequency swing? [4] (19–4)
14. What is the frequency swing of an FM broadcast transmitter when modulated 60 per cent? [4] (19–4)
15. Why is frequency modulation undesirable in the standard broadcast band? [4] (19–5)
16. How wide is an FM broadcast channel? [4] (19–5)
17. What is the relationship between the number of sidebands and the amplitude of the modulating voltage in FM? [4] (19–5)
18. What are the criteria for determining bandwidth of emission in FM? (19–5)
19. What is the relationship between the spacing of the sidebands and the modulating frequency of FM? (19–5)
20. What is the relationship between the number of sidebands and the bandwidth of emission in FM? (19–5)

Commercial License Questions

21. What is the relationship between modulation index or deviation ratio and the number of sidebands in FM? (19–5)
22. What is the relationship between percentage of modulation and the number of sidebands in FM? (19–5)
23. What is the relationship between the number of sidebands and the amplitude of the modulating voltage in FM? (19–5)
24. What is the relationship between the number of sidebands and the modulating frequency in FM? (19–6)
25. What is meant by *preemphasis* in an FM broadcast transmitter? (19–6)
26. What are the reasons for preemphasis and deemphasis in FM? (19–6)
27. What is the purpose of a reactance tube in an FM broadcast transmitter? (19–7)
28. How is good stability of a reactance-tube modulator achieved? (19–7)
29. What is a common method of obtaining frequency modulation in an FM broadcast transmitter? (19–7, 19–8)
30. What is the difference between frequency and phase modulation? (19–9)
31. Explain briefly what occurs in a waveform if it is phase modulated. (19–9)
32. Describe briefly the operation of the Armstrong and the phasitron method of obtaining phase modulation. (19–10)
33. Explain in a general way how radio signals are transmitted through the use of frequency modulation. (19–11)
34. Why are high-gain antennas used in FM broadcast stations? (19–11)
35. What is the frequency tolerance of an FM broadcast station? (19–11)
36. Discuss wideband and narrowband FM systems with respect to frequency deviations and bandwidth. (19–12)
37. Why is narrowband rather than wideband FM used in radio communication systems? (19–12)
38. In communication services such as the Public Safety Radio Service (a) what percentage of modulation is normally required when amplitude modulation is used for radiotelephony, and (b) what maximum frequency deviation arising from modulation is permitted when phase or frequency modulation is used for radiotelephony? (19–12)
39. What are the authorized bandwidth and frequency deviation of public-safety stations operating at about 30 MHz? At 160 MHz? (19–12)
40. The carrier frequency of a public-safety transmitter operating at 160 MHz with a license power of 50 W must be maintained within what percentage of the licensed value? (19–12)
41. What legal requirements must be met before installing and operating a radio station in services such as the public safety? (19–12)

42. Where may standard forms applicable to the Public Safety Radio Service be obtained? (19–12)
43. What notification must be forwarded to the engineer-in-charge of the FCC district office prior to testing a new radio transmitter in the Public Safety Radio Service? (19–12)
44. May public-safety stations be operated for short periods of time without a station authorization issued by the Commission? (19–12)
45. Define *control point* as the term refers to transmitters in the Public Safety Radio Service. (19–12)
46. What are the general requirements for transmitting the identification announcements for stations in the Public Safety Radio Service? (19–12)
47. What are the Commission's general requirements for the records or logs that must be kept by stations in the Public Safety Radio Service? (19–12)
48. When a radio operator makes transmitter measurements required by the Commission's rules for a station in the Public Safety Radio Service, what information should be transcribed into the station's records? (19–12)
49. Outline the transmitter measurements required by the Commission's rules for stations in the Public Safety Radio Service. (19–12)
50. What is the maximum percentage of modulation allowed by the Commission's rules for stations in the Public Safety Radio Service that utilize amplitude modulation? (19–12)
51. What are the stereophonic transmission standards provided by the Commission rules? (19–13)
52. What are the transmission standards of subsidiary communications multiplex operations? (19–13)
53. What are some of the possible uses of SCA? (19–13)
54. What items must be included in an SCA operating log? (19–13)
55. Define the term *multiplex transmission* as referred to in FM broadcasting. (19–13)
56. Define the term *crosstalk* as referred to in FM broadcasting. (19–13)
57. Define the term *left signal* as referred to in FM broadcasting. (19–13)
58. Define the term *left sterophonic channel* as referred to in FM broadcasting. (19–13)
59. Define the term *main channel* as referred to in FM broadcasting. (19–13)
60. Define the term *pilot carrier* as referred to in FM broadcasting. (19–13)
61. Define the term *stereophonic separation* as referred to in FM broadcasting. (19–13)
62. Define the term *stereophonic subcarrier* as referred to in FM broadcasting. (19–13)

20
Antennas

20-1 CHARACTERISTICS OF RADIO WAVES

A transmitting antenna converts electrical energy into electromagnetic and electrostatic waves (called radio waves), which radiate from the antenna at the velocity of light. A receiving antenna converts these into electrical energy. When alternating current is fed into a vertical electrical conductor of the proper length, an *electrostatic field* is set up between the conductor and ground. The strength of this field is related to the amount of RF current in the conductor. The conductor acts as one plate of a capacitor and the ground as the other. As the current goes through its cyclical variations, the electrostatic field builds up to a maximum in one direction, then collapses and goes to a maximum in the opposite direction. The rise and fall of the electrostatic field is directly related to the frequency of the current. This is portrayed in Fig. 20-1.

At low frequencies the electrostatic field collapses back into the wire as the current returns to zero. If higher-frequency energy is fed into the conductor (also called an antenna or radiator), all the lines of force in the dielectric field do not get a chance to collapse back into the antenna before the current reverses and another field begins to form. These lines of force *break free* from the antenna and are *radiated into space*. The lowest frequencies useful in radio communications are about 10 kHz.

The lines of force from the first field that did not return to the antenna are pushed out by the second field and become separated from it. This cycle is repeated at the RF rate, and a continuous flow of electrostatic lines of force emanate from the antenna.

The radiated energy spreads out in all directions, just as ripples in a pond when a stone is thrown in. The farther the wave travels, the *weaker* it becomes,

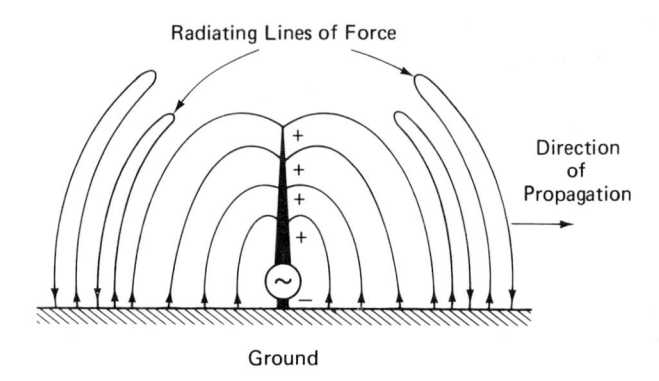

FIG. 20-1 Illustration of how lines of force are radiated from an antenna.

since the total energy in the radiated field is constant. The type of terrain the signal passes over affects its strength. There is more *attenuation* of that portion of the RF wave traveling near the surface of the earth, called the *ground wave*, because the earth absorbs some of the energy. If the earth is *moist*, it is a better conductor than if it is *dry*.

Associated with every moving electrostatic field is an electromagnetic field. These are at *right angles* to each other. This is represented in Fig. 20-2. On the next half-cycle of antenna current the E and H vectors would be pointing in opposite directions.

The velocity of radio waves is *constant* regardless of frequency. Therefore, the *wavelength* (distance traveled by the wave in one cycle) can be determined by the formula

$$\lambda = \frac{3 \times 10^8}{f}$$

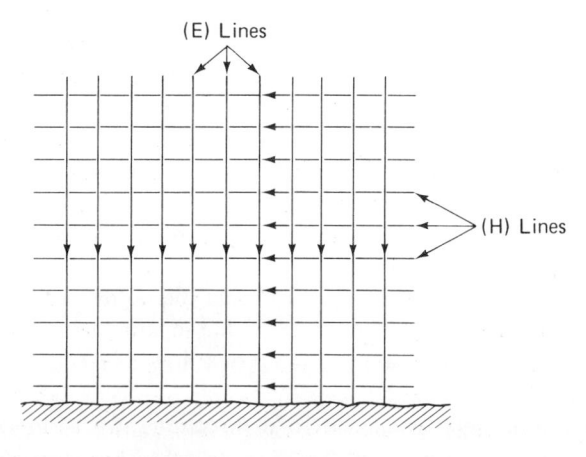

FIG. 20-2 Instantaneous cross section of a radio wave showing electrostatic (E) and electromagnetic (H) components.

Ground Waves

where λ = one wavelength
f = frequency in hertz
3×10^8 = velocity in meters per second

It can also be expressed as

$$f_{\text{MHz}} = \frac{300}{\lambda}$$

Example 20-1

What is the frequency of the current in a transmitting antenna if it is radiating an electromagnetic wave having a wavelength of 2 meters (m)?

Solution

$$f_{\text{MHz}} = \frac{300}{2} = 150$$

Example 20-2

When is the wavelength of a broadcast station operating on 1260 kHz?

Solution

$$\lambda = \frac{300,000}{f_{\text{kHz}}} = \frac{300,000}{1260} = 238 \text{ m}$$

20-2 GROUND WAVES

When a radio wave leaves a vertical antenna, the *field pattern* produced resembles a huge doughnut with the antenna in the center of the hole. Part of the wave moves outward and stays in contact with the ground to form the *ground wave*; the rest moves upward and outward to form the *sky wave*. The ground and sky portions of the electromagnetic wave are responsible for two different modes of transmitting information from the transmitter to the receiver.

Lower-frequency radio waves (long wavelengths) tend to travel along the surface of the earth with little attenuation. These are called ground waves and are usable, day or night, for distances of hundreds to thousands of miles at high power levels. The reception of commercial stations during the daytime is largely by means of ground waves.

The ground wave is made up of two parts: a *surface* and a *space wave*. The surface wave travels along or very near the surface of the earth. The space wave travels in two paths, immediately above the surface of the earth. One is directly from the transmitter to the receiver. The other is a path in which the space wave is reflected from the ground prior to reaching the receiver. These are illustrated in Fig. 20-3. Inasmuch as these two paths are different lengths, the signals arriving at the receiving point may be in or out of

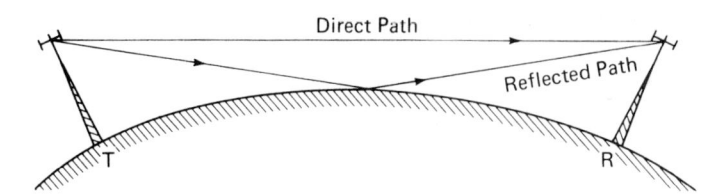

FIG. 20-3 Paths of direct and ground-reflected waves.

phase with respect to each other. The space-wave portion of the ground wave becomes more important as the transmitter antenna height is increased or as the frequency is increased.

As the surface wave passes over the ground it induces a voltage into it setting up eddy currents. This absorbs energy from the surface wave and weakens it as it moves away from the antenna. The attenuation *increases* as the frequency is increased. This phenomenon limits surface-wave communication to relatively low frequencies.

The electrical properties of the earth are relatively constant; therefore, the surface-wave strength from a given station, at a given point, is nearly constant. In regions where there are distinct dry and rainy seasons the amount of moisture in the soil varies, and thus the *conductivity* of the soil changes. This affects the distance that the surface wave may be received. When possible, stations are located close to continually damp ground, or bodies of water, because of the superiority of surface-wave conduction.

20-3 IONOSPHERE

Surrounding the earth is a rather thick layer of ionized gas called the *ionosphere*. This is found in the rarefied atmosphere approximately 40 to 350 miles above the earth. It differs from the other atmosphere in that it contains a much higher number of positive and negative ions. These are produced by the ultraviolet and particle radiations from the sun. The rotation of the earth and the annual course of the earth around the sun, together with the development of sunspots, all affect the number of ions present in the ionosphere. These, in turn, affect the quality and distance of radio transmission.

The ionosphere is composed of three layers, designated, respectively, from lowest to highest as D, E, and F. The F layer is divided into two layers designated F_1 (the lower layer) and F_2. The presence or absence of these layers, and their height, vary with the rotation of the earth on its axis and around the sun, and also with sunspot activity. The relative position of these layers is shown in Fig. 20-4. The D layer ranges from about 40 to 55 miles. This layer has the ability to *refract* signals of low frequency back to the earth. Higher frequencies pass through. After sunset this layer *disappears* because of the rapid recombination of ions.

The E layer ranges from approximately 55 to 90 miles. Sometimes this

Sky Waves

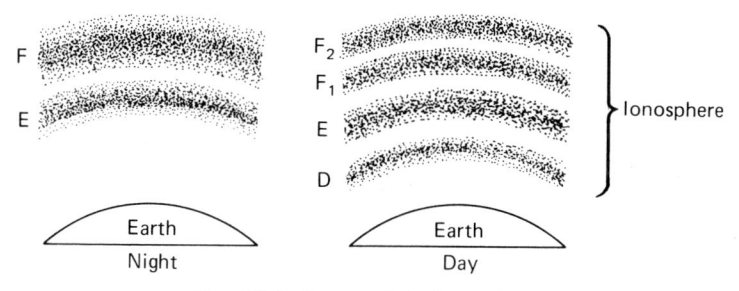

FIG. 20-4 Layers of the ionosphere.

is referred to as the Kennelly–Heaviside layer because these two men were the first to propose its existence. This layer begins to disappear after sunset and is nearly gone by midnight. It has the ability to refract signals of a higher frequency, up to perhaps as high as 20 MHz, for distances up to about 1500 miles.

The F layer exists from about 90 to 350 miles. This splits into two layers designated F_1 and F_2 during the daylight hours. The ionization level in these layers is quite high and varies widely during the course of the day. Since the atmosphere is rarefied at these heights, the recombination of the ions occurs solely after sunset. This results in a reasonably constant ionized layer present at all times. The F layers are responsible for *high-frequency, long-distance* transmission.

20-4 SKY WAVES

The part of the wave that moves upward and is not in contact with the ground is called the *sky wave*. Depending on the frequency, sky waves are refracted by the ionosphere so that they come back to the earth. These waves permit the receiving of strong signals several hundred miles beyond the range of the ground wave. An example of this bending is shown in Fig. 20-5,

The ability of the ionosphere to refract waves depends on the *angle* at which they strike the ionosphere their *frequency*, and the *ion density*. If the frequency is correct, the ionosphere sufficiently dense, and the angle correct, the wave will eventually emerge and return to earth. If the receiver is located at point B or D, it will be received. For convenience in analyzing sky waves, the wave shown in Fig. 20-5 is assumed to be composed of waves that emanate from the antenna in three distinct groups identified according to the angle of elevation. That part of the wave which strikes the ionosphere at a large angle is bent, but not sufficiently to return to earth. That part which strikes the ionosphere at point A is refracted back to earth. The angle θ formed by this wave front is called the *critical angle* for that frequency. Any ray leaving the antenna at an angle greater than this will pass through

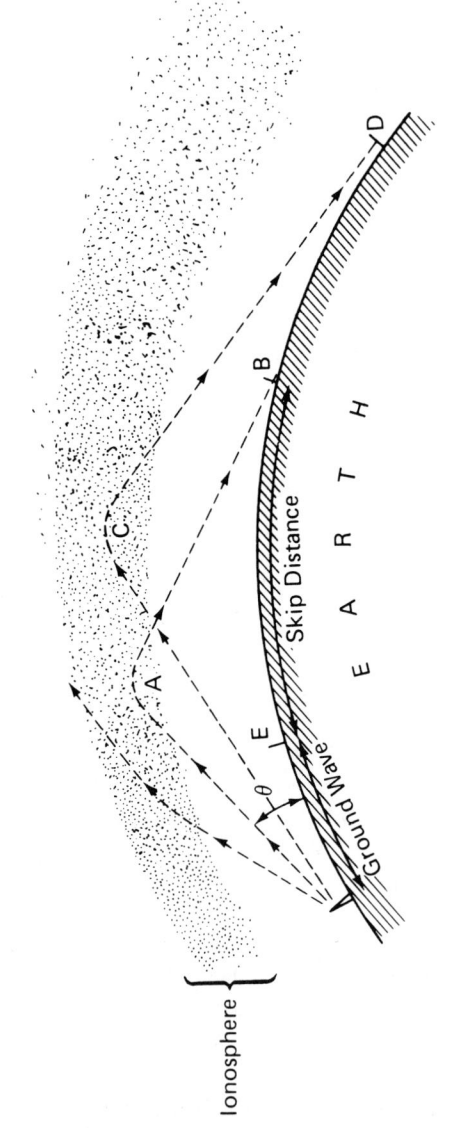

Fig. 20-5 Refraction of skywaves from ionosphere showing skip distance and critical angle θ.

Atmospheric Disturbances 431

the ionosphere and not be refracted back. As the transmitted frequency increases, the critical angle decreases.

Low frequencies can be projected almost straight upward and still be returned to earth. The highest frequency that can be sent upward and still return to earth is called the *critical frequency*. Higher frequencies, regardless of the angle, will not return. The critical frequency is not constant, but seems to vary from one locality to another, with the time of day, the season of the year, and the amount of sunspot activity. In Fig. 20-5 the area between points B and D will receive the transmission by way of refracted sky waves. The area between the transmitter and point E will receive the transmission by means of ground waves. If a receiver is located in the area marked skip distance, between points E and B, it will not receive the transmitted signal because neither the sky wave nor the ground wave reaches this area.

In practice, the wave front intercepts the ionosphere at a great number of angles, and much of this energy is refracted back to earth. Some of this energy strikes the earth and is reflected back up to the ionosphere, where it is bent back to the earth again. Thus a portion of the wave front may be refracted from the ionosphere several times. These are called *two-hop* and *three-hop* waves, and so on. Now, if two parts of the wave front arrive *in phase* at a particular receiving spot, the signal strength will be *increased*. However, if the two waves arrive *out of phase*, they will tend to cancel and the signal will be *weakened or lost*. This is called *fading*.

Fading can also occur if the ground and sky waves come in contact with each other. This type of fading becomes severe if the two waves are approximately equal in amplitude. Variations in the amount of energy absorbed in the ionosphere, and in the length of the path that the wave travels in the ionosphere, are also responsible for fading. Occasionally, disturbances in the ionosphere may cause complete absorption of all sky-wave radiation. Receivers located near the outer end of the skip zone are subject to fading as the sky wave alternately strikes and skips over the area.

20-5 ATMOSPHERIC DISTURBANCES

Many factors other than its strength influence the quality of a signal and the distance at which it can be received.

Lightning is one of the greatest sources of disturbance, or static. Like any spark, it produces RF energy that extends over most of the usable radio spectrum. At lower radio frequencies the energy is great and the interference produced can be considerable, even with storms hundreds of miles away, particularly if a good ground wave is present. In the 5- to 15-MHz range local storms are more troublesome. Above these frequencies the amount of interference produced by lightning diminishes. This is why the higher radio frequencies are more immune to lightning-produced static and generally provide more static-free communications.

The density of the number of layers in the ionosphere is influenced by the amount of ultraviolet radiation from the sun. This radiation varies from night to day as well as daily and seasonally. Very marked changes in ionization also occur throughout the 11-year sunspot cycle.

Not to be ignored is the *aurora borealis*, sometimes called northern lights. These are assumed to be *magnetic storms* caused by changes in the earth's magnetic field. They produce significant changes in the ionosphere and are responsible for the generation of such great amounts of static that lower communications frequencies can be blanked out. This disturbance is greatest near the polar regions and diminishes as the distance from there increases.

20-6 TRANSMISSION LINES

The electrical characteristics of a transmission line are dependent upon such factors as length, spacing, type of dielectric materials, and the amount of L and C present.

If the current flowing in an infinitely long line and the applied voltage are known, the impedance can be determined by using Ohm's law for the ac circuit. This is called the *characteristic impedance* of the line and is the same no matter where it is measured along the line. The characteristic impedance is also called the *surge impedance* and is determined by the equation

$$Z_o = \sqrt{\frac{L}{C}}$$

The formula indicates that the characteristic impedance depends on the ratio of the distributed inductance and capacitance in the line. An increase in the separation of the wires increases the inductance and decreases the capacitance.

Thus, the effect of increasing the spacing of the two wires is to *increase* the characteristic impedance, because the L/C ratio is increased. Similarly, a *reduction* in the diameter of the wires also increases the characteristic impedance, because the wire size affects the capacitance more than the inductance. The impedance can be calculated by the formula

$$Z_o = 276 \log_{10} \frac{d}{r}$$

where d = center-to-center distance between conductors
 r = radius of conductors (measured in same units as d)

Example 20-3

Calculate the impedance of two No. 12 wires (diameter 0.081 in., radius 0.04 in.) held 6 in. apart between centers.

Solution

$$Z_o = 276 \log_{10} \frac{d}{r}$$

$$= 276 \log_{10} \frac{6}{0.04} = 276 \log 150$$

$$= 276 \times 2.1761 = 600 \, \Omega$$

The characteristic impedance of a coaxial line also varies with L and C, but in a different manner. Hence, the following formula must be used:

$$Z_o = 138 \log_{10} \frac{D}{d}$$

where D = inner diameter of outer conductor
d = outer diameter of inner conductor

If a line is terminated in its characteristic impedance, the same amount of current would flow as if the line were infinitely long. This permits a maximum transfer of power between the generator and load.

The characteristics of a theoretically infinite line may be summarized as follows:

1. The voltage and current are *in phase* throughout the line.
2. The ratio of the voltage to the current is constant over the entire line and is known as the characteristic impedance.
3. The input impedance is equal to the characteristic impedance.
4. Since the voltage and current are in phase, the line operates at maximum efficiency.
5. Any length of line can be made to appear like an infinite line if it is terminated in its characteristic impedance.

When a transmission line properly terminated, *reflections* do not occur. However, if there is an *abrupt discontinuity* in the line, such as an open or short circuit, complete reflection will occur. This means that as the current and voltage waves progress down the line they will reflect back to the generator end. Here they encounter the energy being put in by the generator. They are reflected back and forth again, ad infinitum, and cause *standing waves* of current and voltage to be established on the line. When the line is open, the voltage and current relationships shown in Fig. 20-6a exist. When the line is shorted, the e and i relationships are as shown in Fig. 20-6b.

From this we may conclude that a resonant line will present to its source a high or low impedance at multiples of a quarter-wavelength, depending on how the output is terminated. At points that are not exact multiples of a quarter-wavelength, the line acts as a capacitor or inductor.

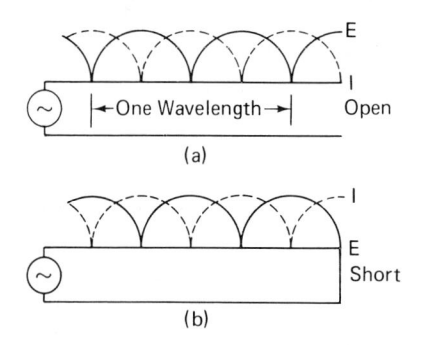

FIG. 20-6 Standing waves of voltage and current: (a) line terminated in an open; (b) line terminated in a short.

If a transmission line one quarter-wavelength long is terminated in an impedance equal to its surge impedance, the ratio between currents at the opposite ends is 1:1. Since the line is properly terminated, it becomes nonresonant or behaves as an infinite-length line. There will be no standing waves, and the current has the same amplitude all along the line.

A transmission line, such as a *coaxial cable*, must have sufficient spacing and insulation between conductors to withstand the voltage between them. For example, suppose that it is necessary to determine the rms voltage between the inner conductor and metal sheath of a 72-Ω coaxial cable when 5 kW is applied. Assuming that the line is nonresonant, the voltage can be determined as

$$e_{rms} = \sqrt{WZ} = \sqrt{5000 \times 72} = 600 \text{ V}$$

The peak value would be 1.414 times this.

In the above example, what is the rms value of current in the transmission line?

$$i_{line} = \sqrt{\frac{W}{Z_L}} = \sqrt{\frac{5000}{72}} = 8.33 \text{ A}$$

20-7 HALF-WAVE ANTENNA

The simplest type of antenna is a single wire, properly suspended above ground, whose length is approximately equal to half the transmitted wavelength. Such an arrangement is shown in Fig. 20-7 where the generator represents the transmitter output. Each wire is one quarter of the wavelength of the generator's output. This is known as a *dipole antenna*.

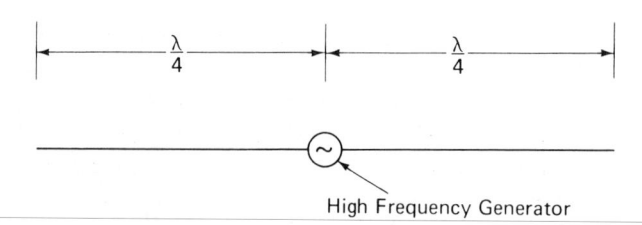

FIG. 20-7 Dipole antenna.

Half-Wave Antenna

Consider that the instantaneous generator polarity is as shown in Fig. 20-8a. Electrons are forced out of the negative terminal toward the left end of the wire. The positive terminal attracts the electrons from the right-hand

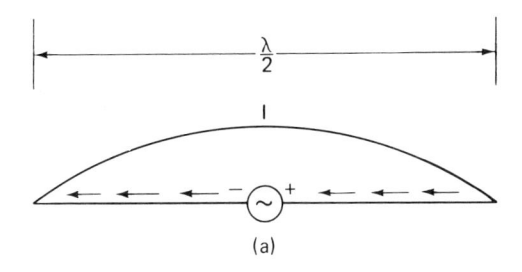

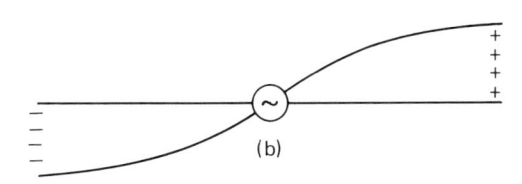

FIG. 20-8 Dipole antenna showing distribution of (a) current and (b) voltage.

wire. The result is a current flow from right to left. The current distribution curve indicates that the current flow is greatest at the center of the dipole and zero at the ends. On the alternate half-cycle the current curve would be on the bottom and the arrows pointing in the opposite direction. Assuming the generator output to be sinusoidal, the relative current distribution will also be sinusoidal. Thus, an RF ammeter inserted near the center of the antenna will indicate a *relatively large current*, and one near either end a *small current*. Although the relative current distribution over the antenna will always be the same, no matter how much or how little current is flowing, the amplitude at any given point will vary directly with the amount of voltage developed at the generator output.

An antenna has *distributed capacitance* and *inductance* and acts like a *resonant circuit*. The applied voltage will not be in phase with the resulting current at all points along the antenna. At the center the current and voltage are in phase. At the ends of the dipole they are out of phase. The voltage distribution of a dipole driven from a sinusoidal generator can be seen in Fig. 20-8b. The voltage, like the current, varies sinusoidally with respect to time and antenna length. If an RF voltmeter is connected between one end of the antenna and ground, it will indicate a *relatively large voltage*. If it is moved toward the center, the voltage reading drops to a *low value*. On the alternate half-cycle the voltage waveform and polarity would be reversed.

Standing waves are *undesirable* on transmission lines; however, on an antenna they are *required*, as evidenced by the low voltage point at the center of a dipole and high voltage at the ends.

The formulas in Section 20-1 give the electrical length of an antenna presumed to be in *free space*. From a practical standpoint antennas are not mounted in free space, as they must be supported by insulators attached to some form of support. The dielectric constant of these insulators is greater than 1. Consequently, the velocity of the wave along the antenna is always slightly less than the velocity in free space. Therefore, the *physical length of the antenna is correspondingly less (about 5 per cent) than the theoretical length in free space*. This is called *end effect*.

To find the length of a half-wave antena in feet, the following formula can be used, which is sufficiently accurate for wire antennas up to 30 MHz:

$$L = \frac{492 \times 0.95}{f_{MHz}} = \frac{468}{f_{MHz}}$$

Example 20-4

Find the length of a half-wave antenna, in feet, designed to operate at 7150 kHz.

Soultion

$$L = \frac{468}{7.15} = 65.45 \text{ ft}$$

At times it is desirable to express the height or length of an antenna in wavelengths.

Example 20-5

A vertical antenna is 405 ft high and is operated at 1250 kHz. What is its physical height, expressed in wavelengths (1 m = 3.28 ft)?

Solution

The first step is to convert the antenna height, in feet, to meters.

$$\text{height} = \frac{405}{3.28} = 123.4 \text{ m}$$

Next determine the wavelength of an RF signal at 1250 kHz:

$$\lambda = \frac{3 \times 10^8}{1.25 \times 10^6} = 240 \text{ m}$$

The antenna height, in wavelengths, is then

$$\frac{123.44}{240} = 0.5143$$

Wave Polarization

20-8 ANTENNA IMPEDANCE

Antennas have resistance as well as inductive and capacitive reactance. This impedance is

$$Z_o = \frac{e}{i} \quad \text{or} \quad Z_o = R \pm jX$$

The impedance varies along the length of the antenna, being *highest* where the current is the *lowest*, and vice versa. At the center of a half-wave antenna the impedance is approximately 73 Ω and increases to about 2500 Ω at either end.

The input impedance is affected by the presence of nearby conductors, such as guy wires and metal objects. This makes calculations difficult, and exact impedance matching is usually done by trial and error.

The value of resistance that would dissipate the same power that the antenna radiates is called *radiation resistance* and is calculated by

$$R_r = \frac{P}{I^2}$$

where R_r = radiation resistance in ohms
 P = total power radiated from the antenna
 I = effective value of antenna current at the feed point in amperes

The radiation resistance varies with antenna length. For a half-wave antenna it is approximately 73 Ω and for a quarter-wave about 36.6 Ω when measured at the current maximum. These are *free-space* values that would exist if the antenna were completely isolated. In practice, the height of the antenna and ground reflections affect the radiation resistance. Generally speaking, as the radiation resistance is reduced, the field intensity increases.

20-9 WAVE POLARIZATION

The electric field, or E component, of the electromagnetic wave determines the *plane of polarization*. This component is in the *same plane* as the antenna. Hence, a vertical antenna radiates a vertically polarized wave, and a horizontal antenna, a horizontally polarized wave.

At low frequencies the polarization is not disturbed, and the radiation field has the same polarization at the receiving station that it had when transmitted. However, at higher frequencies the plane of polarization usually varies, rapidly at times, because the wave splits into several components that travel different paths. Because the lengths of these various paths differ, the recombined electric field will have either a *circularly* or *elliptically* polarized field.

Transmissions below 2 MHz are generally vertically polarized as with broadcast stations. In the 3- to 30-MHz range either type of polarization may be used with varying degrees of success, depending upon such factors as distance and time of day. Above 30 MHz vertical polarization is generally used, except for TV and FM.

Horizontally polarized antennas are used with TV and FM broadcasting because they produce *better ground waves*. The receiving antenna must therefore be mounted in the same plane as the polarization of the transmitted signal. Hence, TV and FM antennas are horizontally mounted.

The sky wave usually has its plane of polarization *reversed* with respect to the incident wave. Consequently, the vertically polarized wave may sometimes be best received on a horizontal antenna, and vice versa.

20-10 HERTZ ANTENNAS

An antenna that is one half-wavelength long, or any multiple thereof, is a *Hertz* antenna. A dipole is therefore a Hertz antenna and can be mounted either horizontally or vertically. At low and medium frequencies they are quite long, and are usually only found on ground installations. They are more likely to be used in medium- or high-frequency installations where operation is not required on a large number of frequencies.

Figure 20-9 shows two different methods of connecting feedlines to

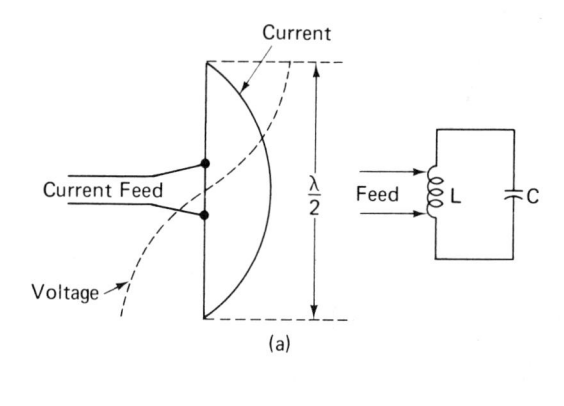

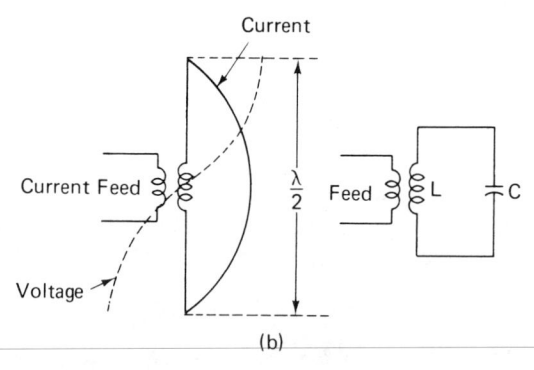

FIG. 20-9 Two methods of current-feeding a Hertz antenna and their equivalent circuits.

Marconi Antennas 439

Hertz antennas. In Fig. 20-9a the feedline is spread out just far enough at the antenna end to match the antenna's impedance. The current is nearly maximum and the voltage minimum at this point. The equivalent circuit of this arrangement is shown to the right.

An alternative method of current feeding a Hertz antenna appears in Fig. 20-9b. In this case an RF transformer is used to provide the impedance match. The equivalent of this circuit is seen at the right of the drawing. Figure 20-10 shows a method of coupling an RF amplifier to a Hertz antenna.

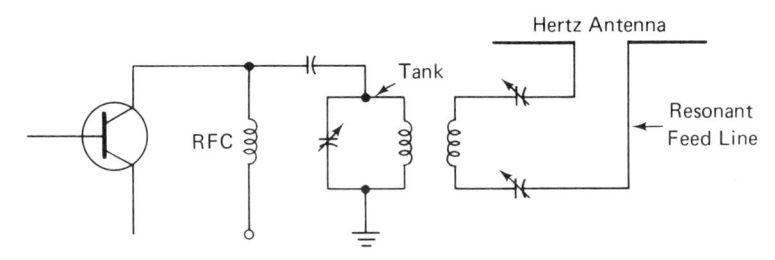

Fig. 20-10 Method of coupling an RF amplifier to a Hertz antenna.

20-11 MARCONI ANTENNAS

The *Marconi* antenna is physically a quarter-wavelength long, but it operates as a half-wave antenna. This is possible because of its connection to ground, which is used as the *other quarter-wavelength*. Thus, the antenna is electrically a half-wavelength. Because the earth is considered to be a good RF conductor, it provides a reflection equivalent to a quarter-wavelength. All the voltage, current, and impedance relationships characteristic of a half-wave antenna will also exist in this antenna. The only exception is that the input impedance is approximately 36.6 Ω.

A Marconi antenna, with its current and voltage distribution, is shown in Fig. 20-11. The transmitter output is connected between the antenna and ground. The image antenna of the upper quarter-wave is represented by the straight dotted line. The grounded Marconi antenna can be any odd multiple of a quarter-wavelength.

Marconi antennas are used more frequently for low- and medium-frequency applications. When a quarter-wave whip antenna is installed on top of an automobile, the roof becomes the ground plane. When mounted on the back bumper, the rear of the car and the road act as the ground plane.

When the conductivity of the soil on which the antenna is mounted is very low, the reflected wave from the ground may be greatly attenuated. To overcome this disadvantage, the site could be relocated to a place where the soil has high conductivity. If this is not practical, provisions must be made to improve the reflecting characteristics of the ground. A Marconi antenna radiates omnidirectionally in the horizontal plane.

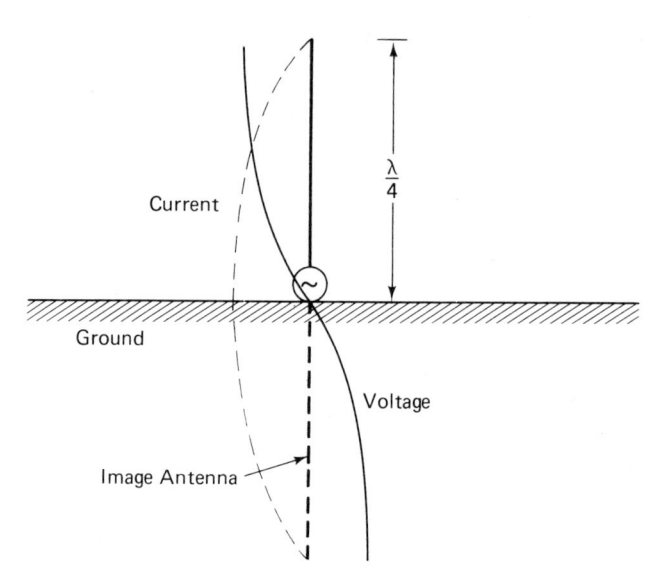

FIG. 20-11 Current and voltage distribution in a Marconi antenna.

20-12 COUNTERPOISE

When the soil conductivity is very poor for the operation of a Marconi antenna, a large buried ground screen can be installed. This must be of *conductive* material and should extend out in all directions at least the physical length of the antenna. If any of the radials become corroded, or broken, radiation will be impaired in that direction. If a ground screen, or system of radials, is not practical, a *counterpoise* may be used to replace the usual direct ground connection. The counterpoise consists of a structure made of heavy wire, or metal framework, erected a short distance above the ground and insulated from ground. The size of the counterpoise should be at least equal to and preferably larger than the size of the antenna.

The counterpoise and the surface of the earth form a large capacitor. Antenna current is collected in the form of charge and discharge currents on this capacitor. The end of the antenna normally connected to ground is connected through the large capacitance formed by the counterpoise. If the counterpoise is not well insulated from ground, *leakage currents* will flow with a resultant loss greater than if no counterpoise is used.

The geometry and size of the counterpoise are not critical, but it should extend *equidistant* in all directions. Normally, it should be about 8 to 12 ft above ground. Typically shaped counterpoises are represented in Fig. 20-12. It must be of such physical dimensions that it *will not resonate* at the operating frequency. A Marconi antenna using a counterpoise operates as effectively as one with a good ground system.

For VHF and UHF operation vertical antennas use a counterpoise

Long-Wire Antennas

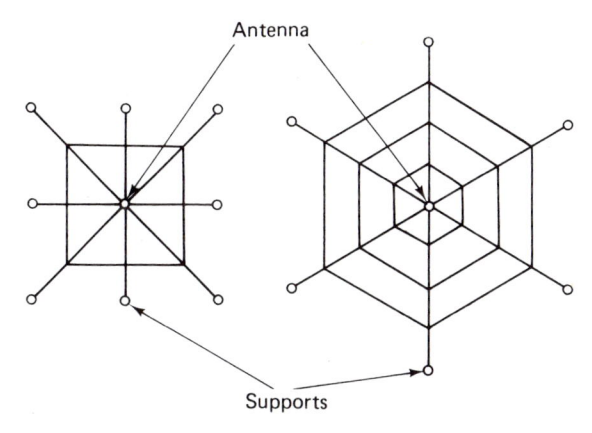

FIG. 20-12 Two different types of counterpoise.

consisting of at least four horizontal quarter-wave radials. Sometimes these radials are pointed downward at about 45°, resulting in what is referred to as a *drooping* ground plane. This raises the input resistance of the antenna to about 50 Ω, which provides an easier impedance match to a coaxial transmission line. In some installations the radials are bent straight down and connected together to form a half-wave coaxial antenna. A metal sleeve or cylinder can be used in place of the radials.

20-13 LONG-WIRE ANTENNAS

An antenna is known as long wire only when it is long in regard to the *wavelength* of the frequency fed to it. If the length is long enough so that an integral number of standing waves of voltage and current can exist along its length, it will radiate effectively.

The distribution of current and voltage waves along an antenna operated at various harmonics of its fundamental resonant frequency is shown in Fig. 20-13. In a, the antenna is one half-wavelength long; in b, c, and d the same length antenna is shown operating as a full wave, one and one half, and two wavelengths, respectively. Therefore, in the last three examples the antenna is operating on the *second*, *third*, and *fourth harmonics* of the fundamental.

In practice, the physical length is not an exact multiple of a half-wave because of end effects. To determine the length of a long-wire antenna, the following formula can be used:

$$L_{ft} = \frac{492(N - 0.05)}{f_{MHz}}$$

where N is the number of half-waves on the antenna.

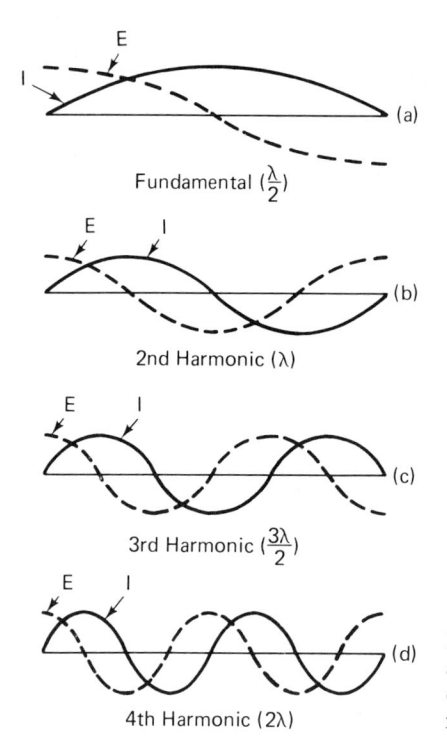

FIG. 20-13 Standing waves of current and voltage on a long-wire antenna operated at various harmonics of its fundamental resonant frequency.

(a) Fundamental ($\frac{\lambda}{2}$)

(b) 2nd Harmonic (λ)

(c) 3rd Harmonic ($\frac{3\lambda}{2}$)

(d) 4th Harmonic (2λ)

Example 20-6

How long would an antenna three half-wavelengths long be at 21.3 MHz?

Solution

$$\frac{492(3 - 0.05)}{21.3} = \frac{492(2.95)}{21.3} = 68.14 \text{ ft}$$

As the wire is lengthened, the direction in which maximum radiation occurs tends to approach the direction in which the antenna is pointing. This is the result of less radiation occurring from the sides of the antenna. The antenna can be fed at either end or at any current loop.

20-14 FOLDED DIPOLES

A single dipole has an impedance of about 73 Ω when measured at the center. This makes feeding relatively easy with low-impedance transmission lines. In other cases, however, when it is desired to use an open-wire line, a higher impedance feed point is preferable. If two dipoles are connected as shown in Fig. 20-14, the input resistance can be increased, because the total length of the radiating element is now approximately one wavelength. This allows the antenna to be fed with *high voltage and low current*.

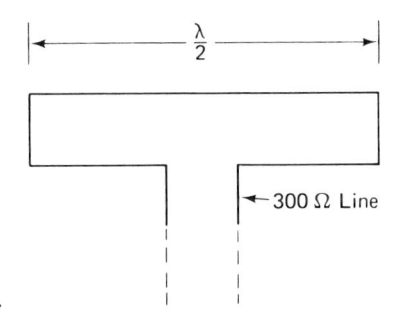

FIG. 20-14 Folded dipole.

An antenna of this type is frequently called a *half-wave doublet* and consists of two parallel, closely spaced half-wave wires connected together at the ends. Very small spacing, on the order of 0.01 wavelength, can be used for wavelengths above about 0.5 m. With the antenna in free space, the input impedance is about 300 Ω. Thus, it is possible to connect an open-wire transmission line of this impedance directly to the terminals of the antenna without a matching tranformer. This type of antenna is commonly used for TV and FM reception, and 300-Ω TV cable is used for the connection between antenna and receiver.

20-15 PARASITIC ARRAYS

A *parasitic array* is an antenna system that consists of two or more elements in which all are not driven. The elements that are not driven are excited by induction and radiation fields produced by the driven element. The advantage of a parasitic array is that it is possible to obtain a *highly directional* radiation pattern. The phase relationship between elements varies according to the spacing between them. They are usually spaced a fraction of a wavelength apart.

A two-element array is shown in Fig. 20-15. The driven element is one half-wavelength long and fed by a low-impedance transmission line. A parasitic element located about 0.15 wavelength behind the driven element serves as a *reflector*. It is about 5 per cent longer than the driven element. The current induced in the parasitically excited element is 180° out of phase with that in the driven element.

The radiated wave produced by the induced current in the parasitic element tends to cancel the original wave, reducing the radiation on the reflector side of the array and reinforcing it on the driven element side. Because of losses, the reradiated wave cannot be as strong as the driven wave, and there is no complete null in any direction. The net result is that most of the radiation is on the side of the driven element, while very little occurs on the reflector side.

The radiated field from the reflector induces a voltage back into the driven element. This causes a change in the input current to the antenna.

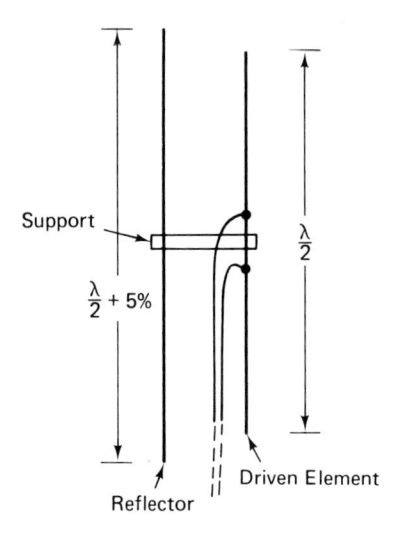

FIG. 20-15 Half-wave antenna with parasitically excited reflector.

Because the input current is changed, the *input impedance drops* from 73 to about 50 Ω.

When a parasitic element is made about 5 per cent shorter than the driven element, it becomes a *director*. If it is spaced about 0.1 wavelength from the driven element, most of the radiated energy is on the director side of the array. This arrangement is shown in Fig. 20-16.

When an antenna array uses both a reflector and a director, it is called a *Yagi*, after its inventor. As in the previously described arrays, the driven element is insulated, but the reflector and director are *electrically connected* (sometimes welded) to a piece of metal tubing that runs parallel to the direc-

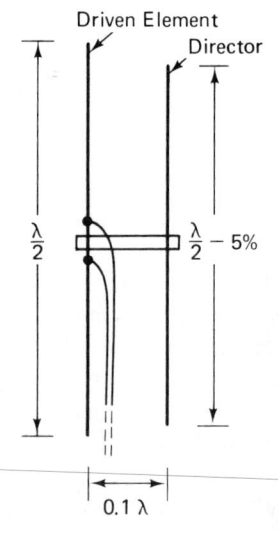

FIG. 20-16 Half-wave antenna with parasitically excited director.

Parasitic Arrays 445

tion of propagation. A three-element Yagi is represented in Fig. 20-17. If additional directivity is desired, more directors can be added.

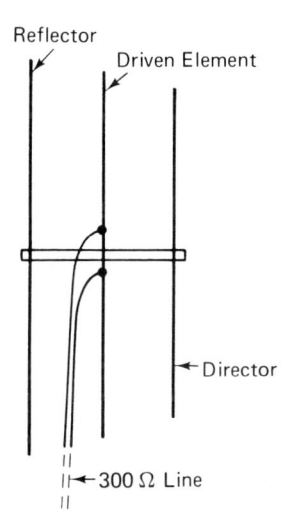

FIG. 20-17 Three-element Yagi antenna.

The directivity of an antenna array can be expressd in one of two ways. One is to determine the ratio of the power in the major lobe to the power radiated by a simple half-wave antenna. This is a measurement of *antenna gain*. The other involves measuring the power in the direction of the major lobe to the power in the opposite direction. This is called the *front-to-back ratio*.

Comparison between the more directive types of arrays is made in terms

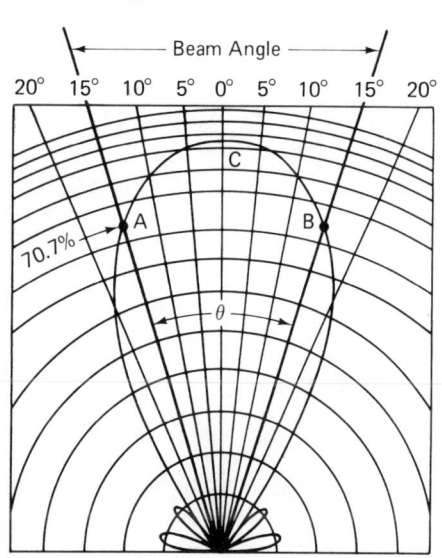

FIG. 20-18 Measurement of beam angle.

of the *beam angle*. This is a measurement between the *half-power points* to the major lobe. These points are where the electric field strength (measured in volts per meter) is 0.707 times as great as that along the axis of the beam. An example of this can be seen in Fig. 20-18. In this illustration the angle between points A and B (angle θ) is the beam angle. At each of these points the signal strength is 70.7% of its value at point C. This means that the power along lines A and B is one half that at corresponding points on line C.

20-16 COLLINEAR ARRAYS

A *collinear array* is formed when two half-wave elements are placed end to end and excited in phase. There is no directivity in the plane perpendicular to the antenna, but there is a strong pattern in any plane containing the

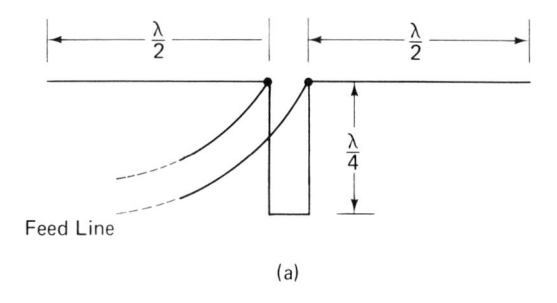

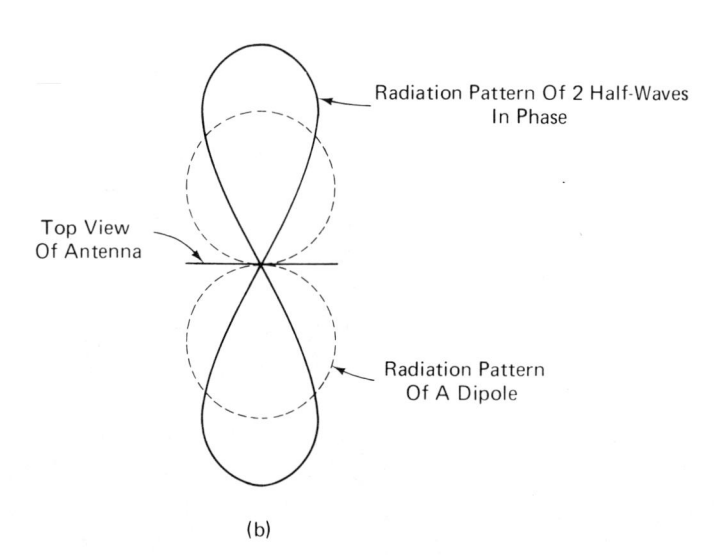

FIG. 20-19 (a) Collinear array consisting of two half-waves in phase; (b) comparison of collinear array and dipole radiation patterns.

Loop Antennas 447

antenna. By increasing the number of half-wave elements, the directivity is increased. A typical collinear array and its radiation pattern are seen in Fig. 20-19.

It is important that the current being fed to the antenna be in the *correct phase relationship*. Since the current direction changes for each half-wavelength of an antenna, it is not possible to connect the two sections directly together. Instead, the section that carries the current in the wrong direction is folded to form a quarter-wave section of RF line. This brings together the ends of the sections in which current flows in the same direction. Note how the RF feedline is connected to the antenna. The connection is made at a point of *high impedance*. This requires the use of a resonant line. If a non-resonant line is used, it must be connected somewhat down from the top of the quarter-wave stub to a point where the impedances will match.

20-17 LOOP ANTENNAS

Loop antennas are used principally in radio direction finder applications. The loop consists of a relatively few turns of wire, closely spaced, on a vertical form 1 to 3 ft in diameter. The form may be either round or square. By rotating the loop so that its plane is perpendicular to the direction of travel of the radio wave, equal voltages will be induced in each vertical section, if the wave is vertically polarized. If the loop is properly balanced, these induced voltages will cancel and no output results. The loop is adjusted for *minimum* rather then maximum response, because it is much more sensitive in the null position.

A simple loop operating in this fashion gives the bearing angle of the passing wave, but leaves an *ambiguity* of 180°. This uncertainty as to the exact position of the transmitting station can be eliminated by using a *vertical sense antenna* in conjunction with the loop.

The accuracy with which a loop can be electrostatically balanced to ground is enhanced by enclosing it in an *electrostatic shield*. This ensures that all parts of the loop will have the same capacity to ground, regardless of loop orientation or proximity to surrounding objects. The shield is usually a metallic housing in the form of a pipe provided with a small insulated joint so that it will not act as a short-circuit turn. This joint should be equidistant from the output of the loop. The presence of this shield has practically no effect upon the performance of the loop.

If a loop antenna is mounted in a horizontal plane, it loses its directional characteristics and becomes omnidirectional. If a radio wave transmitted from a vertical antenna is refracted back to earth from the ionosphere, its plane of polarization changes. Under these conditions the loop will have voltages induced in the horizontal members of the antenna that do not give zero resultant loop output when the plane of the loop is perpendicular to the bearing of the wave. In fact, if no vertical component is present, the null position will occur when the plane of the loop is parallel to the oncoming wave.

For optimum receiving efficiency the loop operates as a *tuned circuit*. The loop inductance is resonated by an external tuning capacitor, as shown in Fig. 20-20a, and its directional characteristics are as shown in Fig. 20-20b.

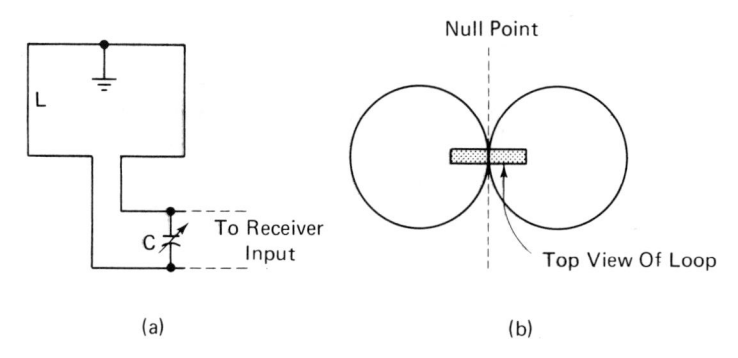

FIG. 20-20 (a) Loop antenna used as part of tuned circuit; (b) directional characteristics of loop antenna.

20-18 METHODS OF ANTENNA FEED

Radio-frequency energy can be fed to an antenna in many different ways. In Fig. 20-9 two methods of current feeding a half-wave antenna were shown. This is also known as *center feeding*. If energy is applied to the end of an antenna, it is known as *end fed* or voltage feeding, because (assuming a half-wave antenna) the voltage is at a maximum.

It is seldom possible to connect the output of an RF power amplifier directly to an antenna. Some form of transmission line is needed to transfer the RF energy from the FPA to the antenna. These may be resonant, nonresonant, or a combination of both. Resonant feedlines are not as commonly used, because they tend to be inefficient and are *very critical* with respect to length for a particular operating frequency.

One method of feeding a half-wave antenna, using resonant lines, is illustrated in Fig. 20-21. The feedline is one half-wavelength long with current and voltage distribution represented by the dotted and solid curved lines. Each end is a low impedance. Therefore, by connecting one end of the line to the center of the half-wave antenna (the low-impedance and high-current point), a proper impedance match is effected.

The transmission line in Fig. 20-22 is a quarter-wavelength, two-wire line. Its characteristics are such that its impedance is high at the antenna end and low at the sending end. Customarily, each leg of the secondary winding of the RF transformer would have a series capacitor installed (not shown) before it connects to the transmission line. These, in conjunction with the

Methods of Antenna Feed

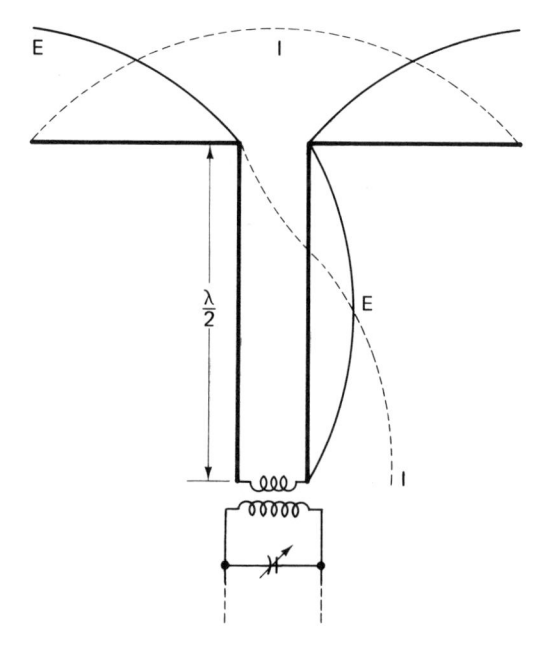

Fig. 20-21 Current-feeding a half-wave antenna with a half-wave resonant feedline.

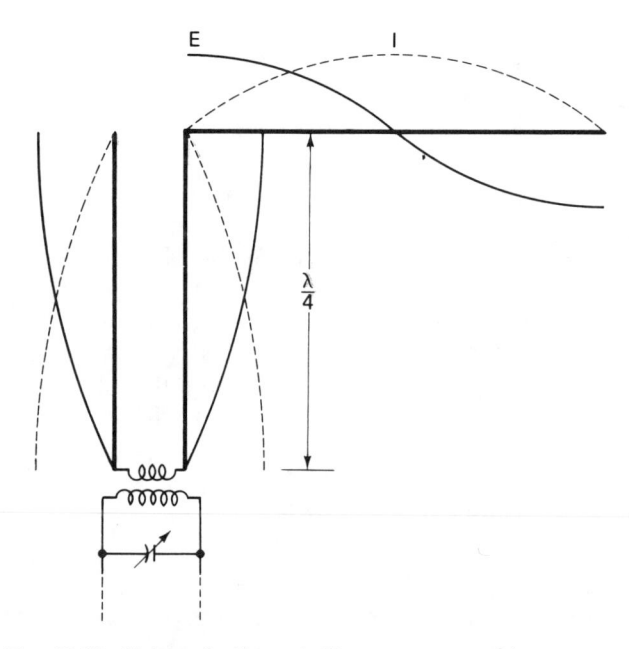

Fig. 20-22 Voltage-feeding a half-wave antenna with a quarter-wave resonant feedline.

secondary winding, are tuned to form a series resonant circuit, which develops a large current to feed the transmission line. Adjusting these capacitors compensates for irregularities in line and antenna length. By leaving one end of the feedline unconnected, we have what is known as a *zepp antenna*.

An alternative method of feeding a half-wave antenna is to use a *quarter-wave matching stub*, as illustrated in Fig. 20-23. In this case the high impedance at the end of the antenna matches the open end of the stub. The

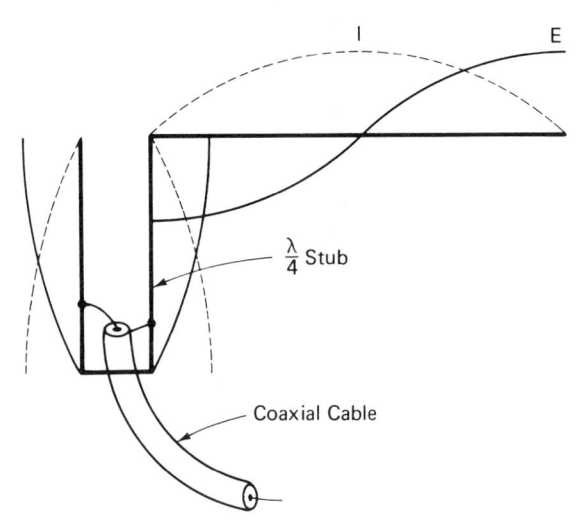

FIG. 20-23 Half-wave antenna, end-fed, using a quarter-wave shorting stub.

impedance on the stub varies from zero at the short-circuit end to several thousand ohms at the open end. This makes it possible to connect a 70-Ω coaxial cable a short distance up from the shorted end to the 70-Ω point. Almost any impedance transmission line can be matched at an appropriate point on the stub.

Another impedance-matching device is the *quarter-wave transformer* or matching transformer shown in Fig. 20-24. This matches the low Z of the antenna to a line of higher impedance. To determine the characteristic impedance (Z_o) of the quarter-wave section, the following formula is used:

$$Z_o = \sqrt{Z_s Z_r}$$

where Z_s = impedance of the feedline
Z_r = impedance of the radiating element

In the illustration Z_o has a value slightly more than 191 Ω. Standing waves will exist on the quarter-wave section, but not on the 600-Ω line.

Nonresonant feedlines are most frequently used. These may be the open-wire line, the shielded pair, the coaxial line, and the twisted pair. They have

Methods of Antenna Feed

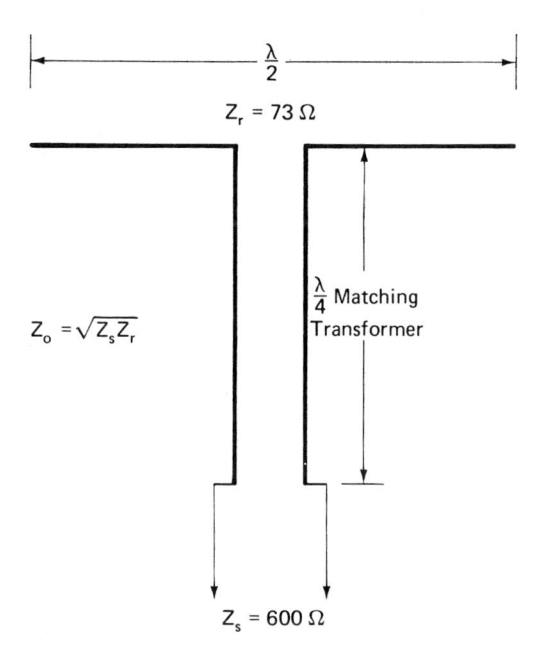

FIG. 20-24 Quarter-wave matching transformer.

great advantage over the resonant line in that their operation is practically independent of length.

A unique way of terminating a low-impedance, nonresonant transmission line to a half-wave antenna is illustrated in Fig. 20-25a. A feedline of about 70 Ω is connected directly to the center of a half-wave antenna. Actually, the outer conductor connects to the center of the antenna and the inner conductor connects to a point where the impedance is equal to that of the line. This is known as a *gamma-match* feed system, and it is slightly unbalanced. An improved method is through the use of two parallel coaxial cables, as illustrated in Fig. 25-25b. This is known as a *T match*. The purpose of the capacitors in series with the inner conductors is to cancel their inductive reactance.

Another common transmission line is the *twisted-pair* illustrated in Fig. 20-26. It is an untuned line and works well on lower radio frequencies. Due to excessive losses occurring in the insulation, it is not suitable for higher frequencies. The characteristic impedance is about 70 Ω.

When a feedline does not match the impedance of a center-fed antenna, an impedance-matching device, such as shown in Fig. 20-27, must be used. This is known as a *delta match*. As the end of the line (nearest the antenna) is spread, its characteristic impedance increases and will equal the antenna impedance at some critical point.

The delta section becomes part of the antenna and, consequently,

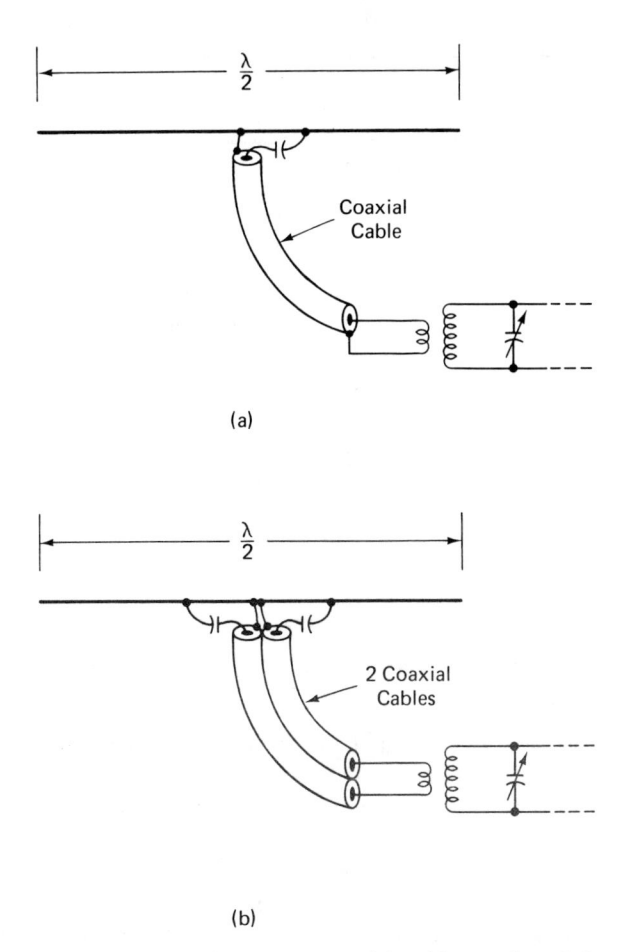

FIG. 20-25 (a) Low-impedance coaxial cable terminated in a gamma match; (b) two coaxial cables terminating in a T-match.

introduces radiation loss. Another disadvantage is that some trial-and-error methods are usually required to determine the optimum dimensions of the *A* and *B* sections. This can be simplified by using the following formulas to determine the lengths of the A and B sections:

$$A_{ft} = \frac{118}{f_{MHz}}$$

$$B_{ft} = \frac{148}{f_{MHz}}$$

20-19 TUNING OR LOADING THE ANTENNA

If the physical length of an antenna is approximately equal to its electrical length, it will resonate at the required frequency. Under these conditions

Tuning or Loading the Antenna

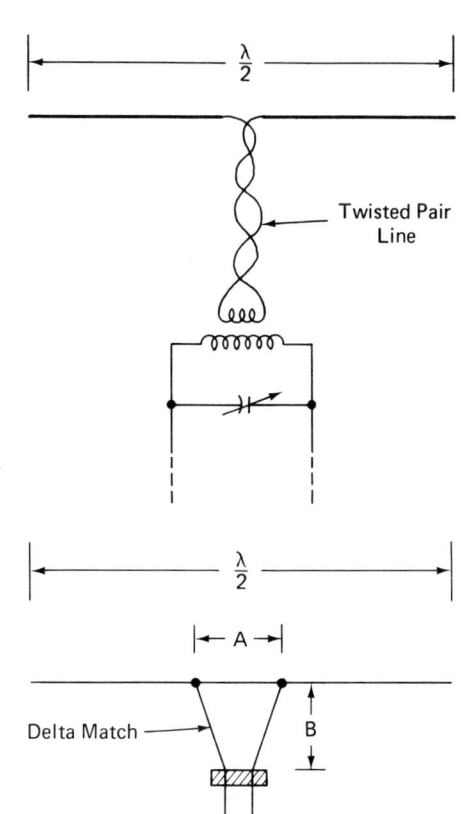

FIG. 20-26 Half-wave antenna fed by twisted-pair line.

FIG. 20-27 Delta-matched half-wave antenna.

the antenna will present to the input power source an impedance that is all resistive. However, if the antenna is not of the proper length, the source will see an opposition other than the pure resistance offered under perfect conditions. This will either be inductive or capacitive, depending on whether the antenna is longer or shorter than the specified wavelength. A Hertz antenna slightly longer than a half-wavelength will act like an *inductive circuit*; a shorter one will appear as a *capacitive circuit* to the source.

Compensation can be made for an antenna that is electrically and physically too long, for a half-wave length, by cutting it to proper length, or by *tuning out* the inductive reactance by adding the proper amount of series capacitance. When an antenna is shorter than the required length, it may be brought into resonance by adding an appropriate amount of inductance in series. If both the inductance and the capacitance inserted in the antenna are variable, it is possible to resonate the antenna over a wide range of frequencies.

Marconi antennas can also be electrically lengthened or shortened by adding series inductance or capacitance. *Top loading* of broadcast-station antennas is commonly encountered, particularly at the lower frequencies. At the low end of the broadcast frequency spectrum a quarter-wave antenna has to be very high and is costly. By adding a series inductance near the top of the vertical antenna, it becomes top loaded, and its electrical length is increased so that it will resonate at the required frequency.

Top loading can also be accomplished by adding a metallic, wheel-like structure to the top of a vertical antenna, called a *top hat*, as shown in Fig. 20-28a. This materially increases the capacitance of the antenna to ground.

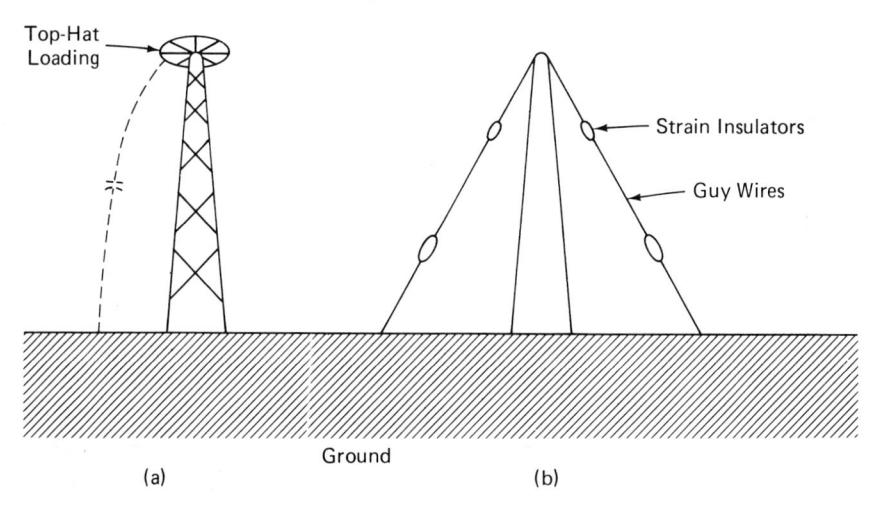

FIG. 20-28 Methods of top-loading a marconi antenna: (a) using top-hat, (b) using top portion of guy wires.

An alternative method of top loading can be accomplished by using the top part of the several guy wires as the top hat, as in Fig. 20-28b. The supporting guy wires must be broken up into lengths that are not harmonically related to the transmitted frequency to prevent pickup and reradiation of RF energy. The *strain insulators* are usually porcelain, which are strong and have a hard, smooth surface that minimizes the accumulation of dirt and moisture, and thereby reduces losses across the insulators. In coastal regions salt encrustation may be built up on the insulators, causing leakage losses due to the relatively high conductance. Because of the egg shape of the strain

Tuning or Loading the Antenna 455

insulators, if they should break, the two wire loops are still joined and the guy wire does not part.

Antennas are often fed by a *pi network*. The simplified schematic for a shunt-fed circuit appears in Fig. 20-29. The pi network, composed of L, C_1,

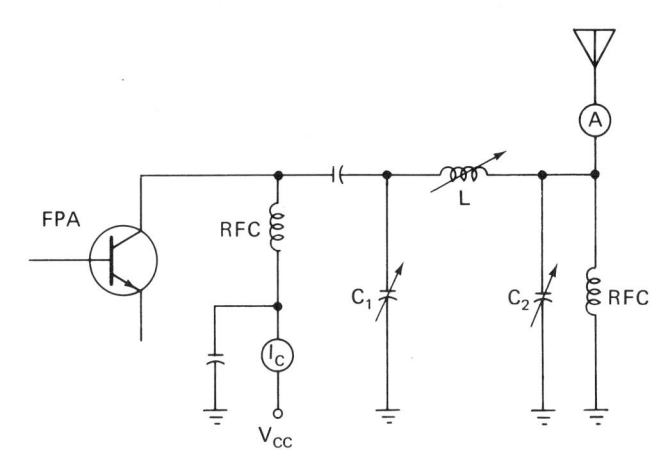

FIG. 20-29 A pi network used to couple the FPA to the base of an antenna.

and C_2, serves as the tank circuit as well as the antenna coupling, and resonates at the operating frequency. The impedance of C_2 must be made to equal the base impedance of the antenna if maximum power is to be drawn. Likewise, C_1 must match the output impedance of the FPA. The correct value of inductance, L, must be used to provide resonance with C_1 and C_2 when they are properly adjusted. The *lowest* capacity setting of C_2 should be used, consistent with normal operation, so that a large reactive voltage will be developed across it. This will cause *maximum* current to flow in the antenna.

The following procedure can be used to tune the pi network: set C_2 to maximum capacity (minimum coupling). Tune C_1 to resonance, as indicated by minimum collector current, and then observe the antenna current. Reduce the capacity of C_2 somewhat and redip the collector current. This should cause the I_c minimum and antenna current to both rise. Continue reducing C_2 in small amounts and retuning C_1 until I_c minimum reads the value recommended by the specifications for the device. Observe the value of antenna current. Repeat the procedure using a reduced value of L. Next repeat the entire tuning procedure using more L. The stage will be properly tuned when maximum antenna current flows at the rated value of collector current (or I_p, if a vacuum tube is used).

An advantage of pi-network tuning is that the output capacitor presents a *low reactance to harmonics* of the operating frequency, thus bypassing them to ground. The *RFC* between antenna and ground provides a discharge path for electrostatic charges picked up by the antenna. It does not adversely affect antenna tuning.

A method of shunt feeding a vertical quarter-wave, grounded antenna is shown in Fig. 20-30. Irrespective of the transmission-line impedance, there

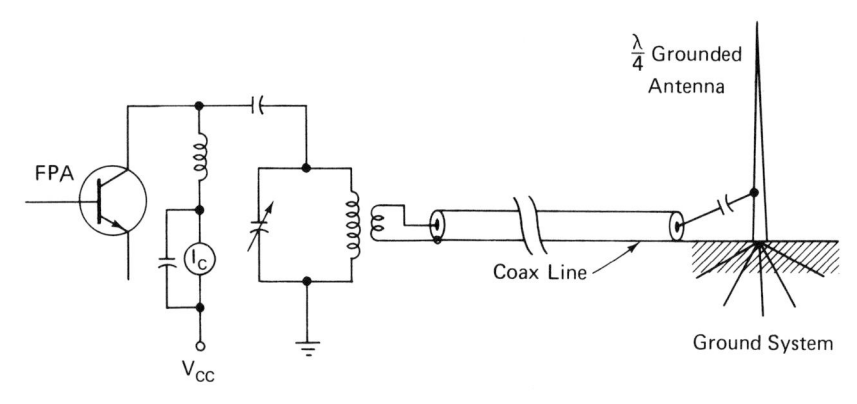

FIG. 20-30 Shunt-fed, quarter-wave, grounded antenna.

will be some place on the antenna having the same impedance. If a 72-Ω line is used, it would have to connect to a point on the antenna where the impedance is 72 Ω. In this illustration a *shunt-fed, half-delta* feed system is used. Some inductance exists on the center conductor of the coaxial cable between the end of the cable and the point where it connects on the antenna. This is canceled by a capacitor, as illustrated. Usually some experimentation is required to find the point of minimum SWR (standing-wave ratio).

It is desirable to use a *dummy antenna* when tuning a transmitter. It contains lumped constants, usually noninductive resistance and capacitance, in a relatively small metal enclosure designed to prevent radiation of RF energy. The dummy antenna should closely approximate the electrical characteristics of the original antenna. It enables the transmitter to be tested and adjusted while causing a minimum of interference due to radiation. Naturally, the regular antenna must be disconnected during such testing operations.

20-20 HARMONIC SUPPRESSION

A transmitter should radiate only on its carrier frequency with no harmonics present, although experience indicates they often are. There are circuits that will filter out these undesirable frequencies.

A method of coupling a two-wire transmission line to the output of an RF amplifier with suppression of second and third harmonic energy is illustrated in Fig. 20-31. The Faraday screen serves as an *electrostatic shield* between the tank and the transmission-line pickup coil. This effectively

Harmonic Suppression

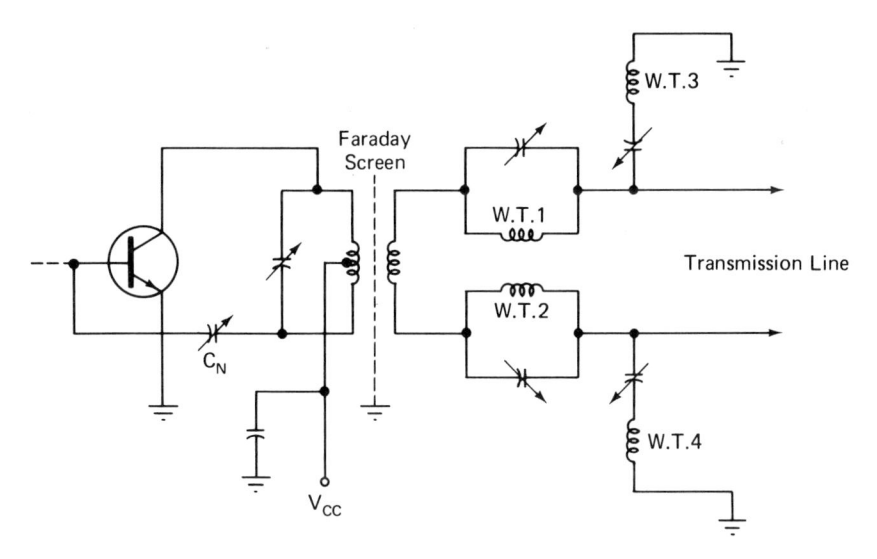

FIG. 20-31 Two-wire transmission line coupled to an RF amplifier showing method of suppressing second and third harmonics.

eliminates the transfer of harmonic energy from the tank to the pickup coil through electrostatic coupling. The parallel-tuned wave traps, designated W. T. 1 and W. T. 2, are resonant at the second harmonic, thus blocking them. The two series traps, designated W. T. 3 and W. T. 4, are tuned to resonate at the third harmonic and bypasses them to ground.

The transmission line can be coupled to the antenna by means of a T-network, as shown in Fig. 20-32. This arrangement can provide a step-up or step-down of impedance ratios, if necessary, to match transmission line to antenna. Inductor L_1 is tuned for minimum dip in the output current of the

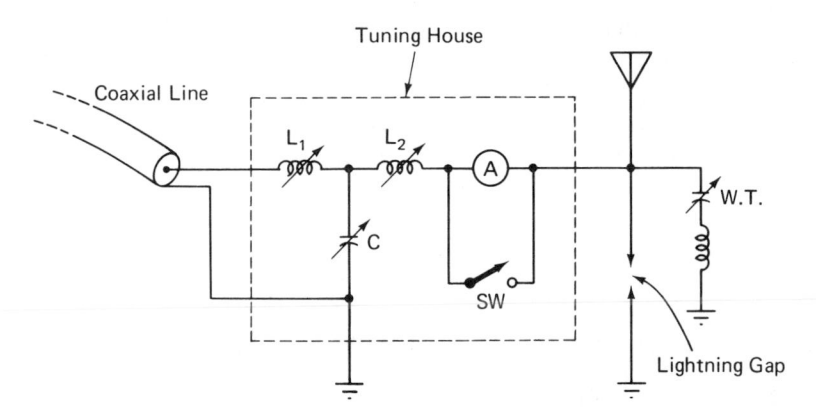

FIG. 20-32 T-network coupling between transmission line and antenna with trap for harmonic attenuation and lightning gap.

FPA; L_2 is tuned for maximum antenna current. Some experimentation with capacitor C will be required to find the optimum setting that will cause proper dipping of the FPA with L_1 and maximum antenna current with L_2. These components are located in the *tuning house* at the base of the antenna. This should be well shielded and grounded to minimize the radiation of harmonics.

The wave trap is an optional item to be used in the event a particular harmonic needs to be suppressed. The antenna ammeter may be switched out of the circuit, except when making adjustments, to prevent burnout due to lightning strikes near, or on, the antenna. A *lightning gap* is normally located between the base of the antenna and ground. It must be adjusted so that it is just wide enough to not arc over with normal transmitter operation, but close enough to break down with a lightning strike, thus protecting the transmitter.

20-21 ANTENNA RADIATION PATTERNS

The strength of a transmitted RF signal can be determined by how many volts it develops in a receiving antenna 1 m long. For example, if the receiving antenna is 25 miles away from the transmitter, the signal may develop 45 mV in it. Therefore, the signal strength is 45 mV/m. The signal strength varies *inversely* with distance.

Commercial stations are required, as part of their licensing, to take a large number of field-strength measurements in all directions around the transmitting antenna. Points of like intensity are then joined together to form a pattern showing how the radiated signal strength varies in the geographic area surrounding the transmitter. From these field-strength patterns the FCC, and the station engineers, can determine the *service area* of the station.

The radiation patterns of a Marconi antenna are shown in Fig. 20-33. In a can be seen the *omnidirectional* characteristics of the antenna. The circular plot might represent a field strength of 850 μV/m. Similar plots could be made showing signal strengths of 750, 650 μV/m, and so on. These would be concentric plots at greater distances from the transmitting antenna.

In Fig. 20-33b is shown a plot of signal strengths in the vertical plane for vertical Marconi antennas one quarter- and one half-wavelength long. Notice that the $\lambda/4$ wave antenna produces a stronger signal in the vertical plane than the $\lambda/2$ wave. As the wavelength of the antenna is increased, the radiation is concentrated increasingly along the horizontal until it reaches a little more than one half-wavelength. Beyond this, high-angle lobes of increasing amplitude begin to appear until at a height of one wavelength none of the energy is radiated along the horizontal. These latter conditions are not shown.

The radiation pattern of a full-wave horizontal antenna in free space is shown in Fig. 20-34. Because the antenna has equal currents flowing in opposite directions at the same instant, the radiation pattern of each half will be the same. This results in zero effective radiation at right angles to the antenna.

Antenna Radiation Patterns

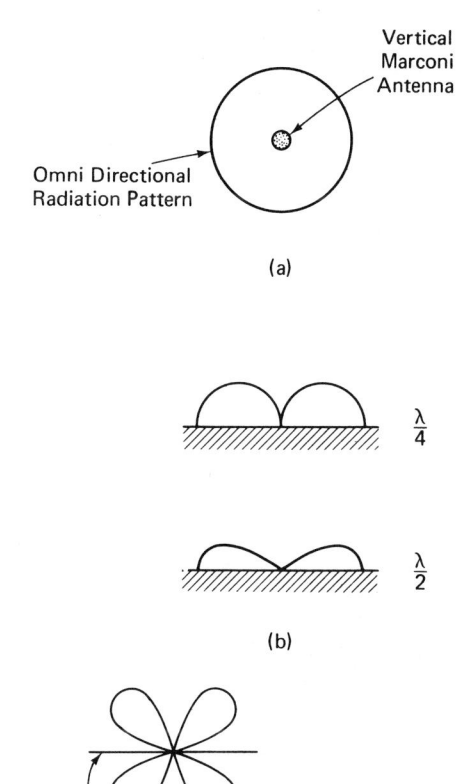

FIG. 20-33 (a) Omnidirectional radiation pattern of a vertical Marconi antenna; (b) vertical radiation patterns of $\lambda/4$ and $\lambda/2$ Marconi antennas.

FIG. 20-34 Radiation pattern produced by a horizontal antenna one wavelength long in free space.

From a study of Fig. 20-34 it can be seen that maximum radiation occurs at angles of 45° with respect to the direction of the antenna.

Consider the direction of maximum radiation from two vertical antennas spaced 180° and having equal currents in phase. The signal radiating outward from one antenna (A) will arrive at the other antenna (B) at the moment its exciting current is changing direction. This means that the wave radiated by antenna B is exactly 180° out of phase with the wave arriving from antenna A. Thus, the two waves cancel in the direction leading away from antenna B. A similar action takes place in the direction from antenna B to antenna A. From this we may conclude that all radiation along a line connecting the two antennas is zero. However, on a line that perpendicularly bisects the line between the two antennas, the two radiated waves arrive in phase and are additive. This causes maximum radiation to occur along the perpendicular bisector.

20-22 CALCULATING THE POWER IN AN ANTENNA

Part of the records that a station must maintain concern the amount of RF power delivered to the antenna. Calculations are generally necessary to determine this. Some examples will be given of how the losses in a feeder line and power delivered to an antenna can be determined.

Example 20-7

A long transmission line delivers 10 kW into an antenna; at the transmitter end of the line current is 5 A and at the coupling house it is 4.8 A. Assuming the line to be properly terminated and negligible losses in the coupling system, what power is lost in the line?

Solution

First, determine the impedance of the line as follows:

$$Z_L = \frac{W}{I_L^2} = \frac{10,000}{4.8^2} = 435 \, \Omega$$

This is also the antenna input impedance. Next, calculate the input power to the line:

$$P_i = I_L^2 \times Z_L = 5^2 \times 435 = 10,875 \text{ W}$$

The power loss is the difference between the line input and antenna input:

$$P_{\text{loss}} = 10,875 - 10,000 = 875 \text{ W}$$

Example 20-8

Suppose that it is desired to find the power delivered to a properly terminated 500-Ω transmission line when the input current is 3 A.

Solution

The power input is

$$P = I^2 Z = 3^2 \times 500 = 4.5 \text{ kW}$$

Example 20-9

The FCC requires that the nighttime power of some stations be reduced to eliminate interference with other distant stations. Suppose that the daytime transmission-line current of a 10-kW transmitter is 12 A, and the transmitter is required to reduce to 5 kW at sunset. What is the new value of transmission-line current?

Solution

First, calculate the transmission-line impedance.

$$Z = \frac{P}{I^2} = \frac{10,000}{12^2} = 695 \, \Omega$$

Calculating the Power in an Antenna 461

Knowing the impedance of the line, we may now determine the input current:

$$I^2 = \frac{P}{Z} = \frac{5000}{695} = 7.2$$

Therefore,

$$I = \sqrt{7.2} = 2.685 \text{ A}$$

Example 20-10

If the resistance at the base of a Marconi antenna is 36 Ω and the input current is 11.8 A, what is the power input to the antenna?

Solution

$$P = I^2 R$$
$$= 11.8^2 \times 36 \cong 5 \text{ kW}$$

Example 20-11

What is the antenna current when a transmitter is delivering 900 W into an antenna having a resistance of 16 Ω?

Solution

$$P = I^2 R$$
$$I = \sqrt{\frac{P}{R}} = \sqrt{\frac{900}{16}} = \sqrt{56.85} \cong 7.55 \text{ A}$$

Example 20-12

If the antenna current of a station is 9.7 A for 5 kW, what is the current necessary for a power of 1 kW?

Solution

The power is proportional to the square of the current.

$$\frac{P_1}{P_2} = \frac{I_1^2}{I_2^2}, \quad \frac{1}{5} = \frac{I_1^2}{9.7^2}$$

Therefore

$$I_1 = \frac{9.7}{\sqrt{5}} = 4.33 \text{ A}$$

Example 20-13

If the day input power to a certain broadcast-station antenna having a resistance of 20 Ω is 2000 W, what would be the night input power if the antenna current were cut in half?

Solution

First, the daytime antenna current must be determined.

$$I = \sqrt{\frac{P}{R}} = \sqrt{\frac{2000}{20}} = 10 \text{ A}$$

Half of this current is 5 A. Therefore, the night input power becomes
$$P = 5^2 \times 20 = 500 \text{ W}$$

Example 20-14

The dc input power to the FPA is exactly 1500 V at 700 mA. The antenna resistance is 8.2 Ω and the antenna current is 9 A. What is the plate efficiency of the final?

Solution

The dc power into the final is
$$P = EI = 1500 \times 0.7 = 1050 \text{ W}$$

The power into the antenna is
$$P = I^2 R = 9^2 \times 8.2 \cong 664 \text{ W}$$

This problem assumes no loss in the transmission line; therefore, the efficiency of the final is

$$\% \text{ efficiency} = \frac{P_o}{P_i} \times 100 = \frac{664}{1050} \times 100 \cong 63$$

Example 20-15

The ammeter connected to the base of a Marconi antenna has a certain reading. If this is increased 2.77 times, what is the increase in output power?

Solution

The output power is proportional to the square of the current. Hence, the power output will be increased by the square of 2.77, or
$$2.77^2 = 7.67$$

Example 20-16

If the power output of a broadcast station has been increased so that the field intensity at a given point is doubled, what increase has taken place in antenna current?

Solution

The field intensity varies as the *square root* of the radiated power. Hence, to double the field intensity, it is essential to *quadruple* the antenna power. The antenna current will be proportional to the square root of the antenna power. Consequently, quadrupling the antenna power will result in a *doubling* of antenna current. From this we see that as the field intensity at a given point is doubled, the antenna current must be doubled.

Antenna Gain 463

Example 20-17

How does the field strength of a standard broadcast station vary with distance from the antenna?

Solution

The field strength varies inversely as the distance from the antenna.

20-23 ANTENNA GAIN

A directional antenna is capable of delivering a stronger signal, at a given point and power input, than a dipole antenna. The ratio of the two signal strengths is a measure of the *antenna gain*. Expressed another way, the antenna gain is generally taken as the *ratio* of the power that must be supplied to some standard-comparison antenna (usually a dipole) to the power that must be supplied to the antenna under test in order to produce the same field strengths in the desired direction at a given receiving antenna. This amounts to a ratio of the respective radiated fields.

Example 20-18

One kilowatt is supplied to a directional antenna, which produces a field strength of 20 μV/m at the receiving station. A dipole antenna located near the directional antenna and properly oriented must be supplied 4 kW to produce the same signal strength. What is the gain of the antenna?

Solution

$$\mathrm{dB} = 10 \log \frac{P_1}{P_2} = 10 \log \frac{4}{1}$$
$$= 10 \times 0.6021 \approx 6$$

The gain of a directional antenna can also be expressed in terms of its *effective radiated power*, abbreviated *ERP*.

Example 20-19

What is the ERP of a television broadcast station if the output of the transmitter is 1 kW, antenna transmission line loss if 50 W, and the antenna power gain is 3?

Solution

The power delivered to the antenna is the difference between the transmitter output and the line loss, or

$$1000 - 50 = 950 \text{ W}$$

With an antenna gain of 3, the ERP is

$$\text{ERP} = 950 \times 3 = 2850 \text{ W}$$

It is necessary to use *high-gain antennas* for TV and FM stations because there is no useful sky wave at these frequencies. Also, the ground wave is attenuated very rapidly as it travels outward from the transmitting antenna. Most of the useful radiation is confined to line of sight. At these high frequencies it is practical to construct directional antennas to concentrate all the radiated RF energy to low vertical angles, which is where the receiving antennas are located.

20-24 PHASE MONITORS

When a directional antenna system consists of two or more radiating elements, the correct radiation pattern of the system depends upon the *phase relationships* of the currents in the various elements. The FCC requires that the currents in the elements be maintained within 5 per cent of their licensed values. To determine the relative phases, RF pickups are installed in the base of each antenna to sample the currents. Leads from these are connected to meters at the operating position so that constant indication of the relative current amplitudes is available at the phase monitor. The monitor requires about 0.2 to 4 W of RF power per element. Its input impedance is about 65 Ω, which permits the use of ordinary coaxial cable.

20-25 MISCELLANY

Some stations use the *direct method* of computing their radiated power output. In these cases the antenna current must be measured at the point in the antenna system where the radiation resistance is measured. This is usually a maximum current point.

The effective height of a receiving antenna can be determined by the amount of induced voltage in it from the passing wave of a radio station.

Example 20-20

If the field intensity of a station is 25-mV/m and this causes 2.7 V to be developed in a certain antenna, what is its effective height?

Solution

The effective height in meters is equal to the ratio of the two voltages. Thus,

$$\text{height}_{\text{eff}} = \frac{2.7}{0.025} = 108 \text{ m}$$

Expressed in feet, this amounts to

$$108 \times 3.28 = 354 \text{ ft}$$

Practice Problems

Example 20-21

In a certain directional antenna system the two towers must be separated by 120 electrical degrees. What is the distance between them in feet if the carrier frequency is 950 kHz?

Solution

It is necessary to first determine the wavelength of a 950-kHz signal.

$$\lambda = \frac{3 \times 10^2}{f_{MHz}} = \frac{3 \times 10^2}{0.95} = 315.8 \text{ m}$$

The tower separation is 120 electrical degrees, or one third of a wavelength. Hence,

$$\frac{315.8}{3} = 105.27 \text{ m}$$

The distance between the towers is therefore

$$105.27 \times 3.28 = 345.3 \text{ ft}$$

Practice Problems

1. What is the frequency of an antenna current that radiates a 0.66-m electromagnetic wave?
2. Calculate the surge impedance of a transmission line having an inductance of 0.5 μH and 2 pF/ft.
3. Determine the characteristic impedance of a coaxial cable when the inside diameter of the outer conductor is 0.120 in. and the outer diameter of the inner conductor is 50 mils.
4. A two-wire transmission line is made up of two No. 14 copper wires (diameter 64 mils) spaced 2 in. apart. What is the surge impedance of the line?
5. What is the peak value of the voltage between the inner and outer conductors of a 58-Ω coaxial cable that is carrying 800 W?
6. What is the rms value of the transmission-line current in problem 5?
7. Calculate the radiation resistance of a half-wave antenna radiating 500 W when the antenna current is 2.6 A.
8. What should the length in feet be of a long-wire antenna that is four half-wavelengths long and operates at 149 MHz?
9. Calculate the power loss on a transmission line that delivers 50 kW into an antenna when the line current is 10 A at the antenna end and 10.2 A at the transmitter end.
10. The license of a certain station stipulates that its daytime power of 2.5 kW must be reduced to 1 kW at night. If the daytime transmission-line current is 5.78 A, what is the value of the nighttime current?

Commercial License Questions

Sections in which answers to questions are given appear in parentheses. A bracketed number following a question implies that it applies only to that element.

1. What kinds of fields emanate from a transmitting antenna? What are their relationships to each other? (20–1)
2. Can either of the fields radiated from an antenna produce a voltage in a receiving antenna? (20–1, 20–2)
3. What are the lowest useful frequencies in radio communications? (20–1)
4. If the period of one complete cycle of a radio wave is 0.000001 s, what is the wavelength? (20–1, 20–7)
5. What is the velocity of propagation of RF waves in space? (20–1)
6. What is the formula for determining wavelength when the frequency, in kilocycles, is known? (20–1, 20–7)
7. If the two towers of a 950-kHz directional antenna are separated by 120 electrical degrees, what is the tower separation in feet? [4] (20–1, 20–7)
8. What factors determine the resonant frequency of any particular antenna? (20–1, 20–7, 20–8)
9. Which radio frequencies provide more reliability for long-distance communications? (20–2)
10. How is ground-wave coverage affected by operating frequency? (20–2)
11. What frequencies are unaffected by the ionosphere and have line-of-sight characteristics? (20–3)
12. Discuss the effects the ionosphere has on different radio frequencies. (20–3)
13. What effect does the frequency of emission and density of the ionosphere have on the length of the skip zone? (20–3, 20–4)
14. How is the length of the skip zone affected by the angle of radiation of the RF wave? (20–4)
15. Discuss the various causes of fading. (20–4, 20–5)
16. What effect do sunspots and aurora borealis have on RF communications? (20–5)
17. Discuss the importance of a transmission line matching the impedance of a transmitter to an antenna. (20–6)
18. What is meant by the *surge*, or *characteristic impedance* of a transmission line? What factors control this? (20–6)
19. What happens to the impedance of a two-wire transmission line if the center-to-center spacing is unchanged but large diameter conductors are used? (20–6)

Commercial License Questions 467

20. What change takes place in the characteristic impedance of a two-wire transmission line if the conductor spacing is doubled? (20–6)
21. Discuss the relative merits of a solid-dielectric cable over a hollow, pressurized cable for use as a transmission line. (20–6)
22. What is the meaning of *standing waves* and *standing-wave ratio* (SWR)? How can these be kept to a minimum? (20–6)
23. The power input to a 72-Ω concentric transmission line is 5000 W. What is the rms voltage between the inner conductor and sheath? [4] (20–6)
24. What is the ratio between the currents at the opposite ends of a transmission line one quarter-wavelength long and terminated in an impedance equal to its surge impedance? [4] (20–6)
25. Explain the properties of a quarter-wave section of an RF transmission line. [4] (20–6)
26. The power input to a 72 Ω concentric line is 5000 W. What is the current flowing in it? [4] (20–6)
27. What is the primary reason for terminating a transmission line in an impedance equal to the characteristic impedance of the line? [4] (20–6)
28. Why are standing waves desirable on an antenna but not on a transmission line? (20–7)
29. What is the current and voltage relationship in a half-wave antenna? (20–7)
30. If a vertical antenna is 405 ft high and is operated at 1250 kHz, what is its physical height expressed in wavelengths (1 m equals 3.28 ft)? (20–8)
31. What is meant by *antenna impedance*? How is it determined? (20–8)
32. What is meant by the term *radiation resistance*? (20–8)
33. What is meant by horizontal and vertical *polarization* of a radio wave? (20–9)
34. How should a transmitting antenna be designed if a vertically polarized wave is to be radiated, and how should the receiving antenna be designed for best performance in receiving the ground wave from this transmitting antenna? (20–9)
35. Show by a diagram how a two-wire RF transmission line may be connected to feed a Hertz antenna. (20–10)
36. What is the relationship between the electrical and physical length of a Hertz antenna? (20–10)
37. Draw a simple schematic diagram of a system of coupling a single electron tube employed as an RF amplifier to a Hertz-type antenna. (20–10)
38. What will be the effect upon the resonant frequency if the physical length of a Hertz antenna is reduced? (20–10)
39. What is the radiation pattern of a vertical antenna? (20–11)

40. What are the voltage and current relationships in a quarter-wave antenna? (20–11)

41. Which type of antenna has a minimum of directional characteristics in the horizontal plane? (20–11)

42. Draw a simple schematic diagram of a push–pull, neutralized RF amplifier stage coupled to a Marconi-type antenna system. (20–11)

43. Discuss the various types of ground planes that are encountered in mobile stations for VHF and UHF. (20–12)

44. What effect do broken ground conductors have on a standard broadcast antenna? [4] (20–12)

45. Describe the current and voltage relationships in a full-wave antenna. (20–13)

46. Describe the physical and electrical characteristics of a folded dipole antenna. (20–14)

47. What is meant by a *parasitic array*? (20–15)

48. Describe the function of directors in parasitic arrays. (20–15)

49. What is a reflector's length and spacing with respect to the driven element? (20–15)

50. Describe the function of a reflector in a parasitic array. (20–15)

51. What is a collinear array and what are its directional characteristics? (20–16)

52. Describe the directional characteristics of a vertical loop antenna. (20–17)

53. What is meant by *stub tuning*? (20–18)

54. Explain the characteristics of a quarter-wave section of an RF transmission line. (20–18)

55. What should be the approximate surge impedance of a quarter-wave matching line used to match a 600-Ω transmission line to a 70-Ω antenna? (20–18)

56. Draw a simple schematic diagram of a T-type coupling network suitable for coupling a coaxial line to a standard broadcast antenna. Include means for harmonic attenuation. [4] (20–18, 20–20)

57. How is the degree of coupling varied in a pi network used to transfer RF energy from a plate circuit to an antenna? (20–19)

58. Discuss series and shunt feeding of quarter-wave antennas with regard to impedance matching. (20–19)

59. If you desire to operate on a frequency lower than the resonant frequency of an available Marconi antenna, how may this be accomplished? (20–19)

60. What is the effect on the resonant frequency of connecting an inductor in series with an antenna? (20–19)

Commercial License Questions 469

61. What is the effect on the resonant frequency of adding a capacitor in series with an antenna? (20–19)

62. What material is best suited for use as an antenna strain insulator that is exposed to the elements? (20–19)

63. Why do some standard broadcast stations use top-loaded antennas? [4] (20–19)

64. Why are insulators sometimes placed in antenna guy wires? (20–19)

65. What is the purpose of a dummy antenna? (20–19)

66. Draw a simple schematic diagram showing a method of coupling the RF output of the final power amplifier stage of a transmitter to a two-wire transmission line, with a method of suppression of second and third harmonic energy. [4] (20–20)

67. Why is it that sometimes harmonic radiation can cause interference at distances from a transmitter where the fundamental signal could not be heard? (20–20)

68. What is the purpose of a Faraday screen between the final tank of a transmitter and the antenna inductance? (20–20)

69. Draw a schematic diagram of a T-type coupling network between a coaxial transmission line and an antenna, showing an antenna ammeter and method harmonic suppression. (20–20)

70. How may a standard broadcast antenna ammeter be protected from lightning? [4] (20–20)

71. Describe the directional characteristics of the following types of antennas:
    (a) Horizontal Hertz antenna.
    (b) Vertical Hertz antenna.
    (c) Vertical loop antenna.
    (d) Horizontal loop antenna.
    (e) Vertical Marconi antenna. (20–21)

72. What is the direction of maximum radiation from two vertical antennas spaced 180° and having equal currents in phase? [4] (20–21)

73. If the resistance and the current at the base of a Marconi antenna are known, what formula could be used to determine the power in the antenna? (20–22)

74. If the antenna current of a station is 9.7 A for 5 kW, what is the current necessary for a power of 1 kW? [4] (20–22)

75. What is the antenna current when a transmitter is delivering 900 W into an antenna having a resistance of 16 Ω? [4] (20–22)

76. An antenna is being fed by a properly terminated two-wire transmission line. The current in the line at the input end is 3 A. The surge impedance of the line is 500 Ω. How much power is being supplied to the line? [4] (20–22)

77. If the daytime transmission-line current of a 10-kW transmitter is 12 A and the transmitter is required to reduce to 5 kW at sunset, what is the new value of transmission-line current? [4] (20–22)

78. A long transmission line delivers 10 kW into an antenna; at the transmitter end the line current is 5 A and at the coupling house it is 4.8 A. Assuming the line to be properly terminated and the losses in the coupling system negligible, what is the power lost in the line? [4] (20–22)

79. If the day input power to a certain broadcast station antenna having a resistance of 20 Ω is 2000 W, what would be the night input power if the antenna current were cut in half? [4] (20–22)

80. The dc input power to the final amplifier stage is exactly 1500 V and 700 mA. The antenna resistance is 8.2 Ω and the antenna current is 9 A. What is the plate efficiency of the final amplifier? [4] (20–22)

81. If the power output of a broadcast station is quadrupled, what effect will this have upon the field intensity at a given point? [4] (20–22)

82. The ammeter connected at the base of a Marconi antenna has a certain reading. If this reading is increased 2.77 times, what is the increase in output power? [4] (20–22)

83. If the power output of a broadcast station has been increased so that the field intensity at a given point is doubled, what increase has taken place in antenna current? [4] (20–22)

84. How does the field strength of a standard broadcast station vary with distance from the antenna? [4] (20–22)

85. Explain why high-gain antennas are used at FM broadcast stations. [4] (20–23)

86. What is the ERP of a TV broadcast station if the output of the transmitter is 1000 W, antenna transmission loss is 50 W, and the antenna power gain is 3? (20–23)

87. What is meant by *antenna field gain* of a TV broadcast antenna? (20–23)

88. What is an antenna phase monitor? (20–24)

89. If the two towers of a 950-kHz directional antenna are separated by 120 electrical degrees, what is the tower separation in feet? (20–25)

90. At broadcast stations using the direct method of computing output power, at what point in the antenna system must the antenna current be measured? (20–25)

91. The currents in the elements of a directive broadcast antenna must be held to what percentage of their licensed value? [4] (20–25)

92. If the field intensity of 25 mV/m develops 2.7 V in a certain antenna, what is its effective height? [4] (20–25)

21

Receivers

21-1 BASIC FUNCTIONS OF A RECEIVER

A receiver must be able to perform certain functions. These are, in order of their performance,

1. *Reception.* The antenna system of the receiver has minute voltages inducted into it by the many RF waves passing through it.
2. *Selection.* The ability of the receiver through its tuned circuits to pick out or select the desired signal from among all those developed in the antenna.
3. *Detection.* The recovering of the original intelligence sent out by the transmitter. A reverse process of modulation.
4. *Audio-frequency amplification.* Building up the detected signal to a level required to drive the reproducer.
5. *Reproduction.* A loudspeaker or earphone that converts the electrical signals to sound waves.

A block diagram of a simple receiver is shown in Fig. 21-1, where each of these functions is indicated.

The quality of any receiver can be measured by three basic characteristics. These are *sensitivity*, *selectivity*, and *stability*. The first relates to the ability of the receiver to pick up weak signals. The second is the capability to select the desired signal and reject all undesirable ones. The latter characteristic relates to the ability of a receiver to remain tuned to a particular frequency over a long period of time without drifting.

471

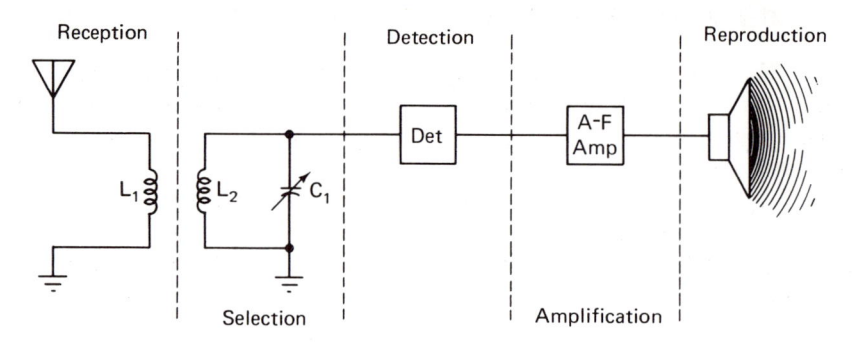

FIG. 21-1   Diagram showing basic functions of a receiver.

21-2   CRYSTAL DETECTORS

The simplest type of receiving circuit is the *crystal detector* shown in Fig. 21-2. No transistor or vacuum tubes are used to amplify the signals;

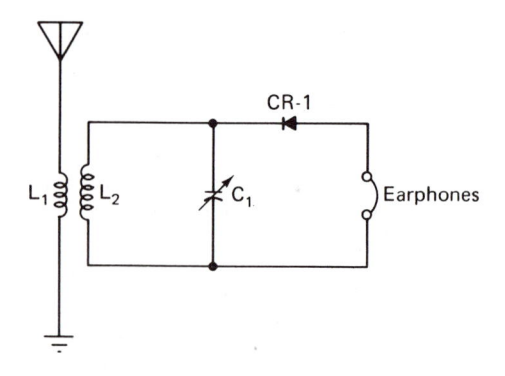

FIG. 21-2   Circuit diagram of a simple receiver.

therefore, no external power is required. The RF currents generated in L_1 by the passing waves induce voltages in L_2 of the antenna transformer. The resonant voltage across L_2C_1 is fed to the series combination of CR_1 and the earphones. The crystal diode rectifies the RF and causes a series of pulses to flow through the phones. These are varying in amplitude according to the modulation envelope of the carrier. Consequently, the pulse strength through the phones is varied at a rate that causes the diaphragm to move back and forth, thereby reproducing the sound.

Although the design of this circuit is very simple, it has several major drawbacks. First, the circuit has little selectivity. Second, the sensitivity is very poor. Generally speaking, only the stronger stations in the area are capable of developing enough voltage across the tuned circuit to provide sufficient energy to actually drive the earphones. There is not enough power developed to drive a small loudspeaker.

Diode Detectors 473

21-3 TUNED RADIO-FREQUENCY RECEIVER

The serious limitations of the crystal detector were partly overcome when the advent of vacuum tubes made it possible to amplify the weak signals. By using tuned circuits between stages, the selectivity and sensitivity can be increased. Such a circuit is shown in block diagram form in Fig. 21-3 and is called a *tuned radio-frequency* (TRF) receiver. The detector recovers the modulation component of the RF wave, which is then amplified in the AF stage and fed to the loudspeaker. A power supply (not shown) is required to operate the tubes.

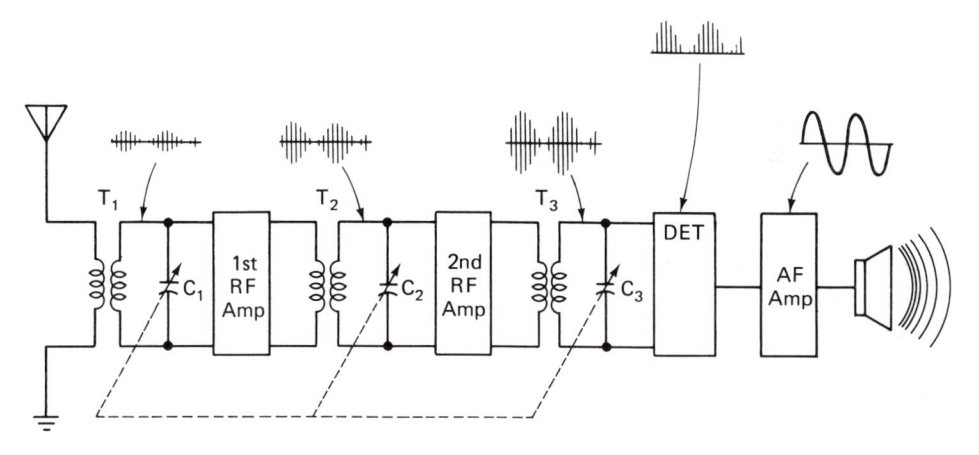

FIG. 21-3 Block diagram of a TRF receiver showing waveforms.

One disadvantage of the TRF receiver is that the selectivity is not constant but *decreases* as it is tuned toward the high end of its tuning range. Also, the amplification is not constant. The *superheterodyne* receiver (discussed later in this chapter) overcomes these disadvantages.

21-4 DIODE DETECTORS

The detection of an AM wave can be accomplished by a *diode detector*, as shown in Fig. 21-4. It has a nearly linear *IV* characteristic curve when operated above the heel. It can handle large input signals with minimum distortion and is sometimes called a *power detector*. Basically, it is a half-wave rectifier. The modulated signal voltage developed across the tuned circuit made up of L_2C_1 is applied to CR_1 and its load resistor R (Fig. 21-4a). On each positive half-cycle the diode conducts and develops a voltage across its load. The rectified signal flowing through the diode actually consists of a series of RF pulses and is not a smooth outline of the modulation envelope. The average of these pulses increases and decreases at the AF rate.

The detector output after rectification is a dc voltage that varies at an

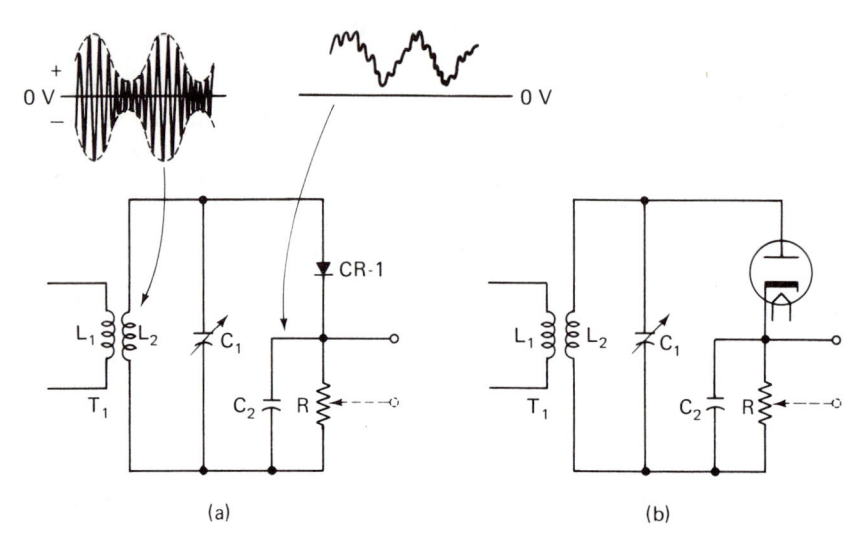

Fig. 21-4 Diode detector circuits using (a) solid state diode, (b) vacuum diode.

audio rate. This is shown by the waveform whose arrow is pointing to the top of C_2. The curve of the output voltage across the capacitor is shown somewhat jagged. Actually, the RF component of this voltage is negligible, and after amplification it is filtered out by the passband characteristics of the AF amplifier.

The correct choice of R and C_2 in the diode detector circuit is very important if optimum sensitivity and fidelity are to be obtained. The load resistor and the resistance of the diode act as a voltage divider to the detected signal. Load resistor R should be high compared to the resistance of the diode so that most of the voltage will be across it. The value of C_2 should be such that its RC time constant is *long* compared with the time of one RF cycle, but *short* compared with the time of one AF cycle. In many diode detector circuits it is customary to use a potentiometer as the load resistor so that varying levels of audio voltage can be taken from the load to drive the following stage. This provides *volume-control* action.

The circuit has the disadvantage of *drawing power* from its driving source. This reduces the circuit Q, sensitivity, and selectivity.

21-5 GRID-LEAK DETECTORS

A triode can be used as a detector, as shown in Fig. 21-5. The circuit is called a *grid-leak detector* and functions like a diode detector combined with a triode amplifier. Detection and amplification can be considered as separate functions.

The RF voltage applied to the grid causes pulses of grid current to flow each time it is driven positive with respect to the cathode. Hence, the pulses

Plate Detectors 475

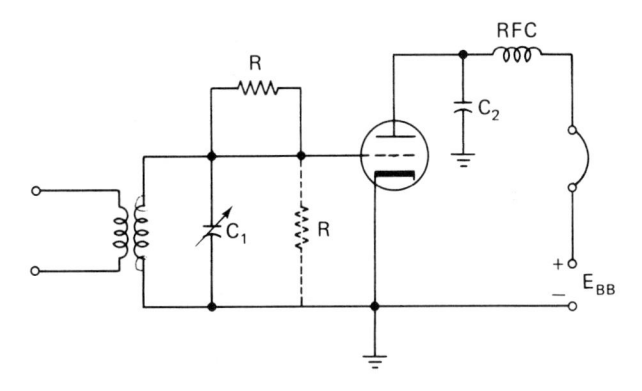

FIG. 21-5 Grid-leak detector.

of grid current vary in accordance with the modulated carrier. This causes a negative charge to appear on the right side of C_1. Grid-leak resistor R permits this charge to leak off. Otherwise, the tube would be blocked. This resistor can also be connected, as shown by the dotted lines, and provides the same function.

By selecting the correct values of R and C, a short time constant for RF and a long time constant for AF can be achieved. This means that the charge appearing across C will vary in accordance with the modulation of the RF carrier.

The charge on C is also controlling the grid bias. Consequently, the plate current is caused to vary at an audio rate. This part of the circuit operates as a triode amplifier. Any tendency for RF to pass through the tube is bypassed to ground via the RF filter made up of C_2 and the *RFC*.

The plate voltage of the grid-leak detector is usually less than 75 V. This causes the tube to operate on the *curved portion* of the characteristic curve, which means that the variation in plate current is essentially proportional to the *square* of the grid voltage. Therefore, grid circuit detectors are frequently referred to as *square-law detectors*. Grid detection is characterized by *high distortion* and *high sensitivity*. This type of detector is infrequently used in modern receivers designed for broadcast reception.

21-6 PLATE DETECTORS

With *plate detection*, Fig. 21-6, the RF signal is amplified and then detected in the plate circuit. Cathode bias resistor R_1 is chosen so that the tube is nearly at cutoff. Hence, pulses of plate current flow only during the positive half-cycles of the input signal. By proper design the peak value of the RF input signal is limited to slightly less than the cutoff bias, thus preventing the grid being driven positive. As a consequence, plate detectors do not load the input circuit. Capacitor C_2 is sufficiently large to hold the voltage across R_1 steady at the lowest audio frequency to be detected. R_2 is the plate

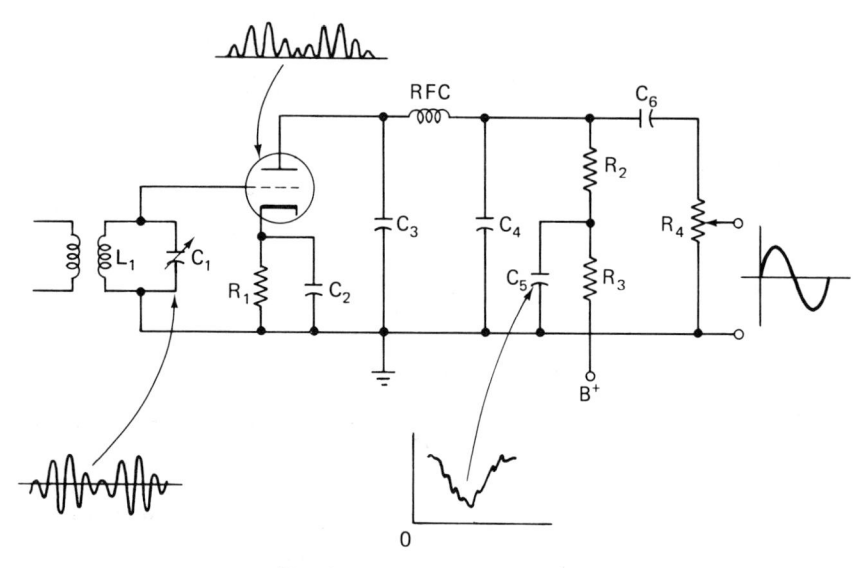

FIG. 21-6 Plate-detector circuit.

load resistor across which the AF component is developed. C_4 in conjunction with R_2 has a long time constant with respect to RF but a short time constant as far as AF is concerned. The *RFC* and capacitor C_3 provide a filter to keep RF energy out of the audio circuit.

The plate detector has *excellent selectivity*. Its sensitivity is also greater than that of the diode detector. However, it is inferior to the diode detector in that *it is unable to handle strong signals without overload*. Another disadvantage is that the operating bias varies according to the strength of the RF signal and causes distortion, unless provision is made to maintain the signal input at a somewhat constant level. This can be accomplished by *automatic volume control* circuits (discussed in Section 21-15).

21-7 REGENERATIVE DETECTORS

A typical *regenerative detector* using a JFET is shown in Fig. 21-7. The principle of operation is the same as one using a vacuum tube. The circuit is essentially a *gate-leak detector* (same as grid-leak detector) and an AF amplifier. The gate current that flows as a result of the rectifying action charges C_2 and leaks off through gate-leak resistor R_1. By properly selecting the values of C_2 and R_1, the voltage at the gate of the JFET will vary at a rate corresponding to the modulation of the carrier. These audio frequencies are amplified by the JFET.

The drain load circuit consists of inductor L_3 and the primary winding of the AF output transformer. The circuit is *regenerative* because some of the signal from the drain circuit is fed back to the gate by inductive coupling

Superregenerative Detectors 477

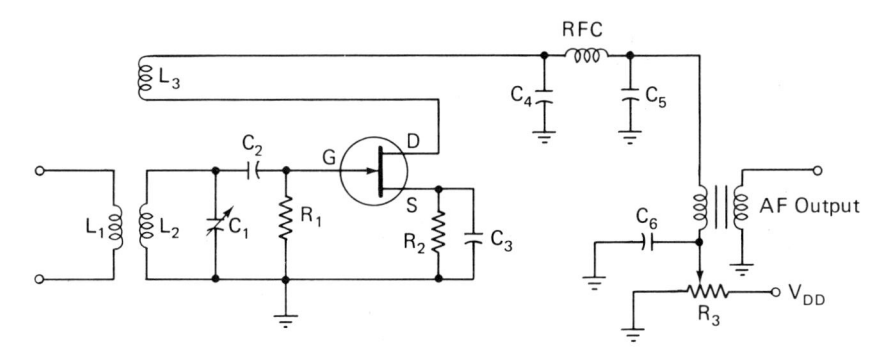

FIG. 21-7 A regenerative detector using a JFET.

(between L_3 and L_2). The amount of regeneration must be controllable, because maximum regenerative amplification is obtained at the *critical point* where the circuit is just about to go into oscillation. This depends on the type of signal being received. Regeneration is controlled by potentiometer R_3, which varies the voltage to the drain of the JFET.

Although this detector is more sensitive than most other types, it has several disadvantages. The linearity is poor and the signal-handling capability is limited. The degree of antenna coupling is also critical. These limitations restrict its use to the simplest types of receivers. When receiving A3 emission, it is preferable to reduce the regeneration to the point just before the receiver goes into oscillation. This provides the most sensitive operating point. For CW reception the regeneration control should be advanced until the circuit just breaks into a hiss, which indicates the circuit is oscillating.

21-8 SUPERREGENERATIVE DETECTORS

Of all detectors the *superregenerative* detector provides the highest sensitivity. It is characterized by simplicity, and is particularly applicable to circuits where weight is a factor. These advantages are offset to some degree by poor selectivity, the inability to demodulate other than AM and wideband FM signals, and high noise level. The oscillatory nature of the circuit will cause *radiation* of RF energy from the antenna unless some form of isolation is used between the antenna system and the detector.

The basic circuit of a superregenerative detector using a JFET appears in Fig. 21-8. The great sensitivity results from the use of an *ac quench voltage*, which is beyond the audible range, generally between 25 and 350 kHz. This quench frequency should be approximately 100 times lower than the frequency at which the circuit is designed to operate. This means that the superregenerative detector is frequently used as a simple type of VHF receiver.

Operation is controlled by R_2, which is a *regeneration control*. This establishes a mode of operation such that the detector goes into oscillation on each positive peak of the quench voltage and is cut off by negative swings. By

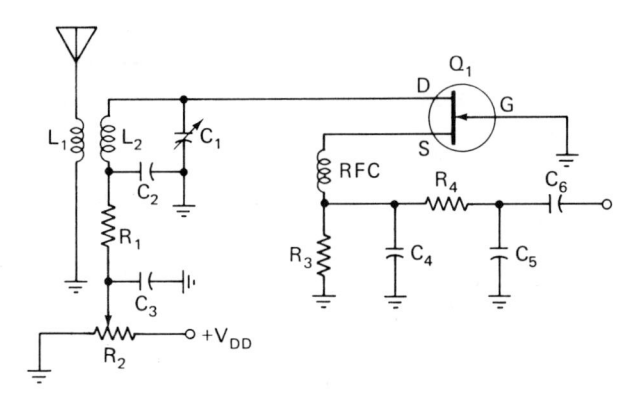

FIG. 21-8 Typical superregenerative detector using a JFET as the detector.

cycling the superregenerative detector on and off at the quench frequency, the circuit is permitted to regenerate to a point far beyond that which would be available in a straight regenerative detector. The result is excellent sensitivity. Even though the quench frequency is well above the audio spectrum, it is essential that it be filtered out of the detector output to prevent reaching subsequent audio stages. The quench filter is made up of R_4 and C_5. The drain circuit will have pulses of current at the quench frequency. When no signals are present, the quench frequency establishes a certain average current in the drain circuit. When a signal is received, the amplitude of the quench pulses changes, causing the average drain current to vary in accordance with the amplitude of the intelligence that is modulated in the carrier. Hence, demodulation occurs.

The selectivity of this circuit is limited by the Q of the tuned circuit comprised of L_2 and C_1. The selectivity is also affected by the quench frequency, with better selectivities at lower quench frequencies. The quenching frequency is determined by R_3 and C_4. R_3 also establishes the operating bias for the detector. The *RFC* maintains the source above ground at the signal frequency. R_1 and C_3 form a decoupling network to prevent the circuit from *motor boating* when audio stages are added to the circuit. A JFET should be selected that has a high transconductance and is designed for VHF use. If a vacuum tube is used, one should be selected having a high mu.

21-9 SUPERHETERODYNE RECEIVERS

The essential difference between TRF and superheterodyne receivers is that in the former the RF amplifiers preceding the detector are tunable over a band of frequencies, whereas in the latter the corresponding amplifiers are tuned to one fixed frequency called the *intermediate frequency* (IF). The principle of *frequency conversion* (heterodyne action) is employed to convert any desired frequency within the receiver range to this intermediate frequency before detecting the AF component. The IF amplifier thus provides *optimum selectivity*, *voltage gain*, and *bandwidth* to contain all of the desired sideband components associated with the AM carrier.

The block diagram and typical waveforms of a superheterodyne receiver are shown in Fig. 21-9. The RF signal from the antenna passes through an RF amplifier or *preselector*. A *local oscillator* generates an unmodulated RF signal of constant amplitude, which is mixed with the carrier in the *mixer stage*. The mixing or heterodyning of these two frequencies produces an intermediate frequency that contains all the modulation characteristics of the original signal. The IF is equal to the difference between the station frequency and the local oscillator. This frequency is then amplified in one or more stages called the IF amplifier. It is then detected (second detector) and fed to the AF amplifier.

21-10 RADIO-FREQUENCY AMPLIFIER OR PRESELECTOR

Better superheterodyne receivers use an RF amplifier or preselector stage. The advantages are increased signal amplification, increased selectivity (due to an additional tuned circuit), isolation of the local oscillator from the antenna, and improved *image rejection*. If the antenna is connected directly to the mixer stage, there is a possibility that a part of the local oscillator frequency may be radiated into space, causing interference in nearby electronic equipment.

A typical RF amplifier using permeability tuning is shown in Fig. 21-10. Radio-frequency transformers T_1 and T_2 have their tuning slugs ganged together as indicated by the dotted lines, so they will *track* (tune to the same frequency) as the tuning dial is rotated. Trimmer capacitors C_1 and C_5 are used for alignment purposes. Capacitors C_2 and C_6 bypass the RF to ground to keep it out of the power supply.

The sensitivity of a receiver is a measure of the *minimum signal input required to produce a specified output*. Various noises are also present that will be amplified with the signal unless means are taken to minimize them. Two sources of noise must be considered: *external* and *internal*. External noise results from such things as electrical storms, sunspots and aurora borealis, electrical arcs, neon signs, motors, and fluorescent lights. Internal noise results from the passing of the electric current through circuit components such as resistors, coils, and the active devices being used. The ratio of the desired signal amplitude to the noise voltage amplitude is known as the *signal-to-noise ratio*. The effect of noise voltages, whose frequencies do not lie within the bandwidth of the desired signal, can be minimized by decreasing the bandwidth (increase selectivity) of the input circuits as much as possible without decreasing the overall bandwidth below the required amount. Therefore, decreasing the bandwidth results in an increased signal-to-noise ratio.

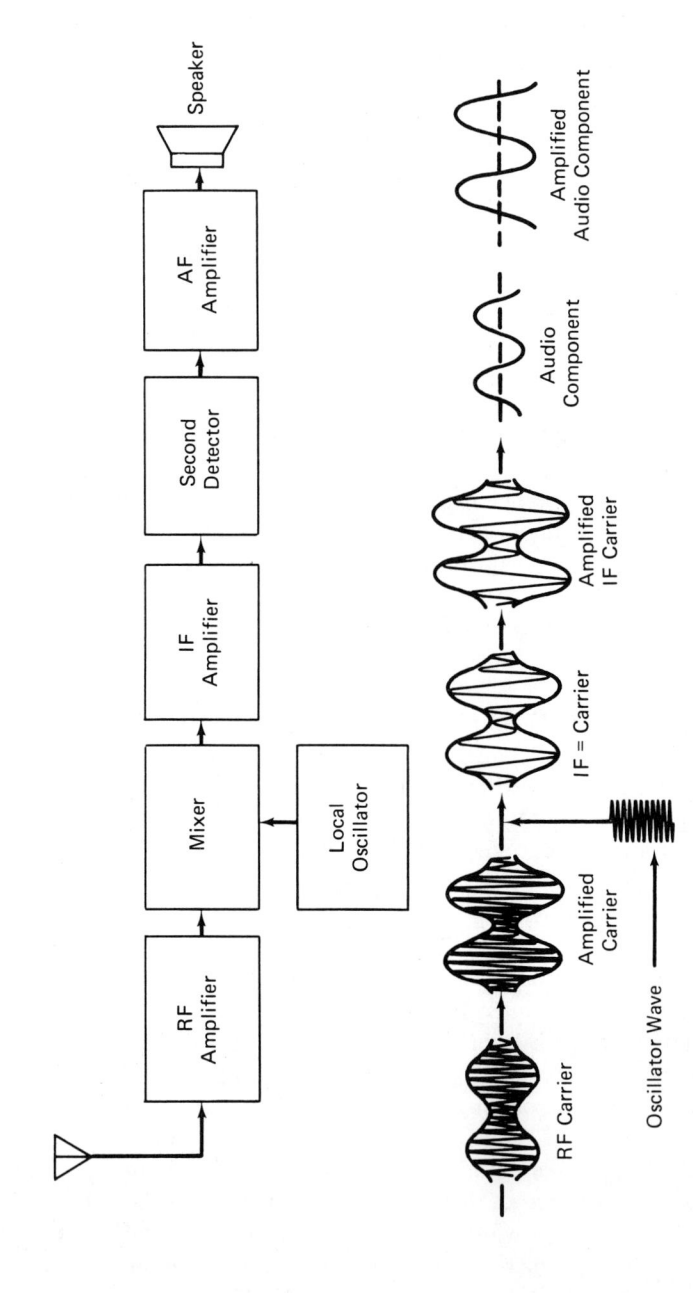

Fig. 21-9 Block diagram of a superheterodyne receiver showing typical waveforms.

Mixers and Converters 481

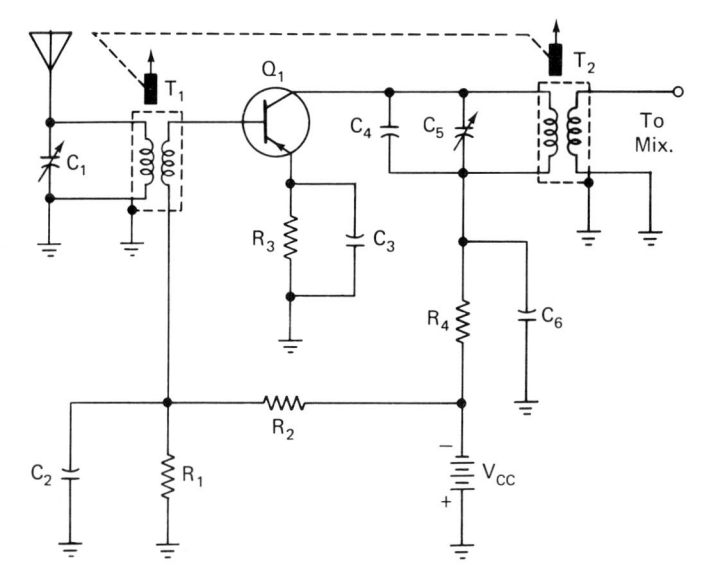

FIG. 21-10 Typical solid state RF amplifier using permeability tuning.

21-11 MIXERS AND CONVERTERS

The characteristics of a superheterodyne receiver are *uniform gain and selectivity* over its tunable range. These are possible because the incoming signals are converted to an intermediate frequency having a constant center frequency. This is achieved in the frequency conversion stage of the receiver. There are two basic types of frequency conversion stages used in the *superhet* receiver, one type being the *mixer* and the other the *converter*. The frequency conversion process is the heart of the superheterodyne principle.

If two different frequencies are fed to a *nonlinear device*, such as the *IV* characteristic curve of a transistor, a *heterodyne* action occurs. This generates *two new frequencies;* one is the *sum* of the original two, and the other is their *difference*. These exist in the output circuit along with the two original ones. This action is graphically illustrated in Fig. 21-11. Assume that an RF signal of 1000 kHz has been tuned in and fed to the mixer stage. Within the receiver (not shown) is a local oscillator designed to operate on 1450 kHz. Its output is also connected to the mixer. The nonlinearity of the mixer causes these two frequencies to beat together. This results in the four frequencies shown in the output. Only the lowest of these, 450 kHz, is used in a typical broadcast receiver.

In a practical situation the received AM signal contains upper and lower sidebands. These would also beat against the local oscillator, and a multitude of significant output frequencies would result.

Since the output of the mixer contains many frequencies, a tuned circuit, acting as a filter, selects the desired ones. The characteristic of this filter

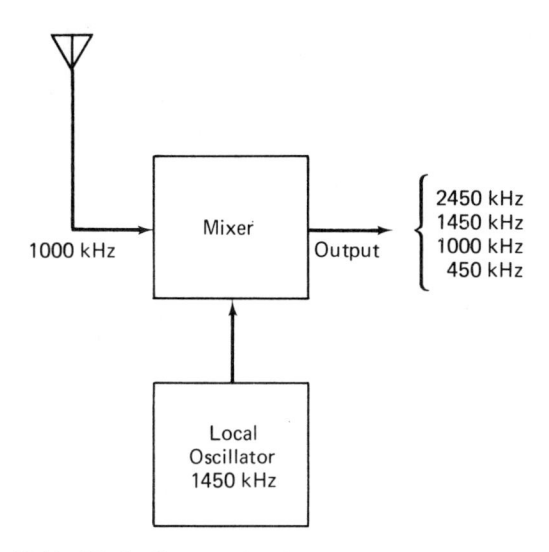

FIG. 21-11 Block diagram showing heterodyne action between two different frequencies.

is shown in Fig. 21-12. The gain is relatively uniform within the 0.707 points. Therefore, only frequencies centered around this value will develop sufficient voltage to be passed on to the RF amplifier. This curve represents the bandwidth of a complete receiver. The half-power points have been chosen arbitrarily, and are dependent on receiver circuit design and characteristics.

If the mixer is capable of delivering a relatively large amount of IF output voltage with respect to the RF signal input, it is said to have a high *conversion efficiency*. A high ratio is desirable. The noise generated by the mixer should also be low if a good signal-to-noise ratio is to be achieved. This is particularly true if the mixer is the first stage in the receiver.

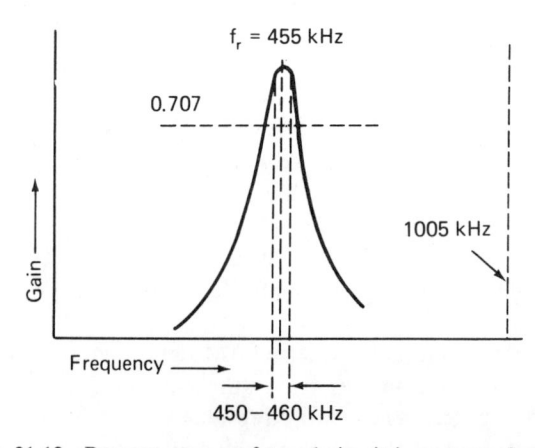

FIG. 21-12 Response curve of tuned circuit in output of mixer.

Mixers and Converters 483

If the mixer and local oscillator are separate devices, the converter portion is called a *mixer*. If the two are contained in one active device, which is frequently the case, the stage is referred to as a *converter*.

The local oscillator and the received signal may be applied to the mixer in a number of different ways. One typical circuit using a *pentagrid mixer* is shown in Fig. 21-13. This provides good isolation between the tuned RF input and the local oscillator, and thus prevents *oscillator pulling* when the mixer grid circuit is tuned. The local oscillator output is fed to the mixer via an *injection grid*.

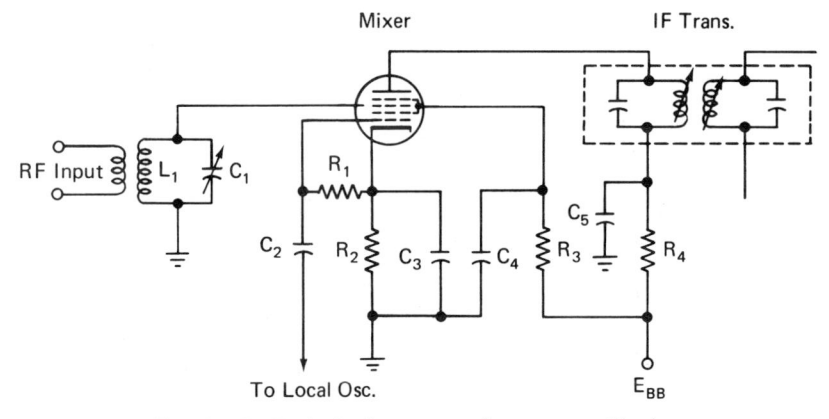

Fig. 21-13 Typical mixer stage using a pentagrid tube.

A transistorized mixer is shown in Fig. 21-14. Output of the local oscillator is injected into the base via C_2. If emitter injection were used, bypass

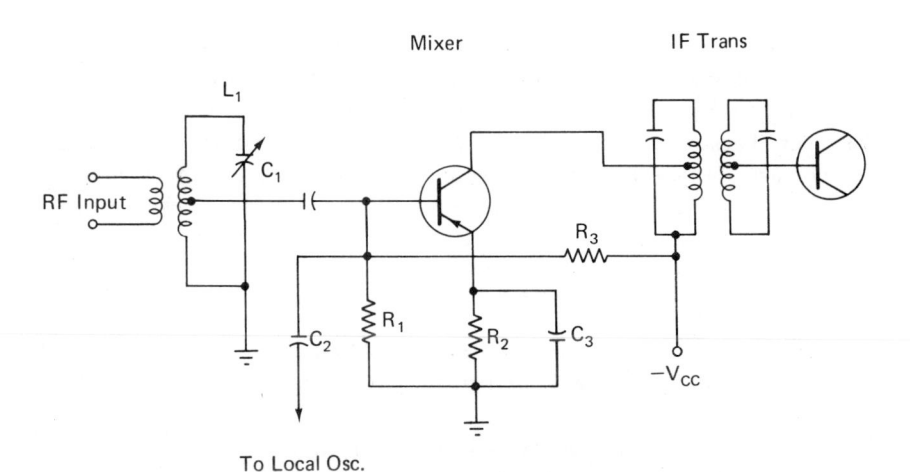

Fig. 21-14 Typical transistorized mixer stage.

capacitor C_3 would have to be removed. Large RF signals cannot be successfully handled due to the limitations of the dynamic characteristics of most transistors.

Mixing action may be better understood by considering the transistor as being controlled simultaneously by the oscillator and the incoming signals. The oscillator output is usually much larger (10 or more times) in amplitude than the incoming RF. For this reason the collector current of the mixer is controlled primarily by the oscillator signal.

A pentagrid converter circuit is shown in Fig. 21-15. The function of

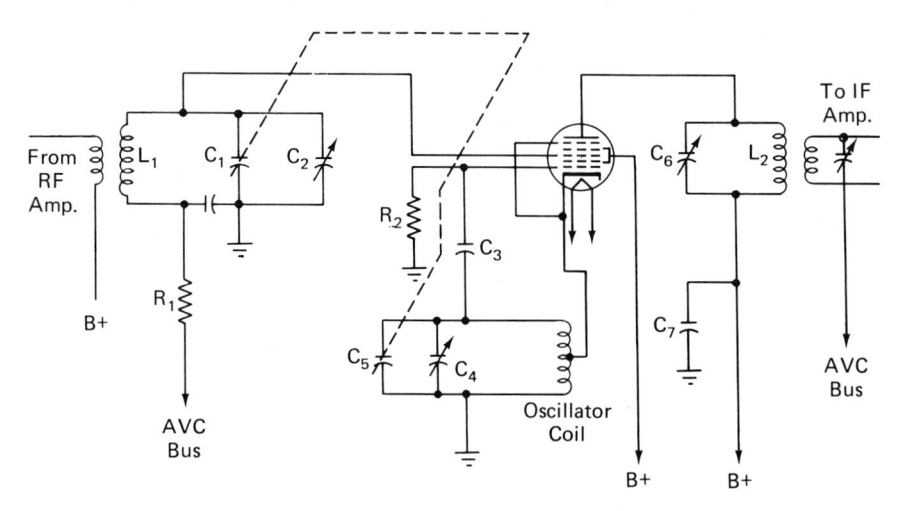

FIG. 21-15 First detector employing a pentagrid converter.

oscillator and mixer are combined in one tube. Pentagrid converters have the advantage of requiring fewer parts. However, less stability results, particularly at the higher frequencies. Some oscillator voltage is coupled to the signal grid via the space charge. The coupling increases with frequency. Radio-frequency tuning capacitor C_1 and oscillator tuning capacitor C_5 are ganged together. These two circuits are designed to track so that the oscillator frequency will always differ from the incoming signal by an amount equal to the IF. Generally, but not always, the oscillator operates above the incoming frequency. The function of the AVC circuit is discussed in Section 12-15.

21-12 IMAGE FREQUENCY

An *image frequency* is defined as an interfering transmitted signal whose frequency differs from the desired one by twice the IF. Expressed mathematically,

$$\text{image frequency} = \text{station frequency} \pm 2 \text{ IF}$$

The plus sign is used if the local oscillator operates above the desired station frequency, and minus if below. Generally, the local oscillator operates below

Intermediate-Frequency Amplifier

the incoming signal on higher frequency bands and above on lower bands.

If a receiver having an IF of 455 kHz is tuned to a frequency of 1500 kHz, its local oscillator will be at 1955 kHz. If a strong station is operating at 2410 kHz, it too will beat with the 1955-kHz oscillator signal and produce a difference frequency of 455 kHz. Even though this station is out of the broadcast range, *it can cause interference*. Both stations when mixed with the local oscillator will produce the correct IF. Thus, the IF amplifier cannot separate the desired station from the image signal.

Even though a superhet does not depend exclusively on the input circuit for selectivity, the Q of the input circuit must be high enough to provide good *image rejection*.

21-13 INTERMEDIATE-FREQUENCY AMPLIFIER

Since the IF amplifier operates at a fixed frequency, it can be designed to provide optimum gain and bandwidth. The choice of the IF is usually a compromise between several factors. For example, the use of a low IF results in slightly better gain, stability, and selectivity. However, these advantages are offset by an increased susceptibility to image-frequency reception.

Figure 21-16 illustrates a typical IF amplifier. The input circuit (the

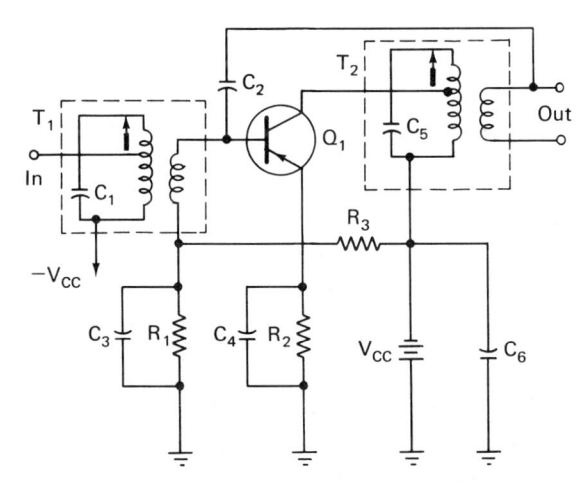

FIG. 21-16 Single stage IF amplifier.

primary of T_1 and C_1) is actually the output circuit of the mixer stage and is tuned to the receiver's IF. Tuning is by adjustment of a powdered iron core (indicated by the arrow and slug above the primary coil), while the secondary is untuned.

The turns ratio and the coefficient of coupling are chosen so that the impedance of the secondary will match the input impedance of Q_1. Transformer T_2 is also tuned to the IF. Its secondary may feed either the detector or another IF stage. In inexpensive receivers a single IF stage is used, whereas in communications and other high-quality receivers the gain, bandwidth, and selectivity requirements are satisfied by *cascading* two or three IF stages.

Although the Q of the IF tuned circuits should be high for good selectivity, it must not be so high as to reduce the overall bandwidth to the point where the higher sideband frequencies are attenuated. This is important for hi-fi receivers, but not for communications receivers where selectivity it more important.

Some applications require that all sideband frequencies within the band pass receive equal amplification. Others may require a wide band pass, but also a high degree of selectivity for frequencies immediately outside the band pass. These requirements can be satisfied by the use of *double tuning* in the IF stages. This refers to an interstage transformer in which both the primary and secondary contain resonant circuits.

The band-pass characteristics of a double-tuned stage depend on several things, such as coefficient of coupling (k) between the primary and secondary windings, the Q's of the windings, and the mutual inductance. The effect of varying the coupling is shown in Fig. 21-17.

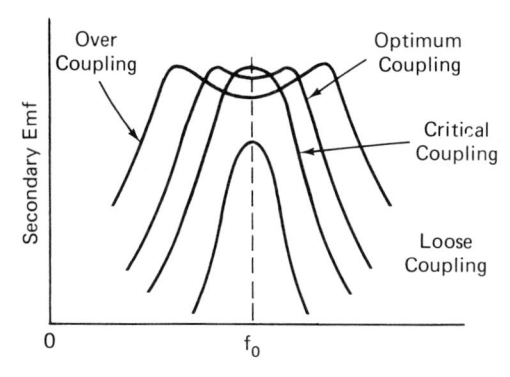

FIG. 21-17 Effects of varying the coupling between primary and secondary of an IF transformer.

Intermediate-frequency stages can also be *stagger tuned*, where the two tuned circuits (one in each stage) are tuned to a different resonant frequency, one slightly above and the other slightly below the receiver's IF. The net result is an IF response that is broadened compared to the characteristics of one tuned circuit alone.

21-14 CRYSTAL FILTERS

In communication receivers it is often desirable to have much higher selectivity than is possible by the use of tuned transformers. A method of achieving this is by the use of a piezoelectric crystal in the IF amplifier. Because the Q of a quartz crystal is extremely high, it can easily limit the passband of the amplifier to several kilohertz. The crystal acts as a high Q tuned circuit, which is many times more selective than conventional tuned circuits. The crystal must resonate at the intermediate frequency. This high selectivity affords good discrimination against adjacent signals and also tends to reduce the noise.

An IF amplifier containing a crystal filter is shown in Fig. 21-18a. The crystal is located in one arm of a bridge circuit, and phasing capacitor C_4 is in the other. The crystal acts as a high Q series resonant circuit permitting signals only within the immediate vicinity of resonance to pass. The secondary, L_2, of the input transformer is center tapped to provide a balance to ground.

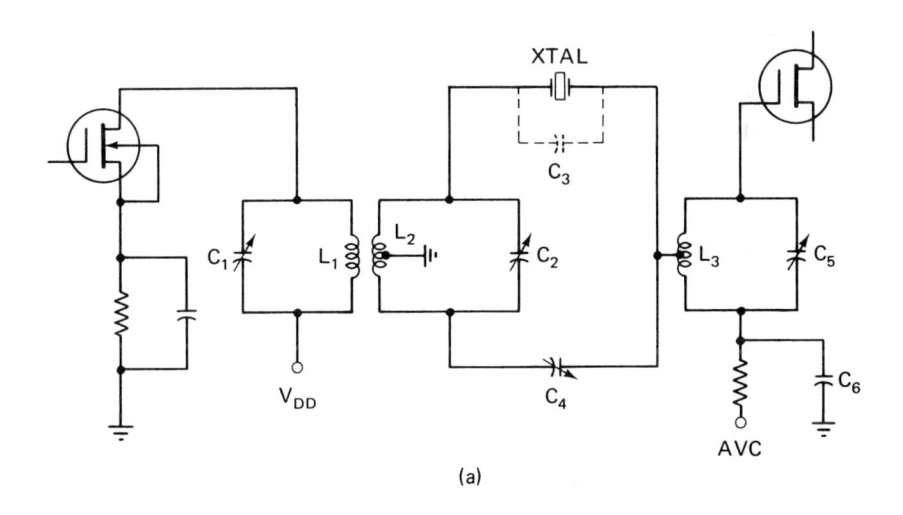

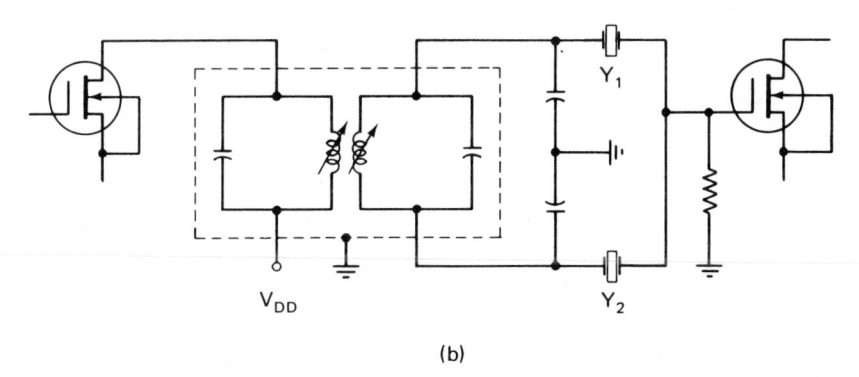

FIG. 21-18 Crystal filters used in the IF section of a superheterodyne receiver.

Because of the capacity between the crystal holder plates, undesirable signals may be bypassed around the crystal. Consequently, neutralizing capacitor C_4 is necessary.

A more effective circuit utilizes two crystal filters, such as shown in Fig. 21-18b. The two crystals must operate at slightly different frequencies. If the frequencies are only several hundred hertz apart, the IF characteristics will be excellent for CW reception. If they are about 2 kHz apart, reasonably good speech characteristics are obtainable. More elaborate circuits are available using four and six crystals, which give further reduced bandwidth.

Mechanical filters are finding widespread application in the IF circuits of receivers. These are made up of three sections: an input transducer, a mechanically resonant filter section, and an output transducer. The transducers are basically *magnetostrictive* devices that convert electrical signals to mechanical energy and then back again. The mechanically resonant sections are made up of precisely machined metal discs supported and coupled by thin rods. Each disc has its own resonant frequency, dependent upon the material and its dimensions. The effective Q of a single disc may be in excess of 2000.

21-15 SECOND DETECTOR AND AUTOMATIC VOLUME CONTROL CIRCUIT

The function of the second detector in a superhet receiver is to demodulate or rectify the AM signal. Either half of the wave can be rectified, as they both contain the same intelligence. Most circuits demodulate the positive half of the modulated wave, as shown by the schematics in Figs. 21-19a and b. A vacuum diode is used in a. Each time the plate is driven positive relative to the cathode a pulse of plate current flows through load resistor R_2, which is in series with the secondary winding of the IF transformer. The emf developed across R_2 varies with the modulation component. R_2 also serves as the volume control, with maximum output occurring with the wiper positioned at the negative end. C_3 couples the audio component to the first AF stage. Capacitor C_2 places the bottom end of the secondary winding at RF ground, but is essentially open to the AF present.

A similar circuit using solid-state components is shown in Fig. 21-19b. CR_1 demodulates the carrier, and the rectified output is across R_2 and potentiometer R_3 which serves as the volume control.

As we have learned, it is desirable for a receiver to have high sensitivity. However, when strong signals are received, *overloading* of the RF and IF sections may result, causing the audio output to become distorted. To overcome this problem, a method of automatically controlling the gain of these sections has been developed that will cause the sensitivity to vary *inversely* with signal strength. With this arrangement it is possible to set the manual volume control to the desired level and then, as the receiver is tuned across the band, the strong signals will not produce appreciably more audio output than weak signals. This is called *automatic volume control* (AVC).

Second Detector and Automatic Volume Control Circuit

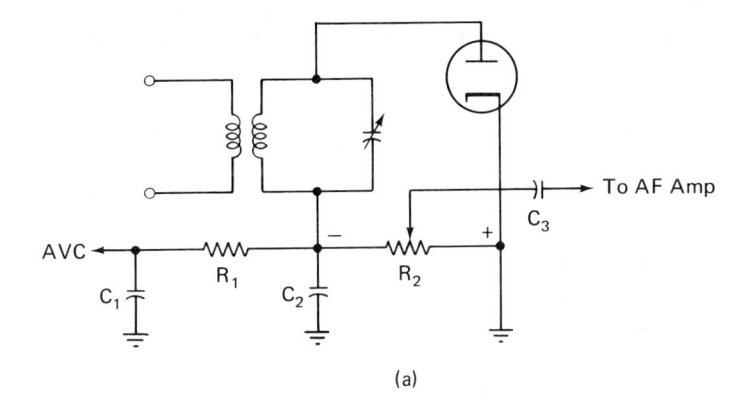

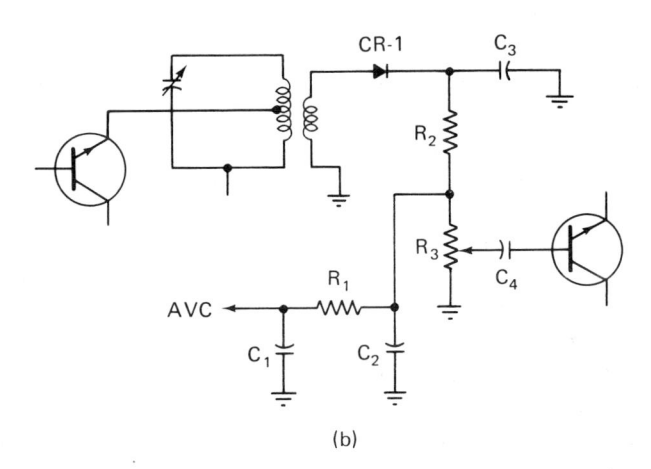

FIG. 21-19 Second detector and AVC circuits for (a) vacuum tube and (b) transistor configurations.

To provide AVC action, a voltage that will control the bias of the RF and IF stages is needed so that with strong signals the bias will be increased and the gain reduced. The load resistor of a diode detector is an excellent source of this voltage, since the rectified voltage appearing across it will increase and decrease with signal strength. A suitable filter is used to remove the AF component of the signal and at the same time to isolate the AVC circuit from the audio output.

The voltage appearing at the junction of R_1 and R_2 (Fig. 21-19a) is varying at an audio rate. This varying negative voltage is filtered by the combination of C_1, R_1, and C_2 and is fed as AVC voltage to the appropriate RF and IF stages of the receiver. The time constant of this combination is long with respect to the AF signal, so that only variations in signal strength are available as AVC voltage. Figure 21-19b shows the equivalent transistorized circuit. Radio-frequency energy is filtered out of the audio circuit by C_3. The

AF component of the rectified signal is removed by the filter circuit made up of C_1, R_1, and C_2.

21-16 DELAYED AUTOMATIC VOLUME CONTROL

A disadvantage of AVC circuits is that even the weakest signals produce some AVC bias, and, as a consequence, the receiver suffers some reduction of gain. This is overcome by the use of a *delayed AVC* circuit, as shown in Fig. 21-20. The AVC diode, plate 2, is separated from the detector diode, plate 1, and both are contained in the same envelope with the triode. This type of tube is called a duodiode high-mu triode.

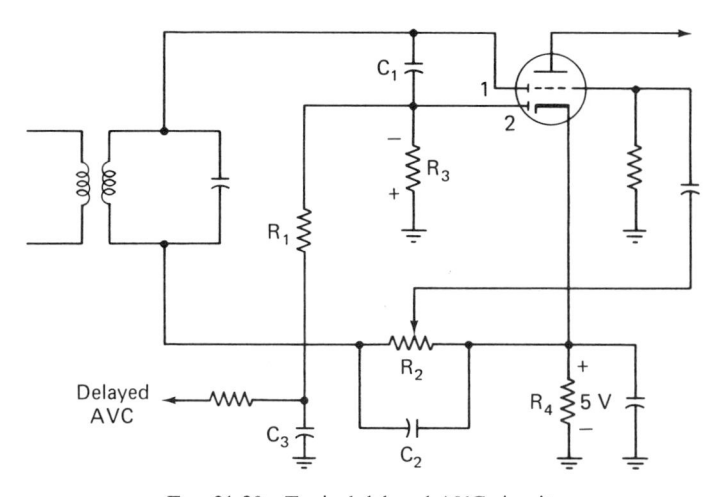

FIG. 21-20 Typical delayed AVC circuit.

In this circuit the cathode of the triode is maintained at an average potential of 5 V above ground by virtue of the voltage drop across R_4. This places a bias of 5 V on the delayed AVC diode plate 2, inasmuch as its load resistor R_3 is returned directly to ground. The signal across the secondary of the IF transformer is coupled to this diode by C_1. When the IF signal is less than 5 V peak, no delayed AVC voltage is developed across R_3. Under these conditions no delayed AVC voltage will be supplied to the RF or IF stages. Therefore, maximum gain will be provided on weak signals. When the signal exceeds the 5-V bias, the AVC diode conducts and supplies delayed AVC.

21-17 BEAT-FREQUENCY OSCILLATORS

A CW signal will not produce an audible-tone output in a superhet receiver. To receive these signals, it is necessary to generate a frequency that can be heterodyned or beat against the IF. This can be accomplished by a *beat-frequency oscillator* (BFO) whose output is coupled into the second

Signal-Strength Meters 491

detector. The BFO is adjusted by a front panel control, so that it is 1 or 2 kHz above or below the IF. The difference frequency is detected and amplified in the AF amplifier, resulting in an audible tone.

The BFO should be shielded to prevent its output from being radiated and interfering with other signals in the receiver. If AVC voltage is used, it should be obtained from a separate diode isolated from the second detector. One way is to couple the output of an IF amplifier stage ahead of the second detector to the AVC diode. Otherwise, the output of the BFO would be rectified by the second detector and would develop an AVC voltage, even on no signal.

21-18 SIGNAL-STRENGTH METERS

Communications receivers customarily have some means by which the relative signal strength may be determined. A *signal-strength meter*, or S meter, provides this indication. Calibration is in numerical units of from 1 to 9, with 9 being approximately middle scale. The numbers indicate the relative signal strength. Table 21-1 defines the 1-to 9 rating. The calibrations on the

TABLE 21-1 THE R–S SYSTEM

Readability
1. Unreadable
2. Barely readable, some words distinguishable
3. Readable with difficulty
4. Readable with little difficulty
5. Perfectly readable

Signal Strength
1. Faint signals, barely perceptible
2. Very weak signals
3. Weak signals
4. Fair signals
5. Fairly good signals
6. Good signals
7. Moderately strong signals
8. Strong signals
9. Very strong signals

meter face usually vary 6 dB/calibration unit. On this basis an S9 reading is equivalent to approximately 100 μV of received signal strength. Few signal-strength meters are accurate across the entire indicating range. Each S meter should be calibrated with the receiver in which it is intended to be used. Usually they are only good as relative indicating instruments when comparing the strength of signals at a given time at a given frequency. By having S9 approximately at midscale, extremely strong signals will be crowded at the far end of the meter scale. Weak signals can then be spread out over the lower half of the dial scale, making for more accurate readings.

Two typical S-meter circuits are shown in Fig. 21-21. In Fig. 21-21a the second detector is comprised of CR_2 and R_3. Output from the last IF transformer also drives CR_1, whose load is R_1 and R_2 together with the S meter.

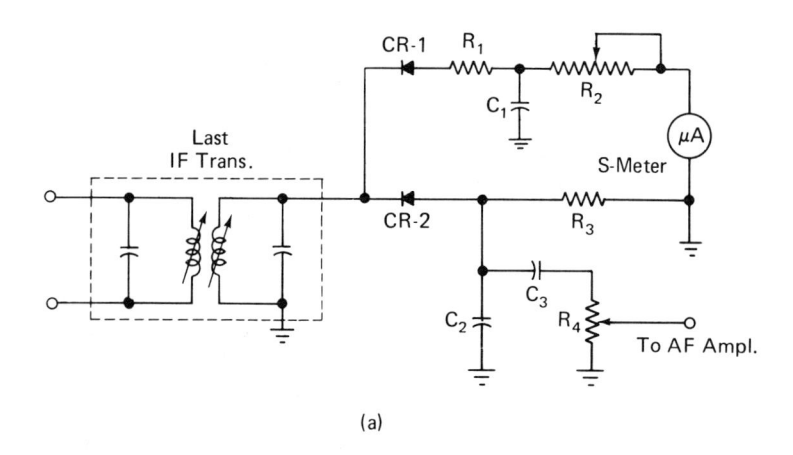

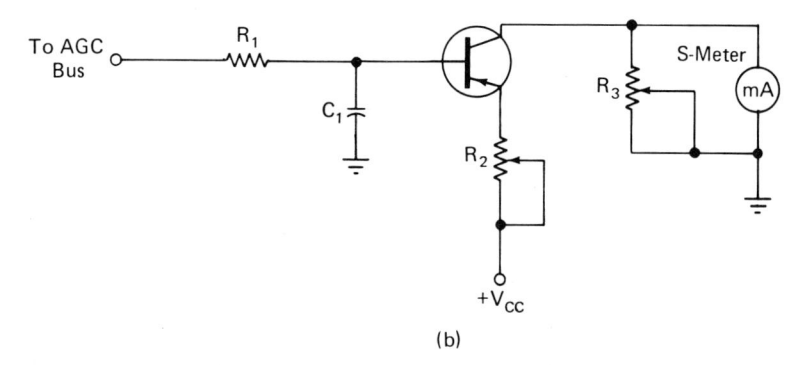

FIG. 21-21 Examples of S-meter circuits: (a) using diodes, (b) using transistor to drive S-meter.

Filtering is by means of R_1 and C_1. Potentiometer R_2 serves to linearize the circuit and thus compensate for the nonlinear characteristics of CR_1. It is also used as a means of setting the meter to zero reading in the absence of a signal. The meter has a microampere movement and draws little current; hence, the loading of the last IF transformer is insignificant.

In Fig. 21-21b a transistor is used to drive the S meter. Its input comes from the AGC bus. As the AGC voltage increases with signal strength, the forward bias increases, resulting in a rise of collector current and causing the S meter to increase.

21-19 NOISE LIMITERS

The performance of a receiver can be seriously impaired by the presence of noise. Good circuit design, particularly in the first stage, can greatly reduce this. Other types of noise that must be contended with are those generated by automobile ignition systems, lightning, and switch and key clicks. These generally appear in the form of short-duration, high-energy pulses having strengths that may be 100 times greater than that of the signal being received. If these are not eliminated, intelligibility may be lost. This noise affects AM reception, but not FM.

Several circuits have been developed to eliminate this noise. The simplest consists of two Zener diodes back to back, as shown in Fig. 21-22a.

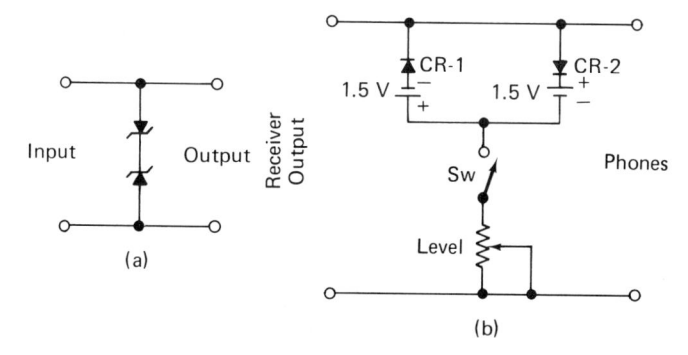

FIG. 21-22 Audio shunt limiters using: (a) zener diodes, (b) diodes.

They would normally be connected between the output of the receiver and the reproducer. Whenever noise levels exceed the voltage of either diode, the diode breaks down and the output is shorted to ground via the other diode.

An alternative circuit is shown in Fig. 21-22b. The circuit is activated by closing SW_1. The two diodes are connected in parallel with polarity reversed which will clip both positive and negative peaks. Both diodes are back biased by their respective batteries, which permits the level control to serve as a clipping level adjustment. Any audio appearing above the threshold setting of the level control will not be reproduced.

A more effective limiter is the series type shown in Fig. 21-23. Two diodes are used, one serving as the second detector and the other as the noise limiter. The demodulated signal appears across diode load resistors R_1 and R_2. Assume that these are of equal ohmic value and that the amplitude of the average signal appearing across this load is -8 V. This places the anode of CR_2 at -4 V with respect to ground. The -8 V appearing at the left of resistor R_1 is connected via R_3 and R_4 to the cathode of CR_2. This places approximately 4 V across CR_2 with the cathode being negative. Conduction

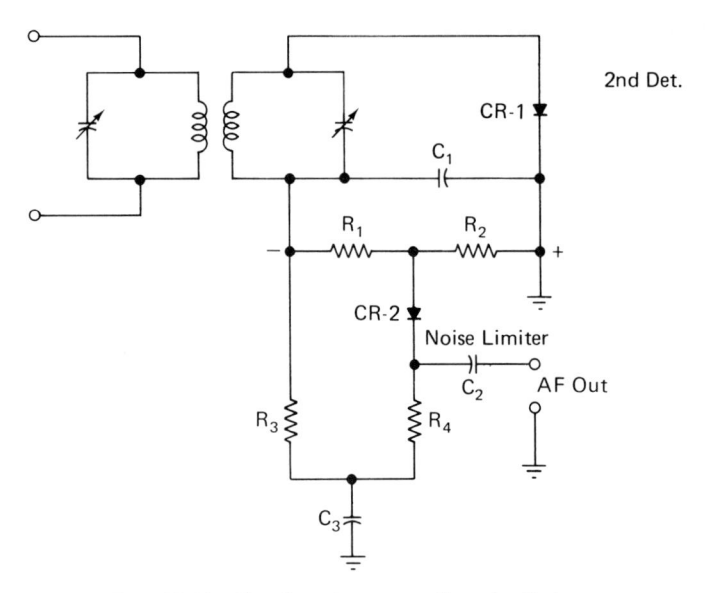

FIG. 21-23 Simple series-type audio noise-limiter.

occurs and the audio signal is developed across resistor R_4. When a noise pulse is received, the instantaneous voltage appearing across R_1 and R_2 will increase to perhaps several times the former value. If the voltage is assumed to increase instantaneously to -16 V across these two resistors, the anode of CR_2 will be momentarily -8 V below ground. The time constant of R_3, C_3, and R_4 is such that audio signals can easily be passed. However, for the extremely short duration of noise pulses the time constant is too long, and capacitor C_3 prevents pulses of noise from reaching the cathode. Hence, the cathode remains at a potential higher than its anode. Consequently, conduction ceases and there is no output. By proper choice of circuit values, audio output will remain uninterrupted for modulation levels of approximately 90 per cent. Anything in excess of this value will cause the noise-limiter diode to stop conducting, thus limiting the output.

21-20 SQUELCH CIRCUITS

When a receiver employing AVC is tuned from one signal to another, there is an undesirable increase in noise output between stations. This can be overcome by silencing the receiver in the absence of a signal. Circuits to accomplish this are known by such names as *muting*, *quiet AVC*, or *squelch systems*.

A typical silencer circuit appears in Fig. 21-24. One section of the twin triode, V_1, is connected as a diode, which connects the output of the first AF stage to the input of the second AF amplifier. Silencer amplifier V_2 serves as

Squelch Circuits

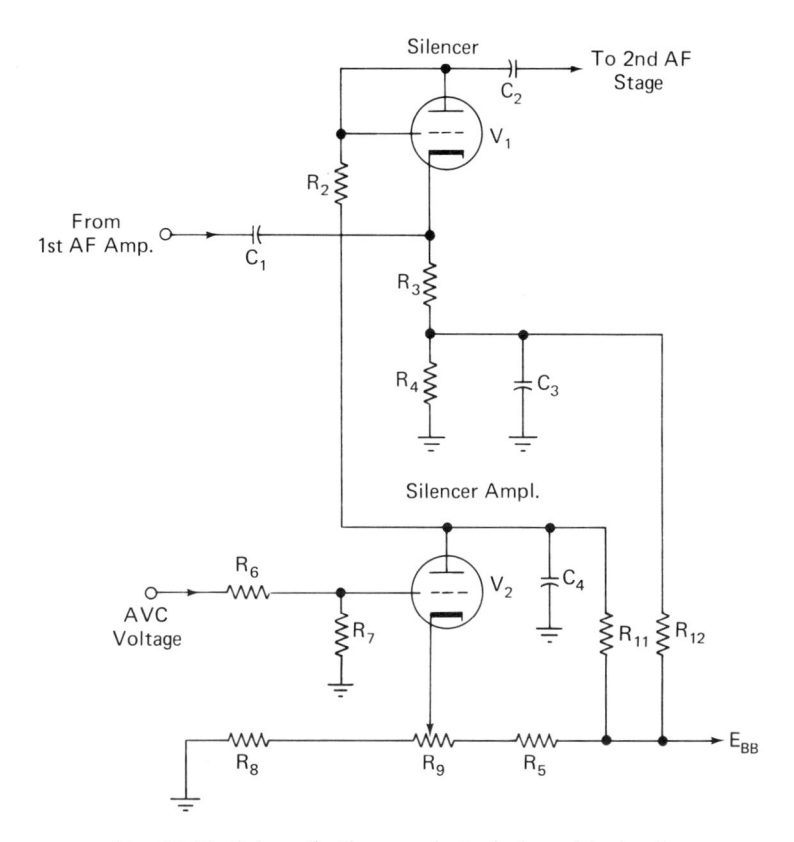

FIG. 21-24 Schematic diagram of a typical squelch circuit.

the control tube for the silencer. The plate voltage of V_1 is supplied via R_2 from the plate of V_2 (which is supplied from E_{bb} via R_{11}), and is positive with respect to ground. The cathode voltage of V_1 is also positive with respect to ground, since it is connected to the B+ supply through a voltage-divider network made up of R_{12} and R_4. In the absence of an input signal R_9 is adjusted until V_2 draws enough plate current to reduce its plate voltage and that of V_1 to a value below the voltage of the cathode of V_1. Thus, the silencer plate voltage is negative with respect to the cathode, conduction ceases, and the silencer cuts off. The output is reduced to zero and the receiver is mute. The grid of V_2 is connected to the AVC line.

The AVC voltage produced by a received signal is applied to the grid of V_2, thereby reducing the plate current and increasing the plate voltage on both V_2 and V_1. When the plate of V_1 becomes positive with respect to the cathode, the tube conducts and the signal is passed to the second AF amplifier.

21-21 DOUBLE-CONVERSION SUPERHETERODYNE RECEIVERS

At very high frequencies it is difficult to obtain an adequate image-rejection ratio when intermediate frequencies of the order of 455 kHz are used. By using a much higher IF, say 2 MHz, it is impossible to obtain the necessary selectivity required for a communications receiver. The solution to this dilemma lies in a superheterodyne receiver employing *double-conversion* techniques. To minimize image response, the incoming signal is first converted to a high IF (i.e., 2 MHz), and then reconverted to a lower IF where the required selectivity may be obtained.

The operation of a double-conversion superhet receiver is shown in block diagram form in Fig. 21-25. The output of the RF amplifier is fed to the

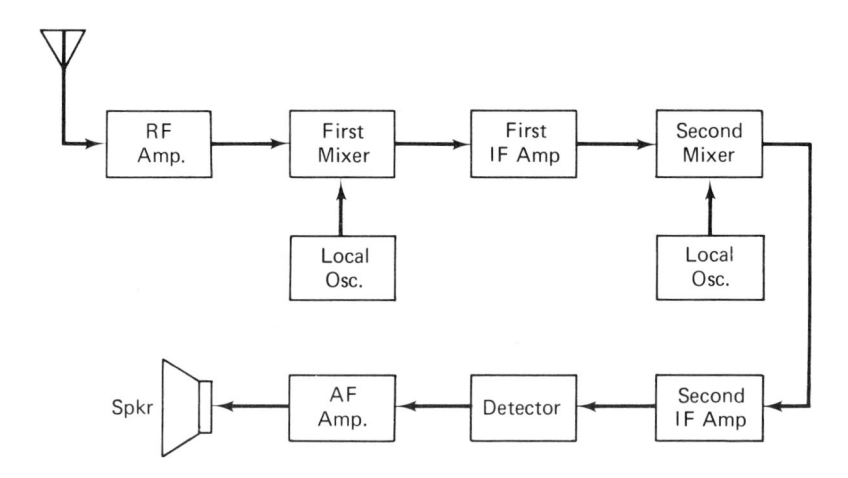

FIG. 21-25 Block diagram of a double-conversion superheterodyne receiver.

first mixer stage. The local oscillator associated with this stage usually operates below the incoming signal by an amount equal to the first IF (typically, 2 MHz). The first IF amplifier may consist of two or three stages to provide the necessary amplification.

The output of the first IF amplifier is fed to the second mixer where another local oscillator beats with it to produce a second IF. This provides additional selectivity. After amplification the signal is connected to the detector stage for demodulation.

The alignment of a receiver of this type is considerably more difficult than a conventional superhet because of the necessity to maintain tracking between the RF amplifier and the two local oscillators. In a good communications receiver, using this principle, we would expect to find an AVC circuit, a BFO, a noise limiter, and perhaps a crystal filter, none of which are shown in the simplified block diagram.

Diversity Receiving System　　　　　　　　　　　　　　　　　　　　　497

21-22　DIVERSITY RECEIVING SYSTEM

Signal fading is a common problem in receiving systems. Although the AVC circuit tends to partially overcome this, its effect is limited when the signal momentarily fades out. This problem can usually be overcome by using *two receiving antennas* spaced at least 5 to 10 wavelengths apart. At this distance fading rarely occurs on both antennas at the same time. By connecting a separate receiver to each of the two antennas and feeding them to a common second detector, the fading problems can largely be overcome.

The block diagram of a typical *diversity receiver system* is shown in Fig. 21-26. Each antenna feeds into its own RF amplifier. These stages then feed

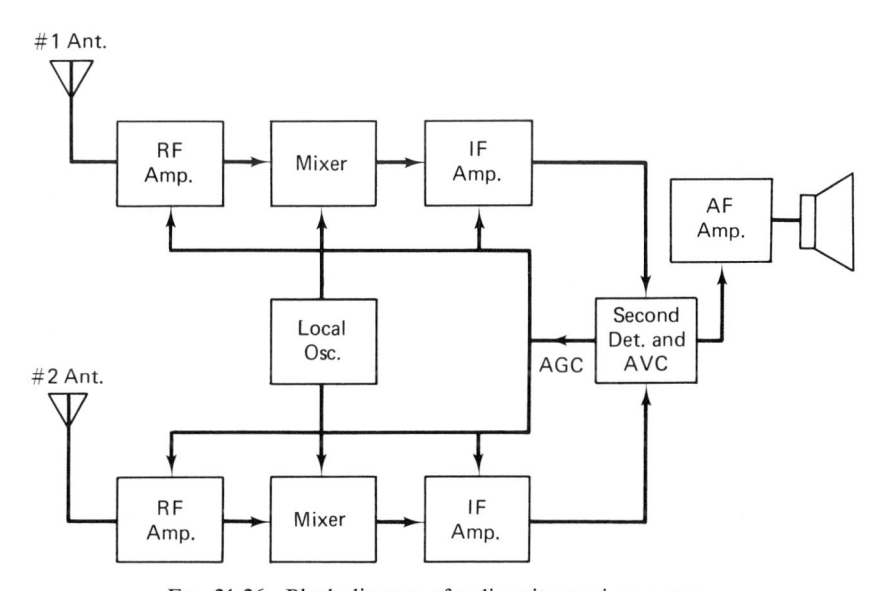

FIG. 21-26　Block diagram of a diversity receiver system.

into mixer circuits. From the block diagram it is apparent that a single local oscillator is used to inject the signal into both mixers. This ensures that the IF output of each mixer will be the same. The output of the two IF amplifiers feeds into the common second detector and AVC circuit. A common AGC circuit is used for each receiving circuit. If each receiver had its own AGC circuit, the effect of the system would be largely nullified. By using a common AGC circuit, the receiver having the strongest signal develops the most AGC. Thus, the output is derived principally from the one receiver. Hence, the AGC system permits the receiver receiving the largest signal to be operative while it biases the other one off. This produces on output signal of somewhat uniform intensity, irrespective of which receiver is picking up the stronger signal.

21-23 WAVE TRAPS

It is possible for certain transmitted frequencies to cause interference in receivers. This can be the result of *overloading* due to extremely strong signals, *beat frequencies* produced with the local oscillator, or *spurious radiations* from a transmitter. These troublesome frequencies can be filtered out by the appropriate use of *wave traps*. These are either series or parallel resonant circuits tuned to the frequency that is to be eliminated.

Some possible uses of parallel resonant circuits as wave traps are shown in Fig. 21-27. A trap can be connected in series with the antenna circuit.

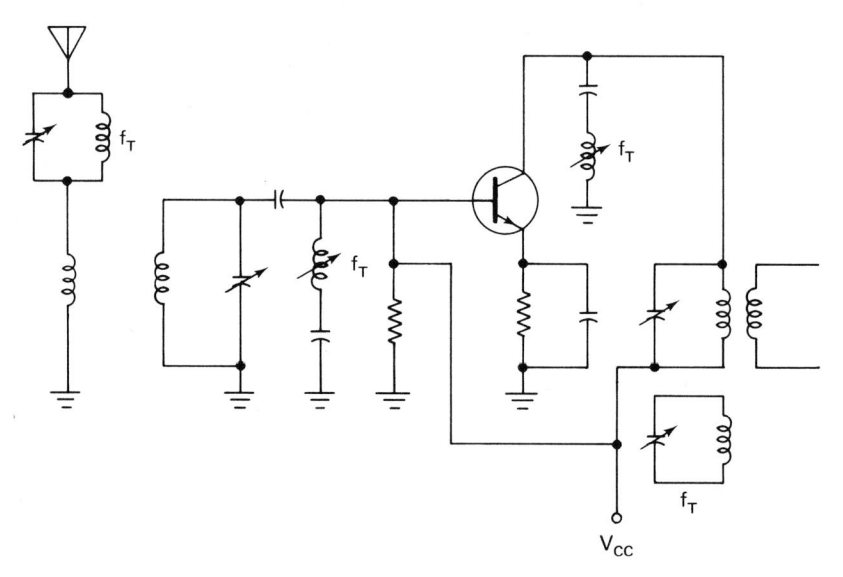

FIG. 21-27 Possible locations of wave or frequency traps (f_t) in an RF amplifier and antenna circuit.

By tuning this to the *undesired frequency*, a high impedance results and virtually no current flows through the primary of the antenna coil. A parallel resonant trap can also be inductively coupled in the collector of the transistor. The undesired frequency induced into the wave trap causes it to oscillate. The RF field built up around this coil induces a voltage back into the original primary winding of the same frequency, but 180° out of phase. This cancels the signal in the primary winding to which the wave trap is tuned.

Figure 21-27 also shows some possible applications of series resonant wave traps. One is connected in the base circuit of the transistor. By tuning this trap to the undesirable frequency, an effective short circuit is placed across the tuned circuit, thus bypassing to ground the unwanted signal. A similar filter could be installed in the collector circuit as shown. In this case the undesired frequency is bypassed to ground and kept out of the tuned cir-

Frequency-Modulated Receivers

cuit in the collector. Traps may cause some detuning of the associated resonant circuits.

21-24 FREQUENCY-MODULATED RECEIVERS

An FM receiver has several differences compared to an AM receiver. The greatest is in the method of detection. Also, the tuned circuits of the FM receiver have a *wider band pass*, and the last IF stage is especially adapted for *limiting* the amplitude of the incoming signal.

The block diagram of an FM receiver is shown in Fig. 21-28. The RF

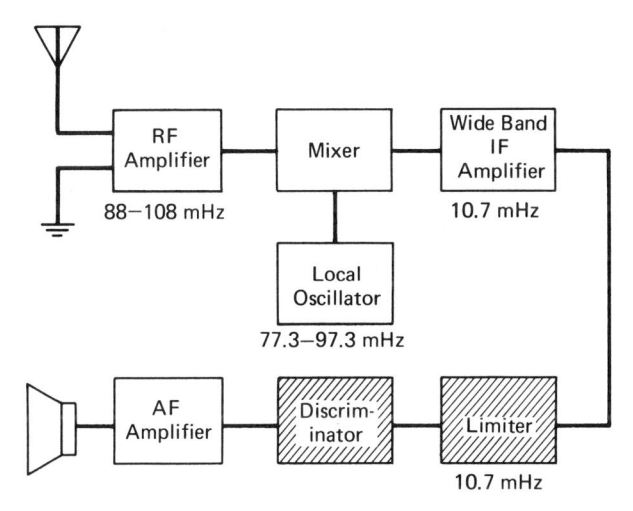

FIG. 21-28 FM receiver block diagram.

amplifier, mixer, and local oscillator perform the same functions as in the AM receiver. However, they are designed to operate in the 88- to 108-MHz band. At these frequencies there are some problems, such as the stability of the local oscillator. It has a tendency to synchronize with the incoming signal and thus to lose the IF output entirely. This is because the station and oscillator are relatively closer together than with AM. Therefore, for maximum stability, a separate oscillator is frequently used. Especially designed pentagrid converters that have reasonably good frequency stability, high conversion transconductance, and oscillator transconductance are employed in some less expensive commercial sets.

Even in well-designed FM receivers such factors as a change in *internal capacitance* of the oscillator tube and *expansion* of coil windings and capacitor plates during warm up may cause the local oscillator to drift. This may be overcome in several ways. For example, the second harmonic of the local oscillator is sometimes used because of the increased stability attainable at a lower frequency. Capacitors are used having *negative temperature coeffi-*

cients. Proper voltage regulation, as well as the choice of oscillator tubes having low internal capacitances, will also increase stability.

The local oscillator usually operates below the incoming signal (see Fig. 21-28) to ensure more stability. The IF amplifier is tuned to 10.7 MHz and has a band pass of about 200 kHz. This wider band pass provides less gain per stage than the IF of an AM receiver; therefore, more IF stages are required.

21-25 LIMITERS

Frequency-modulated signals suffer some amplitude modulation as they travel from transmitter to receiver. This is the result of natural and man-made static combining to produce variations in the amplitude of the modulated signal. Other variations are caused by fading. These are amplified as the signal passes through successive stages of the receiver up to the input of the *limiter*, whose function is to remove any AM components of the wave. The signal before and after limiting is shown in Figs. 21-29a and b.

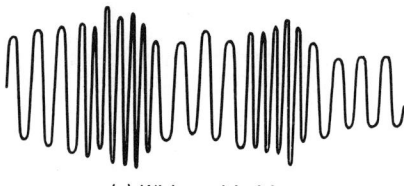

(a) Without Limiting

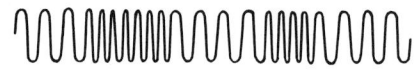

(b) With Limiting

FIG. 21-29 Waveforms showing FM signal before and after limiting.

A limiter using grid-leak bias is shown in Fig. 21-30. The tube is a sharp cutoff pentode with plate and screen voltages purposely made low. That means that the tube will go into *saturation* and *cutoff* with input signals having a magnitude of only a few volts. Thus, the limiter output is approximately constant for all signals having an amplitude great enough to develop a grid-leak bias voltage that is greater than the cutoff voltage. The frequency variations in the FM signal are maintained in the output, because the plate current pulses are produced at the signal frequency and excite the tuned circuit, which has a relatively low Q and a wide band pass. Thus, because of the flywheel effect, a complete ac waveform is passed to the discriminator stage.

When the peak amplitude of the grid signal is less than the cutoff voltage, the limiting action fails, because the stage is practically a class A amplifier for such signals, and the average plate current varies as the grid-leak bias changes with varying signal amplitudes. For this reason the stages preceding the limiter must have sufficient gain to provide satisfactory limiting action on the weakest signal to be received.

Foster–Seeley Discriminator

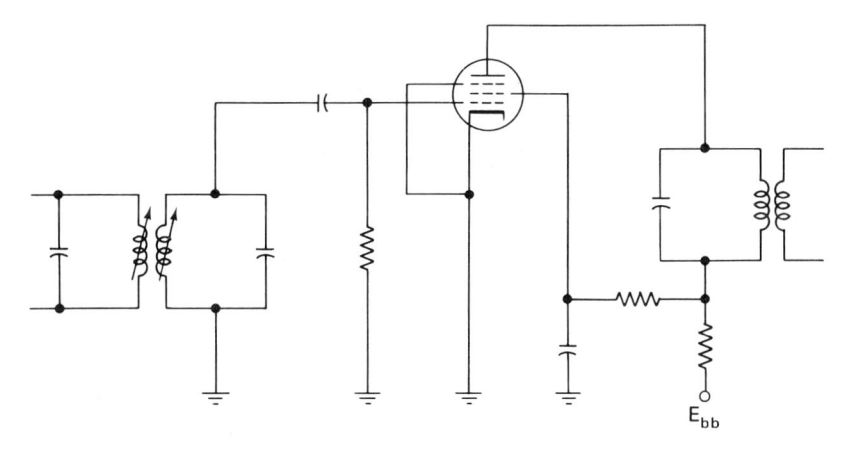

FIG. 21-30 Limiter using grid-leak bias.

21-26 FOSTER–SEELEY DISCRIMINATOR

The function of the *discriminator* is to convert the frequency deviations of the FM signal into the intelligence added at the transmitter. The circuit of a typical Foster–Seeley discriminator is shown in Fig. 21-31. Before analyzing

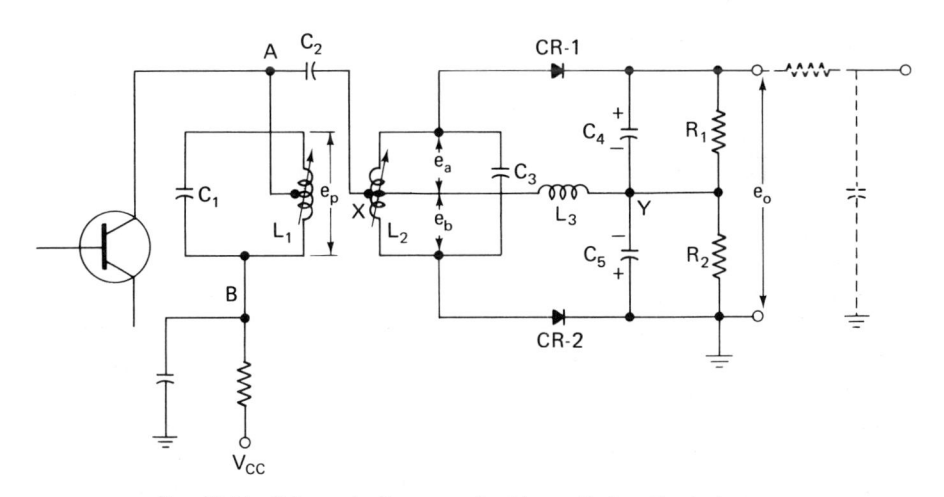

FIG. 21-31 Schematic diagram of a Foster–Seeley discriminator.

the circuit three point should be brought out: (1) The two tank circuits made up of L_1, C_1, and L_2C_3 must be tuned to the resting frequency of the FM signal. (2) The primary tank must have sufficient bandwidth to maintain essentially constant output for at least the full swing of the input signal. (3) Point B in the primary tank is at RF ground, and capacitors C_2, C_4, and C_5 all have low reactance at the signal frequency.

The low reactance of C_4 and C_5 places point Y at RF ground. At the same time points A and X are at the *same* RF potential. As a consequence, the full primary voltage e_p is connected across choke coil L_3. By transformer action a secondary voltage e_s is induced into the secondary winding, developing voltages e_a and e_b. Diode CR_1 conducts because of the vector sum of voltage e_a and e_p. During this conduction, capacitor C_4 charges with the polarity as shown. Conduction of CR_2 is dependent on the vector sum of e_b and e_p. Observe that the charge developing on C_5 is opposite in polarity to that of C_4. Therefore, if the diodes conduct equally, the voltages across R_1 and R_2 are equal and opposite and the output voltage e_o is zero. If the FM signal is not on its resting frequency, the output will swing either negative or positive, depending upon which diode is conducting most heavily. The dotted resistor and capacitor shown in the output circuit are the *deemphasis* circuit necessary in all FM receivers.

From these basic facts we may now proceed to analyze the operation of the circuit with the aid of the vectors shown in Fig. 21-32. Drawing a shows the conditions that exist when the incoming signal is at its resting frequency. Using the primary voltage e_p as the reference vector, we observe that the induced secondary voltage is 180° out of phase by standard transformer action.

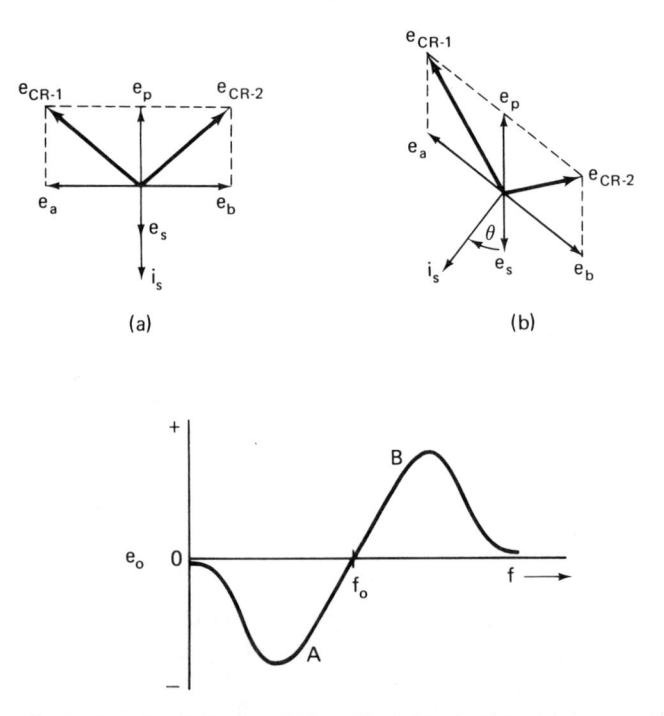

FIG. 21-32 Discriminator phase relationships: (a) signal at resting frequency; (b) signal above resting frequency; (c) response curve of discriminator.

Ratio Detector 503

Because the secondary circuit is at resonance, the circulating tank current is in phase with the secondary voltage e_s. This current flowing through the inductance of the secondary winding produces a voltage drop $(e_a + e_b)$ that lags the current by 90°. The center-tap action of L_2, however, produces voltages across the extremes of the inductor that are 180° out of phase. The vector sum of the voltages supplied to CR_1 and CR_2 are equal in magnitude; consequently, the diodes conduct equally and the output voltage is zero.

Next, let us consider a situation in which the modulation of the signal has caused the incoming frequency to swing above its resting value. The voltages that result from this are shown in Fig. 21-32b. Voltages e_p and e_s remain as before. Because the frequency is higher, the induced voltage is in an inductive circuit and the secondary current is caused to lag by an angle θ. Because of the center-tap action of L_2, voltages e_a and e_b will remain at right angles to i_s. From the vector drawing it is observed that these voltages swing clockwise with respect to e_p. The resultant voltage e_{CR1} is now greater than e_{CR2}. Consequently, CR_1 conducts harder, resulting in a positive output. If the frequency deviation increased even more, the angle θ would also increase. The resultant vector e_{CR1} becomes even larger, and the output of the discriminator stage further increases. This type of discriminator circuit is frequently referred to as a *phase discriminator* because of its dependence on the phase relationships of the voltages involved.

If the incoming FM signal had been deviated below the resting frequency, the vectors would have swung in a counterclockwise direction with respect to e_p. This would have resulted in vector e_{CR2} being larger, and the net result would have been an output voltage that was negative. A plot of the voltage versus frequency response of a discriminator circuit such as the one being discussed results in a curve like the one shown in Fig. 21-32c. Because of its shape, it is frequently called an *S curve*. The design of a demodulator must be such that operation is restricted to the linear portions of the curve shown between the points A and B in the drawing.

21-27 RATIO DETECTOR

Another commonly used FM detector is the *ratio detector* shown in Fig. 21-33. At first glance it seems as though this is essentially the same circuit as the Foster–Seeley discriminator. However, there are three significant differences between the two circuits. One of the diodes, CR_2, is reversed. A large capacitor, C_6, of approximately 10 μF is connected across the load resistors R_1 and R_2. The output is taken between the junction of the load resistors R_1 and R_2 (point Z) and the series capacitors C_4 and C_5 (point Y). The deemphasis circuit, not shown for simplicity, would normally be connected between points Y and Z.

At the resting frequency f_0 the voltage vector shown in Fig. 21-32a applies. Under these conditions both diodes conduct equally. As the frequen-

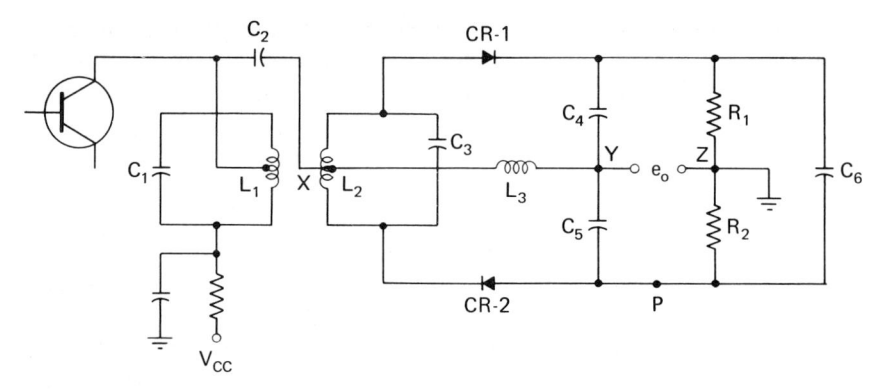

FIG. 21-33 Schematic diagram of a ratio detector.

cy is deviated upward, the vector representation would be as shown in Fig. 21-32b, which causes diode CR_1 to conduct more. Because one diode is reversed, current can now flow in the overall circuit, that is, from the bottom of secondary L_2 through CR_2, to the long time constant circuit comprised of R_1, R_2, and C_6, and back through CR_1 to the top of L_2. After several RF cycles have been applied to the circuit, C_6 charges to nearly the peak value of the voltage across L_2. Because of the relatively long time constant of this circuit, any amplitude variation of the FM signal will have little effect on the charge of C_6. Consequently, the circuit has great immunity to noise and other forms of amplitude modulation.

In some receiving circuits using ratio detectors noise-limiter circuits are not employed. Some manufacturers will nevertheless include a limiter stage, because large variations in amplitude can cause some noise output in the circuit.

As stated before, at the resting frequency f_0 both diodes are conducting equally, and the voltages across C_4 and C_5 are equal. Because of the parallel nature of C_4 and C_5 and R_1 and R_2, points Y and Z are at the same potential, and zero output results. Assume now that as the carrier is deviated CR_2 conducts more than CR_1. Therefore, e_{c5} increases while e_{c4} decreases. However, the sum of their two voltages must always equal the supply voltage across L_2. This results in a positive output. This action may possibly be better understood by giving an example. Suppose that e_{c6} is 8 V. Then e_{c5} could be 6 V and e_{c4}, 2 V. Compared to point P the potential of point Y is $+6$ V. At the same time the voltage at point Z (compared to point P) is $+4$ V. This makes the output value $+2$ V. Conversely, when CR_1 conducts more, e_{c4} exceeds e_{c5} and the output voltage goes negative. Now it is seen that the sum of e_{c4} and e_{c5} remains constant, but that their ratio changes depending upon the signal frequency. Because of this action, the circuit is called a ratio detector.

Other demodulator circuits have been developed that can be used with

FM receivers. Some of these are the *gated-beam detector*, the *slope detector*, the *cycle-counting detector*, and the *double-tuned detector*. However, none of these circuits have received much popularity and will not be discussed in this section.

21-28 MAINTENANCE PROCEDURES

Receivers are not immune to troubles. Any attempt to list them all would be futile. A good communications receiver is a sophisticated piece of electronic equipment. An operator or technician should not attempt any alignment without being thoroughly familiar with the circuit and having available the necessary test equipment. Certain routine maintenance procedures can and should be followed, however, that do not require more than a cursory knowledge of the equipment.

Poor antenna or ground connections can be an annoying source of noise and cause poor reception. They should be occasionally inspected to determine that they are free of corrosion and making good connection. Particularly is this true in mobile equipment, where much vibration may be encountered.

In vacuum-tube equipment it is necessary to perform periodic checks on the tubes, as the emission falls off over a period of time, and a loss of transconductance results. Also, *interelement shorts* occasionally develop, which may cause intermittent reception, complete loss of signal, and sometimes the burning out of screen-grid or plate resistors. Tubes also are prone to become *gassy*, and this can cause serious distortion.

Solid-state receivers are not subject to many of the problems of the vacuum-tube receiver. However, voltage transients can cause transistors to short, although good design minimizes this.

Mobile equipment is often plagued with ignition noise. Noise is also produced by the commutation process in generators and alternators. Blinking lamps (turn signals) and electric motors (starter, windshield wiper, air conditioners, etc.) are other potential causes. In fact, anything that produces small sparks is a noise source. Even a slightly dragging brake shoe on a brake drum can cause static.

These noises can be eliminated, or at least greatly reduced, by proper *shielding* of antenna lead-in cables and good ground connections. The use of spark suppressors in the ignition system is very important. Most new automobiles use a special ignition wire that has a thin graphite or carbon filament as the conductive element. This replaces the spark suppressor and is presumably more effective.

It is customary to place a capacitor between the hot lead and ground of the various electric motors to suppress the small sparks caused by commutation. For mobile receivers it is important that the input dc power be passed through a pi-type filter to bypass any noise on the dc line. This is generally mounted in a small metal compartment that is a part of the chassis. This shields the inductor in the filter and prevents radiation of stray magnetic fields.

The local oscillator is generally mounted in a shielded compartment to prevent spurious radiation. In some poorly designed receivers part of the oscillator energy feeds back to the antenna and is radiated. Beat-frequency oscillators should be shielded. If the receiver uses a *dc-to-dc converter* in the power supply high-energy, short-duration pulses are created that can be very troublesome. Vacuum-tube voltage regulators are another source of trouble. Electrostatic and magnetic shields should be maintained in their proper positions.

Commercial License Questions

Sections in which answers to questions are given appear in parentheses. A bracketed number following a question implies that it applies only to that element.

1. Describe the operation of a crystal detector (rectifier). (21–2)
2. What type of radiotelephone receiver using vacuum tubes does not require an oscillator? (21–3)
3. Draw a diagram of a TRF-type radio receiver. (21–3)
4. Compare the selectivity and sensitivity of the following types of receivers:
   (a) TRF receiver.
   (b) Superregenerative receiver.
   (c) Superheterodyne receiver. (21–3, 21–8, 21–9)
5. Explain the operation of a diode detector. (21–4)
6. Draw a simple schematic diagram of a diode vacuum tube connected for diode detection, and show a method of coupling to an audio amplifier. (21–4)
7. What is the principal advantage in the use of a diode detector instead of a grid-leak-type triode detector? (21–4, 21–5)
8. Explain the operation of a grid-leak detector. (21–5)
9. Draw a simple schematic diagram of a triode vacuum tube connected for grid-leak condenser detection. (21–5)
10. List and explain the characteristics of a square-law type of vacuum-tube detector. (21–5)
11. What effect does the reception of modulated signals have on the plate current of a grid leak–grid condenser type of detector? On a grid-bias type of detector? (21–5, 21–6)
12. Is a grid-leak type of detector more or less sensitive than a power detector (plate rectification)? Why? (21–5, 21–6)
13. Draw a simple schematic diagram of a triode vacuum tube connected for plate or power detection. (21–6)

Commercial License Questions

14. What operating conditions determine that a tube is being used as a power detector? (21-6)
15. Explain the operation of a power or plate-rectification type of vacuum-tube detector. (21-6)
16. What are the characteristics of plate detection? (21-6)
17. Draw a simple schematic circuit of a regenerative detector. (21-7)
18. Explain what circuit conditions are necessary in a regenerative receiver for maximum response to a modulated signal. (21-7)
19. What feedback conditions must be satisfied in a regenerative detector for most stable operation of the detector circuit in an oscillating condition? (21-7)
20. What might be the cause of low sensitivity of a three-circuit regenerative receiver? (21-7)
21. What effects might be caused by a shorted grid condenser in a three-circuit regenerative receiver? (21-7)
22. Describe the operation of a regenerative-type receiver. (21-7)
23. How may a regenerative-type receiver be adjusted for maximum sensitivity? (21-7)
24. How does the value of resistance in the grid leak of a regenerative-type detector affect the sensitivity of the detector? (21-7)
25. What feedback conditions must be satisfied in a regenerative detector to obtain sustained oscillations? (21-7)
26. What would be the effect upon a receiver if the vacuum-type plate potential were reversed in polarity? (21-8, 21-9)
27. If a frequency of 500 Hz is beat with a frequency of 550 kHz, what will be the resultant frequencies? [4] (21-9, 21-11)
28. Explain the relation between the signal frequency, the oscillator frequency, and the image frequency in a superheterodyne receiver. (21-9, 21-11, 21-12)
29. Draw a block diagram of a superheterodyne receiver capable of receiving AM signals and indicate the frequencies present in the various stages when the receiver is tuned to 2450 kHz. What is the frequency of a station that might cause image interference to the receiver when tuned to 2450 kHz? (21-9, 21-11, 21-12)
30. What is the principal advantage of the tetrode as compared to the triode when used in a radio receiver? (21-9, 21-10)
31. What type of radio receivers contain intermediate frequency transformers? (21-9, 21-24)
32. What type of radio receiver is subject to image interference? (21-9, 21-24)
33. If a tube in the only RF stage of your receiver burned out, how could temporary repairs or modifications be made to permit operation of the receiver if no spare tube is available? (21-9, 21-10, 21-11)

34. What effect do sunspots and aurora borealis have on radio communications?
(21–10)

35. What are the advantages to be obtained from adding a TRF amplifier stage ahead of the first detector (converter) stage of a superheterodyne receiver?
(21–10)

36. Discuss methods whereby interference to radio reception can be reduced.
(21–10)

37. Explain the purpose and operation of the first detector in a superheterodyne receiver. (21–11)

38. If a superheterodyne receiver is tuned to a desired signal at 1000 kHz, and its conversion oscillator is operating at 1300 kHz, what would be the frequency of an incoming signal that would possibly cause image reception?
(21–12)

39. Draw a diagram showing how AVC is accomplished in a standard broadcast receiver. [4] (21–15)

40. How is AVC accomplished in a radio receiver? (21–15)

41. What is the purpose of an oscillator in a receiver operating on a frequency near the intermediate frequency of the receiver? (21–17)

42. What is the purpose of a squelch circuit in a radio communication receiver?
(21–20)

43. What is meant by *double detection* in a receiver? (21–21)

44. What is the purpose of a diversity antenna receiving system? (21–22)

45. What is the purpose of a wave trap in a radio receiver? (21–23)

46. Show by a diagram how to connect a wave trap in the antenna circuit of a radio receiver to attenuate an interfering signal. (21–23)

47. Draw a block diagram of an FM receiver and explain its principle of operation.
(21–24)

48. How wide a frequency band must the IF amplifier of an FM broadcast receiver pass? [4] (21–24)

49. What types of radio receivers do not respond to static interference?
(21–24)

50. What is the purpose of a limiter stage in an FM broadcast receiver? [4]
(21–24, 21–25)

51. What is the purpose of a discriminator in an FM broadcast receiver? [4]
(21–24, 21–26)

52. What is the purpose of a deemphasis circuit in an FM broadcast receiver? [4]
(21–24, 21–26)

53. Draw a diagram of a limiter stage in an FM broadcast receiver. [4]   (21–25)

54. Draw a diagram of an FM broadcast receiver detector circuit. [4]
(21–26, 21–27)

Commercial License Questions 509

55. What is a ratio detector? [4] (21–27)
56. What is the purpose of shielding in a multistage radio receiver? (21–28)
57. Discuss the cause and prevention of interference to radio receivers installed in motor vehicles. (21–28)

22

Broadcast Stations

22-1 STANDARD BROADCAST STATION

The broadcast industry accounts for a large part of the field of communications. The aim of this chapter is to acquaint the novice with the FCC rules that apply to this specialized field. Also, it is hoped that a basic understanding of the duties of operators and technicians can be obtained.

A *standard broadcast station*, as defined by the FCC, is one that operates in the 535- to 1605-kHz spectrum. There are 106 channels, each with an allowed 10 kHz of bandwidth. The stations are licensed to operate on 540, 550, 560 kHz, and so on, through 1600 kHz. The carrier is not allowed to drift more than ± 10 Hz from its assigned frequency. The FCC allows these stations to operate with a power level of 100 W to 5 kW, 1 kW to 5 kW, and 10 kW to 50 kW, depending upon the class assignment for the proposed station. The station is allowed to use only AM (A3), and modulating frequencies must not exceed 15 kHz to prevent sidebands from interfering with adjacent stations. Many stations must *reduce* their power at night to prevent interference with other stations sharing the same channel, but in a different geographical location.

The FCC states that it licenses broadcast stations for the purpose of serving *public interest*, *convenience*, and *necessity*, and may revoke the station's license if it is shown that these requirements are not being met.

Service areas of broadcast stations fall into three categories: *primary*, *secondary*, and *intermittent*. The FCC defines each as follows:

1. Primary. No fading of the signal or objectional interference.
2. Secondary. Some fading but no objectional interference.
3. Intermittent. Signal is received with some interference and fading.

The FCC breaks down each day into three parts:

1. Daytime, from local sunrise to local sunset.
2. Nighttime, from local sunset to midnight.
3. Experimental period, from midnight to local sunrise.

These are *local* times. By FCC definition, the term *broadcast day* refers to the period between sunrise and midnight.

The station studios are usually located within the business area of a city; the transmitter is generally located near the edge or outside the city. It is desirable to locate the antenna (and consequently the transmitter) on flat, moist land to take advantage of the good grounding conditions necessary to produce a good ground wave. This is not true for FM broadcasting or TV stations. The spectrum allocated for FM broadcasting is 88 to 108 MHz, and 54 to 890 MHz for TV stations. Since both of these bands fall in the VHF to UHF category, transmission is line of sight. Therefore, FM and TV stations try to locate their transmitter and antenna at a high point overlooking the intended service area to prevent signal loss due to natural or man-made obstructions.

22-2 FREQUENCY-MODULATED BROADCAST STATIONS

The FCC licenses FM stations to operate from 88.1 to 107.9 MHz, channels 201 to 300, respectively. The stations are placed at 200-kHz intervals throughout the spectrum (i.e., 88.1 MHz, 88.3 MHz, etc.) with a total of 100 channels available. The 88.1- to 91.9-MHz part of the band is reserved for use only by noncommercial, educational broadcast facilities. The FCC states that FM transmitters operated within this spectrum must have a carrier stability of ± 2 kHz. The FCC set a carrier deviation of ± 75 kHz as being the equivalent of 100 per cent modulation in an AM transmitter. Stations operate with power levels as low as 10 W for educational stations to 100 kW for commercial stations.

Many definitions used in AM broadcasting also apply to FM. It is not necessary for an FM station to drop power or go off the air at night to protect stations in adjacent cities, due to the fact that the transmission is usually line of sight and there is very little skipping to distant areas.

22-3 COMPONENTS OF A BROADCAST SYSTEM

A possible block diagram of a broadcast station is shown in Fig. 22-1. Programs originate in the centrally located studio from the turntables, tapes, microphones, and so forth. They are mixed together into one output by means of an *audio mixing console*. This is also used to control the volume or levels of each individual signal source feeding the console. To help establish audio levels, the operator is aided by the *volume-units meter*.

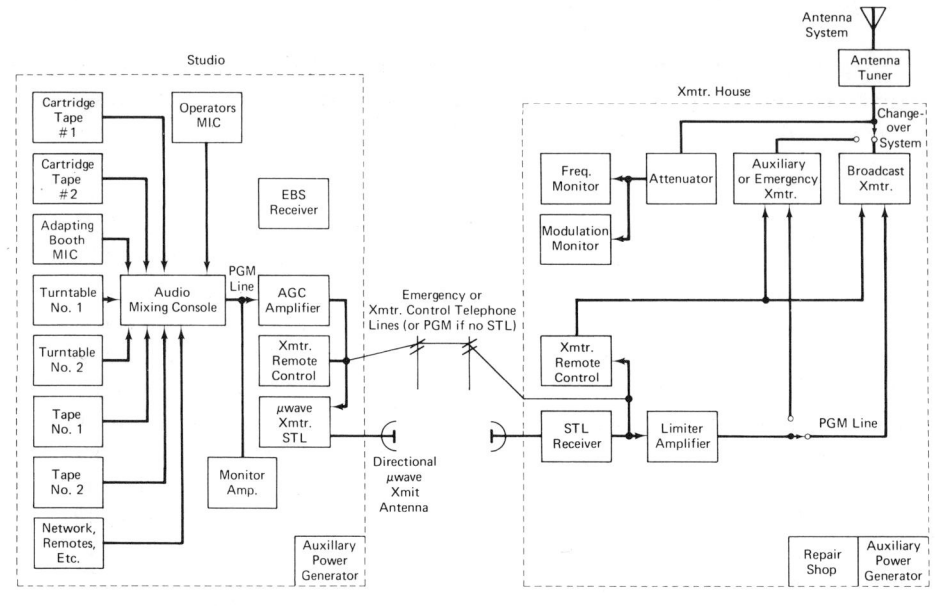

Fig. 22-1 Block diagram of remote-controlled broadcast station (medium size).

The output of the audio console is then fed to the AGC amplifier or limiter, which helps to keep audio levels more constant and limits peaks. The output of the limiter then feeds a transmission line that links the studio to the transmitter. This can either be telephone lines or a microwave *studio-to-transmitter link* (STL) operating in the 942- to 952-MHz region. The STL link utilizes FM. The FCC rules concerning the use of STLs state that the transmitter must have a frequency tolerance of 0.005 per cent and no more power than that needed to secure an adequate signal at the receiving point. Intercity relay stations also operate within this spectrum and usually carry network radio and television programs.

If telephone lines are used to link the studio to the transmitter, they have to be leased from the local phone company. Often many miles of cable link the studio and transmitter, so there may be many booster amplifiers along the cable to overcome the losses. Due to interconductor capacitances of the cable, high-frequency information tends to be lost in the cable. To compensate for this, *equalizers* are placed in the line.

Often stations maintain telephone lines as well as an STL in case one fails. Since the transmitter is at a remote location, there must be some means for the operator to control the transmitter and be able to read the meters. The usual procedure is to use a remote-control unit that controls the transmitter by means of dc pulses or tones, if telephone lines are used, or subcarriers on the STL, which are above the human hearing response. Using these methods the transmitter can be turned on and off, power can be controlled, and meter readings can be sent back to the studio.

Components of a Broadcast System

At the transmitter building, the telephone line is fed to a common audio (line) amplifier and then to the transmitter modulator. In rare instances a second peak limiter is used for increasing levels of modulation; however, such practice serves to almost completely destroy the *dynamic range* of the material being broadcast.

Many stations maintain two transmitters: one *main* and one *auxiliary* or *emergency*. An auxiliary transmitter, by FCC definition, is one used in the event the main transmitter fails. An auxiliary transmitter is usually of a lower output power than the main and cannot be used in place of the main transmitter except when the main fails. The FCC requires that the auxiliary transmitter be tested weekly. An *alternate* transmitter can be used if it is capable of the same power output as the main transmitter and can be used at will in place of the main. The *operating log* must always indicate which transmitter is in service. *Proofs of performance* must be made annually of main and alternate transmitter; however, such proofs are not required for the auxiliary transmitters.

Modulation monitors and *frequency-deviation meters* are used at the transmitter to monitor the transmitter's operation. A sampling device is usually placed at the transmitter output and fed to the monitors to reduce the power to a level that the monitors can handle (usually 2 to 10 W).

Stations operating in the standard broadcast band use *antenna tuners* at the base of their tower to resonate the antenna to the operating frequency. The tuner may also reject harmonics that the transmitter failed to eliminate. An RF ammeter is usually contained within the tuning unit to measure antenna current. If it is not contained within the tuning unit, one is placed between the tuning unit and the antenna so that power output can be monitored and logged as required.

Many stations utilize directional antenna arrays to beam their signals to the desired service area and also to eliminate interference with stations operating on the same frequency in adjacent cities. Two or more towers are usually used for directional stations, and each tower is fed with RF energy that is advanced or retarded in phase from that of the other towers. To accomplish this, a *phaser* is placed between the transmitter and antenna. Directional stations are required by the FCC to monitor the phase of each antenna during broadcast and to log the results.

The heart of every broadcast station is the audio console, control console, or mixer. The purpose of the mixer is to switch the various sources of audio, control the levels of audio, amplify signals when necessary, and combine them all into one output. A simplified block diagram of an audio console appears in Fig. 22-2. It shows a control room, microphone, turntable, and tape machine entering the console. A monitor speaker allows the operator to hear what is being broadcast.

The operator is aided in determining the output level of the audio console by the VU or *volume-units meter*. The meter uses an attenuator so that

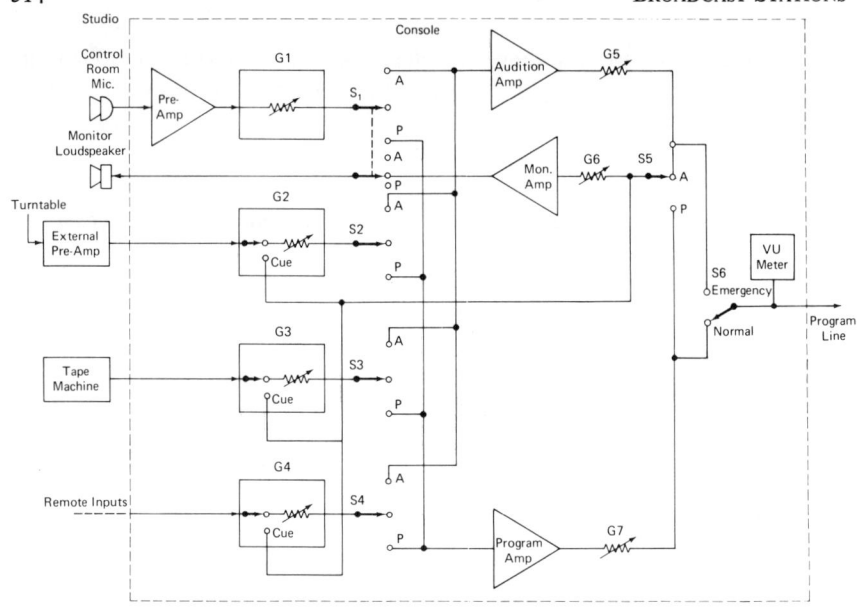

FIG. 22-2 Block diagram of simplified audio console.

the output level of the console will read 0 dB, or 100 per cent modulation, when an appropriate level has been reached. The operator tries to keep the levels of each source as high as possible without exceeding 100 per cent on the meter.

22-4 REMOTE BROADCAST FACILITIES

Often in broadcasting it is necessary to originate a program from some point remote from the studio. Examples would be sports events, live reporting from news scenes, and so on. The most common method of handling these is by the use of a small remote audio console at the scene and a pair of rented telephone lines to link the remote console to the main studio. The telephone lines used for remote broadcasts are commonly referred to as *nemo lines*. A pair of nemo lines are used so that one line can carry the program material while the other can be held open for cuing and communicating back to the studio. Also, if something happened to the main line, the second line could be used to complete the link so that the program would not go off the air.

For on-the-spot remote broadcasts, such as coverage of news events and on-the-street interviews, a small transmitter is usually used. These are licensed on special remote broadcast pickup frequencies allocated to the broacdasting industry and occasionally in commercial business bands. The latter bands are fraught with peril to the broadcaster from nonbroadcast-type interference. These transmissions are picked up on receivers at the studio. If there is a large distance between the remote transmitter and studio, an intermediate location is used for the receiver. The signal is then fed to

Broadcast Automation 515

telephone lines, which link the remaining distance to the studio. The only time radio-link remotes are used is if telephone lines are inaccessible or the remote is in motion, such as with traffic reports from an airplane.

Many larger radio stations have remote broadcast trucks equipped with an audio console similar to that of the main control room. These contain turntables, tape recorders, remote transmitter, and other facilities. Frequently, equalizer amplifiers are included. These are designed to boost the higher audio frequencies, which are attenuated on long lines. Use of low-impedance lines helps to overcome this problem, but often the equalizer amplifier is necessary to restore full frequency response. A diagram of a possible equalizer amplifier is shown in Fig. 22-3. Also, it is not uncommon for the equalizer amplifier to be used in the main studio to overcome losses between studio and transmitter.

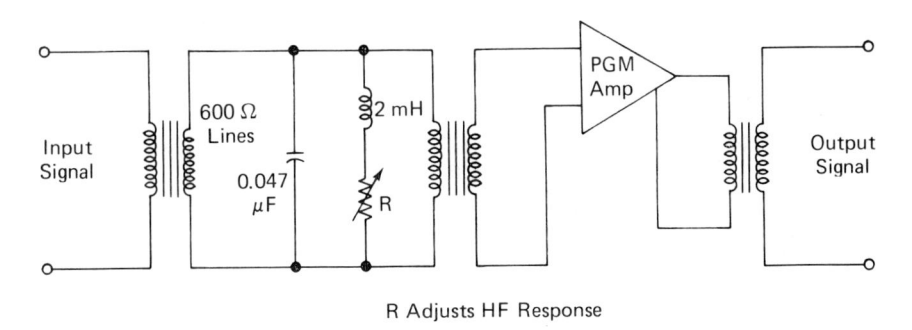

R Adjusts HF Response

FIG. 22-3 Possible equalizer amplifier circuit.

22-5 BROADCAST AUTOMATION

Many AM and FM stations now program by means of *automation*. Most automation systems employ *reel-to-reel tape machines*. Each tape contains about 10 to 50 musical selections and appropriate announcements, depending on the tape size and length of recorded material. Three to four tape decks are used with usually two to three cartridge machines for commercial spots or other material. A master programmer programs the specific machine that is to play at a particular time. When a selection is finishing on one tape, an inaudible tone is placed on the tape. This is the cue for the next tape machine to start running. The machine just finishing its run stops at a cue tone inserted just before the next material on the tape. By proper insertion of tones, tapes can be made to overlap if so desired. The programmer can select from any of the machines or cartridges, all of which are cued from tones at beginning and ending.

It is required by the FCC to keep *program* and *operation logs* for all programs aired. To do this in an automation system, since these installations usually run for many hours without any operator present, various recording systems are used. Probably the simplest, and at the same time the most sophisticated, is a typewriter of the input–output variety that prints out a proper description of what was scheduled to be played, what actually played, and at what time. Also, data regarding type of program, sponsors, and other pertinent facts are logged.

22-6 VOLUME-UNIT METERS AND AUDIO LEVELS

The operator at the control console is aided in maintaining his audio levels by means of a VU meter. These are used in broadcasting as well as all types of audio work. The meter is calibrated in decibels (or volume units) from −20 to +3 dB. When connected across an audio line, the VU meter will read relative power levels. The meters are supplied with type A and type B scales, as represented in Fig. 22-4. The 0 to 100 percent scale is used to correspond to transmitter modulation.

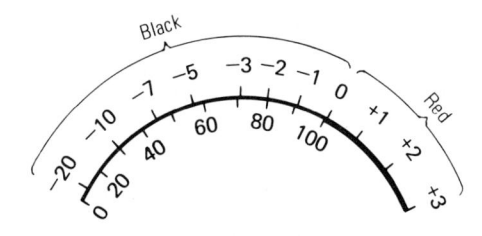

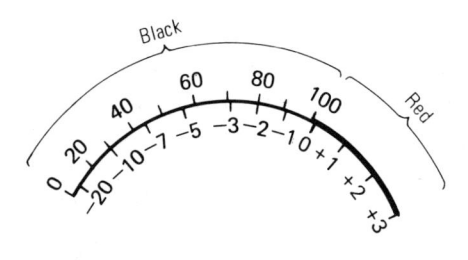

FIG. 22-4 Two types of VU meter scales. Graduations are in plus and minus volume units and 0–100 percent modulation.

In 1940 it was decided to set a standard for VU meters, so that all meters would read the same when connected across an audio line. The standard was set so that the meter would indicate 0 VU or dB, or 100 percent, when a power level of 1 mW across a 600-Ω line was applied to the meter. The damping was standardized so that the meter would swing *sharply upward* when audio was applied, and then swing back to zero *slightly slower* when audio was removed. The reference level of 1 mW across a 600-Ω line is referred to as 0 dBm.

Attenuator Networks 517

Often in broadcast work levels higher than 0 dBm are required, such as +5 to +15 dBm. If the operator desired to have a level of +10 dBm correspond to 0 VU on the meter, an appropriate *attenuator pad* could be placed on the meter (discussed later).

The average broadcast microphone has an output of about −50 to −75 dBm. If the audio console is to produce an output level of 0 dBm with a microphone level of −70 dBm, the microphone preamplifiers will have to have a gain of at least 70 dB to bring up the level. Preamplifiers with more than 70-dB gain would be desired, since a weaker sound striking the microphone could produce an output level as low as −100 dBm, and usually 6 to 10 dB of gain is lost in the level pots. If the mircophone has an output level of −100 dBm and 8 dB is lost in the level control, the microphone preamplifier will have to have a gain of 108 dB to correspond to an output level of 0 dB. If an output of +10 dBm were desired at the audio console, the preamplifiers would have to have a gain of 118 dB.

Telephone companies also reference audio on the lines to 0 dBm as the maximum signal level to prevent *crosstalk* between different lines.

22-7 ATTENUATOR NETWORKS

It is often necessary to reduce a signal to a certain level. To do this an *attenuator pad* or *attenuator network* is used. The following example illustrates this. Suppose that an engineer were setting up a remote broadcast. Upon tesing the remote audio console, he finds that 0 VU on the VU meter corresponds to an output level of +10 dBm. Since the maximum input level to the telephone line is 0 dBm, the engineer must insert a 10-dB attenuator network between the output of the audio console and the phone lines.

The four major types of attenuators used in broadcast work are the T-pad, H-pad, L-pad, and U-pad. Attenuator pads consist of a series of resistors placed in various configurations to act as voltage dividers, while maintaining a *constant input and output impedance* (usually 600 Ω for broadcast use). Since every broadcast station has many attenuator networks in both studio and transmitter, a good understanding of their operation is essential.

The T-pad attenuator is designed for use with an *unbalanced line*, which is one having a common ground point. The T-pad gets its name from the fact that the resistors are placed in such a manner that they form the letter T. Figure 22-5 shows a T-pad attenuator used to attenuate the voltage 6 dB in a 600-Ω unbalanced line. If 10-dB attenuation is desired, R_1 and R_2 are approximately 300 Ω, and R_3 is approximately 400 Ω. For 20-dB attenuation, R_1 and R_2 could be approximately 500 Ω and R_3 approximately 120 Ω. Formulas for calculating the values of R_1, R_2, and R_3 can be found in most engineering handbooks. When the input impedance is equal to the output impedance, R_1 and R_2 will always be equal in value. The 600-Ω secondary of transformer

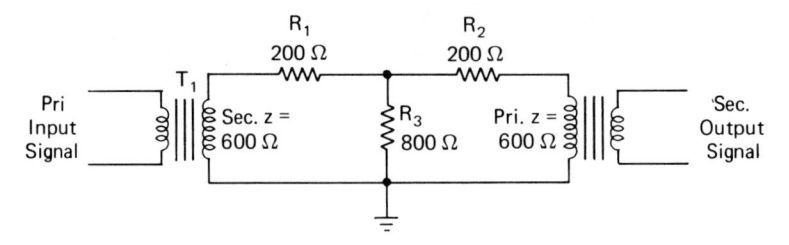

FIG. 22-5 6-db T-pad attenuator used in 600Ω unbalanced line.

T_1 looks into R_1 (200 Ω) in series with two parallel impedances of R_2 (200 Ω) plus the primary impedance of transformer T_2 (600 Ω), which is in parallel with R_3 (800 Ω), which equals 600 Ω of the transformer. If T_1 produces an output voltage of 1 V, the ratio of R_1 to the equivalent value of R_2 is 200 to 400 Ω, which produces a voltage drop across the equivalent R_3 of 0.67 V. Since the 200 Ω of R_2 and the 600 Ω of impedance of T_2 are in series across the 0.67 V, the primary of T_2 "sees" $\frac{600}{800}$ of the 0.67 V, or 0.5 V. The T-pad attenuator is usually used to attenuate signals for VU meters.

For *balanced audio lines* (neither side of the line grounded), the H-pad

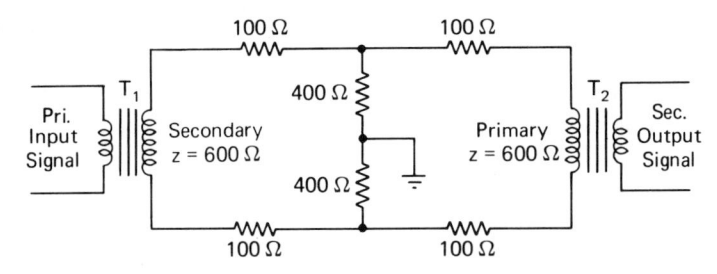

FIG. 22-6 6-db H-pad attenuator used in 600Ω balanced line.

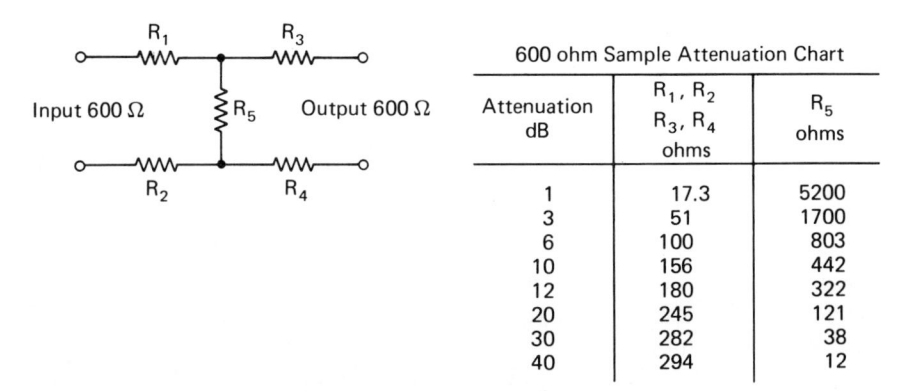

600 ohm Sample Attenuation Chart

Attenuation dB	R_1, R_2 R_3, R_4 ohms	R_5 ohms
1	17.3	5200
3	51	1700
6	100	803
10	156	442
12	180	322
20	245	121
30	282	38
40	294	12

FIG. 22-7 H-pad attenuator with chart showing resistor values necessary to provide indicated db attenuation.

Attenuator Networks 519

attenuator is used. Balanced lines are used for long lines where hum would likely be introduced by currents circulating through different ground paths. In this manner *ground-loop hum* from double grounding can be eleminated.

An example of an H-pad attenuator across 600-Ω balanced lines is shown in Fig. 22-6. Note that the values of the series resistors are exactly one half those shown for the T-pad attenuator. This drops the voltage by one half, or 6 dB. The resistances across the line are the same as that of a T-pad attenuator, except that it has a center-tap ground. An example of a floating ground H-pad attenuator for a balanced line, along with a typical chart of attenuation, is shown in Fig. 22-7. The H-pad attenuator gets its name from the fact that resistors are placed in such a manner that they form the letter H.

When *unequal impedances* in an unbalanced line are to be matched, either a *coupling transformer* or an *L-pad* can be utilized. Figure 22-8 shows an

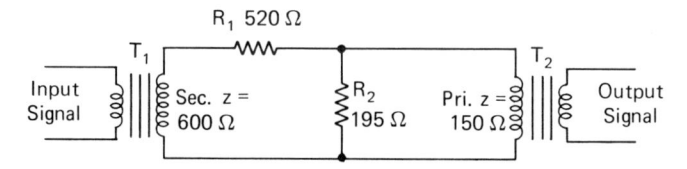

FIG. 22-8 L-pad used to match 600Ω to 150Ω line.

L-pad used to match a 600-Ω line to a 150-Ω line. It is not possible to vary the amount of loss in an L-pad, but the average loss is about 10 dB.

For balanced lines the U-pad is used. A typical U-pad impedance-matching network is shown in Fig. 22-9. The series resistances are halved

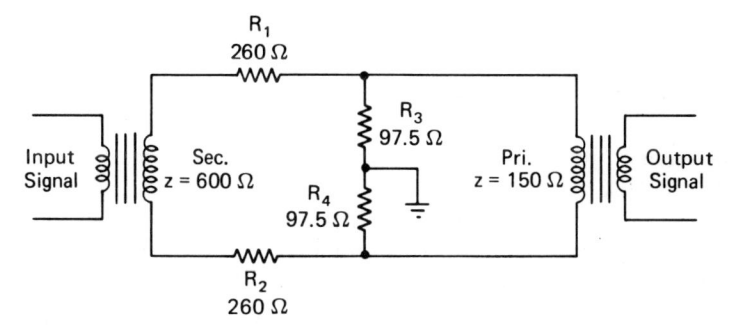

FIG. 22-9 U-pad used on balanced line to match impedances.

as in the H-pad, and the shunt resistance is center tapped. Often the pickup end transformer is center tapped on the primary for grounding. This tends to balance out any hum or noise that the lines might have picked up. Occasionally, the shunt or center-tapped resistor is replaced by a potentiometer to allow a more accurate balance of the lines.

Often it is desired to vary the amount of attenuation, such as in the pots on an audio console. To do this, a *variable T-pad* is often utilized. The variable T-pad attenuator is shown in Fig. 22-10. The arms of the pots vary at the same time (ganged) so that the input and output impedance is held essentially constant, regardless of the setting of the wiper on the pots. This is far superior to the potentiometer, which varies its input and output impedance with the setting of the wiper. At the high and low settings of the attenuator, T_1 and T_2 "see" a slight impedance mismatch, but the closer to center the operation of the attenuator, the closer the impedance match.

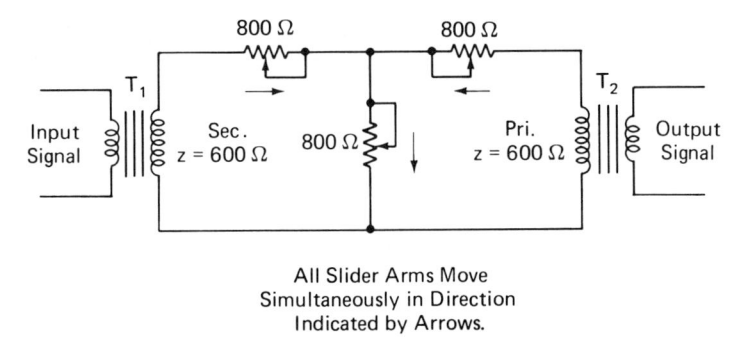

FIG. 22-10 Variable T-pad attenuator.

For use in gain controls in audio consoles, a 24-position step switch is used instead of potentiometers for purposes of reliability. The steps of the switches usually change levels at about 2 dB.

22-8 AUTOMATIC GAIN CONTROL AMPLIFIERS

Operators try to keep their audio levels as high as possible without causing overmodulation. Often there are unpredictable peaks in music or speech that could cause overmodulation before the operator could turn the level down. To prevent this kind of problem, the *peak-limiting amplifier* is used. In the AGC (or compressor) amplifier, as the incoming audio level gets higher, the amplification gets lower. Therefore, the gain will be reduced on high passages, thereby keeping the output level relatively constant. The AGC amplifier has very fast gain reduction (attack), but slow increase back to normal (release), so background noise will not be amplified during pauses in speech or music. The amplifier is usally utilized in the control studio to keep levels feeding the STL or telephone lines relatively constant. It is not effective in controlling all audio peaks, but is just used to compress dynamic range and keep levels constant. To eliminate overmodulation, the audio is then fed to another amplifier with fast attack and variable release, which has a peak-limiting circuit. This is known as a *limiter* or *peak-limiting amplifier*. The speech clipper used in speech-type communications equipment is not adequate for broadcast work due to the high distortion that clipping creates.

Frequency Monitors 521

With the use of AGC and peak-limiting amplifiers, a station can maintain a high average modulation, while the peaks are smoothed out so overmodulation does not occur. The average amount of limiting used by broadcast stations ranges from about 5 to 10 dB; stations playing classical music in which there are a number of low passages use only about 5-dB limiting, to prevent destroying the dynamic range.

22-9 FREQUENCY MONITORS

The FCC requires that AM and FM stations continuously monitor their carrier frequency. This is done with *frequency monitors* that operate independently of the transmitter frequency control. All monitors must be FCC type approved.

A simplified block diagram of a broadcast frequency monitor is shown in Fig. 22-11. The input signal is tapped off the transmitter's output through

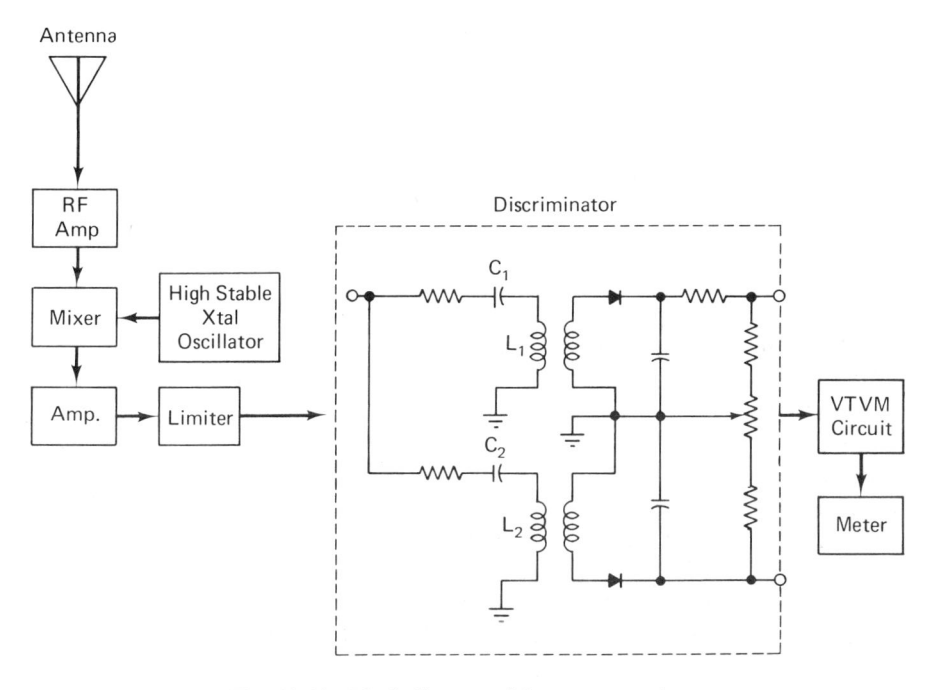

FIG. 22-11 Block diagram of frequency monitor.

an appropriate attenuator pad to give an input of about 5 V to the RF amplifier. The output of the RF amplifier and a highly stabilized crystal oscillator, using precise temperature control, are heterodyned together in the mixer stage. The crystal oscillator usually runs on a frequency that is precisely 990 Hz below the assigned carrier frequency. If the transmitter is exactly on frequency, there will be a 990-Hz output from the mixer. If the transmitter is not on frequency, there will be a slightly higher or lower frequency, depending on the exact frequency of the transmitter. The output is then fed to the amplifier stage. The amplifier output then goes to the limiter stage, where the sine wave is squared and fed to the discriminator circuit.

The discriminator has two tuned circuits consisting of C_1 and L_1 and C_2 and L_2. One circuit is tuned slightly above and the other slightly below 990 Hz. If the transmitter's carrier is exactly on frequency, the 990 Hz is rectified, and the net output of the discriminator is 0 V. If the transmitter is operating off frequency, there will be more output voltage in the tuned circuit *closest* to the new frequency, resulting in either a positive or negative output voltage, depending on whether the transmitter's carrier is positive or negative. The output voltage of the discriminator is then fed to a VTVM, which reads out on a meter calibrated in plus or minus cycles of deviation. Some types of frequency monitors are designed to give an *audible warning* when the carrier deviation is in excess of FCC specifications.

The accuracy of the frequency monitor must be checked periodically by having the station's frequency checked by a commercial monitoring service and compared to the reading of the monitor. For example, if a commercial monitoring service measures the frequency as being 5 Hz higher than the assigned carrier frequency, and the monitor in the station reads minus 5 Hz, the frequency monitor is 10 Hz off, and should be recalibrated.

If the frequency monitor becomes defective, the station must replace or repair it within 60 days and also notify the FCC engineer-in-charge. *Appropriate entries* must be made in the log as to the date and time the monitor was taken out and restored to service. Also, the station's carrier frequency must be measured once every 7 days by a commercial monitoring service and this information entered into the log.

22-10 MODULATION MONITORS

The FCC requires that all AM and FM broadcast stations continuously monitor their modulation. As with the frequency monitor, the *modulation monitor* must be FCC type approved. The requirements are that all modulation monitors be capable of

1. Reading on a dc meter movement the *average carrier level* that will display carrier shift during modulation.
2. By means of a rapidly moving meter, indicate the *percentage of positive and negative modulation*, selectable by a switch.
3. Have a peak-indicating device, either visual or audible, that can be set at any value between 50 to 120 per cent modulation on positive peaks or 50 to 100 per cent on negative peaks, or both.

The basic block diagram of a modulation monitor appears in Fig. 22-12. Input is obtained by tapping off part of the transmitter's output. The

Broadcast Transmitters

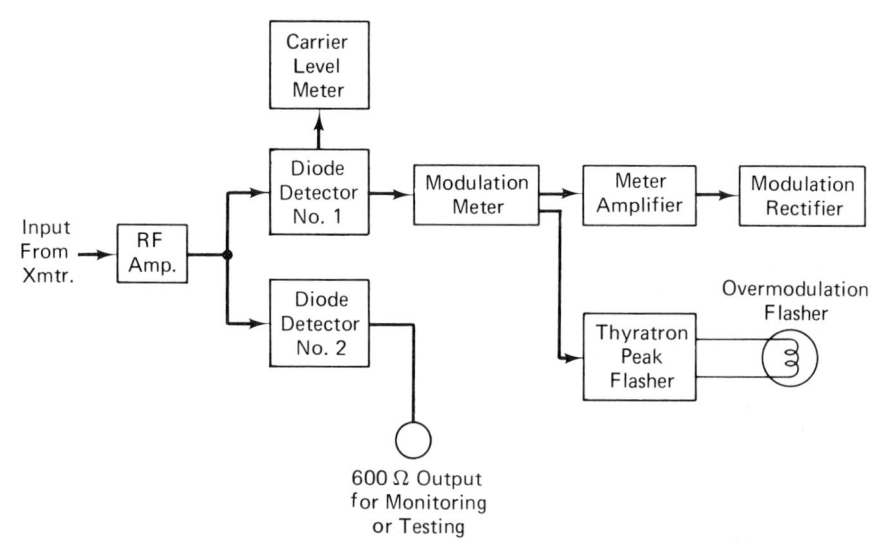

FIG. 22-12 Modulation monitor block diagram.

output of the RF amplifier feeds two very linear diode detectors. The output of detector 2 is fed to an output jack for use as an off-the-air monitor or for test measurements. Detector 1 provides a dc reference level that feeds the carrier level meter. A second output is fed to the modulation rectifier, which converts the detected signal to a pure audio signal whose level corresponds to the percentage of modulation of the transmitter. One output of this stage is fed to a calibration amplifier, which converts the various levels of modulation to a corresponding reading on the meter. The other output feeds a flasher circuit, which will indicate peaks on an adjustable overmodulation lamp.

The accuracy of the modulation monitor can be checked by displaying the modulation as either an envelope or *trapezoidal* waveform on an oscilloscope. Many stations use oscilloscopes as well as modulation monitors, because the oscilloscope can detect an instantaneous peak that the damped meter of the monitor could not.

If the modulation monitor should become defective, the FCC allows the station a period of 60 days to either replace or repair the monitor, provided that the engineer-in-charge of the district is notified. Appropriate entries must be made in the log as to the date and and time when the monitor was removed and restored to service, and that modulation is displayed by means of an oscilloscope.

22-11 BROADCAST TRANSMITTERS

All transmitters used for broadcast must meet certain FCC specifications and be FCC *type approved*. The transmitter must be crystal controlled, and the crystal must have a low temperature coefficient. The crystal must be used inside a temperature-controlled chamber or crystal oven. The transmitter must be capable of producing 85 to 95 per cent modulation when running at the authorized power. The carrier shift (current) at any percentage of modulation must not exceed 5 per cent. Voltmeters and ammeters that are associated with the FPA must have an accuracy of at least 2 per cent of the full-scale reading.

Adequate margin must be provided for all component parts to avoid overheating at the maximum rated power output. Any emission appearing on a frequency removed from the carrier by between 15 and 30 kHz, inclusive, must be attenuated at least 25 dB *below* the level of the *unmodulated carrier*. Compliance with the specification will be deemed to show the occupied bandwidth to be 30 kHz or less. Any emission appearing on a frequency removed from the carrier by more than 30 kHz and up to and including 75 kHz, inclusive, must be attenuated at least 35 dB below the level of the unmodulated carrier. Any emission appearing on a frequency removed from the carrier by more than 75 kHz must be attenuated at least $43 + 10 \log_{10}$ (power in watts) dB below the level of the unmodulated carrier, or 80 dB, whichever is the lesser attenuation.

In general, the transmitter shall be constructed either on racks or panels or in totally enclosed frames protected as required by article 810 of the National Electrical Code [section 8192 (a), (b), and (c)]. Adequate provision must be made for varying the transmitter power output between sufficient limits to compensate for excessive variations in line voltage or other factors that may affect the power output.

22-12 OPERATING POWERS

When the FCC licenses a broadcast station, they state the operating power in the license. It is referred to as the *licensed power* or the *authorized operating power*, and is the actual power that is being fed to the antenna. Since antenna gain is not a part of standard broadcasting, the radiated power is a function of the antenna's efficiency. The FCC sets the power limits to 5 per cent above the licensed power or 10 per cent below the licensed power. Each station is required to stay within this tolerance except in times of emergency, when a lower power may be utilized for a limited period of time. The operating power is sometimes referred to as the *carrier power*, and it should be noted that this is always the power of the *unmodulated carrier*.

The maximum rated carrier power is defined as being the maximum amount of power that a specific transmitter can produce. For example, if a station is licensed to operate with 500 W, and the transmitter is rated for 1 kW, the maximum rated carrier power of that transmitter is 1 kW.

The plate input power of a broadcast transmitter is determined by multiplying the plate voltage by the plate current of the last radio stage ($P_i = E_p I_p$).

The operating power, or output power, is equal to $P = I^2 R$, where I is the antenna current in amperes with no modulation, and R is the antenna resistance *at the same point* where antenna current is measured. This is known as the *direct method* for determining the unmodulated carrier power of a broadcast station. This is the FCC approved method for determining operating power.

The *indirect method* of determining operating power is used primarily for FM broadcast stations and the aural section of TV transmitters. It can be used only temporarily for standard broadcast stations under the following conditions:

1. In case of an emergency where the licensed antenna system has been damaged by causes beyond the control of the licensee.
2. Pending completion of authorized changes in the antenna system.
3. If any change is made that may affect the antenna system.

The indirect method for determining unmodulated operating power is as follows: operating power $= E_p \times I_p \times F$, where E_p is the dc plate voltage applied to the last radio stage, I_p is the unmodulated value of plate current applied to the last radio stage, and F is the *efficiency factor* derived from Table 22-1.

TABLE 22-1

Factor (F)	Method of Modulation	Maximum Rated Carrier Power	Class of Amplifier
0.70	Plate	0.1–1.0 kW	–
0.80	Plate	5 kW and over	–
0.35	Low level	0.1 kW and over	B
0.65	Low level	0.1 kW and over	BC*
0.35	Grid	0.1 kW and over	–

*All linear amplifiers where efficiency approaches that of class C operation.

22-13 REMOTE-CONTROL SYSTEMS

It is often desirable to operate the transmitter by remote control from the main studio and eliminate the transmitter operator. This permits the operator at the main studio to control the operation of the transmitter, as well as to take the required meter readings. A station can utilize remote control if the transmitter's power ouput is 10 kW or less; the station does not use a directional antenna array; the transmitter is inaccessible to unauthorized personnel; the transmitter ceases operation if trouble should occur; and the operator is able to turn on and off the transmitter and take readings for logs. Remote-control operation can take place in stations with a power in excess of 10 kW, or with a directional antenna system, if the station can prove that the transmitter and/or antenna system is stable enough to operate without periodic adjustment.

Should failure of any part of the remote-control system occur, causing inaccurate or loss of meter readings, or loss of transmitter control, the station must cease operation by remote control, and a licensed operator will then be required at the transmitter site.

22-14 PROOF-OF-PERFORMANCE TESTS

During the construction and testing periods of a broadcasting station, a station must prove that it can meet several technical standards. The FCC permits testing of the transmitting equipment between midnight and local sunrise for short periods of time. These are known as *equipment tests.* When construction is completed, the FCC permits service or program tests before the actual license is issued. After the station is in normal operation, periodical tests are required by the FCC to assure proper performance of the equipment. These are known as *proof-of-performance* tests and must be performed on an annual basis. The experimental period is usually used to make such tests, provided that the tests do not interfere with other stations operating on such frequency at that time.

The specifications that must be met during proof-of-performance tests are that the transmitter must be capable of modulating 85 to 95 per cent with a total harmonic distortion to the audio frequencies not to exceed 7.5 per cent. For modulation percentages under 85 per cent the total harmonic distortion must not exceed 5 per cent. Distortion tests must be made at 50, 100, and 400 Hz, and 1, 5, and 7.5 kHz, plus any other frequencies found necessary. During normal operation modulation must be maintained as high as possible.

A frequency response graph must be made from 50 Hz to 7.5 kHz at 25, 50, 85, 95, and 100 per cent (if possible) modulation. The hum level, carrier frequency, carrier shift, spurious emission, and harmonic content must be recorded and submitted to the FCC 4 months prior to the filing for license renewal. Occasional field intensity measurements in all directions of azimuth are required for directional stations. The carrier hum and extraneous noise (exclusive of microphone and studio noises) level must be at least 45 dB below 100 per cent modulation for the frequency band of 50 to 20,000 Hz. Periodical tests of the modulation monitor and frequency monitor are also required.

The above specifications are for standard broadcast stations only; FM and aural TV transmitters are required to meet more precise specifications.

22-15 LOG REQUIREMENTS

The FCC requires that each broadcast station retain a *program log*, an *operating log*, and a *maintenance log*. Each program log must contain the following information:

1. Entry of each time a station identification is given (must include call letters and location and be given at least once every half hour).

2. A brief program description (speech, music, drama, etc.), the name of the program, program's sponsor, start and finish times, any mechanical recordings used, and political affiliation of political candidate speakers.
3. Entry showing that each sponsored program has been aired as agreed upon by the sponsor, and paid for or furnished by sponsor.
4. An entry indicating source of program (live, tape, record, network and name, etc.).

Each operating log must contain the following information:

1. Entry indicating times and dates at which the station begins supplying power to the antenna and the time it ceases.
2. Times indicating when program begins and ends.
3. An entry of the time at which an interruption to the carrier wave occurs, its cause and duration.
4. Entry regarding antenna structure illumination, such as time when lights are turned on and off (only if required by the FCC).
5. An entry of the following must be made each 30 min:
    a. Plate voltage and plate current of the last radio stage.
    b. Antenna current.
    c. Carrier frequency readings (these must be made prior to making any adjustment).

Each maintenance log must contain the following information:

1. Entries on a weekly basis of the following: base currents of the antenna; times and results of auxiliary transmitter tests; results of carrier frequency tests made by commercial monitoring services (if made); results of calibration checks made on monitors or chart recording devices (if made).
2. Entries of times and dates indicating the removal of and restoration dates of any of the following equipment: modulation monitor and frequency monitor; final-stage plate voltmeter or ammeter; antenna ammeter or common point ammeter (if used).
3. An entry indicating any inspections made of tower lights or replacement of tower lights.
4. An entry describing any experimental operations made during experimental periods.

Logs are kept on a daily basis and are required to be retained for a period of *2 years*, unless they contain information regarding complaints, distress communications or other claims. The FCC requires that logs be kept in ink, that each log be signed by the appropriate engineer, and that times be noted for going on and off duty. Only *licensed operators* are permitted to sign the log. Abbreviations may be used in a log only if there is a key on the log as to the meaning of the abbreviations. If an error is made in log information, a single line must be struck through the erroneous information and the correct information entered. The operator must initial the correction. The FCC has no prescribed form for a log other than it (1) contain the necessary information, (2) be kept in a neat form, and (3) have all information printed and clearly readable.

22-16 EMERGENCY BROADCAST SYSTEM

If a state of emergency exists (caused by war, public disaster, or other national emergency), all but a few stations will go off the air and observe radio silence. Before leaving the air, the station will give instructions as to where to tune for stations remaining in service. The stations (AM, FM, or TV) that stay on the air will be under the National Defense Emergency Authorization (NDEA). They will broadcast information to the public according to an Emergency Broadcast System (EBS) plan.

When a state of emergency is declared, all stations not authorized to remain on the air will indicate that such a condition exists. They will transmit the *Emergency Attention Signal*, which is an announcement of the emergency. The carrier is then turned off for 5 s, then on for 5 s, then off again for 5 s. The carrier is then put on the air and a 1-kHz tone is broadcast for 15 s. Then the state of emergency will again be announced and instructions will be given on where the EBS plan will be broadcast. If the station is not an NDEA station, it must then cut the carrier and observe radio silence for the duration of the emergency.

All radio and television stations are required to give periodical EBS tests. The procedure followed is the same as if an emergency existed, except that it must be announced before and after the test that it is only a test. A short announcement is then made indicating the procedure to follow if an actual emergency did exist. Each station is required to make an EBS test at least once a week on an unscheduled basis between the hours of 8:30 A.M. and local sunset.

22-17 OPERATOR LICENSE REQUIREMENTS

Any person operating any equipment that is placed on the air is *required to be licensed* by the FCC. It would be desirable for all personnel in a broadcast station to have first-class radiotelephone licenses. Often the announcer or "combo" operator does not have sufficient background to warrant a first-class license. There are lower-class licenses that include the second- and third-class radiotelephone licenses. Operators with *second- or third-class radiotelephone licenses* may operate the following types of stations:

1. Any station that has a power output of 10 kW or less.

2. The station utilizes a nondirectional antenna.
3. All transmitting equipment is inspected by an operator with first-class license on a daily basis.
4. A first-class licensed operator is at the studio or transmitter for 5 days each week at intervals of no less than 12 h.

The second- and third-class operators may make the following adjustments:

1. Those necessary to turn the transmitter on and off.
2. External controls to maintain proper power-supply voltages.
3. External controls to maintain proper modulation percentages.
4. External controls to correct operating power requirements.
5. External controls to effect operation during an emergency action condition.

Operators with *third-class licenses* must have a *broadcast endorsement* (Element 9) to make any of the above adjustments. All other adjustments must be made by an operator with a first-class license.

For noncommercial educational FM stations with a transmitter output of no more than 10 W a *second-class operator* can make any necessary adjustments on the transmitter or associated equipment. At least a second-class license is required by a technician before he can service any portion of non-broadcast communications transmitting equipment.

22-18 BROADCAST MICROPHONES

The general types of microphones used for broadcast application are the *dynamic, velocity or ribbon*, and *condenser* types. Microphones used in the control room when combo operation (operator spins records as well as announces) is used are of the *unidirectional* type to help cancel out the surrounding noises of cuing records, loading tape machines, and so on. All microphones used in the broadcast field are *low-impedance* types to prevent the low-level signals in the microphone lines from picking up hum and other noises. *Balanced lines* are also used to help eliminate the hum pickup from ground loops. The average output impedance of broadcast microphones ranges from 50 to 250 Ω. Microphones and cables are also carefully positioned so that they are out of strong magnetic fields created by power transformers and other equipment.

22-19 TRANSMISSION LINES

The four important characteristics that must be known for broadcast-station transmission lines are

1. Impedance.
2. Power-handling capability.

3. Attenuation.
4. Ability to withstand weather extremes.

The average impedance used in broadcast-station transmission lines is 50 Ω. Four types of transmission lines are used. They are

1. Flexible coaxial cable, RG/U type, using a solid polyethylene dielectric. This is used in nondirectional stations employing powers up to 1 kW.
2. The semiflexible-type transmission line is used in many stations using directional antenna arrays and in many medium power nondirectional stations. Semiflexible transmission line is made of a center conductor and a tubular outer conductor. The center conductor is insulated from the outer conductor by small rings placed at intervals throughout the line. The dielectric is air.
3. One recent type of cable consists of a copper inner conductor and an aluminum outer conductor. The inner conductor is suspended from the outer conductor by a continuous swirling insulator around the inner conductor. The dielectric is air.
4. Rigid line is widely used. Both inner and outer conductors are tubular and are suspended from each other by Teflon discs or pegs. This type of line uses air as the dielectric and has excellent power-handling capabilities and low loss at high frequency.

To prevent moisture from entering into the line, it is *pressurized with nitrogen*. Since all the connections and flanges are made airtight, once the cable is pressurized refilling is rarely needed.

At modulation peaks of 100 per cent the transmitter's power peaks have *four times* the average power. The peak-to-peak power that will be present at 100 per cent modulation is $4E_p \times I_p$, where E_p is the final plate voltage and I_p is the final plate current.

22-20 ANTENNA SYSTEMS

Antennas used in standard broadcasting are of the *vertical* type. The radiated pattern of a single vertical antenna is essentially nondirectional and has a low radiation angle. The tower itself is used as the radiating element. Any *guying wires* that support the structure must be broken up with insulators to prevent any part of the lines from becoming resonant to the carrier or its harmonics.

Antenna towers are used in both the base-grounded and the base-ungrounded mode. The most popular is the ungrounded type where the base of the tower is insulated from ground and power is fed across the insulator. Since the ground system is a part of the antenna system, long copper wires

Antenna Systems 531

are buried under the tower to establish better grounding conditions. The height of antenna is usually a quarter-wavelength, and the ground acts like an "image" quarter-wavelength giving an effective antenna length of a half-wavelength.

The radiation resistance of the vertical antenna is theoretically 36 Ω, but, depending on the ground characteristics, this changes quite substantially. To tune the antenna to resonance and match it to the impedance of the transmission line, a *tuning unit* is used at the base of the antenna structure. A simplified unit is shown in Fig. 22-13. Basic tuning is accomplished by the T-section

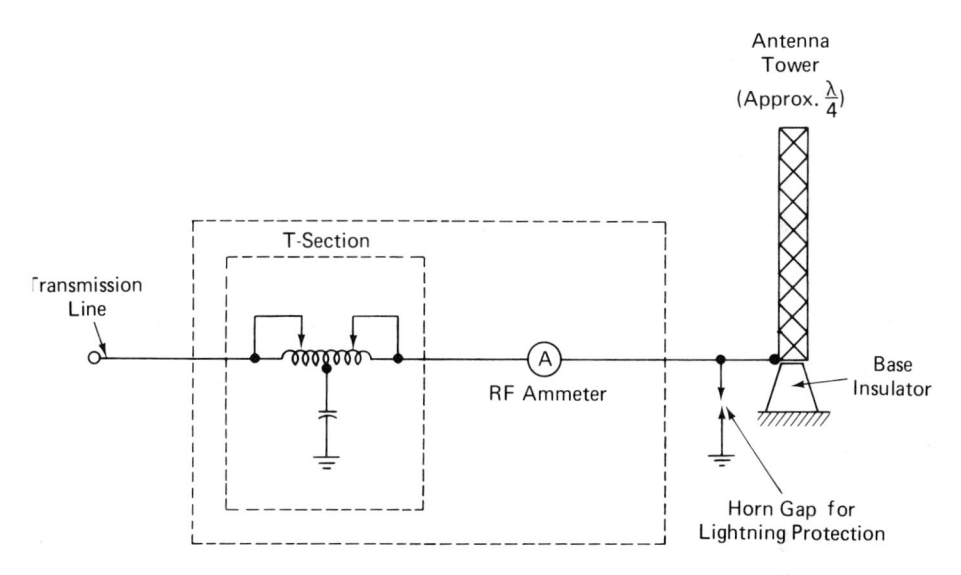

FIG. 22-13 Basic antenna tuning unit.

matching circuit. The RF then passes through the ammeter, which reads the antenna current to the base of the antenna. A *horn gap* is placed from the antenna base to ground so that there will be a path to ground in the event the antenna is struck by lightning. The horn gap is spaced far enough apart so that 100 per cent peak modulation will not cause arc-over.

Often high-power stations use top loading on their antennas to increase the equivalent electrical wavelength without increasing antenna height. To do this, an umbrella-like structure is placed on the top of the tower. This method substantially increases the ground and sky wave.

For towers above 150 ft in height there must be two 500- or 620-W code *beacon lights* at the top of the tower, as well as one or more side lights on the tower, depending on the height. The lights must burn continuously from sunset to sunrise or be controlled by a photoelectric switch that turns on the lights when an intensity of 35 footcandles (fc) is reached. It must turn off the lights when an intensity of 58 fc is reached. Spare lamps must be kept on hand at all times, and in case of failure lamps should be replaced as soon as practical. If a tower light should fail, the Federal Aviation Administration (FAA) must be notified immediately as to the nature of the failure. They must also be notified when repairs have been made. All information regarding checks and services to tower lighting must be entered in the station's maintenance log.

Occasional checks of *field intensity* are necessary to assure that the antenna efficiency has not decreased. Field-intensity tests are measured in microvolts per meter at specific distances from the antenna.

The antenna structure must be painted with alternate bands of aviation service orange and white, terminating with aviation orange at both top and bottom. The antenna structure must be cleaned and painted as often as necessary to maintain good visibility.

22-21 MISCELLANEOUS REQUIREMENTS

The FCC also licenses the operation of international broadcast stations for operation on shortwave frequencies. These must maintain a carrier frequency tolerance of 0.005 per cent.

Should any indicating instrument fail and no substitute meter be immediately available, the station must notify the engineer-in-charge of the radio district at the time the meter is found defective. He must also be notified at the time it is replaced. Appropriate entries must be made in the maintenance log. An alternative method of measurement must be devised until the meter is replaced. The station has a period of 60 days in which to replace the defective meter.

All meters associated with transmitting and monitoring equipment must have a full-scale accuracy of 2 per cent or better. Indicating instruments must also meet the following specifications: the length of the scale shall not be less than $\frac{2}{10}$ in.; the maximum rating of the meter shall be such that it does not go off the scale during modulation; the scale shall have at least 40 divisions; and full-scale reading shall not be greater than five times the minimum normal reading. These specifications must be met for the final amplification stages of broadcast transmitters.

The following specifications must be met for antenna current, base current, or common-point current meters: instruments must have logarithmic or square-law scales; full-scale readings shall not be greater than three times the minimum normal indication; and no scale division above one third full-scale reading (in amperes) shall be greater that one thirteenth of the full-scale reading.

Commercial License Questions

Sections in which answers to questions are given appear in parentheses. A bracketed number following a question implies that it applies only to that element.

Commercial License Questions

1. What is the frequency tolerance for a standard broadcast station? [4]
   (22–1)
2. With reference to broadcast stations, what is meant by the *experimental period*? [4]
   (22–1)
3. What is the frequency tolerance for a broadcast STL station? [4]   (22–3)
4. Define *auxiliary broadcast transmitter* and state the conditions under which it may be used. [4]
   (22–3)
5. In what part of a broadcast-station system is a phase monitor sometimes found? What is the function of this instrument? [4]
   (22–3)
6. How frequently must the auxiliary transmitter of a standard broadcast station be tested? [4]
   (22–3)
7. For what purpose is an auxiliary transmitter maintained? [4]   (22–3)
8. Why are preamplifiers sometimes used ahead of mixing systems? [4]
   (22–3)
9. What is the purpose of a variable attenuator in a speech input system? [4]
   (22–3)
10. What is the purpose of a line equalizer? [4]   (22–4)
11. Draw a diagram of the equalizer circuit most commonly used for equalizing wire-line circuits. [4]
    (22–4)
12. If a preamplifier having a 600-Ω output is connected to a microphone so that the power output is -40 dB, and assuming the mixer system to have a loss of 10 dB, what must be the voltage amplification necessary in the line amplifier to feed $+10$ dB into the transmitter line? [4]
    (22–6)
13. What unit has been adopted by leading program transmission organizations as a volume unit and to what power is this unit equivalent? [4]   (22–7)
14. Why is a high-level amplifier feeding a program transmission line generally isolated from the line by means of a pad? [4]
    (22–7)
15. Why are grounded center-tap transformers frequently used to terminate program wire lines? [4]
    (22–7)
16. What are the purposes of H- or T-pad attenuators? [4]   (22–7)
17. What is the purpose of a line pad? [4]   (22–7)
18. In what part of a broadcast-station system are limiting devices usually employed? What are their functions? [4]
    (22–8)
19. What are the results of using an audio peak limiter? [4]   (22–8)
20. What is the purpose of using a frequency standard or service independent of the transmitter frequency monitor or control? [4]
    (22–9)
21. What is the reason that certain broadcast-station frequency monitors must receive their energy from an unmodulated stage of the transmitter? [4]
    (22–9)
22. If a broadcast station receives a frequency measurement report indicating that the station frequency was 45 Hz low at a certain time, and the transmitter log

for the same time shows the measured frequency to be 5 Hz high, what is the error in the station frequency monitor? [4] (22–9)

23. What is the required full-scale accuracy of the plate ammeter and plate voltmeter of the final radio stage of a standard broadcast transmitter? [4] (22–11)

24. What portion of the scale of an antenna ammeter having a square-law scale is considered as having acceptable accuracy for use at a broadcast station? [4] (22–11)

25. What is the required full-scale accuracy of the ammeters and voltmeters associated with the final radio stage of a broadcast transmitter? [4] (22–11)

26. What is the maximum carrier shift permissible at a standard broadcast station? [4] (22–11)

27. In accordance with the Commission's Standards of Good Engineering Practice, what determines the maximum permissible full-scale reading of indicating instruments required in the last radio stage of a standard broadcast transmitter? [4] (22–11)

28. What percentage of modulation capability is required of a standard broadcast station? [4] (22–11)

29. Define *maximum rated carrier power* of a broadcast-station transmitter. [4] (22–12)

30. Define *plate input power* of a broadcast-station transmitter. [4] (22–12)

31. What is the power that is actually transmitted by a standard broadcast station termed? [4] (22–12)

32. Are the antenna current, plate current, and so forth, as used in the rules and regulations of the Commission with reference to radiotelephone transmitters, modulated or unmodulated values? [4] (22–12)

33. Describe the various methods by which a standard broadcast station may compute its operating power, and state the conditions under which each method may be employed. [4] (22–12)

34. What factors enter into the determination of power of a broadcast station that employs the indirect method of measurement? [4] (22–12)

35. Under what conditions may a standard broadcast station be operated at a power lower than specified in the station license? [4] (22–12)

36. When the authorized nighttime power of a standard broadcast station is different from the daytime power and the operating power is determined by the indirect method, which of the efficiency factors established by FCC rules is used? [4] (22–12)

37. What is the power specified in the instrument of authorization for a standard broadcast station called? [4] (22–12)

38. What are the permissible tolerances of power of a standard broadcast station? [4] (22–12)

Commercial License Questions 535

39. Under what conditions may a standard broadcast station be operated by remote control? [4] (22–13)
40. What is the frequency tolerance allowed an international broadcast station? [4] (22–13)
41. If the plate ammeter in the last stage of a broadcast transmitter burned out, what should be done? [4] (22–13)
42. What are meant by *equipment, program,* and *service tests* mentioned in the rules and regulations of the Commission? [4] (22–14)
43. For what purpose may a standard broadcast station, licensed to operate daytime or specified hours, operate during the experimental period without specific authorization? [4] (22–14)
44. How frequently must a remote reading ammeter be checked against a regular antenna ammeter? [4] (22–15)
45. Why should impedances be matched in speech-input equipment? [4] (22–18)
46. What is the ratio of unmodulated carrier power to instantaneous peak power at 100 per cent modulation at a standard broadcast station? [4] (22–19)
47. In what units is the field intensity of a broadcast station normally measured? [4] (22–20)

23

Television

23-1 BASIC PRINCIPLES

Television means "to see from a distance." It is a system of transmitting and receiving *visual* scenes in motion by RF broadcasting techniques. The sound associated with the *picture or video information* is transmitted simultaneously. Hence, a TV transmitter is basically two transmitters, *video* and *audio*, coupled to a single antenna.

A television system requires a *camera tube* and an *image-reproducing tube*. The camera tube is a photoelectric device that produces electrical signals which correspond to the various levels of light intensity of the picture. At the receiving end a picture tube converts these signals back to the visual information detected by the camera tube.

A camera tube, such as the *image orthicon* or *vidicon*, is used to convert the visual information into electrical signals. It does so by *scanning* the picture, which is focused on the face of the camera tube, with an electron beam in a sequential order. Scanning is accomplished by starting at the top and moving from left to right and from top to bottom, one line at a time. This is called *horizontal scanning*. At the end of each horizontal line the beam is returned very quickly to the left side to commence scanning the next horizontal line. During retrace, no picture information is scanned, as both camera and receiving picture tube are blanked out during this period.

When the beam is returned to the left side, its vertical position is slightly lowered so that it will not scan the same line again. This is accomplished by a *vertical scanning* circuit.

The total number of scanning lines for a complete picture should be large so as to provide the maximum amount of resolution or detail. However, the greater the number of scanning lines, the greater is the RF bandwidth

Magnetic Deflection and Focusing 537

requirement. As a compromise, a complete picture containing 525 scanning lines has been chosen. This is the optimum number of *lines per frame* for the standard 6-MHz bandwidth of the television broadcast channels.

23-2 MAGNETIC DEFLECTION AND FOCUSING

A beam of electrons can be deflected by a magnetic field at right angles to it. The device used to do this in a picture tube or CRT is a *yoke*. It has two pairs of coils symmetrically displaced around the neck of the tube. The vertical pair deflect the beam *horizontally*, while the horizontal pair provide *vertical* deflection.

The scanning beam can also be *focused* by a properly designed and positioned coil. Providing the right amount of current through the coil, and adjusting its position, will provide a sharp focus for the beam and hence give a sharper picture. A CRT with its associated deflection coils is shown in Fig. 23-1.

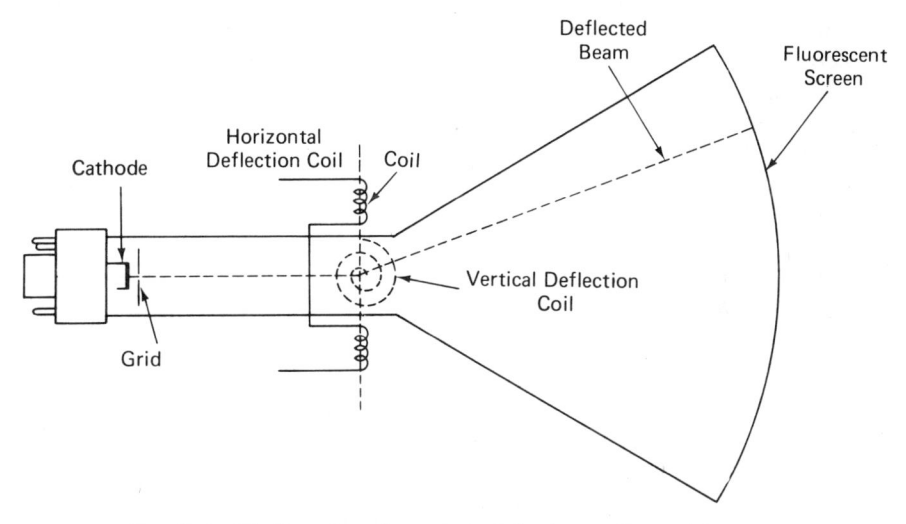

FIG. 23-1 Horizontal and vertical deflection coils positioned around the neck of a TV picture tube.

Because of the high vacuum inside picture tubes, they must be handled with *extreme care*. Do not allow any part of the glass surface to become scratched, as the glass may crack due to the enormous air pressure pressing against it. This may result in an *implosion*, which can be very hazardous to surrounding persons. Likewise, no unnecessary pressure should be brought against the tube, particularly the stem and bell parts. Manufacturers recomment that persons handling picture tubes wear leather gloves and jacket and protective glasses. Never pick up the tube by the neck, only from the face.

23-3 INTERLACED SCANNING

Not only is the picture broken down into its numerous picture elements, but the complete picture is scanned 30 times or frames per second to provide the illusion of motion. The slight flicker observed at this rate was solved by a technique known as *interlaced scanning*. This divides the 525 picture lines into two groups of lines called *fields*. Each field contains 262.5 lines with one scanning the odd-numbered lines and the other the even. This provides a repetition rate of 60 fields/s and was designed to synchronize with the 60-Hz/s. power-line frequency. This minimizes any tendency for the picture to roll. Because of the *persistence of vision* of the human eye and the *persistence of fluorescence* of the CRT screen, no flicker is apparent to the observer.

23-4 ICONOSCOPE

One of the first successful television camera tubes was the *iconoscope*. Although not highly sensitive, it may still be found in use as the pickup tube for motion pictures and slide films. It consists of an electron gun, deflection coils, a *mosaic picture screen* on an insulating sheet, a conductive plate, and a collector anode, as shown in Fig. 23-2. Microscopic globules of photoelectric silver are deposited on one side of a thin mica insulating sheet. Each globule is electrically insulated from all those adjacent to it. The other side of the mica sheet is coated with a conductive material. Thus, each globule forms a tiny capacitor between itself and the back plate, with the mica as the dielectric.

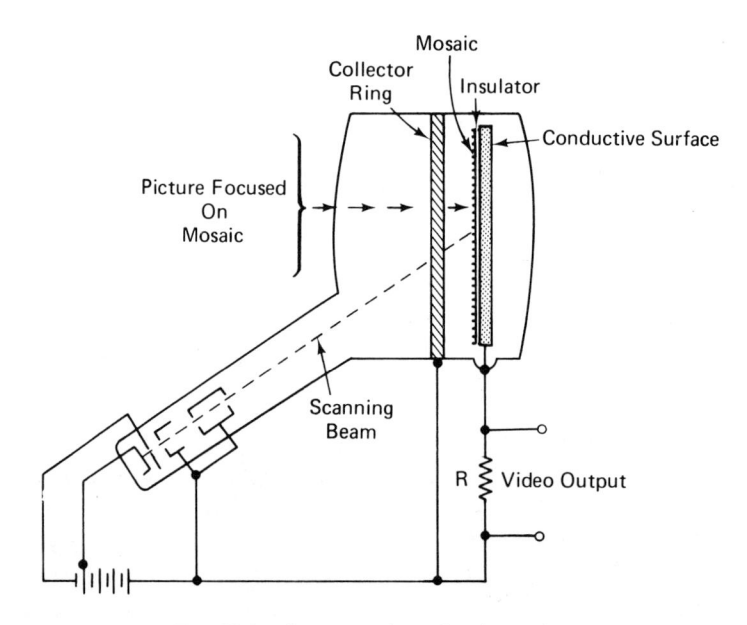

FIG. 23-2 Cut-away view of an iconoscope.

Image Orthicon 539

When a scene is focused on the mosaic screen, *photons* (bundles of light energy) strike the photoelectric globules, releasing electrons and leaving the light areas positive and the darker areas less positive. The electrons released from the globules are drawn to the collector-anode ring. This leaves the mosaic screen with areas of electric charge that conform to the light and dark areas of the scene focused on it. Each tiny capacitor is charged in proportion to the light striking it. Since only a few electrons reach the collector ring, the difference in potential between the mosaic and collector ring is but a few volts.

Most of the emitted electrons collect around the mosaic and form a space charge. When the *scanning electron beam* strikes a globule, the number of secondary electrons emitted is much greater than the number emitted due to light rays from the picture. This greatly increased supply of electrons increases the difference of potential between the collector ring and the mosaic and causes current flow through the load resistor, which is in *synchronism* with the scanning beam. This current is modulated by the amount of light striking each globule and produces a video output voltage across load resistor R, which conforms to the picture projected on the mosaic.

23-5 IMAGE ORTHICON

The *image orthicon* is intended for use in black and white cameras for outdoor and studio pickup. It is very stable in performance at all incident light levels. To understand the operation of the image orthicon, refer to Fig. 23-3 while studying the following paragraphs.

Light from the scene being televised is focused on the semitransparent photocathode and causes electrons to be emitted proportional to the light striking the area. These electrons are accelerated toward the target by grid 6 and focused by the magnetic field produced by an external coil. The target

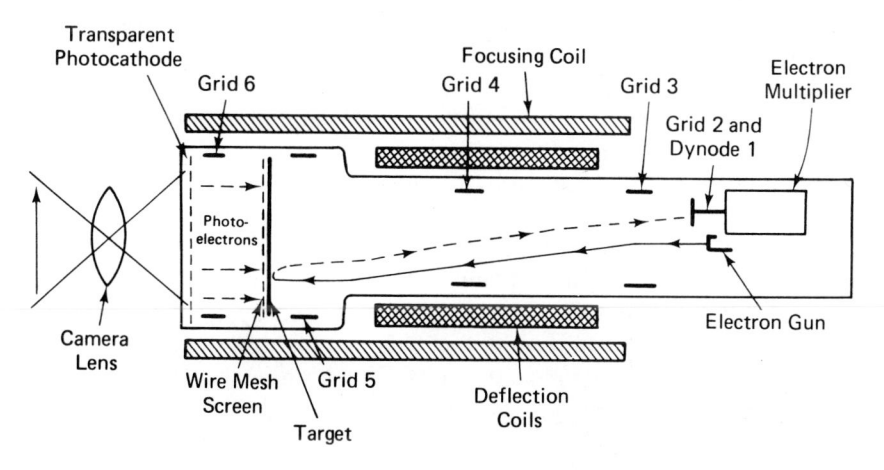

FIG. 23-3 Cut-away view of image orthicon tube.

consists of a special thin glass disc with a fine-mesh screen on the photocathode side. Focusing is also accomplished by varying the potential of the photocathode.

When the electrons strike the target, secondary emission from the glass takes place. These secondary electrons are collected by the wire mesh, which is maintained at a constant potential of approximately 1 V. This limits the potential of the glass disc and accounts for its stability in changing intensities of light. As electrons are emitted from the photocathode side of the glass disc, positive charges are built up on the other side, which vary with the amount of electrons emitted. Thus, a pattern of positive charges corresponding to the intensities of light of the scene being televised is set up. This constitutes the image section of the image orthicon. The action described is completely independent of the electron beam and scanning circuits of the tube.

The backside of the target is scanned with a low-velocity beam from the electron gun. The beam is focused by the magnetic field generated by an external coil and by the electrostatic field of grid 4. The potential applied to grid 5 adjusts the decelerating field between grid 4 and the target. As the low-velocity beam strikes the target, it is turned back and focused on dynode 1, the first element of an *electron multiplier*. When the beam is turned back from the target, however, some electrons are taken from the beam to neutralize the charge on the glass. The greater the charge on the glass, the more electrons are taken from the beam, leaving fewer electrons to return to dynode 1. This action leaves the scanned side of the target negatively charged, while the opposite side is positively charged. Because the glass disc target is extremely thin, these charges neutralize themselves by conduction through the glass. This neutralization takes place in less than the time of one frame.

As the amplitude-modulated stream of electrons strikes dynode 1, several secondary electrons are emitted for each primary electron striking it. This electron multiplication takes place from dynode 1 through 5, which is located inside the box labeled electron multiplier. Thus, the electrons are multiplied many times before reaching the anode of the multiplier. The approximate gain of the multiplier section of this tube is 1000.

Darker picture elements produce less positive areas on the target, and need fewer deposited scanning-beam electrons to neutralize the charge. Under these conditions a larger number of electrons from the scanning beam are returned to dynode 1. White picture elements cause more positive areas on the target and fewer electrons are turned back to the electron multiplier. From this we observe that the returning beam of electrons is modulated by the charge distribution of the target plate. This load resistor in the image-orthicon output is connected between the anode and the power supply.

If the image orthicon remains focused on a bright, stationary image for several minutes, a *sticking picture* results. This is an image of the televised scene in which the black and white areas are reversed. This is more apt to happen if the image orthicon is operated without having sufficient warm-up

Vidicon 541

time. This situation can usually be corrected by focusing on a clear white wall or screen. This may require a few minutes or several hours depending on how old the tube is. After the image orthicon has been in operation for 200 to 300 h, it should remain idle for 3 to 4 weeks to recover its original sensitivity and resolution.

23-6 VIDICON

The vidicon is a small camera tube suitable for use in black and white and color television cameras for either broadcasting or closed-circuit applications. Its structural arrangement, shown in Fig. 23-4, consists of a target,

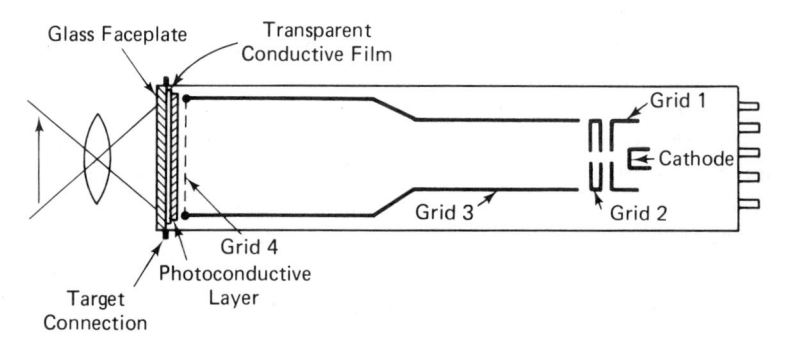

FIG. 23-4 Cut-away view of vidicon camera tube.

fine-mesh screen (grid 4), a beam-focusing electrode (grid 3), and an electron gun. The target is composed of a transparent conducting film on the inner surface of the faceplate. Grid 4 is adjacent to the photoconductive layer, and grid 3 is connected to grid 4. Each small portion of the photoconductive layer is an insulator when there is no light on the faceplate, but becomes slightly conductive when illuminated.

The gun side of the photoconductive layer is scanned by a low-velocity beam produced by the electron gun. The gun consists of a cathode, a control grid (grid 1), and an accelerating grid (grid 2). The beam is focused by the magnetic field of an external coil and by the electrostatic field of grid 3. Grid 4 provides a uniform decelerating field between itself and the photoconductive layer so that the beam will approach the layer perpendicular to it, a condition necessary for linear scanning.

When the gun side of the photoconductive layer, with its positive-potential pattern, is scanned by the beam, electrons are deposited from the beam until the surface potential is reduced to that of the cathode. Thereafter, the electrons are turned back to form a return beam, which is not used. Electron deposits on the scanned surface of any portion of the photoconductive layer change the difference of potential between the two surfaces of the photoconductive layer. When the two surfaces of this layer (which in effect is a charged capacitor) are connected to a load through the target connection, power supply, and the scanning beam, a capacitive current flows, which constitutes the video signal.

23-7 TELEVISION STATION

The basic functions of a TV station are shown in Fig. 23-5. The transmitting system is divided into three basic parts: the studios, the monitor and control facility, and the transmitter.

Most stations would likely have several studios, each equipped with at least one TV camera, even though the block diagram shows but one. This permits camera crews and other studio personnel to prepare and record programs, commercials, and so forth, without going directly on the air. If some scenes do not come out as desired, they can be rerecorded.

The monitor and control-room facilities are complex and can only be given the most superficial explanation. Perhaps the heart of this operation is the *sync generator*, which produces the necessary pulses to *synchronize* all units. This includes the camera control consoles, the video tape recorders, the slide and motion picture film projectors, and the various video monitors that allow studio personnel to monitor the program. Notice that part of the sync generator output is fed to a unit called the *sync pulse adder*. This unit adds the sync pulses to the output of the *video switcher* before the picture is fed to the transmitter.

The output of all cameras, video tape recorders, and film projection equipment feed into the video switcher. This unit enables the operator to select the desired scene(s) and at the same time keep all equipment synchronized so that no vertical roll or horizontal jitter will occur.

The audio part of the program is picked up by appropriate studio microphones and amplified before going to the audio mixer console. Part of the output from this console goes to the monitor speaker(s) and part to the FM transmitter, where it modulates the oscillator. The FM transmitter signal is fed to the antenna through a *diplexer*, which prevents any of the FM signal from being coupled back into the AM(video) transmitter, and vice versa.

The complex nature of the video signal and sync pulses makes it difficult to develop the large amounts of power that would be required to amplitude modulate the final power amplifier. Consequently, one of the earlier stages is amplitude modulated by using grid-modulation techniques, where the modulator power requirements are small. The function of the *vestigial sideband filter* will be discussed in Section 23-13.

The transmitted signal is approximately 6 MHz wide. To transmit such a wide band at TV frequencies, either a slotted-cylinder or turnstile-type antenna is needed. The FCC requires that the output be *horizontally polarized*.

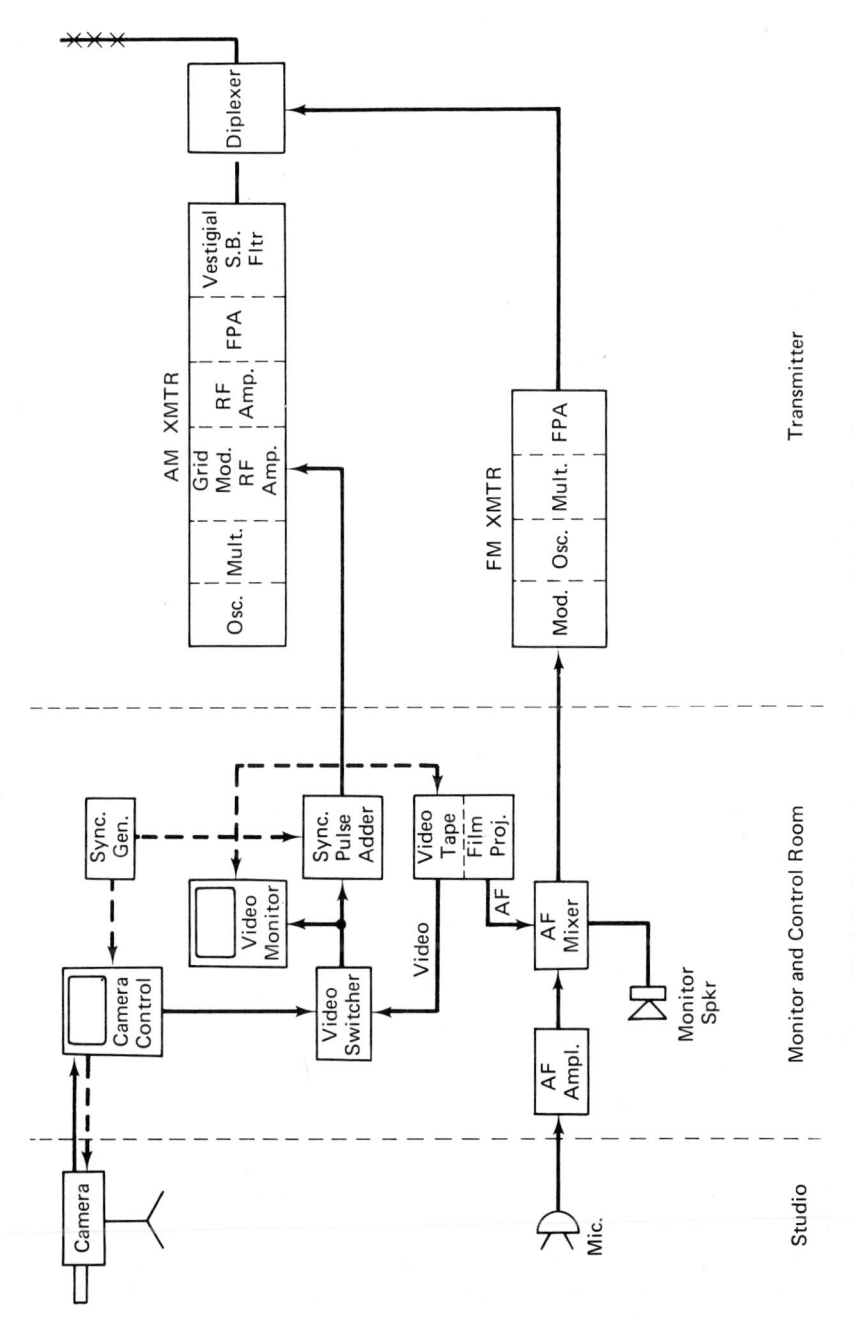

FIG. 23-5 Simplified block diagram of a TV transmitting station.

23-8 SYNCHRONIZING-PULSE GENERATOR

To reproduce an intelligible signal, the camera must be scanned in synchronization with the receiver. The synchronizing-pulse generator, or sync-pulse generator, is a complex piece of equipment requiring many components. Basically, the sync generator consists of an oscillator having a frequency of twice the horizontal sweep rate, or 31,500 Hz. These pulses are used as *equalizing pulses* and to *serrate* the vertical sync pulses. These serrations in the vertical sync pulse help keep the horizontal oscillator on frequency during vertical sync. The horizontal sync pulses are derived by taking the half-frequency of 31,500 Hz.

In the receiver the vertical sync pulses are fed through a low-pass filter and trigger the vertical sweep oscillator. The horizontal sync pulses are fed through a high-pass filter and are used to trigger the horizontal circuits. A partial train of sync pulses is shown in Fig. 23-6. Detailed views of the *blanking*, *synchronizing*, and *equalizing pulses* are shown in Fig. 23-7.

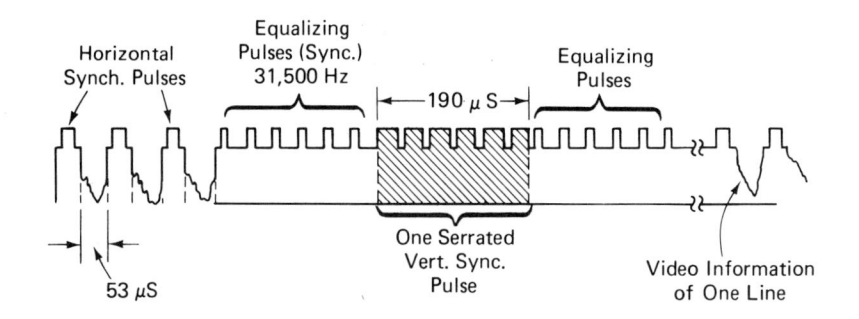

FIG. 23-6 Synchronizing, blanking, and equalizing pulses used in TV.

Before the horizontal sync pulse is produced, a blanking pulse must appear to cut off the electron beam in the camera tube, the transmitter, and the picture tube of the receiver. During the interval of the blanking pulse, the horizontal oscillator returns the electron beam to the starting condition for the beginning of the next line. The sync pulses are transmitted as part of the complete television signal, and are used to make any necessary correction to the oscillator to assure the proper starting time of each line after the blanking pulse drops off. Thus, the horizontal and vertical sync pulses are only added to the transmitted television signal as an aid to synchronize the horizontal and vertical circuits of the receiver.

The vertical sync pulse is synchronized to the power-line frequency. This prevents any power-supply ripple beating with the vertical oscillator to cause *vertical jitters or hum bars* (dark horizontal bars) on the screen.

Video Modulation 545

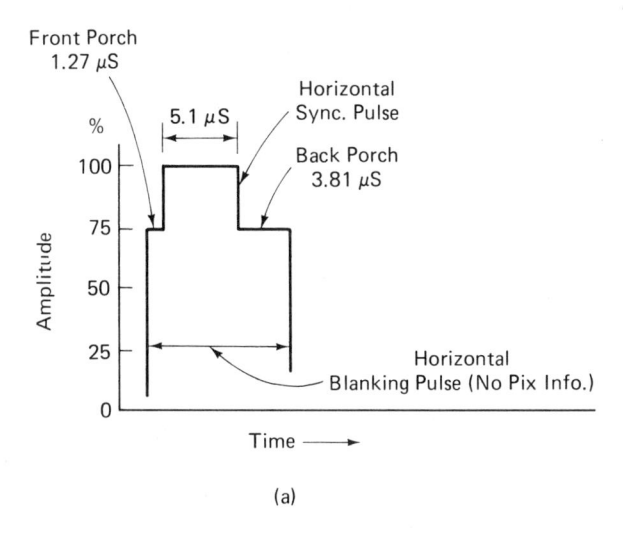

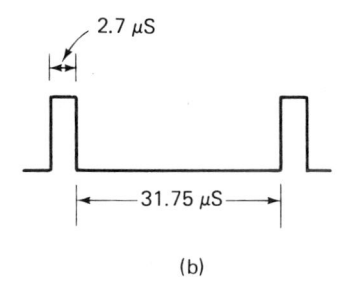

FIG 23-7 (a) Details of a horizontal blanking and sync pulse; (b) details of an equalizing pulse.

Most sync generators are capable of producing their own controlled 60-Hz signal when a local source is not available. When remote or field installations are set up, the remote sync generator develops the pulses for the main transmitter. The sync generator at the station has an input circuit to allow it to lock in on signals originating from remote sync generators, network programs, or field equipment. Transmission to the main transmitter from a studio or field pickup is usually made by an FM klystron transmitter. A high-gain, parabolic, directional transmitting antenna is used to beam the signals in the 6.875- to 7.125-GHz or 12.75- to 13.25-GHz bands.

23-9 VIDEO MODULATION

Since the video signal is amplitude modulated, the carrier amplitude varies from scene to scene. However, the amplitude of the sync and blanking pulses remains constant. The top of each sync pulse is the *maximum emitted carrier* and is considered the 100 *per cent level* (see Fig. 23-7). The top of the sync pulse is called the *blacker-than-black level*.

The top of the blanking or pedestal level is 75 per cent of total modulation (± 2.5 per cent). The front of the pedestal is called the *front porch*; the back is called the *back porch*.

The reference black level, which will produce the darkest black in the transmitted picture, is at approximately 70 per cent. The reference white level, or the whitest picture, is 12.5 per cent. If the white level were allowed to fall to zero, the power emission of the transmitter would be so low that the signal-to-noise ratio would allow local disturbances to affect the receiving picture quality. Another effect would be that there would not be sufficient signal to beat against the aural signal to produce a 4.5-MHz sound IF.

23-10 SOUND TRANSMITTER

The *aural*, or sound, transmitter is essentially the same as an FM broadcast station except that 100 per cent modulation is represented by only a 25-kHz frequency swing of the carrier. The same 30- to 15,000-Hz audio capability of the amplifiers is required. The sound signals are amplified, controlled by an operator who monitors and mixes them, and then feeds them to an FM modulator in the transmitter. The FM transmitter signal is monitored and fed to the antenna. The aural signal is 4.5 MHz away from the video carrier.

23-11 OPERATING POWER OF THE TRANSMITTER

The peak RF power output of a typical TV transmitter varies between 1 and 50 kW. Because of the gain of the antenna system, the effective radiated power (ERP) will be somewhat higher. The FCC specifies that the minimum ERP of a TV transmitter shall be 50 kW for an area whose population is 1 million or more, with an antenna height of 500 ft. For populated areas under 50,000 people, the minimum ERP shall be 1 kW with an antenna height of 300 ft.

For black and white transmission the radiated power of the aural signal shall not be less than 50 per cent or more than 150 per cent of the radiated power of the picture carrier. To minimize sound interference in the picture, the sound power is limited to 50 to 70 per cent of the picture power for color transmission.

The output power is calculated in the same manner as any broadcast transmitter. This can be either by the direct or indirect method. In either case no modulation should be present when the readings are taken. The indirect method uses the formula

$$P_o = e_p i_p F$$

where F is the efficiency of the final stage as determined by the manufacturer and approved by the FCC.

The picture transmitter's output is measured at the output of the *vestigial sideband filter*. Measurements are made with the transmitter connected to a dummy load whose resistance is equal to the transmission-line surge impedance. A black and white picture should be transmitted during these measurements. When the transmitter is connected to the antenna, the average power reading must be essentially the same as with the dummy load. Television transmitters are rated in peak power output, which can be determined by multiplying the average power output by 1.68.

To determine the ERP of a TV transmitter, the peak power output is multiplied by the square of the antenna gain. For example, if the antenna has a field gain of 4 and a peak power output of 10 kW, the ERP is 10 kW times 4^2, or 160 kW.

The frequency tolerance for the picture or sound carrier is ± 0.002 per cent. The center frequency of the aural transmitter must always remain 4.5 MHz, ± 1000 Hz, above the visual carrier frequency.

23-12 CAMERA CHAINS

A single camera chain consists of one camera and its associated control unit. The camera has a small CRT built in as a view finder so that the operator can see the same scene that the camera is picking up. Within the camera is the image orthicon tube (or vidicon in some cases), deflection and blanking circuits for the camera tube, and a preamplifier that supplies the signal to the control unit. The control unit provides remote control of gain, black level, beam current, target voltage, and *electrical focus* of the camera tube. *Optical focus* is controlled by the cameraman.

Camera chains may consist of several cameras and their control units. They may also consist of one or more film projectors and a slide projector. A block diagram of a typical film multiplexer and camera is shown in Fig. 23-8. Projectors 1 and 3 are typically motion picture units, while 2 would be for slides. The film multiplexer includes the projectors as part of the installation.

23-13 VESTIGIAL SIDEBAND TRANSMISSION

If both sidebands were transmitted, a channel 9 MHz wide would be needed to transmit both the video and sound carriers. This is undesirable because of the limited amount of space available in the frequency spectrum available for TV use. Also, it would be necessary to provide amplifiers in both the transmitter and receiver capable of passing this wideband signal. Since both video sidebands contain identical information, it is possible to filter or suppress one sideband at the transmitter and still obtain all the picture information needed by the receiver.

Filters that can completely cut off the undesired sideband without distorting the remaining 4-MHz channel are difficult to construct. To overcome

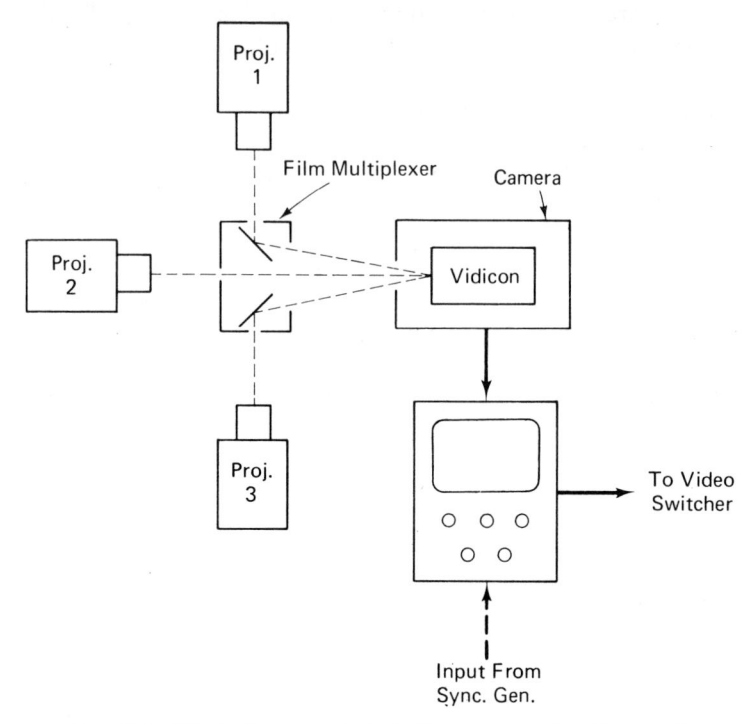

FIG. 23-8 Block diagram of single film camera with film multiplexer for three-projector pick up.

this problem, a part of the lower sideband is attenuated. This results in what is known as *vestigial sideband transmission*. This allows a television channel to be reduced to 6 MHz and still contain all the necessary information for a good picture.

A graphical analysis of the effect that vestigial sideband filter has on the output of the picture transmitter operating on channel 3 appears in Fig..23-9. The lower sideband is allowed to extend only about 1 MHz below the picture carrier frequency of 61.25 MHz. If the lower sideband were completely suppressed, serious phase distortion of the low frequencies would result, causing a blurred picture. The upper sideband is 4 MHz wide and extends to about 65.25 MHz. The sound carrier is 65.75 MHz, which is 4.5 MHz above the picture carrier. The FM sound carrier is deviated 25 kHz above and below its center frequency (100 per cent modulation) and occupies 50 kHz of bandwidth. This leaves a guard band between the upper end of the channel and the beginning of the next channel.

Vestigial sideband transmission presents a problem. A signal transmitted with only a single sideband and carrier represents 50 per cent modulation in comparison with a normal double sideband signal with 100 per cent

Channel Allocations

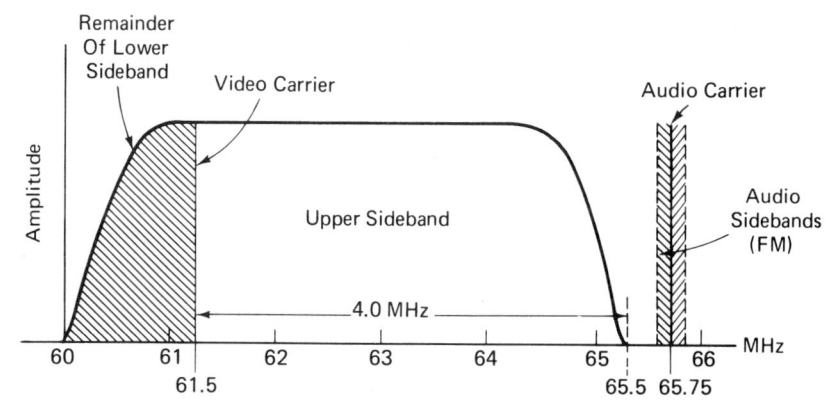

FIG. 23-9 Frequency distribution of signals associated with channel 3.

modulation. The *higher modulating video frequencies* are suppressed in the lower sideband, whereas the *low modulating video frequencies* are not appreciably suppressed in this band. This means that the lower video modulating frequencies are essentially transmitted as *double sideband*. Therefore, the higher video frequencies provide signals with one half the effective carrier modulation produced by the lower video frequencies. This amounts to a *low-frequency boost* in the video signal. This is corrected in the IF amplifier of the receiver by *deemphasizing* the low video frequencies.

23-14 LOG REQUIREMENTS

The FCC's requirements for TV station logs are similar to those for standard broadcast and FM stations. This includes program, operating, and maintenance information. In the case of TV, the log must include entries from both the picture and aural transmitters, plus transmission-line readings. The maintenance log must also include any service work performed on the modulation- and frequency-monitoring equipment, Technicians employed at the TV station transmitter must possess radiotelephone first-class licenses.

23-15 CHANNEL ALLOCATIONS

Originally, the FCC planned for 13 channels for commercial television. These were divided into a *low band* and a *high band*. The low band (channels 2 to 6) covers a frequency range of 54 to 88 MHz with the 72- to 76-MHz band being reserved for other services. The high band covers the 174- to 216-MHz spectrum and is represented by channels 7 to 13. In 1952 the UHF band was opened, including an additional 70 channels (14 to 83) between 470 and 890 MHz. Table 23-1 shows the different channel allocations.

TABLE 23-1 TELEVISION CHANNEL ALLOCATIONS

Channel Number	Frequency Band (MHz)	Channel Number	Frequency Band (MHz)
1*	—	42	638–644
2	54–60	43	644–650
3	60–66	44	650–656
4	66–72	45	656–662
5	76–82	46	662–668
6	82–88	47	668–674
7	174–180	48	674–680
8	180–186	49	680–686
9	186–192	50	686–692
10	192–198	51	692–698
11	198–204	52	698–704
12	204–210	53	704–710
13	210–216	54	710–716
14	470–476	55	716–722
15	476–482	56	722–728
16	482–488	57	728–734
17	488–494	58	734–740
18	494–500	59	740–746
19	500–506	60	746–752
20	506–512	61	752–758
21	512–518	62	758–764
22	518–524	63	764–770
23	524–530	64	770–776
24	530–536	65	776–782
25	536–542	66	782–788
26	542–548	67	788–794
27	548–554	68	794–800
28	554–560	69	800–806
29	560–566	70	806–812
30	566–572	71	812–818
31	572–578	72	818–824
32	578–584	73	824–830
33	584–590	74	830–836
34	590–596	75	836–842
35	596–602	76	842–848
36	602–608	77	848–854
37	608–614	78	854–860
38	614–620	79	860–866
39	620–626	80	866–872
40	626–632	81	872–878
41	632–638	82	878–884
		83	884–890

*The 44- to 50-MHz band was television channel 1 but is now assigned to other services.

Receiving Functions—The Antenna–Tuner Circuits 551

23-16 RECEIVING FUNCTIONS—THE ANTENNA–TUNER CIRCUITS

A receiver must perform many functions, as shown in Fig. 23-10, which is a block diagram of a black and white TV receiver.

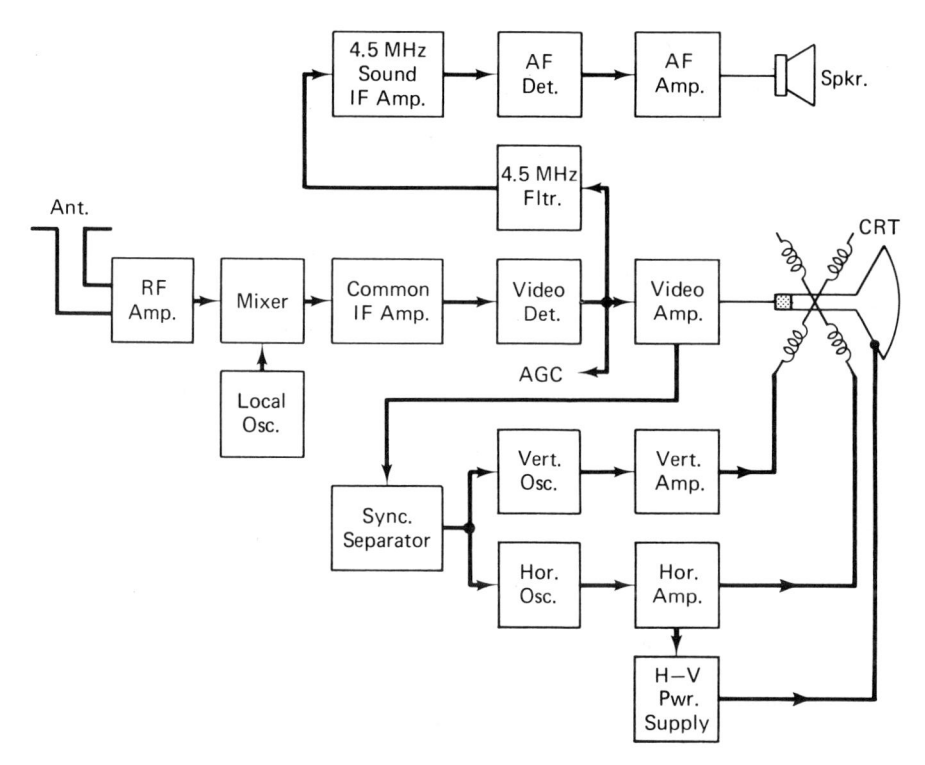

Fig. 23-10 Block diagram of a typical black and white television receiver.

Antennas vary from elaborate parasitically excited Yagi beams on high masts to simple dipoles on the back of the set. Many persons are subscribing to a community antenna TV system (CATV), which brings a cable into the home carrying all the channels available to the area.

The RF amplifier must be capable of amplifying the incoming signal as much as possible above the noise levels that are present. Noise, appearing on the screen of the CRT, shows up as a salt-and-pepper pattern popularly referred to as snow. The circuit of a typical VHF tuner is shown in Fig. 23-11. The amplified signal is coupled through T_2 to the base of mixer Q_2. The local oscillator also feeds its signal to the base of the mixer. An AGC voltage controls the gain of the RF amplifier.

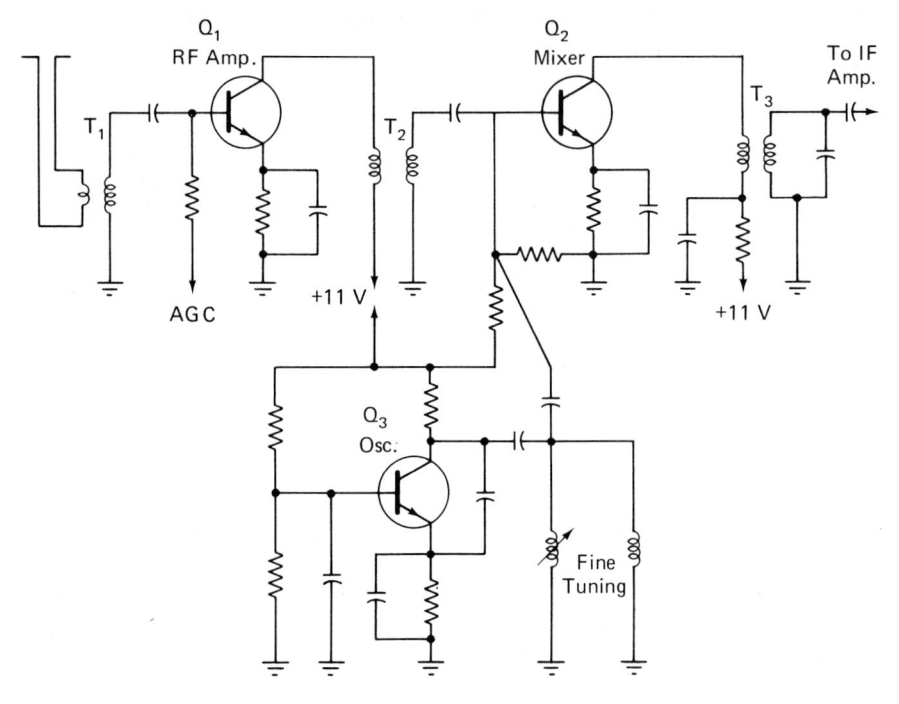

FIG. 23-11 Circuit of a typical VHF tuner.

23-17 INTERMEDIATE-FREQUENCY AMPLIFIER

The performance of the IF amplifier is very critical. Practically all the receiver *gain* and *selectivity* are obtained in this section of the receiver. It must have a band pass in excess of 4 MHz and a frequency response such as to correct for the deemphasis of high frequencies caused by the vestigial sideband transmission.

The ideal response curve is represented in Fig. 23-12. The picture IF is 45.75 MHz and the sound IF is 4.5 MHz below this. Observe that the sound and video IFs have *reversed* relationship in the receiver as compared to the transmitter (refer to Fig. 23-9). This frequency inversion is the result of the local oscillator operating above the incoming signal. The heterodyned frequencies that are produced are inverted, as represented in Fig. 23-12. The higher radio frequencies are closer to the oscillator's frequency, resulting in lower values for the IF difference frequencies.

The picture IF carrier is at the 50 per cent point of the response curve. This reduced response for the picture carrier and side frequencies close to it is *opposite* to the effect of vestigial sideband operation. Even though the sound IF at 41.25 MHz is amplified with the picture signal, its relative gain is only 5 to 10 per cent.

Figure 23-13 is a typical IF response curve of a receiver when swept by

Intermediate-Frequency Amplifier 553

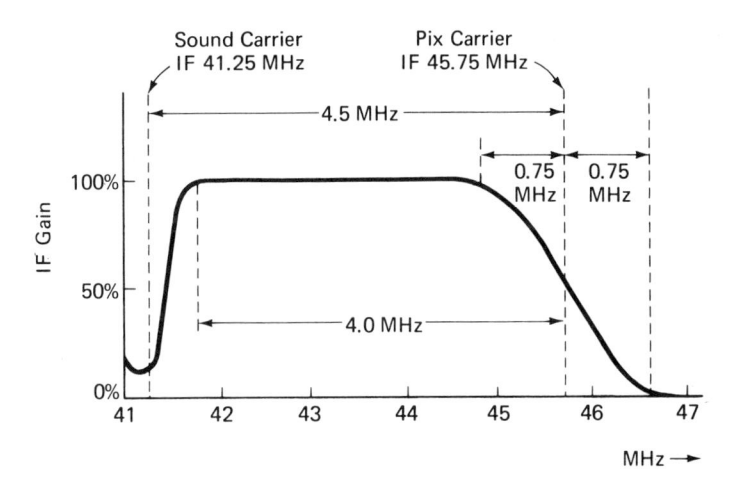

FIG. 23-12 Ideal response curve for intercarrier sound receiver with picture IF carrier of 45.75MHz.

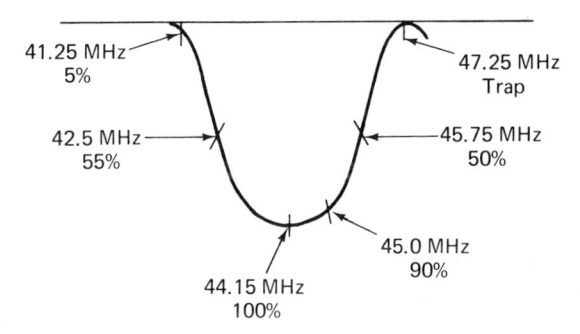

FIG. 23-13 Typical IF response curve produced on an oscilloscope by a TV alignment signal generator. Small vertical lines are alignment markers.

a TV *alignment generator* and presented on an oscilloscope. As can be seen, the actual response curve differs somewhat from the ideal. The fact that the curve is inverted is of no consequence. Notice that the video carrier of 45.75 MHz sits at 50 per cent of maximum response. The small vertical lines shown around the response curve are produced by marker generators in the alignment generator. These markers are very accurate and allow the technician to correctly set all IF transformers and wavetraps to their correct frequencies.

Automatic gain control is applied to the IF strip to control the gain of the amplifier to compensate for strong signals, which may drive the amplifier into cutoff. This would result in a *flattening* of sync tips and a loss of sync. On weak signals the AGC voltage increases the gain of the IF to ensure proper signal-to-noise ratio.

23-18 VIDEO DETECTOR

The video carrier IF is at 45.75 MHz and sound is at 41.25 MHz. When these two signals are passed through the detector diode, a nonlinear device, they beat together and produce a difference frequency of 4.5 MHz with the characteristics of the sound carrier. This is picked off the collector of the video amplifier through the 4.5-MHz pass filter (see Fig. 23-10), and amplified by the sound IF amplifier. A discriminator tuned to 4.5 MHz detects the FM signal.

The video detector also rectifies the picture carrier, filters it, and feeds it to the base of the first video amplifier as a varying dc voltage.

23-19 AUTOMATIC GAIN CONTROL

There are two types of AGC systems: *keyed* and *simple*. The latter method is the least expensive and less effective way of controlling gain. The control voltage is obtained by sampling the signal at the output of the video detector and feeding it to the base of the AGC control transistor.

With keyed AGC, voltage pulses are used to key the circuit on so it will conduct only during a small part of each signal cycle. The advantage is that *noise pulses* in the signal have little effect on the AGC voltage. Keying is done by flyback pulses from the horizontal output transformer. Then the AGC transistor produces collector current only during horizontal sync-pulse time. This results in an output that is free from noise, because the circuit is not operating between keying pulses.

A simple keyed AGC circuit is shown in Fig. 23-14. A pulse is picked up from the horizontal output transformer and coupled to the collector via C_2. The diagram shows the approximate shape of this pulse and its coinci-

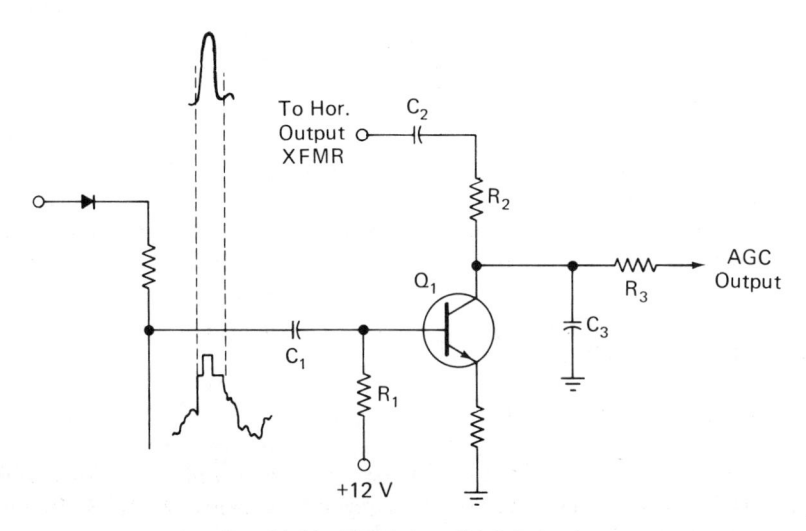

Fig. 23-14 Simple keyed AGC circuit.

Vertical Deflection 555

dence with the blanking pulse. Resistor R_1 biases the transistor slightly on. Superimposed in this bias voltage is the blanking pulse and video signal. The amount of transistor conduction is determined principally by the *amplitude, or level, of the blanking pulse* and not the video signal level, which is constantly changing. The amplitude of the blanking pulse is a function of the *strength* of the received signals. Therefore, weak stations will not cause Q_1 to conduct as much as strong stations.

The dc potential on the collector of Q_1 varies inversely with signal strength. Therefore, strong signals cause the collector voltage to drop. This places a lower charge on C_3. Between pulses, when Q_1 does not conduct, C_3 discharges. The circuit resistance on discharge is high enough to make the discharge time constant much longer than the 65.5 μs between pulses. Therefore, the voltage across C_3 is a relatively steady dc potential, which is proportional to the peak value of the video signal input (the top of the blanking pulse).

23-20 VIDEO AMPLIFIER

The *video amplifier* amplifies the output of the video detector so that it can drive or vary the *kinescope* (picture tube) beam current. The modulation of this current reproduces the various degrees of light and dark shades of the picture elements on the kinescope screen.

The input frequency to the video amplifier may be nearly direct current, when the camera is scanning a scene that has no change of illumination, to about 4.5 MHz when scanning a scene with many contrasts.

23-21 SYNCHRONIZING-PULSE SEPARATION

If the received picture is to remain steady on the screen, it must remain synchronized with the transmitter. This will only occur if the horizontal and vertical synchronizing pulses can be recovered and synchronize their respective oscillators.

On the top of every blanking pulse is a *synchronizing pulse*. Since this pulse occurs during the time the kinescope is blanked out, it has no effect on the picture. A special circuit, called a *sync separator*, is designed to recover these pulses and is shown in Fig. 23-15. The composite video signal is fed to the base of Q_1, which is biased to cutoff. The sync pulses riding on top of the blanking pulses are of sufficient amplitude to drive Q_1 into conduction. The signal at the collector will only contain the sync pulses. These sync pulses are then fed to each oscillator.

23-22 VERTICAL DEFLECTION

Part of the output of the sync separator goes to the *vertical oscillator*, which is designed to free run at some frequency just below 60 Hz. By feeding vertical sync pulses to it, its frequency will be *increased* so that it synchronizes

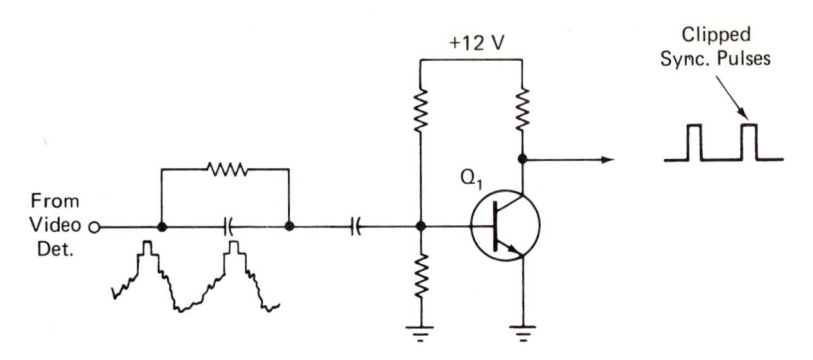

FIG. 23-15 Simplified sync separator circuit.

with the vertical pulses. These are the same pulses that are controlling the vertical sweep of the camera at the TV station. Therefore, the vertical sweep of the kinescope should be in exact synchronism with the transmitted picture. As long as this condition remains, there will be no rolling of the picture.

A typical vertical oscillator and amplifier circuit are represented in Fig. 23-16. Transistor Q_1 is the oscillator and receives its positive feedback from the collector of Q_3, via C_1 and R_1. The output of Q_2 is not sufficiently large to drive the base of the vertical output stage (Q_3), so vertical amplifier Q_2 is used. This stage operates as an emitter follower.

The circuit generates a sawtooth waveform. If the rate of rise is not *absolutely linear*, the picture will be distorted. Objects that are supposed to be round will appear oblong. To compensate for any nonlinearity, a *vertical linearity control* is provided. The *height control* determines the amount of drive to Q_3, which controls the height of the picture displayed on the kinescope. The *vertical hold control* also sets the approximate operating frequency. The output of Q_3 is connected to output inductor L_1, whose reactive voltage is capacitively coupled to the vertical deflection coils.

To keep the oscillator synchronized, the vertical sync pulses from the sync separator are fed through an *integrator circuit* to the base of Q_1. The integrator is a resistor–capacitor network with a long time constant relative to the horizontal pulses. Since the vertical sync pulses are of longer duration than the horizontal, the only time the integrator is allowed to charge to a level high enough to trigger the multivibrator is during these pulses.

23-23 HORIZONTAL DEFLECTION

The *horizontal oscillator* generates the 15,750-Hz signal to drive the horizontal output stage. A typical circuit is shown in Fig. 23-17. When the receiver is turned on, forward bias causes collector current to flow through the collector winding of T_1. The field established by this winding induces a voltage into the base winding, which causes additional increase in collector current. As the regenerative drive to the base of Q_1 increases, a point is

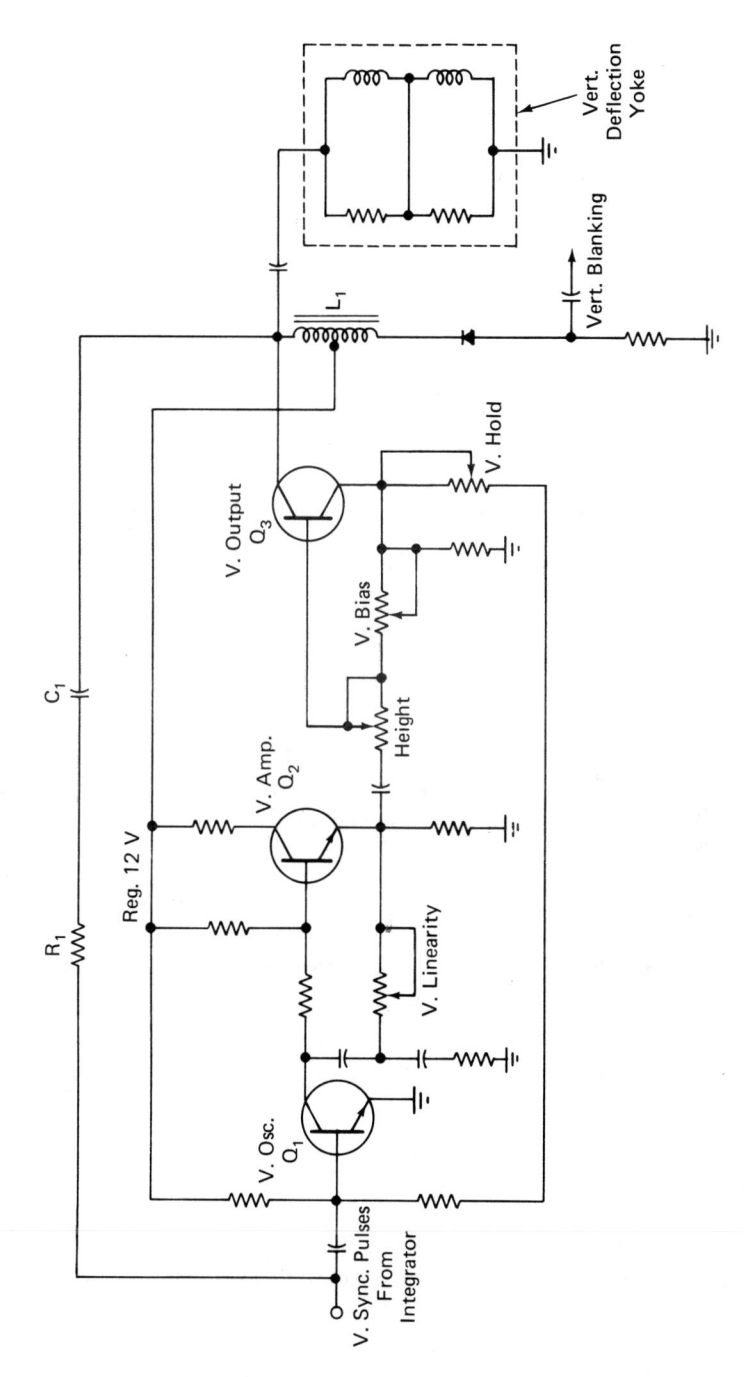

FIG. 23-16 Typical vertical oscillator and output amplifier.

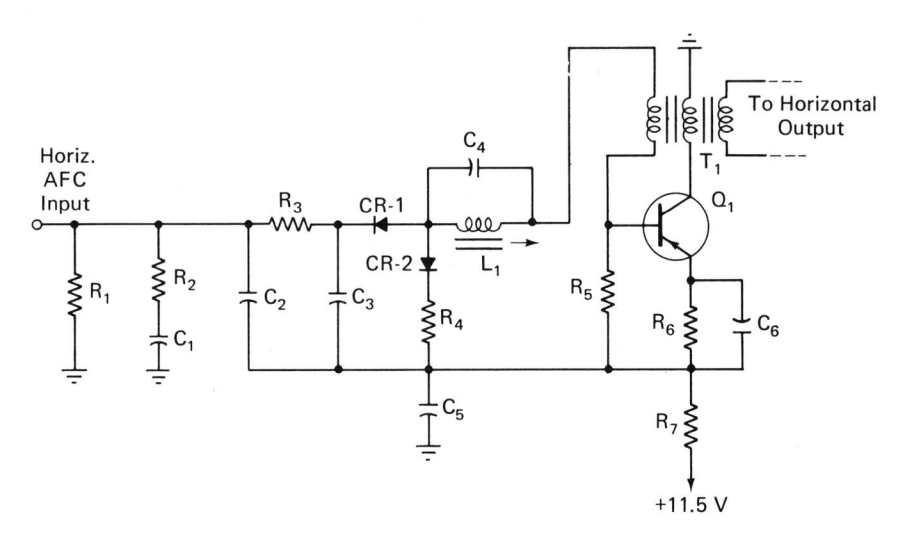

FIG. 23-17 Horizontal oscillator circuit.

reached where CR_1 becomes reversed biased as a result of the charge built up on C_3. Now the only forward bias remaining for Q_1 comes from the magnetic coupling between collector and base windings of T_1. Blocking oscillator transformer T_1 soon reaches saturation, and base drive ceases. Q_1 remains cut off until C_3 discharges enough to forward bias CR_1. The cycle then repeats itself.

The basic oscillator frequency is determined by the constants of L_1 and C_4. If it deviates from its assigned frequency, the input from the horizontal AFC will sync it to the correct frequency. The oscillator drives the horizontal output circuit, which provides three basic functions:

1. Supplies current to the horizontal deflection coils.
2. Produces the high voltage for the kinescope or CRT.
3. Develops a boosted voltage for the horizontal output amplifier and for the first anode of the picture tube.

A simplified circuit of this stage is seen in Fig. 23-18. The input is essentially a sawtooth waveform applied to the base of amplifier Q_1. The output of this stage is transformer coupled to the base of horizontal output transistor Q_2. Collector current of this stage flows through part of the winding of the horizontal output transformer T_2. This same pulse causes current to flow through the horizontal deflection coils, which are connected, via C_1, across the collector-emitter circuit of Q_2. Because of the inductance of the horizontal deflection coils, the current increases in a linear, sawtooth fashion. This causes the electron beam to move at a uniform rate as it travels horizontally across the screen.

When the collector current reaches a peak value, it falls back to zero

Color

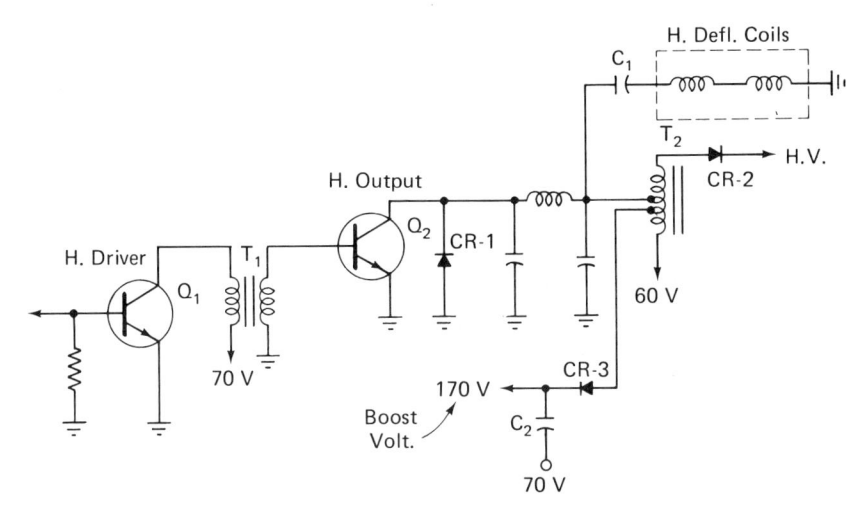

FIG. 23-18 Simplified horizontal output circuit.

and awaits the next pulse. The collapsing field around the lower half of T_2 causes a very high voltage to be developed across the entire winding of T_2. CR_2 provides half-wave rectification of this voltage and feeds it to the high-voltage anode of the kinescope where it is filtered by the capacitance of the CRT. The dc value of this voltage will be 15 to 25 kV or more, depending on circuit constants. This voltage is developed during the retrace time and is referred to as the *flyback voltage*. The transformer is called the flyback transformer.

During the flyback time, the energy stored in the magnetic field around the horizontal deflection coils would tend to produce damped oscillations in the *LC* circuit consisting of the output transformer and the distributed capacitance that is present. This *ringing* would interfere with the generation of the next sawtooth wave and is prevented by *damper diode CR_1* (see Fig. 23-18).

During the flyback interval the collapsing magnetic field induces a relatively large emf in the bottom part of T_2. This is rectified by CR_3 and added to the dc charge on C_2, which in this case is 70 V_{dc}. In the illustration shown, 100 V_{dc} is developed by this rectifying action. The net result is a 170-V_{dc} boosted emf that supplies certain output and CRT circuits.

When a TV set is serviced, care must be exercised to discharge this circuit after the power has been turned off. Unless discharged, this high voltage may exist for several hours before finally leaking off.

23-24 COLOR

Light waves are a form of *radiant energy* that travels through space. They differ from radio waves in frequency or wavelength. Light waves are measured in *millimicrons* [1 micron (μ) equals 1×10^{-6} m]. Therefore, 1 millimicron (mμ) is one thousandth of a millionth of a meter. Red light, for example, has a wavelength of about 700 mμ, green light about 530 mμ, and blue light about 450 mμ.

These are the *primary colors* (hues) used for television. If these colors exist in their pure, or saturated, form and are mixed together in the correct proportions, any desired color can be produced. For example, if blue and green are added, with no red, a blue-green, called cyan, will be produced. Green and red, with no blue, will result in yellow. With proper levels of blue, green, and red displayed simultaneously, all color perception in the eye is canceled and only white is seen.

To transmit a picture in full color, it is necessary to scan it for its *blue*, *green* and *red* content. When the same percentage of each color is projected on a screen at the same time, and in the correct places, the original picture is reproduced. Bright white areas in the picture will be projected as strong pure green, red, and blue colors. Black (dark) areas will be projected with no green, red, or blue. By proper combinations of hues and intensities, any desired color can be reproduced, as well as white, grays, and black.

The drawing in Fig. 23-19 represents an approximation of the color distribution that can be produced with an electron beam striking the colored

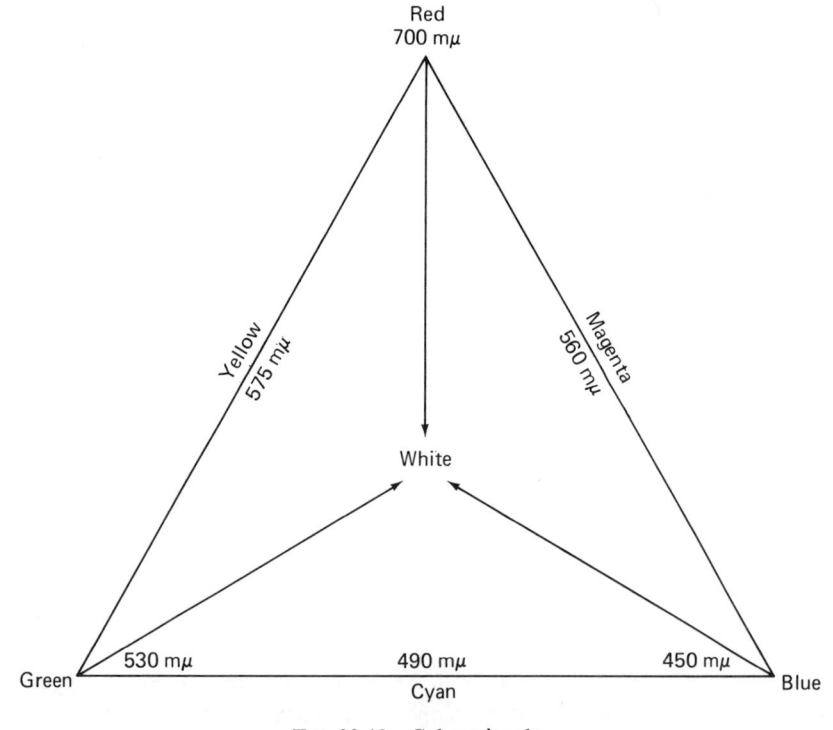

FIG. 23-19 Color triangle.

Transmitting Color Signals

phosphors of a TV kinescope. The amount of color saturation possible is equal to or better than that attainable with printing inks. When the transmitting and receiving equipment are properly adjusted, lifelike color pictures can be received.

23-25 TRANSMITTING COLOR SIGNALS

There are several stringent requirements of the composite color signal. For example, it must not only carry all the color information, but it must be *compatible* with black and white (monochrome) television. In other words, the composition of the color signal must be such that it can be received in a monochrome receiver in black and white without any receiver modification. Also, the *chrominance* (color) part of the signal must be transmitted in such a manner that no degradation of the quality of the picture will occur when a monochrome receiver is tuned to a TV station transmitting a color picture.

The chroma information must be transmitted in terms of the three primary colors: red, green, and blue. The three physical aspects of these colors—*brightness*, *hue*, and *saturation*—must also be transmitted, because the eye sees color in terms of these aspects.

To maintain compatibility with black and white TV, the monochrome signal specifications have to be retained. To these must be added the chrominance and *luminance* (brightness) information. This latter signal is much the same as the video signal used in standard monochrome transmission.

A typical tricolor camera is shown in block diagram form in Fig. 23-20. Three camera tubes, each sensitive to one of the three primary colors, are used in conjunction with two *dichroic mirrors*. A dichroic mirror permits all

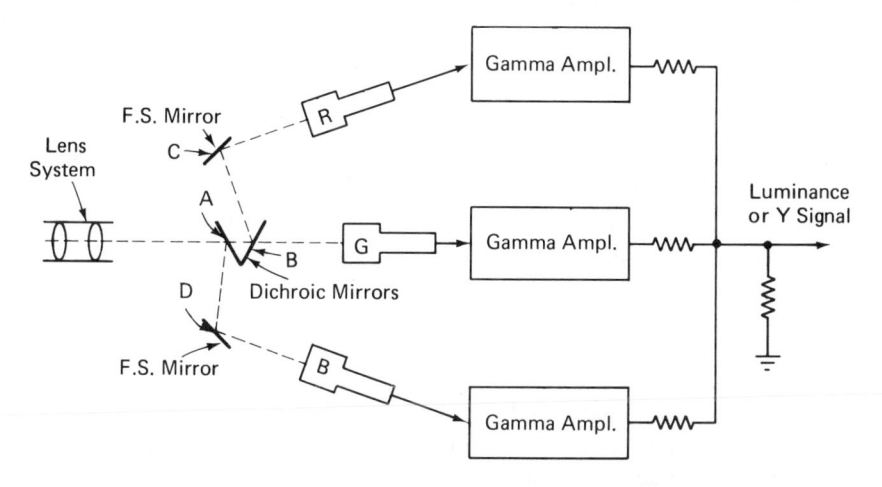

FIG. 23-20 Block diagram of color camera and associated circuits used to produce luminance dry signal.

the light frequencies to pass through except the particular primary color it is designed to reflect. By using this type of mirror, white light can be separated into the three primary colors. The blue dichroic mirror, position A, passes all light frequencies but blue. The tilt angle of the mirror at position A is such that the blue light is reflected to a *front-surface* (F.S.) mirror at point D, which reflects the light onto the face of the blue camera tube.

The light that passes through the dichroic mirror positioned at A goes on to the second dichroic mirror positioned at B. This mirror passes all light frequencies except red, which is reflected to a front-surface mirror at position C, which in turn reflects the light onto the face of the red camera tube.

With both the red and blue colors removed from the incoming light only green remains, which passes directly to the face of the green camera tube. By this scheme, the incoming light is broken up into the three primary colors.

The output of the camera tubes is amplified and *gamma corrected*. Gamma correction is necessary to compensate for the nonlinear operation of the picture tubes. That is, the camera tubes produce a brightness output that does not correspond to the brightness levels recognized by the human eye. The gamma-corrected video signals are then fed to a *matrix* represented by the resistive network shown in Fig. 23-20. This combined signal constitutes the luminance, or Y, output.

Definite proportions of the color signals from each camera are used to form the luminance signal. These are 59 per cent of the green signal, 30 per cent of the red signal, and 11 per cent of the blue signal. These percentages also provide a good reproduction of white, grays, and black when viewed on a monochrome receiver. It is this luminance signal that is used to amplitude modulate the TV transmitter carrier directly.

To transmit a color picture (chrominance information) without increasing the bandwidth, it is necessary to use a *subcarrier*. The frequency of this subcarrier, 3.579545 MHz, was carefully selected so that its *sideband energy would interleave with the sideband energy of the luminance signal*. The selection of this frequency made it necessary to slightly change the horizontal and vertical scanning frequencies for color transmission and reception to 15,734.26 and 59.94 Hz, respectively. These frequencies are so close to the ones used for standard monochrome that the TV receiver oscillators will hold synchronism without adjustment of the sync controls.

Another matrix in the transmitter combines the three color signals in the proportions of 60 parts red, 32 parts blue, and 28 parts green. This composite signal is known as the $B-Y$, the blue-chrominance, or the I (in-phase) signal.

A third signal is generated within the transmitter, which is known as the $R-Y$, red-chrominance, or Q (quadrature) signal. This is made up of 54 parts green, 31 parts blue, and 21 parts red.

The subcarrier is modulated by both the $R-Y$ and $B-Y$ signals. The $R-Y$ signal is first fed through a 0- to 0.5-MHz low-pass (LP) filter to remove all

Transmitting Color Signals

higher-order frequencies. The signal is then fed to a *balanced modulator* that also has the 3.579545-MHz subcarrier fed to it. Because of the balanced modulator action, the carrier is canceled while the sidebands remain. These sidebands range from 3.08 to 4.08 MHz (the approximate subcarrier frequency of $3.58 + 0.5$ MHz and $3.58 - 0.5$ MHz). These sidebands are then passed through a 3- to 4.2-MHz band-pass filter to remove any undesirable modulation products.

The $B-Y$ signal also modulates the subcarrier. However, it is first passed through a 0- to 1.5-MHz LP filter to remove higher frequencies. Also, the 3.58-MHz subcarrier is shifted 90° before it is modulated. This creates a *quadrature relationship* with the carrier modulated by the $R-Y$ signal. When these signals ($R-Y$ and $B-Y$) are added to the Y signal, the two sets of sidebands are 90° out of phase. A form of *phase modulation* is produced as a result of this. The phase angle is determined by the relative amplitude of the two signals. *At the receiver, these out-of-phase signals will represent the various colors or hues* to be displayed. The amplitudes of these signals will vary the saturation, or intensity, of the color.

Little reference has been made to the green signal. It might seem that it has been forgotten, but this is not so. Since the $R-Y$, $B-Y$, and Y signals contain components of all three colors, the receiver has circuitry that combines these signals to produce $G-Y$ information (green chrominance), even though this is not transmitted as such.

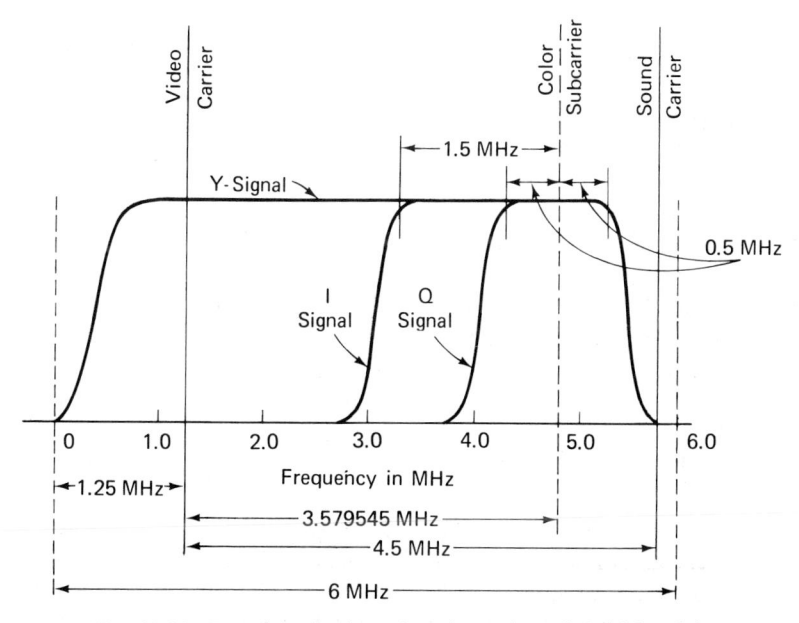

FIG. 23-21 Location of color subcarrier and I and Q sidebands in a 6-MHz TV channel.

The relative positions of the video carrier and its sidebands, the subcarrier and the I and Q sidebands, and the sound carrier can be seen in Fig. 23-21. The subcarrier is shown by a dotted line to imply that it is suppressed and not transmitted. All signals fall within the FCC allotted 6-MHz channel width.

23-26 COLOR SYNCHRONIZATION

The chrominance signal is *constantly changing phase* with every change of the hue of the color it is representing. The particular hue being transmitted at a given moment is represented by the *phase difference* between the chrominance signal and the output of the subcarrier generator. For the receiver to detect these signals it must have some method of comparing the phase of the chrominance signal with a fixed reference phase that is identical to that of the subcarrier at the transmitter. To accomplish this, a *reference phase* is provided in the receiver by an oscillator synchronized to the subcarrier generator. Synchronization is accomplished by transmitting a *colorburst signal* during the horizontal blanking period. The burst consists of a minimum of eight cycles at 3.579545 MHz.

The colorburst is located on the *back porch* of the horizontal blanking pulse, as seen in Fig. 23-22. By placing the colorburst at this position, it will

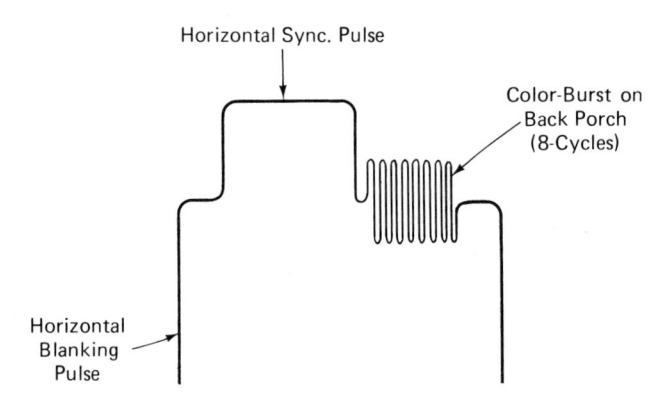

FIG. 23-22 Location of color-burst or back porch of horizontal blanking pulse.

not affect the operation of the horizontal oscillator circuits, because the horizontal circuits used in receivers are designed for immunity against noise or pulses for a short time after being triggered. The voltage level of the colorburst is the same as that of the blanking level, so it does not produce any signal on the picture tube during retrace time.

Circuits in the color receiver are designed to detect the colorburst signal from the other signals being transmitted. The burst is used to synchronize the color section of the receiver in much the same manner as the horizontal and vertical pulses are used to sync the horizontal and vertical sweep circuits.

23-27 COLOR TELEVISION RECEIVER

A color TV receiver includes all the circuits of a monochrome receiver plus the chrominance signal circuits. However, in the color receiver the *bandwidth* of the RF circuits is of greater importance. This is because the sideband frequencies around 3.58 MHz, which contain the higher video frequencies, are important for the chrominance information in the picture. If the fine-tuning control on the RF tuner is not exactly tuned, these higher frequencies will be partially discriminated against, and the color will be impaired.

A partial block diagram of a color receiver is shown in Fig. 23-23. Only the basic chrominance and luminance blocks are indicated. The upper part of the drawing indicates those circuits essential to provide luminance information; the lower part represents the chrominance section. The function of these blocks will be explained in the following paragraphs.

The output of the video detector is the composite signal sent from the transmitter. This includes the luminance signal, two sets of chrominance sidebands, and a colorburst of at least eight cycles of a 3.58-MHz subcarrier. These signals are all passed on to a video amplifier. Part of the output of this stage is fed through a 1-μs *delay line* to the luminance amplifier. The delay line is usually a short length of special coaxial cable. The purpose of the delay line is to introduce a delay approximately equal to the transit time of the chrominance signals through their respective narrow passband filters, which have considerable capacitance and inductance. This assures that the luminance signal will be in the proper phase relationship with the chrominance signals at the kinescope.

The luminance amplifier is wideband, passing all frequencies from 0 to 4.2 MHz. This Y signal produces the black and white pictures when only monochrome signals are being received. The Y signal also is fed to the red, green, and blue matrix and their adders (not shown). The intensity or strength of this signal varies the brightness of the picture display on the screen.

Part of the output of the video amplifier is fed to the 3.58-MHz chrominance band-pass amplifier. This stage passes all frequencies between 2 and 4.2 MHz, which contain the $R-Y$ and $B-Y$ sidebands of the subcarrier. The output of this amplifier is fed to the I and Q color demodulators. The signal level coming out of this stage controls the saturation of the color picture. It is sometimes called the *color amplifier*.

Observe that another part of the video amplifier output (Fig. 23-23) is fed to the *burst separator*. Remember that the burst signal is riding on the back porch of every horizontal sync pulse. This stage is a 3.58-MHz amplifier tuned to the color subcarrier frequency, but keyed on only during the hori-

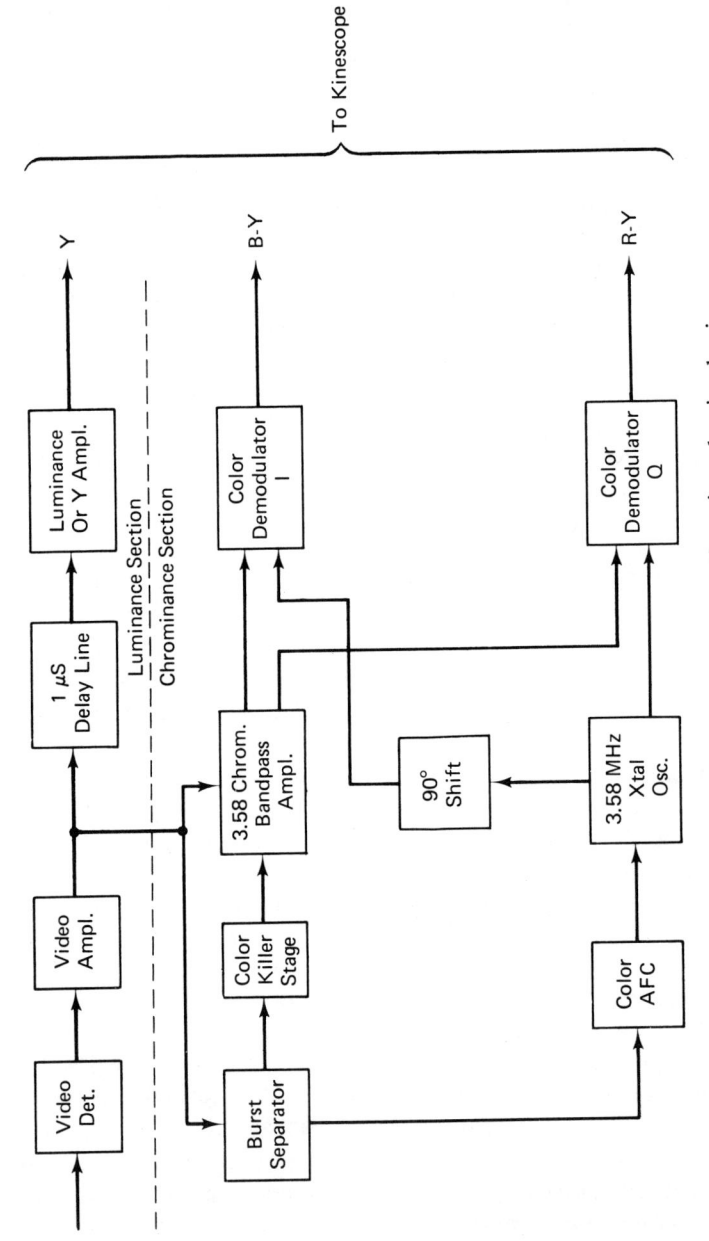

FIG. 23-23 Partial block diagram of a color TV receiver showing luminance and chrominance sections; RF, sync, sound, deflection and power supplies not shown.

zontal flyback time. During the time the beam is scanning each trace, the burst separator is cut off. Therefore, output is realized at 3.58 MHz only for the color sync burst during the flyback interval.

Even though the burst separator is a tuned amplifier at 3.58 MHz, like the chrominance amplifier, it conducts at a different period of time. As previously stated, the burst separator is on during flyback time and off during trace time. The chrominance amplifier is only on during trace time.

Part of the burst separator output goes to the color AFC circuit to furnish the 3.58-MHz synchronizing voltage for the *crystal subcarrier oscillator*. Although the oscillator is accurately controlled by its crystal for stability, its *phase* must be precisely controlled for correct hues in the picture. Hence, the crystal oscillator is locked in the correct phase by the burst signal.

The burst separator also supplies a signal to the *color-killer stage*. The name describes its function, that is, to cut off the chrominance amplifier when monochrome programs are received. With the chrominance amplifier disabled, there is no input to the color demodulators, and consequently no color signals to the picture tube. The purpose is to prevent noise generated in the color amplifier from appearing as colored snow, or confetti, on the screen. For a color program the color-killer stage does not function.

It is necessary to use two color demodulators to detect the two different phases of the modulated 3.58-MHz chrominance signal. Output from the chrominance band-pass amplifier feeds the two demodulators, which have different phases of input voltage from the 3.58-MHz crystal oscillator. This injected oscillator voltage beats with the sideband frequencies in the modulated chrominance signal. The demodulators must be able to detect both the *amplitude* and *phase* of the color signal. To reproduce the original colors, the demodulator must convert the chrominance signal into the I and Q signals.

The I and Q signals from the demodulators are mixed in *matrices* (not shown in Fig. 23-23) in proper proportions to produce a result equivalent to the green, red, and blue signals picked up by the camera tubes. These three signals can now be used to key the three separate scanning beams of the kinescope.

23-28 THREE-GUN COLOR PICTURE TUBE

The color picture tube, or kinescope, must be able to produce an image in accordance with the color variations and brightness of the original televised scene. Some of the characteristics of such a tube are three beams originating from three electron guns, which must strike only the correct set of phosphor dots; a phosphor-dot combination screen, consisting of three colors that cover the entire rear surface of the screen; and a shadow mask designed to allow each beam to only strike the correct set of colored phosphor dots.

The three electron guns are identical. Their principle of operation is the same as any CRT, except that extra elements have been added to achieve proper focus and control of the beams. Usually the three guns are symmetrically displaced around a central axis of the gun assembly. One of the critical requirements of the color tube is that the beams must be accurately aligned so that each beam will only strike its respective colored phosphor dot. Another requirement is that all three beams must be made to pass simultaneously through any given hole in the shadow mask and strike their respective phosphors. A simplified sketch showing the relative positions of the electron guns, shadow mask, and phosphor dots is given in Fig. 23-24. Not shown are the convergence coils, magnets, and deflection coils necessary for proper operation.

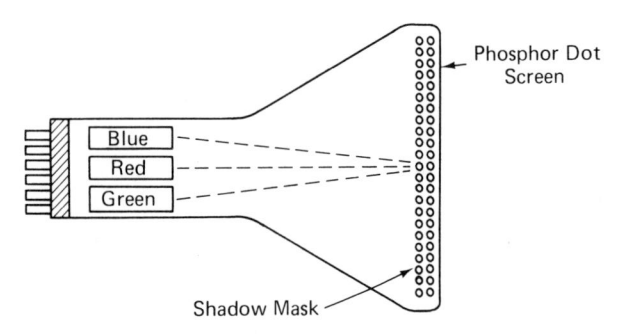

FIG. 23-24 Sketch of three-gun color kinescope.

The phosphor dots are deposited on the rear surface of the viewing screen in a very precise manner. They are arranged in groups of *three dots*, one for each primary color. An enlargement of a very small part of the screen is shown in Fig. 23-25. Notice the triangular pattern of the three color-dot grouping. These are referred to as *trios* or *triads*. The dotted triangles have been added to aid the reader in identifying the three color groups, It must be mentioned that even though the dots are placed very close together, they do not overlap or touch. Each dot glows individually when struck by its respective beam. The characteristics of the human eye are such that the light emitted from the three dots (in a given triad) cannot be distinguished individually at normal viewing distances. When properly viewed, the light from the three dots blends together to give the appearance of a single hue or color.

The *shadow mask* is a thin metal sheet with numerous tiny holes each in line with a dot triad. For a 21-in. picture tube there are about 400,000 holes. This means that there are three times that number of phosphor dots, as there are three dots per triad. The mask enables each beam to strike its own respective color dots, when the beam converges at the proper angle, without striking the other dots in the triad. This action takes place over the entire screen as the deflection yoke deflects all three beams to produce the

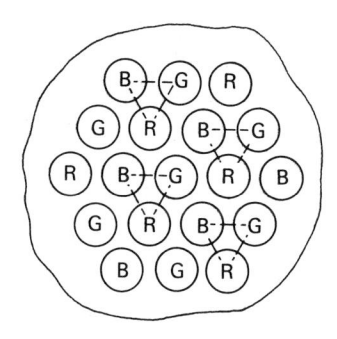

FIG. 23-25 Enlargement of small portion of phosphor dot screen. Note arrangement of triads.

raster. Any electrons that have not been properly converged are blocked by the shadow mask.

Commercial License Questions

Sections in which answers to questions are given appear in parentheses. A bracketed number following a question implies that it applies only to that element.

1. What is meant by *horizontal scanning*? *Vertical scanning*? (23–1)
2. Explain what is meant by *interlaced scanning*. Why is it used? (23–3)
3. Briefly describe the operation of an iconoscope. [4] (23–4)
4. What is the principle of operation of an image orthicon? [4] (23–5)
5. What causes a sticking picture in an image orthicon? How is it corrected? [4] (23–5)
6. Briefly explain the operation of a vidicon. [4] (23–6)
7. Draw a simple block diagram of a monochrome TV transmitter and briefly explain the function of each block. (23–7)
8. What is the function of a sync generator in a TV station? [4] (23–7)
9. Is the video output of a TV transmitter AM or FM? (23–7, 23–9)
10. Explain the function of a diplexer. [4] (23–7)
11. Describe what is meant by vestigial sideband. Why is this used in TV? [4] (23–7, 23–13)
12. What is the function of blanking pulses? (23–9)
13. What is the frequency deviation of the aural carrier corresponding to 100 per cent modulation? (23–10)
14. How can the ERP of a TV transmitter be determined if the antenna gain is known? [4] (23–11)

15. What is the meaning of *camera chain*? (23–12)
16. What is the function of a film multiplexer in a TV studio? [4] (23–12)
17. What are the FCC requirements about keeping a station log? (23–14)
18. What are the bandwidth requirements of a TV channel? (23–15)
19. Describe the functions performed by the video detector. (23–18)
20. Explain the operation of a keyed AGC circuit. (23–19)
21. Explain the purpose of the sync separator in a TV receiver. (23–21)
22. How is boosted voltage obtained in a TV receiver? (23–23)
23. What is the function of dichroic mirrors in a color TV camera? [4] (23–25)
24. What is gamma correction in a TV color camera? [4] (23–25)
25. Explain the function of a subcarrier in color transmission. [4] (23–25)
26. What is the colorburst signal used for? (23–26)

24

Generators, Motors, and Batteries

24-1 BASIC DIRECT-CURRENT GENERATOR

Many radio stations have a generator to furnish auxiliary power in the event the main electrical service fails. Electrical motors are used in a wide variety of applications from driving large radar antennas to small units for operating band-switching mechanisms. Batteries are used as a source of standby power for many kinds of communication equipment. This chapter will cover the basic operation of these devices and essential maintenance procedures.

A dc generator is a rotating machine that converts *mechanical energy* into *electrical energy*. This is accomplished by rotating a coil(s) through a magnetic field. A voltage is induced into the coil that is proportional to the

1. Number of turns in the coil.
2. Strength of the magnetic field.
3. Speed of the coil cutting through the magnetic flux.
4. Angle at which the coil passes through magnetic lines of force.

A simple generator is shown in Fig. 24-1. The north and south poles of a permanent magnet are arranged so that the coil may rotate through the flux established by them. These are referred to as *field poles*. The coil is mounted on a shaft, supported by bearings at each end, that is connected to some prime mover. The two ends of the coil are fastened to curved copper strips called *commutator segments*. Riding on these are *carbon brushes* that provide the electrical connections to the load.

When the rotating coil is parallel to the field poles, the largest emf is generated in it, as it is *cutting through* the greatest number of magnetic lines

FIG. 24-1 Simple dc generator.

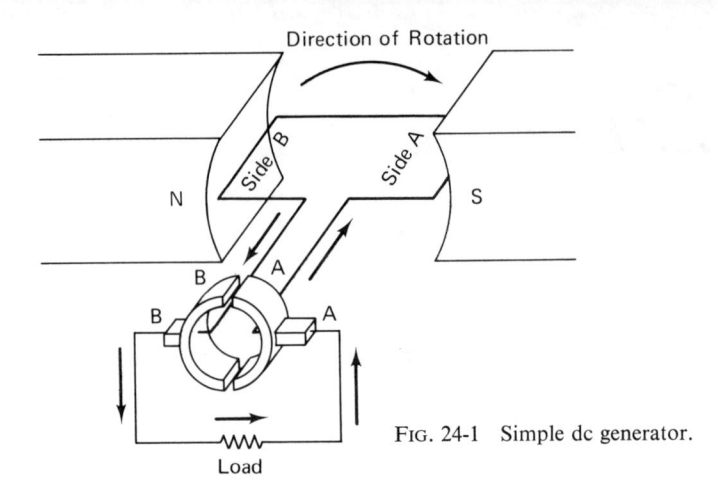

of force. The direction of the resultant current is dependent upon the direction of rotation of the coil.

When the coil has rotated 90° from the position shown in the illustration, it is momentarily parallel to the magnetic flux, and *no lines of force* are being cut. Hence, no emf is induced in the coil. When the coil rotates another 90° (in the same direction), maximum voltage is again induced in it, but of the *opposite polarity*. Hence, the brushes are now in contact with the opposite commutator segments. This causes the generated current in side *A* of the coil to flow *in the same direction* through the load. From this we can see that the commutator is a form of *rotating switch* connecting the output of the coil to the load at precisely the right moment, so that current always flows in one direction through the load.

The current produced by a dc generator is not constant, even though it always flows in the same direction. This is because the magnetic flux between the poles is strongest in the center and gets weaker near the edges. The graph in Fig. 24-2 shows this variation as the coil rotates through 360°.

Since the current increases and decreases, it is called a *pulsating current*.

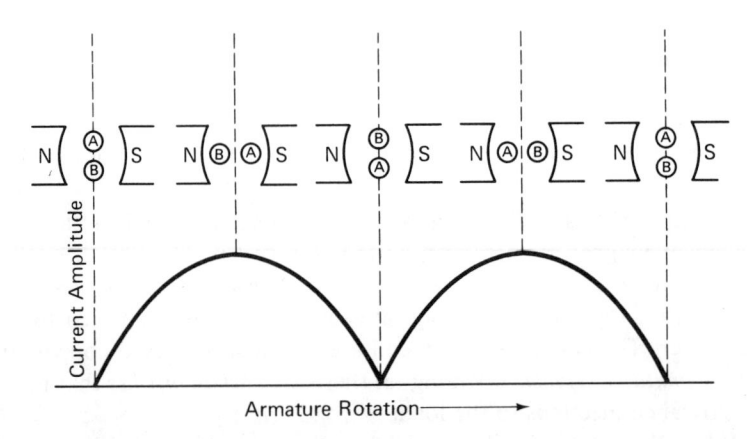

FIG. 24-2 Variation of output current from dc generator with only one pair of poles.

Practical Generators 573

However, it is still direct current because it always flows in the same direction. The number of pulses that occur every second is called the *ripple frequency*.

24-2 PRACTICAL GENERATORS

Practical generators have more than one pair of poles and many turns on each coil. Also, instead of permanent magnets, *electromagnets* are used. By controlling the current through them, the output voltage can be established. A *four-pole* generator, with armature removed, is shown in Fig. 24-3.

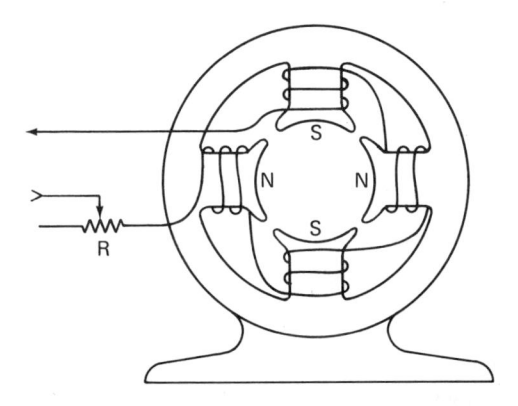

FIG. 24-3   Four-pole generator with armature removed.

The rheostat R controls the current flowing through the field coils. A part of the generator's output can be connected to the field windings, providing *self-excitation*. This is possible because a certain *residual magnetism* remains in the field poles even when it is not operating. This causes a weak initial emf to be generated, which establishes some current flow through the windings. The strengthened field induces more emf in the rotating coils. The cycle continues until the full generator output voltage is reached.

Sometimes self-excited generators will not build up. This may be due to insufficient residual magnetism. By momentarily connecting an external voltage of correct polarity across the field windings, the residual magnetism will be strengthened and the generator will build up. This is called *flashing the field*.

Generators (as well as motors) have the following losses:

1. I^2R in the windings.
2. Eddy currents in the armature and field poles.
3. Hysteresis.

It is necessary to make the armature and poles of *thin laminated pieces of soft steel* coated with a thin layer of lacquer to minimize eddy currents. Otherwise, the currents generated in these members, as a result of the armature turning at high speeds, would cause considerable heat.

24-3 GENERATOR TYPES

Generators are classified according to the way their fields and armatures are connected. They are *series, shunt,* and *compound.* The schematic of a series-wound dc generator is shown in Fig. 24-4. The field coils consist of a

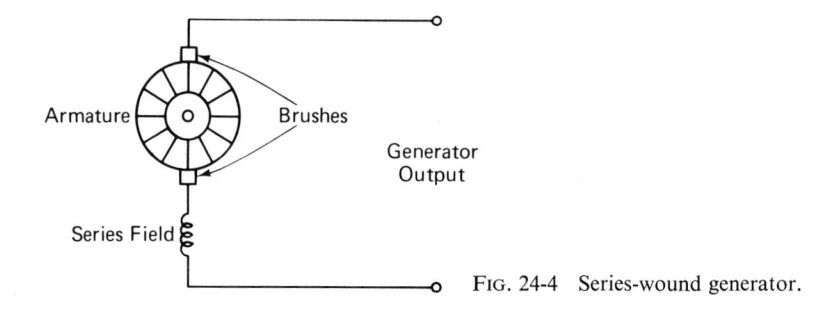

FIG. 24-4 Series-wound generator.

few turns of large wire in series with the armature. The output voltage is largely dependent upon load, being low when the load is low. Because of this undesirable characteristic, series-wound generators can only be used when the *load is fairly constant.*

The shunt generator shown in Fig. 24-5 has field coils consisting of many turns of small wire connected in shunt with the armature and load. The relatively large resistance of these turns results in a field current approximately 5 to 10 per cent of the total armature current. The field current is *essentially constant* for normal variations of load. Hence, the generator is essentially a *constant potential* machine that delivers current according to load.

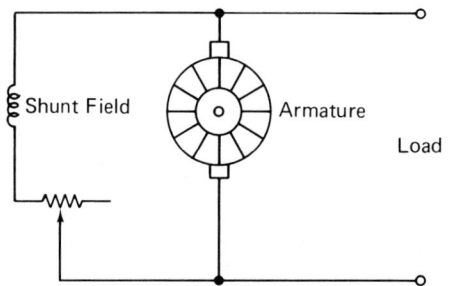

FIG. 24-5 Shunt-wound generator.

Compound generators have *series* and *shunt* field windings on the same poles. The schematic diagram is shown in Fig. 24-6. This technique incorporates the desirable features of the series and shunt generators.

The effect of the series winding is to increase the field flux with increased load. The extent of the increased flux depends on the degree of saturation of the magnetic circuit as determined by the shunt field current. Thus, the terminal voltage of the generator may increase or decrease with load, depending

Direct-Current Motors 575

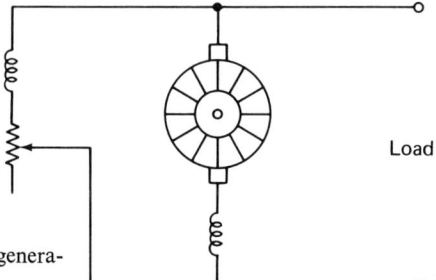

FIG. 24-6 Compound-wound generator.

upon the influence of the series field coils. This influence is referred to as the *degree of compounding*.

Generators require some method to control their output voltage. When the load is relatively constant, a *manual* control is usually adequate, as represented by the rheostat in Figs. 24-5 and 24-6. In old installations a third brush was sometimes used. Its position on the commutator determined the amount of current flow through the field windings.

When the load changes are frequent and large, it is necessary to use some form of automatic voltage regulation, such as the carbon pile regulator.

24-4 DIRECT-CURRENT MOTORS

The operation of a dc motor depends on the principle that a current-carrying conductor in a magnetic field tends to move at *right angles* to the direction of the field. This is illustrated in Fig. 24-7a, where the field around the conductor is aiding or strengthening the field at the bottom of the pole pieces and opposing it at the top. This creates a *torque*, forcing the conductor

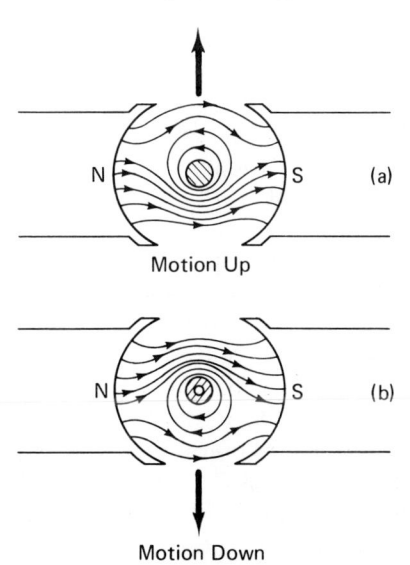

FIG. 24-7 Motion of a current-carrying conductor in a magnetic field.

up. If the current is reversed through the conductor, an opposite condition results, and the conductor is forced down, as shown in Fig. 24-7b. This is the underlying principle of motor action.

The essential parts of a simple dc motor are shown in Fig. 24-8. A loop

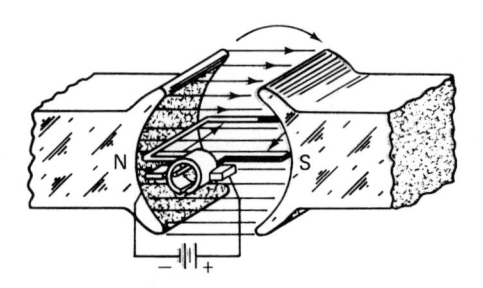

FIG. 24-8 Basic dc motor action.

of wire acting as an *armature* (bearings not shown) is designed to rotate in the magnetic flux established by the field poles. The ends of the loop are connected to a commutator–brush assembly. The battery forces current through the armature, via the brushes and commutator. The flux created by the armature current creates a torque, and the armature begins to rotate. It would seem that the armature would stop when it rotated 90° from the indicated starting position. Its inertia carries it beyond this point and causes each commutator segment to contact the other brush. Current will still flow into the left-hand brush and out the right, resulting in a magnetic flux about the armature that continues the rotation.

A practical motor has many coils of wire in the armature winding. The armature has many slots into which are inserted many turns of wire. This increases the number of armature conductors and thus produces a greater and more constant torque.

In each motor there is some *generator action*. As the armature revolves through the field flux, a voltage is induced in it opposite to the impressed emf. This *counter electromotive force* (cemf) is directly proportional to the number of turns on the armature and its speed, together with the strength of the field between the poles.

Normally, the pole pieces of a motor contain field coils whose flux is determined by the number of turns and strength of current through the windings. By reversing the connections of either the field winding or armature, the motor will turn in the opposite direction.

A series-wound dc motor is shown schematically in Fig. 24-9. The rheostat serves as both a *speed control* and a *starting control* device.

When a series motor is started, some means must be used to *limit the starting current*. At zero speed there is no cemf, and since the ohmic resistance of the motor is low, the inrush current would be enormous as soon as it was connected to the line. By setting the rheostat or *starting resistance* to its maximum value, the current can be limited during the starting operation. As

Direct-Current Motors

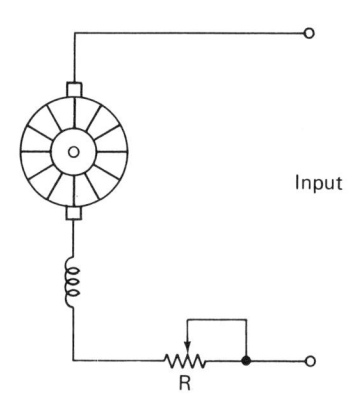

FIG. 24-9 Schematic of series-wound dc motor with speed and starting control.

the motor picks up speed, the starting resistance can be reduced. This operation continues until the motor is up to speed. If the motor is under a heavy load and is started too slowly, the starting resistance may burn out. If the load is removed completely, the speed will become dangerously high and the motor will race to destruction. Series motors, therefore, should *never* be used in any application where it is possible for it to operate without load or with a belt drive, as belt failure would remove the load.

Since the speed of the series motor changes with load, it is not suitable for applications where the load is variable or constant-speed drive is essential. This disadvantage is offset, at least partially, by the unusually good torque characteristics of this type of motor.

A series-wound motor can also be used on alternating current if there are sufficient turns on the armature and field coils. For this reason series-wound motors are frequently called *universal motors*. Most small electrical appliances have this kind of motor.

The field circuit of a *shunt motor* is connected across the line and is thus in parallel with the armature, as shown in Fig. 24-10. It consists of many

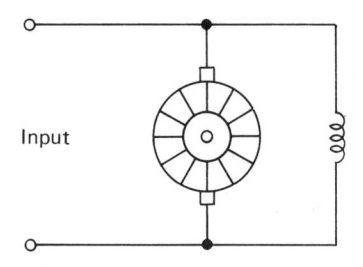

FIG. 24-10 Schematic diagram of shunt-wound motor.

turns of small wire whose current is about 5 per cent of the total motor current.

If the supply voltage is constant, the current through the field coils, and consequently the field flux, will be constant. When there is no load on the shunt motor, the only torque necessary is that required to overcome bearing friction and windage loss. The rotation of the armature coils through the field flux establishes a cemf that limits the armature current to the relatively small value required to establish the necessary torque to run the motor on no load.

When a load is applied to the shunt motor, it tends to slow down slightly. The slight decrease in speed causes a corresponding decrease in cemf. Since the armature resistance is low, the resulting increase in armature current and torque is relatively large. Therefore, the torque is increased until it matches the resisting torque of the load. The speed of the motor then remains constant at the new value as long as the load is constant. The shunt motor is considered as nearly a *constant-speed motor*.

The schematic diagram of a compound-wound dc motor is the same as Fig. 24-6, and has two separate field windings on each of its field poles. One is connected in parallel with the armature and the other is in series with it. This gives the motor the characteristics of both the series- and shunt-wound motor. The operating characteristics of the compound motor lie between those of the series and shunt motors. By properly proportioning the two field windings, the motor can be made to approach the characteristics of either the series or shunt motor.

The commutation process in motors and generators is likely to cause some sparking and thus generate RF interference. By using suitable RFCs in series with the device and appropriate bypass capacitors, the interference can be kept out of electronic equipment.

24-5 MOTOR GENERATORS

In some communications installations it is necessary to convert electrical energy from one value to another, such as 12 V_{dc} to a high dc voltage to supply the plate potentials of transmitting and receiving tubes. Such a device is called a *dynamotor*. In another case it may be necessary to convert alternating current to direct current to operate the filaments of large transmitting tubes. Customarily, this would be accomplished by transformer rectifiers. In older installations, when high current was needed, this was achieved by an ac motor driving a dc generator. This kind of unit is called a *rotary converter*. Some applications exist where an inverter is needed. This is the reverse of the rotary converter. Direct current is fed into the device and alternating current is taken out. These units are all forms of motor–generator sets. Generally speaking, these are large, heavy units that are noisy and costly to purchase and install. As with any other piece of rotating machinery, a certain amount of maintenance is necessary.

In many older mobile installations dynamotors were used to provide the necessary dc-to-dc conversion. These are generally small units providing relatively low power output. They are essentially a motor and generator built into one unit having two or more windings on one armature. There is one commutator for each winding. The commutators are usually located at opposite ends of the shaft, and the windings are customarily located in the same armature slots. Since both input and output windings are on the same armature, a *single field* common to both motor and generator is used.

A disadvantage of the dynamotor is that it is nearly impossible to change the value of the output voltage. True, a speed-control device, such as a rheostat, can be installed in series with the motor and thus change the motor speed and generator output voltage but optimum efficiency would be sacrificed and poor voltage regulation would result.

24-6 ALTERNATORS

Alternating-current generators are divided into three main types, the *single-phase*, *two-phase*, and *three-phase* generator. Commercial ac generators are usually called *alternators*, since they generate alternating current.

If the commutator of a dc generator were removed and replaced with two *slip rings*, it would supply alternating rather than direct current. A sinusoidal output would result as the coil was continuously rotated through 360°. This generator is a basic single-phase alternator.

The rotating part of a dc generator is called the armature. In an ac generator this rotating part is called the *rotor*. The stationary part of an ac generator is called the *stator*. It compares to the field of a dc generator. The brushes ride on the slip rings and are held in brush holders. The brush holders are fastened to a part of the frame.

The magnetic field of an ac generator results from the flow of direct current furnished by an outside source. The field is said to be *separately excited*.

In the revolving-armature ac generator, the stator provides the electromagnetic field. The rotor, acting as the armature, revolves in the field, cutting the lines of force and producing the desired output voltage. The armature output is taken from the slip rings.

The revolving-armature alternator is seldom used, because its output power is conducted through slip rings and brushes. These are subject to frictional wear and sparking and liable to arc-over at high voltages. Consequently, they are limited to low-power, low-voltage applications.

The *revolving-field* alternator is the most widely used. Direct current from a separate source is passed through windings on the rotor by means of slip rings and brushes. This maintains a rotating electromagnetic field of fixed polarity (similar to a rotating bar magnet). As the rotor revolves, alternating voltages are induced in the stator windings, since magnetic fields of one polarity and then another cut through them. The output is connected through fixed terminals directly to the external loads. This is advantageous in that there are no sliding contacts, and the whole output circuit is continuously insulated, thus minimizing the danger of arc-over.

Alternators are rated in terms of armature load current and voltage output, or *kilovolt–ampere* (kVA) output, at a specified frequency and power factor (usually 80 per cent lagging). For example, a single phase designed to deliver 100 A at 1000 V is rated at 100 kVA. This would supply 100 kW at unity power factor or an 80-kW load at 80 per cent power factor.

A simple two-phase alternator is illustrated in Fig. 24-11a. Two identical pairs of coils are wound on the stator poles. The coils on the horizontal poles are series connected as are those on the vertical poles. Leads from the horizontal poles are arbitrarily designated ϕA and the others as ϕB. By connecting one lead from ϕA to one from ϕB, such as at point X, it is only necessary to have *three leads* coming out of the alternator.

The most common type of *polyphase* (more than one phase) system is

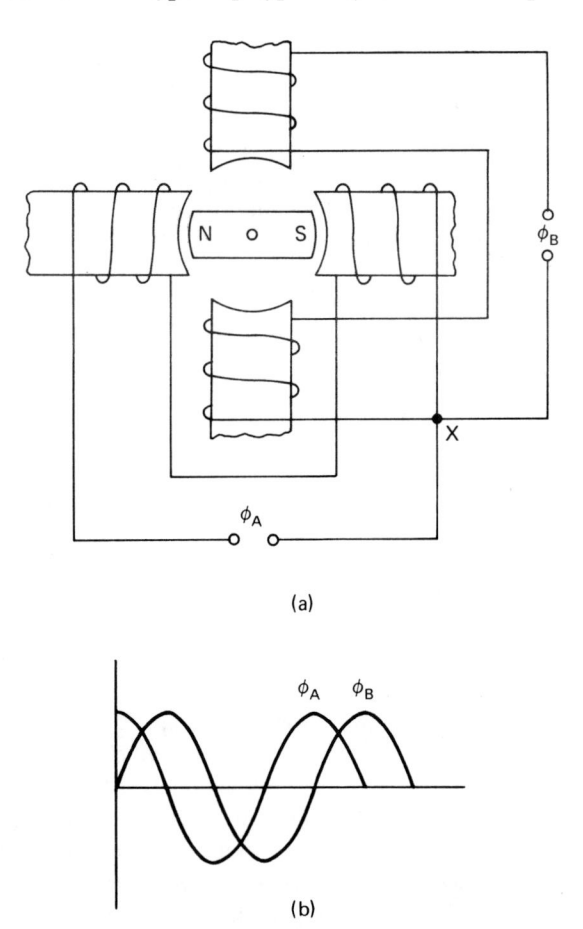

FIG. 24-11 (a) Simple two-phase alternator; (b) waveforms showing 90° displacement.

Alternators

the three-phase system. A simple three-phase alternator has three coils equally spaced (120° apart) around the stator. If each end of each coil goes to a separate terminal, it is a six-wire three-phase alternator. This cumbersome system has been replaced in practice by either a three- or four-wire system. One method of accomplishing this consists of connecting one terminal of each coil at a common point and leaving the other terminal of each coil free. If the three free ends are brought out so that loads may be connected to them, it is called a three-wire system. The schematic diagram of this arrangement is shown in Fig. 24-12a, and is called a *wye* (Y) connection.

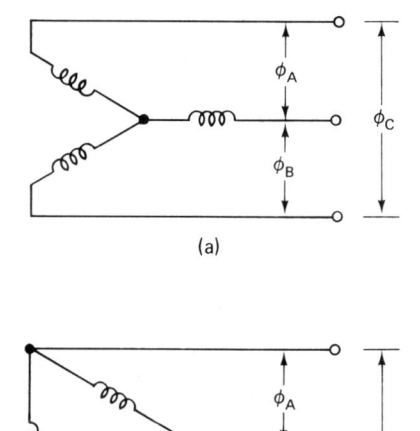

FIG. 24-12 Schematic diagrams of: (a) 3ϕ Y-wound alternator, (b) 3ϕ Δ-wound alternator. Phase representations are arbitrary.

An alternative method of connecting the coils of a 3ϕ alternator is shown in Fig. 24-12b and is called a *delta* connection (Greek letter Δ). No matter how the coils are connected, however, the voltages (and currents) must be 120° out of phase with each other. Either sine waves or vectors may be used to show this phase relationship.

If it is ever necessary to parallel two alternators to accommodate a larger load, it is very important that they be *properly phased*. Suppose that both alternators are single-phase machines. Not only must their *output frequencies* be the same, but they must be in the *same phase relationship*. If the machines are approximately in phase before being paralleled, they will *lock in* when connected to the same load. If the phasing should be far off (i. e., 180°), the alternators will short circuit each other when paralleled. If they are not properly protected by circuit breakers, they may burn out.

24-7 INDUCTION AND SYNCHRONOUS MOTORS

The most widely used polyphase motor is the *squirrel-cage induction motor*. It derives its name from the fact that the rotor is shaped somewhat like a squirrel cage, as illustrated in Fig. 24-13. It consists of a laminated iron

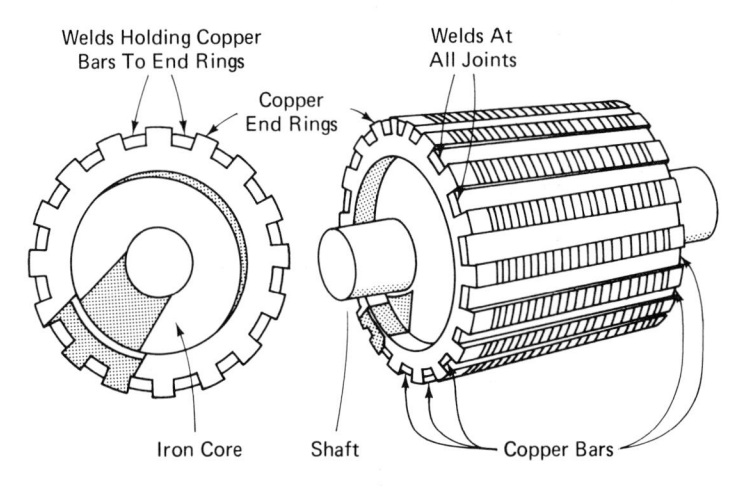

FIG. 24-13 Squirrel-cage rotor.

core mounted on a shaft. Copper bars are laid in slots the length of the core and are welded to copper end rings. Thus, the rotor is an absolute *short circuit*. It is important that these rotors be good short circuits, because currents are produced in the rotor only through mutual induction.

The rotor turns because of the mutual induction that occurs between it and the rotating magnetic field of the stator windings. The currents produced in the rotor conductors set up magnetic fields whose poles are opposite to that of the rotating magnetic field. The rotor therefore is attracted by the rotating magnetic field and rotates in the *same direction* as the field. There is no electrical connection between rotor and stator.

Although the rotor and the stator field rotate in the same direction, there is *relative motion* between the two, because the speed of rotation of the field is greater than that of the rotor. The difference in speed is called *slip*. At no load the rotor runs at almost the same speed as the magnetic field rotates; therefore, the slip is approximately zero. At full load, however, the rotor must run at a speed that is from 3 to 20 per cent slower than the magnetic field rotation to develop the necessary torque.

Squirrel-cage motors are rugged and safe, rugged because of the rotor design and the fact there are no rotating contacts, and safe because there are no sparks that may set fire to flammable fuels.

The *synchronous* speed of an induction motor is directly proportional to the frequency of the input voltage and inversely proportional to half the

Single-Phase Motors

number of magnetic poles around the stator. Expressed mathematically,

$$S = \frac{f \times 60}{0.5P} \quad \text{or} \quad \frac{f \times 120}{P}$$

where S = synchronous speed
 f = frequency of input voltage in hertz
 P = number of poles

Example 24-1

What is the synchronous speed of a motor that has four poles and is fed from a 60-Hz line?

Solution

$$S = \frac{f \times 120}{P} = \frac{60 \times 120}{4} = 1800 \text{ rpm}$$

If the motor has 5 per cent slip under load, the rotor speed would be 90 rpm less.

24-8 SINGLE-PHASE MOTORS

Motors run from a single-phase ac source are called *single-phase motors*. The series universal motor is one type of such motor. Single-phase motors operate on the principle of mutual induction. The difficulty with a single-phase input to an ac motor is that it is impossible to set up a rotating magnetic field with a voltage of only one phase. If the rotor is made to turn by some means, the attraction between the magnetic poles of the rotor and the opposing poles of the stator exerts a torque, and the rotor continues as long as voltage is applied to the stator windings.

Note that a single-phase induction motor cannot start automatically but continues to run if started by some means. The problem, therefore, is to find methods of starting it. For this reason single-phase motors are often classified as induction motors of three general types: *repulsion–induction*, *split phase*, and *shaded pole*.

A repulsion single-phase motor resembles a dc motor in that it has a commutator, brushes, and field windings. The two brushes are placed 180° apart and are short circuited, as shown in Fig. 24-14.

Because of the expanding and contracting single-phase field of the stator, current flows in the rotor through induction. Since the short-circuited brushes furnish a path for the induced current, magnetic poles are created around the rotor. Because this rotor is repelled by the stator, the motor is called a *repulsion motor*. Note that, as the rotor turns, the magnetic poles on the rotor constantly develop a torque due to repulsion by the stator field, and the motor rotates. The speed of a repulsion motor varies inversely with the load. If the load is removed, the motor races.

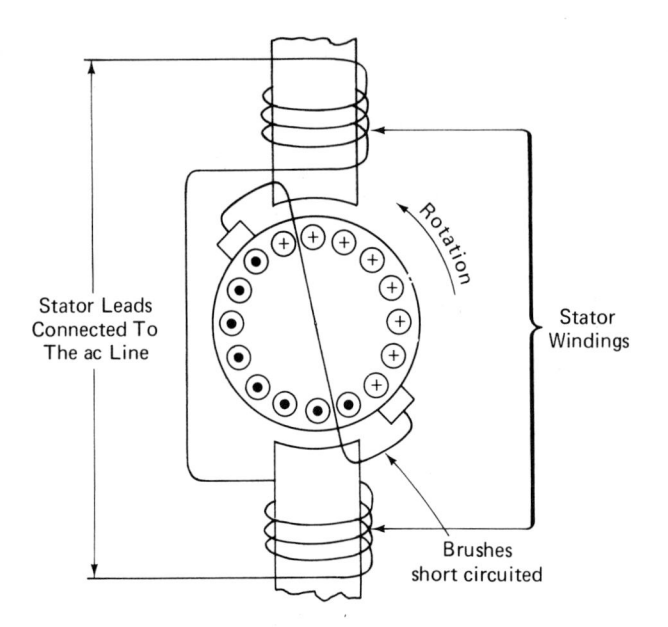

Fig. 24-14 Schematic diagram of single-phase repulsion motor.

If a centrifugal switch is provided that will short circuit the commutator segments after the motor has attained a speed of approximately 75 per cent of rated speed, the motor will continue to run as an induction motor. A repulsion motor with a centrifugal switch is called a *repulsion–induction* motor because it starts as a repulsion motor and runs as an induction motor. Such a motor has a constant speed and is not affected by load as much as the straight repulsion type.

Split-phase motors are motors that start by splitting the single-phase applied voltage into two voltages 90° out of phase with each other. The two phases produce a rotating magnetic field and the motor starts. After the motor has attained sufficient speed, one phase is cut out by a centrifugal switch, and the motor continues to run as an induction motor.

Another type of split-phase motor is the *capacitor* motor in which the two-phase voltages necessary for developing a starting torque are produced by means of a capacitor placed in series with an auxiliary winding. In this type of motor the phase difference between the currents in the two windings is almost 90°, since the capacitor causes the current to lead the voltage, and the inductance of the other winding causes the current to lag the voltage. A greater starting torque is therefore produced by the rotating magnetic field. After the motor reaches about 75 per cent of rated speed, a centrifugal switch cuts out the capacitor and auxiliary winding, and the motor runs as a single-phase induction motor.

Another type of single-phase induction motor is a *shaded-pole motor*. A schematic diagram of such a motor is shown in Fig. 24-15. A slot is cut into

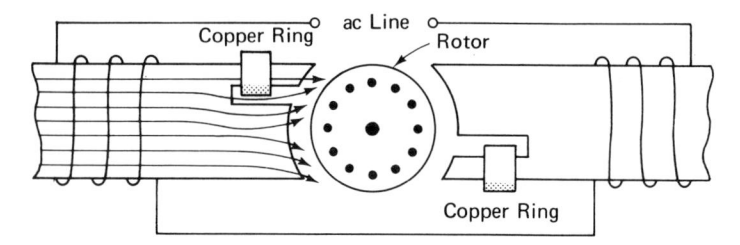

Fig. 24-15 Schematic diagram of a shaded-pole motor.

each pole piece, and short-circuited copper rings are placed over the slotted sections thus formed. The main field flux causes an induced current in the shorted rings that sets up a flux which opposes the main field flux. Thus, most of the main field flux passes through the unshaded portion of the pole piece. The flux produced in the shaded portion lags behind the flux produced in the unshaded portion because of the inductance of the copper rings. This causes a magnetic field, which effectively rotates in the direction of the shaded portion. This rotating magnetic field is very weak and develops very little torque. Shaded-pole stators are therefore used for very small motors [up to about $\frac{1}{20}$ horsepower (hp)] and operate only on alternating current. They are very simple in construction, low in cost, and very reliable.

24-9 MAINTENANCE OF MOTORS AND GENERATORS

There are a few maintenance procedures that, if followed, will provide trouble-free operation. The most important is *lubrication*. The bearings of all rotating machines must be adequately lubricated. An overheated bearing will soon cause serious trouble. If a bearing becomes overheated because of a lack of lubrication, it should be flushed with a light oil until it has cooled down. The machine should be kept in operation during the flushing procedure. Lubrication should then be followed with a normal-weight oil. When lubricating bearings, caution should be exercised so that excessive oil does not drip down on the insulation, which would likely cause it to become soft. Disintegration of the insulation will eventually cause serious trouble. Some small machines may have special bearings that do not require lubrication. These contain a high percentage of oil that is forced out of the pores of the metal when the bearing becomes heated by rotation.

Next to bearings, commutators and brushes are the chief source of trouble in rotating machines. The continual sliding of the brushes against the commutator wears both the brushes and the commutator. The brushes generally wear more rapidly than the commutator or slip rings. Excessive wear, of the brushes can cause sparking during commutation, which may result in burning and pitting of the commutator. If the commutator becomes grooved pitted, or out of round, it is necessary to remove the armature and turn the commutator down on a lathe. Before the armature is installed, make certain that all metal particles that may have been deposited in the mica insulation between segments and around the commutator are removed. It is necessary to *undercut* the mica insulation between commutator segments, and this should be done prior to turning down the armature.

If any of the coils on the armature become shorted, the armature will heat, the brushes will spark, and permanent damage will result. For satisfactory commutation, continuous contact must be maintained between the brushes and the commutator or slip rings. When correct commutation is taking place, the commutator takes on a dark chocolate color. The brushes must be free to slide up and down in their holders and are made to bear upon the commutator or slip rings with a spring. Proper spring pressure is important, as too little pressure causes poor brush contact and unnecessary sparking. Too much pressure causes excessive brush and commutator wear. Excessive sparking at the brushes may indicate shorted turns on the armature or too great a load on the machine.

Motors and generators should always be kept clean and free of dust. If the commutator or slip rings become dirty, they can be cleaned with a heavy piece of canvas. It is also possible to use commercial cleaners, such as trichlorethylene or carbon tetrachloride. The latter is *very toxic* and should only be used in an emergency. Avoid contact with the skin and do not breathe the fumes. If for any reason the commutator or slip rings have become slightly rough, it is possible to sand them smooth with a very fine grade of sandpaper. *Never use emery paper or steel wool*, as the particles that break loose may become lodged between sections of the commutator or other parts of the machine and cause severe shorting. If a motor does not start when connected to the line, check the line voltage, circuit breakers, or fuses. If the line protective devices do not remain closed, it is possible that the rotor or armature is frozen and will not rotate. Also, some of the field or armature windings may be shorted. Poor brush contact may be another reason that the machine is inoperative.

24-10 PRIMARY CELLS

Batteries have played a vital role in communications for decades. Since the advent of transistors, batteries have been used more than ever.

Batteries may be classified according to the design of their cells, that is *primary* or *secondary*, or whether a *dry* or *wet* electrolyte is used. Primary cells *can not be successfully recharged*. Flashlight batteries are examples of these. Secondary cells can be recharged. Automobile batteries are typical of this kind.

A *primary cell* is one in which the chemical action eats away one of the electrodes, usually the negative terminal. This may occur in a relatively short

Primary Cells

or long time, depending upon the amount of current the battery is called upon to supply. When the negative terminal is pretty well eaten away, it and the electrolyte must be replaced or the cell thrown away.

The voltage across the electrodes depends upon the materials comprising them and the composition of the electrolyte. The difference of potential between carbon and zinc electrodes in a dilute solution of sulfuric acid and water is about 1.5 V.

The amount of current a primary, or any, cell can deliver depends not only upon the resistance of the external circuit, but the *internal resistance* of the cell itself. This in turn is a function of the size of the electrodes, the distance between them, and the resistance of the electrolyte. The larger the electrodes and the closer they are together (without touching), the lower will be the internal resistance and the more current it can supply.

The chemical action that takes place in the cell while the current is flowing causes hydrogen bubbles to form on the surface of the positive carbon electrode in great numbers until the entire surface is surrounded. This action is called *polarization*. The hydrogen sets up an emf in the opposite direction to that of the cell, thus increasing the effective internal resistance, reducing the output current, and lowering the terminal voltage. When a cell becomes heavily polarized, it has no useful output.

The common dry-cell battery is a commercial form of the voltaic cell and employs a substance rich in oxygen as a part of the positive carbon electrode, which combines chemically with the hydrogen to form water. One of

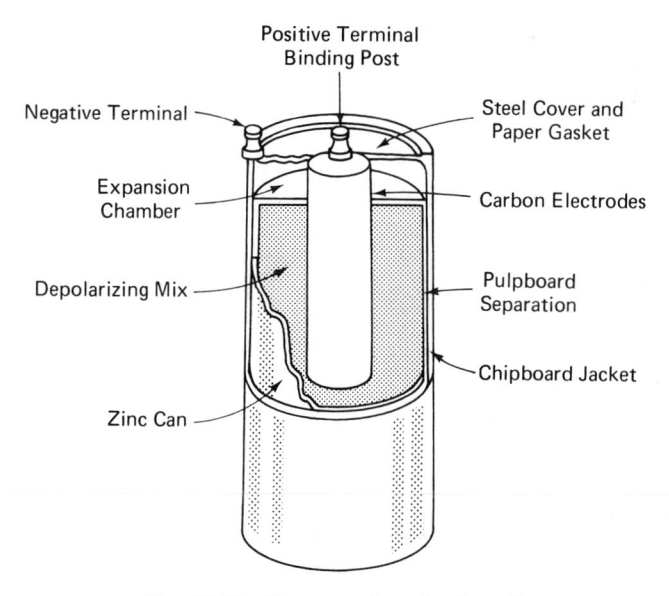

FIG. 24-16 Cut-away view of a dry cell.

the best *depolarizing* agents used is *manganese dioxide*, which supplies enough free oxygen to combine with all the hydrogen so that the cell is practically free from polarization.

When chemically pure zinc is used as an electrode of a primary cell, the zinc is not consumed by chemical action when no external circuit is connected. In short, the cell does not wear out when not in use. However, in most cases zinc electrodes are not chemically pure, but have tiny impurities. These enter into chemical action with the electrolyte to produce small electrical currents around the zinc plate, even when not under load. As a result, chemical energy is consumed and the life of the cell is decreased.

To prevent local action, zinc electrodes are usually *amalgamated with mercury*. When this is done, the zinc dissolves in the mercury, but the impurities do not. As a result, the zinc is free to enter chemical action with the electrolyte during normal operation of the cell (when an external circuit is connected), but the impurities are covered with mercury and do not enter into chemical action with the electrolyte to produce local action. A typical dry-cell battery is shown in Fig. 24-16.

24-11 CELL CONNECTIONS

If a circuit requires more voltage than one cell can provide, several can be connected in series to obtain the desired potential. Figure 24-17a shows four 1.5-V dry cells connected in series to provide 6 V. Observe that the negative terminal of one connects to the positive of the other. The schematic of these cells and a resistive load is shown in Fig. 24-17b.

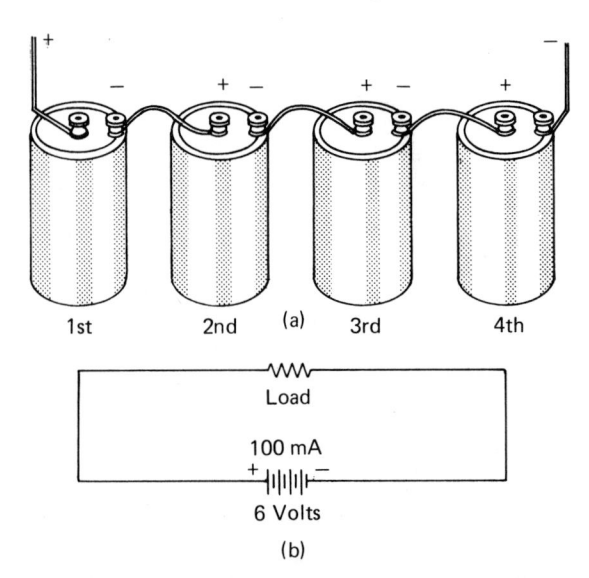

FIG. 24-17 (a) Pictorial view of four series-connected cells. (b) Schematic diagram of same circuit connected to 100mA load.

Cell Connections

Suppose that an electrical load requires 1.5 V at 400 mA. Assume that a number of 1.5-V 100-mA cells are available. To meet this requirement, four of the cells must be connected in parallel. Hence, the output voltage will not be greater than any one cell, but each can deliver 100 mA. Therefore, the four cells can provide 400 mA at 1.5 V, as shown in Fig. 24-18.

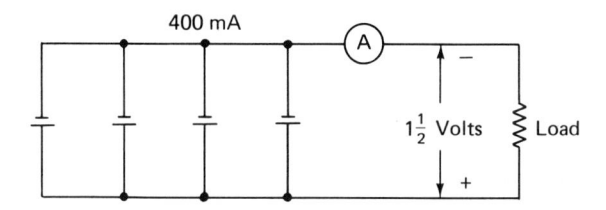

FIG. 24-18 Schematic of four parallel connected cells to 400 mA load.

A *battery* consists of two or more cells connected together, usually in series. Hence, a 9-V transistor battery consists of six 1.5-V cells connected in series. Having all the cells in one container makes for more convenience in handling. An automobile battery contains six cells in series, each supplying 2.1 V when fully charged. Therefore, the battery supplies 12.6 V.

Figure 24-19 shows a battery network supplying power to a load requiring both a voltage and current greater than one cell can provide. To provide the required 4.5 V, groups of three 1.5-V cells are connected in series. To provide the required 0.4 A, four series groups are connected in parallel, each supplying 100 mA of current.

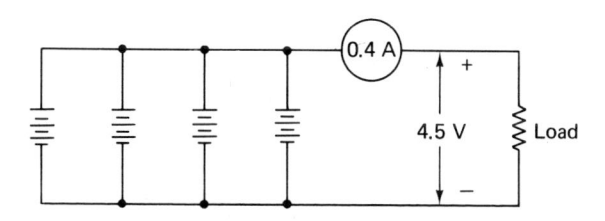

FIG. 24-19 Series-parallel connected cells.

The capacity of a battery is measured in *ampere-hours*. The ampere-hour capacity is equal to the product of the current in amperes and the time in hours during which the battery is supplying this current. The ampere-hour capacity varies inversely with the discharge current. The size of a cell is determined generally by its ampere-hour capacity. The capacity of a cell depends upon many factors, the most important of which are

1. Area of the plates in contact with the electrolyte.
2. Quantity and specific gravity of the electrolyte.

3. General condition of the battery (degree of sulfating, plates buckled, separators warped, sediment in bottom of cells, etc.).

24-12 SECONDARY CELLS

Secondary cells function on the same basic chemical principles as primary cells. They differ mainly in that they may be *recharged*. Some of the materials of a primary cell are consumed in the process of changing chemical energy to electrical energy. In the secondary cell the materials are merely transferred from one electrode to the other as the cell discharges. Discharged secondary cells may be recharged to their original state by forcing an electric current from some other source through the cell in the *opposite direction* to that of the discharge.

The most common type of storage battery is the *lead–acid* type, such as used with automobiles. When fully charged, each cell delivers 2.1 V. The positive plate of the cell is *lead peroxide* and the negative is sponge lead. The electrolyte is a mixture of *sulfuric acid and water*, whose strength is measured in terms of *specific gravity*. A fully charged cell has an electrolyte with a specific gravity of about 1.270. As the battery discharges, the sulfuric acid is depleted and the electrolyte is gradually converted to water. The specific gravity of a particular battery at any one time reflects the state of charge or discharge.

The specific gravity of the electrolyte is measured with a *hydrometer* such as shown in Fig. 24-20; part of the electrolyte is drawn up into a glass tube by means of a rubber bulb at the top.

In a fully charged battery all the acid is in the electrolyte, so that the specific gravity is at its *maximum* value. The active materials of both the positive and negative plates are porous and have absorptive qualities similar to a sponge. The pores are therefore filled with the electrolyte in which they are immersed. As the battery discharges, the acid separates from the electrolyte and forms a chemical combination with the plate's active material, changing it to *lead sulfate*. Thus, as the discharge continues, lead sulfate forms on the plates, and more acid is taken from the electrolyte. The electrolyte's water content becomes progressively higher. As a result, the specific gravity of the electrolyte will gradually decrease during discharge.

When the battery is being charged, the reverse takes place. The acid held in the sulfated plate material is driven back into the electrolyte; additional charging cannot raise its specific gravity any higher. When fully charged, the material of the positive plates is again pure lead peroxide and that of the negative plates is pure lead.

Separators of wood, rubber, or glass are placed between the positive and negative plates to act as insulators. These are grooved vertically on one side and are smooth on the other. The grooved side is placed next to the positive plate to permit free circulation of the electrolyte around the active material.

Battery Charging 591

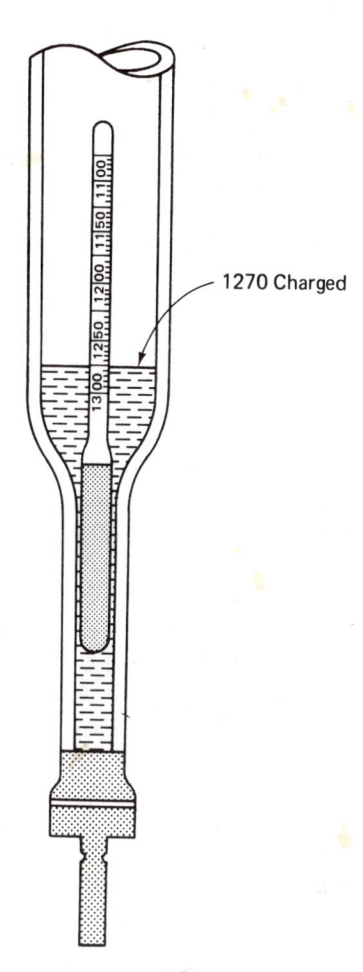

FIG. 24-20 Typical syringe-type hydrometer.

1270 Charged

The assembled lead–acid cell has the positive and negative terminals projecting through a cell cover. A hole fitted with a filler cap is provided in each cell cover to permit filling and testing. The filler cap has a vent hole to allow gas that forms in the cell during charge to escape. Hydrogen gas forms at the negative terminal and oxygen at the positive. These gases are *very flammable;* therefore, the battery should be properly *vented*. Never allow sparks near the battery, lest they explode.

24-13 BATTERY CHARGING

A secondary cell or battery must be recharged periodically for it to maintain maximum output. If it becomes discharged, it should be recharged immediately or permanent damage may result. Only direct current can be used to charge a battery. The positive and negative terminals of the charging source must be connected to the *corresponding* battery terminals. Furthermore, it must supply a potential *slightly higher* than that of the battery or no recharging will occur. Batteries may be recharged by either the *constant-current* or *constant-voltage* method.

The constant-current method, shown in Fig. 24-21, is used to charge single batteries or a number of batteries in series. The rheostat and ammeter are included to adjust and observe the charging rate. The rheostat is adjusted to supply a voltage slightly more than the internal voltage of the battery with a charging current of from 5 to 10 A. As the battery voltage rises, the rheostat is set to an increased voltage to keep the charging current constant. This is slow, requiring from 20 to 30 h. If properly watched, this method has the advantage of not overheating the battery, thus reducing the danger of *buckling the plates*.

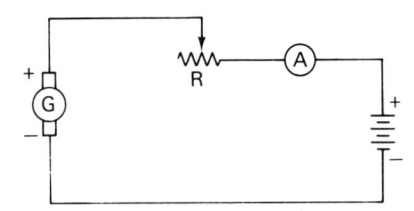

Fig. 24-21 Constant-current charging circuit.

The constant-voltage charger provides a voltage that is slightly higher than the voltage of the battery. When more than one battery is to be charged, the batteries are connected in *parallel*. As the battery becomes charged and its voltage increases, the amperage from the charger reduces. This is called the *taper-charge* method. This type of charging is relatively fast, taking 6 to 8 h. One drawback to this method is that it may cause *overheating* and *buckling* of plates if used on a completely discharged battery.

Two specialized types of charging are *fast charging* and *trickle charging*. A fast charger delivers an extremely high current (100 A or more) for a very short time and then drops to a normal value. This may cause overheating of the battery and buckled or shorted plates.

Trickle charging is the continuous application of low-charging current and is usually a constant-voltage method. It maintains the electrolyte at maximum specific gravity and the batteries remain at full charge for long periods of time. This is generally used for stationary batteries.

There are other methods by which the charge of a storage battery may be determined if no hydrometer is available. Perhaps the best way is to place a *normal load* on the battery and read the terminal voltage. The terminal voltage under load should be approximately that of the voltage with no load.

Suppose that a 6-V storage battery is to be charged at a 3-A rate from 110-V dc source. What value of series dropping resistance must be used to prevent charging at a rate greater than that specified? By Ohm's law it can be determined that the dropping resistor must drop 104 V (110 − 6 = 104). Dividing this voltage by 3 A gives a resistance of approximately 34.7 Ω.

Battery Charging

Suppose that a storage battery with a terminal voltage of 12 V is to be trickle charged at a 0.5-A rate. What value of series resistance is needed if the charging source is a 120-V dc line? In this instance the series dropping resistor must drop 108 V. Dividing this by 0.5 A gives a 216-Ω resistor.

Many battery chargers use step-down transformers to provide the necessary voltage for charging, as shown in Fig. 24-22. A half-wave gas-filled rectifier tube called a *Tungar bulb* is frequently used. The tap switch at the bottom end of the transformer secondary permits varying the charging rate.

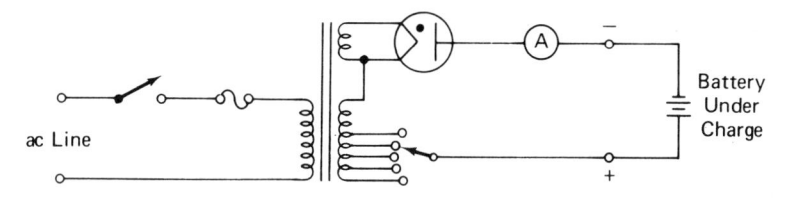

FIG. 24-22   Battery charger using rectifier tube.

If a discharged storage battery having three cells has an open circuit voltage of 1.8 V per cell and an internal resistance of 0.1 Ω per cell, how much charging voltage is required to produce an initial charging rate of 10 A? The terminal voltage of the battery must be 5.4 V (1.8 × 3 = 5.4). If each cell has an internal resistance of 0.1 Ω, the total internal resistance must be 0.3 Ω. The voltage required to force 10 A through 0.3 Ω is 3 V. This must be added to the 5.4-V battery potential. Consequently, the charging force must supply 8.4 V to assure an initial charging rate of 10 A.

If a lead–acid cell is found to have a specific gravity of 1.180, it can be assumed that the battery is not fully charged. The battery should be placed on a charger and left there until the specific gravity is brought up to normal. Sulfuric acid *should not be added* to bring the specific gravity up to the required value.

If a battery is placed on a charger and the overload circuit breaker continues to trip, it generally indicates that the internal resistance of the battery is very low and excessive current is flowing. This would usually indicate one or more *shorted cells* within the battery.

In some battery-powered installations it is necessary to have a second battery standing by, on charge, so that when the output of the one connected to the load begins to drop the other may be switched on. This can be accomplished by the use of a four-pole double-throw switch like the one shown in Fig. 24-23.

In installations where batteries are used to supply power to a transmitter, it is necessary that the leads between battery and transmitter be kept as short as possible. The presence of RF currents in the various transmitter circuits may get into the battery leads and cause undesirable radiation.

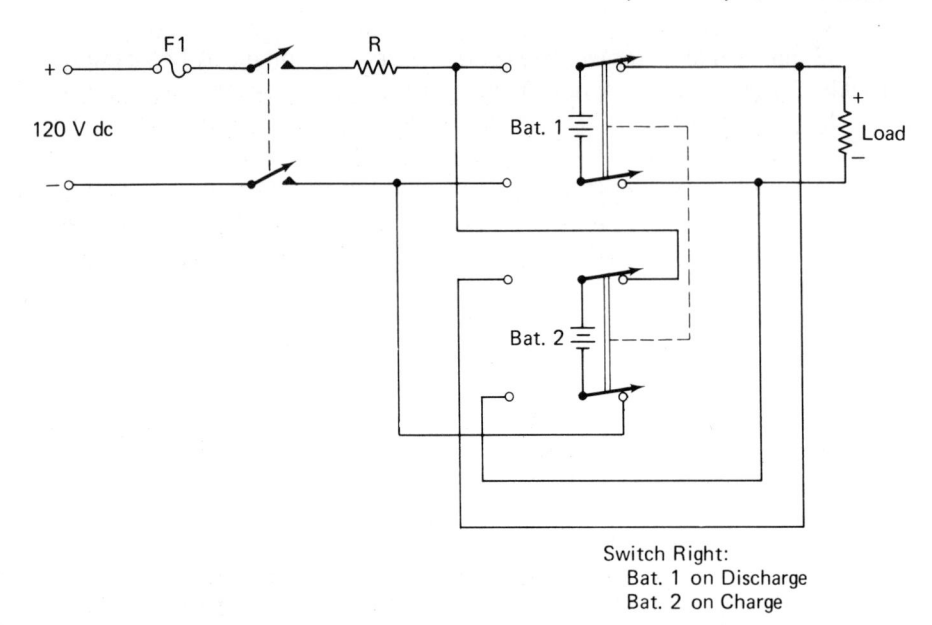

FIG. 24-23 Circuit showing how one battery can be connected to a load while the other is on charge.

Switch Right:
Bat. 1 on Discharge
Bat. 2 on Charge

24-14 LEAD–ACID BATTERY MAINTENANCE

To guarantee long life and trouble-free operation, the following items of maintenance are recommended.

1. Maintain the electrolyte at the proper level.
2. Gassing results when batteries are either charged or discharged. They must be properly vented.
3. Do not permit any flame or sparks near batteries that are being charged or discharged.
4. Avoid spilling the acid as it may cause burns and will eat holes in clothing. (If acid comes in contact with the skin, the affected area should be washed as soon as possible with large quantities of fresh water, after which a salve such as vaseline, boric acid, or zinc ointment should be applied. Acid spilled on clothing may be neutralized with *dilute ammonia* or a solution of *baking soda and water*.)
5. Keep the top of the battery clean to prevent leakage between the cells.
6. If corrosion develops around the electrodes, scrape them clean and then apply a solution made of soda and water to neutralize the acidity. Wash away the residue.
7. Coat the battery terminals with some form of petroleum jelly or other similar material to prevent formation of corrosion.

Edison Storage Battery

8. If it becomes necessary to add new electrolyte, use only *chemically pure* sulfuric acid properly diluted.
9. If the electrolyte is low, add only distilled water to bring the level up to the proper point.
10. Never allow a battery to remain discharged for any appeciable time. If allowed to remain discharged for a long time, the lead sulfate will grow into a hard, white crystalline formation. This is known as *sulfation*, and closes the pores of the active material and destroys the plates.
11. Test operate batteries at least once a month.
12. Discharged batteries must always be brought back to full charge.
13. Do not overcharge.
14. Prevent contaminants from entering the battery cell when adding distilled water.
15. If battery boxes are used, make sure that they are properly ventilated so that explosive gases are not permitted to accumulate. Clean the inside surfaces occasionally to remove any condensation of electrolyte.

24-15 EDISON STORAGE BATTERY

The *Edison storage cell* or *nickel–iron–alkaline* battery is not as common as the lead–acid storage cell, but finds some use.

The positive plate consists of rows of nickel-coated tubes filled with alternate layers of nickel hydroxide and very thin flakes of pure nickel. The negative plate consists of a grid of cold-rolled nickeled steel having a number of rectangular pockets that are filled with an iron oxide. The separators between the plates are made of hard rubber. The electrolyte is a 21 per cent solution of potassium hydroxide in distilled water to which is added a small amount of lithium hydrate. The electrolyte has a specific gravity of approximately 1.200 at 60°F and does not vary appreciably during the charge and discharge cycles. The container is made of nickeled sheet steel.

The active material of the positive plate is an oxide of nickel and that of the negative plate is pure iron. On discharge the pure iron of the negative plate is oxidized, and the nickel dioxide of the positive plate is changed to nickel oxide. On charge the reverse action takes place.

As in the case of the lead–acid cell, charging requires direct current. If by accident the cell should be charged in the wrong direction, no permanent damage will result, as long as the temperature does not rise above 115°F. The state of charge *cannot be determined with a hydrometer*, since the density of the electrolyte does not change between the charge and discharge cycles. Testing the condition of the battery is done by measuring the full-load voltage which should be 1.37 V for a fully charged cell and 1 V for a discharged one.

Even though Edison batteries are lighter in weight, require less attention, and have a longer life (from 15 to 20 years) than lead–acid batteries, their use is limited because of their higher initial cost and their reduced efficiency at low temperatures.

Other characteristics of the Edison cell are its higher internal resistance compared to the lead–acid cell. Therefore, this type of battery cannot supply the same amount of current, under load, as a lead–acid cell of the same capacity.

Commercial License Questions

Sections in which answers to questions are given appear in parentheses. A bracketed number following a question implies that it applies only to that element.

1. What is the purpose of a commutator on a dc motor? A generator? (24–1)
2. Why is laminated iron or steel generally used in the construction of the field and armature cores of motors and generators instead of solid metal? (24–2)
3. Describe the action and list the main characteristics of a series dc generator. (24–2)
4. Draw the schematic diagram of a compound-wound dc generator and explain its operation. (24–3)
5. Explain the principle of operation of a shunt-wound generator. (24–3)
6. State the principal advantage of a third-brush generator for radio power supply in automobiles. (24–3)
7. How may RF interference, often caused by sparking at the brushes of a high-voltage generator, be minimized? (24–3)
8. Describe the action and list the main characteristics of a series dc motor. (24–4)
9. If the field of a shunt-wound dc motor were opened while the machine was running under no load, what would be the probable result(s)? (24–4)
10. What is meant by *counter emf* in a dc motor? (24–4)
11. What determines the speed of a series dc motor? (24–4)
12. What is the disadvantage of using a series motor to drive a radio power-supply generator? (24–4)
13. What may happen if a series dc motor is operated without a load? (24–4)
14. Why is it sometimes necessary to use a starting resistance when starting a dc motor? (24–4)
15. What might be the result of starting a motor too slowly using a hand starter? (24–4)
16. Describe the action and list the main characteristics of a shunt-wound dc motor. (24–4)

Commercial License Questions

17. Draw a diagram of a shunt-wound dc motor. [4] (24-4)
18. Explain why RF chokes are sometimes placed in the power leads between a motor–generator power supply and a high-powered radio transmitter. (24-4)
19. Why are bypass condensers often connected across the brushes of a high-voltage dc generator? (24-4)
20. What controls the output voltage of a dynamotor? (24-5)
21. How can the output voltage of a motor–generator set be increased? (24-5)
22. Discuss the construction and operation of dynamotors. (24-5)
23. What is the advantage of a dynamotor over a motor–generator set in supplying power to a mobile transmitter? (24-5)
24. List the comparative advantages and disadvantages of motor–generator and transformer–rectifier power supplies. (24-5)
25. In what unit is the output of an alternator generally rated? (24-6)
26. How may the output voltage of a separately excited ac generator, at constant output frequency, be varied? (24-6)
27. What conditions must be satisfied in order to parallel two alternators? (24-6)
28. What is meant by *power factor* as regards an alternator? (24-6)
29. What is the approximate speed of a 220-V, 60-Hz, four-pole, three-phase induction motor? [4] (24-7)
30. What controls the speed of an induction motor? (24-7)
31. Describe the operation of an induction motor. (24-7)
32. What determines the speed of a synchronous motor? (24-7)
33. What effect do shorted turns on an armature have on the performance of a dc motor? (24-9)
34. List three causes of sparking at the commutator of a dc motor. (24-9)
35. What materials should be used to clean the commutator of a motor or generator? (24-9)
36. What procedure should be followed if a motor fails to start when the start button is depressed? (24-9)
37. What may cause a motor–generator bearing to overheat? (24-9)
38. Explain why emery cloth or steel wool should never be used to clean a commutator. (24-9)
39. Name four causes of excessive sparking at the brushes of a dc motor or generator. (24-9)
40. In what form is the energy stored in a lead–acid battery? (24-10)
41. What may cause sulfation of a lead–acid storage cell? (24-10, 24-12)

42. What are the effects of sulfation? (24–10, 24–12, 24–13)
43. Describe an electrolyte. (24–10, 24–12)
44. How does a primary cell differ from a secondary cell? (24–10, 24–12)
45. What is polarization as applied to a primary cell and how may its effect be counteracted? (24–10)
46. How may a dry cell be tested to determine its condition? (24–10, 24–11)
47. Name the materials used as electrodes in the dry-cell battery. (24–10)
48. Show by a diagram how to connect battery cells in series (24–11)
49. Show by a diagram how to connect battery cells in parallel. (24–11)
50. What method of connection should be used to obtain the maximum short-circuit current from a group of similar cells in a storage battery? (24–11)
51. What method of connection should be used to obtain the maximum no-load output voltage from a group of similar cells in a storage battery? (24–11)
52. In what unit is the capacity of a battery rated? (24–11, 24–12)
53. What is the effect of local action in a lead–acid storage cell and how may it be compensated? (24–12)
54. Why is low internal resistance desirable in a storage cell? (24–12)
55. What is the approximate fully charged voltage of a lead–acid cell? (24–12)
56. What form of energy is stored in lead-type storage batteries? (24–12)
57. What causes secondary cells to heat up when being rapidly charged or discharged? (24–12, 24–13)
58. What is the chemical composition of the active material composing the negative plate of a lead–acid-type storage cell? (24–12)
59. What is the chemical composition of the active material composing the positive plate of a lead–acid-type storage cell? (24–12)
60. What is the chemical composition of the electrolyte of a lead–acid storage cell? (24–12)
61. Sketch the construction of a lead–acid cell. (24–12)
62. Define *specific gravity* as used in electrolytic solutions. (24–12)
63. What is the cause of the heat developed within a storage cell under charge or discharge condition? (24–13)
64. How may the state of charge of a lead–acid storage cell be determined?
65. What should be done to a lead–acid cell if some of the electrolyte has evaporated? (24–13, 24–14)
66. What will be the result of discharging a lead–acid storage cell at an excessively high current rate? (24–13)
67. If the charging current through a storage battery is maintained at the normal rate, but its polarity is reversed, what will result? (24–13)

Commercial License Questions 599

68. What should be done to a lead–acid cell having a specific gravity of 1.180? (24–13, 24–14)

69. What may cause the plates of a lead–acid storage cell to buckle? (24–13, 24–14)

70. How may the polarity of the charging source to be used with a storage battery be determined? (24–13)

71. Describe the characteristics of a battery-charging rectifier tube. (24–13)

72. How may the approximate state of charge of a storage battery be determined if no hydrometer is available? (24–13)

73. Draw the schematic diagram of a battery charger using a rectifier tube. (24–13)

74. What value of resistance must be connected in series with a 6-V battery that is to be charged at a 3-A rate from a 110-V dc source? (24–13)

75. A storage battery with a terminal voltage of 12 V is to be trickle charged at a 0.5-A rate. Indicate the value of series resistance needed if the charging source is a 120-V dc line. (24–13)

76. Draw the diagram of a charging circuit for two batteries using a four-pole double-throw switch so that while one battery is on discharge the other is being charged. Show the dc charging source, voltage dropping resistors, and connections to the battery load. (24–13)

77. A discharged storage battery, having three cells, has an open circuit voltage of 1.8 V per cell and an internal resistance of 0.1 Ω per cell. What charging emf is required to produce an initial charging rate of 10 A? (24–13)

78. If a battery is placed on charge and the overload circuit breakers will not stay closed, what may be the cause of trouble? (24–13)

79. Why is it necessary to keep the unshielded leads between a transmitter and battery as short as possible? (24–13)

80. Why does the rate of charge of a storage battery decrease as the charging progresses when being charged from a fixed voltage source? (24–13)

81. What steps may be taken to prevent corrosion of lead–acid storage-cell terminals? (24–14)

82. What chemical may be used to neutralize a storage-cell acid electrolyte? (24–14)

83. Why should adequate ventilation be provided in the room housing a large group of storage cells? (24–14)

84. When should distilled water be added to a lead–acid storage cell and for what purpose? (24–14)

85. Describe the care that should be given a group of storage cells to maintain them in good operating condition. (24–14)

86. Why should the top of a storage battery be kept clean and dry? (24–14)

87. How does a lead–acid battery differ from an Edison battery? (24–14, 24–15)

88. How may the condition of charge of an Edison cell best be determined?
(24–15)

89. What is the approximate fully charged voltage of an Edison storage cell?
(24–15)

90. Name three causes of a decrease in capacity of an Edison storage battery.
(24–15)

91. What is the chemical composition of the acitve material composing the negative plate of an Edison-type storage cell? (24–15)

92. What is the chemical composition of the active material composing the positive plate of an Edison-type storage cell? (24–15)

93. What is the chemical composition of the electrolyte used in an Edison-type storage cell? (24–14)

25

Federal Communications Commission Licenses and Laws, Element 1

25-1 ROLE OF THE FEDERAL COMMUNICATIONS COMMISSION

Element 1 is an integral part of all the FCC commercial license examinations. This chapter deals with the essential information needed to pass this element.

As the art of electrical and electronic communication developed, it became necessary to establish laws to provide for the regulation of interstate and foreign communication by wire or radio. The communications Act of 1934 set up general laws to be followed in the United States that would also be agreeable with the communications agreements established with foreign countries. The Act provides for the creation of a Federal agency, called the *Federal Communications Commission*, to carry out and enforce its provisions. It is located in Washington, D.C., and has district offices located in a number of cities throughout the country (see Appendix E). The commission supervises the transmission of *all forms of communication*, whether by wire or radio.

Since its inception, the FCC has created a series of rules and regulations to govern the many different types of communications facilities. These are known as *Rules and Regulations* and constitute a number of volumes. Revisions occur from time to time, as the state of the art changes or as new conditions arise. Hence, it is incumbent on the owner and operators of these radio services to keep abreast of the changes affecting their operation.

⟨The FCC has the authority to inspect all radio stations that are licensed to ascertain that their construction, installation, and operations meet the requirements of the rules and regulations of the commission. In addition, it has the authority to issue station and operator licenses to qualified applicants.⟩ This necessitates maintaining a number of examination facilities throughout the United States.

25-2 TYPES OF FEDERAL COMMUNICATIONS COMMISSION LICENSES AND PERMITS

To ensure proper maintenance and operation of radio and other electronic transmitting equipment all technician personnel must be licensed. Licenses are only issued by the FCC in this country to applicants who have successfully passed the appropriate written examination(s).

The FCC issues two basic classes of licenses: *commercial* and *amateur*. A commercial license is required by an operator of any transmitting installation used as a business enterprise. If the radio equipment is used strictly as a hobby, on frequencies assigned for this purpose, an amateur license is required.

The FCC issues several different commercial licenses and permits that are made up of one or more of the following nine elements:

Element 1: Basic laws. Consists of 20 questions covering laws, regulations, and treaties that the applicant must be familiar with (covered in this chapter).

Element 2: Basic operating practice. Consists of 20 questions covering radio operating procedures and practices in radiotelephone communication (covered in Chapter 27).

Element 3: Basic radiotelephone. Consists of 100 questions covering radio theory, testing, and operational procedures for radiotelephone stations (most chapters).

Element 4: Advanced radiotelephone. Consists of 50 questions dealing with advanced radio theory, testing, and regulations, mostly applicable to the broadcast industry (most chapters).

Element 5: Radiotelegraph operating practice. Fifty questions dealing primarily with operating practices and procedures for radiotelegraph stations other than commercial telegraph or maritime (not covered in this text).

Element 6: Advanced radiotelegraph. Consists of 100 questions on radio theory, testing, and other relative subjects applicable to the operation of radiotelegraph stations, radio navigational aids, emergency messages, and message handling (not covered in this text).

Element 7: Aircraft telegraph (seldom used anymore).

Element 8: Ship radar techniques. Consists of 50 questions specifically dealing with theory, installation, and maintenance of ship radar gear (not covered in this text).

Element 9: Broadcast endorsement. Consists of 20 questions relating to elementary theory and operational practices of standard AM and FM broadcast stations (Chapter 22).

All questions are of the multiple-choice type for which the candidate is instructed to choose the best answer of five given. A minimum grade of 75 per cent is required to successfully pass any of the examinations.

The FCC issues four types of operator's licenses. Each license consists of one or more elements as shown below:

RADIOTELEPHONE—First Class

> Elements 1, 2, 3, 4 (No code). The holder can, in general, operate, repair, and maintain any transmitting equipment not transmitting Morse code.

RADIOTELEPHONE—Second Class

> Elements 1, 2, 3 (No code). The holder can maintain, repair, and operate any transmitting equipment except AM, FM, and TV broadcast stations and stations transmitting Morse code.

RADIOTELEGRAPH—First Class

> Elements 1, 2, 5, 6. Code test: 20 code groups/min and 25 words/min plain language. The holder can operate the same equipment identified in the radiotelegraph second-class license. He can serve as chief operator and must have at least 1 year of experience as a second-class radiotelegraph operator and be at least 21 years old.

RADIOTELEGRAPH—Second Class

> Elements 1, 2, 5, 6. Code test: 16 code groups/min and 20 words/min plain language. The holder can, in general, operate any transmitter, voice or code, except TV, AM, and FM broadcast stations. He cannot act as chief operator on most passenger vessels and there are no experience or age requirements.

In addition to the licenses mentioned above, the FCC also issues *permits* that enable nontechnical personnel to only operate certain transmitters. They are not permitted to make any adjustments that may cause improper transmitter operation unless they are made under the direct supervision of a licensed operator. The permits are

RADIOTELEPHONE—Third-Class Operator Permit

> Elements 1, 2, 9 (No code). Persons who wish to obtain this permit with the broadcast endorsement as authority to routinely operate the transmitters at certain standard AM and FM broadcast stations may so specify.

RADIOTELEGRAPH—Third-Class Operator Permit

> Elements 1, 2, 5. Code test: 16 code groups/min and 20 words/min plain language.

RESTRICTED Radiotelephone Operator Permit

> This permit requires only a written certification that the applicant is familiar with the appropriate rules and regulations, knows how to keep a station log, and can understand and speak messages in English. This permit is used by the Public Safety Radio Services, aircraft pilots, policemen, taxi drivers, and so forth.

Element 8 is a radar endorsement that can be used with any license.

Element 9 is the broadcast endorsement for use with the radiotelephone third-class operator permit.

25-3 LICENSE AND PERMIT REQUIREMENTS

All commercial and amateur radio stations operating in the United States *must be licensed by the FCC. Operators* of these stations must also be properly licensed by this same agency. Exceptions to this are government and military-service stations and certain citizen-band equipment.

Applicants for commercial licenses or permits must be citizens or nationals of the United States, except in the case of aliens who hold Aircraft Pilot Certificates issued by the Federal Aviation Administration. The age and fee requirements for each license or permit are indicated in Table 25-1. When only U.S. citizenship is required (no age limit), U.S. appears under the age column.

TABLE 25-1 AGE AND FEE REQUIREMENTS FOR FCC LICENSES AND PERMITS

License or Permit	Age	Fee ($)
Radiotelephone: first class	U.S.	5.00
Radiotelephone: second class	U.S.	4.00
Radiotelephone: third class	U.S.	3.00
Restricted radiotelephone	14	8.00
Radiotelegraph: first class	21	5.00
Radiotelegraph: second class	U.S.	4.00

As a convenience, the FCC will issue a *verification card* to license or permit holders if they submit FCC Form 758-F properly filled out. This is a wallet-sized card that can be carried by the individual in lieu of the original license or permit when operating a station that does not require the posting of an operator's license. However, the license or permit must be reasonably available upon request. The fee for a verification card is $2.00.

License examinations are given at the various FCC field offices located in many of the major cities in the United States. Contact the nearest field office to determine which day of the week the desired examination is to be given. The examination schedule is not the same throughout the field offices. Applicants for licenses must first fill out an application form before the examination will be given.

There is no time limit on any of the examinations. Consequently, it is wise to start the examination as early in the day as possible. This allows time to relax and consider each question with care. If some questions are difficult, pass them by and complete the rest of the examination. Then go back to the questions that gave you trouble. This time your memory may be refreshed and they may be easier. If more than one element is to be taken, it may be helpful to leave the room (this is permitted) for some relaxation before beginning the next element.

Commercial operator licenses are normally issued for a period of 5 years from the date of issuance. Application for renewal of a license may be filed at any time during the final year of the license term or during a 1-year grace period after the date of expiration of the license sought to be renewed. During this 1-year grace period, an expired license is not valid. A renewed license issued upon the basis of an application filed during the grace period will be dated currently and will not be backdated to the date of expiration of the license being renewed. A renewal application must accompany the license to be renewed.

If the holder of a license qualifies for a higher class in the same group, the license held will be canceled upon the issuance of the new license. Similarly, if the holder of a restricted operator permit qualifies for a first- or second-class license of the corresponding type, the permit will be canceled upon issuance of the new license.

If an operator is working at two different stations, he must have his license *conspicuously posted* at one of them. At the other he may post a verified statement obtainable from the FCC, Form 759.

If a license or permit has been lost, mutilated, or destroyed, the operator should *immediately* notify the FCC. A properly filled out application form for a duplicate should be submitted to the office that issued the original, stating the circumstances involved in the loss of the license. The applicant must state that a reasonable search has been made for the lost license, and, if found at a later date, will be returned for cancellation. It is incumbent on the applicant to submit documentary evidence of the service that has been performed under the original license, or a statement under oath affirming that information. It is permissible to operate under these circumstances if the operator posts a signed copy of the application for duplicate.

25-4 TYPES OF COMMUNICATION

Radio stations are normally involved in routine or general communications. However, in times of emergency a system of priorities has been established to ensure the flow of essential communications. This involves three categories of transmissions, which are, according to their priorities, (1) *distress*, (2) *urgency*, and (3) *safety*.

For mobile radiotelephone operation the distress signal is the word MAYDAY (from the French "m'aider"), which is spoken three or more times to alert listening stations. This means the station is in need of immediate assistance. The words THIS IS . . . should then be spoken along with the name of the station. The distress message, which follows, should contain the exact position (if possible) of the vessel or aircraft, the nature of the distress, the kind of assistance needed, along with other pertinent information that might facilitate rescue, such as true heading and airspeed, altitude, type of aircraft, and the intention of the person in command (i.e., ditching at sea or forced landing).

The mobile station in distress is in control of distress-message traffic. If it is unable to do this, because of the nature of the emergency, it may delegate control of the distress traffic to another station. Distress messages are *not* subject to the secrecy provisions of the law. The transmission of false or fraudulent distress signals is prohibited by law and punishable by a fine of not more than $10,000 and not more than 1 year in prison. The distress signal for mobile stations using radiotelegraph is SOS ($\cdots---\cdots$) sent as one character with no spacing between letters.

The urgent signal for radiotelephonic communication is the word PAN, spoken three times. This should precede any message concerning a situation that requires immediate attention and may become distress in nature. In radiotelegraphy the urgent signal consists of the transmission of the three letters XXX.

The safety signal for radiotelephone communication is the word SECURITY (from the French sécurité) spoken three times before the message. This signal is used when the station is about to transmit a message concerning the safety of navigation or giving important meteorological warnings. In radiotelegraphy the safety signal consists of the three letters TTT.

25-5 SUSPENSION OF OPERATOR LICENSES

Whenever grounds exist for suspension of an operator license as provided by the Communications Act, the Chief, Safety and Special Radio Services Bureau, with respect to amateur operator licenses, or the Chief, Field Engineering Bureau, with respect to commercial operator licenses, may issue an order suspending the operator license. "No order of suspension of any operator's license shall take effect until 15-days' notice in writing of the cause for the proposed suspension has been given to the operator licensee, who may make written application to the Commission at any time within said 15 days for a hearing upon such order. The notice to the operator licensee shall not be effective until actually received by him, and from that time he shall have 15 days in which to mail the said application. In the event that physical conditions prevent mailing of the application before the expiration of the 15-day period, the application shall then be mailed as soon as possible thereafter, accompanied by a satisfactory explanation of the delay. Upon receipt by the Commission of such application for hearing, said order of suspension shall be designated for hearing by the Chief, Safety and Special Radio Services Bureau or the Chief, Field Engineering Bureau, as the case may be, and said order of suspension shall be held in abeyance until the conclusion of the hearing. Upon the conclusion of said hearing, the Commission may affirm, modify, or revoke said order of suspension."[1] If the license is ordered suspended, the operator shall send his operator license to the office of the Commission in Washington, D.C., on or before the effective date of the order, or, if the effective date has passed at the time notice is received, the license shall be sent to the Commission forthwith.

25-6 NOTICE OF VIOLATIONS

Except in cases of willfulness or those in which public health, interest, or safety require otherwise, any licensee who appears to have violated any provision of the Communications Act will, before revocation, suspension, or cease and desist proceedings are instituted, be served with a written notice calling these facts to his attention and requesting a statement concerning the matter; FCC Form 793 may be used for this purpose.

Within 10 days from receipt of notice, or such other period as may be specified, the licensee shall send a written answer in duplicate direct to the office of the Commission originating the official notice. If an answer cannot be sent nor an acknowledgment made within such 10-day period, by reason of illness or other unavoidable circumstances, acknowledgment and answer shall be made at the earliest practicable date with a satisfactory explanation of the delay.

The answer to each notice shall be complete in itself and shall not be abbreviated by reference to other communications or answers to other notices. In every instance the answer shall contain a statement of action taken to correct the condition or omission complained of and to preclude its recurrence. In addition,

1. If the notice relates to violations that may be due to the physical or electrical characteristics of transmitting apparatus and any new apparatus is to be installed, the answer shall state the date such apparatus was ordered, the name of the manufacturer, and the promised date of delivery. If the installation of such apparatus requires a construction permit, the file number of the application shall be given, or, if a file number has not been assigned by the Commission, such identification shall be given as will permit ready identification of the application.
2. If the notice of violation relates to lack of attention to or improper operation of the transmitter, the name and license number of the operator in charge shall be given.

"No person within the jurisdiction of the United States shall knowingly utter or transmit, or cause to be uttered or transmitted, any false or fraudulent signal of distress, or communication relating thereto, nor shall any broad-

[1] Extracts from the Communications Act of 1934, as amended. Sec. 303

casting station rebroadcast the program or any part thereof of another broadcasting station without the express authority of the originating station."[2]

"Any person who willfully and knowingly does or causes or suffers to be done any act, matter, or thing, in this Act prohibited or declared to be unlawful, or who willfully or knowingly omits or fails to do any act, matter, or thing in this Act required to be done, or willfully and knowingly causes or suffers such omission or failure shall, upon conviction thereof, be punished for such offense, for which no penalty (other than a forfeiture) is provided in this Act, by a fine of not more than $10,000 or by imprisonment for a term not exceeding 1 year, or both, except that any person, having been once convicted of an offense punishable under this section, who is subsequently convicted of violating any provision of this Act punishable under this section, shall be punished by a fine of not more than $10,000 or by imprisonment for a term not exceeding 2 years, or both."[3]

"Any person who willfully and knowingly violates any rule, regulation, restriction, or condition made or imposed by the Commission under authority of this Act, or any rule, regulation, restriction, or condition made or imposed by any international radio or wire communications treaty or convention, or regulations annexed thereto, to which the United States is or may hereafter become a party, shall, in addition to any other penalties provided by law, be punished, upon conviction thereof, by a fine of not more than $500 for each and every day during which such offense occurs."[4]

The Communications Act and the treaties to which the United States is a party may revoke the license of any operator who is guilty of the following:

1. Has failed to carry out a lawful order of the master or person lawfully in charge of the ship or aircraft on which he is employed.
2. Has willfully damaged or permitted radio apparatus or installations to be damaged.
3. Has transmitted superfluous radio communications or signals or communications containing profane or obscene words, language, or meaning, or has knowingly transmitted
   a. False or deceptive signals or communications.
   b. A call signal or letter that has not been assigned by proper authority to the station he is operating.
4. Has willfully or maliciously interfered with any other radio communications or signals.
5. Has obtained or attempted to obtain, or has assisted another to

[2] Extracts from the Communications Act of 1934, as amended, Sec. 325
[3] Extracts from the Communications Act of 1934, as amended. Sec. 501
[4] Extracts from the Communications Act of 1934, as amended. Sec. 502

obtain or attempt to obtain, an operator's license by fraudulent means.

25-7 RESTRICTIONS ON OPERATING, SERVICING, AND ADJUSTING TRANSMITTERS

Holders of Restricted Radiotelephone Permits and Radiotelephone Third-Class Permits are *not allowed* to make adjustments to transmitting equipment. Any needed adjustments that may affect the proper operation of the equipment must be made by or under the immediate supervision of a person holding the proper class of license required for the equipment involved. The licensed operator performing the repairs is responsible for the proper functioning of the station equipment.

Any radio station requiring a licensed operator must be a licensed station, and it is the responsibility of the operator to determine that the required license is held by the station before he operates the equipment.

Almost all Public Safety Radio Services equipment such as police, fire, aircraft, railroad, petroleum, taxicab, and special-emergency, require only a Restricted Radiotelephone Permit to operate. This permit only allows the individual to turn the equipment on or off and to speak over the microphone.

Transmitting equipment designed for radiotelegraph operation (Morse code) can only be operated by persons holding a suitable radiotelegraph license or permit.

Radio announcers, communicators, and others in similar work require at least Radiotelephone Third-Class Permits with broadcast endorsement (Element 9).

Holders of radio amateur licenses are not authorized to make adjustments to commercial radio transmitting equipment (except limited operation of disaster communication services stations). Likewise, holders of commercial licenses only are not authorized to operate or adjust any radio amateur transmitting equipment.

25-8 LOGS

The FCC requires that *accurate logs* be kept for all licensed radio tranmissions. Log entries must show the date, time, nature of the communication, and operator in charge. If the transmission was with another station, that station must be indicated.

Logs can only be made out by persons having authority to do so. If an error is made in an entry, it *must not be erased*. Any necessary correction must be made by the person originating the entry. This is done by striking out the erroneous portion, initialing the correction, and dating it. Logs must be kept for a period of 2 years for standard broadcast stations. If they contain distress traffic, or information relative to any legal proceedings or investigations by the FCC, they must be kept until written authority to destroy them is received from the FCC.

The following entries shall be made in the program log:

1. For each program: an entry identifying the program by name or title, an entry of the time each program begins and ends, an entry classifying each program as to type (news, religious, entertainment, sports), an entry classifying each program as to source, an entry for each program presenting a political candidate showing the name and political affiliation of such candidate.
2. For commercial matter: an entry identifying the sponsor(s) of the program, the person(s) who paid for the announcement, or the person(s) who furnished materials or services of any kind (such as records, transcriptions, talent, scripts), an entry showing the total amount or commercial continuity within each commercially sponsored program, an entry showing the duration of each commercial announcement, an entry that shows either the beginning time of each such announcement or divides the log to show the 15-min time segment within which the announcement was broadcast, and an entry showing that the appropriate announcements (sponsorship, furnishing material, or services) have been made.
3. For public-service announcements: an entry showing that a public-service announcement has been broadcast together with the name of the organization or interest on whose behalf it has been made.

Commercial License Questions

Applicants for all commercial licenses must know the answers to all the following questions relating to Element 1. Sections in which answers to questions are given appear in parentheses.

1. Where and how is an operator license or permit obtained? (25–1, 25–3)
2. Are radio stations subject to inspection by the FCC? (25–1)
3. Must a person designated to operate a radiotelephone station post his operator license or permit and, if so, where? (25–3)
4. What must a person do whose operator license or permit has been lost, mutilated, or destroyed? (25–3)
5. In applying for a duplicate operator license or permit, what documentary evidence must be submitted along with an application? (25–3)
6. Is it permissible to operate pending receipt of a duplicate operator license or permit after application has been made for reissue? (25–3)
7. What provision is made for operation without an actual operator license or permit, pending receipt of a duplicate? (25–3, 25–5)

Commercial License Questions 611

8. How soon before expiration of an operator license or permit should application be made for renewal? (25–3)
9. Is it prohibited by law to transmit unnecessary and superfluous signals? (25–4)
10. Are communications bearing upon distress situations subject to the secrecy provisions of law? (25–4)
11. In radiotelephony what are the distress, urgency, and safety signals? (25–4)
12. In radio communication, what does the transmission of distress, urgency, and safety signals signify, respectively? (25–4)
13. What information must be contained in a distress message? (25–4)
14. Under what conditions may a mobile radio station send a distress message for another mobile station in distress? (25–4)
15. In the case of a mobile radio station in distress what station is responsible for the control of distress message traffic? (25–4)
16. What does the distress call consist of when sent by radiotelephony? (25–4)
17. What is the priority of the urgency signal? (25–4)
18. May the FCC suspend an operator license or permit for due cause? (25–5)
19. Can suspension of an operator license or permit take effect prior to notification? (25–5)
20. How soon after receiving notification of suspension of an operator license or permit does a suspension order become effective? (25–5)
21. May a person who has received an order of suspension of operator license or permit request a hearing? (25–5)
22. How must a person who receives a Notice of Violation from the FCC reply? (25–6)
23. How soon does the FCC require a response to a Notice of Violation? (25–6)
24. If a person cannot respond to a Notice of Violation in the time prescribed by the FCC, is it necessary to explain the reason for any delay? (25–6)
25. Should the answer to each Notice of Violation be complete and should reference be made to remedial action if any specific remedial steps are necessary? (25–6)
26. To whom is a response to a Notice of Violation addressed? (25–6)
27. Does the Government have authority to impose fines for failure to comply with the rules and regulations governing the use of radio on compulsorily equipped ships? (25–6)
28. What penalty is provided by law for willful and knowing violation of regulations imposed by the FCC and radio treaties? (25–6)
29. What penalty is provided by law for willful and knowing violation of the radio laws? (25–6)

30. Is it prohibited by law to transmit false or fraudulent signals of distress? (25–6)
31. Is the holder of a Radiotelephone Third-Class Operator Permit authorized to make technical adjustments to the transmitter he operates? (25–7)
32. Should a radio station that is required to be operated by a licensed radio operator be a licensed radio station? (25–7)
33. How may necessary corrections to the log record be made? (25–8)

26

Basic Operating Practice, Element 2

26-1 MESSAGE PRIORITY FOR MOBILE SERVICE

This chapter contains all the necessary information needed to pass Element 2 of the FCC examination. Applicants who wish to obtain a Radiotelephone Third-Class Operator Permit as authority to operate equipment on board small boats may so request when applying for this permit. The larger part of this chapter deals with information needed to answer these questions. Those sections of the chapter dealing specifically with questions for the Radiotelephone Third-Class Permit for broadcast work are designated by (G), implying general questions. Persons desiring the latter permit, with broadcast endorsement, must also take FCC Element 9, which is covered in the next chapter.

All licensed stations are given a call by which they can be identified. No two stations have the same call. The call may consist of letters only or a combination of letters and numbers. Because of the very great number of licensed stations, operators must give their calls with clarity to eliminate the possibility of confusion.

Because there are so many stations, both domestic and international, it is imperative that procedures be followed to reduce interference and provide for the orderly flow of emergency traffic. This has been discussed in some detail in Section 25-4, but will be elaborated upon in Section 26-2.

The order of priority for communications in the mobile service is

1. Distress calls, distress messages, and distress traffic.
2. Communications preceded by the urgency signal.
3. Communications preceded by the safety signal.
4. Communications relating to radio direction finding.

5. Communications relating to the navigation and safe movement of aircraft.
6. Communications relating to the navigation, movements, and needs of ships, and weather observation messages destined for an official meteorological service.
7. Government communications for which priority has been requested.
8. Service communications relating to the working of the radio communications previously exchanged.
9. All other routine traffic.

26-2 DISTRESS, URGENT, AND SAFETY MESSAGES

Distress messages have priority over all other transmissions. Any station hearing a distress signal or message must *immediately* cease any transmission that may possibly cause interference and must listen on the frequency being used for the distress call. Upon hearing a distress call the operator must immediately acknowledge receipt of the distress message from the mobile station if he is in the vicinity. Other stations should stand by in the event they may be of assistance. If stations are in a position to be of assistance, they should transmit as soon as possible, on the order of the person in charge of the ship or aircraft, their (1) name, (2) present position, and (3) speed toward the station in distress. In these emergency conditions an operator can increase his power beyond normal limits and operate the station without regard to certain provisions of his license.

If an operator is not in position to render direct assistance, he must do all in his power to attract the attention of stations that might possibly assist. He must follow the distress traffic even if he does not participate in it.

No station must use the distress frequency for other types of calls or traffic during the time the distress messages are being transmitted. If an operator is told he is interfering with distress traffic, he must cease transmission immediately and listen to find out if he can be of assistance.

Land stations receiving a distress call must immediately notify the proper authorities (usually the Coast Guard) who are in position to handle rescue operations. Unless involved in the distress traffic, land stations must not use the distress frequency. Stations not involved in the distress traffic may continue their operation on frequencies that will not interfere after it is established that the distress traffic is being adequately handled.

When the distress traffic is completed, the station handling the traffic should make an announcement on the distress frequency such as the following:

MAYDAY ALL STATIONS (three times)
THIS IS (call letters of the station transmitting the message)
STATION (name of station in distress)
DISTRESS TRAFFIC ENDED. OUT.

Stations receiving the urgency signal (PAN) must continue to listen for at least 3 min. If no urgency message is received during this period, they may resume normal operation.

When an operator hears the safety signal (SECURITY), he must continue to listen on the frequency on which it is being transmitted until he is assured that the message is of no interest to him. He has an obligation to make no transmission that will cause interference with the message. Commercial stations and any others cannot charge for handling safety messages.

26-3 MICROPHONE TECHNIQUES (G)

To achieve high-quality voice and music reproduction, it is necessary to use a good microphone. However, if proper microphone techniques are not employed, the results will be unsatisfactory due to the distortion generated. This, when added to other possible distortion factors, such as a malfunctioning speech amplifier, overmodulation, poor power supply regulation, static, fading, and so forth may cause the signal to become distorted or even unreadable. Several important microphone—and speech—techniques will enhance voice communications. These are

1. Speak in a well-modulated voice.
2. Do not place the microphone too close to the mouth (unless special differential microphones are used).
3. If the microphone is too far away, its output will be weak.
4. If the ambient noise level is high, it may be necessary for the operator to cup his hands over the microphone to exclude the noise.
5. Pronounce words clearly.
6. Use well-known words and phrases.
7. If the receiving operator is having difficulty with his reception, important words should be repeated or spelled out.

26-4 PHONETIC ALPHABET (G)

Many speech sounds in the English language sound almost alike, particularly when transmitted over a radiotelephone circuit if a little distortion is present. The problem is compounded in such industries as international air transportation where pilots from all over the world communicate with airport control towers. To ensure that letters or words will be understood, an *international phonetic alphabet* has been adopted. Each letter is represented by a well-known easy-to-pronounce word. Thus the word "say" would be spelled out as sierra, alfa, yankee. The receiving operator writes down the first letter of each word and receives "say."

The International Civil Aviation Organization (ICAO) phonetic alphabet is shown in Table 26-1.

TALLE 26-1 INTERNATIONAL CIVIL AVIATION ORGANIZATION PHONETIC ALPHABET

A	— ALFA	N	— NOVEMBER
B	— BRAVO	O	— OSCAR
C	— CHARLIE	P	— PAPA
D	— DELTA	Q	— QUEBEC
E	— ECHO	R	— ROMEO
F	— FOXTROT	S	— SIERRA
G	— GOLF	T	— TANGO
H	— HOTEL	U	— UNIFORM
I	— INDIA	V	— VICTOR
J	— JULIETT	W	— WHISKEY
K	— KILO	X	— X-RAY
L	— LIMA	Y	— YANKEE
M	— MIKE	Z	— ZULU

26-5 COMMONLY USED RADIOTELEPHONE ABBREVIATIONS (G)

It is important in radiotelephone communications that operators use familiar and well-known words and phrases to ensure accuracy and save time from undue repetition of words. A list of the more commonly used ones is:

Words	Meaning
Roger	I have received all of your last transmission.
Wilco	Your last message received, understood, and will be complied with.
Out or Clear	This conversation is ended and no response is expected.
Over	My transmission is ended, and I expect a response from you.
Speak slower	Speak slowly.
Say again	Repeat.
Words twice	Give every phrase twice.
Break	I am changing from one part of the message to another (i.e., address or preamble to text). Also used to allow the receiving operator to indicate if he has received all parts of the message thus far transmitted.
Repeat back	Repeat the message back to me.
Stand by	Wait for another call or further instructions.

26-6 CALLING ANOTHER STATION

When one station wants to contact another for routine communication purposes, the following procedure should be followed. Suppose that station WABC wishes to contact station WXYZ. The calling station would say, "WXYZ, WXYZ, WXYZ, this is WABC, WABC, WABC, over." Station WXYZ, upon hearing the call, would reply, "WABC, this is WXYZ, over." It is not necessary for the called station to repeat its call three times, nor that of the calling station.

An example of a message from WABC to WXYZ is as follows:

Calling and Working Frequencies 617

WXYZ THIS IS WABC. MESSAGE NUMBER THREE FROM FREIGHTER WHITE CLOUD AUGUST EIGHTH TEN FIFTEEN AM. BREAK. ESTIMATED TIME ARRIVAL NEW YORK TWELVE THIRTY-FIVE PM TODAY. BREAK. SIGNED CAPTAIN ANDREWS. BREAK. THIS IS WABC. OUT.

The operator at WXYZ would likely copy the message as follows:

WABC NR 3 FREIGHTER WHITECLOUD AUGUST 8, 1970 10:15 A. M.	Preamble
ESTIMATED TIME ARRIVAL NEW YORK 12:35 P.M. TODAY	Text
CAPT. ANDREWS	Signature
10:20 AM LA	Service

The service indicates the time the operator received the message followed by his initials.

The operator at WXYZ would acknowledge receipt of the message and sign out as follows:

WABC THIS IS WXYZ. ROGER YOUR MESSAGE NUMBER THREE. OUT.

If several messages are to be transmitted between stations that have already established contact, it is unnecessary for each station to identify itself until the traffic has terminated (unless of 15-min or more duration).

When a mobile station is calling another station by radiotelephone, it must not continue for a period of more than 30 s in each instance. If the called station is not heard to reply, it shall not be called again until after an interval of 1 min (emergencies excepted).

Radiotelephone traffic in the 2000- to 3000-kHz band should be limited to 5 min to allow other stations to communicate (excepting emergency communications).

26-7 CALLING AND WORKING FREQUENCIES

The international radiotelephone distress and calling frequency for the maritime mobile service 2182 kHz. All vessels should have a receiver tuned to this frequency throughout the night and day. If one station wants to contact another, it may do so on this frequency. Having established contact with each other, they may then shift to a working frequency to carry on their traffic. This reduces congestion on this frequency.

This frequency can also be used for the following:

1. The international urgency signal, and urgent messages concerning the safety of a ship, aircraft, or persons aboard same.
2. The international safety signal and messages preceded by this signal.
3. Brief radio operating signals.

4. Brief signal tests to determine that the station's transmitting equipment is in good working condition on this frequency.

There is another international radiotelephone frequency for calling, safety, intership, and harbor-control purposes. This is 156.8 MHz, located in the 156.25-to 162.05-MHz band. Its use is basically the same as the 2182-kHz frequency.

Transmissions by stations using the calling frequencies shall be kept to a minimum, and in general any one exchange of communications shall not exceed 3 min in duration (excepting distress or emergency signals).

If an operator is required to "stand watch" on the international distress frequency, he must continue until relieved by another operator, or he may stop listening if the particular emergency is ended.

26-8 STANDARD OPERATING PRACTICES (G)

It is bad practice to initiate a call or test before listening on the frequency to be used. Failure to do so may interfere with someone else's transmission already on the frequency. Normally, an operator should not transmit on a frequency he cannot monitor.

Be brief in your conversations. Think out what you want to say so that the message will be coherent and to the point.

If transmitting and receiving conditions are poor due to static, interference, or fading, it may be desirable to let routine messages wait until conditions improve. It may be possible to shift to a better frequency. It is poor practice to tie up a channel with slow-moving nonemergency traffic.

An operator is responsible for all signals transmitted over his station. If he allows someone else to speak over the microphone, he must be certain that the person does not violate any of the FCC rules and regulations.

Avoid long conversations. Allow for frequent breaks so that the receiving station may come in to advise you if he is not copying you for some reason.

A radiotelephone transmitter should be tested at least once each day to determine that it is in proper working condition. This can be done by making a test transmission to an appropriate station. If a radiotelephone transmitter is in normal use during the day, it is not necessary to make any special test.

The governments of countries that a mobile station visits may require the operator to produce his and the station's licenses for examination. The operator must permit this course of action. The license should be kept so that it can be produced upon request. It should be permanently exhibited in the station. This is part of the Geneva Treaty of 1959.

If a radio transmitter is in a public place, it should be attended at all times by, or supervised by, a licensed operator, or the transmitter should be made inaccessible to unauthorized persons.

No person receiving, or assisting in receiving, or transmitting any

Miscellaneous (G)

interstate or foreign communication by wire or radio shall divulge or publish the existence, contents, or meaning of the message to any person other than the addressee or his authorized agents. This does not apply to the receiving, divulging, or publishing of contents of any radio communication brodacast, radio amateur transmission, or others for the general public use. Likewise, it does not apply to ships or aircraft in distress.

26-9 MISCELLANEOUS (G)

Coastal stations have been established to provide essential communication with surface vessels and/or aircraft. This includes the reception and transmission of distress, alarm, urgent, or safety signals in radiotelephone service that have a bearing on the safety of life or property. These stations must also keep accurate logs of all their communications. Time notations (for transmissions, message receptions, equipment failures, etc.) are usually in Greenwich Mean Time (except on inland waters where local time is used).

Public coastal stations are not permitted to charge for the transmission, receipt, or relaying of information concerning dangers to navigation.

The licensee of any radio station having an antenna structure requiring illumination to minimize aircraft hazards must do the following:

1. Make an observation of the tower lights as least once each 24 h either visually or by observing automatic equipment designed to register any failure of such lights.
2. Report immediately by telephone or telegraph to the nearest Airways Communication Station or FAA office any failure of a code or rotating beacon light or top light not corrected within 30 min, regardless of the cause of failure.
3. Provide immediate notification, by telephone or telegraph, to the appropriate agency when the required illumination is resumed.
4. Inspect, at intervals not to exceed 3 months, all automatic or mechanical control devices associated with the tower lighting to ensure that the equipment is functioning properly.
5. The licensee must also ensure that the following information is entered in the station log:
   a. Time the tower lights are turned off and on and the daily inspection made, if manually controlled.
   b. The nature of any failure and the date and time observed. The time and date the appropriate agency was notified (more than 30-min failure)
   c. Date, time, and nature of adjustments or repairs that are made and when appropriate agency was notified.
   d. Dates and times of all inspections and adjustments.

Radio transmitters should be kept off the air when signals are not being transmitted so that the carrier radiation will not cause interference. The operator should not press the push-to-talk button except when he intends to transmit.

Commercial License Questions

Applicants for this element have the choice of answering 20 questions in either the general or marine category. Sections in which answers to questions are given appear in parentheses.

General

1. Why is the station's call sign transmitted? (26–1)
2. When may an operator use his station without regard to certain provisions of his station license? (26–2)
3. What may happen to the received signal when an operator has shouted into a microphone? (26–3)
4. How should a microphone be treated when used in noisy locations? (26–3)
5. What is meant by a phonetic alphabet in radiotelephone communications? (26–4)
6. What are the meanings of clear, out, over, roger, words twice, repeat, and break? (26–5)
7. Why should an operator use well-known words and phrases? (26–5)
8. What should an operator do when he leaves a transmitter unattended? (26–8)
9. What should an operator do if he hears profanity being used at his station? (26–8)
10. Who bears the responsibility if an operator permits an unlicensed person to speak over his station? (26–8)
11. How does the licensed operator of a station normally exhibit his authority to operate the station? (26–8)
12. What precautions should be observed in testing a station on the air? (26–8)
13. Why should radio transmitters be off when signals are not being transmitted? (26–9)
14. Where does an operator find specifications for obstruction marking and lighting (where required) for the antenna towers of a particular radio station? (26–9)

Marine

1. Describe completely what actions should be taken by a radio operator who hears a distress message; a safety message. (26–2)

Commercial License Questions 621

2. What information must be contained in distress messages? What procedure should be followed by a radio operator in sending a distress message? What is a good choice of words to be used in sending a distress message? (26-2)

3. What do distress, safety, and urgency signals indicate? What are the international urgency, safety, and distress signals? (26-2)

4. In the case of a mobile radio station in distress, what station is responsible for the control of distress-message traffic? (26-2)

5. In the mobile service, why should radiotelephone messages be as brief as possible? (26-5, 26-8)

6. What are the meanings of clear, out, over, roger, words twice, repeat, and break? (26-5)

7. Why are call signs sent? Why should they be sent clearly and distinctly? (26-6)

8. How long may a radio operator in the mobile service continue attempting to contact a station that does not answer? (26-6)

9. Under what circumstances may a coast station contact a land station by radio? (26-6, 26-7)

10. What is the importance of the frequency 2182 kHz? (26-7)

11. What are the requirements for keeping watch on 2182 kHz? If a radio operator is required to "stand watch" on an international distress frequency, when may he stop listening? (26-7)

12. In regions of heavy traffic, why should an interval be left between radiotelephone calls? Why should a radio operator listen before transmitting on a shared channel? (26-7, 26-8)

13. How often should the station's call sign be sent? (26-7, 26-8)

14. What is the difference between calling and working frequencies? (26-7)

15. Why are test transmissions sent? How often should they be sent? (26-8)

16. Does the Geneva 1959 treaty give other countries the authority to inspect U.S. vessels? (26-8)

17. How does the licensed operator of a ship station exhibit his authority to operate a station? (26-8)

18. When may a coast station *not* charge for messages it is requested to handle? (26-9)

27

Logarithms and Decibels—
Microphones and Loudspeakers

27-1 LOGARITHMS

Decibels are extensively used in electronics to measure differences in sound, voltage, current, and power levels. They are *logarithmic* in nature, and therefore a brief review of this branch of mathematics is in order.

From our previous studies involving the use of powers of 10, we found that $10^3 = 1000$. That is, the *exponent* 3 indicated the *power* that 10 must be raised to in order to equal 1000. This is illustrated by the following:

$$10^2 = 100, \quad 10^1 = 10, \quad 10^0 = 1, \quad 10^{-1} = 0.1, \quad 10^{-2} = 0.01$$

However, exponents can be decimal fractions as well as integers, such as $10^{0.3}$. If $10^0 = 1$ and $10^1 = 10$, it follows that $10^{0.3}$ must be some number between 1 and 10. (In these examples 10 is the base.) By referring to a table of logarithms, such as in Appendix H, we can find the number that $10^{0.3}$ equals. Search the vertical columns containing four digits each until you find the number closest to 3000. This is 3010. Opposite this in the N (number) column we find 20. (As we shall see, the table gives the same number for 20, 0.2, 2, or any multiple thereof.) Therefore, $10^{0.3}$ is approximately equal to 2.

Because common logarithms are so universally used, the 10 is usually omitted. The logarithm of a number is divided into two parts: the *integral* and the *decimal*. The integral part is known as the *characteristic* and the decimal part is called the *mantissa*.

The characteristic of a number can be determined by the following rules:

1. *The characteristic of a number larger than 1 is positive and is 1 less than the number of digits to the left of the decimal point.*

Decibels   623

2. *The characteristic of a positive number less than 1 is negative and is 1 more than the number of zeros immediately to the right of the decimal point.*

Therefore, the number 47 has a characteristic of 1; 6281 has a characteristic of 3; 0.0033 has a characteristic of -3.

The mantissa is always the same for a given sequence of integers, regardless of where the decimal point appears among them. For example, the mantissa is the same for 1570, 157, 15.7, 1.57 0.157, and so on. The logarithms of these numbers differ only in their characteristics. Hence, their logarithms are, respectively, 3.1959, 2.1959, 1.1959, 0.1959, and -1.1959. The mantissa is *always positive*, even when the characteristic is negative.

Although the log of $0.157 = -1.1959$, it is not customary to use the negative characteristic. Instead, 10 is added to the characteristic and the same number is subtracted from the entire logarithm. Thus, the log of 0.157 is written as $9.1959 - 10$, and the log of 0.0157 as $8.1959 - 10$.

27-2 DECIBELS

The human ear has a nonlinear or logarithmic response to sound energy. The unit selected to express the ratio between sound levels is the *bel* or its subunit the *decibel* (dB). Expressed mathematically,

$$dB = 10 \log_{10} \frac{P_2}{P_1}$$

where P_1 = power input in watts
       P_2 = power output in watts

Because common logarithms are always used, the 10 is omitted.

Example 27-1

How many decibels correspond to a power ratio of 100?

Solution

$$dB = 10 \log \frac{P_2}{P_1}$$
$$= 10 \log 100$$
$$= 10 \times 2 = 20$$

Example 27-2

What is the decibel gain of an amplifier if it produces 70-W output with 5.3-mW input?

Solution

$$dB = 10 \log \frac{70}{0.0053}$$
$$= 10 \log 13207 = 10(4.1206) = 41.2$$

Example 27-3

An amplifier provides a gain of 65 dB with an input of 20 μW. What is its power output?

Solution

$$65 \text{ dB} = 10 \log \frac{P}{2 \times 10^{-5}}$$

$$6.5 = \log \frac{P}{2 \times 10^{-5}}$$

Therefore, the log of $P/(2 \times 10^{-5})$ is 6.5.

The characteristic is 6 and the mantissa is 5000. Searching through the mantissas in Appendix H, 5000 is found in the 31-6 line and column. Therefore, the power ratio $P/(2 \times 10^{-5}) = 3{,}160{,}000$. Then $P = 3.16 \times 10^6 \times 2 \times 10^{-5} = 6.28 \times 10^1$ W.

The dB is used as a unit of measurement when power decreases to a threshold value before it reaches an absolute zero. This value has been arbitrarily designated as the *zero* or *reference* level. If power levels increase above this, they are considered positive, and minus when below.

In the past several reference levels have been used as 0 dB. Today, the most commonly used one is 1 mW. When this reference is used, the resulting power level is abbreviated dBm, and means decibels above or below 1 mW. Thus, 0 dBm = 1 mW; 10 dBm = 10 mW; 30 dBm = 1 W. One decibel is equal to a power increase (or decrease) of 1.26 or 26 per cent.

Decibels can be used to express the ratio between two voltages or currents. Because power can also be calculated by E^2/R or I^2R, the basic formulas shown below can be used for making these calculations.

$$dB = 10 \log \frac{e_o^2/r_o}{e_i^2/r_i}$$

$$= 10 \log \frac{i_o^2 \, r_o}{i_i^2 \, r_i}$$

where e_o = output voltage
e_i = input voltage

When the input and output impedances of an amplifier are equal, the resistance or impedances cancel, and these formulas reduce to

$$dB = 10 \log \frac{e_o^2}{e_i^2}$$

and

$$dB = 10 \log \frac{i_o^2}{i_i^2}$$

Inasmuch as doubling the value of a logarithm is equivalent to squaring,

Decibels

these formulas can be simply written as:

$$dB = 20 \log \frac{e_o}{e_i}$$

and

$$dB = 20 \log \frac{i_o}{i_i}$$

Example 27-4

If an amplifier has an output of 23 V when its input is 2 mV, what is its decibel gain? (Assume equal input and output impedance.)

Solution

$$dB = 20 \log \frac{e_o}{e_i}$$

$$= 20 \log \frac{23}{2 \times 10^{-3}} \quad \text{or} \quad 20 \log 1.15 \times 10^4$$

$$= 20 \times 4.0607 = 81.2$$

Example 27-5

If an amplifier has a 30-dB gain, what voltage ratio does this gain represent (assume equal resistances)?

Solution

Voltage ratio is $e_o/e_i = x$.

$$dB = 20 \log x$$
$$30 = 20 \log x$$
$$\log x = 1.5$$
$$x = 31.6$$

Therefore, the voltage ratio is 31.6 to 1.

When the input and output voltage of an amplifier are known but the input and output impedances are different it is necessary to *first* calculate the power developed across the input impedance. The power developed across the output impedance is calculated in the same way. Knowing now the input and output power, the gain is determined by the use of the formula $dB = 10 \log P_o/P_i$.

Example 27-6

If a 1-V signal is applied across the 600-Ω input impedance of an amplifier and 500 V is developed across the 5-kΩ output impedance, what is the decibel power gain?

Solution

$$P_o = \frac{e_o^2}{z_o} = \frac{500^2}{5000} = 50 \text{ W}$$

$$P_i = \frac{e_i^2}{z_i} = \frac{1^2}{600} = 1.66 \text{ mW}$$

$$\text{dB} = 10 \log \frac{50}{1.66 \times 10^{-3}}$$

$$= 10 \log 3 \times 10^4$$

$$= 10(4.477) \cong 44.8$$

Practice Problems

1. Find the logarithms of the following numbers (use base 10): (a) 47,100, (b) 126, (c) 0.0891.
2. An amplifier has an input power of 100 μW and a gain of 55 dB. What is the power output?
3. A broadcast station increases its power from 1.5 to 25 kW. What decibel gain does this represent?
4. How many decibels correspond to a power gain of 1000?
5. How many decibels correspond to a voltage gain of 1000?
6. A transistorized repeater amplifier, having 600-Ω input and output impedances, produces 21.9 V$_{ac}$ when its input is 8.2 mV$_{ac}$. What is its decibel gain?
7. A high-frequency attenuator has an output of 0.6 V across 600 Ω when fed with a 6-V signal across its 600-Ω input. What is the decibel loss of the attenuator?
8. An amplifier has an input and output impedance of 5 kΩ and 8 Ω, respectively. If 5-mV input produces 17-V output, what is its gain?

27-3 MICROPHONES

A microphone *converts sound energy to electrical energy*. The sound waves cause a diaphragm to vibrate, which causes current to flow in proportion ot the instantaneous pressure applied to it.

Most microphones are relatively inefficient, producing a very small electrical output for the amount of acoustical energy input.

A microphone has a sensitivity of 0 dB (the level of comparison) if a force of 1 dyn/cm² on the diaphragm produces an output of 1 V. The usual method, however, is to assume that the 0-dB level represents an input of 1 dyn/cm² and an output of 1 mW. If it is assumed that the 1 mW is developed across an impedance of 600 Ω, then dBm or volume units (VU) may be used. Suppose that a microphone is rated at -80 dB. This means that the energy

Microphones

output is much less than the energy input. Actually, the output is 10^{-8} mW for an equivalent input of 1 mW, and this is equivalent to -80 dB.

It is important to have the microphone sensitivity as high as possible. High microphone output levels require less gain in the amplifiers used with them and thus provide a greater margin over thermal noise, amplifier hum, and noise pickup in the line between the microphone and the amplifier.

Carbon microphones are the most commonly used. They were the first developed and are still used for telephone communications everywhere, because they are inexpensive, rugged, and have high sensitivity. However, their frequency response is poor (limited essentially to voice frequencies), they have relatively high inherent noise levels, and require a source of direct current for excitation.

The principle of operation is simple. Tiny *carbon granules* are enclosed in a small cup or button with a diaphragm at the open end. A typical cross-sectional view of a single-button carbon microphone appears in Fig. 27-1.

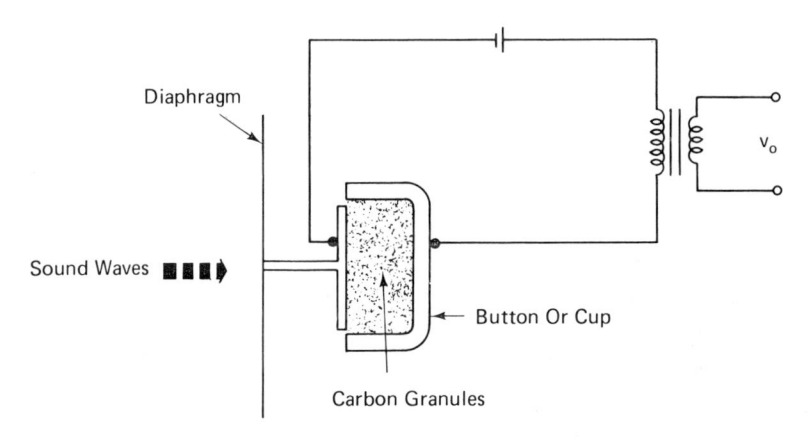

FIG. 27-1 Single-button carbon microphone.

The diaphragm is connected to a small piston or electrode. Sound vibrations striking the diaphragm vary the pressure on the carbon granules, thereby increasing and decreasing the electrical resistance between diaphragm and button. This constantly changing resistance, in proportion to the sound waves striking the diaphragm, varies the battery current flowing through the primary of the transformer. The induced voltage in the secondary is connected to the input of an amplifier. Because this kind of microphone has a low impedance, the transformer also serves to match it to the input of an amplifier.

In the *double-button* microphone, there is more of a push–pull action. Any movement of the diaphragm increases the pressure in one cup while decreasing the pressure in the other cup by approximately the same amount. The word approximately is used here, because it is highly unlikely that the two cups could be filled with exactly the same amount of carbon granules.

In the circuit shown in Fig. 27-2 the microphone button is placed in series with a battery. Any current flowing in the microphone circuit will also flow through the primary of the transformer. With no movement of the diaphragm, there will be no change in the resistance of the cup. Under these conditions, the direct current flowing through the cup will be constant, and the resulting magnetic field about the primary of the transformer will not fluctuate. There will be no voltage induced in the secondary of the transformer. However, there is a constant "hiss" in the output due to minute currents flowing through the carbon granules. The frequency response of a carbon microphone is approximately 100 to 5 kHz.

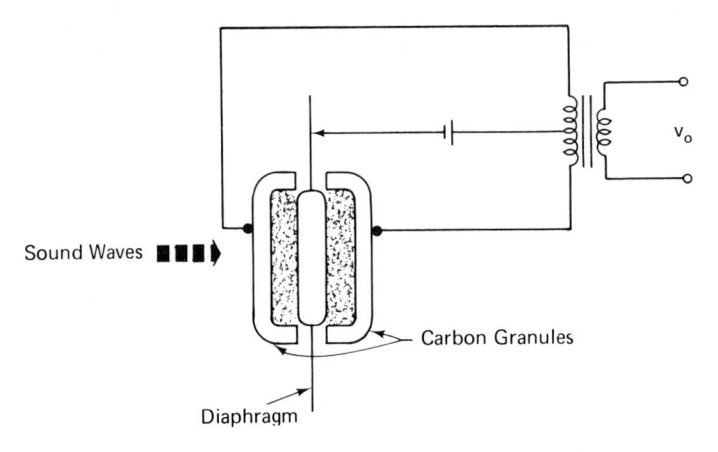

FIG. 27-2 Double-button carbon microphone.

A type of microphone that uses the principles of inductance is the *moving coil* or *dynamic* microphone, shown diagrammatically in Fig. 27-3. The coil that is wound around the pole piece is able to move back and forth

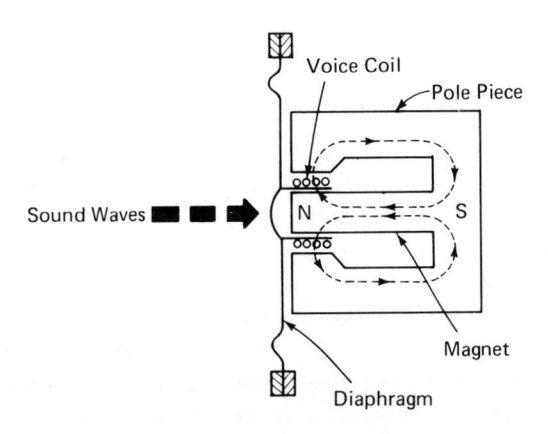

FIG. 27-3 Dynamic microphone.

Loudspeakers 629

along the pole. It is attached to a flexible diaphragm and is caused to move by sound waves striking it. As it moves, it passes between the poles of the magnet. The voltage induced into the coil when it moves will also vary at an audio rate.

The dynamic microphone has an impedance between 50 and 100 Ω. The frequency response is from about 40 to 18,000 Hz and has an output of about −45 to −65 dBm. It is light, rugged, moisture proof, small, and not subject to the effects of temperature and humidity.

Crystal microphones make use of the *piezoelectric* effect. Sound waves strike the diaphragm, as shown in Fig. 27-4, and cause a voltage to appear across the crystal that is proportional to the frequency and intensity of the incident wave. The type of crystal most widely used is Rochelle salt because of its sensitivity. The disadvantages of the crystal microphone are its sensitivity to temperature and humidity changes, and that it can be easily damaged by rough handling.

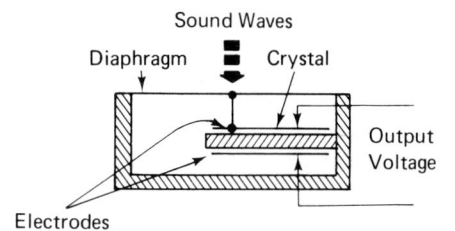

FIG. 27-4 Schematic diagram of crystal microphone.

27-4 LOUDSPEAKERS

The purpose of a loudspeaker is to convert electrical energy into acoustical energy. There are basically two types of dynamic loudspeakers: *permanent magnet* and *electromagnet*. In principle, the operation is the same. The permanent magnet or pm speaker is the most commonly used, as only two leads need be connected to it from the output of an amplifier. The electromagnetic speaker needs two additional leads to excite the electromagnet from some external source of power.

The construction of an elementary dynamic loudspeaker is shown in Fig. 27-5. The coil that is shown wrapped around the base of the cone of the loudpeaker is called a *voice coil*. The AF output current from the amplifier flows through it. Since the voice coil is located between the poles of the permanent magnet, the magnetic field created by the voice coil will interact with the stationary field of the permanent magnet. The voice coil is wound around a Bakelite or aluminum form and has the ability to move back and forth. Since the voice coil is physically connected to the paper cone, any movement of the voice coil will cause a corresponding movement of the cone. The higher the value of current, the louder the sound will be. The paper cone and voice coil are suspended from a metal frame by a flexible device called a *spider*.

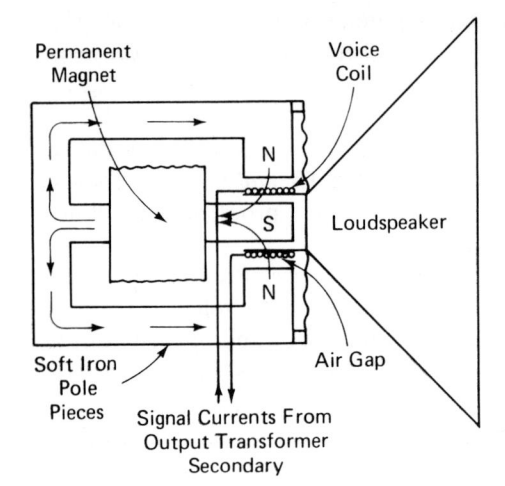

FIG. 27-5 Basic construction of a pm dynamic speaker.

The spider ensures that the voice coil remains centered around the permanent magnet, but allows back and forth movement of the voice coil and cone. It also returns the coil to its original position when no current is flowing through the coil.

A voice coil is a low-impedance device (usually 3.2, 6, 8, or 16 Ω) and is connected to the output of a power amplifier that has a high output impedance. An impedance-matching transformer is used to match the output of the amplifier to the input of the speaker.

27-5 EARPHONES

The basic components of earphones are shown in Fig. 27-6. When no signal currents are present, the permanent magnets exert a *steady pull* on the

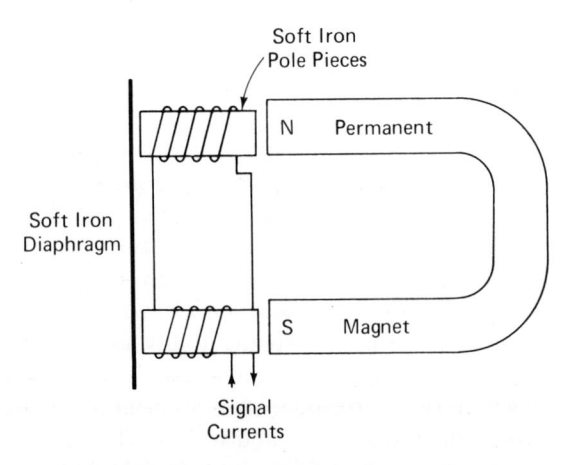

FIG. 27-6 Basic construction of an earphone.

soft-iron diaphragm. Signal currents flowing through the coils mounted on the soft-iron pole pieces develop a magnetomotive force that either adds to or subtracts from the pull of the permanent magnets. This causes the air around the diaphragm to vibrate at the frequency of the signal currents.

If earphones are connected directly in the output circuit of an amplifier, care must be taken to connect the leads in such a way that the direct current flowing through them will *add* to the pull of the magnets and not tend to demagnetize them. The manufacturer usually provides some form of identification on the leads, such as a red tracer, to indicate which one connects to the positive side of the power supply. It is less hazardous to the operator if the earphones are connected to the output of the receiver through an *output transformer*.

Both high- and low-impedance earphones are available. The latter kind have fewer turns and require more signal current to drive them. An impedance-matching transformer is necessary to provide a proper match.

Commercial License Questions

Sections in which answers to questions are given appear in parentheses. A bracketed number following a question implies that it applies only to that element.

1. Define the term *decibel*. (27–2)
2. What is the formula for determining the decibel loss or gain in a circuit? [4] (27–2)
3. If a certain audio-frequency amplifier has an overall gain of 40 dB and the output is 6 W, what is the input? [4] (27–2)
4. Draw a diagram of a single-button carbon microphone circuit, including the microphone transformer and source of power. (27–3)
5. Describe the construction and characteristics of a carbon button-type microphone. (27–3)
6. What is the most serious disadvantage of using carbon microphones with high-fidelity amplifiers? [4] (27–3)
7. What type of microphone employs a coil of wire, attached to a diaphragm, which moves in a magnetic field as the result of the impinging of sound waves? [4] (27–3)
8. What precaution should be observed when using and storing crystal microphones? (27–3)
9. Describe the construction and characteristics of a crystal-type microphone. (27–3)
10. Why are high-reactance head telephones generally more satisfactory for use with radio receivers than low-reactance types? (27–5)
11. Why should polarity be observed in connecting head telephones directly in the plate circuit of a vacuum tube? (27–5)

12. If low-impedance head telephones of the order of 75 Ω are to be connected to the output of a vacuum-tube amplifier, how may this be done to permit most satisfactory operation? (27–5)

Appendix A
Copper-Wire Table, American Wire Gauge (B. and S.)

B & S Gauge No.	Diam. in Mils at 20°C	Area Circular Mils	Ohms per 1,000 Ft. 25°C, 77°F.	Approx. Pounds per 1,000 Ft.
1	289.3	83,690.0	0.1264	253
2	257.6	66,370.0	0.1593	201
3	229.4	52,640.0	0.2009	159
4	204.3	41,740.0	0.2533	126
5	181.9	33,100.0	0.3195	100
6	162.0	26,250.0	0.4028	79
7	144.3	20,820.0	0.5080	63
8	128.5	16,510.0	0.6405	50
9	114.4	13,090.0	0.8077	40
10	101.9	10,380.0	1.018	31
11	90.74	8,234.0	1.284	25
12	80.81	6,530.0	1.619	20
13	71.96	5,178.0	2.042	15.7
14	64.08	4,107.0	2.575	12.4
15	57.07	3,257.0	3.247	9.8
16	50.82	2,583.0	4.094	7.8
17	45.26	2,048.0	5.163	6.2
18	40.30	1,624.0	6.510	4.9
19	35.89	1,288.0	8.210	3.9
20	31.96	1,022.0	10.35	3.1
21	28.46	810.1	13.05	2.5
22	25.35	642.4	16.46	1.9
23	22.57	509.5	20.76	1.5
24	20.10	404.0	26.17	1.2
25	17.90	320.4	33.00	0.97
26	15.94	254.1	41.62	0.77
27	14.20	201.5	52.48	0.61
28	12.64	159.8	66.17	0.48
29	11.26	126.7	83.44	0.38
30	10.03	100.5	105.2	0.30
31	8.93	79.70	132.7	0.24
32	7.95	63.21	167.3	0.19
33	7.08	50.13	211.0	0.15
34	6.31	39.75	266.0	0.12
35	5.62	31.52	335.5	0.095
36	5.00	25.00	423.0	0.076
37	4.45	19.83	533.4	0.060
38	3.96	15.72	672.6	0.048
39	3.53	12.47	848.1	0.038
40	3.14	9.89	1,069.0	0.030
41	2.80	7.84	1,323.0	0.0229
42	2.50	6.22	1,667.0	0.0189
43	2.22	4.93	2,105.0	0.0153
44	1.98	3.91	2,655.0	0.0121

Appendix B
Natural Sines, Cosines, and Tangents

Degs.	Function	0.0°	0.1°	0.2°	0.3°	0.4°	0.5°	0.6°	0.7°	0.8°	0.9°
0	sin	0.0000	0.0017	0.0035	0.0052	0.0070	0.0087	0.0105	0.0122	0.0140	0.0157
	cos	1.0000	1.0000	1.0000	1.0000	1.0000	1.0000	0.9999	0.9999	0.9999	0.9999
	tan	0.0000	0.0017	0.0035	0.0052	0.0070	0.0087	0.0105	0.0122	0.0140	0.0157
1	sin	0.0175	0.0192	0.0209	0.0227	0.0244	0.0262	0.0279	0.0297	0.0314	0.0332
	cos	0.9998	0.9998	0.9998	0.9997	0.9997	0.9997	0.9996	0.9996	0.9995	0.9995
	tan	0.0175	0.0192	0.0209	0.0227	0.0244	0.0262	0.0279	0.0297	0.0314	0.0332
2	sin	0.0349	0.0366	0.0384	0.0401	0.0419	0.0436	0.0454	0.0471	0.0488	0.0506
	cos	0.9994	0.9993	0.9993	0.9992	0.9991	0.9990	0.9990	0.9989	0.9988	0.9987
	tan	0.0349	0.0367	0.0384	0.0402	0.0419	0.0437	0.0454	0.0472	0.0489	0.0507
3	sin	0.0523	0.0541	0.0558	0.0576	0.0593	0.0610	0.0628	0.0645	0.0663	0.0680
	cos	0.9986	0.9985	0.9984	0.9983	0.9982	0.9981	0.9980	0.9979	0.9978	0.9977
	tan	0.0524	0.0542	0.0559	0.0577	0.0594	0.0612	0.0629	0.0647	0.0664	0.0682
4	sin	0.0698	0.0715	0.0732	0.0750	0.0767	0.0785	0.0802	0.0819	0.0837	0.0854
	cos	0.9976	0.9974	0.9973	0.9972	0.9971	0.9969	0.9968	0.9966	0.9965	0.9963
	tan	0.0699	0.0717	0.0734	0.0752	0.0769	0.0787	0.0805	0.0822	0.0840	0.0857
5	sin	0.0872	0.0889	0.0906	0.0924	0.0941	0.0958	0.0976	0.0993	0.1011	0.1028
	cos	0.9962	0.9960	0.9959	0.9957	0.9956	0.9954	0.9952	0.9951	0.9949	0.9947
	tan	0.0875	0.0892	0.0910	0.0928	0.0945	0.0963	0.0981	0.0998	0.1016	0.1033
6	sin	0.1045	0.1063	0.1080	0.1097	0.1115	0.1132	0.1149	0.1167	0.1184	0.1201
	cos	0.9945	0.9943	0.9942	0.9940	0.9938	0.9936	0.9934	0.9932	0.9930	0.9928
	tan	0.1051	0.1069	0.1086	0.1104	0.1122	0.1139	0.1157	0.1175	0.1192	0.1210
7	sin	0.1219	0.1236	0.1253	0.1271	0.1288	0.1305	0.1323	0.1340	0.1357	0.1374
	cos	0.9925	0.9923	0.9921	0.9919	0.9917	0.9914	0.9912	0.9910	0.9907	0.9905
	tan	0.1228	0.1246	0.1263	0.1281	0.1299	0.1317	0.1334	0.1352	0.1370	0.1388
8	sin	0.1392	0.1409	0.1426	0.1444	0.1461	0.1478	0.1495	0.1513	0.1530	0.1547
	cos	0.9903	0.9900	0.9898	0.9895	0.9893	0.9890	0.9888	0.9885	0.9882	0.9880
	tan	0.1405	0.1423	0.1441	0.1459	0.1477	0.1495	0.1512	0.1530	0.1548	0.1566
9	sin	0.1564	0.1582	0.1599	0.1616	0.1633	0.1650	0.1668	0.1685	0.1702	0.1719
	cos	0.9877	0.9874	0.9871	0.9869	0.9866	0.9863	0.9860	0.9857	0.9854	0.9851
	tan	0.1584	0.1602	0.1620	0.1638	0.1655	0.1673	0.1691	0.1709	0.1727	0.1745
10	sin	0.1736	0.1754	0.1771	0.1788	0.1805	0.1822	0.1840	0.1857	0.1874	0.1891
	cos	0.9848	0.9845	0.9842	0.9839	0.9836	0.9833	0.9829	0.9826	0.9823	0.9820
	tan	0.1763	0.1781	0.1799	0.1817	0.1835	0.1853	0.1871	0.1890	0.1908	0.1926
11	sin	0.1908	0.1925	0.1942	0.1959	0.1977	0.1994	0.2011	0.2028	0.2045	0.2062
	cos	0.9816	0.9813	0.9810	0.9806	0.9803	0.9799	0.9796	0.9792	0.9789	0.9785
	tan	0.1944	0.1962	0.1980	0.1998	0.2016	0.2035	0.2053	0.2071	0.2089	0.2107
12	sin	0.2079	0.2096	0.2113	0.2130	0.2147	0.2164	0.2181	0.2198	0.2215	0.2232
	cos	0.9781	0.9778	0.9774	0.9770	0.9767	0.9763	0.9759	0.9755	0.9751	0.9748
	tan	0.2126	0.2144	0.2162	0.2180	0.2199	0.2217	0.2235	0.2254	0.2272	0.2290
13	sin	0.2250	0.2267	0.2284	0.2300	0.2318	0.2334	0.2351	0.2368	0.2385	0.2402
	cos	0.9744	0.9740	0.9736	0.9732	0.9728	0.9724	0.9720	0.9715	0.9711	0.9707
	tan	0.2309	0.2327	0.2345	0.2364	0.2382	0.2401	0.2419	0.2438	0.2456	0.2475
14	sin	0.2419	0.2436	0.2453	0.2470	0.2487	0.2504	0.2521	0.2538	0.2554	0.2571
	cos	0.9703	0.9699	0.9694	0.9690	0.9686	0.9681	0.9677	0.9673	0.9668	0.9664
	tan	0.2493	0.2512	0.2530	0.2549	0.2568	0.2586	0.2605	0.2623	0.2642	0.2661
Degs.	Function	0'	6'	12'	18'	24'	30'	36'	42'	48'	54'

Natural Sines, Cosines, and Tangents

Degs.	Function	0.0°	0.1°	0.2°	0.3°	0.4°	0.5°	0.6°	0.7°	0.8°	0.9°
15	sin	0.2588	0.2605	0.2622	0.2639	0.2656	0.2672	0.2689	0.2706	0.2723	0.2740
	cos	0.9659	0.9655	0.9650	0.9646	0.9641	0.9636	0.9632	0.9627	0.9622	0.9617
	tan	0.2679	0.2698	0.2717	0.2736	0.2754	0.2773	0.2792	0.2811	0.2830	0.2849
16	sin	0.2756	0.2773	0.2790	0.2807	0.2823	0.2840	0.2857	0.2874	0.2890	0.2907
	cos	0.9613	0.9608	0.9603	0.9598	0.9593	0.9588	0.9583	0.9578	0.9573	0.9568
	tan	0.2867	0.2886	0.2905	0.2924	0.2943	0.2962	0.2981	0.3000	0.3019	0.3038
17	sin	0.2924	0.2940	0.2957	0.2974	0.2990	0.3007	0.3024	0.3040	0.3057	0.3074
	cos	0.9563	0.9558	0.9553	0.9548	0.9542	0.9537	0.9532	0.9527	0.9521	0.9516
	tan	0.3057	0.3076	0.3096	0.3115	0.3134	0.3153	0.3172	0.3191	0.3211	0.3230
18	sin	0.3090	0.3107	0.3123	0.3140	0.3156	0.3173	0.3190	0.3206	0.3223	0.3239
	cos	0.9511	0.9505	0.9500	0.9494	0.9489	0.9483	0.9478	0.9472	0.9466	0.9461
	tan	0.3249	0.3269	0.3288	0.3307	0.3327	0.3346	0.3365	0.3385	0.3404	0.3424
19	sin	0.3256	0.3272	0.3289	0.3305	0.3322	0.3338	0.3355	0.3371	0.3387	0.3404
	cos	0.9455	0.9449	0.9444	0.9438	0.9432	0.9426	0.9421	0.9415	0.9409	0.9403
	tan	0.3443	0.3463	0.3482	0.3502	0.3522	0.3541	0.3561	0.3581	0.3600	0.3620
20	sin	0.3420	0.3437	0.3453	0.3469	0.3486	0.3502	0.3518	0.3535	0.3551	0.3567
	cos	0.9397	0.9391	0.9385	0.9379	0.9373	0.9367	0.9361	0.9354	0.9348	0.9342
	tan	0.3640	0.3659	0.3679	0.3699	0.3719	0.3739	0.3759	0.3779	0.3799	0.3819
21	sin	0.3584	0.3600	0.3616	0.3633	0.3649	0.3665	0.3681	0.3697	0.3714	0.3730
	cos	0.9336	0.9330	0.9323	0.9317	0.9311	0.9304	0.9298	0.9291	0.9285	0.9278
	tan	0.3839	0.3859	0.3879	0.3899	0.3919	0.3939	0.3959	0.3979	0.4000	0.4020
22	sin	0.3746	0.3762	0.3778	0.3795	0.3811	0.3827	0.3843	0.3859	0.3875	0.3891
	cos	0.9272	0.9265	0.9259	0.9252	0.9245	0.9239	0.9232	0.9225	0.9219	0.9212
	tan	0.4040	0.4061	0.4081	0.4101	0.4122	0.4142	0.4163	0.4183	0.4204	0.4224
23	sin	0.3907	0.3923	0.3939	0.3955	0.3971	0.3987	0.4003	0.4019	0.4035	0.4051
	cos	0.9205	0.9198	0.9191	0.9184	0.9178	0.9171	0.9164	0.9157	0.9150	0.9143
	tan	0.4245	0.4265	0.4286	0.4307	0.4327	0.4348	0.4369	0.4390	0.4411	0.4431
24	sin	0.4067	0.4083	0.4099	0.4115	0.4131	0.4147	0.4163	0.4179	0.4195	0.4210
	cos	0.9135	0.9128	0.9121	0.9114	0.9107	0.9100	0.9092	0.9085	0.9078	0.9070
	tan	0.4452	0.4473	0.4494	0.4515	0.4536	0.4557	0.4578	0.4599	0.4621	0.4642
25	sin	0.4226	0.4242	0.4258	0.4274	0.4289	0.4305	0.4321	0.4337	0.4352	0.4368
	cos	0.9063	0.9056	0.9048	0.9041	0.9033	0.9026	0.9018	0.9011	0.9003	0.8996
	tan	0.4663	0.4684	0.4706	0.4727	0.4748	0.4770	0.4791	0.4813	0.4834	0.4856
26	sin	0.4384	0.4399	0.4415	0.4431	0.4446	0.4462	0.4478	0.4493	0.4509	0.4524
	cos	0.8988	0.8980	0.8973	0.8965	0.8957	0.8949	0.8942	0.8934	0.8926	0.8918
	tan	0.4877	0.4899	0.4921	0.4942	0.4964	0.4986	0.5008	0.5029	0.5051	0.5073
27	sin	0.4540	0.4555	0.4571	0.4586	0.4602	0.4617	0.4633	0.4648	0.4664	0.4679
	cos	0.8910	0.8902	0.8894	0.8886	0.8878	0.8870	0.8862	0.8854	0.8846	0.8838
	tan	0.5095	0.5117	0.5139	0.5161	0.5184	0.5206	0.5228	0.5250	0.5272	0.5295
28	sin	0.4695	0.4710	0.4726	0.4741	0.4756	0.4772	0.4787	0.4802	0.4818	0.4833
	cos	0.8829	0.8821	0.8813	0.8805	0.8796	0.8788	0.8780	0.8771	0.8763	0.8755
	tan	0.5317	0.5340	0.5362	0.5384	0.5407	0.5430	0.5452	0.5475	0.5498	0.5520
29	sin	0.4848	0.4863	0.4879	0.4894	0.4909	0.4924	0.4939	0.4955	0.4970	0.4985
	cos	0.8746	0.8738	0.8729	0.8721	0.8712	0.8704	0.8695	0.8686	0.8678	0.8669
	tan	0.5543	0.5566	0.5589	0.5612	0.5635	0.5658	0.5681	0.5704	0.5727	0.5750
Degs.	Function	0'	6'	12'	18'	24'	30'	36'	42'	48'	54'

Natural Sines, Cosines, and Tangents

Degs.	Function	0.0°	0.1°	0.2°	0.3°	0.4°	0.5°	0.6°	0.7°	0.8°	0.9°
30	sin	0.5000	0.5015	0.5030	0.5045	0.5060	0.5075	0.5090	0.5105	0.5120	0.5135
	cos	0.8660	0.8652	0.8643	0.8634	0.8625	0.8616	0.8607	0.8599	0.8590	0.8581
	tan	0.5774	0.5797	0.5820	0.5844	0.5867	0.5890	0.5914	0.5938	0.5961	0.5985
31	sin	0.5150	0.5165	0.5180	0.5195	0.5210	0.5225	0.5240	0.5255	0.5270	0.5284
	cos	0.8572	0.8563	0.8554	0.8545	0.8536	0.8526	0.8517	0.8508	0.8499	0.8490
	tan	0.6009	0.6032	0.6056	0.6080	0.6104	0.6128	0.6152	0.6176	0.6200	0.6224
32	sin	0.5299	0.5314	0.5329	0.5344	0.5358	0.5373	0.5388	0.5402	0.5417	0.5432
	cos	0.8480	0.8471	0.8462	0.8453	0.8443	0.8434	0.8425	0.8415	0.8406	0.8396
	tan	0.6249	0.6273	0.6297	0.6322	0.6346	0.6371	0.6395	0.6420	0.6445	0.6469
33	sin	0.5446	0.5461	0.5476	0.5490	0.5505	0.5519	0.5534	0.5548	0.5563	0.5577
	cos	0.8387	0.8377	0.8368	0.8358	0.8348	0.8339	0.8329	0.8320	0.8310	0.8300
	tan	0.6494	0.6519	0.6544	0.6569	0.6594	0.6619	0.6644	0.6669	0.6694	0.6720
34	sin	0.5592	0.5606	0.5621	0.5635	0.5650	0.5664	0.5678	0.5693	0.5707	0.5721
	cos	0.8290	0.8281	0.8271	0.8261	0.8251	0.8241	0.8231	0.8221	0.8211	0.8202
	tan	0.6745	0.6771	0.6796	0.6822	0.6847	0.6873	0.6899	0.6924	0.6950	0.6976
35	sin	0.5736	0.5750	0.5764	0.5779	0.5793	0.5807	0.5821	0.5835	0.5850	0.5864
	cos	0.8192	0.8181	0.8171	0.8161	0.8151	0.8141	0.8131	0.8121	0.8111	0.8100
	tan	0.7002	0.7028	0.7054	0.7080	0.7107	0.7133	0.7159	0.7186	0.7212	0.7239
36	sin	0.5878	0.5892	0.5906	0.5920	0.5934	0.5948	0.5962	0.5976	0.5990	0.6004
	cos	0.8090	0.8080	0.8070	0.8059	0.8049	0.8039	0.8028	0.8018	0.8007	0.7997
	tan	0.7265	0.7292	0.7319	0.7346	0.7373	0.7400	0.7427	0.7454	0.7481	0.7508
37	sin	0.6018	0.6032	0.6046	0.6060	0.6074	0.6088	0.6101	0.6115	0.6129	0.6143
	cos	0.7986	0.7976	0.7965	0.7955	0.7944	0.7934	0.7923	0.7912	0.7902	0.7891
	tan	0.7536	0.7563	0.7590	0.7618	0.7646	0.7673	0.7701	0.7729	0.7757	0.7785
38	sin	0.6157	0.6170	0.6184	0.6198	0.6211	0.6225	0.6239	0.6252	0.6266	0.6280
	cos	0.7880	0.7869	0.7859	0.7848	0.7837	0.7826	0.7815	0.7804	0.7793	0.7782
	tan	0.7813	0.7841	0.7869	0.7898	0.7926	0.7954	0.7983	0.8012	0.8040	0.8069
39	sin	0.6293	0.6307	0.6320	0.6334	0.6347	0.6361	0.6374	0.6388	0.6401	0.6414
	cos	0.7771	0.7760	0.7749	0.7738	0.7727	0.7716	0.7705	0.7694	0.7683	0.7672
	tan	0.8098	0.8127	0.8156	0.8185	0.8214	0.8243	0.8273	0.8302	0.8332	0.8361
40	sin	0.6428	0.6441	0.6455	0.6468	0.6481	0.6494	0.6508	0.6521	0.6534	0.6547
	cos	0.7660	0.7649	0.7638	0.7627	0.7615	0.7604	0.7593	0.7581	0.7570	0.7559
	tan	0.8391	0.8421	0.8451	0.8481	0.8511	0.8541	0.8571	0.8601	0.8632	0.8662
41	sin	0.6561	0.6574	0.6587	0.6600	0.6613	0.6626	0.6639	0.6652	0.6665	0.6678
	cos	0.7547	0.7536	0.7524	0.7513	0.7501	0.7490	0.7478	0.7466	0.7455	0.7443
	tan	0.8693	0.8724	0.8754	0.8785	0.8816	0.8847	0.8878	0.8910	0.8941	0.8972
42	sin	0.6691	0.6704	0.6717	0.6730	0.6743	0.6756	0.6769	0.6782	0.6794	0.6807
	cos	0.7431	0.7420	0.7408	0.7396	0.7385	0.7373	0.7361	0.7349	0.7337	0.7325
	tan	0.9004	0.9036	0.9067	0.9099	0.9131	0.9163	0.9195	0.9228	0.9260	0.9293
43	sin	0.6820	0.6833	0.6845	0.6858	0.6871	0.6884	0.6896	0.6909	0.6921	0.6934
	cos	0.7314	0.7302	0.7290	0.7278	0.7266	0.7254	0.7242	0.7230	0.7218	0.7206
	tan	0.9325	0.9358	0.9391	0.9424	0.9457	0.9490	0.9523	0.9556	0.9590	0.9623
44	sin	0.6947	0.6959	0.6972	0.6984	0.6997	0.7009	0.7022	0.7034	0.7046	0.7059
	cos	0.7193	0.7181	0.7169	0.7157	0.7145	0.7133	0.7120	0.7108	0.7096	0.7083
	tan	0.9657	0.9691	0.9725	0.9759	0.9793	0.9827	0.9861	0.9896	0.9930	0.9965
Degs.	Function	0'	6'	12'	18'	24'	30'	36'	42'	48'	54'

Natural Sines, Cosines, and Tangents

Degs.	Function	0.0°	0.1°	0.2°	0.3°	0.4°	0.5°	0.6°	0.7°	0.8°	0.9°
45	sin	0.7071	0.7083	0.7096	0.7108	0.7120	0.7133	0.7145	0.7157	0.7169	0.7181
	cos	0.7071	0.7059	0.7046	0.7034	0.7022	0.7009	0.6997	0.6984	0.6972	0.6959
	tan	1.0000	1.0035	1.0070	1.0105	1.0141	1.0176	1.0212	1.0247	1.0283	1.0319
46	sin	0.7193	0.7206	0.7218	0.7230	0.7242	0.7254	0.7266	0.7278	0.7290	0.7302
	cos	0.6947	0.6934	0.6921	0.6909	0.6896	0.6884	0.6871	0.6858	0.6845	0.6833
	tan	1.0355	1.0392	1.0428	1.0464	1.0501	1.0538	1.0575	1.0612	1.0649	1.0686
47	sin	0.7314	0.7325	0.7337	0.7349	0.7361	0.7373	0.7385	0.7396	0.7408	0.7420
	cos	0.6820	0.6807	0.6794	0.6782	0.6769	0.6756	0.6743	0.6730	0.6717	0.6704
	tan	1.0724	1.0761	1.0799	1.0837	1.0875	1.0913	1.0951	1.0990	1.1028	1.1067
48	sin	0.7431	0.7443	0.7455	0.7466	0.7478	0.7490	0.7501	0.7513	0.7524	0.7536
	cos	0.6691	0.6678	0.6665	0.6652	0.6639	0.6626	0.6613	0.6600	0.6587	0.6574
	tan	1.1106	1.1145	1.1184	1.1224	1.1263	1.1303	1.1343	1.1383	1.1423	1.1463
49	sin	0.7547	0.7559	0.7570	0.7581	0.7593	0.7604	0.7615	0.7627	0.7638	0.7649
	cos	0.6561	0.6547	0.6534	0.6521	0.6508	0.6494	0.6481	0.6468	0.6455	0.6441
	tan	1.1504	1.1544	1.1585	1.1626	1.1667	1.1708	1.1750	1.1792	1.1833	1.1875
50	sin	0.7660	0.7672	0.7683	0.7694	0.7705	0.7716	0.7727	0.7738	0.7749	0.7760
	cos	0.6428	0.6414	0.6401	0.6388	0.6374	0.6361	0.6347	0.6334	0.6320	0.6307
	tan	1.1918	1.1960	1.2002	1.2045	1.2088	1.2131	1.2174	1.2218	1.2261	1.2305
51	sin	0.7771	0.7782	0.7793	0.7804	0.7815	0.7826	0.7837	0.7848	0.7859	0.7869
	cos	0.6293	0.6280	0.6266	0.6252	0.6239	0.6225	0.6211	0.6198	0.6184	0.6170
	tan	1.2349	1.2393	1.2437	1.2482	1.2527	1.2572	1.2617	1.2662	1.2708	1.2753
52	sin	0.7880	0.7891	0.7902	0.7912	0.7923	0.7934	0.7944	0.7955	0.7965	0.7976
	cos	0.6157	0.6143	0.6129	0.6115	0.6101	0.6088	0.6074	0.6060	0.6046	0.6032
	tan	1.2799	1.2846	1.2892	1.2938	1.2985	1.3032	1.3079	1.3127	1.3175	1.3222
53	sin	0.7986	0.7997	0.8007	0.8018	0.8028	0.8039	0.8049	0.8059	0.8070	0.8080
	cos	0.6018	0.6004	0.5990	0.5976	0.5962	0.5948	0.5934	0.5920	0.5906	0.5892
	tan	1.3270	1.3319	1.3367	1.3416	1.3465	1.3514	1.3564	1.3613	1.3663	1.3713
54	sin	0.8090	0.8100	0.8111	0.8121	0.8131	0.8141	0.8151	0.8161	0.8171	0.8181
	cos	0.5878	0.5864	0.5850	0.5835	0.5821	0.5807	0.5793	0.5779	0.5764	0.5750
	tan	1.3764	1.3814	1.3865	1.3916	1.3968	1.4019	1.4071	1.4124	1.4176	1.4229
55	sin	0.8192	0.8202	0.8211	0.8221	0.8231	0.8241	0.8251	0.8261	0.8271	0.8281
	cos	0.5736	0.5721	0.5707	0.5693	0.5678	0.5664	0.5650	0.5635	0.5621	0.5606
	tan	1.4281	1.4335	1.4388	1.4442	1.4496	1.4550	1.4605	1.4659	1.4715	1.4770
56	sin	0.8290	0.8300	0.8310	0.8320	0.8329	0.8339	0.8348	0.8358	0.8368	0.8377
	cos	0.5592	0.5577	0.5563	0.5548	0.5534	0.5519	0.5505	0.5490	0.5476	0.5461
	tan	1.4826	1.4882	1.4938	1.4994	1.5051	1.5108	1.5166	1.5224	1.5282	1.5340
57	sin	0.8387	0.8396	0.8406	0.8415	0.8425	0.8434	0.8443	0.8453	0.8462	0.8471
	cos	0.5446	0.5432	0.5417	0.5402	0.5388	0.5373	0.5358	0.5344	0.5329	0.5314
	tan	1.5399	1.5458	1.5517	1.5577	1.5637	1.5697	1.5757	1.5818	1.5880	1.5941
58	sin	0.8480	0.8490	0.8499	0.8508	0.8517	0.8526	0.8536	0.8545	0.8554	0.8563
	cos	0.5299	0.5284	0.5270	0.5255	0.5240	0.5225	0.5210	0.5195	0.5180	0.5165
	tan	1.6003	1.6066	1.6128	1.6191	1.6255	1.6319	1.6383	1.6447	1.6512	1.6577
59	sin	0.8572	0.8581	0.8590	0.8599	0.8607	0.8616	0.8625	0.8634	0.8643	0.8652
	cos	0.5150	0.5135	0.5120	0.5105	0.5090	0.5075	0.5060	0.5045	0.5030	0.5015
	tan	1.6643	1.6709	1.6775	1.6842	1.6909	1.6977	1.7045	1.7113	1.7182	1.7251
Degs.	Function	0'	6'	12'	18'	24'	30'	36'	42'	48'	54'

Natural Sines, Cosines, and Tangents

Degs.	Function	0.0°	0.1°	0.2°	0.3°	0.4°	0.5°	0.6°	0.7°	0.8°	0.9°
60	sin	0.8660	0.8669	0.8678	0.8686	0.8695	0.8704	0.8712	0.8721	0.8729	0.8738
	cos	0.5000	0.4985	0.4970	0.4955	0.4939	0.4924	0.4909	0.4894	0.4879	0.4863
	tan	1.7321	1.7391	1.7461	1.7532	1.7603	1.7675	1.7747	1.7820	1.7893	1.7966
61	sin	0.8746	0.8755	0.8763	0.8771	0.8780	0.8788	0.8796	0.8805	0.8813	0.8821
	cos	0.4848	0.4833	0.4818	0.4802	0.4787	0.4772	0.4756	0.4741	0.4726	0.4710
	tan	1.8040	1.8115	1.8190	1.8265	1.8341	1.8418	1.8495	1.8572	1.8650	1.8728
62	sin	0.8829	0.8838	0.8846	0.8854	0.8862	0.8870	0.8878	0.8886	0.8894	0.8902
	cos	0.4695	0.4679	0.4664	0.4648	0.4633	0.4617	0.4602	0.4586	0.4571	0.4555
	tan	1.8807	1.8887	1.8967	1.9047	1.9128	1.9210	1.9292	1.9375	1.9458	1.9542
63	sin	0.8910	0.8918	0.8926	0.8934	0.8942	0.8949	0.8957	0.8965	0.8973	0.8980
	cos	0.4540	0.4524	0.4509	0.4493	0.4478	0.4462	0.4446	0.4431	0.4415	0.4399
	tan	1.9626	1.9711	1.9797	1.9883	1.9970	2.0057	2.0145	2.0233	2.0323	2.0413
64	sin	0.8988	0.8996	0.9003	0.9011	0.9018	0.9026	0.9033	0.9041	0.9048	0.9056
	cos	0.4384	0.4368	0.4352	0.4337	0.4321	0.4305	0.4289	0.4274	0.4258	0.4242
	tan	2.0503	2.0594	2.0686	2.0778	2.0872	2.0965	2.1060	2.1155	2.1251	2.1348
65	sin	0.9063	0.9070	0.9078	0.9085	0.9092	0.9100	0.9107	0.9114	0.9121	0.9128
	cos	0.4226	0.4210	0.4195	0.4179	0.4163	0.4147	0.4131	0.4115	0.4099	0.4083
	tan	2.1445	2.1543	2.1642	2.1742	2.1842	2.1943	2.2045	2.2148	2.2251	2.2355
66	sin	0.9135	0.9143	0.9150	0.9157	0.9164	0.9171	0.9178	0.9184	0.9191	0.9198
	cos	0.4067	0.4051	0.4035	0.4019	0.4003	0.3987	0.3971	0.3955	0.3939	0.3923
	tan	2.2460	2.2566	2.2673	2.2781	2.2889	2.2998	2.3109	2.3220	2.3332	2.3445
67	sin	0.9205	0.9212	0.9219	0.9225	0.9232	0.9239	0.9245	0.9252	0.9259	0.9265
	cos	0.3907	0.3891	0.3875	0.3859	0.3843	0.3827	0.3811	0.3795	0.3778	0.3762
	tan	2.3559	2.3673	2.3789	2.3906	2.4023	2.4142	2.4262	2.4383	2.4504	2.4627
68	sin	0.9272	0.9278	0.9285	0.9291	0.9298	0.9304	0.9311	0.9317	0.9323	0.9330
	cos	0.3746	0.3730	0.3714	0.3697	0.3681	0.3665	0.3649	0.3633	0.3616	0.3600
	tan	2.4751	2.4876	2.5002	2.5129	2.5257	2.5386	2.5517	2.5649	2.5782	2.5916
69	sin	0.9336	0.9342	0.9348	0.9354	0.9361	0.9367	0.9373	0.9379	0.9385	0.9391
	cos	0.3584	0.3567	0.3551	0.3535	0.3518	0.3502	0.3486	0.3469	0.3453	0.3437
	tan	2.6051	2.6187	2.6325	2.6464	2.6605	2.6746	2.6889	2.7034	2.7179	2.7326
70	sin	0.9397	0.9403	0.9409	0.9415	0.9421	0.9426	0.9432	0.9438	0.9444	0.9449
	cos	0.3420	0.3404	0.3387	0.3371	0.3355	0.3338	0.3322	0.3305	0.3289	0.3272
	tan	2.7475	2.7625	2.7776	2.7929	2.8083	2.8239	2.8397	2.8556	2.8716	2.8878
71	sin	0.9455	0.9461	0.9466	0.9472	0.9478	0.9483	0.9489	0.9494	0.9500	0.9505
	cos	0.3256	0.3239	0.3223	0.3206	0.3190	0.3173	0.3156	0.3140	0.3123	0.3107
	tan	2.9042	2.9208	2.9375	2.9544	2.9714	2.9887	3.0061	3.0237	3.0415	3.0595
72	sin	0.9511	0.9516	0.9521	0.9527	0.9532	0.9537	0.9542	0.9548	0.9553	0.9558
	cos	0.3090	0.3074	0.3057	0.3040	0.3024	0.3007	0.2990	0.2974	0.2957	0.2940
	tan	3.0777	3.0961	3.1146	3.1334	3.1524	3.1716	3.1910	3.2106	3.2305	3.2506
73	sin	0.9563	0.9568	0.9573	0.9578	0.9583	0.9588	0.9593	0.9598	0.9603	0.9608
	cos	0.2924	0.2907	0.2890	0.2874	0.2857	0.2840	0.2823	0.2807	0.2790	0.2773
	tan	3.2709	3.2914	3.3122	3.3332	3.3544	3.3759	3.3977	3.4197	3.4420	3.4646
74	sin	0.9613	0.9617	0.9622	0.9627	0.9632	0.9636	0.9641	0.9646	0.9650	0.9655
	cos	0.2756	0.2740	0.2723	0.2706	0.2689	0.2672	0.2656	0.2639	0.2622	0.2605
	tan	3.4874	3.5105	3.5339	3.5576	3.5816	3.6059	3.6305	3.6554	3.6806	3.7062
Degs.	Function	0'	6'	12'	18'	24'	30'	36'	42'	48'	54'

Natural Sines, Cosines, and Tangents

Degs.	Function	0.0°	0.1°	0.2°	0.3°	0.4°	0.5°	0.6°	0.7°	0.8°	0.9°
75	sin	0.9659	0.9664	0.9668	0.9673	0.9677	0.9681	0.9686	0.9690	0.9694	0.9699
	cos	0.2588	0.2571	0.2554	0.2538	0.2521	0.2504	0.2487	0.2470	0.2453	0.2436
	tan	3.7321	3.7583	3.7848	3.8118	3.8391	3.8667	3.8947	3.9232	3.9520	3.9812
76	sin	0.9703	0.9707	0.9711	0.9715	0.9720	0.9724	0.9728	0.9732	0.9736	0.9740
	cos	0.2419	0.2402	0.2385	0.2368	0.2351	0.2334	0.2317	0.2300	0.2284	0.2267
	tan	4.0108	4.0408	4.0713	4.1022	4.1335	4.1653	4.1976	4.2303	4.2635	4.2972
77	sin	0.9744	0.9748	0.9751	0.9755	0.9759	0.9763	0.9767	0.9770	0.9774	0.9778
	cos	0.2250	0.2232	0.2215	0.2198	0.2181	0.2164	0.2147	0.2130	0.2113	0.2096
	tan	4.3315	4.3662	4.4015	4.4374	4.4737	4.5107	4.5483	4.5864	4.6252	4.6646
78	sin	0.9781	0.9785	0.9789	0.9792	0.9796	0.9799	0.9803	0.9806	0.9810	0.9813
	cos	0.2079	0.2062	0.2045	0.2028	0.2011	0.1994	0.1977	0.1959	0.1942	0.1925
	tan	4.7046	4.7453	4.7867	4.8288	4.8716	4.9152	4.9594	5.0045	5.0504	5.0970
79	sin	0.9816	0.9820	0.9823	0.9826	0.9829	0.9833	0.9836	0.9839	0.9842	0.9845
	cos	0.1908	0.1891	0.1874	0.1857	0.1840	0.1822	0.1805	0.1788	0.1771	0.1754
	tan	5.1446	5.1929	5.2422	5.2924	5.3435	5.3955	5.4486	5.5026	5.5578	5.6140
80	sin	0.9848	0.9851	0.9854	0.9857	0.9860	0.9863	0.9866	0.9869	0.9871	0.9874
	cos	0.1736	0.1719	0.1702	0.1685	0.1668	0.1650	0.1633	0.1616	0.1599	0.1582
	tan	5.6713	5.7297	5.7894	5.8502	5.9124	5.9758	6.0405	6.1066	6.1742	6.2432
81	sin	0.9877	0.9880	0.9882	0.9885	0.9888	0.9890	0.9893	0.9895	0.9898	0.9900
	cos	0.1564	0.1547	0.1530	0.1513	0.1495	0.1478	0.1461	0.1444	0.1426	0.1409
	tan	6.3138	6.3859	6.4596	6.5350	6.6122	6.6912	6.7720	6.8548	6.9395	7.0264
82	sin	0.9903	0.9905	0.9907	0.9910	0.9912	0.9914	0.9917	0.9919	0.9921	0.9923
	cos	0.1392	0.1374	0.1357	0.1340	0.1323	0.1305	0.1288	0.1271	0.1253	0.1236
	tan	7.1154	7.2066	7.3002	7.3962	7.4947	7.5958	7.6996	7.8062	7.9158	8.0285
83	sin	0.9925	0.9928	0.9930	0.9932	0.9934	0.9936	0.9938	0.9940	0.9942	0.9943
	cos	0.1219	0.1201	0.1184	0.1167	0.1149	0.1132	0.1115	0.1097	0.1080	0.1063
	tan	8.1443	8.2636	8.3863	8.5126	8.6427	8.7769	8.9152	9.0579	9.2052	9.3572
84	sin	0.9945	0.9947	0.9949	0.9951	0.9952	0.9954	0.9956	0.9957	0.9959	0.9960
	cos	0.1045	0.1028	0.1011	0.0993	0.0976	0.0958	0.0941	0.0924	0.0906	0.0889
	tan	9.5144	9.6768	9.8448	10.02	10.20	10.39	10.58	10.78	10.99	11.20
85	sin	0.9962	0.9963	0.9965	0.9966	0.9968	0.9969	0.9971	0.9972	0.9973	0.9974
	cos	0.0872	0.0854	0.0837	0.0819	0.0802	0.0785	0.0767	0.0750	0.0732	0.0715
	tan	11.43	11.66	11.91	12.16	12.43	12.71	13.00	13.30	13.62	13.95
86	sin	0.9976	0.9977	0.9978	0.9979	0.9980	0.9981	0.9982	0.9983	0.9984	0.9985
	cos	0.0698	0.0680	0.0663	0.0645	0.0628	0.0610	0.0593	0.0576	0.0558	0.0541
	tan	14.30	14.67	15.06	15.46	15.89	16.35	16.83	17.34	17.89	18.46
87	sin	0.9986	0.9987	0.9988	0.9989	0.9990	0.9990	0.9991	0.9992	0.9993	0.9993
	cos	0.0523	0.0506	0.0488	0.0471	0.0454	0.0436	0.0419	0.0401	0.0384	0.0366
	tan	19.08	19.74	20.45	21.20	22.02	22.90	23.86	24.90	26.03	27.27
88	sin	0.9994	0.9995	0.9995	0.9996	0.9996	0.9997	0.9997	0.9997	0.9998	0.9998
	cos	0.0349	0.0332	0.0314	0.0297	0.0279	0.0262	0.0244	0.0227	0.0209	0.0192
	tan	28.64	30.14	31.82	33.69	35.80	38.19	40.92	44.07	47.74	52.08
89	sin	0.9998	0.9999	0.9999	0.9999	0.9999	1.000	1.000	1.000	1.000	1.000
	cos	0.0175	0.0157	0.0140	0.0122	0.0105	0.0087	0.0070	0.0052	0.0035	0.0017
	tan	57.29	63.66	71.62	81.85	95.49	114.6	143.2	191.0	286.5	573.0
Degs.	Function	0′	6′	12′	18′	24′	30′	36′	42′	48′	54′

Appendix C

Greek Alphabet

A	α	Alpha
B	β	Beta
Γ	γ	Gamma
Δ	δ	Delta
E	ϵ	Epsilon
Z	ζ	Zeta
H	η	Eta
Θ	θ	Theta
I	ι	Iota
K	κ	Kappa
Λ	λ	Lambda
M	μ	Mu
N	ν	Nu
Ξ	ξ	Xi
O	o	Omicron
Π	π	Pi
P	ρ	Rho
Σ	σ	Sigma
T	τ	Tau
Υ	υ	Upsilon
Φ	ϕ	Phi
X	χ	Chi
Ψ	ψ	Psi
Ω	ω	Omega

Appendix D
Classification of Emissions

Type of Modulation or Emission	Type of Transmission	Supplementary Characteristics	Symbol
1. Amplitude	Absence of any modulation		A0
	Telegraphy without the use of modulating audio frequency (on–off keying)		A1
	Telegraphy by the keying of a modulating audio frequency or audio frequencies or by the keying of the modulated emission (special case: an unkeyed modulated emission)		A2
	Telephony	Double sideband, full carrier	A3
		Single sideband, reduced carrier	A3a
		Two independent sidebands, reduced carrier	A3b
	Facsimile		A4
	Television		A5
	Composite transmission and cases not covered by the above		A9
	Composite transmissions	Reduced carrier	A9c
2. Frequency (or phase) modulated	Absence of any modulation		F0
	Telegraphy without the use of modulating audio frequency (frequency-shift keying)		F1
	Telegraphy by the keying of a modulating audio frequency or audio frequencies or by the keying of the modulated emission (special case: an unkeyed emission modulated by audio frequency)		F2
	Telephony		F3
	Facsimile		F4
	Television		F5
	Composite transmissions and cases not covered by the above		F9

641

Type of Modulation or Emission	Type of Transmission	Supplementary Characteristics	Symbol
3. Pulsed emissions	Absence of any modulation intended to carry information		P0
	Telegraphy without the use of modulating audio frequency		P1
	Telegraphy by the keying of a modulating audio frequency or audio frequencies, or by the keying of the modulated pulse (special case: an unkeyed modulated pulse)		
		Audio frequency or audio frequencies modulating their pulse in amplitude	P2d
		Audio frequency or audio frequencies modulating the width of the pulse	P2e
		Audio frequency or audio frequencies modulating the phase (or position) of the pulse	P2f
	Telephony	Amplitude-modulated pulse	P3d
		Width modulated pulse	P3e
		Phase (or position) modulated pulse	P3f
	Composite transmissions and cases not covered by the above		P9

Appendix E
Federal Communications Commission Field Offices

Applicants for any of the FCC licenses should contact the nearest field office of the FCC to determine the exact place and time the examination will be given and what forms and fees, if any, are required. Street address can be found in local telephone directories under the heading "United States Government." Address all communications to the Engineer-in-Charge, Federal Communications Commission.

Mobile	Alabama	36602
Anchorage	Alaska	99501
Los Angeles	California	90012
San Diego	California	92101
San Francisco	California	94111
San Pedro	California	90731
Denver	Colorado	80202
Washington	D.C.	20554
Miami	Florida	33130
Tampa	Florida	33602
Atlanta	Georgia	30303
Savannah	Georgia	31402
Honolulu	Hawaii	96808
Chicago	Illinois	60604
New Orleans	Louisiana	70130
Baltimore	Maryland	21201
Boston	Massachusetts	02109
Detroit	Michigan	48226
St. Paul	Minnesota	55101
Kansas City	Missouri	64106
Buffalo	New York	14203
New York	New York	10014
Portland	Oregon	97204
Philadelphia	Pennsylvania	19106
San Juan	Puerto Rico	00903
Beaumont	Texas	77701
Dallas	Texas	75202
Houston	Texas	77002
Norfolk	Virginia	23502
Seattle	Washington	98104

Appendix F
Commonly Used Q Signals

QRA	What is the name of your station?	The name of my station is....
QRG	Will you tell me my exact frequency (or that of...)?	Your exact frequency (or that of...) is... kHz (or MHz).
QRI	How is the tone of my transmission?	The tone of your transmission is... (good, variable, bad).
QRK	What is the readability of my signals (or those of...)?	The readability of your signals (or those of...) is... 1. Unreadable 2. Readable now and then 3. Readable, but with difficulty 4. Readable 5. Perfectly readable
QRM	Are you being interfered with?	I am being interfered with.
QRN	Are you troubled by static?	I am troubled by static.
QRQ	Shall I send faster?	Send faster (... words per minute).
QRS	Shall I send more slowly?	Send more slowly (... words per minute).
QRT	Shall I stop sending?	Stop sending.
QRU	Have you anything for me?	I have nothing for you.
QRV	Are you ready?	I am ready.
QRX	When will you call me again?	I will call you again at... hours (on... kHz or MHz).
QSA	What is the strength of my signals (or those of...)?	The strength of your signals (or those of...) is... 1. Scarcely perceptible 2. Weak 3. Fairly good 4. Good 5. Very good
QSB	Are my signals fading?	Your signals are fading.
QSY	Shall I change to transmission on another frequency?	Change to transmission on another frequency (or on... kHz or MHz).
QSZ	Shall I send each word or group more than once?	Send each word or group twice (or... times).
QTE	What is my true bearing from you? *or* What is my true bearing from... (call sign)? *or* What is the true bearing of... (call sign) from... (call sign)?	Your true bearing from me is... degrees (at... hours) *or* Your true bearing from... (call sign) was... degrees (at... hours) *or*

QTF	Will you give me the position of my station according to the bearings taken by the direction-finding stations which you control?	The true bearing of ... (call sign) from ... (call sign) was ... degrees at ... hours. The position of your station according to the bearings taken by the direction-finding stations which I control was ... latitude ... longitude, class ... at ... hours.
QTH	What is your position in latitude and longitude (or according to any other indication)?	My position is ... latitude ... longitude (or according to any other indication).
QTR	What is the correct time?	The correct time is ... hours.

Appendix G
Miscellaneous Abbreviations and Signals—General

Abbreviation or Signal	Definition
AA	All after . . . (used after a question mark to request a repetition).
AB	All before . . . (used after a question mark to request a repetition).
ABV	Repeat (*or* I repeat) the figures in abbreviated form.
ADS	Address (used after a question mark to request a repetition).
AR	End of transmission ($\cdot - \cdot - \cdot$ to be sent as one signal).
AS	Waiting period ($\cdot - \cdot \cdot \cdot$ to be sent as one signal).
BK	Signal used to interrupt a transmission in progress.
BN	All between . . . and . . . (used after a question mark to request a repetition).
BQ	A reply to an RQ.
C	Yes.
CFM	Confirm (*or* I confirm).
CL	I am closing my station.
COL	Collate (*or* I collate).
CP	General call to two or more specified stations.
CQ	General call to all stations.
CS	Call sign (used to request a call sign).
DB	I cannot give you a bearing, you are not in the calibrated sector of this station.
DC	The minimum of your signal is suitable for the bearing.
DF	Your bearing at . . . (time) was . . . degrees, in the doubtful sector of this station, with a possible error of . . . degrees.
DG	Please advise me if you note an error in the bearing given.
DI	Bearing doubtful in consequence of the bad quality of your signal.
DJ	Bearing doubtful because of interference.
DO	Bearing doubtful. Ask for another bearing later [*or* at . . . (time)].
DP	Possible error of bearing may amount to . . . degrees.
DS	Adjust your transmitter, the minimum of your signal is too broad.
DT	I cannot furnish you with a bearing; the minimum of your signal is too broad.
DY	This station is not able to determine the sense of the bearing. What is your approximate direction relative to this station?
DZ	Your bearing is reciprocal. (To be used only by the control station of a group of direction-finding stations when it is addressing stations of the same group.)
DE	Used to separate the call sign of the station called from the call sign of the calling station.
ER	Here . . . .

Abbreviation or Signal	Definition
ETA	Estimated time of arrival.
ITP	The punctuation counts.
JM	Make a series of dashes if I may transmit. Make a series of dots to stop my transmission (not to be used on 500 kHz except in cases of distress).
K	Invitation to transmit.
MN	Minute (*or* Minutes).

Appendix H
Common Logarithm Table

N	0	1	2	3	4	5	6	7	8	9
10	0000	0043	0086	0128	0170	0212	0253	0294	0334	0374
11	0414	0453	0492	0531	0569	0607	0645	0682	0719	0755
12	0792	0828	0864	0899	0934	0969	1004	1038	1072	1106
13	1139	1173	1206	1239	1271	1303	1335	1367	1399	1430
14	1461	1492	1523	1553	1584	1614	1644	1673	1703	1732
15	1761	1790	1818	1847	1875	1903	1931	1959	1987	2014
16	2041	2068	2095	2122	2148	2175	2201	2227	2253	2279
17	2304	2330	2355	2380	2405	2430	2455	2480	2504	2529
18	2553	2577	2601	2625	2648	2672	2695	2718	2742	2765
19	2788	2810	2833	2856	2878	2900	2923	2945	2967	2989
20	3010	3032	3054	3075	3096	3118	3139	3160	3181	3201
21	3222	3243	3263	3284	3304	3324	3345	3365	3385	3404
22	3424	3444	3464	3483	3502	3522	3541	3560	3579	3598
23	3617	3636	3655	3674	3692	3711	3729	3747	3766	3784
24	3802	3820	3838	3856	3874	3892	3909	3927	3945	3962
25	3979	3997	4014	4031	4048	4065	4082	4099	4116	4133
26	4150	4166	4183	4200	4216	4232	4249	4265	4281	4298
27	4314	4330	4346	4362	4378	4393	4409	4425	4440	4456
28	4472	4487	4502	4518	4533	4548	4564	4579	4594	4609
29	4624	4639	4654	4669	4683	4698	4713	4728	4742	4757
30	4771	4786	4800	4814	4829	4843	4857	4871	4886	4900
31	4914	4928	4942	4955	4969	4983	4997	5011	5024	5038
32	5051	5065	5079	5092	5105	5119	5132	5145	5159	5172
33	5185	5198	5211	5224	5237	5250	5263	5276	5289	5302
34	5315	5328	5340	5353	5366	5378	5391	5403	5416	5428
35	5441	5453	5465	5478	5490	5502	5514	5527	5539	5551
36	5563	5575	5587	5599	5611	5623	5635	5647	5658	5670
37	5682	5694	5705	5717	5729	5740	5752	5763	5775	5786
38	5798	5809	5821	5832	5843	5855	5866	5877	5888	5899
39	5911	5922	5933	5944	5955	5966	5977	5988	5999	6010
40	6021	6031	6042	6053	6064	6075	6085	6096	6107	6117
41	6128	6138	6149	6160	6170	6180	6191	6201	6212	6222
42	6232	6243	6253	6263	6274	6284	6294	6304	6314	6325
43	6335	6345	6355	6365	6375	6385	6395	6405	6415	6425
44	6435	6444	6454	6464	6474	6484	6493	6503	6513	6522
45	6532	6542	6551	6561	6571	6580	6590	6599	6609	6618
46	6628	6637	6646	6656	6665	6675	6684	6693	6702	6712
47	6721	6730	6739	6749	6758	6767	6776	6785	6794	6803
48	6812	6821	6830	6839	6848	6857	6866	6875	6884	6893
49	6902	6911	6920	6928	6937	6946	6955	6964	6972	6981
50	6990	6998	7007	7016	7024	7033	7042	7050	7059	7067
51	7076	7084	7093	7101	7110	7118	7126	7135	7143	7152
62	7160	7168	7177	7185	7193	7202	7210	7218	7226	7235
53	7243	7251	7259	7267	7275	7284	7292	7300	7308	7316
54	7324	7332	7340	7348	7356	7364	7372	7380	7388	7396

Common Logarithm Table

N	0	1	2	3	4	5	6	7	8	9
55	7404	7412	7419	7427	7435	7443	7451	7459	7466	7474
56	7482	7490	7497	7505	7513	7520	7528	7536	7543	7551
57	7559	7566	7574	7582	7589	7597	7604	7612	7619	7627
58	7634	7642	7649	7657	7664	7672	7679	7686	7694	7701
59	7709	7716	7723	7731	7738	7745	7752	7760	7767	7774
60	7782	7789	7796	7803	7810	7818	7825	7832	7839	7846
61	7853	7860	7868	7875	7882	7889	7896	7903	7910	7917
62	7924	7931	7938	7945	7952	7959	7966	7973	7980	7987
63	7993	8000	8007	8014	8021	8028	8035	8041	8048	8055
64	8062	8069	8075	8082	8089	8096	8102	8109	8116	8122
65	8129	8136	8142	8149	8156	8162	8169	8176	8182	8189
66	8195	8202	8209	8215	8222	8228	8235	8241	8248	8254
67	8261	8267	8274	8280	8287	8293	8299	8306	8312	8319
68	8325	8331	8338	8344	8351	8357	8363	8370	8376	8382
69	8388	8395	8401	8407	8414	8420	8426	8432	8439	8445
70	8451	8457	8463	8470	8476	8482	8488	8494	8500	8506
71	8513	8519	8525	8531	8537	8543	8549	8555	8561	8567
72	8573	8579	8585	8591	8597	8603	8609	8615	8621	8627
73	8633	8639	8645	8651	8657	8663	8669	8675	8681	8686
74	8692	8698	8704	8710	8716	8722	8727	8733	8739	8745
75	8751	8756	8762	8768	8774	8779	8785	8791	8797	8802
76	8808	8814	8820	8825	8831	8837	8842	8848	8854	8859
77	8865	8871	8876	8882	8887	8893	8899	8904	8910	8915
78	8921	8927	8932	8938	8943	8949	8954	8960	8965	8917
79	8976	8982	8987	8993	8998	9004	9009	9015	9020	9025
80	9031	9036	9042	9047	9053	9058	9063	9069	9074	9079
81	9085	9090	9096	9101	9106	9112	9117	9122	9128	9133
82	9138	9143	9149	9154	9159	9165	9170	9175	9180	9186
83	9191	9196	9201	9206	9212	9217	9222	9227	9232	9238
84	9243	9248	9253	9258	9263	9269	9274	9279	9284	9289
85	9294	9299	9304	9309	9315	9320	9325	9330	9335	9340
86	9345	9350	9355	9360	9365	9370	9375	9380	9385	9390
87	9395	9400	9405	9410	9415	9420	9425	9430	9435	9440
88	9445	9450	9455	9460	9465	9469	9474	9479	9484	9489
89	9494	9499	9504	9509	9513	9518	9523	9528	9533	9538
90	9542	9547	9552	9557	9562	9566	9571	9576	9581	9586
91	9590	9595	9600	9605	9609	9614	9619	9624	9628	9633
92	9638	9643	9647	9652	9657	9661	9666	9671	9675	9680
93	9685	9689	9694	9699	9703	9708	9713	9717	9722	9727
94	9731	9736	9741	9745	9750	9754	9759	9763	9768	9773
95	9777	9782	9786	9791	9795	9800	9805	9809	9814	9818
96	9823	9827	9832	9836	9841	9845	9850	9854	9859	9863
97	9868	9872	9877	9881	9886	9890	9894	9899	9903	9908
98	9912	9917	9921	9926	9930	9934	9939	9943	9948	9952
99	9956	9961	9965	9969	9974	9978	9983	9987	9991	9996

Appendix I
Standard Fixed Resistor Values

The following information concerns fixed composition resistors. The figures were standardized by the Electronics Industries Association (EIA) so that successive values cover all intermediate values within the tolerance of that column.

For example: A 330-Ω ±10 per cent resistor may have an actual value anywhere in the range from 297 to 363 Ω. Note that a 270-Ω ±10 per cent resistor may have an actual high value of 297 Ω and a 390-Ω ±10 per cent resistor may have an actual value as low as 351 Ω.

Resistors beyond the 10- to 100-Ω range of values are merely values from the chart multiplied by an appropriate power of 10 (0.1, 10, 100, etc.), such as the 330-Ω resistor in the example above.

±20%	±10%	±5%
10	10	10
		11
	12	12
		13
15	15	15
		16
	18	18
		20
22	22	22
		24
	27	27
		30
33	33	33
		36
	39	39
		43
47	47	47
		51
	56	56
		62
68	68	68
		75
	82	82
		91
100	100	100

Index

A

Abbreviations, signal, 646-647
Absorption frequency meters, 275-277
Acorn tubes, 164
Admittance, 96
Alarm messages, 619
Alternating current, 36-45
   average value, 39
   effective or root-mean-square value, 40
   frequency, 41
   generators, 36-37, 579-581
   peak and instantaneous values, 38-39
   phase relationships, 41-43
   radians, 43
   rectifier meters, 267-268
   sine wave, 37-38
   vacuum-tube voltmeters, 265-267
   vectors, 44
   waveshapes, 36, 37-38
Alternating-current circuits, 84-103
   complex, 99-101
   J operator, 90
   parallel, 96-99
   power in, 90-95
   power factor, 95
   series inductance-capacitance-resistance circuit, 88-90
   series resistance-capacitance circuit, 86-88
   series resistance-inductance circuit, 84-86
Alternators, 37, 579-581
AM, *see* Amplitude modulation

Ammeters, 5, 256-258, 270-272, 387-388
Ampere turns, 27-29
Amperes, 3
Amplifiers
   audio, *see* Audio amplifiers
   automatic gain control, 520-521
   buffer, 304, 334-335
   intermediate-frequency, 226, 485-486, 552-553
   peak-limiting, 520, 521
   power, 332-338
   radio-frequency, *see* Radio-frequency amplifiers
   two-stage, using regenerative feedback, 295-296
   video, 555
Amplitude modulation, 362-399
   antenna ammeter indications, 387-388
   balanced modulator circuits, 393-394
   bandwidth of wave, 365
   base modulation, 381
   basic concepts, 362-364
   basic system, 365-367
   carrier shift, 386-387
   checking percentage of, 381-384
   collector modulation, 380-381
   control-grid modulation, 365, 375-377
   Heising modulation, 378-380
   level of, 374
   linear radio-frequency amplifiers, 384-386
   percentage of, 369-371

Amplitude modulation *(cont.)*
    plate modulation, 367-369, 374, 375
    radiotelephone transmitters, 388-391
    sideband
        filters, 394-395
        power, 371
        single, 391-393, 395-396
    suppressor-grid modulation, 365, 377-378
Anodes, 129, 159-160
Antennas, 359, 425-470
    ammeter indications, 387-388
    atmospheric disturbances, 431-432
    broadcast station, 530-532
    calculating power in, 460-463
    collinear arrays, 446-447
    counterpoise, 440-441
    dummy, 338-339, 456
    Federal Communications Commission rules and regulations for, 619
    feed, methods of, 448-452
    folded dipoles, 442-443
    gain, 463-464
    ground waves, 426, 427-428
    half-wave, 434-436
    harmonic suppression, 456-458
    Hertz, 438-439, 453
    impedance, 437
    ionosphere, 428-429
    long-wire, 441-442
    loop, 447-448
    Marconi, 439, 454, 458
    parasitic arrays, 443-446
    phase monitors, 464
    radiation patterns, 458-459
    radio waves, characteristics of, 425-427
    sky waves, 427, 429-431
    television, 551
    transmission lines, 432-434
    tuning or loading, 452-456
    wave polarization, 437-438
Apparent power, 92, 94
Armstrong FM system, 411-412
Armstrong oscillator, 296-298
Atmospheric disturbances, 431-432
Attenuator networks, 517-520
Audio amplifiers, 199-225
    biasing, 204-207
    cathode and emitter followers, 221-222
    Class A, 199-201
    Class AB, 202-203
    Class B, 201-202

Class C, 203-204
complementary, 218-219
coupling circuits, 207-211
distortion, 211, 219
gain, 211-212, 219
inverse feedback, 219-221
paraphase, 213-214
push-pull power amplifiers, 215-217
transistor, 217-219
transistor configurations, comparison of, 213
Audio levels, broadcast stations, 516-517
Audio mixing console, 511-512
Audio oscillators, 314-315
Automatic gain control, 520-521, 554-555
Automatic volume control, 488-490
Automation, broadcast, 515-516
Autotransformers, 59
Avalanche voltage, 135-136

B

Balanced modulator, 393-394, 411-412
Band pass, 112-113
Band-pass filters, 117-119
Band-rejection filters, 117-119
Bandwidth, 112-114, 381, 405-407
Batteries
    cell connections, 588-590
    charging, 591-594
    Edison storage, 595-596
    lead-acid, 590-591, 594-595
    primary cells, 586-588
    secondary cells, 590-591
Beam power, 164
Beam-forming plates, 164
Beat-frequency oscillators, 490-491
BH magnetization curves, 29
Biasing
    to detect oscillation, 325
    field-effect transistors, 204-207
    grid-leak, 298-299, 339-341
    oscillator, 298-299
    protective, 341
    rectifying junction, 126-128
    transistors, 145-149
    transmitters, 339-341
    vacuum tubes, 204-207
Bleeder resistors, 185-187
Blocked-grid keying, 344-346
Blocking capacitors, 208

Index

Bridge rectifier, 170-172
Broadcast stations, 510-535
    antenna systems, 530-532
    attenuator networks, 517-520
    automation, 515-516
    automatic gain control amplifiers, 520-521
    components of, 511-514
    Emergency Broadcast System, 528
    frequency monitors, 521-522
    frequency-modulated, 511
    log requirements, 526-528
    microphones, 529
    modulation monitors, 513, 522-523
    operating powers, 524-525
    operator license requirements, 528-529
    proof-of-performance tests, 526
    remote broadcast facilities, 514-515
    remote control systems, 525-526
    standard, 510-511
    transmission lines, 529-530
    transmitters, 523-524
    volume-unit meters and audio levels, 516-517
Buffer amplifiers, 304, 334-335
Bulkhead feed-through capacitors, 355

C

Capacitors, 68-83
    basic, 68-69
    bulkhead feed-through, 355
    capacitive reactance, 78-81
    charging of, 70-71
    color codes, 76-77
    coupling, 207-209, 234-236, 296-299
    factors affecting, 71-72
    inductance-capacitance-resistance circuit, series, 88-90
    input filtering, 173-174
    losses in, 81
    motor, 584
    in parallel and series, 75-76
    resistance-capacitance circuit, series, 86-88
    resistance-capacitance time constant, 78-79
    screen bypass, 162-163
    series-connected filter, 184-185
    split tank, 302-304
    types of, 72-75

    unit of capacitance, 69-70
    variable, 306
    voltage ratings of, 81-82
Carrier
    frequency, 331
    power, 524
    shift, 289, 386-387
Cathodes, 129-131, 159-160
    biasing resistor, 205
    followers, 221-222
    keying, 344, 347-348
Cathode-ray oscilloscope, 273-275
Cavity resonators, 318
Channel allocations, television, 549-550
Chirping signal, 342
Chokes, filter, 174-178
Circuit breakers, 19-20, 355-356
Circuits review, 1-16
Collector, 139
    family of curves, 141-142
    modulation, 380-381
Collinear arrays, 446-447
Color codes, capacitor, 76-77
Color television, 559-569
Colpitts oscillator, 302-304
Common-base amplifier, 149-150
Common-collector amplifier, 150
Common-emitter amplifier, 149
Communication, types of, 605-606
Community antenna TV system (CATV), 551
Complementary amplifiers, 218-219
Complex waves, 373
Compound generators, 574
Condensers, *see* Capacitors
Conduction band, 123
Conductors, 17-23
Constant impedance, 202
Constant-speed motor, 578
Continuous wave transmitters, 330, 356-358
Control-grid modulation, 365, 375-377
Converters, 481-484
Copper-wire table, 633
Corona discharge, 20-21
Cosines, 634-639
Coulomb, 69
Counter emf, 46
Counterpoise, 440-441
Coupling, 207-211, 226-229, 234-236, 247-248, 296-299
Cross neutralization, 240

Crossover distortion, 219
Crystals, 307-310
   -controlled oscillators, 305-307, 313-314
   detectors, 472
   filters, 486-488
   lattice, 123
   lattice filter, 395
   microphones, 629
   ovens, 310-313
Current, 2-3
   See also Alternating current; Direct current
Current feedback, 220-221
Cutoff frequency, 114-115

D

Damping, 255
Decibels, 622-626
Decoupling circuits, 325
Deemphasis, 407-408
Degenerative feedback, 206-207, 238
Delta match feed system, 451-452
Dichroic mirrors, 561-562
Dielectric strength, 20
Difference of potential, 1
Diodes, 121, 123, 128
   detectors, 473-474
   as rectifiers, 129
   solid-state, 121, 168-169
   tunnel, 156, 319-320
   vacuum-tube, 121, 129-132, 168-169
   volt-ampere characteristic, 132
   Zener, 135-136, 187
Dipole antenna, 434, 442-443
Direct-current
   ammeter connections and shuts, 256-258
   generators, basic, 571-573
   meter movement, basic, 253-256
   motors, 575-578
   vacuum-tube voltmeters, 263-265
   voltmeters, 259-262
Discriminator, 501-503
Displacement current, 70
Distortion, 211, 219, 232, 236, 386, 387, 407
Distress message, 605-606, 614, 617, 619
Diversity receiving system, 497
Doping, 121, 123
Driver, 365
Driving power, 232-233, 241

Dummy antennas, 338-339, 456
Dynamic microphone, 628-629
Dynamic transfer characteristic curves, 199-200
Dynamotor, 578
Dynatron oscillators, 318-320

E

Earphones, 33, 630-631
Eddy currents, 57-58, 178
Edison storage battery, 595-596
Electrodynamometers, 268-269
Electrolytic capacitors, 74-75
Electromagnetic field, 109
Electromagnets, 24, 28, 573
Electromechanical filter, 395
Electromotive force, 2
Electron multiplier, 540
Electron-coupled oscillators, 304
Electron-pair bond, 123
Electrons, 1, 121-125, 163
Electrostatic energy, 68
Electrostatic field, 1, 425
Electrostatic shield, 59, 162, 447, 456
Emergency Attention Signal, 528
Emergency Broadcast System, 528
Emissions, classification of, 641-642
Emitter, 139
Emitter followers, 221-222
Equalizers, 512
Equalizing resistors, 76, 182, 184-185
Exciting power, 232-233

F

Fading, 350-351, 431
Faraday shield, 355
Farads, 69-70
Federal Communications Commission
   field offices, 643
   licenses and laws
      element 1: basic laws, 601-612
      element 2: basic operating practice, 613-621
Feedback, 206-207, 219-221, 238, 245, 295-296, 300, 302, 304
Field poles, 571
Field-effect transistors, 154-155
Filament, 129, 130

Index 655

FM, *see* Frequency modulation
Focusing, television, 537
Forward current transfer ratio, 151
Foster-Seeley discriminator, 501-503
Free-running oscillators, 322
Frequency meters, 275-277, 283-289, 513
Frequency modulation, 400-424
   broadcast stations, 511
   fundamental principles, 401-403
   modulation index, 403-404
   percentage of modulation, 404
   phase modulation, 410-413
   preemphasis and deemphasis, 407-408
   public safety radio service, 414-417
   reactance-tube modulators, 408-409
   reasons for, 400-401
   receivers, 499-505
   sidebands and bandwidth, 405-407
   solid-state modulator, 409-410
   stereo multiplex, 417-421
   transmitter, block diagram of, 413-414
Frequency monitors, 521-522
Frequency multipliers, 142-145
Frequency standards, 279-283
Frequency-deviation limits, 403-404
Frequency-shift keying, 349-350
Full-wave rectifier circuit, basic, 169-170
Fuses, 19-20, 355

G

Gain
   antenna, 463-464
   control, 520-521, 554-555
   audio amplifier, 211-212, 219
Galvanometer, 31, 33, 253-255
Gamma-match feed system, 451
Gas-filled cold cathode glow tube, 187-188
Gauss, 25, 29
Generators
   alternating current, 36-37, 579-581
   basic direct-current, 571-573
   maintenance of, 585-586
   motor, 578-579
   practical, 573
   types of, 574-575
Gilbert, 29
Greek alphabet, 640
Grid, 159-160, 162-164
   amplifier, grounded, 245-246
   current loading, 232-233

-dip meters, 278-279
keying, blocked, 344-346
-leak, 298-299, 339-341, 474-475
modulation, 365, 374-378, 387
neutralization, 238-239
tuned-plate tuned-grid oscillator, 299-301
voltage, screen, 348
Ground base amplifier, 149-150
Ground waves, 426, 427-428

H

Half-wave antenna, 434-436
Half-wave rectifiers, 167-169
Harmonic radiation, 201-202, 351-354
Harmonic suppression, 456-458
Harmonics, 242-245
Hartley oscillator, 301-302
Hazeltine neutralization system, 236-238
Heising modulation, 378-380
Henry, 48
Hertz, 37
Hertz antennas, 438-439, 453
Heterodyne frequency meters, 283-289
Heterodyning, 393
High pass filters, 115-116
High-gain antennas, 464
Horizontal deflection, 556, 558-559
Hum voltages, 337
Hydrometer, 590
Hysteresis, 29-30, 57, 81, 178

I

Iconoscope, 538-539
Image frequency, 484-485
Image orthicon, 539-541
Image-reproducing tube, 536
Impedance coupling, 207, 209, 234-236, 296-298
Impedance matching, 60-61
Induced current, 31-33
Inductance, 46-53, 65-67
   -capacitance-resistance circuit, series, 88-90
   input filter, 174-175
   -resistance circuit, series, 84-86
Induction motors, 582-583
Inductive coupling, 207

Insulators, 20-21
Interference, 400
Interlaced scanning, 538
Intermediate-frequency amplifier, 226, 485, 552-553
Internal impedance, 232
Internal resistance, 9, 168
Interstage coupling techniques, 234-236
Inverse feedback, 219-221
Ionosphere, 428-429

J

J operator, 90
Junction, rectifying, 125-129

K

Key clicks, 342
Keying relay, 334
Keying transmitters, 342-350
Kinescope, 567-569

L

Lag, 41-42
Lead-acid batteries, 590-591, 594-595
Lecher lines, 318
Lecher wires, 290
Licenses, Federal Communications Commission, 602-609
Lighthouse tubes, 164, 246
Lightning, 431
Limiters, 500
Linear radio-frequency amplifiers, 384-386
Linear voltage regulator, 189
Liquids, conduction in, 20
Load lines, transistor, 143-145
Loftin-White amplifier, 211
Log requirements
  broadcast stations, 526-528
    Federal Communications Commission, 609-610
  television stations, 549
Logarithms, 622-626, 648-649
Long-wire antennas, 441-442
Loop antennas, 447-448
Loudspeakers, 629-630
Low pass filters, 114

M

Magnetic deflection, 537
Magnetic field, 50
Magnetism review, 24-35
  ampere turns, 27-29
  BH curve, 29
  earphones, 33
  hysteresis loss, 29-30
  induced current, 31-33
  magnetic circuit, 29
  magnetic flux, 25
  permanent magnetism, 24
  permeability, 26-27
Magnetomotive force, 29
Magnetostrictive transducer, 395
Marconi antennas, 439, 454, 458
Master oscillator power amplifiers, 332-334
Maximum inverse voltage rating, 180
Maximum plate dissipation rating, 161
Maximum-power-transfer theorem, 10
Maxwell, 25
Measuring instruments, 253-293
  ammeters, 5, 256-258, 270-272, 387-388
  basic direct-current meter movement, 253-256
  carrier-shift detector, 289
  electrodynamometers, 268-269
  frequency meters, 275-277, 283-289
  frequency standards, 279-283
  grid-dip meters, 278-279
  Lecher wires, 290
  ohmmeters, 262-263
  oscilloscopes, 37, 273-275, 325, 381-384
  rectifier meters, 267-268
  voltmeters
    alternating current vacuum-tube, 265-267
    direct current, 259-262
    direct current vacuum-tube, 263-265
    peak-reading, 268
  volt-ohm-milliammeters, 263
  volume-unit meters, 272, 511, 513-514, 516-517
  watthour meters, 270
  wattmeters, 269
Mechanical filters, 488
Mechanical resonators, 395
Mercury-vapor rectifiers, 180-182
Meters, *see* Measuring instruments
Mho, 8

Index

Microfarad, 70
Microphones, 529, 615, 626-629
Mil, 17-18
Mil-foot, 17
Mixers, 481-484
Modulation, *see* Amplitude modulation; Frequency modulation
Modulation monitors, 513, 522-523
Motors, 575-578, 582-586
Multielectrode tube, 304
Multigrid tubes, 164
Multiplex system, stereo, 417-421
Multipliers, frequency, 242-245
Multirange voltmeter, 261
Multivibrators, 322-324
Mutual conductance, 161-162
Mutual inductance, 32-33, 48

N

Narrowband FM, 415
National Bureau of Standards, 279-281
National Defense Emergency Authorization, 528
Natural sines, 634-639
Negative feedback, 206-207, 219-221
Negative resistance, 319
Nemo lines, 514
Neon lamp to detect oscillation, 325
Neutralization, 236-241
Noise limiters, 493-494
Noninductive capacitors, 73
Null point, 48

O

Oersteds, 26, 29
Ohm, 3
Ohmmeters, 262-263
Ohm's law, 3, 5, 29, 133
Operator license requirements, broadcast stations, 528-529
Opposite polarity, 33
Oscillators, 37, 294-329, 358
    Armstrong, 296-298
    audio, 314-315
    beat-frequency, 490-491
    biasing, 298-299
    Colpitts, 302-304
    crystal subcarrier, 567
    crystal ovens, 310-313
    crystal-controlled, 305-307, 313-314
    crystals, 307-310
    dynatron, 318-320
    electron-coupled, 304
    Hartley, 301-302
    high-frequency, 316-318
    horizontal, 556
    keying, 344
    master, power amplifiers, 332-334
    methods used to detect oscillation, 325-326
    multivibrators, 322-324
    parasitic oscillations, 324-325
    phase-shift, 315-316
    Pierce, 313-314
    requirements for oscillation, 294-295
    resistance-capacitance, 320-321
    stabilization, 326
    tuned-plate tuned-grid, 299-301
    two-stage amplifier using regenerative feedback, 295-296
    unijunction transistor, 321-322
    variable-frequency, 351
    vertical, 555-556
Oscilloscopes, 37, 273-275, 325, 381-384
Output conductance, 151
Output transformers, 62
Overload relay, 233
Overloads, 19
Oxide-coated cathode, 130

P

Parallel resonant circuits, 108-111
Parallel-connected inductors, 49-50
Paraphase amplifiers, 213-214
Parasitic arrays, 443-446
Parasitic oscillations, 324-325
Peak clipping, 232
Peak inverse voltage, 168, 183-184
Peak-limiting amplifiers, 520, 521
Peak-reading voltmeters, 268
Pentavalent elements, 123
Pentodes, 163-164
Permeability, 26-27
Permits, Federal Communications Commission, 604-605
Phase angle, 85

Phase discriminator, 503
Phase distortion, 211
Phase modulation, 410-413
Phase monitors, 464
Phase relationships
 alternating current, 41-43
 inductance, 53
Phase-shift method of generating single sidebands, 395-396
Phase-shift oscillator, 315-316
Phasitron, 412-413
Phonetic alphabet, 615-616
Pi network, 455-456
Picofarad, 70
Picture tube, color, 567-569
Pierce oscillator, 313-314
Piezoelectric effect, 305
Pi-section filter, 175-176
Plate
 circuit, 132
 current, to detect oscillation, 325
 detectors, 475-476
 keying, 348
 modulation, 367-369, 374, 375, 386-387
 neutralization, 236-238
 resistance, 133-135
Polarization, 587-588
Polyphase rectifiers, 194-196
Power, 5-6
Power amplifiers, 332-338
Power detectors, 473-474
Power factor, 95
Power supplies, 167-198
 basic full-wave rectifier circuit, 169-170
 bleeder resistors, 185-187
 bridge rectifier, 170-172
 capacitance input filter, 173-174
 filter chokes, 177-178
 half-wave rectifiers, 167-169
 inductance input filter, 174-175
 mercury-vapor rectifiers, 180-182
 peak inverse voltage, 183-184
 pi-section filter, 175-176
 regulated, 189-192
 series-connected filter capacitors, 184-185
 three-phase, 194-196
 vacuum rectifiers, 178-180
 vibrators, 192-194
 voltage dividers, 185-187
 voltage regulator, 187-189

 voltage-doubler circuits, 192
 waveform analysis, 172-173
Preemphasis, 407-408
Prefixes used in electronics, 10-11
Primary cells, 586-588
Proof-of-performance tests, 526
Public safety radio service, 414-417
Pulsating current, 572
Pulse repetition rate, 322
Push-pull amplifiers, 202, 215-217, 242

Q

Q signals, commonly used, 644-645
Quarter-wave matching stub, 450
Quarter-wave transformer, 450

R

Radiation, 432
 harmonic, 351-354
 patterns, antenna, 458-459
 resistance, 437
Radio waves, 425-427
Radio-frequency ammeters, 270-272
Radio-frequency amplifiers, 226-252, 359, 479
 Class C, 229-232
 coupling to the load, 247-248
 frequency multipliers, 242-245
 grid current loading, 232-233
 grounded-grid, 245-246
 interstage coupling techniques, 234-236
 linear, 384-386
 neutralization, 236-241
 parallel operation of, 241-242
 push-pull operation of, 242
 series and shunt feed, 233-234
 transformer coupling, 226-229
 troubleshooting procedures, 249-250
 very high and ultrahigh, 246
Radio-frequency indicator to detect oscillation, 325
Radio-frequency receivers, 473
Radio-frequency transformers, 62-63
Radiotelephone transmitters, 388-391
Ratio detector, 503-505
Reactance-tube modulators, 408-409

Index 659

Receivers, 471-509
  automatic volume control circuit, 488-490
  basic functions of, 471
  beat-frequency oscillators, 490-491
  color television, 565-567
  crystal filters, 486-488
  detectors
    crystal, 472
    diode, 473-474
    grid-leak, 474-475
    plate, 475-476
    ratio, 503-505
    regenerative, 476-477
    second, 488-490
    superregenerative, 477-478
  diversity receiving system, 497
  frequency modulated, 499-505
  image frequency, 484-485
  intermediate-frequency amplifier, 485-486
  maintenance procedures, 505-506
  mixers and converters, 481-484
  noise limiters, 493-494
  radio-frequency amplifier or preselector, 479
  signal-strength meters, 491-492
  squelch circuits, 494-495
  superheterodyne, 478-479, 496
  wave traps, 498-499
Rectifiers, 121-137
  basic full-wave circuit, 169-170
  biasing, 126-128
  bridge, 170-172
  capacitance input filter, 173-174
  diodes as, 129
  half-wave, 167-169
  inductance input filter, 174-175
  junctions, 125-129
  mercury-vapor, 180-182
  meters, alternating current, 267-268
  pi-section filter, 175-176
  plate resistance, 133-135
  polyphase, 194-196
  semiconductor materials, 121-125
  vacuum, 178-180
  vacuum-tube diode, 129-132
  volt-ampere characteristic, 132
  waveform analysis, 172-173
Regenerative detectors, 476-477
Regenerative feedback, 295-296, 298, 323

Regulators, 189-192
Remote-control broadcast systems, 525-526
Repulsion motor, 583
Residual magnetism, 24
Resistance, 3, 18-19
  -capacitance circuit, series, 86-88
  -capacitance coupling, 207-209
  -capacitance oscillators, 320-326
  -capacitance time constant, 78-79
  -inductance circuit, series, 84-86
  measurement, 262-263
  negative, 319
Resistors
  bleeder, 185-187
  color code, 3
  cathode biasing, 205
  equalizing, 76, 182, 184-185
  fixed, values, 650
  grid-leak, 299
  source biasing, 205
  voltage-dropping, 6-7
Resonant circuits, 104-120
  band rejection, 117-119
  band-pass filters, 117-119
  bandwidth, 112-114
  high-pass filters, 115-116
  low-pass filters, 114-115
  parallel, 108-111
  resonance, 104-105
  series, 106-108, 111-112
Reverse bias, 169
Reverse transfer voltage ratio, 152
Rice neutralization system, 238-239
Rotary converter, 578

S

Safety signal, 606, 613, 615, 617, 619
Scanning, 536-538
Screen bypass capacitor, 162-163
Screen grid, 162, 348, 365, 374, 375
Secondary cells, 590-591
Secondary emission, 163, 181, 274, 319
Semiconductor materials, 121-125
Series generators, 574
Series motors, 576-577
Series-connected inductors, 49-50
Shaded-pole motor, 583, 584-585
Shielding, transmitter, 354-356
Short-circuit, 191, 505

Shorted turns, 64-65
Shorting bars, 355
Shunt feeding, 233-234, 302, 303
Shunt generators, 574-575
Shunt motors, 577-578
Shunting capacitance, 209
Sidebands, 351, 364, 365, 369-370, 391-396, 405-407
Siemen, 8, 96
Signals, 646-647
Signal-strength meters, 491-492
Signal-to-noise ratio, 479
Simplex operation, 334
Sine wave, 37-38
Single-phase motors, 583-585
Skin effect, 21, 316
Sky waves, 427, 429-431
Soft signal, 342
Soldering, 21-22
Solenoid, 27
Solid state diode, 121, 168-169
Solid-state modulator, 409-410
Sound transmitter, television station, 546
Source biasing resistor, 205
Spark-gap oscillators, 295
Spatial effect, 417
Specific resistance, 17, 18-19
Speech amplifier, 365
Splatter effect, 370
Split tank capacitor, 302-304
Split-phase motors, 583, 584
Square mil, 18
Square wave, 37
Square-law detector, 475
Squelch circuits, 494-495
Static, 400
Step-up transformer, 55-56
Stereo multiplex, 417-421
Store energy, 109
Strain insulators, 454
Studio-to-transmitter link, 512
Subsidiary Communication Authorization (SCA), 420
Superheterodyne receivers, 478-479, 496
Superregenerative detectors, 477-478
Suppressor grid, 163-164, 365, 377-378
Surge impedance, 432
Susceptance, 96
Swinging choke, 178
Switching regulator, 189

Synchronization, color, 564-565
Synchronized-pulse generator, 544-545
Synchronizing-pulse separation, 555
Synchronous motors, 582-583

T

Tapped tank inductance, 302
Television, 536-570
    basic principles, 536-537
    color, 559-569
    iconoscope, 538-539
    image orthicon, 539-541
    interlaced scanning, 538
    magnetic deflection and focusing, 537
    receiving functions, 551-559
    stations, 542-550
    vidicon, 541-542
Tetrode transistor, 154
Tetrodes, 162-163
Thermal runaway, 153
Thermionic cathodes, 130
Thermionic emission, 129
Thermocouple ammeters, 270-272
Thoriated tungsten cathode, 129
Three-gun color picture tube, 567-569
Three-phase transformers, 63-64
Transconductance, 161-162
Transducer, magnetostrictive, 395
Transformer coupling, 207, 226-229, 296-298
Transformers, 53-65
    autotransformers, 59
    basic, 53-55
    impedance matching, 60-61
    losses, 57-59
    modulation, 365, 367
    output, 62
    quarter-wave, 450
    radio-frequency, 62-63
    shorted turns, 64-65
    three-phase, 63-64
    turns ratio, 55-57
Transient, 64
Transistor fundamentals, 138-158
    basic transistor, 138-139
    biasing techniques, 145-149
    circuit configurations, 149-150
    collector family of curves, 141-142

Index

Transistor fundamentals *(cont.)*
  equivalent circuit, 152
  hybrid parameters, 151-152
  leakage currents, 152-153
  load lines, 143-145
  types of transistors, 152-156
Transmission lines, 432-434, 529-530
Transmitters
  basic, 330-361
  bias methods, 339-341
  break-in operation, 334
  broadcast, 523-524
  buffer amplifiers, 334-335
  continuous wave, 330, 356-358
  dummy antennas, 338-339
  frequency-modulated, 413-414
  fundamental concepts, 330-331
  harmonic radiation, reducing, 351-354
  key-click filters, 343-344
  keying, 342-350
  locating troubles in, 358-359
  master oscillator power amplifier, 332-334
  modulated continuous wave signals, 350-351
  power amplifiers, 335-338
  radiotelephone, 338-391
  restrictions on operating, servicing, and adjusting, 609
  shielding, 354-356
  single-stage, 331-332
  television, 546-547
  variable-frequency-oscillator operation, 351
Triangular wave, 37
Triggered oscillators, 322
Trigonometric functions, 39
Triodes, 159-162
Trivalent elements, 124
Tuned-plate tuned-grid oscillator, 299-301
Tunnel diodes, 156, 319-320
Tuner circuits, television, 551
Turns ratio, 55-57

U

Ultra-audion, 303
Ultrahigh-frequency amplifiers, 246
Unijunction transistor oscillator, 321-322

Unijunction transistors, 154
Universal motors, 577
Urgent signal, 606, 613, 617, 619

V

Vacuum rectifiers, 178-180
Vacuum tubes, 159-166
  biasing, 204-207
  diodes, 121, 129-132, 168-169
  keying, 346-347
  voltmeters, 263-267
Valence bond, 123
Valence electrons, 121-122
Varactors, 409
Variable capacitors, 72-73, 306
Variable-frequency-oscillator operation, 351
Varicaps, 409
Vectors, 44
Vertical deflection, 555-556
Very high frequency amplifiers, 246
Vestigial sideband transmission, television, 547-549
Vibrator power supplies, 192-194
Video amplifier, 555
Video detector, 554
Video modulation, 545-546
Video switcher, 542
Vidicon, 541-542
Voltage
  amplifiers, *see* Amplifiers
  avalanche, 135-136
  breakdown rating, 20
  dividers, 147, 185-187
  -double circuits, 192
  -dropping resistors, 6-7
  feedback, 220
  feeding, 448
  gradient, 154
  measurement, *see* Voltmeters
  peak inverse, 168, 183-184
  ratings, 19, 81-82
  regulation, 187-189
  screen, 368
Voltmeters
  direct current, 259-262
  peak-reading, 268
  vacuum-tube, 263-267
Volt-ohm-milliammeters, 263

Volts, 2
Volume-unit meter, 272, 511, 513-514, 516-517

W

Watthour meters, 270
Watthours, 5
Wattmeters, 269
Watts, 5
Wave polarization, 437-438
Wave traps, 498-499

Waveform analysis, 172-173
Wavelength, 331, 426-427
Wavemeter, 275-277
Waveshapes, 36, 37-38
Wire gauge, American, 633
Wire sizes, 19-20

Z

Zener diodes, 135-136, 187
Zeppo antenna, 450

Review Chap 13
14